INTRODUCTION TO MOLECULAR BIOLOGY

S. E. BRESLER

HEAD, DEPARTMENT OF BIOPOLYMERS
PHYSICO-TECHNICAL INSTITUTE
ACADEMY OF SCIENCES OF THE U.S.S.R.
LENINGRAD, U.S.S.R.

AND

PROFESSOR OF BIOPHYSICS
POLYTECHNIC INSTITUTE OF LENINGRAD
LENINGRAD, U.S.S.R.

Translation Editor: Robert A. Zimmermann
DEPARTMENT OF BACTERIOLOGY AND IMMUNOLOGY
HARVARD MEDICAL SCHOOL
BOSTON, MASSACHUSETTS

ACADEMIC PRESS   New York and London

Copyright © 1971, by Academic Press, Inc.
ALL RIGHTS RESERVED
NO PART OF THIS BOOK MAY BE REPRODUCED IN ANY FORM,
BY PHOTOSTAT, MICROFILM, RETRIEVAL SYSTEM, OR ANY
OTHER MEANS, WITHOUT WRITTEN PERMISSION FROM
THE PUBLISHERS.

ACADEMIC PRESS, INC.
111 Fifth Avenue, New York, New York 10003

United Kingdom Edition published by
ACADEMIC PRESS, INC. (LONDON) LTD.
Berkeley Square House, London W1X 6BA

LIBRARY OF CONGRESS CATALOG CARD NUMBER: 76-91429

PRINTED IN THE UNITED STATES OF AMERICA

Translated from the Second Russian Edition
Published under the title: Vvedenie v Molekulyarnuyu
Biologiyu, Moscow, 1966.

MICHIGAN STATE UNIVERSITY
LIBRARY
JUL 31 2020
WITHDRAWN

PLACE IN RETURN BOX to remove this checkout from your record.
TO AVOID FINES return on or before date due.
MAY BE RECALLED with earlier due date if requested.

DATE DUE	DATE DUE	DATE DUE
1 2 0 1 0 0 DEC 1 6 2000		
JUL 0 5 2003 08 13 03		

1/98 c:/CIRC/DateDue.p65-p.14

Introduction to Molecular Biology

Contents

Preface ix

Introduction xi

I. Protein Structure

1. Introduction 1
2. The Chemical Structure of Proteins 2
3. Analysis of Amino Acid Composition 7
4. Determination of Amino Acid Sequence in the Protein Chain 10
5. The Types of Macromolecular Structure 23
6. Synthetic Polypeptides as Model Proteins 26
7. Secondary Structure of Polypeptides and Proteins 31
8. Optical Properties of Polypeptides and Proteins 50
9. Theory of the Order–Disorder Transition 65
10. Tertiary Structure of Proteins 69
11. X-Ray Structure Analysis of Proteins 79
12. Physicochemical Techniques for Measuring Molecular Size and Shape 101
13. Purification and Identification of Proteins 115
    References 124

II. Protein Function

1. Introduction 127
2. Enzymatic Catalysis 129

	3. Kinetics of Enzymatic Catalysis	142
	4. The Mechanism of Enzymatic Reactions	153
	5. Allosteric Regulation of Enzymatic Activity	165
	6. Active Transport	174
	7. The Mechanochemical Function of Proteins	186
	References	198

III. Nucleic Acid Structure

	1. Introduction	200
	2. Chemical Structure of the Nucleic Acids	204
	3. Macromolecular Structure of DNA	213
	4. Thermodynamics and the Mechanism of Phase Transition in Polynucleotides	231
	5. Synthesis and Properties of Model Polynucleotides	236
	6. Enzymatic Synthesis of Nucleic Acids	247
	7. The Synthesis of Unusual Nucleic Acids	259
	8. The Use of Isotopes in Studies on DNA Replication	261
	9. Molecular Heterogeneity, Reversible Denaturation, and the Formation of Molecular Hybrids	269
	10. Macromolecular Structure of RNA	284
	References	295

IV. DNA Function

	1. Introduction	299
	2. Bacterial Genetics	301
	3. Mutation and Mutagenesis	304
	4. Selection of Bacterial Mutants	308
	5. The Luria-Delbrück Fluctuation Experiment	312
	6. Sex in Bacteria; Conjugation	318
	7. Replication of the Bacterial Chromosome during Vegetative Growth and during Conjugation	328
	8. Mechanism of Recombination	334
	9. Genetic Mapping by Conjugation	340
	10. Genetic Mapping by Recombination Frequencies	344
	11. The Physical Scale of the Genetic Map; Experiments with Radioactive Cells	352
	12. Bacterial Transformation	355
	13. Genetic Processes in Viruses and Bacteriophages	366
	14. Lysogeny and Transduction	391
	15. The Chemistry of Mutagenesis	403
	16. Mutation and the Genetic Code	419
	17. Direct Methods for Deciphering the Genetic Code; General Properties of the Code	433
	References	443

V. RNA Function

	1. Introduction	449
	2. Transfer RNA and the Activation of Amino Acids	451
	3. Ribosomes and the Synthesis of Proteins	457
	4. The Rate and Direction of Protein Synthesis	469

Contents

- 5. Messenger RNA and Protein Synthesis   477
- 6. Regulation of Protein Synthesis   486
  - References   510

Author Index   513
Subject Index   525

Preface

This text was written to acquaint biologist and nonbiologist alike with the remarkable advances made in molecular biology, molecular genetics, and molecular biophysics during the 1950's and 1960's. The discipline of biology has been radically altered in these past two decades by investigators from all over the world. Beyond their role in the laboratory, biologists have also expended their efforts in the dissemination of newly acquired information to their colleagues in other countries. Professor Bresler's book represents a further example of the continuing exchange within the international scientific community.

Because his professional career has been devoted to the study of high-molecular-weight compounds, Professor Bresler is eminently qualified to apply the principles of polymer physics and chemistry to fundamental phenomena in the biological sciences. The main body of the text is focused upon the structure, synthesis, and function of nucleic acids and proteins, with a full discussion of the physicochemical techniques necessary for the determination of macromolecular structure, the kinetics and mechanism of enzyme action, the genetics of bacteria and their viruses, and the genetic code. Topics to which the author gives particular stress are the importance of precise quantitative analysis in biochemistry and biophysics, the intimate relationship between the configuration of biological macromolecules and their function, and the unique mechanisms that regulate biological activity within the cell.

The first edition of this book was published in the Soviet Union in 1963 and a second edition, incorporating considerable new data, appeared in 1966. This English translation is based on the second Russian edition and includes supplementary material which reviews the literature to early 1969.

It is my hope that students of biology as well as scientists in related disciplines will find this volume a stimulating introduction to current research in this rapidly expanding field.

ROBERT A. ZIMMERMANN*

* *Present Address:* Institut de Biologie Moléculaire, Université de Genève, Quai de l'Ecole de Médecine, Genève, Switzerland.

Introduction

The discipline of biology, one of the most important of the natural sciences, has undergone enormous qualitative changes in recent years. The far-reaching implications of new discoveries have led to a revolution in our conception of life processes. In its significance and content, contemporary biology resembles physics at the beginning of this century. That was the era in which the fundamental constituents of matter—electrons, nuclei, atoms, and molecules—became the objects of intensive study. Physics of the nineteenth century was based on the mathematical formulation of numerous empirical laws. Its subdivisions such as mechanics, thermodynamics, and electrodynamics provided adequate quantitative descriptions of the behavior of macroscopic bodies. But not until the advent of the new physics, which sought laws governing the interactions of elementary particles, was it possible to draw hitherto uncoordinated physical phenomena together under a single unified system of concepts. Significantly, the new discipline did not abandon the achievements of classical physics, but, instead, broadened and expanded them to an enormous extent, opening up a vast world of undiscovered phenomena. A new method of thinking became firmly established in the physics of that time; by searching for the atomic and molecular mechanisms of physical phenomena, it was possible to relate outwardly dissimilar processes with one another. It is not necessary to mention the colossal scientific and technical advances wrought by the new physics in order to realize that this was the first revolution in the natural sciences during the twentieth century.

The second revolution, still in full swing, has resulted from the study of biological processes at the molecular level. This new branch of science has grown up straddling molecular physics, organic chemistry, and biology, and the name by which it is currently known—molecular biology—reflects its diverse heritage. Before attempting to define this field of inquiry any further, it is essential to indicate the experimental discoveries to which the new science owes its existence.

The principal levels of biological organization treated by the classical science were either the structure and function of complex whole organisms or the microscopic structure of the cell, although it was long recognized that cells must ultimately be comprised of molecules and atoms. Nonetheless, most biological phenomena occur on the level of the whole organism or, at most, of the cell. The scale of these phenomena exceeds molecular dimensions by a large factor and there was no direct and obvious link between them and the properties of molecules which constituted cellular matter. Biology is no exception in this respect. There exist many macroscopic physical processes in which molecular structure is immaterial. Hydrodynamics and aerodynamics, for instance, are not based on the molecular nature of liquids or gases, but rather on their properties as continuous media. Hydrodynamic motion differs greatly in scale from molecular motion, and only in very rarified gases is this distinction obliterated. For similar reasons, it did not seem profitable to reduce to molecular terms those biological phenomena whose mechanisms belong wholly to the sphere of cellular or multicellular physiology. Any attempt to do so would have appeared to contradict scientific methodology. How then did molecular biology come into being as a separate science?

The origin of molecular biology was partly a response to new empirical discoveries, partly a result of fresh ideas different in principle from those current in classical biology. In the past fifteen years, great strides have been made in determining the physical and chemical structure of two extremely important classes of biological compounds— proteins and nucleic acids. These substances possess molecular weights ranging from ten or twenty thousand to many hundred million. High-molecular-weight compounds of this type are commonly referred to as macromolecules. The successful investigation of the most complicated biological macromolecules was in large measure predetermined by advances in the physics and chemistry of high-molecular-weight polymers.

The most important functions of proteins have been shown to occur on the level of individual molecules and they must hence be investigated as molecular phenomena.

The first major function of proteins is catalytic. Specialized proteins, called enzymes, catalyze all metabolic, synthetic, and degradative reactions in living organisms. Many proteins of multicellular organisms and practically all proteins of single-celled organisms such as bacteria possess enzymatic activity. Catalytic or enzymatic capacity is inherent in the individual protein macromolecules, and the biochemical properties of purified proteins prepared from living tissues have long been the subject of intensive study.

Contractile proteins constitute a second major class. These proteins, of which actomyosin, the active component of muscle, is one example, behave as mechanochemical machines. They transform the chemical energy stored in specific labile compounds—adenosine triphosphate (ATP), in particular—into mechanical work. Contractile proteins are found not only in the muscles of higher organisms, but also in the locomotor flagellae of microbes, and even in the contractile "tail" of certain viruses. Experiments have shown that flagellar locomotion occurs on the molecular level. The contractile reaction is also an inherent property of individual actomyosin molecules and does not depend on the integrity of muscle tissue as was thought at one time.

The third principal group of proteins consists of structural proteins. The cell cannot be likened to a vessel in which all the metabolites and enzymes are simply mixed up together. On the contrary, the cellular interior is divided into many organelles which are protected by protein and lipoprotein membranes that prevent the free diffusion of soluble substances. Similarly, the cell wall also consists of a lipoprotein membrane with a very selective permeability. The majority of cellular enzymes are located within one or another organelle, and for this reason most biochemical processes are localized at specific points in the cell. Mitochondria, rather large oblong particulates approximately 0.5μ in length, contain the enzymes of oxidation and oxidative phosphorylation. These are the enzymes catalyzing reactions which store up energy for later use by the cell. Ribosomes or microsomes are small, almost spherical particles, 150 to 200 Å in diameter, whose structural proteins are essential for the synthesis of all cellular proteins. In some cases, structural proteins play a strictly architectural role, providing the material from which the cell synthesizes various morphological structures. In other cases, they regulate the passage of chemical substances into organelles, sometimes participating in active transport of metabolites against a concentration gradient and thus counter to diffusion. In the differentiated and highly specialized cells of higher organisms, certain structural proteins are present in considerable quantities, providing substance for the formation of specialized tissues. Examples are collagen in bone and connective tissue, fibrinogen in the blood, and scleroproteins in the cornea. The distinctive structural properties of these proteins are closely linked to the function that they fulfill. Thus, in the case of structural proteins, too, we have reason to speak of functional activity on the molecular level.

Among the proteins of higher organisms we encounter two additional types of functional activity. These are the transport proteins responsible for reversibly binding and transporting specific compounds of vital importance to the organism. Hemoglobin, the major protein of red blood cells, transports molecular oxygen throughout the bodies of vertebrates and is perhaps the prime example of this kind of protein. Serum albumin, with its exceptional ability to bind fatty acids, steroids, and other lipids, is most likely another example of a transport protein. The capacity to bind and transport various substances is exhibited by the protein macromolecules as such and it appears to result from a reversible chemical reaction.

Finally, many organisms contain proteins whose function is defense against foreign chemical compounds, macromolecules, and cells. Immunological reactions mediated by γ-globulins can be studied with the same techniques applicable to any other molecular interaction.

To summarize, it is now evident that all types of functional activity exhibited by proteins—enzymatic catalysis, contractility, formation of selectively permeable membranes, reversible binding and transport, and immunological activity—are related to molecular interactions and therefore fall within the purview of molecular biology.

Let us now consider the functional activity of nucleic acids. It has been firmly established over the past several years that nucleic acids perform the task of storing and transmitting information whether in viruses, microbes, plants, or animals. All the genetic information necessary to specify the proteins needed by the organism is fixed in the cellular deoxyribonucleic acid (DNA) or, in the case of some viruses, in ribonucleic acid (RNA). Direct experiments on the transformation of bacteria with purified DNA preparations, on the infection of bacteria with DNA extracted from viruses, and on the infection of plant cells with viral RNA all show that genetic information is carried specifically by macromolecules of DNA and RNA. To borrow an analogy from the realm of computers, we might, along with von Neumann, liken the cell to a machine which contains all the information necessary to reproduce itself. In such a machine there must be working elements—in cells, the enzymes organized into three-dimensional organelles—and memory elements in which all the details of design and construction are specified by means of a code. This latter function, performed by a magnetic tape in a computer, is the task of the nucleic acids in living cells. The longer the nucleic acid polymer is, the more information can be stored in it.

It will be instructive to examine some of the magnitudes involved. An autonomous bacterial cell contains, on the average, 2000 different proteins. The overall length of DNA in the bacterial chromosome is of the order of 0.2 cm. This is not very great if we were to compare it with a long magnetic tape. But as we shall see in what follows, the symbols in which the genetic code is recorded consist of just a few neighboring monomer units in a long polymer chain. If we assume (1) that three successive nucleotide residues represent one bit of information capable of specifying an amino acid, (2) that one nucleotide residue occupies 3.4 Å along the chain, and (3) that the molecule is double-stranded, we discover that 0.2 cm of DNA contains 3×10^6 nucleotide pairs or 10^6 coding symbols. This is just what is necessary to specify the structure of 2000 proteins if each one contains an average of 500 amino acid residues with an aggregate molecular weight of 50,000. The DNA segment carrying information for the synthesis of one protein has a molecular weight of the order of one million. In the old terminology this would have been called a gene, but it is currently referred to as a cistron, the unit of functional activity in a DNA molecule. In the primitive bacterial virus, bacteriophage T2, there is a single molecule of DNA which contains about 200 cistrons. The overall length of the phage DNA is about 0.1 mm. By contrast, in the cells of higher organisms there is several hundred times—and in man, a thousand times—more DNA than in the bacterial cell. The combined length of all the DNA molecules would reach several meters. This should not be at all surprising since the nucleus of a gamete stores information not only for its own structure and metabolism, but for the entire future development and differentiation of the cell into a complex multicellular organism. The supply of information required for this task is hundreds or thousands of times greater than in a bacterial cell.

Nucleic acids thus provide the material substrate of heredity. Their most important property is a capacity for reduplication or self-replication, which stems from a distinctive molecular structure. The hereditary variability of organisms is related to chemical changes in their DNA molecules, the basis of mutation. In a chemical sense, it has been possible to direct mutagenesis; mutations have been produced by various chemical agents that selectively attack certain of the nucleic acid components.

Spontaneous and induced mutations have both been subjected to detailed investigations in microorganisms and in viruses. Powerful methods of genetic analysis have established that each mutation corresponds to an alteration in the genetic material of the cell; the size of the affected region falls within the range of atomic dimensions. Since genetic information is stored in the polymer chain of DNA, it is inferred that mutation results from the alteration of a single nucleotide residue in that chain. Within the functional region of the genetic material, the cistron, there are many potentially mutable sites called mutons. Let us assume that we can isolate the protein whose synthesis is specified by a particular cistron and, furthermore, that we have cells which contain mutationally altered variants of that cistron in addition to normal cells. We can now directly determine the effect of mutations on the products of the cistron by comparing the chemical structure of normal and mutant proteins. Such studies have established that in each case a mutational alteration in the DNA molecule corresponds to a well-defined change in a specific amino acid residue of the protein chain. This provides proof that all the genetic information necessary for the synthesis of a protein is stored in the DNA molecule and that biological variability is a chemical process occurring on the molecular level.

Synthesis of the other cellular nucleic acid, RNA, requires the participation of DNA, whose chemical structure it reflects. One function of RNA is to carry genetic information from the nucleus to the cytoplasm and, in particular, to the cytoplasmic granules called ribosomes where protein synthesis takes place. At the same time, another specialized type of RNA, characterized by low molecular weight and high solubility, transports amino acids, the building blocks of protein, to the ribosomal particles. Thus we see once again that biological molecules, here RNA of various kinds, play a direct role in a vital cellular function, protein synthesis.

The discoveries enumerated above created the experimental basis of molecular biology. They reflect the general tendency in contemporary natural science to regard life processes in terms of molecular phenomena. In particular, this approach focuses on the structure, function, and biosynthesis of the most important macromolecular constituents of living matter—proteins and nucleic acids. The sources of molecular biology are to be found in a wide variety of scientific disciplines: organic chemistry has provided the methods for analyzing the chemical structure of proteins and nucleic acids, their chemical reactions, and their chemical synthesis; biochemistry has made possible the detailed study of metabolic reactions in biological systems, including intermediates and mechanisms; cytology has contributed to the understanding of cellular ultrastructure and physiology; genetics has defined our knowledge of hereditary processes; microbiology and virology have revealed basic life processes in the simplest of organisms. From the physical sciences, molecular biology has drawn on the ideas and methods of molecular physics, the physical chemistry of polymers, spectroscopy, and x-ray structure analysis.

The task of this book is to acquaint the reader with the most important facts and ideas at the disposal of contemporary molecular biology.

S. E. BRESLER

Chapter I

Protein Structure

1. Introduction

The sum of our knowledge about protein structure has increased enormously over the last 10 to 15 years. It may safely be said that just a few years ago no one could have dreamed of the extraordinarily powerful methods of analysis that are now available to us, nor could they have imagined the comprehensive results which have been obtained. At the present time we know the structural formulas of several dozen proteins, and in the cases of myoglobin, hemoglobin, ribonuclease, chymotrysin, carboxypeptidase, papain, subtilisin, and lysozyme, the complete three-dimensional structure has been determined down to the coordinates of their individual component atoms. In regard to this last statement, it will be necessary to touch for a moment on one very important point.

Although it has been known for a long time that proteins are high-molecular-weight polymers made up of hundreds, if not thousands, of amino acid residues, the nature of the polymer has not always been properly understood. Proteins could either be individual chemical species consisting of mathematically identical molecules or else they could consist of complex mixtures of homologous polymers characterized by an average composition but with statistical fluctuations among individual macromolecules. This is far from an idle distinction. When we consider the properties of ordinary linear polymers, we always deal with statistical ensembles of molecules. Thus the concept of polymer molecular weight is a statistical one. Indeed, it is more correct to speak of a distribution function over molecular weights or to discuss this property in terms of various kinds of averages such as the number average or weight average. Even the chemical structure of ordinary polymers is not a constant: chains may become branched or cross-linked and there are often alternative ways of joining successive monomeric units. The frequency of these events is determined by the laws of probability.

Some time ago it would not have been hard to believe that proteins too were inherently statistical in nature. The techniques in protein chemistry necessary to resolve this fundamental question were slow to evolve. The inertia of traditional concepts had to be overcome as did the numerous experimental difficulties which plagued the early biochemists who devoted themselves to this field. The first major advance was the development of methods for fractionating and purifying individual protein species. These have presently been perfected to such a degree that it is possible to obtain chemically pure protein preparations with a contamination less than 0.01 %. This compares favorably with the purity of any organic compound. Only purified proteins could be subjected to rigorous chemical studies, the most important of which was the deduction of their structural formulas.

It was precisely the investigation of protein structural formulas, that is, of the sequence of amino acid residues in the protein chain, which proved that all molecules of a given protein are mathematically identical to one another. In general, a cell can synthesize altered proteins only as a result of mutations in its hereditary constitution. Even at that, all molecules of an altered protein produced by the mutant cell are identical. It was Ingram, working with hemoglobin, who first showed that a simple genetic mutation leads to the replacement of a unique amino acid residue in the polypeptide chain. His work has been subsequently confirmed by many other investigators. In such cases, the properties of the affected protein may undergo a large change even though the chemical "damage" appears insignificant. This arises from the following circumstances. Protein macromolecules assume a helical secondary structure which directly results from the formation of intramolecular hydrogen bonds; the helical regions themselves are folded and packed into a compact tertiary structure determined by an extremely delicate balance between different kinds of attractive and repulsive forces within the molecule. A change in the nature of a single amino acid residue can thus evoke tremendous alterations in the structural integrity of the protein.

In this chapter we shall approach the problem of protein structure in the following way. First the chemical structure, or amino acid sequence, of the protein molecule will be considered. Next, the helical secondary structure of the polypeptide chain will be discussed and it will be shown that the investigation of synthetic model peptides has been of enormous value in understanding the structure of proteins themselves. The third topic concerns protein tertiary structure, in particular how the exact three-dimensional configuration of a protein molecule can be determined from x-ray structure analysis, the most highly perfected technique applicable. It will be demonstrated that numerous physicochemical properties of proteins—optical, electrical, and hydrodynamic—directly depend on macromolecular configuration. Finally, we shall take up modern methods for the fractionation, purification, and identification of individual proteins.

2. The Chemical Structure of Proteins

Proteins are linear polymers or copolymers composed of amino acids joined together by peptide bonds. That the backbone of all protein molecules consists of a *polypeptide chain* has been proved without a doubt. Its general formula is

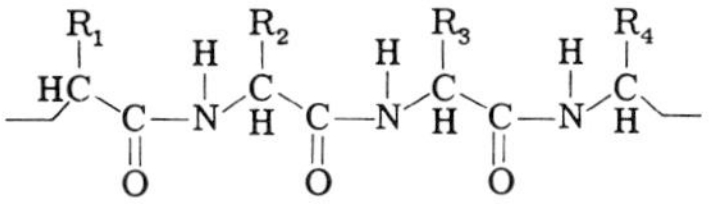

2. The Chemical Structure of Proteins

$$\text{---HC}\underset{\underset{O}{\|}}{\overset{R_1}{\underset{}{C}}}\text{---N}\overset{H}{\underset{H}{}}\underset{\underset{O}{\|}}{\overset{R_2}{\underset{}{C}}}\text{---N}\overset{H}{\underset{H}{}}\underset{\underset{O}{\|}}{\overset{R_3}{\underset{}{C}}}\text{---N}\overset{H}{\underset{H}{}}\overset{R_4}{\underset{}{C}}\text{---}$$

One can imagine that such a polymer is the product of a series of condensation reactions between various amino acids in which the amino (—NH$_2$) and carboxyl (—COOH) groups of successive residues are united to form peptide bonds (—CO—NH—) with the exclusion of a water molecule:

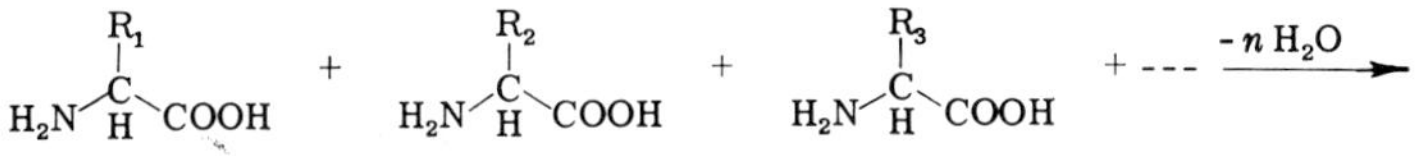

$$\text{H}_2\text{N}\overset{R_1}{\underset{H}{C}}\text{COOH} + \text{H}_2\text{N}\overset{R_2}{\underset{H}{C}}\text{COOH} + \text{H}_2\text{N}\overset{R_3}{\underset{H}{C}}\text{COOH} + \cdots \xrightarrow{-n\,\text{H}_2\text{O}}$$

As a rule, there is a free or acylated amino group at one end of each polypeptide chain (the N-terminus) and a free or amidated carboxyl group at the other end (the C-terminus).

All amino acids contain in common both an amino and a carboxyl group attached to the α-carbon atom. They differ in the nature of their side chains, designated by $R_1, R_2, R_3, \ldots, R_n$ in the above reaction diagram. There are twenty common amino acids which are found in practically all proteins and a few more which occur only rarely. The structural formulas of the twenty universal amino acids are presented in Table 1-1, classified according to the properties of their side chains, R. We see that a number of amino acids are electrochemically active by virtue of acidic (—COOH, —SH) or basic (—NH$_2$, —NH) radicals in their side groups. Even when present in a polypeptide chain, such groups retain their acidic or basic properties and become ionized at certain pH's. A considerable number of the side chain carboxyl groups are amidated in conformity with the following reaction:

$$\text{—COOH} + \text{NH}_3 \longrightarrow \text{—CONH}_2 + \text{H}_2\text{O}.$$

Thus, in addition to aspartic acid residues, we also find the corresponding amide known as asparagine:

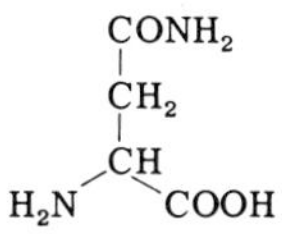

$$\begin{array}{c}\text{CONH}_2\\|\\\text{CH}_2\\|\\\text{H}_2\text{N}\diagup\text{CH}\diagdown\text{COOH}\end{array}$$

and in addition to glutamic acid, its amide, glutamine:

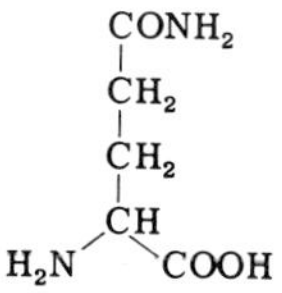

$$\begin{array}{c}\text{CONH}_2\\|\\\text{CH}_2\\|\\\text{CH}_2\\|\\\text{H}_2\text{N}\diagup\text{CH}\diagdown\text{COOH}\end{array}$$

Among the rarer amino acids which do not belong to the twenty most common compounds of this type is phosphoserine, found in phosphoproteins such as casein and pepsin:

$$O=P(OH)(OH)-O-CH_2-CH(NH_2)-COOH$$

and thyroxine, a hormone and constituent of the protein thyroglobulin elaborated by the thyroid gland:

$$HO-C_6H_2I_2-O-C_6H_2I_2-CH_2-CH(NH_2)-COOH$$

TABLE 1-1
The Twenty Amino Acids Commonly Found in Proteins

Aliphatic hydrocarbon side chains				
Glycine (Gly)	Alanine (Ala)	Valine (Val)	Leucine (Leu)	Isoleucine (Ile)
H$_2$N–CH$_2$–COOH	CH$_3$–CH(NH$_2$)–COOH	(H$_3$C)$_2$CH–CH(NH$_2$)–COOH	(H$_3$C)$_2$CH–CH$_2$–CH(NH$_2$)–COOH	CH$_3$–CH$_2$–CH(CH$_3$)–CH(NH$_2$)–COOH

Acidic side chains		Side chains containing hydroxyl groups	
Aspartic acid (Asp)	Glutamic acid (Glu)	Serine (Ser)	Threonine (Thr)
HOOC–CH$_2$–CH(NH$_2$)–COOH	HOOC–CH$_2$–CH$_2$–CH(NH$_2$)–COOH	HO–CH$_2$–CH(NH$_2$)–COOH	CH$_3$–CH(OH)–CH(NH$_2$)–COOH

Basic side chains		
Lysine (Lys)	Arginine (Arg)	Imidazole group in side chain: histidine (His)
H$_2$N–CH$_2$–CH$_2$–CH$_2$–CH$_2$–CH(NH$_2$)–COOH	H$_2$N–C(=NH)–NH–CH$_2$–CH$_2$–CH$_2$–CH(NH$_2$)–COOH	imidazole–CH$_2$–CH(NH$_2$)–COOH

2. The Chemical Structure of Proteins

TABLE 1-1 (Continued)

Aromatic side chains		Heterocyclic side chain containing indole group: tryptophan (Trp)
Phenylalanine (Phe)	Tyrosine (Tyr)	

(Structures of Phenylalanine, Tyrosine, and Tryptophan)

Sulfur-containing side chains		
Cysteine (Cys)	Cystine (Cys-S-S-Cys)	Methionine (Met)

(Structures of Cysteine, Cystine, and Methionine)

Imino acids containing pyrrolidine ring as core	
Proline (Pro)	Hydroxyproline (Hypro)

(Structures of Proline and Hydroxyproline)

Certain amino acids are never found as constituents of proteins. Some are intermediates in metabolic reactions while others are found in various low-molecular-weight peptides including certain antibiotics. Widely distributed amino acid intermediates are α-aminovalerianic acid or norvaline:

$$CH_3-CH_2-CH_2-C{\overset{H}{\underset{NH_2}{\diagdown COOH}}}$$

and ornithine:

$$H_2N-CH_2-CH_2-CH_2-C{\overset{H}{\underset{NH_2}{\diagdown COOH}}}$$

As examples of amino acids from polypeptide antibiotics we may cite *N*-hydroxylysine:

$$\text{HO—NH—CH}_2\text{—CH}_2\text{—CH}_2\text{—CH}_2\text{—C}\begin{smallmatrix}\text{H} & \text{COOH}\\ & \\ & \text{NH}_2\end{smallmatrix}$$

and lanthionine:

$$\begin{smallmatrix}\text{HOOC}\\ \\ \text{H}_2\text{N}\end{smallmatrix}\text{CH—CH}_2\text{—S—CH}_2\text{—C}\begin{smallmatrix}\text{H} & \text{COOH}\\ & \\ & \text{NH}_2\end{smallmatrix}$$

All amino acids except glycine contain asymmetric carbon atoms, namely, the α-carbon to which the —COOH, —NH$_2$, and —R groups are attached. There are two stereoisomers of each amino acid, the left (L-) and the right (D-) configurations, which are determined by the way in which the various groups are distributed around the α-carbon atom. An example of an L-amino acid, α-aminobutyric acid, is illustrated in Fig. 1-1. Of the two possible amino acid enantiomorphs (stereoisomers), the left

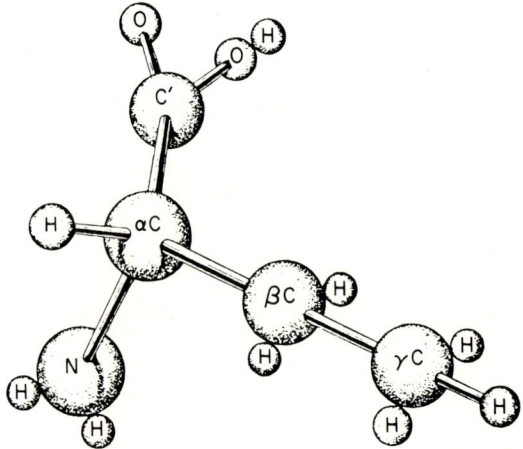

Fig. 1-1. Model of an L-amino acid molecule.

configuration is by far the most common in nature, particularly in proteins, where the right form is all but unknown. How and why the complete separation of stereoisomers in living tissue was started remains an enigma.

Nonetheless, D-amino acids have been discovered in the cell walls of bacteria as well as among their metabolic products. Furthermore, D-amino acids are found in antibiotics which constitute a cellular defense directed against other microorganisms. Antimetabolites containing these stereoisomers can bind to certain enzymes so as to inhibit their metabolic activity.

It is difficult to understand the way in which a cell selects L-amino acids as constituents for its proteins, since both stereoisomers are thermodynamically and kinetically identical. Many attempts have been made to achieve an asymmetric synthesis of polymers *in vitro* from racemic mixtures of monomeric stereoisomers. The desired product would be enriched for one of the enantiomorphs. Among the techniques used were photochemical synthesis with circularly polarized light and catalytic synthesis on complexes devoid of symmetry planes. The most successful experiment of this type was performed by Natta and associates (1), who used the second procedure to obtain synthetic polymers in which one stereoisomer was significantly more frequent than the other, comprising up to 70% of the total residues. By contrast, the proportion of L-amino acids in proteins is 100%. We can only speculate that this remarkable phenomenon originally occurred as the result of very rare large-scale fluctuations associated with the origin of life.

Proteins frequently consist of a single polypeptide chain. However, it is not uncommon to encounter macromolecular proteins containing several polypeptides, such as hemoglobin, which possesses four individual chains. Even insulin, the smallest protein known with only 51 amino acid residues, contains two chains in the active state. In many cases, the multiple chains are joined together by covalent or electrostatic bonds between their side groups. The most widely distributed type of "molecular bridge" is the disulfide bond that forms between cysteine residues. Oxidation of two cysteine residues through their —SH groups produces cystine:

$$2 \; \underset{HOOC}{\overset{H_2N}{>}}CH-CH_2-SH \xrightarrow[-H_2O]{+O} \underset{HOOC}{\overset{H_2N}{>}}CH-CH_2-S-S-CH_2-\underset{COOH}{\overset{H \;\; NH_2}{C<}}$$

For the most part, any polypeptide chains containing free cysteine can be cross-linked or "vulcanized" in this way with the aid of an oxidizing agent.

Another type of bridge which is specific for phosphorproteins results from the formation of a double ester of phosphoric acid with serine:

$$\underset{COOH}{\overset{H_2N}{>}}CH-CH_2-O-\underset{OH}{\overset{\overset{O}{\|}}{P}}-O-CH_2-\underset{COOH}{\overset{H \;\; NH_2}{C<}}$$

This kind of cross-linking occurs in casein and pepsin.

3. Analysis of Amino Acid Composition

A fundamental analytical parameter for any protein is its *amino acid composition*, the number of amino acid residues of each type that comprise the macromolecule (2–4). Elementary analysis, a more usual means of characterizing organic compounds, yields relatively little information in this case since most proteins contain about the same amounts of nitrogen (15–18%), hydrogen (6.5–7.3%), carbon (50–55%), oxygen (21–24%), and sulfur (0–2.4%).

Standard methods are presently available for determining the composition of proteins with a precision approaching 0.1% for most amino acids. The protein is first completely hydrolyzed to amino acids, usually by heating in a solution of 6 N hydrochloric acid for 10 to 15 hours. After hydrolysis the amino acid mixture is introduced into a chromatographic column where all the amino acids can be fully resolved from one another as they are eluted. The chromatographic analysis of protein hydrolyzates has been highly automated and can be completed by a special machine called an amino acid analyzer in 2 to 4 hours. Owing to the development of these methods and instruments, the amino acid analysis of proteins, extremely time-consuming and inexact just a few years ago, has become a commonplace and precise procedure. The method and the apparatus were devised mainly by Spackman, Stein, and Moore (5).

The principle underlying the amino acid analyzer is the use of *ion-exchange resins*. Amino acids are separated by chromatography on polystyrene resins containing —SO_3H groups. This resin is a strong electrolyte since its acidic groups are dissociated to as great an extent as those of sulfuric acid. At any pH the resin behaves as a macromolecular anion (polyanion) because it bears —SO_3^- groups; the counterions, H^+ in this case, are dissolved in the surrounding medium. The resin is normally used in the sodium form, however, so that it does not alter the pH of the solution. To change the resin from the H^+ to the Na^+ form, it is washed with NaOH and the functional groups are converted to —$SO_3^-Na^+$.

When an amino acid residue is adsorbed to the resin, it replaces a Na^+ ion, but in order for this to occur, the amino acid must be positively charged. This is achieved by allowing adsorption to take place from a buffer at pH 2. At this pH, all the carboxyl groups are undissociated according to the following reaction:

$$-COO^- + H^+ \longrightarrow -COOH$$

while all the amino groups are charged:

$$-NH_2 + H^+ \longrightarrow -NH_3^+$$

Thus, all the amino acids behave as cations under these conditions and are adsorbed to the resin in exchange for Na^+ ions. Amino acids with two amino groups will bear two positive charges in acid medium and are therefore adsorbed more strongly. Elution of amino acids from the column is begun by passing buffer through it at a rate of from 50 to 100 ml per hour per cm^2 of cross section. Amino acids are separated from one another as they move down the column because of differences in their electrochemical characteristics.

Since amino acids are weak electrolytes, each containing at least two ionizable groups, they continually oscillate between their charged and uncharged forms. The equilibrium for the electrolytic dissociation:

$$-COOH \rightleftharpoons -COO^- + H^+$$

is determined by the constant:

$$K_1 = \frac{C_{COO^-} \cdot C_{H^+}}{C_{COOH}} \quad (1\text{-}1)$$

For the sake of convenience, this parameter is usually expressed in terms of its negative logarithm, pK_1:

$$pK_1 = -\log_{10} K_1 \quad (1\text{-}2a)$$

3. Analysis of Amino Acid Composition

one of the principal defining parameters of a weak electrolyte. A second dissociation is associated with the process:

$$-NH_3^+ \rightleftharpoons -NH_2 + H^+$$

to which another constant, K_2, and the corresponding pK_2

$$pK_2 = -\log_{10} K_2 \tag{1-2b}$$

apply. The values of both pK_1 and pK_2 vary somewhat from amino acid to amino acid, depending on the nature of the side group, —R:

$$R-C\begin{matrix}H\\ \\ \end{matrix}\begin{matrix}COOH\\ \\ NH_2\end{matrix}$$

attached to the amino acid. Moreover, the side chains of certain amino acids contain dissociable groups characterized by a third constant, pK_3.

The value of pK_1 lies between pH 3.5 and 5.5, while pK_2 falls between pH 7 and 9. At pH values less than pK_1, the amino acid is positively charged (cationic), since dissociation of the —COOH group is prevented. At pH's greater than pK_2, the amino acid is an anion, since the amino group is not charged. When the pH lies between pK_1 and pK_2, the amino acid simultaneously bears two opposite charges:

$$R-C\begin{matrix}H\\ \\ \end{matrix}\begin{matrix}COO^-\\ \\ NH_3^+\end{matrix}$$

and is called a *zwitterion* or dipolar ion. Understandably, an amino acid can no longer remain adsorbed to the polyanion in its zwitterion form, since the anionic group is repulsed from the surface of the resin. Consequently, as the pH of the solution nears the pK_1 of an amino acid, it will start to elute from the column. Since the pK_1 of amino acids varies within a range of 2 pH units, it is possible to wash off the entire series consecutively with two buffers, one at pH 3.5 and a second at pH 5.5. The basic amino acids lysine and arginine, which contain two amino groups, must be washed off separately because they are adsorbed through the ionized ε-amino or guanidino groups, respectively, located in their side chains. When the carboxyl groups of these two compounds become ionized they are not eluted from the resin unless basic groups in their side chains are also discharged. The use of a buffer at pH 11 accomplishes this, resulting in the removal of both basic amino acids from the polyanionic resin.

In the amino acid analyzer, very precise micropumps force buffer through the column during elution. The transition from one buffer to another takes place automatically at predetermined intervals. After leaving the column, the eluant containing amino acids is mixed with a solution of ninhydrin (triketohydrindene hydrate), a specific reagent for the detection of amino acids. The amino acid solution containing ninhydrin at a final concentration of 0.1% passes through a fine (1-mm diameter) plastic tube 30 m in length which is maintained at a temperature of 100°C. At this stage ninhydrin reacts with amino acids to form a derivative that is usually violet in color. A yellowish product results in the case of the imino acids proline and hydroxyproline, however. The colored solution next passes through the cuvette of a colorimeter whose output is connected to a recorder. The recorder pen draws a curve on a paper chart corresponding to the optical density of the eluate, thereby marking the positions and quantities of the separated amino acids. The amount of color developed is practically the same per mole of any amino acid with the exception of proline and hydroxyproline. The colorimeter is therefore calibrated with just one of the amino acids and it is necessary to introduce only small corrections on the order of a few percent to quantitate each of the other amino acids. Proline and hydroxyproline must be assayed separately.

Since the rate at which amino acids elute from the column and pass through the colorimeter is constant from run to run, all that one must do to identify the amino acids in an unknown mixture is to compare their order of elution with that of known amino acids. Their relative amounts are determined from the area of the colorimeter tracings. The amino acid analyzer of Stein and Moore has thus resolved one of the fundamental difficulties of protein chemistry.

4. Determination of Amino Acid Sequence in the Protein Chain

The structural formula, or *primary structure*, of a protein is much more difficult to establish than its amino acid composition. This task entails the determination of the sequence of all amino acids along each of the component polypeptide chains as well as the location of all chemical "bridges" which connect the chains together. At present we know the structural formulas of many proteins, including insulin, pancreatic ribonuclease, the coat protein of tobacco mosaic virus (TMV), cytochrome c, trypsinogen, hemoglobin, and lysozyme, and a large number of others (6).

In order to understand the principles underlying sequence analysis, imagine the polypeptide chain to be a succession of similar repeated elements such as letters in a very long word. Some of the letters, or amino acids, are repeated very often, some are encountered only rarely, but we can identify them only individually, not when they are part of the word. The polypeptide must therefore be broken down into shorter peptides containing three, four, or five amino acids by partial hydrolysis with acid or enzymes. These small peptides, or syllables, to pursue the analogy, are much easier to deal with. More important than the precise method of fragmenting the polypeptide is that the procedure be reproducible and that the resulting peptides be neither too large nor too small. The mixture of small fragments must now be separated by chromatography, either on paper or on ion-exchange resins.

Once the peptides have been purified, their amino acid composition can be determined, but the problem of establishing their amino acid sequence remains. One approach is to label the free amino group at one end of the peptide, or the free carboxyl at the other, with specific reagents. The peptide is then hydrolyzed to amino acids and the labeled component is identified. This establishes which amino acids are

4. Determination of Amino Acid Sequence in the Protein Chain

located at the ends of the sequence. Suppose that a hypothetical peptide contains the three amino acids B, C, and G, and that B is found to have a free reactive amino group. Furthermore, if the dipeptides BC and CG are found when the tripeptide is partially hydrolyzed under mild acid conditions, its sequence must be BCG (Fig. 1-2a). Peptides consisting of five or six amino acids can also be split up into two or

```
1. -D-D-A-            -B-A-C-D-
2. -B-F-B-C-          -C-D-D-D-
3. -B-C-G-            -D-D-A-
4. -B-A-C-D-          -E-B-
5. -A-J-F-D-          -B-F-B-C-
6. -G-G-              -B-C-G-
7. -G-H-I-A-J-        -G-G-
8. -E-B-              -G-H-I-A-J-
9. -C-D-D-D-          -A-J-F-D-
       (a)                (b)
```

H₂N- B - A - C - D - D - D - A - E - B - F - B - C - G - G - H - I   A - J - F - D - COOH

(c)

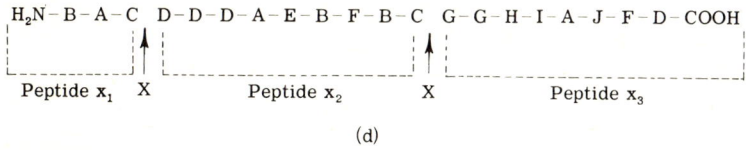

(d)

Fig. 1-2. Scheme illustrating the principles of protein sequence analysis. Each letter stands for one kind of amino acid and dashes represent peptide bonds. Where not indicated, the left end of the peptide is N-terminal and the right end is C-terminal. (a) The sequences of amino acid residues in nine peptides of a hypothetical polypeptide chain. (b) Order of peptides deduced from overlapping sequences. (c) Sequence of amino acid residues. (d) Cleavage of the chain at C residues by enzyme X. See text for further explanation.

three fragments by partial hydrolysis, and these smaller peptides can then be analyzed as described.

When the sequences of all the short peptide fragments have been determined, they must be fit together in their proper order relative to one another. Imagine the polypeptide to consist of twenty amino acid residues and the sequences of its nine component peptides to be those shown in Fig. 1-2a. In addition, B is known to occupy the amino-terminal end of the chain and D the carboxyl-terminal. The entire sequence may now be generated by aligning the short peptides according to their overlapping segments (Fig. 1-2b). The analysis in the illustration is quite straightforward and leads to a unique answer (Fig. 1-2c). Fitting the small peptides together is not always so easy, however, and there frequently remain ambiguous solutions for one or another region of the chain. The proper alignment of peptides 1 and 8, for instance, is only inferred from our knowledge of the N- and C-terminal residues since there are no peptides which overlap the AE sequence. Overlaps between peptides can often be elucidated by hydrolyzing the protein with an enzyme that cleaves it at specific points. In Fig. 1-2d, enzyme X is shown to hydrolyze the chain following C residues. Analysis of the peptide x_2 gives us the AE overlap between peptides 1 and 8. The use of two or more enzymes that split the protein at different but specific amino acid residues is of great assistance in aligning long polypeptide fragments.

The specific chemical methods used to decipher protein structure are quite interesting in themselves. The usual starting point is to determine the number of polypeptide chains in the protein macromolecule. One way to accomplish this is to find the number of different N-terminal or C-terminal amino acids in a purified preparation of the protein.

There are excellent methods available for chemically marking the N-terminal amino acid residues. In one of these, developed by Sanger (7; see also ref. 4), the terminal amino group is reacted with 1-fluoro-2,4-dinitrobenzene (FDNB):

$$\sim\!\!\sim\!\!\sim\! NH_2 + F\!\!-\!\!\!\bigcirc\!\!(NO_2)\!\!-\!\!NO_2 \longrightarrow \sim\!\!\sim\!\!\sim\! HN\!\!-\!\!\!\bigcirc\!\!(NO_2)\!\!-\!\!NO_2 + HF$$

The amino and hydroxy groups in the side chains of lysine, tyrosine, and other amino acids also react with this substance. After complete hydrolysis of the protein in 6 N HCl, the bright yellow FDNB-amino acids, which have the structure,

$$R\!\!-\!\!CH(COOH)\!\!-\!\!HN\!\!-\!\!\!\bigcirc\!\!(NO_2)\!\!-\!\!NO_2$$

can be isolated chromatographically. Since these derivatives are acidic, they can be adsorbed at neutral pH to an anion exchange resin containing quaternary ammonium groups of the type:

$$-\!CH_2\!-\!N^+(CH_3)_3 \cdot Cl^-$$ (repeated)

Internal residues of the polypeptide chain which reacted through their side chains will behave as zwitterions at neutral pH and will not adsorb to the ion exchange resin. The N-terminal residue, or residues if there is more than one polypeptide in the protein, is thus selectively adsorbed.

After eluting the FDNB-amino acid from the resin with dilute acid, the colored compound can be identified by any of several methods. One of the simplest ways is to use paper chromatography to measure the R_f of the compound. The R_f is defined as the ratio of the rate of movement of the compound on the chromatogram to that of the solvent front. Each FDNB-amino acid is characterized by a different R_f so that

4. Determination of Amino Acid Sequence in the Protein Chain

known FDNB derivatives can be used to identify the amino acid in question. Sanger's method requires the presence of 0.01 μmole or about 1 μg of the amino acid per spot on the chromatogram.

Another reaction has been recently employed to label the N-terminal amino acid using 1-dimethylaminonaphthalene-5-sulfonyl chloride (DNSCl):

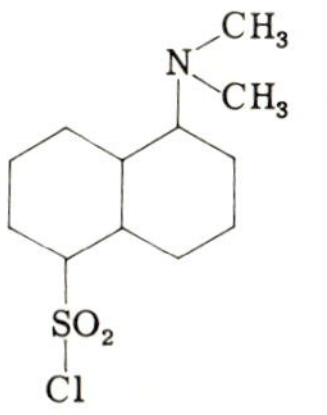

[Gray and Hartley (8)]. This compound reacts with the amino groups of proteins, peptides, amino acids, and also with the phenol group of tyrosine:

$$R-SO_2Cl + R'NH_2 \xrightarrow{-HCl} R-SO_2-NH-R'$$

The products formed possess an intense yellow fluorescence in ultraviolet light, permitting the detection of as little as 10^{-5} μmole or 10^{-3} μg of any one amino acid. With such phenomenal sensitivity, extremely small amounts of material can be used for the analysis. This is particularly important in the investigation of rare and hard-to-prepare proteins. Moreover, the sulfonilamide derivatives are more stable during acid hydrolysis than are FDNB-amino acids.

A third means of identifying N-terminal residues is known as the phenylthiohydantoin method and was worked out by Edman. (9; see also ref. 4). Here, the N-terminal amino acid reacts with phenylisothiocyanate:

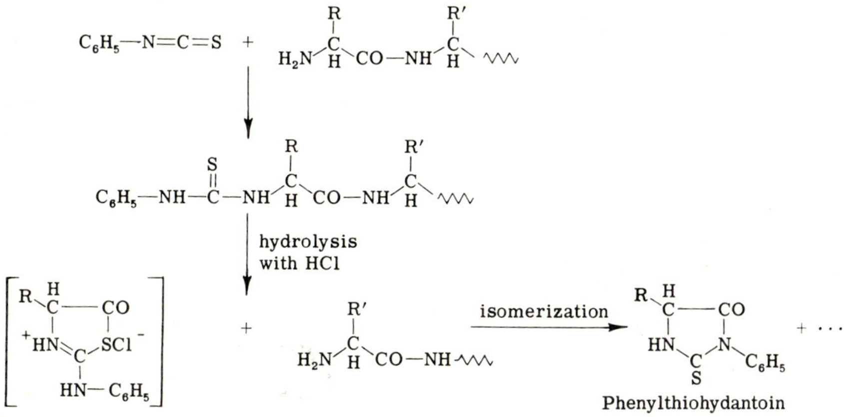

After reaction, the N-terminal phenylthiohydantoin derivatives are selectively cleaved from the end of the polypeptide chain and then chromatographed on paper. The positions of the spots are located either with ultraviolet light or after coloration with iodine. They are then compared with the positions of controls prepared from known amino acids. Under favorable conditions it is possible to carry out the stepwise removal of up to ten to twelve residues from the N-terminus of a protein molecule by the Edman method without hydrolyzing sufficient internal peptide bonds to confuse the results. If a peptide is not too long, the sequence of amino acids in it can be determined fairly easily with the Edman reaction alone. A machine has been designed by Edman to automate sequence analysis by the phenylisothiocyanate technique.

Methods for determining the number of N-terminal groups and for identifying these residues have proved to be quite satisfactory. But techniques are not as straightforward for measuring the number of C-terminal amino acids. One of the methods used to investigate the C-terminal residue is to cleave it enzymatically with purified pancreatic carboxypeptidase. This enzyme can remove amino acids one by one from the C-terminal end of the polypeptide chain. Certain amino acids, however, such as cysteine, proline, and hydroxyproline, are not attacked by this enzyme. Although the carboxypeptidase method is not a universal one, it is nonetheless used more than any other. If a single residue is removed from the end of the chain by keeping incubation times with the enzyme short, it is possible to isolate and identify it by paper or ion-exchange chromatography.

When the number of terminal residues in a protein preparation has been determined, we know how many polypeptide chains comprise the macromolecule, provided that the molecular weight of the protein is also known.

The next question is how to separate polypeptide chains that are linked by disulfide bridges. As Sanger showed in the case of insulin, which is made up of two polypeptides designated A and B, this may be accomplished with performic acid, a strong oxidant (10). Oxidation of the disulfide bridges proceeds according to the following reaction:

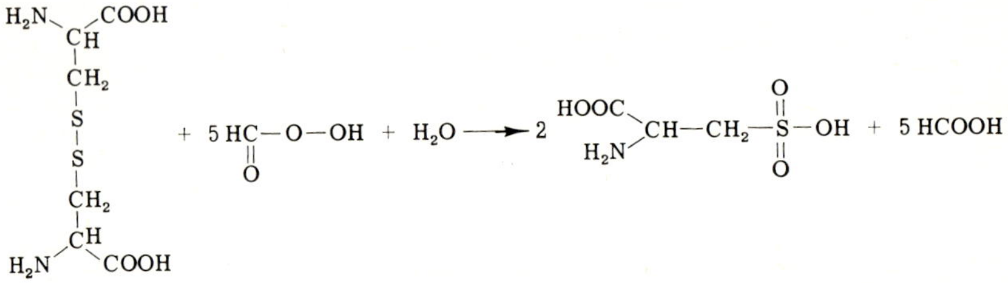

$$\underset{\substack{H_2N \\ }}{\overset{}{\text{CH}}}\overset{COOH}{\underset{\substack{| \\ CH_2 \\ | \\ S \\ | \\ S \\ | \\ CH_2 \\ | \\ CH \\ H_2NCOOH}}{}} + 5\,HC(=O)-OH + H_2O \longrightarrow 2\,\underset{H_2N}{\overset{HOOC}{\diagdown}}CH-CH_2-\underset{O}{\overset{O}{\overset{\|}{S}}}-OH + 5\,HCOOH$$

As a result, the polypeptide chains acquire strongly acidic —SO$_3$H groups which completely dissociate in the manner of sulfuric acid. The individual chains can be separated and purified by ion-exchange chromatography.

A drawback in this method of disrupting disulfide bonds lies in the fact that the amino acid tryptophan is irreversibly destroyed in the oxidation. Hence, —S—S— bridges are frequently broken by reduction, with mercaptoethanol, for instance,

$$R—S—S—R' + 2HS—C_2H_4—OH \rightarrow R—SH + R'SH + HO—C_2H_4—S—S—C_2H_4—OH$$

rather than by oxidation. Under these conditions, cystine is reduced to two cysteine residues. Since the sulfhydryl groups are extremely reactive, they are immediately treated with iodoacetamide,

$$R—SH + R'SH + 2\,ICH_2CONH_2 \xrightarrow{-2HI} R—S—CH_2CONH_2 + R'—S—CH_2CONH_2$$

4. Determination of Amino Acid Sequence in the Protein Chain

in order to prevent their oxidation. While insulin is characterized by two interchain disulfide bonds which join the A and B polypeptides, ribonuclease, consisting of just one polypeptide chain, is drawn tightly together by four *intrachain* disulfide bridges. In the latter case, the disulfides must be disrupted prior to sequence analysis. At a later stage in the analysis, however, the disulfides must be maintained intact so that the specific cysteine residues which participate in each one can be located.

Hemoglobin, a protein with a total molecular weight of 66,000, provides an example of a different kind of subunit binding. In hemoglobin, there are two subunits, weakly held together by what appear to be salt bonds. Each of the subunits contains two different polypeptide chains, the α-chain and the β-chain, also joined by ionic bonds. In order to separate the subunits into individual α- and β-chains, the protein is first adsorbed on the anion exchanger, carboxymethylcellulose, and then eluted with a formic acid–pyridine buffer (pH 2). During elution, the concentration of the buffer, initially 0.02 M formic acid and 0.02 M pyridine, is gradually increased in strength until it reaches ten times the starting concentration. This technique is called gradient elution. The α- and β-chains are washed from the resin at different buffer concentrations and each of them may thus be obtained in pure form.

Now that the problem has been reduced to the study of separate polypeptide chains, the method of limited hydrolysis is used to cleave the macromolecule into peptide fragments. Of the proteolytic or digestive enzymes which can be used for this purpose, *trypsin* is particularly suitable, since it is distinguished by a very strict specificity. Trypsin splits peptide bonds only where the basic amino acids lysine and arginine occur, and the reaction is such that these amino acids become the C-terminal residues of the resulting peptides:

$$\begin{array}{c} \sim\sim\sim \\ CO \\ | \\ NH \\ | \\ CH-(CH_2)_4-NH_2 \\ | \\ CO \\ | \\ NH \\ | \\ CH-R \\ \sim\sim\sim \end{array} + H_2O \longrightarrow \begin{array}{c} \sim\sim\sim \\ CO \\ | \\ NH \\ | \\ CH-(CH_2)_4-NH_2 \\ | \\ COOH \end{array} + \begin{array}{c} \sim\sim\sim \\ NH \\ | \\ CO \\ | \\ CH-R \\ | \\ NH_2 \end{array}$$

Limited digestion with trypsin reaches its maximum when all peptide bonds in the protein involving either lysine or arginine, but no others, have been hydrolyzed. This treatment usually produces ten to twenty different polypeptide fragments, depending on the size of the protein and the distribution of basic amino acids in it. One of the main characteristics of trypsin hydrolysis is its high degree of reproducibility. If the initial protein preparation and the trypsin are both quite pure, then the same fragments are obtained in identical proportions every time the hydrolysis is carried out.

The mixture of tryptic peptides must now be fractionated so as to obtain each in a pure state. Once again ion-exchange chromatography is used, this time with a resin containing weakly acidic carboxyl groups. A portion of the peptides in the trypsin hydrolysate may be di- and tripeptides, but free amino acids are never found. This is because trypsin is an *endopeptidase* which hydrolyzes a peptide bond only if there is another peptide bond located next to it, that is, only when the sensitive bond is within the chain. Older work indicating the presence of free amino acids in trypsin hydrolyzates resulted from the use of impure hydrolytic enzymes.

Rather long peptide fragments containing on the order of 20 to 30 amino acids may also be found in the hydrolyzate. The length of any given fragment is of course determined by the position of lysine and arginine residues along the protein backbone. Carboxy resins are appropriate for separating the longer peptides since they frequently adsorb irreversibly to sulfo resins and thus cannot be eluted quantitatively. Carboxy resins lose their acidic characteristics between pH 3 and 4 because their carboxyl groups are discharged. Under these conditions, practically all peptides can be eluted from the resin. Thus, when polypeptides are chromatographed on carboxy resins and eluted by a pH gradient running from pH 7 to pH 2, all fragments from the hydrolyzate are obtained in quantitative yield.

Another method for the separation of peptides is called *fingerprinting* and involves the use of both electrophoresis and chromatography to achieve this end (11). In this technique, a small amount of the peptide mixture in solution is applied to the corner of a large sheet of chromatography paper, previously moistened with buffer containing 1.5% pyridine, 2.5% acetic acid, and 5% n-butanol, pH 4.5. Up to 3 mg of the peptide mixture can be separated on chromatography paper 0.5 mm thick. The peptides are first subjected to electrophoresis along the sheet. When an electric field of from 15 to 25 V/cm is applied for several hours, the initial polypeptide spot is fractionated into several spots according to the charge they bear. If the peptide contains amino groups in its side chains and thus carries a net positive charge, it will move more rapidly toward the cathode and will be located farther from the origin than less highly charged peptides.

Electrophoresis does not result in complete fractionation of the polypeptides, however. Therefore, after electrophoresis, but before developing the spots, chromatography is carried out in a direction perpendicular to electrophoresis. When the paper sheet is dry, its edge is next immersed in a bath containing the chromatography solvent, a mixture of 5 volumes n-butanol, 2 volumes of glacial acetic acid, and 2 volumes of H_2O, pH $\approx$ 3. This mixture must be prepared 2 weeks in advance to allow the formation of an equilibrium concentration of the ester. Due to the movement of the eluting solvent along the paper, which requires about 24 hours for a sheet 80 cm long, each electrophoretic spot is resolved into several different spots in the second dimension. Separation by paper chromatography does not result from charge differences, but from the varied affinity of the peptides for the organic solvent. Those peptides which dissolve best in the butanol–acetic acid buffer move most rapidly as the solvent proceeds along the paper and thus have a larger R_f.

After the paper sheet is dried once again, it is sprayed with a weak solution of ninhydrin (0.1%). The colored spots which mark the position of each peptide on the paper are developed by heating at low temperatures for 10 hours in the dark. Under the gentle conditions used to develop the spot, only the terminal amino group of each peptide is chemically destroyed. Hence, each spot may be cut out and the peptide eluted from it with hot water; the purified peptides are thus obtained with only minimal losses.

For a given protein hydrolyzed in a specific manner the resulting pattern of peptide

4. Determination of Amino Acid Sequence in the Protein Chain

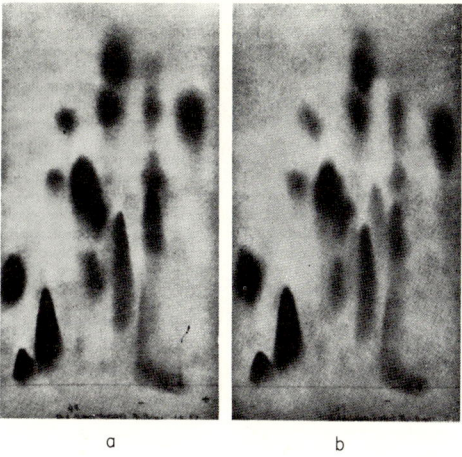

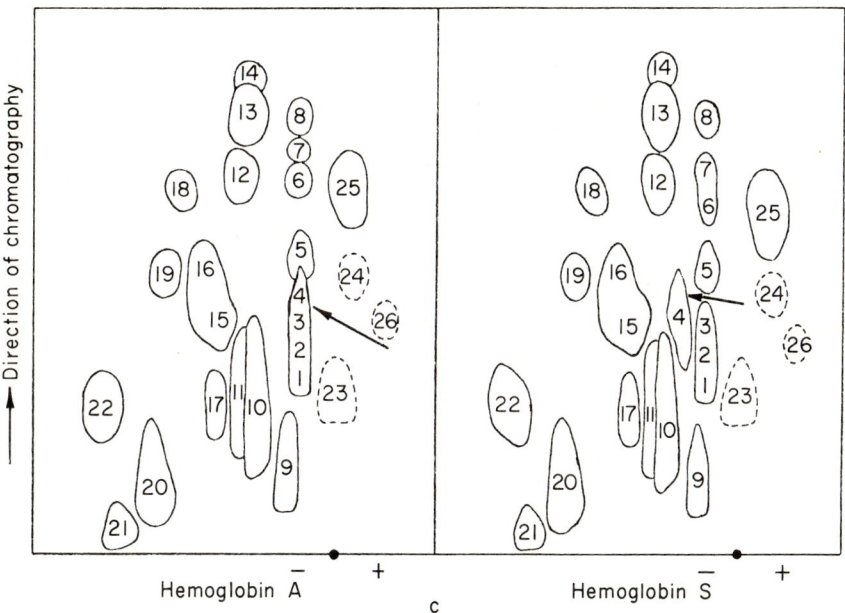

Fig. 1-3. Fingerprints of trypsin digests of hemoglobin A and S. (a) Normal hemoglobin A. (b) Abnormal sickle-cell hemoglobin S. (c) Tracings showing the positions of 26 peptides on each fingerprint. The protein hydrolyzate was pipetted onto the paper at the black dot on lower margin, and an electric field was applied in the direction indicated. After electrophoresis, the peptides were further separated by ascending chromatography in the second dimension and the fingerprints were then developed with ninhydrin. Arrows point to the single peptide difference between the normal (A) and abnormal (S) hemoglobins. This difference was later found to result from the replacement of a single amino acid residue in hemoglobin S. From Ingram (11).

spots, called a fingerprint, is extremely distinctive. Only closely related proteins will have similar fingerprints and even at that, differences in a single amino acid residue can be detected. As an example, Fig. 1-3 presents a fingerprint of hemoglobin after hydrolysis with trypsin. If measures are taken to ensure that procedures and materials do not vary the method of fingerprinting yields strikingly reproducible results. The positions of the respective spots fall within 3 to 5 mm of each other from experiment to experiment.

Now the task becomes one of investigating each of the peptides in the enzymatic hydrolyzate. First, an amino acid analysis of the fragments is performed and then either the Sanger or Edman technique is used to determine the N-terminal amino acids. The sequence of short peptides, containing up to five residues, may be deciphered with relative ease by combining sequential application of the Edman reaction to the N-terminus with carboxypeptidase digestion of the C-terminus. Several color reactions for specific amino acids are also useful at this stage. With the help of such reactions it is possible to discover which polypeptides contain some of the rarer or more complex amino acids and sometimes this leads to further structural information. Thus, for example, if cysteine is identified in a peptide, the presence of a disulfide bond is strongly indicated.

The longer tryptic peptides require further fragmentation by partial acid hydrolysis or with a second proteolytic enzyme such as chymotrypsin, thermolysin, pepsin, subtilisin, or papain. Unfortunately, this last group of proteases lacks the specificity of trypsin. Chymotrypsin, for instance, attacks aromatic amino acids and tryptophan most readily of all, but it also hydrolyzes bonds between other amino acids at a lesser rate. In such cases, it is necessary to severely restrict the extent of the reaction by proper choice of incubation time, pH, temperature, and enzyme concentration. Besides the use of proteolytic enzymes, polypeptide chains can also be cleaved by purely chemical methods. Particularly useful is reaction with cyanogen bromide, $N{\equiv}C{-}Br$, which specifically attacks methionine residues and converts them to sulfonium derivatives (12). Under acidic conditions the modified methionine spontaneously decomposes to homoserine lactone via an iminolactone intermediate with the rupture of an adjacent peptide bond:

$$\begin{array}{c}|\\CHR\\|\\NH\\|\\Br^-\quad O{=}C{-}CH{-}NHR'\\H_3C{-}\overset{+}{S}{-}CH_2{-}CH_2\\|\\C{\equiv}N\end{array}\quad\xrightarrow{+H_2O}\quad\begin{array}{c}|\\CHR\\|\\{}^+NH_3\cdot Br^-\\\\O\quad NHR'\\\diagdown\;\diagup\\C{-}CH\\O\quad|\\\diagdown\;\\H_2C{-}CH_2\end{array}\quad+\;H_3C{-}S{-}C{\equiv}N$$

Once small peptide fragments have been obtained from the initial large polypeptide, they can be separated and the N-terminal amino acid of each one can be determined. By examining the overlapping sequences of various sets of peptides, the full sequence of the original tryptic peptide can be established (Fig. 1-2b).

But even after finding the amino acid sequence in each of the tryptic peptides, we cannot immediately write down the structural formula of the protein from which

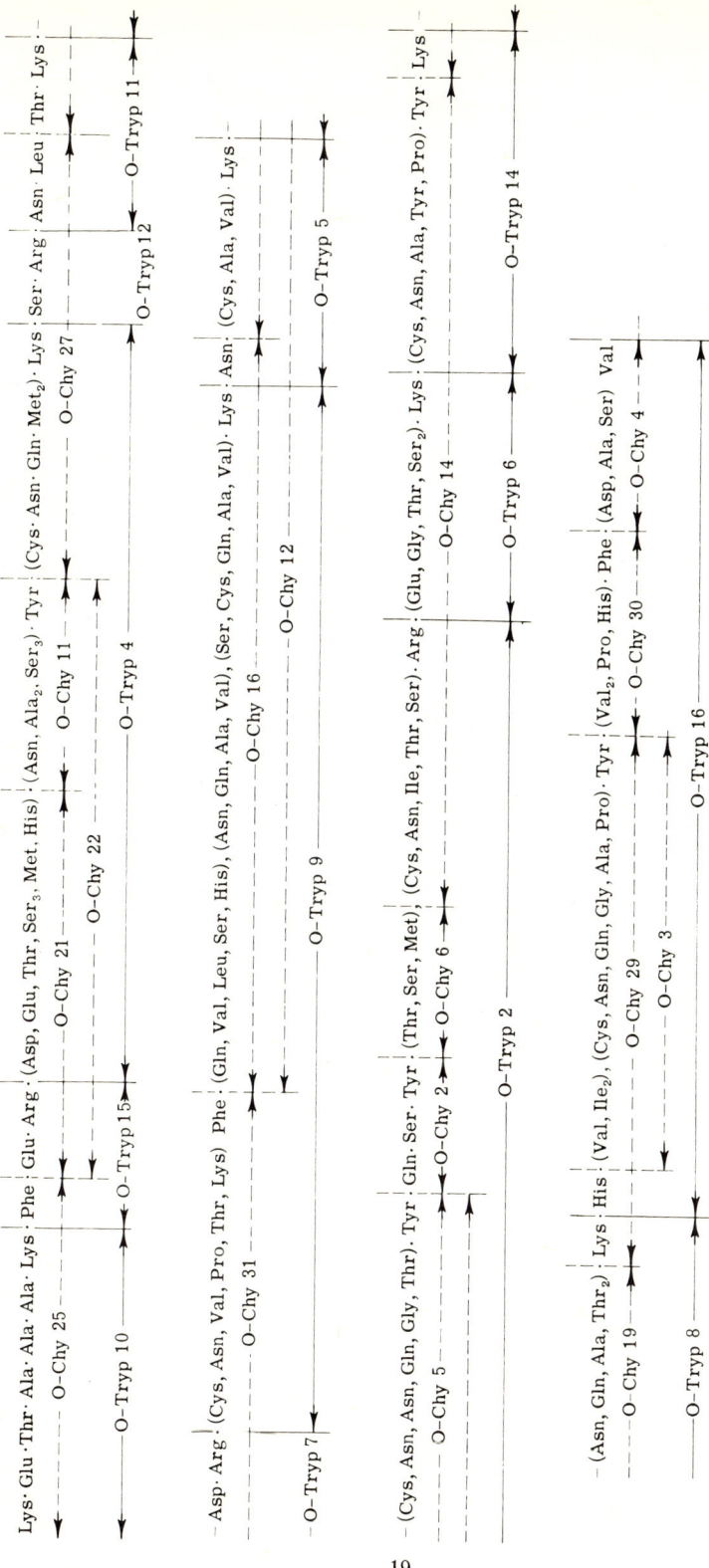

Fig. 1-4. A preliminary primary structure for pancreatic ribonuclease demonstrating the way in which overlapping peptides from tryptic and chymotryptic hydrolyzates were used to reconstruct the sequence. The points of cleavage by trypsin (Tryp) and chymotrypsin (Chy) are indicated by solid and dashed lines, respectively. The sequence of amino acids within parentheses has not been elucidated at this stage of the analysis. Abbreviated symbols for amino acids generally correspond to the first three letters of name (see Table 1-1). Adapted from Hirs *et al.* (13).

Fig. 1-5. The sequence of amino acid residues in bovine pancreatic ribonuclease. Asn stands for asparagine and Gln for glutamine. From Smyth *et al.* (14).

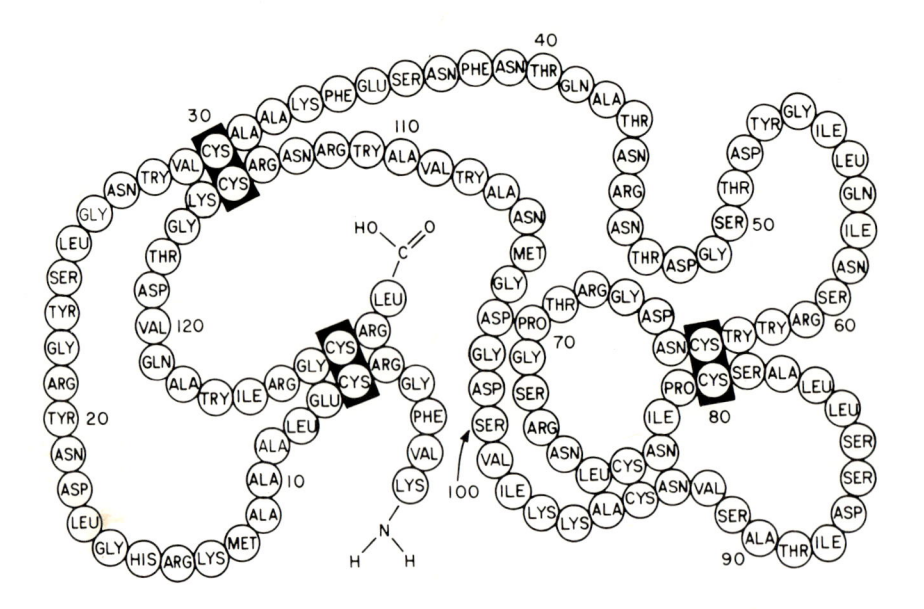

Fig. 1-6. The primary structure of egg-white lysozyme indicating the positions of the four disulfide bonds. Asn and Gln denote asparagine and glutamine, respectively. From Canfield and Liu (15).

A-Chain

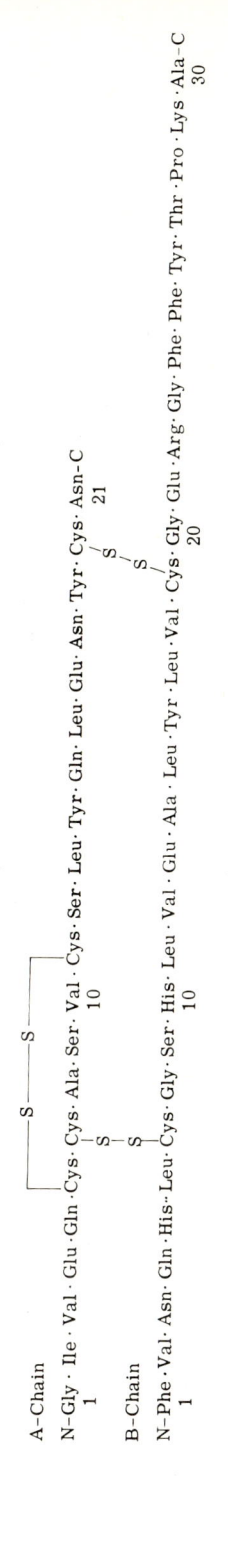

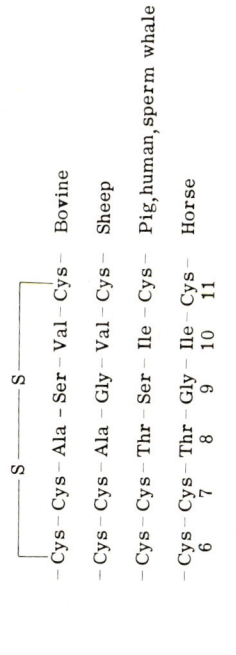

```
                    ┌──S─────S──┐
N-Gly·Ile·Val·Glu·Gln·Cys·Cys·Ala·Ser·Val·Cys·Ser·Leu·Tyr·Gln·Leu·Glu·Asn·Tyr·Cys·Asn-C
 1                            │                 10                                  S  21
                              S                                                     │
                              │                                                     S
B-Chain                       S                                                     │
N-Phe·Val·Asn·Gln·His·Leu·Cys·Gly·Ser·His·Leu·Val·Glu·Ala·Leu·Tyr·Leu·Val·Cys·Gly·Glu·Arg·Gly·Phe·Phe·Tyr·Thr·Pro·Lys·Ala-C
 1                      10                                       20                                              30
```

```
    ┌──S─────S──┐
   -Cys- Cys- Ala - Ser- Val- Cys-    Bovine
   -Cys- Cys- Ala - Gly- Val- Cys-    Sheep
   -Cys- Cys- Thr - Ser- Ile- Cys-    Pig, human, sperm whale
   -Cys- Cys- Thr - Gly- Ile- Cys-    Horse
     6    7    8     9   10   11
```

Number of residue in A-chain

Fig. 1-7. The amino acid sequence of bovine insulin (above) as worked out by Sanger and associates (16). The structures of insulins from several other mammalian sources (below) are identical, with the exception of residues within the cystine loop of the A chain shown in lower part of the illustration (17). In addition, the C-terminal alanine residue of the B chain is replaced by threonine in human insulin.

they were derived. The difficulty is that we do not know how the various peptides are arranged relative to one another along the chain. We must therefore resort to fragmenting the protein with cyanogen bromide or with a second protease whose specificity differs from that of the first, so as to to yield a new and independent set of peptides in which we can find the sequences overlapping two adjacent tryptic peptides (Fig. 1-2c). Although the entire amino acid sequence of the protein can often be established after examining two independent sets of peptides, there generally remain a certain number of ambiguities which can be resolved only by further varying the methods used to cleave the protein into peptides. Figure 1-4 shows the juxtaposition of peptides obtained from ribonuclease after digestion by trypsin and chymotrypsin. When this experiment was performed in 1956, the amino acid sequence of the individual peptides had not been completely worked out; where only the amino acid composition was known, the symbols for the residues are enclosed in brackets. The complete amino acid sequence of ribonuclease as it finally emerged, including location of its four disulfide bridges, is illustrated in Fig. 1-5.

The general approach to determining the structural formula of any protein molecule is similar to the one outlined here. The task is not an easy one; despite the use of modern analytical techniques, it may take years to decipher the sequence of one protein. Naturally there exist many modifications of the procedures described here for digesting the protein and for separating its fragments by chromatography and

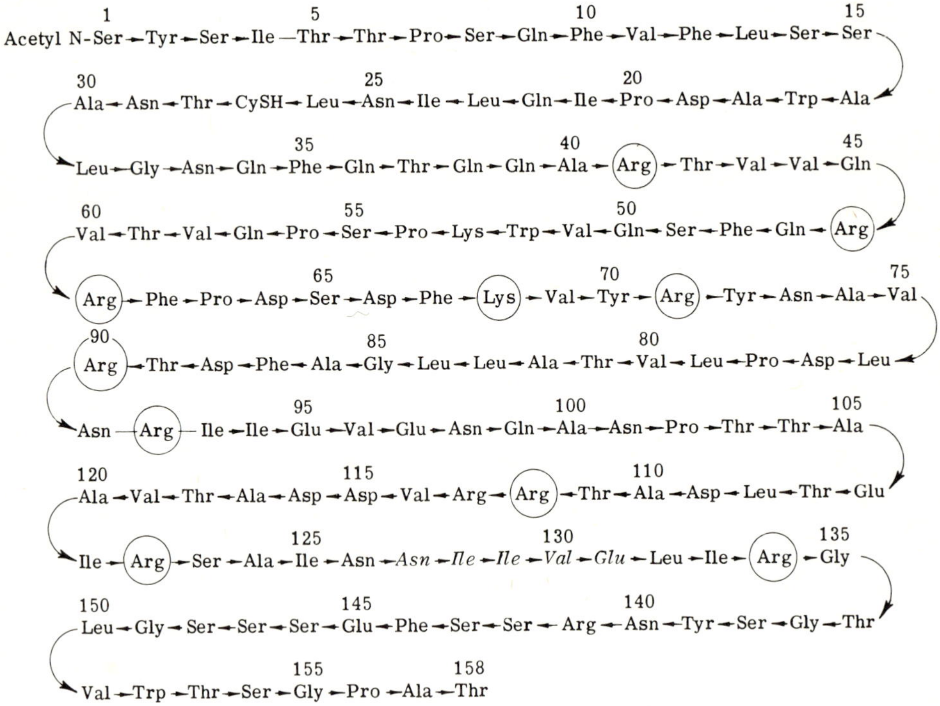

Fig. 1-8. Sequence of the 158 amino acid residues in the coat protein of tobacco mosaic virus. Note the absence of disulfide bonds in this protein. From Tsugita *et al.* (18) as modified by Anderer *et al.* (19).

electrophoresis. Although these methods have already proved extremely effective, the entire process will undoubtedly be simplified and accelerated in the future. Nonetheless, determination of protein structure has proved to be a fruitful endeavor, as evidenced by the many structural formulas presently known. In addition to ribonuclease, illustrated in Fig. 1-5, the structures of three other proteins of biological interest are presented in Figs. 1-6, 1-7, and 1-8.

5. The Types of Macromolecular Structure

Even though proteins are high-molecular-weight polymers, they possess a relatively compact structure. In this regard, proteins are quite sharply distinguished from ordinary linear polymers. The latter generally consist of loose random coils of considerable spatial extension in which the solvent occupies up to 99% of the total volume. This structure accounts for the rather large linear dimensions and for the significant viscosity which solutions of these compounds display. Soluble proteins are quite different. They are globular, or tightly packed, and contain a minimal amount of bound water, on the order of 20 to 30%. Hence protein molecules occupy much less space and increase the viscosity of solutions to a much smaller extent than do linear polymers of the same molecular weight. Another difference between proteins and ordinary linear polymers is that the former, by virtue of the complexity of their structure and of the variety of molecular forces acting within them, exhibit several types of spatial organization, some of which are illustrated in Fig. 1-9.

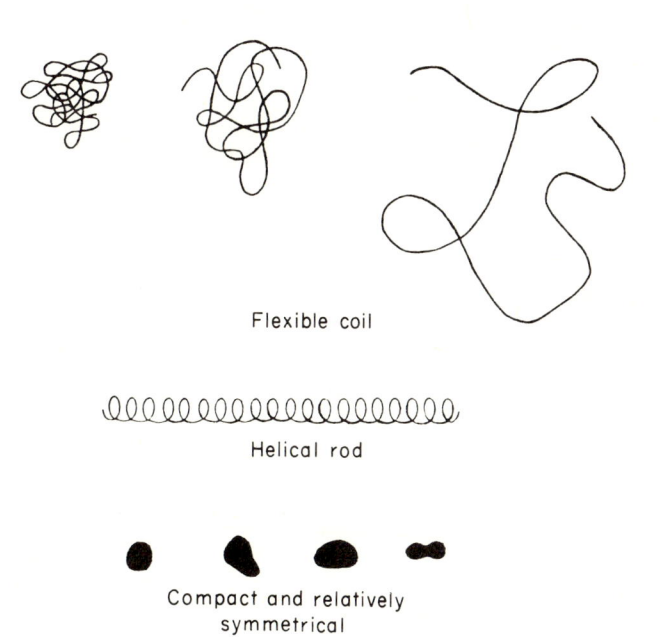

Flexible coil

Helical rod

Compact and relatively symmetrical

Fig. 1-9. Schematic diagrams of some possible configurations of macromolecules in solution. The fully extended length of each chain is the same and the relative sizes of the illustrations correspond to the relative space occupied by the molecules in solution. From Tanford (20), p. 127.

We have already considered one type of organization, the linear polypeptide chain itself, which is held together by covalent chemical bonds. Using the terminology of Linderström-Lang, we shall call this type of organization the *primary structure* of the protein. Further, *hydrogen bonds* are formed between residues of the polypeptide chain and these are not disrupted even by water. In this regard the peptide groups themselves are especially important since each of them can form two hydrogen bonds with other similar groups:

$$\begin{array}{c} \text{H} \\ | \\ \sim\!\!\sim\!\!\sim\text{C}-\text{N}\!\sim\!\!\sim\!\!\sim \\ \parallel \\ \text{O} \\ \vdots \\ \text{H} \\ | \\ \sim\!\!\sim\!\!\sim\text{C}-\text{N}\!\sim\!\!\sim\!\!\sim \\ \parallel \\ \text{O} \\ \vdots \\ \text{H} \\ | \\ \sim\!\!\sim\!\!\sim\text{C}-\text{N}\!\sim\!\!\sim\!\!\sim \\ \parallel \\ \text{O} \end{array}$$

The polypeptide chains seek to align themselves in such a way that the maximum number of hydrogen bonds are formed. Since about 1400 cal/mole are released upon formation of such bonds, the more of them that form within the molecule, the lower is its energy. Thus, the residues of the polypeptide chain in proteins do not freely participate in rotatory Brownian movement, as do those of simple polymers. Rather, proteins tend to form rigid, ordered helices, or flat sheets, constructed in such a way that all peptide groups take part in two hydrogen bonds.

If covalent bonds along the peptide chain are first by magnitude of their binding energy, then hydrogen bonds occupy second place in the determination of protein structure. The helical or folded conformation which they evoke is called the *secondary structure* of the protein. Thus in proteins as well as in nucleic acids, we encounter a special class of macromolecules in which large intramolecular forces of a nonchemical nature play an important role. The specific spatial organization of the protein brought about by hydrogen bonds is frequently referred to as intramolecular crystallization, and each macromolecule of this type evidences some of the characteristics of a true crystal. For example, the helical structure which occurs in many proteins with significant hydrogen bonding imparts to the molecule both a short- and long-range order. Furthermore, the mechanical characteristics of this structure prove to be quite unusual: a tightly coiled helical chain is rigid, not flexible as are ordinary linear polymers. As we shall see in what follows, the analogy with a crystal may be carried even further, since the secondary structure is characterized by a melting point.

Since there are a number of factors which disrupt the regularity of secondary structure in proteins, the effects of intrachain hydrogen bonding can be studied best in simple model polypeptides. Some of the reasons for this are enumerated below.

First, residues of the imino acids proline and hydroxyproline are interspersed along the peptide chain. At these residues, the chain takes a sharp turn of about 130°. Such a change of direction is clearly incompatible with the continuation of the helix and at these points the helical structure of the chain is interrupted. In fact, there is a correlation between the proportion of proline and hydroxyproline residues in a protein and the degree of regularity of its secondary structure.

A second reason for irregularity is the presence of disulfide bonds. Disulfide bridges may either join distant points of one helical polypeptide chain or they may join two separate chains together. At the places where these bridges occur, the polypeptide is strained and the secondary structure cannot be maintained.

The free energy of interaction among the side groups of different amino acid residues may be counted as a third factor which sometimes diminishes the regularity of secondary structure. Many side groups contain a considerable number of hydrocarbon radicals whose interactions are accompanied by large increases in entropy and relatively small changes of energy. Just as hydrocarbons suspended in water tend to coalesce so as to decrease their surface free energy, the hydrocarbon radicals of proteins merge and form symmetrical hydrophobic "droplets". A competition exists between the interaction of amino acid side groups and the tendency of the polypeptide chain to form a helix or a pleated sheet.

Thus, the secondary structure of a protein molecule must be pictured as a series of ordered hydrogen-bonded segments connected by irregular regions in which the system of hydrogen bonds between peptide groups is disrupted. The *tertiary structure* of the protein refers to the way in which ordered and irregular segments of the polypeptide chains are folded into a compact globule. The forces which determine the tertiary structure of a protein are, on the one hand, the van der Waals forces among amino acid side groups and, on the other hand, disulfide bridges and other covalent bonds which impart a certain rigidity to the molecular structure.

Proteins are differentiated from one another primarily by tertiary structure, for it has been shown that the primary and secondary structures of all proteins are fundamentally similar. The two major classes of tertiary structure are exemplified by *fibrous* or *filamentous* proteins, and by *globular* proteins. This distinction is to a certain extent arbitrary, since there is a whole gamut of intermediate types that falls between the two classes. Certain water-insoluble structural proteins such as silk fibroin and keratin, found in hair, fingernails, and bird feathers, are indeed distinguished by a fibrous, oriented configuration. At the opposite end of the scale are the extremely soluble proteins of blood serum—serum albumin and serum globulin. Molecules of the latter type have roughly spherical or ellipsoidal tertiary structures. In addition, serum albumin is extensively "vulcanized" with —S—S— disulfide bonds and as a result, its tertiary structure is quite rigid. Many soluble proteins are intermediate between these extremes. Myosin, fibrinogen, and procollagen, for instance, have a compact macromolecular structure but are drawn out into long asymmetrical rods. In order to perform their function these proteins must be part of a fibrous structure such as muscle, blood clot, or skin. But by many criteria, such as their folded, topologically complex tertiary structure, they must be assigned to the globular proteins.

In the next section we shall consider quantitative data relating to the second and third levels of macromolecular organization in proteins. It is necessary first, however, to say a few words about the fourth level of organization, the *quaternary structure*. Quaternary structure arises as the result of the association of two or more globular molecules, called subunits, into a composite or complex globule. The reversible dissociation of protein complexes into subunits is frequently observed. Dissociation or association of subunits usually has significant effects on the functional characteristics of the protein. The enzyme alkaline phosphatase, for instance, loses activity when its two subunits are dissociated. Phosphoglyceraldehyde dehydrogenase, however, loses its activity when two protein molecules are joined together. The association of proteins into complexes sometimes constitutes the basis of their biological activity. Thus after activation, fibrinogen molecules coalesce to form a filament of fibrin. This is the material that closes off disrupted blood vessels after traumatic or inflammatory damage, thereby preventing excessive loss of blood. Fibrin is also deposited on serous membranes, inhibiting the spread of inflammation. The polymerization reaction between the proteins actin and myosin leads to the formation of actomyosin fibers which are responsible for muscle contraction.

All four levels of structural organization are important for the functional activity of the protein molecule. Although these types of structure all influence each other to a certain extent, they are nonetheless distinct, determined in the main by different types of molecular interactions: primary structure by covalent bonds along the polypeptide backbone; secondary structure by hydrogen bonds between neighboring peptide groups of the helical chain; tertiary structure by van der Waals interactions between the amino acid side groups and by chemical bridges such as disulfide bonds. Finally, the quaternary structure results from localized forces between functional groups distributed on the surface of the protein globule, one example of which is the electrostatic attraction of oppositely charged ionic groups.

The destruction of a protein's native secondary and tertiary structure is referred to by the general term denaturation, regardless of the means used to bring this about. Among the many causes of denaturation are high temperature, sudden and large changes in pH, or the presence of certain agents which break hydrogen bonds between —CO and —NH groups such as urea, guanidine salts, trifluoroacetic acid, and dichloroacetic acid.

6. Synthetic Polypeptides as Model Proteins

The secondary structure of proteins was understood in detail only after the study of model polypeptides synthesized from a single type of amino acid (21a,b). In particular, the use of such compounds permitted the development of methods to measure quantitatively the degree of regularity in a protein molecule. The first major success in synthesizing high-molecular-weight polypeptides was achieved in 1947 by Woodward and Schramm, who performed the polycondensation of amino acid carboxyanhydrides in nonaqueous solutions (22).

That amino acids could not be reacted with each other by simple condensation was apparent *a priori*, because the only solvent which could be used for this is water, and in water the equilibrium of the reaction is strongly displaced toward hydrolysis, that is, toward cleavage of the peptide bond. Under these conditions, peptide bond formation requires the expenditure of energy so as to displace the chemical reaction toward synthesis. This can be achieved if a chemically active derivative is used from which a small molecule other than water is excluded upon condensation. Conditions would be even more favorable if the condensation reaction could be performed in nonaqueous medium.

6. Synthetic Polypeptides as Model Proteins

At present, the best general method for obtaining high-molecular-weight polypeptides consists in the condensation of amino acid carboxyanhydrides:

$$n \begin{array}{c} \text{HN} - \text{CH(R)} \\ | \\ \text{O=C} - \text{O} \end{array} \text{C=O} \longrightarrow \cdots -\text{NH}-\text{CH(R)}-\text{CO}-\text{NH}-\text{CH(R)}-\text{CO}-\text{NH}- \cdots + n\,CO_2$$

The carboxyanhydrides themselves are produced by reacting amino acids with phosgene:

$$H_2NCHRCOOH + COCl_2 \xrightarrow{-HCl} \begin{array}{c} H \\ RC-COOH \\ | \\ HN-COCl \end{array} \xrightarrow{-HCl} \begin{array}{c} H \\ RC-CO \\ | \quad \quad \backslash \\ \quad \quad \quad O \\ HN-CO \end{array}$$

The amino acid derivatives are easily crystallized from chloroform, ether, tetrahydrofuran, and various solvent mixtures. The crystalline solids have melting points between 50° and 130°C and they do not decompose at these temperatures.

The polycondensation of carboxyanhydrides can be carried out in nonaqueous solvents such as nitrobenzene, dimethylformamide, tetrahydrofuran, and dioxane. The principal demand placed on the solvent is that both monomer (carboxyanhydride) and polymer (polypeptide) remain in solution, even if the latter is in the form of a gel. Moreover, the solvent molecules must not themselves be capable of initiating the condensation reaction, a condition that eliminates substances containing hydroxyl and amino groups. The only active center of polymer chain growth during the process of polycondensation must be the terminal amino group to which the monomeric units are successively added. Of course it is essential that the terminal amino group be available for reaction with the monomer throughout the incubation. If optimal conditions for polycondensation are maintained, it is possible to obtain polypeptides with molecular weights of 300,000 and higher.

In polypeptide synthesis, as in every polycondensation or polymerization, the initiating reaction must be distinguished from that which extends the length of the chain. The chain-terminating reaction can be ignored in the present case since all chains continue to grow until the supply of monomer in the medium is exhausted. To initiate the polycondensation of carboxyanhydrides, small amounts of H_2O, aqueous NaOH, amines, or sodium methoxide or ethoxide are added to the reaction mixture. Water served as initiator in the pioneering studies of Woodward and Schramm. In subsequent work, primary and secondary amines have become the favored initiators for in the case of these compounds there is a close resemblance between the reactions of chain initiation and chain growth.

The various carboxyanhydrides differ greatly in their reactivity, depending on the amino acid from which they were derived. A qualitative study of carboxyanhydride reactivities was performed by measuring the rates of condensation of several carboxyanhydrides with initiators consisting of amino acid dimethylamides:

$$\underset{H_2N}{\overset{R'}{\underset{|}{C}}}\underset{H}{\overset{}{}}CON\overset{CH_3}{\underset{CH_3}{}}$$

A scale of reactivities is obtained which is independent of the particular dimethylamide, that is, of the particular radical, R′-, present in the initiating substance (Table 1-2). The nature of the peptide with whose amino group the carboxyanhydrides react

TABLE 1-2
Relative Reactivity of Amino Acid Carboxyanhydrides[a]

N-Carboxyamino acid anhydride $\begin{array}{c}R_1\\ \backslash\\ HC{-}CO\\ |\quad\backslash O\\ R_2{-}N{-}CO\end{array}$	—R₁	—R₂	Relative rate of polymerization in nitrobenzene at 25°C referred to N-carboxysarcosine anhydride
Glycine	—H	—H	3
DL-Alanine	—CH₃	—H	1
DL-Phenylalanine	—CH₂—⟨ ⟩	H	10⁻¹
N-Isopropylglycine	—H	—C(H)(CH₃)(CH₃)	10⁻⁴

[a] Adapted from Bamford, et al. (21a), p. 40.

has relatively little influence on the condensation.

Special reagents have been developed to sequester H₂O released in the condensation of amino acids. One such substance is dicyclohexylcarbodiimide, which participates in peptide bond formation as follows:

$$R_1{-}NH_2 + R_2{-}COOH + \underset{H_2C-CH_2}{\overset{H_2C-CH_2}{\diagup\diagdown}}CH{-}N{=}C{=}N{-}CH\underset{H_2C-CH_2}{\overset{H_2C-CH_2}{\diagup\diagdown}}$$

↓

$$R_1{-}NH{-}CO{-}R_2 + \underset{H_2C-CH_2}{\overset{H_2C-CH_2}{\diagup\diagdown}}CH{-}N{-}\overset{O}{\overset{\|}{C}}{-}N{-}CH\underset{H H_2C-CH_2}{\overset{H_2C-CH_2}{\diagup\diagdown}}$$

This reaction will take place in the solvents nitrobenzene, dimethylformamide, and dioxane. It is interesting to note that a *water-soluble* carbodiimide can be synthesized by replacing one cyclohexyl radical with a group containing a quaternary amine (23). Using this water-soluble carbodiimide it is possible to condense the carboxyl group with the peptide amino group even in aqueous solutions.

6. Synthetic Polypeptides as Model Proteins

In order to obtain a polypeptide chain from α-amino acids with reactive side groups, it is necessary to protect or mask the side groups before polycondensation. After formation of the polymer the protecting substance must be removed under conditions sufficiently gentle to avoid degradation of peptide bonds or racemization of asymmetric carbon atoms. The side groups of lysine, for instance, are carbobenzoxylated in a reaction that simultaneously serves as the first stage in the synthesis of the carboxyanhydride itself:

$$\begin{array}{c} NH_2 \\ | \\ (CH_2)_4 \\ | \\ H_2N-CH-COOH \end{array} \xrightarrow[-HCl]{2\,C_6H_5CH_2OCOCl} \begin{array}{c} NHCOOCH_2C_6H_5 \\ | \\ (CH_2)_4 \\ | \\ C_6H_5CH_2OOC-NH-CH-COOH \end{array} \xrightarrow[\text{in ether}]{PCl_5} \begin{array}{c} NHCOOCH_2C_6H_5 \\ | \\ (CH_2)_4 \\ | \\ HC-CO \\ |\quad\ \ \backslash \\ |\quad\ \ \ O \\ |\quad\ \ / \\ HN-CO \end{array}$$

Removal of the masking group is effected by gentle reducing agents such as anhydrous HBr or phosphonium iodide (PH_4I) in glacial acetic acid.

Protection of the side chain carboxyl groups of aspartic and glutamic acids is accomplished by converting them to their methyl or benzyl esters:

$$\begin{array}{c} COOH \\ | \\ (CH_2)_2 \\ | \\ H_2N-CH-COOH \end{array} \xrightarrow{ROH} \begin{array}{c} COOR \\ | \\ (CH_2)_2 \\ | \\ H_2N-CH-COOH \end{array} \xrightarrow[-HCl]{COCl_2} \begin{array}{c} COOR \\ | \\ (CH_2)_2 \\ | \\ HC-CO \\ |\quad\ \ \backslash \\ |\quad\ \ \ O \\ |\quad\ \ / \\ HN-CO \end{array}$$

Mild alkaline hydrolysis in aqueous ethanol is used to remove the methoxy group from the anhydride, while benzyl alcohol may be cleaved with phosphonium iodide as above.

The hydroxyl groups of serine and tyrosine can be blocked by acetylation or carbobenzoxylation:

$$\begin{array}{c} CH_2OH \\ | \\ H_2N-CH-COOH \end{array} \xrightarrow{Ac_2O/HClO_4} \begin{array}{c} CH_2-OOCCH_3 \\ | \\ H_2N-CH-COOH \end{array} \xrightarrow{COCl_2} \begin{array}{c} CH_2OOCCH_3 \\ | \\ HC-CO \\ |\quad\ \ \backslash \\ |\quad\ \ \ O \\ |\quad\ \ / \\ HN-CO \end{array}$$

$$\begin{array}{c} CH_2OH \\ | \\ H_2N-CH-COOH \end{array} \xrightarrow{2\,C_6H_5CH_2OCOCl} \begin{array}{c} CH_2-OCOOCH_2C_6H_5 \\ | \\ C_6H_5CH_2OOC-NH-CH-COOH \end{array} \xrightarrow[\text{in ether}]{PCl_5} \begin{array}{c} CH_2OCOOCH_2C_6H_5 \\ | \\ HC-CO \\ |\quad\ \ \backslash \\ |\quad\ \ \ O \\ |\quad\ \ / \\ HN-CO \end{array}$$

The acetyl groups are removed from the polymer after polycondensation by mild hydrolysis in aqueous ammonia at room temperature.

In the reactions considered thus far, we have been speaking of the polycondensation of only one type of amino acid monomer or of a random mixture of several different monomers, a process called copolymerization. It is quite another matter to synthesize a polypeptide of defined sequence—in the limiting case, a protein—from a variety of amino acid components. Such a synthesis must be carried out stepwise by adding one amino acid residue after another to the growing chain according to a specific program. With our present knowledge of organic chemistry the stepwise construction of a polypeptide is entirely feasible. To maintain the specificity of the reaction, it is necessary to block all reactive groups of the monomeric amino acids except the α-carboxyl, and all those of the peptide except the terminal α-amino. The condensation of the amino acid —COOH group with the free —NH$_2$ group of the peptide then results in formation of a peptide bond.

For the synthesis of long polypeptide chains with a defined amino acid sequence, an ingenious method has been worked out which markedly simplifies the purification of the product after the addition of each successive residue (24a, b). The polypeptide is attached to a solvent-saturated gel through the carboxyl group of its C-terminal amino acid. The gel is composed of a chlormethylated polystyrene of the following structure:

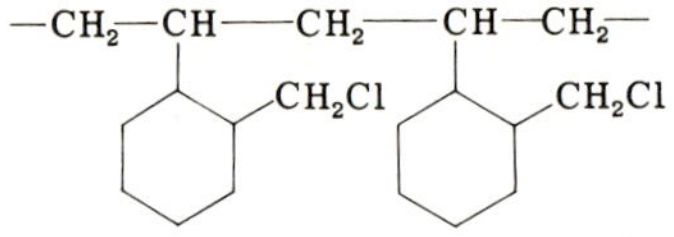

$$-CH_2-CH-CH_2-CH-CH_2-$$
with CH$_2$Cl substituents on phenyl rings

The carboxyl group of the C-terminal residue anchors the growing polypeptide to the sorbent through an ester linkage:

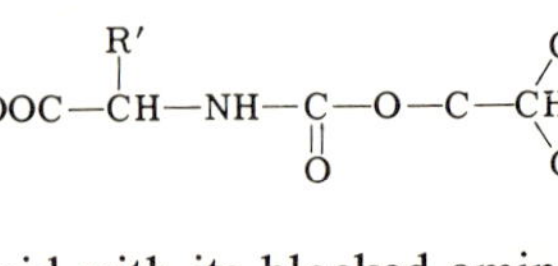

$$CH_2-O-C(=O)-CH(R')-NH_2$$

The amino group of the amino acid is masked beforehand by replacing a hydrogen atom with a tertiary butyloxycarbonyl group:

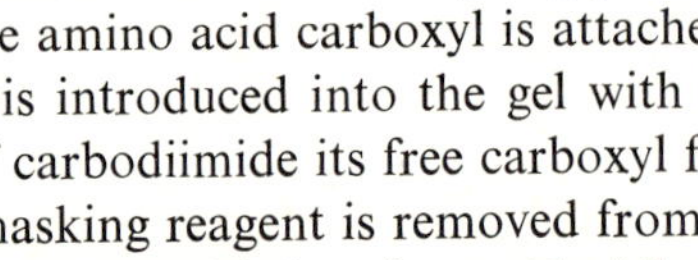

$$HOOC-CH(R')-NH-C(=O)-O-C(CH_3)_3$$

After the C-terminal amino acid with its blocked amino group has been attached to the chlormethylated gel, the amino group is unmasked by mild acid hydrolysis with HCl. Under these conditions the NH—COOC(CH$_3$)$_3$ bond is especially labile, but the ester bond by which the amino acid carboxyl is attached to the gel is unaffected. Now the next amino acid is introduced into the gel with its amino group similarly masked. In the presence of carbodiimide its free carboxyl forms a peptide bond with the first amino acid. The masking reagent is removed from the amino group and the whole process is repeated with the third amino acid. All unreacted compounds and by-products are easily removed by washing the gel between steps. When the polypeptide is finished, it is released from the supporting polystyrene with HBr or HF in the presence of trifluoracetic acid.

In the past few years these new methods in polypeptide chemistry have been used for the complete synthesis of several polypeptide hormones such as the nonapeptide bradykinin (24b), the A and B chains of both sheep and human insulin (25), and ribonuclease (26a, b). In the case of insulin, the synthetic A and B peptides exhibit precisely the same properties as the natural A and B chains produced by reduction of the native protein. Furthermore, molecules with the specific hormonal activity of native insulin can be obtained after the two chains of synthetic insulin are linked with disulfide bonds.

The most significant achievement in the field of synthetic peptides is the complete chemical synthesis of a linear polypeptide with the sequence of bovine pancreatic ribonuclease A [Gutte and Merrifield (26a)]. The 124 amino acids in this enzyme were assembled sequentially from the C-terminal residue by means of a fully automated solid-phase procedure (24a, 26c) in which the 369 necessary chemical reactions were carried out according to a prearranged program without any intermediate isolation steps. After the polystyrene-bound ribonuclease precursor had been cleaved from its solid support and freed of its protective groups, the product was purified chromatographically and oxidized in air to form the four required disulfide bridges. The totally synthetic enzyme that resulted was found to possess the same rigid substrate specificity and the same Michaelis constant as natural ribonuclease A. Synthetic ribonuclease also evidenced the electrophoretic and chromatographic properties of ribonuclease A, and the two proteins gave very similar amino acid analyses and tryptic peptide maps. The assembly of ribonuclease in these experiments represents the first demonstration that an enzyme with normal catalytic activity can be completely synthesized from its constituent amino acids. Ribonuclease S-protein (p. 131) has also been synthesized (26b). In this case, 19 specially prepared peptide fragments were progressively coupled into a polypeptide chain containing 104 amino acids, which displayed characteristic ribonuclease activity after oxidation in the presence of S-peptide.

Since new methods in the chemical synthesis of polypeptides have already proved their great effectiveness, the complete synthesis of a variety of proteins may be expected in the near future. Synthetic polypeptides afford a unique probe for investigating hormone and protein function, owing to the fact that derivatives with specifically altered amino acid residues can be prepared at will and their activity contrasted with that of the naturally occurring material.

7. Secondary Structure of Polypeptides and Proteins

High-molecular-weight synthetic polypeptides possess many of the fundamental properties of protein molecules, though in a much simpler form. These substances have thus provided chemists with an unprecedented opportunity to verify the principal hypotheses about protein structure. The main difference between polypeptides and other linear polymers consists of the extremely strong secondary forces which unite pairs of peptide groups, supplementing the covalent bonds which link adjacent amino acids. The secondary forces arise from hydrogen bonds between the carbonyl oxygen and the imide hydrogen along the polypeptide backbone. Measurements in the laboratory of the author demonstrated that the energy of these bonds reaches 1400 cal/mole (27).

It was earlier supposed that in aqueous solution water molecules could compete with the imide groups of the polypeptide so as to disrupt hydrogen bonds already established, but this view has been proved incorrect. If water is replaced either partially or completely by organic solvents (dioxane, dimethylformamide, chloroethanol), conditions for the formation of hydrogen bonds between peptide groups are even better. Only those solvents which are capable of forming very strong hydrogen bonds with the imide group (concentrated solutions of urea or guanidine, dichloroacetic, trifluoroacetic, or formic acids) can break apart the intramolecular hydrogen bonds in polypeptides at room temperature.

The maximum possible number of hydrogen bonds are formed within the macromolecule when the polypeptide backbone is coiled into a helix. A macromolecular polypeptide thus resembles a rigid hairspring, each turn of which is tightly bound to the preceding and succeeding turns by hydrogen bonds. This helical structure will be maintained in solution as long as the intramolecular hydrogen bonds are unbroken. In contrast, the backbone of ordinary linear polymers assumes the shape of a statistical random coil in which the various residues or segments are distributed around the center of gravity in an approximately Gaussian fashion.

Pauling and Corey were able to predict the structure of the helix with great accuracy from basic principles of molecular physics and from specific data obtained by x-ray crystallography, in particular, the distances and angles between atoms in covalent and hydrogen bonds (28–30). They noted that each peptide group is conjugated to a certain extent, requiring the superposition of two electronic structures, one neutral, the other polar, to adequately describe it:

$$\begin{array}{ccc} \text{H} & \text{H} \;\; \text{H} & \\ | & | \;\; | & \\ -\text{C}-\text{C}-\text{N}-\text{C}- & \text{or} & -\text{C}-\text{C}=\overset{+}{\text{N}}-\text{C}- \\ | \quad \| \quad | & & | \quad | \quad | \\ \text{R} \;\; \text{O} \;\; \text{R}' & & \text{R} \;\; \text{O}^- \;\; \text{R}' \end{array}$$

Experimental evidence for this assertion is provided by an unusually short C—N bond, 1.32 Å in length, instead of the 1.47 Å characteristic of C—N bonds. It follows that the C—N bond of a peptide group has about 40% double bond character. The shortening of the C—N bond is observed in all kinds of peptides and its magnitude falls well outside of the normal error (0.02 Å) in such measurements. A second but less dramatic feature of peptides is a slight stretching of the C=O bonds, indicating partial single bond character. In peptides, the C=O is found to be 1.24 Å compared with 1.21 Å in ordinary carbonyl groups. This difference lies closer to the experimental error, though it clearly exceeds it.

If we accept the fact that the peptide bond is partially conjugated, and that π-electrons are present in it, an important consequence follows from the general principles of quantum mechanics: all four atoms of the peptide group

$$\begin{array}{c} \text{H} \\ | \\ -\text{C}-\text{N}- \\ \| \\ \text{O} \end{array}$$

7. Secondary Structure of Polypeptides and Proteins

lie in one plane, that is, they are *coplanar*. This was the first principle used by Pauling and Corey in postulating a helical structure for the polypeptide chain. The second, or equivalency, principle states that each amino acid residue must be placed in exactly the same spatial relation with respect to its neighbors and to the helix as a whole as are all other residues. The third and last principle recognizes that each peptide group must form two hydrogen bonds and do so in such a way that the helix attains the greatest possible compactness.

The length of these hydrogen bonds, taken as the distance between the atoms of oxygen and nitrogen since crystallographic measurements do not yield the position of hydrogen, is 2.8 Å. Furthermore, the hydrogen bond tends to make the C=O and N—H groups colinear because of polarity. Pauling and Corey evaluated the decrease in energy of the hydrogen bond when the axes of the C=O and N—H groups are not colinear, but at an angle, θ, to one another. The expression for energy as a function of θ was found to equal $U \sin^2 \theta$, where U is the energy of the hydrogen bond. From this, Pauling and Corey concluded that the angle θ could not be large and in building their three-dimensional models, they limited the angle to a maximum of 20°.

Using the three principles enumerated above, it is also possible to determine the three-dimensional configuration of the helical polypeptide chain. Figure 1-10 depicts

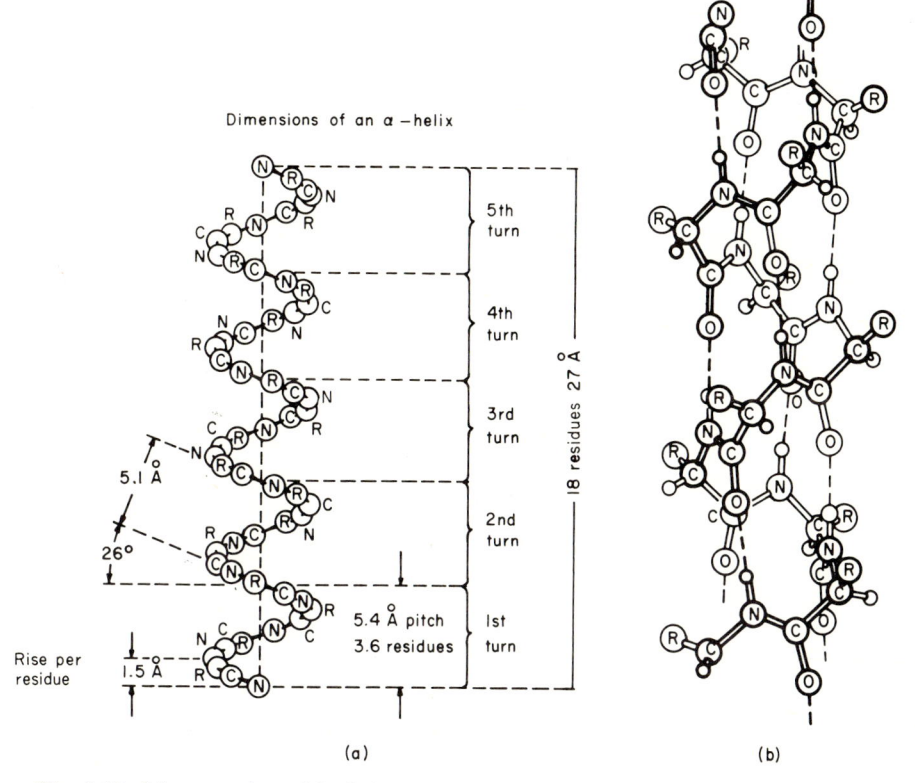

Fig. 1-10. Scheme and model of the Pauling-Corey α-helix. (a) Parameters of the α-helix. (b) Perspective drawing of the α-helix showing configuration of polypeptide backbone and intrachain hydrogen bonds. Adapted from Pauling and Corey (31).

the so-called Pauling-Corey α-*helix* constructed according to these rules. There is room for just 3.6 amino acid residues in one turn of the helix, meaning that one complete period along the axis, that is, the axial translation required to bring the helix back into identity with itself, consists of five turns or eighteen residues. The height of one 360° turn is 5.4 Å; consequently, the axial displacement corresponding to each amino acid residue is 1.5 Å. The diameter of the cylindrical surface on which α-carbon atoms fall is 10.1 Å. From the illustration it is clear that intramolecular hydrogen bonds are formed between every fourth peptide group. The cross section of the α-helix shown in Fig. 1-11 demonstrates how the planar configuration of each peptide bond

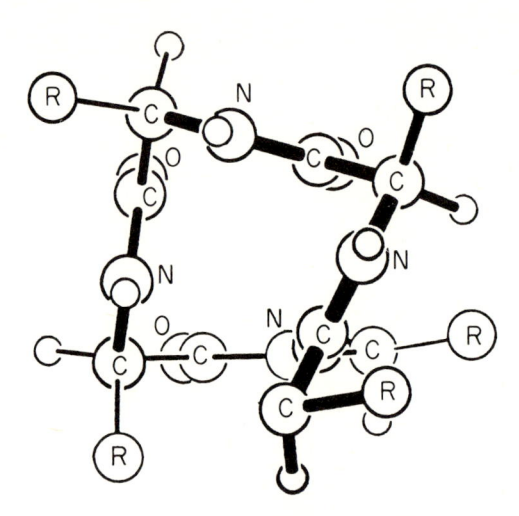

Fig. 1-11. Cross section of the α-helix perpendicular to axis, showing one turn in perspective. From Pauling *et al.* (28).

causes the polypeptide chain to bend by an angle slightly less than 90° at each residue.

x-Ray structure analysis performed by Bamford and co-workers on thin films of the synthetic polypeptides polyleucine, polyphenylalanine, poly-γ-methyl-L-glutamate and poly-γ-benzyl-L-glutamate, confirmed the α-helical structure predicted by Pauling and Corey with remarkable precision (32). The reliability of this conclusion was very great since the complete x-ray analysis of two polypeptides revealed that all the diffraction maxima observed corresponded with those predicted by the theory of the α-helix in both position and intensity.

One question that deserves special attention is what the sense of rotation of the helix should be. Since almost all amino acids in nature are of the L-configuration, and since they contain side groups, it becomes clear that the sense of the helix is not immaterial. To illustrate this, note that the side groups, R, are always on the outside of the helix and that the direction of the chain is designated as that from N-terminal to C-terminal. In a right helix composed of L-amino acids, the side groups will point along the axis in a direction opposite to that of the polypeptide chain (Fig. 1-12). If a left helix is constructed from the same components, the amino acid side groups will extend along the axis in the direction of the polypeptide chain. Since the helix contains 3.6 side groups per turn it is evident that the interactions of these radicals, and hence

7. Secondary Structure of Polypeptides and Proteins

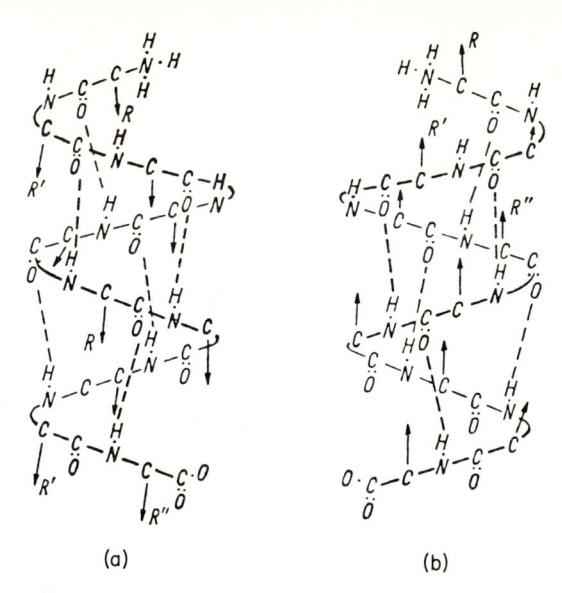

Fig. 1-12. Scheme illustrating the orientation of amino acid side groups in (a) a right and (b) a left α-helix made up of L-amino acid residues.

their packing, will differ greatly in the two structures. Thus the choice between the right and left helices will ultimately be determined by the van der Waals interactions among the side groups as well as by the steric hindrance which they encounter.

In the first phase of the study of polypeptides it appeared that L-amino acids always form right helices. The investigation of proteins also indicated the presence of right-helical segments and attempts were made to generalize these findings. However, Blout and his co-workers succeeded in showing that a polypeptide composed of the benzyl ester of L-aspartic acid (poly-β-benzyl-L-aspartate) forms a left helix although a less stable one than usual (33). We shall return to this problem in a future section (p. 58), but for the time being we shall assume that L-amino acids form a right Pauling-Corey helix, a notion consistent with the bulk of the experimental evidence.

As an alternative to the α-helical structure with its characteristic intramolecular hydrogen bonds, Pauling and Corey considered the structure resulting from the formation of intermolecular hydrogen bonds between different polypeptide chains (30, 34, 35). In this case, the polypeptide chains must be extended to their limit axially and laid either parallel or antiparallel with respect to one another (Fig. 1-13). This conformation is called the *β-structure*, or the *β-pleated sheet*. The side groups, R, project perpendicularly from the plane of the page in which the zigzag polypeptide backbones lie. Contrast this with the α-helix discussed above, where the side chains all fall on a cylindrical surface concentric with the helix axis (Fig. 1-10). A third type of β-structure, called *cross-β*, can be formed intramolecularly if a single polypeptide molecule is folded back on itself so as to permit hydrogen bonding between separate segments of the chain.

The task of differentiating the α- and β-structures from one another experimentally is not an easy one. Optical criteria are in general ambiguous and they lead to a conclusion about the nature of polypeptide secondary structure only if they are used jointly.

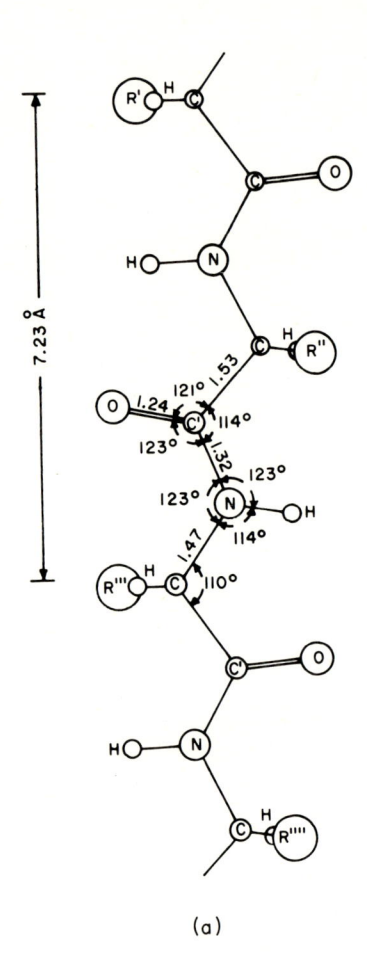

Fig. 1-13a

7. Secondary Structure of Polypeptides and Proteins

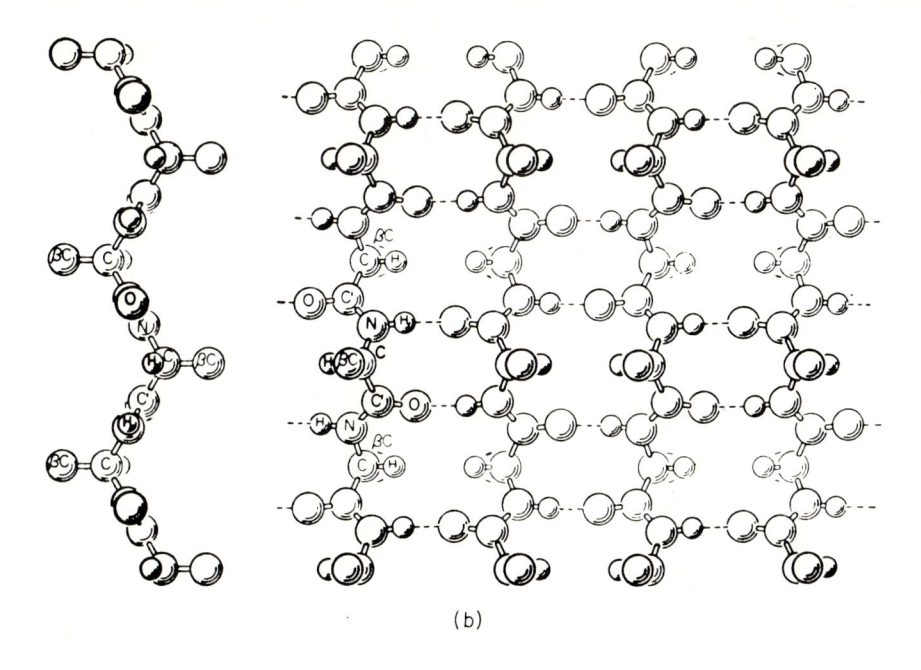

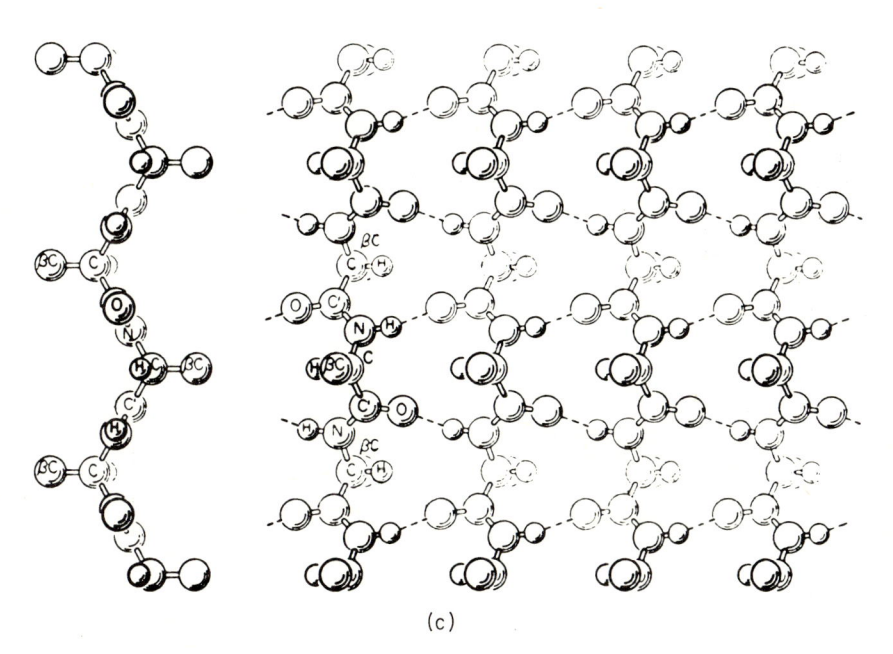

Fig. 1-13. Configuration of fully extended polypeptide chains. (a) Bond lengths and bond angles of the extended chain. From Corey and Pauling (36). (b) The antiparallel β- or pleated sheet structure. (c) The parallel β- or pleated sheet structure. (b) and (c) from Pauling and Corey (35).

X-ray structure analysis is the sole technique which can be used to resolve this question unequivocally, although it is applicable only to crystals or oriented polymer films and not to molecules in solution. The x-ray investigation of films made from synthetic polypeptides has in fact demonstrated that many of these molecules possess regular β-structure under given conditions. Furthermore, in certain of the globular proteins whose detailed structure is known from x-ray analysis, regions of cross-β-conformation have been identified along with α-helical and disordered or amorphous segments. Thus, elements of the cross-β-structure were found in chymotrypsin, while regions with partial β-conformation were observed both in papain and in carboxypeptidase.

The simplicity of synthetic polypeptides makes them particularly suitable for studies on secondary structure. Depending on the nature of their amino acid side groups, polypeptides can assume one of three possible conformations, the α-helix, the β-structure, or the random coil. The interaction of the side groups with each other and with the solvent, together with hydrogen bonds between peptide groups, plays a primary role in establishing the specific equilibrium conformation under a given set of conditions. The contribution of hydrogen bonds to the stabilization of macromolecular structure increases in nonpolar solvents, while strongly polar solvents that are themselves capable of forming hydrogen bonds with CO—NH groups reduce the importance of hydrogen bonding. Water occupies an intermediate position between the two extremes. Furthermore, if the side groups each bear an equivalent charge, the only conformation accessible to the polypeptide chain is the random coil; but if the side groups are uncharged, then, depending on their affinity for one another and for the solvent, one of the two types of ordered structure may be observed. By varying the temperature, the solvent, or the pH, it is very often possible to evoke transitions between ordered and disordered conformations. Moreover, such transitions are strongly cooperative, that is, the majority of residues in the polypeptide macromolecule pass from one state to the other simultaneously.

Some kinds of synthetic polypeptides can even assume all three types of secondary structure in solution under various conditions. The classic example in this case is poly-L-lysine (37). In the charged state, of course, this polypeptide cannot form an ordered structure and it behaves as an ordinary linear polyelectrolyte, but at pH values greater than 10.5, poly-L-lysine is discharged and forms nearly perfect α-helices which are completely stable in aqueous solution as long as the temperature is held below 20°C. In the interval between 20° and 35°, however, a helix–coil transition occurs and the α-structure gives way to a random coil. A second conformational transition takes place above 35°C in which the random coils are converted by a concentration-dependent process to the antiparallel β-pleated sheet structure owing to the ordered aggregation of polypeptide chains. Although intermolecular hydrogen bonds play an important role in the β-form of polylysine, hydrophobic interactions between lysine residues undoubtedly contribute to its stability.

Poly-L-glutamic acid in the charged state is also amorphous, but at pH values lower than 5, it forms α-helical chains which can be transformed back to the random coil conformation in the temperature interval from 20° to 60°C. Esters of polyglutamic acid, such as poly-γ-benzyl-L-glutamate, are soluble in organic solvents (cresol or dichloroethane) and assume the α-helical conformation. Polypeptides with hydrocarbon side groups—poly-L-leucine, for instance—exhibit α-helical structure in aqueous solu-

7. Secondary Structure of Polypeptides and Proteins

tion. Finally, poly-L-serine, poly-L-homoserine, and poly-L-threonine, which all contain hydroxyl groups in their side chains, are highly soluble in water where they form random coils. If the hydroxyl groups of this last set of polymers are esterified, the derivatives are found to dissolve only in organic solvents and they are characterized by a stable cross-β-conformation with wholly intramolecular hydrogen bonds. Poly-O-acetyl-L-serine, for instance, behaves this way in a mixed solvent containing chloroform and dichloroacetic acid. Furthermore, when the concentration of the polar component reaches 30 vol %, a transition from the cross-β to the random coil structure takes place. Poly-O-acetyl-L-threonine also exhibits a stable β-conformation in pyridine or in a mixture of the two solvents, dichloroethane and 2-chloroethanol. The fact that the optical properties of this polymer are independent of concentration indicates that its structure is stabilized by intramolecular hydrogen bonds.

When a macromolecular film is formed from solution, the polypeptide chains frequently retain the secondary structure which they possessed in the corresponding solvent. Thus if a film of poly-γ-benzyl-L-glutamate or poly-L-alanine is made from cresol, the molecules keep their α-helical structure. But if a more polar solvent such as formic acid is employed, films with intramolecular β-structure are obtained. Finally, the formation of films from a solution of the polypeptide in trifluoroacetic acid leads to an amorphous or disordered conformation. That the structure and molecular packing of polypeptides in the solid phase can be predetermined by the characteristics of the solvent from which they were recovered is yet another peculiarity of this class of compounds.

The number of peptide groups participating in hydrogen bonds can be determined quantitatively by studying the kinetics of isotopic exchange between the polypeptide or protein, and water. This method was developed by Linderström-Lang, who used deuterium label for detecting the exchange of hydrogen atoms (38). Tritium has been used for the same purposes in the author's laboratory. The parameter measured here is the rate with which the imide hydrogen of the peptide group exchanges with the solvent. In low-molecular-weight peptides this hydrogen atom exchanges with water so rapidly that the rate can scarcely be measured. In this case, the imide hydrogen atom behaves like the dissociable hydrogen of —NH$_2$, —COOH, —SH, and —OH groups. The imide hydrogens of high-molecular-weight polypeptides, however, exchange with water rather slowly, and the exchange is strongly dependent on temperature, pH, and solvent. The slow exchange of the imide hydrogens with the medium is a sign of intramolecular hydrogen bonding. By measuring the quantity of slowly exchanged hydrogen atoms, presumed to be nonexchangeable owing to their presence in —CO $\cdots$ HN— pairs, we can determine the number of intramolecular hydrogen bonds and, consequently, the proportion of helix or cross-β-structure in the polypeptide or protein. It is impossible to distinguish between the two types of ordered structure by this method.

Figure 1-14 shows the rate at which the imide hydrogens of poly-DL-alanine (containing about 30 residues) are exchanged with water. On the ends of the molecule there are three atoms of hydrogen (in —COOH and —NH$_2$ groups) which exchange instantaneously. Five or six imide hydrogens exchange rapidly as well since hydrogen bonds toward the ends of the chain are weaker and tend to exchange more readily than those in the middle. A greater fraction of the imide hydrogens are exchanged much

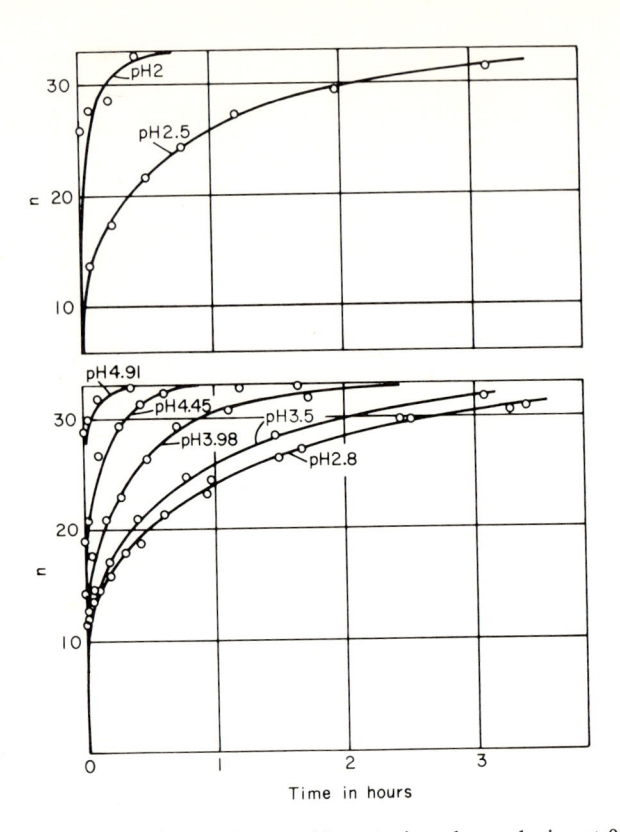

Fig. 1-14. Kinetics of deuterium exchange with water in poly-DL-alanine at 0°C and a variety of different pH values. The number of exchangeable hydrogen atoms belonging to the peptide groups is plotted along the ordinate. Two graphs are shown for clarity. From Berger and Linderström-Lang (39).

more slowly, however, requiring from 2 to 3 hours. The proportion of helix in this polypeptide is apparently close to 100%.

Insulin, which contains a total of 51 amino acid residues in two polypeptide chains and thus possesses 49 peptide bonds, is characterized by about 30 slowly exchanging hydrogen atoms. This indicates that the insulin molecule has about 60% ordered character. Of the 30 atoms which exchange slowly at 0°C, 7 are more weakly bound than the others and begin to exchange rapidly at 20°C. The rest of the hydrogens exchange only after 10 hours at that temperature.

Linderström-Lang suggested that the residues adjacent to —S—S— bridges might be only weakly hydrogen bonded since strains should occur in those regions (Fig. 1-15). When the insulin molecule is split into A and B chains with an oxidant, the A chain exchanges its hydrogens with the medium almost momentarily and is therefore judged to be devoid of ordered regions. The B chain retains most of its initial character under these conditions, exchanging 18 of its 31 imide hydrogens at a very low rate.

The enzyme ribonuclease exhibits an even more varied array of hydrogen-bonded regions than insulin (40). Out of 123 peptide hydrogens, 70 exchange slowly, indicating

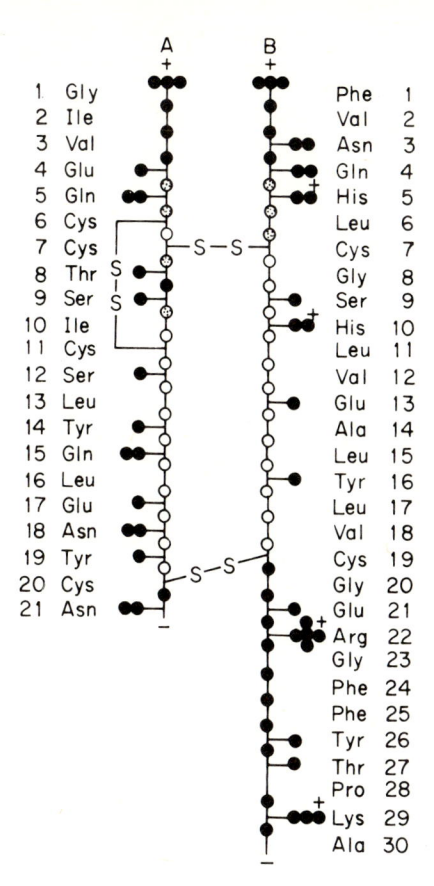

7. Secondary Structure of Polypeptides and Proteins 41

```
           A    B
           +    +
 1  Gly
 2  Ile
 3  Val
 4  Glu
 5  Gln
 6  Cys
 7  Cys      —S—S—              Phe   1
 8  Thr  S                       Val   2
 9  Ser  S                       Asn   3
10  Ile                          Gln   4
11  Cys                          His   5
12  Ser                          Leu   6
13  Leu                          Cys   7
14  Tyr                          Gly   8
15  Gln                          Ser   9
16  Leu                          His  10
17  Glu                          Leu  11
18  Asn                          Val  12
19  Tyr     —S—S—                Glu  13
20  Cys                          Ala  14
21  Asn                          Leu  15
                                 Tyr  16
                                 Leu  17
                                 Val  18
                                 Cys  19
                                 Gly  20
                                 Glu  21
                                 Arg  22
                                 Gly  23
                                 Phe  24
                                 Phe  25
                                 Tyr  26
                                 Thr  27
                                 Pro  28
                                 Lys  29
                                 Ala  30
```

Fig. 1-15. Isotopic exchange in insulin. There are 48 exchangeable hydrogen atoms in the polypeptide backbone and 43 in the end groups and side chains. Of these, 61 exchange instantaneously (solid circles), 7 exchange quite rapidly (shaded circles), and 23 exchange more slowly (open circles). From Linderström-Lang (38).

a content of ordered regions of 57%. Of the 70 peptide groups stabilized by hydrogen bonds, 25 are capable of slow exchange with water at 0°C. An additional 25 groups exchange after a day at 38°C. The remaining 20 do not exchange at all up to the melting temperature of the secondary structure, 60°C. Melting curves for the ordered regions of ribonuclease in both normal and heavy water are presented in Fig. 1-16.

Another means of studying changes in protein conformation is to measure the *specific rotation* $[\alpha]_D$ of the plane of polarization of monochromatic light (usually the sodium D line at 589 mμ) by a protein in solution. This phenomenon is called *optical activity*. Recall that all amino acids with the exception of glycine contain asymmetric α-carbon atoms and therefore rotate the plane of polarized light. The magnitude of the optical rotation is the algebraic sum of two parts: one negative, representing the contribution of all asymmetric carbon atoms, the other positive, resulting from the contribution of α-helix or β-structure (42). This measurement does not distinguish between the ordered α- and β-conformations, however. When α-helices are referred to in the following pages, it will be done for the sake of concreteness

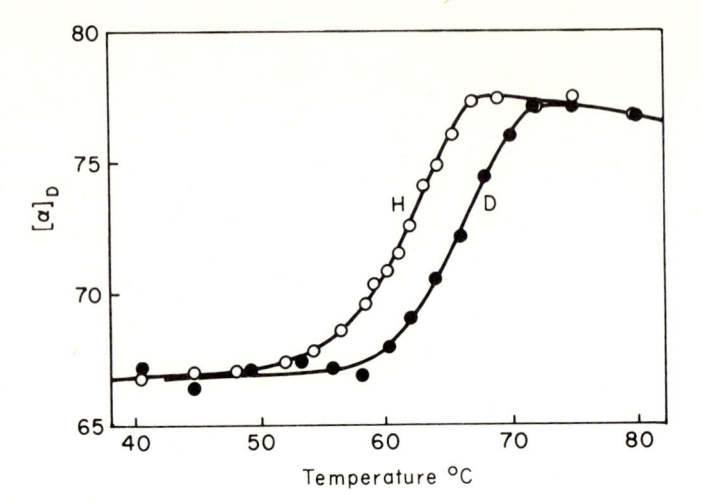

Fig. 1-16. Thermal denaturation of hydrogen-containing ribonuclease in H_2O (H) and of deuterated ribonuclease in D_2O (D). Although the two curves have been shifted vertically by two units relative to one another so as to facilitate comparison, there nonetheless appears to be a real isotope effect. From Hermans and Scheraga (41).

although, in general, examples will be based on polypeptides whose α-helical structure has been established.

Doty, Blout, and others proposed that the proportion of helix in a polypeptide could be assessed quantitatively by determining the positive increment of $[\alpha]_D$ associated with the secondary structure of the polypeptide. When measured in a nonpolar solvent, which does not disrupt the hydrogen bonding necessary to maintain the α-helix, $[\alpha]_D$ relates both to the asymmetric carbon atoms and to the structure as a whole. In order to obtain the desired increment, the negative contribution of the asymmetric carbon atoms must be subtracted. This is accomplished by measuring the optical rotation under conditions which disorganize the α-helix. To this end, Doty either used strongly polar solvents such as dichloroacetic acid or else took advantage of the polyelectrolyte nature of polypeptides such as polyglutamic acid.

Destruction of the α-helix approximates a phase transition, or melting, since formation of the helix is akin to a special kind of intramolecular crystallization. The phase transition analogy is appropriate here because the transition from an ordered α-helical structure to a disordered random coil, called the helix–coil transition, is a cooperative one. In other words, intermediate stages between the completely ordered α-helix and the completely disordered random coil are merely transitional.

Doty found that poly-γ-benzyl-L-glutamate

has its maximum total rotation of $[\alpha]_D = +18°$ in chloroform, a relatively nonpolar solvent, and a rotation of $[\alpha]_D = -30°$ in trifluoroacetic acid, a solvent which completely disrupts hydrogen bonds. The second value reflects the effect of the asymmetric α-carbon atoms alone. Consequently, the increment attributable to the presence of α-helical sections within the molecule is +48°. The α-helix of poly-γ-benzyl-L-glutamate is right-handed and gives a positive specific rotation if it is made out of L-amino acids. If the polypeptide is synthesized from D-amino acids, however, the chain will fold up into a left helix.

In Fig. 1-17a, the specific rotation of poly-γ-benzyl-L-glutamate (molecular

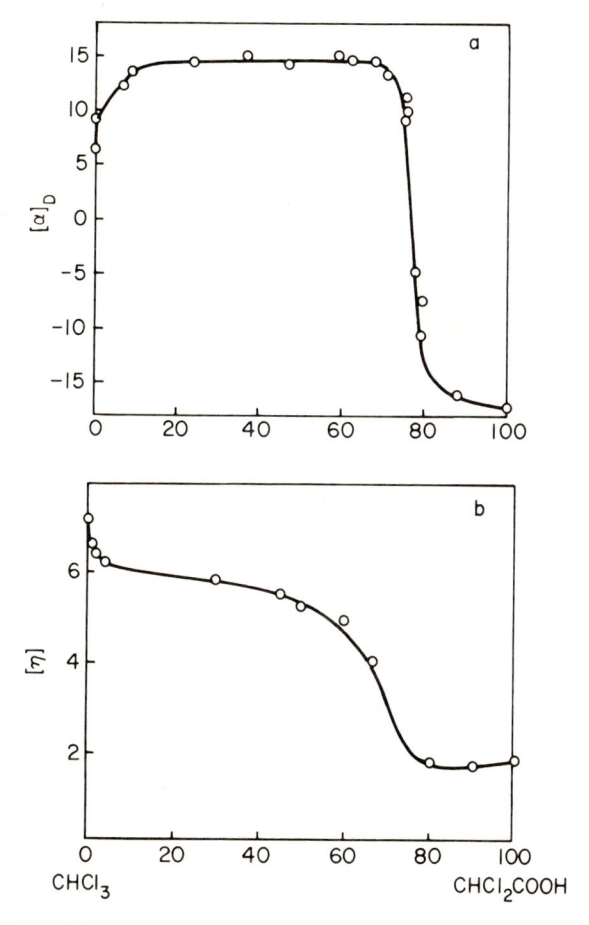

Fig. 1-17. Melting of the helix of poly-γ-benzyl-L-glutamate (PBG) in a mixed solvent composed of chloroform and dichloroacetic acid. (a) Optical rotation and (b) intrinsic viscosity of PBG (molecular weight 350,000) as a function of solvent composition. From Doty (43).

weight 350,000) is presented as a function of solvent composition. The solvent was compounded by mixing chloroform and dichloroacetic acid in different proportions, varying from pure chloroform on the left to pure dichloroacetic acid on the right (43). Measurements were made at room temperature. At a dichloroacetic acid concentration of 75%, the specific rotation suddenly loses its large positive increment, indicating that the α-helix has been melted out.

Simultaneous with the change in optical activity, the hydrodynamic characteristics of the macromolecule also undergo alteration. This may be noted in Fig. 1-17b where the *intrinsic viscosity* $[\eta]$ is plotted as a function of solvent composition for the same polypeptide (43). The transition from a large $[\eta]$, corresponding to a rigid, extended molecule, to a smaller value occurs at approximately the same place (about 75% $CHCl_2COOH$) as the melting of the α-helix. However, the drop in viscosity does not seem to be as sharp as the change in optical rotation, indicating that the melting of the α-helix is preceded by a more gradual breakup of regular structure.

Studies of the way in which the intrinsic viscosity of poly-γ-benzyl-L-glutamate depends on molecular weight have yielded interesting results (44). The molecular weights of a series of polymers were measured by light scattering. The relationship between intrinsic viscosity and molecular weight is always of the form: $[\eta] = kM^v$. According to Simha, $v = 1.7$ for rigid rods (represented as prolate ellipsoids for the purposes of calculation). From Fig. 1-18 it is evident that this theoretically predicted relation is strictly adhered to when the experiment is performed under conditions which favor the α-helical structure (chloroform with a trace of formamide). If the same polypeptide preparations are investigated in dichloroacetic acid, a logarithmic plot of the

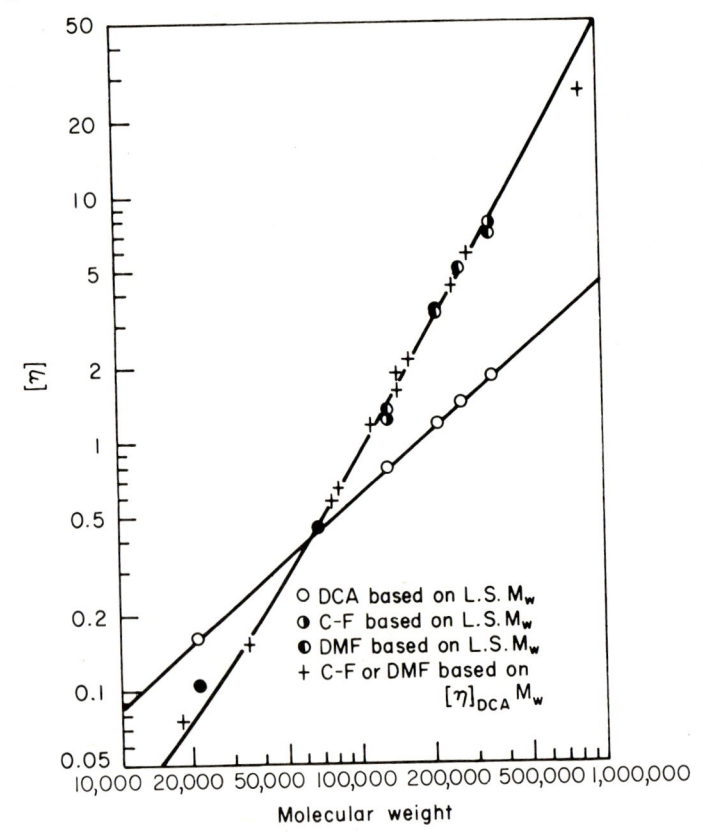

Fig. 1-18. The dependence of the intrinsic viscosity of poly-γ-benzyl-L-glutamate on molecular weight. The half-filled circles correspond to solvents such as chloroform-formamide (C-F) or dimethylformamide (DMF) in which the polypeptide assumes the helical configuration, while the open circles correspond to the randomly coiled configuration in dichloroacetic acid (DCA). From Doty *et al.* (44).

7. Secondary Structure of Polypeptides and Proteins

data results in a straight line whose slope is $v = 0.87$, the value characteristic of a statistical random coil. Many solvents (dimethylformamide, cresol, pyridine, dichloroethane) are just as sparing of the α-helix as chloroform with formamide, yielding identical values for the intrinsic viscosity.

Use of the above formula for the intrinsic viscosity of rigid, rod-shaped molecules in solution permits us to calculate the dimensions of the major and minor axes of the equivalent cylinder. Polypeptides with molecular weights up to 300,000 all turn out to have a diameter of 14.9 Å according to the model. This figure for the outer packing diameter of a Pauling-Corey α-helix is in good agreement with other independent measurements. For example, x-ray crystallographic determinations of the distance between adjacently oriented molecules of poly-γ-benzyl-L-glutamate show the diameter of the helix to be 15 Å. The same value is obtained simply from measurement of the partial specific volume of the polypeptide if it is assumed that each amino acid residue occupies a segment 1.5 Å high along the axis of the helix. Using this figure and the known molecular weight and density of the monomer, a diameter of 15.3 Å is calculated. The agreement of all independently measured values is excellent.

Another method of investigation, measurement of the angular distribution of light scattered from polypeptides in solution, yields additional information about the shape and dimensions of these macromolecules. The length of the particle, expressed in terms of its radius of gyration, was found to be directly proportional to its molecular weight. Using poly-γ-benzyl-L-glutamic acid as an example once again, it was determined that the length of the rod-shaped molecules is equal to 2030 Å when the molecular weight is 358,000. The axial displacement corresponding to one amino acid residue was computed as 1.53, 1.48, and 1.52 Å in three separate determinations, in good agreement with the value predicted from the α-helical structure (44). Finally, the dipole moment of helical poly-γ-benzyl-L-glutamate in dioxane and ethylene dichloride was found to be on the order of a few thousand debyes, depending on molecular weight, as expected from theory (45).

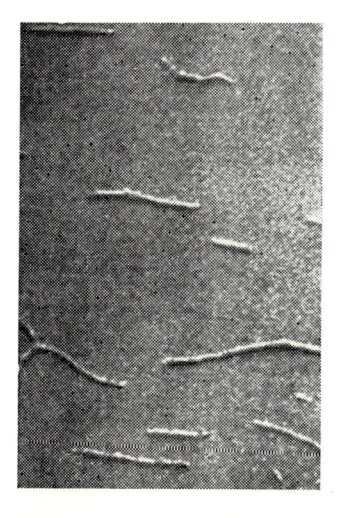

Fig. 1-19. Electron micrograph of poly-L-glutamic acid (molecular weight 43,000) in the form of a helix. From Hall and Doty (46).

An electron photomicrograph of helical poly-L-glutamic acid molecules is presented in Fig. 1-19 (46). The dimensions of the cylindrical molecules coincide within a few percent with those found by other methods. The several structural imperfections that are visible in the photograph most likely arose during preparation of the samples for microscopy. If proper care is not exercised, such regions can lead to an error of up to 30% in the measurement of molecular dimensions by electron microscopy. Polypeptide molecules prepared from solvents which melt the α-helix are observed as disordered clumps in electron micrographs.

All the data presented here serve to convince us that many polypeptides exist as α-helices in a wide variety of solvents and that they behave as rigid rods with a high degree of asymmetry, that is, a large length to diameter ratio.

The behavior of an ionized polypeptide differs from that of a nonpolar one in certain respects as mentioned before. Poly-L-glutamic acid, for instance, consists of rod-shaped macromolecules at acid pH. But when the side chain carboxyls are titrated with base, the uncharged —COOH groups are transformed into charged —COO$^-$ ions which generate strong forces of repulsion along the polypeptide chain. Such electrostatic forces are capable of breaking down α-helical structure (Fig. 1-20).

A very curious yet important feature of the process by which the α-helix becomes disordered is its sharpness: melting occurs only under certain specific conditions. Thus, when a chemical reagent is used to disrupt the hydrogen bonds, the critical concentraition interval is extremely small. If the α-helix is disorganized by heating, then the temperature range in which melting occurs is very narrow. Figure 1-20, for example, illustrates the disorganization of α-helix in poly-L-glutamic acid (molecular weight 34,000) by titration. Within the 0.2 pH units required to increase the degree of ionization of the carboxyl groups from 40 to 80%, the α-helical structure is totally destroyed as evidenced by the drop in specific rotation from −6° to −80°.

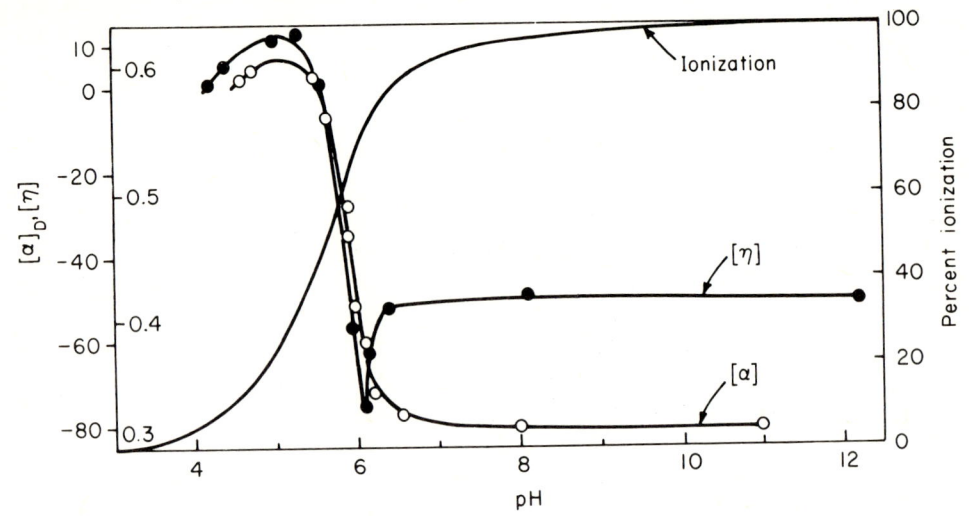

Fig. 1-20. Helix–coil transition in poly-L-glutamic acid as shown by the dependence of intrinsic viscosity and optical rotation on the extent to which the carboxyl groups of the polymer are ionized at different pH values. The solvent employed was a mixture of 0.2 M NaCl and dioxane in the ratio 2:1 at 25°C. From Doty et al. (47).

The change in intrinsic viscosity is particularly revealing here. In the part of the curve corresponding to the α-helical configuration, $[\eta] = 0.65$, quite a large value for this relatively low-molecular-weight polymer. As the α-helix is broken down into a random coil, the intrinsic viscosity drops just as suddenly as does the specific rotation, reaching a value of 0.3. But as the pH rises, the viscosity once again increases to about 0.4 owing to the mutual repulsion of like charges along the chain, a purely polyelectrolytic effect. In order to reduce this effect, the polypeptide solution is maintained at a high ionic strength. In 0.2 M NaCl, the inorganic ions screen the charges on the polypeptide, thereby reducing the repulsion between distant charged groups, but not appreciably altering electrostatic interactions between neighboring charged groups. Poly-L-lysine is subject to the same phenomenon when its amino groups are titrated with acid, starting from pH 10.5, where all the amino groups are uncharged.

The conditions which disrupt the α-helix in synthetic polypeptides are very similar to those which cause protein denaturation, suggesting that helix–coil transitions play a significant role in the latter process. It should be noted, however, that helix–coil transitions in polypeptides are to a large extent reversible. The synthetic compounds lack the structural complexity which, in the case of proteins, both obscures the subtleties of the transition and leads to irreversible denaturation.

Let us return now to the influence of monomer stereoisomerization on the secondary structure of polypeptides. The preceding description bore only on the characteristics of pure enantiomorphs. x-Ray structure analysis shows that racemic polypeptides (i.e., those containing equal proportions of D- and L-stereoisomers) also manifest α-helical structure. These mixed polymers, however, contain both right- and left-helical portions in equal amounts. In order to study the process of helix formation in detail, experiments were performed on a series of poly-γ-benzyl-L-glutamic acid preparations synthesized from mixtures of D- and L-stereoisomers varying from 2 to 50% γ-benzyl-D-glutamic carboxyanhydride.

The behavior of this series of polymers in mixed chloroform-dichloroacetic acid solvents, as determined by their optical activity, is illustrated in Fig. 1-21 (48). Theoretical values of the specific rotation computed according to the principle of additivity are presented to the right of the experimental data for pure dichloroacetic acid (in which α-helix is absent). Upon comparison, we see that there is satisfactory agreement between experiment and theory. For reasons that are apparent, the 50% mixture does not rotate the plane of polarization under any conditions. Considering now the left-hand side of the graph, it is evident that not only helix, but right helix, is formed even when the concentration of the D-isomer is 40%. The contribution of the helix to the optical activity is proportional to the excess of L-isomer over D-isomer.

In a further experiment, L-polypeptides containing about fifteen residues were prepared (49). These short chains were capable of forming a limited amount of right α-helix. Using these molecules as initiators, the chain was extended by polycondensation with (1) pure L-monomers, (2) a mixture of D- and L-monomers, and (3) pure D-monomers. The corresponding changes in the angle of rotation, $\Delta\alpha$, are presented in Fig. 1-22 as a function of reaction time. The results of this experiment are striking. First, when L-monomers are polymerized with the L-polypeptide initiator, $\Delta\alpha$, the difference between the rotations of the polymer and the monomer, increases uniformly with time. This is consistent with the right rotation of polarized light by an α-helix

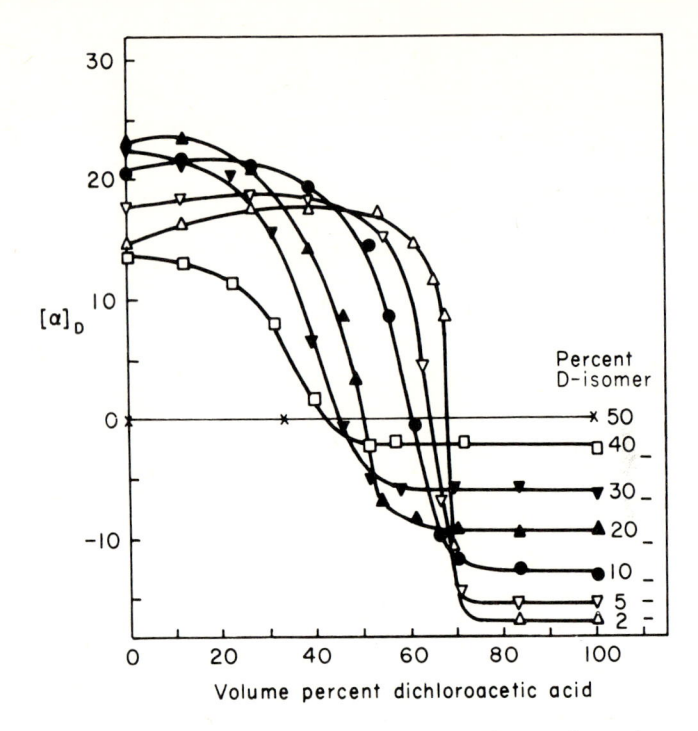

Fig. 1-21. The helix–coil transition in copolymers formed from various mixtures of D- and L-stereoisomers of γ-benzylglutamate as a function of the amount of dichloroacetic acid added to chloroform solutions. The lines on the right indicate the values of $[\alpha]_D$ expected if rotations of D- and L-residues are additive in the random coil configuration. From Blout et al. (48).

composed of L-amino acid residues. When the monomer consists of an equal (racemic) mixture of D- and L-isomers, $\Delta\alpha$ also increases positively with the extent of polymerization. This means that additional helix is formed with the same sense of rotation as the initiator peptide, although the increment in rotation is less for the hybrid polypeptide than for the one composed solely of L-isomers. It is surprising to find that the initial sense of rotation is maintained in this case since it is clear from the reduced overall optical activity that the helix is imperfect.

The third case, in which D-isomers are added to a L-peptide initiator, is much more interesting. The initial rate of polymerization is very small, although $\Delta\alpha$ increases slightly in the initial period. The interpretation is that a few molecules of the D-monomer can attach to the already existing right helix, even though their steric configuration in no way corresponds to it. But when the number of newly attached D-isomers approaches four, a turning point is reached and from there $\Delta\alpha$ starts to fall, eventually becoming negative. After the acquisition of approximately four residues (just a little over one turn of the helix) the sense of the helix apparently changes by 180°, yielding a left helix characteristic of the D-stereoisomer instead of a continuation of the original right helix. The rate of chain growth increases after this change and soon reaches a normal value.

The polymerization of various mixtures of D- and L-anhydrides of γ-benzyl-glutamate, from the pure L-isomer to a racemic mixture of the two, was systematically studied by Doty and Lundberg (50). A peptide composed solely of the L-isomer, and

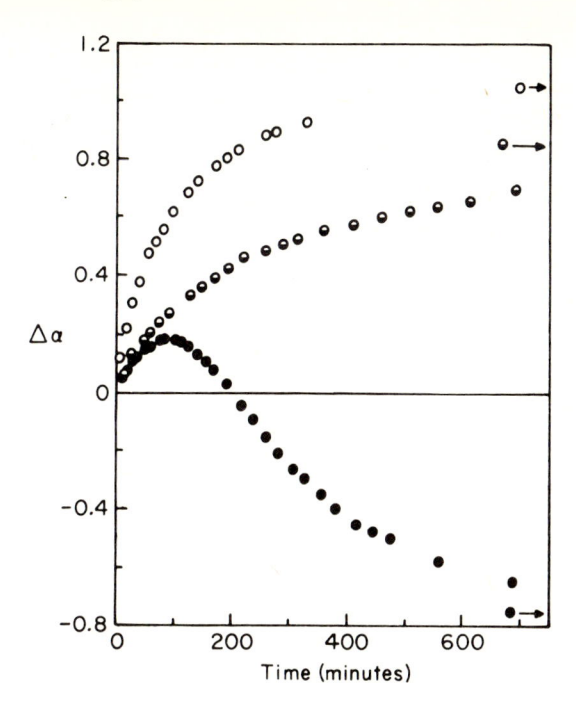

7. Secondary Structure of Polypeptides and Proteins

Fig. 1-22. The change in optical rotation of L- (open circles), DL- (half-solid circles) and D- (solid circles) carboxyanhydrides of γ-benzylglutamate during polymerization initiated by aliquots of an L-polymer. From Lundberg and Doty (49).

therefore in the right helical configuration, was used as an initiator. This peptide contained 15 residues and the secondary polymerization was carried out to a final chain length of 40 residues. The optical rotation of the resulting polymer was analyzed in the following way. The known rotation of the initiating polypeptide was subtracted from the total, as was that of the pure L-polypeptide formed as a result of the excess of L-monomers over D-monomers in the secondary polymerization mixture. This figure was computed by imagining the monomers to consist of the sum of a racemic mixture (in which L-form is equal to the D-form) and an excess of L-form. The latter is treated as if it forms a normal right-helical peptide, the specific rotation of which is known. After subtracting these two contributions, the resulting value of $[\alpha]_D$ should represent the optical rotation of that portion of the helix composed of a racemic mixture of monomers. This parameter, designated $\Delta[\alpha]_D$, is plotted in Fig. 1-23 as a function of the proportion of D-isomer in the secondary reaction mixture. We can see that even with a racemic mixture of monomers (50% D-form) it is primarily a right helix ($\Delta[\alpha]_D = +21°$) which forms. This means that priming with right helical initiator determines the sense of further helix formed, even if the monomers are present in equal proportions. As the fraction of D-form goes to zero, $\Delta[\alpha]_D$ increases toward a limiting value of $+55°$. This value pertains only to the helical increment; the contribution of the asymmetric carbon atoms is not included because $\Delta[\alpha]_D$ is calculated for that part of the mixture in which left and right rotatory forms are present in equal concentrations.

In our previous discussion, the rotation attributable to pure right helix was given as $+48°$. The figure of $+55°$ is a much more exact one. The value $\Delta[\alpha]_D = +21°$

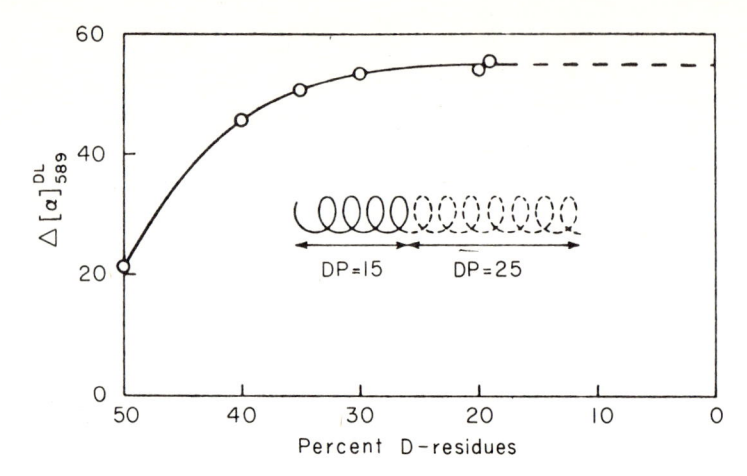

Fig. 1-23. Contribution of the racemic components of the monomer γ-benzylglutamate to the resulting optical rotation of the poly-γ-benzylglutamic acid polymer as a function of the concentration of the D-form of the monomer when synthesis is initiated with pure L-polypeptide. Experiments were carried out in dioxane. Solid helix represents the initiating L-polypeptide, dashed helix the residues added during secondary polymerization, and DP stands for the degree of polymerization of the fragment. From Doty and Lundberg (50).

found when the entire monomer mixture was racemic may also be subjected to theoretical analysis. We have seen that chain growth proceeds in a stereospecific manner; the rate constant for addition of the optical isomer from which the primer is prepared is an order of magnitude greater than that for the other enantiomorph. Thus when a copolymer is formed from a mixture of monomers, one isomer will be preferentially selected. Occasionally there will be a "mistake," and a D-isomer will be added to a right helix. If four D-isomers are added consecutively (representing one complete turn of the helix), the sense of the helix will be reversed and the chain will continue to grow as a left helix by preferential addition of more D-isomers. The polymer chain will thus consist of a series of segments, internally homogeneous as to sense of rotation. If we calculate the probability of inversions resulting from the addition of four D-isomers to a right helix, and of four L-isomers to a left helix, we can determine the theoretical value of the specific rotation of a chain formed from a racemic mixture of monomers. Approximately 30% of the amino acid residues in such a polymer is expected to be in the inverted state. Consequently, the calculated value of the specific rotation is $\Delta[\alpha]_D = 0.7 \times 55° - 0.3 \times 55° = +22°$, in good agreement with the experimental value of $+21°$.

The configurational energy associated with the formation of the α-helix plays an important role in the synthesis of polypeptides. This energy determines the kinetics of formation of all turns of the helix after the first few and forces the growing polymer to pick those stereoisomers out of a mixture which have a natural spatial affinity to the helical structure already present.

8. Optical Properties of Polypeptides and Proteins

Some of the most important methods for investigating secondary structure in polypeptides and proteins are based on their optical properties. We shall first take up

8. Optical Properties of Polypeptides and Proteins

the absorption of ultraviolet and infrared light. If proteins absorb light in the visible region of the spectrum it is wholly attributed to the presence of chromophores or pigments and hence yields little data on structure. Examples of this group are heme-containing proteins such as hemoglobin, myoglobin, the cytochromes, catalase, and peroxidase. All of these proteins are red or bright orange in color owing to the chromophore heme, a porphyrin ring bearing a Fe^{2+} or Fe^{3+} ion.

In the ultraviolet all proteins except protamines have a broad band of absorption near 280 mμ (Fig. 1-24). This band is due solely to excitation of π-electrons in the

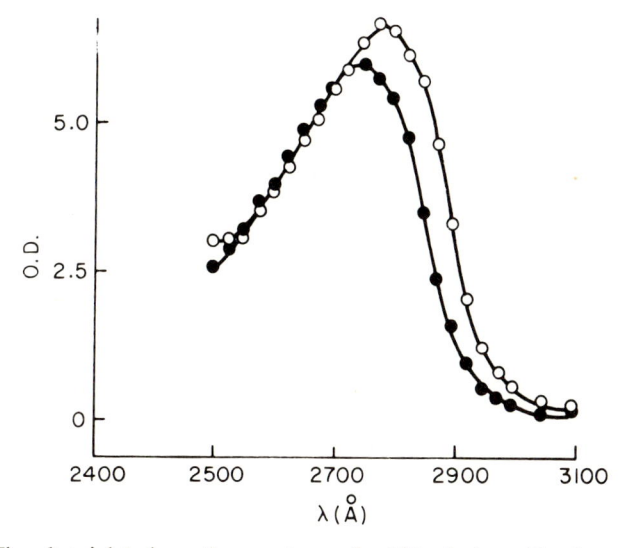

Fig. 1-24. The ultraviolet absorption spectrum of a 1% solution of bovine serum albumin as measured experimentally (O) and as obtained from a mixture of free amino acids taken in the same ratio as in the protein (●). O.D. stands for optical density. From Rideal and Roberts (51).

conjugated rings of tyrosine, phenylalanine, and tryptophan to the π^*-state. The σ-electrons require much larger energies for excitation, and their absorption bands lie in the far ultraviolet as a result.

The fact that absorption bands of various proteins differ in shape and in the position of their maxima is not surprising since the absorption spectra of the aromatic and heterocyclic amino acids are themselves slightly different. Furthermore, the absorption band near 280 mμ is sensitive to a variety of influences which act on the π-electrons of conjugated rings. Among these are various types of complex formation, ionic and dipole interactions, and the formation of hydrogen bonds by functional groups attached to the ring structure.

Especially important is the interaction of tyrosine with the side chain carboxyl groups of glutamic or aspartic acid (52). In insulin and ribonuclease there are, respectively, two and three tyrosine residues associated with carboxylate ions through hydrogen bonds:

The hydroxyl group of tyrosine behaves as a donor and the carboxyl group as acceptor in the formation of this bond. That the hydrogen bond is formed specifically with the carboxylate ion and not with the uncharged carboxyl is evident from the characteristic red shift in the spectrum as the pH is raised from 1.5 (—COOH groups uncharged) to 4.0 (carboxyl groups ionized). This displacement of the absorption maximum is 6 mμ. This effect is most clearly seen when the difference spectrum is plotted, that is, when the optical density measured at pH 1.5 is subtracted from that recorded at pH 4.0.

A typical difference spectrum for insulin, corrected for light scattering by the solution, is presented in Fig. 1-25. There is a maximum at 285 mμ which reveals the

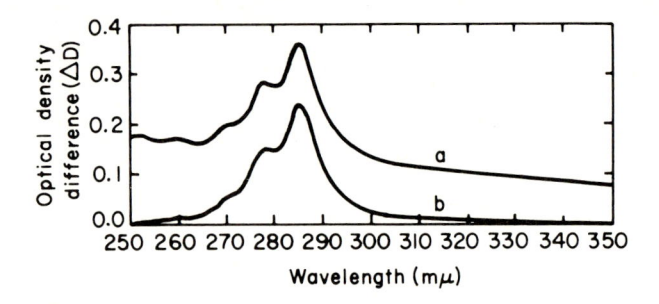

Fig. 1-25. Difference spectrum of a 0.5% solution of bovine insulin measured at pH 3.82 relative to a solution at pH 1.50. (a) Experimental curve. (b) Curve corrected for the effects of light scattering. From Leach and Scheraga (53).

displacement of the absorption band toward the red. The differential optical density at this wavelength, plotted as a function of pH (Fig. 1-26), results in an accurate titration curve for the carboxyl groups of insulin. The pK of this dissociation lies between 3.5 and 3.6. The titration curve of Fig. 1-26 coincides with that measured electrochemically (54). A similar titration curve is obtained for ribonuclease. Moreover,

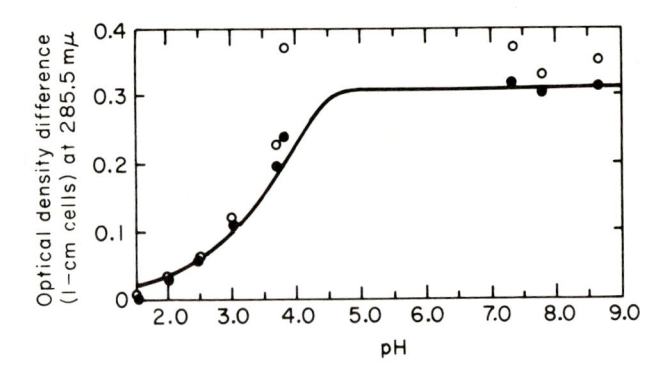

Fig. 1-26. Titration curve of a 0.5% solution of bovine insulin measured by two methods. Open circles correspond to experimental measurements of the difference in optical density at 285.5 mμ and the given pH relative to a solution at pH 1.5. Closed circles represent data corrected for the contribution of light scattering. The solid curve is the electrochemical titration curve of Tanford and Epstein (54) for the —COOH ionization range of insulin. From Leach and Scheraga (53).

8. Optical Properties of Polypeptides and Proteins

its three tyrosine hydroxyls behave anomalously. The hydroxyls become ionized with an apparent pK of 9.5 instead of the pK of 8, normal for free phenol groups. The bond between the —OH group and the —COO⁻ ion evidently impedes ionization of the phenolic hydrogen. Upon denaturation of the protein the hydrogen bond between the tyrosine ring and the carboxylate ion is completely broken. The process of denaturation may thus be studied quantitatively by following the disappearance of the difference peak at 285 mμ.

Let us now consider absorption in the far ultraviolet. The absorption band due to the peptide bond

$$-\underset{\underset{O}{\|}}{C}-\underset{}{\overset{H}{\underset{|}{N}}}-$$

is centered at 190 mμ and has a very special significance (55, 56). Figure 1-27 shows the

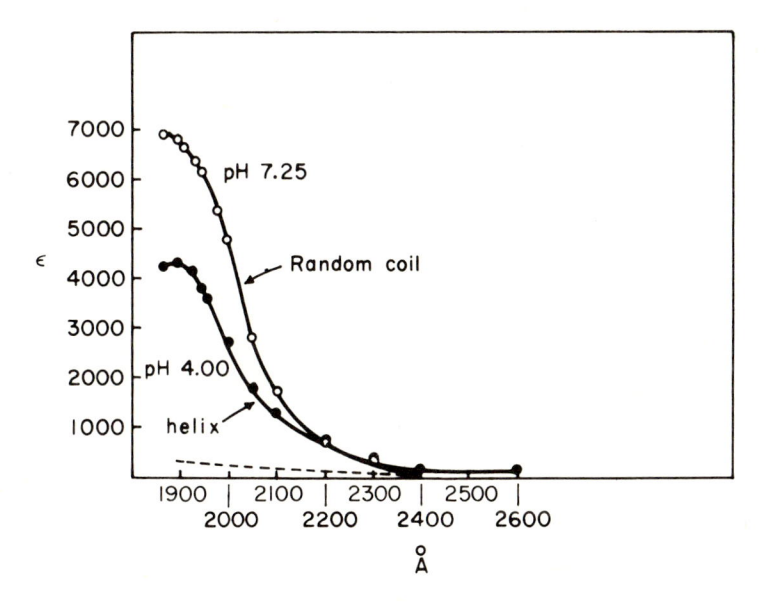

Fig. 1-27. The absorption spectra of poly-L-glutamic acid in the form of a helix (pH 4.0) and of a random coil (pH 7.25) in the far ultraviolet. The broken line shows the scattering of the solution at lower pH. Extinction coefficient given per mole of monomeric units. From Imahori and Tanaka (55).

extinction coefficient of poly-L-glutamic acid in both helical and amorphous configurations as a function of wavelength. The peptide groups of the helical polymer absorb 42% less light at 190 mμ than those of the random coil, although the absorption maximum is not markedly displaced. This phenomenon is called the *hypochromic effect*. It was discovered first in nucleic acids and only later in polypeptides, proteins, and other compounds.

The physical cause underlying the hypochromic effect is the following. The intensity of absorption depends on the dipole transition moment of the electron and the orientation of the dipole transition moments is closely related to the spatial orientation of the peptide groups themselves. When the polypeptide or protein has no secondary structure, the dipole moments are in no way correlated and their electrostatic interaction is neutralized. But when the polypeptide assumes either the α- or β-structure, according to Pauling and Corey, the dipole transition moments are oriented almost parallel to one another, since the peptide groups of adjacent residues in the chain are themselves practically parallel. This spatial correlation of dipole transition moments gives rise to electrostatic forces which decrease ultraviolet absorption. The hypochromic effect can therefore serve as a direct measure of the extent of ordered helical or cross-β-structure in proteins. A practical difficulty with this method, however, is the substantial absorption at 190 mμ by various amino acid side groups. In order to obtain an extinction coefficient for the peptide bond, it is necessary to correct for absorption by side groups, thereby reducing the precision with which the degree of secondary structure can be quantitatively assessed.

The hypochromic effect has been used to study many proteins and some of the results are presented in Table 1-3. Paramyosin, a major constituent of mollusk muscle,

TABLE 1-3

Helix Content of Several Proteins

Protein	Isotopic exchange[a]	Hypochromic effect[b] (average for $\lambda = 190, 197,$ and 205 $m\mu$)	X-ray structure analysis[c]
Paramyosin	—	100	—
Myoglobin	—	82	77
Hemoglobin	—	—	75
Insulin	60	66	—
Ovalbumin	50	—	—
Lysozyme	45	—	42
Chymotrypsin	43	—	3
Bovine serum albumin	35	—	—
Ribonuclease	35	40	16
β-Lactoglobulin, pH 6.4	25	30	—
β-Lactoglobulin, pH 8.7	—	16	—

[a] From Blout et al. (57).
[b] From Rosenheck and Doty (56).
[c] Data on x-ray structure analysis will be found in the following references: 76, 84, 86, 88, and 90.

was used to calibrate the measurement since all data on this protein indicate that its helix content is close to 100%. For comparison, Table 1-3 also presents figures for the helix content determined by isotopic exchange and x-ray structure analysis, the most reliable and precise method available. One advantage of measuring secondary structure by the hypochromic effect is that the protein may be studied in solution. Since the

8. Optical Properties of Polypeptides and Proteins

data for crystalline myoglobin are in satisfactory agreement with those for myoglobin in solution, it is likely that the secondary and tertiary structures of proteins remain essentially the same upon crystallization (58).

Spectroscopy of proteins in the infrared region of the spectrum permits us to discriminate between α- and β-configurations. Although certain of the methods discussed above (isotopic exchange and hypochromic shift) provide us with a means of measuring the number of residues in ordered regions, they do not allow us to decide what the form of the secondary structure is. If the protein molecules under study can be oriented, that is, if samples can be prepared in the form of oriented fibers or films, then infrared spectroscopy—specifically, *infrared dichroism*—can give us the desired information. Infrared dichroism is the difference between the absorption of light polarized parallel to the axis of orientation and that polarized perpendicular to the axis.

Measurement of dichroism yields considerably more than absorption measurements for it is possible to establish the orientation of the most important functional groups (—C=O and —N—H, for example) relative to the molecular axis (59). The principal absorption frequencies for peptide bonds are: 3330 cm^{-1} for the N—H stretching vibration, 1660 cm^{-1} for the C=O stretching vibration and 1540 cm^{-1} for the N—H deformation vibration. In films consisting of α-helical polypeptides, the direction of the stretching vibrations coincide with that of the molecular axes: in the β-structure, however, the stretching vibrations are oriented perpendicular to the axes of the molecules. This exactly corresponds to the models presented in Figs. 1-10 and 1-13. The frequencies themselves are characteristic of hydrogen-bonded

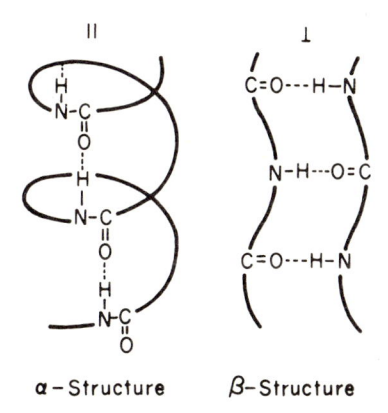

Fig. 1-28. Scheme illustrating the infrared dichroism of polypeptides in the α- and β-configurations. The directions of N—H and C=O stretching vibrations relative to the axis of orientation are shown at the top of the figure. These stretching vibrations are characterized by the following wave numbers and frequencies:

Absorption maxima for hydrogen-bonded groups
  C=O   1660 cm^{-1} ($\lambda = 6.04\ \mu$)
  N—H   3330 cm^{-1} ($\lambda = 3.00\ \mu$)

Absorption maxima for nonhydrogen-bonded groups
  C=O   1695 cm^{-1} ($\lambda = 5.90\ \mu$)
  N—H   3440 cm^{-1} ($\lambda = 2.90\ \mu$)

—C=O and —N—H groups and are displaced from those typical of the free groups (Fig. 1-28).

An interesting sidelight in the behavior of polypeptides is provided by transitions from the α-form to the β-form and vice versa. Naturally occurring fibrous proteins such as hair keratin undergo reversible α → β transitions when stretched to twice their normal length in hot water or water vapor. Alkaline pH further facilitates the α → β transition. Synthetic polypeptides with small side chains such as polyalanine undergo the α → β transition quite readily at room temperatures when the film is placed under tension. This transition can also be reproduced in polypeptides with large hydrophobic side groups by heating the film in water vapor, but the transition is not complete. Many polypeptides assume the β-structure if they are dissolved or even moistened with formic acid. The reverse transition is accompanied by a reduction in length of the molecules and it may be brought about by "annealing" the film at relatively high temperatures (up to 240°C). Alternatively, the α-structure can be restored by moistening the film in one of several solvents (cresol, chloroform) or by completely dissolving the substance in such solvents and then remaking the film.

Infrared spectroscopy also provides a means of determining which of the three basic conformations—random coil, α-helix or β-structure—is present in proteins and polypeptides in solution. Infrared dichroism cannot be used in this case since it is impossible to orient a sample in solution, but the absorption bands themselves undergo a regular, though rather small, displacement depending on polypeptide conformation. Characteristic wave numbers for C=O stretching and N—H deformation bands in the three cases are presented in Table 1-4. The reader should bear in mind that absorp-

TABLE 1-4

Wave Numbers of Infrared Absorption Maxima for Different Polypeptide Chain Configurations[a]

Polypeptide conformation	C=O stretching vibration (cm^{-1})	N—H deformation vibration (cm^{-1})
Random coil	1655–1660	1530–1535
α-Helix	1650–1655	1540–1545
β-Structure	1630–1635	1520–1530

[a] Data from Miyazawa and Blout (59) and Miyazawa (60).

tion differences in proteins are much less than in synthetic polypeptides due to broadening of the bands.

A second important optical property of polypeptides and proteins is their optical activity, the ability to rotate the plane of polarization of light. This phenomenon underlies the most general method for studying the secondary structure of these substances. Optical activity is measured by the *specific rotation* [α], usually expressed in terms of the observed rotation α, the concentration of the solute c in grams per cubic centimeter and the length of the light path d in decimeters:

$$[\alpha] = \alpha/cd \qquad (1\text{-}3)$$

8. Optical Properties of Polypeptides and Proteins

Defined in this manner, the specific rotation is independent of molecular weight.

The optical activity, as noted in a previous section, is composed of two parts, one related to the asymmetric constituents of the polypeptide chain, the second deriving from the asymmetrical secondary structure as a whole. The contribution of the asymmetric carbon atoms of the amino acids can be found by summing the increments due to each residue in the molecule. In natural polypeptide chains, the α-carbons rotate light to the left and thus the configuration of natural amino acids is designated as left. There is an additional complication with free amino acids that arises from the presence of two charged groups, $-NH_3^+$ and $-COO^-$, adjacent to the α-carbon atom. Optical activity is very sensitive to the influence of atoms surrounding the asymmetric center, and charged groups have a particularly marked effect since they tend to polarize the electronic orbitals of the asymmetric atom. This explains the significant differences observed in the optical activities of various amino acids. By changing the nature or pH of the solvent, thereby either curtailing or provoking the electrolytic dissociation of acidic and basic groups, it is possible to alter not only the magnitude, but even the sign of the specific rotation $[\alpha]$. When amino acids are linked together in a polypeptide chain, however, the asymmetric carbon atoms have uncharged $-CO$ and $-NH$ groups as their neighbors so that the problem of polarization disappears. In the latter case, each asymmetric atom makes approximately the same contribution to the optical rotation. Naturally, α-carbon atoms in the terminal amino acids are not taken into account here. These residues all contain one charged group, and the interpretation of their optical activity is quite complicated. Fortunately, the contribution of the terminal amino acid residues is relatively slight when the polymer is of high molecular weight.

For purposes of theoretical analysis, we are primarily interested in the rotation per peptide bond rather than the rotation per unit mass so that the specific optical rotation $[\alpha]$ must be replaced by another parameter, the *mean residue rotation* $[m]$. This is necessary because the mass of the asymmetric centers, that is, the residue mass per peptide bond varies with the nature of the amino acid side chain. Since the specific rotation $[\alpha]$ is defined in terms of total solute mass per volume, it will reflect differences in the composition of the polypeptide chain. Therefore, we shall substitute C, the molar concentration of monomer units, for c, the weight concentration of the polypeptide chains, in Eq. (1-3). If M_0 is the mean residue molecular weight and if C is expressed in moles per 100 cubic centimeters, then $C = 100\, c/M_0$, and the mean residue rotation $[m]$ is

$$[m] = \alpha/dC = (M_0/100)[\alpha] \qquad (1\text{-}4a)$$

The somewhat unusual dimensions for C are chosen so that values of $[m]$ will be of the same order of magnitude as $[\alpha]$. The variation of optical rotation with refractive index, n, must also be taken into account. The required correction can be calculated from electromagnetic theory, permitting us to define an *effective residue rotation* $[m']$:

$$[m'] = \frac{3}{(n^2 + 2)}[m] = \frac{3}{(n^2 + 2)} \times \frac{M_0}{100}[\alpha] \qquad (1\text{-}4b)$$

The multiplicative factor in Eq. (1-4b) effectively reduces optical rotation data to values characteristic of a medium whose index of refraction is unity. The effective residue rotation $[m']$ is therefore the optical rotation parameter generally used in studies on polypeptide secondary structure.

We shall calculate the value of $[m']$ for two water-soluble peptides, poly-L-glutamic acid and poly-L-lysine, under conditions in which the polymers form random coils. Let us first consider the data for poly-L-glutamic acid at alkaline pH, where the polypeptide chain is completely ionized. The specific rotation, $[\alpha]_D$, measured with the sodium D line at 589.3 mμ, is $-107°$, the molecular weight per residue M_0 is 129, and the index of refraction of the medium n is 1.33. From Eq. (1-4b) we compute that $[m']_D$ is $-109°$. For poly-L-lysine in 6 N HCl, the corresponding figures are $[\alpha]_D = -80°$, $M_0 = 130$, $n = 1.38$, and the resulting value of $[m']_D$ is $-82°$. The effective residue rotation thus varies as a function of side-chain structure. Similar variations are observed in denatured protein where the average value of $[m']_D$ per mole of amino acid residues falls between $-85°$ and $-100°$. Such variations reflect differences in the average amino acid composition of individual proteins.

The second part of the optical activity is determined by the structure of the protein macromolecule as a whole. If the Pauling-Corey α-helix is the most important element of the secondary structure of polypeptides and proteins, it should make a large contribution to the optical activity. Calculation of the increment $[m']$ for poly-γ-benzyl-L-glutamate acid yields a figure of $+100°$, while for other peptides it reaches $+105°$. As noted previously (p. 47ff), natural L-amino acids are inclined to form right helices, while D-amino acids form left helices. There we examined the structural peculiarities of left and right helices, especially the differences in packing of amino acid side groups with respect to the sense of the helix.

A very interesting paradox was discovered by Blout and co-workers (33) in their study of poly-β-benzyl-L-aspartate:

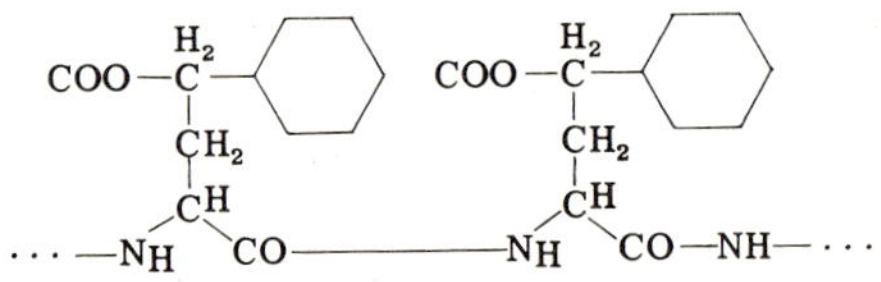

Although the length of the side chain here is shorter than that in poly-γ-benzyl-L-glutamic acid by only one —CH$_2$ group, there was very good evidence that the chain of poly-β-benzyl-L-aspartate was left helical rather than right helical as expected for L-amino acids. In addition, the helical structure of poly-β-benzyl-L-aspartate melts out at a much lower concentration of dichloroacetic acid when chloroform is gradually replaced by the polar solvent (Fig. 1-29). This means that the left rotatory structure, in which the side groups are oriented along the chain, is thermodynamically more advantageous in the case of poly-β-benzyl-L-aspartate, although it is less able to withstand the action of polar solvents and hence, less stable than the right helix of poly-γ-benzyl-L-glutamate. Especially noteworthy is the enormous (negative) increment in optical rotation due to the left helical structure compared to the much smaller (positive) increment of the right helix. This is apparently related to the different orientation of side chains in the two cases.

8. Optical Properties of Polypeptides and Proteins

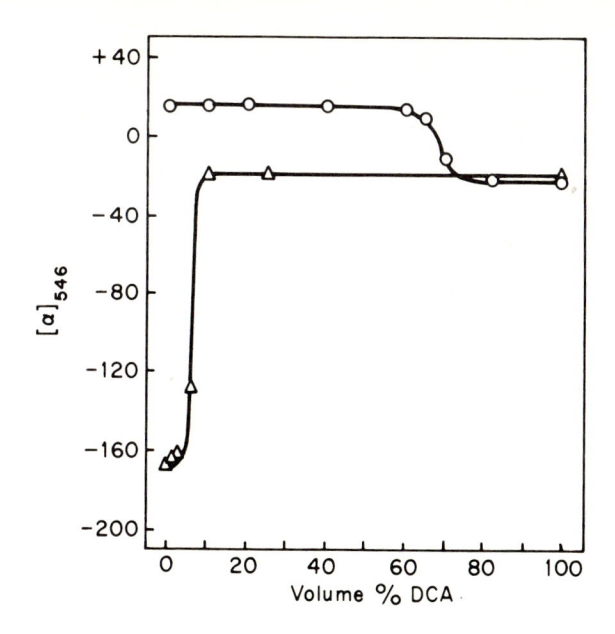

Fig. 1-29. The helix-coil transition in poly-γ-benzyl-L-glutamate (○) and poly-β-benzyl-L-aspartate (Δ) measured by changes in specific optical rotation when dichloroacetic acid is added to a chloroform solution of the polymer. The volume percent of dichloroacetic acid is shown on the abcissa. From Karlson et al. (33).

Blout and Fasman have shown that in thin films, the polypeptide poly-O-acetyl-L-serine:

assumes only the β-structure, as is readily judged from its infrared dichroism (61). Its behavior in solution, however, indicates that it is possible to break down the ordered structure. When the solvent is gradually changed from chloroform to dichloroacetic acid, there is a drop in the optical rotation from positive to negative values, presumably indicating a transition from the β-structure to a random coil. Melting out of the β-structure in this case produces a change in $[m']_D$ of 100°, that is, of about the same value as for helix–coil transitions.

The same type of sharp structural transitions that are encountered in polypeptides are also found in proteins at certain critical temperatures and solvent concentrations. As one example, the change in optical activity of ovalbumin upon irreversible heat denaturation is illustrated in Fig. 1-30 (62, 63). Even more interesting is the reversible heat denaturation of trypsin. Although the functional activity of this proteolytic enzyme is destroyed by boiling, it can be restored simply by cooling to room temperature. A measure of the enzyme's structural integrity is given in Fig. 1-31, where it can

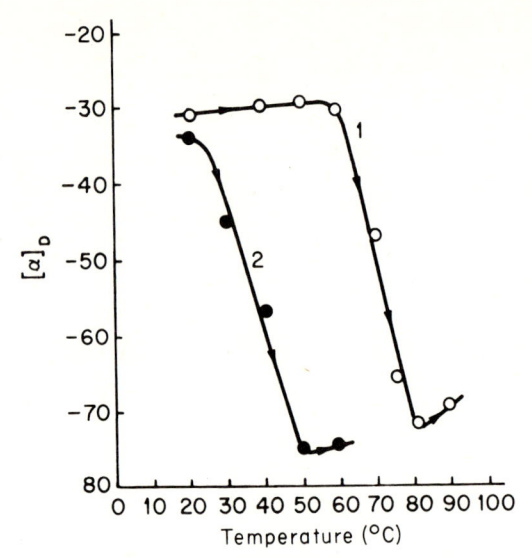

Fig. 1-30. Melting of the secondary structure in ovalbumin at different pH values as measured by changes in specific optical rotation; curve 1: pH 9.4; curve 2: pH 1.8. From Bresler et al. (63).

be seen that the optical activity returns almost to the initial value after a cycle of heating and cooling. The resistance of trypsin to denaturation is attributed to the presence of six disulfide bridges in the molecule.

We cannot *a priori* assert that the ordered right rotatory structure destroyed by heating a protein is α-helix. The presence of structural optical activity in proteins has at times been erroneously interpreted as an indication of helical configuration. However, the absence of a symmetry plane in a chemical structure is the sole necessary and

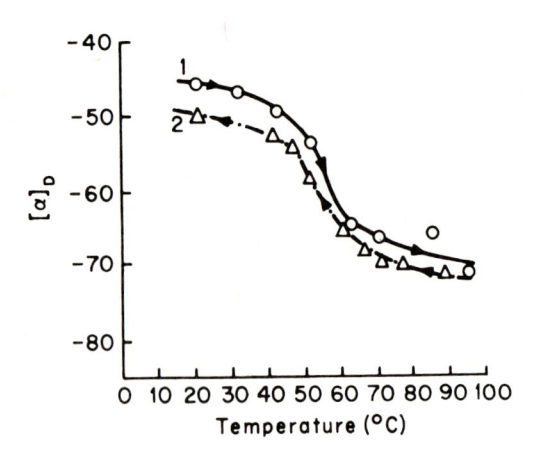

Fig. 1-31. Melting of the secondary structure in trypsin at pH 3.1. Curve 1 shows the changes in optical activity that occur during melting of the original 1% solution. Curve 2 traces the restoration of secondary structure after the solution was heated to 100°C and then cooled. From Kushner and Frenkel' (64).

8. Optical Properties of Polypeptides and Proteins

sufficient condition for the occurrence of optical activity. A quartz crystal serves as a simple example of this. Although the SiO_2 molecule is itself symmetrical, the lattice in which the molecules are packed is devoid of a plane of symmetry. As a result, the quartz crystal is optically active though there is no helical element present. The criterion of optical activity is therefore not sufficient to discriminate between the helix of the α-structure and the layered packing of the β-structure. Both configurations represent internal order in protein molecules and both result in structural optical activity. At present, there exist no unimpeachable physical methods, with the exception of x-ray structure analysis, by which to determine separately the amount of α-helix and of β-structure in a given protein.

Yet another phenomenon which should be considered is *optical rotatory dispersion*, the dependence of optical activity on the wavelength of light used. The general theory of this phenomenon was worked out by Kirkwood, while Moffitt investigated the special case of helical molecules (65). Under conditions in which the optical activity is determined solely by asymmetric carbon atoms, the relationship between effective residue rotation, $[m']$, and wavelength, λ, is given by a Drude equation:

$$[m'] = a_0 \lambda_0^2 / (\lambda^2 - \lambda_0^2) \tag{1-5}$$

in which a_0 is a constant. This equation is consistent with experiment if the wavelengths used are not too close to the absorption bands, λ_0, of atoms or groups adjacent to the asymmetric atoms. An effect known as *anomalous dispersion* occurs near this absorption band. According to the Drude formula, the optical activity goes to infinity at λ_0, which of course cannot happen with real substances. The optical activity does nevertheless become large in this region and evidences a very peculiar dependence on wavelength, which we will discuss further below.

Upon the formation of intramolecular order such as α-helix there is a change in the absorption of light by peptide bonds induced by the interaction between parallel oriented dipole transition moments. The nature of the optical rotatory dispersion undergoes a simultaneous change. Using the theory of Kirkwood, Moffitt obtained a semiquantitative expression for the dispersion of $[m']$ by macromolecules containing helical segments:

$$[m'] = a_0 \lambda_0^2 / (\lambda^2 - \lambda_0^2) + b_0 \lambda_0^4 / (\lambda^2 - \lambda_0^2)^2 \tag{1-6}$$

The second term represents the contribution of helix to the optical rotatory dispersion (Fig. 1-32). The coefficient, b_0, is proportional to the fraction of α-helix present and is negative for right helix, positive for left helix. For random coils, $b_0 = 0$ and the dispersion is subject to the Drude equation. The approximation introduced by Moffitt was to replace several real absorption bands with one effective wavelength, λ_0. According to Moffitt's calculation, $\lambda_0 = 200$ mμ, while the empirical value determined for proteins is $\lambda_0 = 212$ mμ. For paramyosin, which appears to be the most highly helical of all known proteins, the coefficient b_0 is equal to $-630°$.

The proportion of helix in other proteins can thus be determined either by measuring $[\alpha]_D$, or by finding the Moffitt coefficient, b_0, and dividing it by 630 (42). The proportions of helix content calculated from $[\alpha]_D$ and b_0 agree with each other in some cases (Table 1-5), although there is a large discrepancy in the case of insulin and ovalbumin. The value of b_0 for poly-β-benzyl-L-aspartate, is -600, indicating nearly

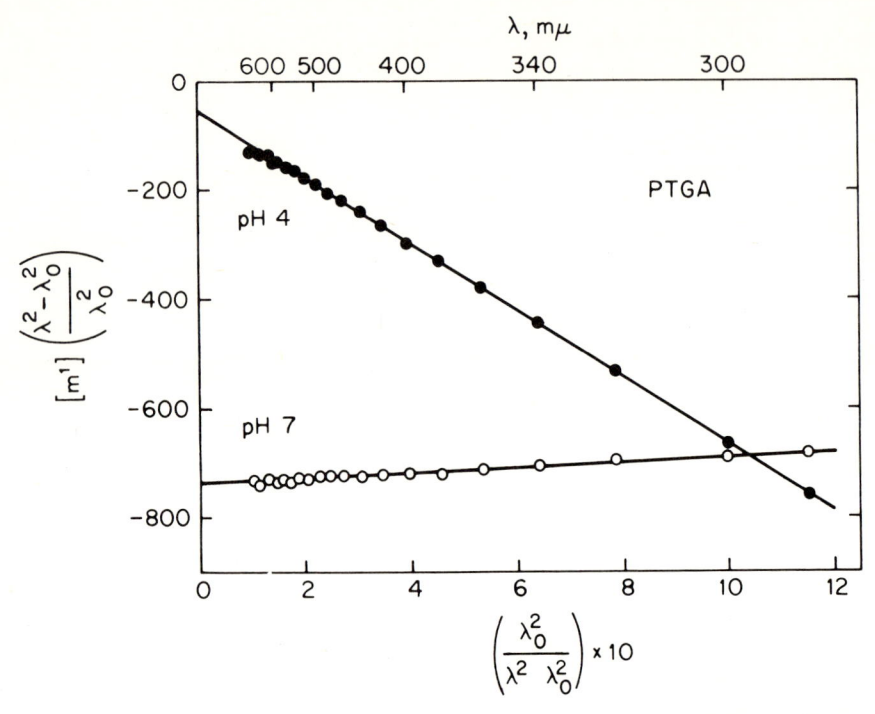

Fig. 1-32. Graphical treatment by the Moffitt equation of the optical rotatory dispersion of a synthetic polypeptide in two configurations. The polypeptide, a copolymer of 5% L-tyrosine with 95% L-glutamic acid, is in the helical configuration at pH 4.0 in 0.1 M phosphate buffer (a_0 is close to zero and $b_0 = -650$) and in the random coil configuration at pH 7.0 in the same solvent ($a_0 = -750$, b_0 close to zero). In these calculations, $\lambda_0 = 212$ mμ. From Urnes and Doty (42).

TABLE 1-5

Helix Content of Native Proteins in Solution[a]

Protein	Percent helix as estimated by Optical rotation	Percent helix as estimated by Optical rotatory dispersion
Paramyosin	100	100
Myoglobin	—	74
Insulin	59	38
Ovalbumin	53	31
Bovine serum albumin	46	46
Lysozyme	37	29
Pepsin	26	31
Ribonuclease	17	16
Chymotrypsin	—	15
β-Lactoglobulin	—	11

[a] Data from Urnes and Doty (42) and Rosenheck and Doty (56.)

8. Optical Properties of Polypeptides and Proteins

100% left helix. Data on the amount of β-structure in polypeptides and proteins are consistent with the Moffitt equation, but attempts to obtain reasonable values of b_0 in this case have been unsuccessful. The magnitude of b_0 appears to vary within a broad range and may even change sign, depending on whether the polymer is characterized by parallel, antiparallel, or cross-β-structure, or by still more subtle variations of conformation. In general, there are so many possible differences in protein secondary structure that optical rotatory dispersion may only be qualitatively interpreted.

Important results have been obtained by Simmons, Blout, Szent-Györgyi, and others by investigating the anomalous dispersion of optical rotation in polypeptides and proteins close to the absorption band of the CO—NH bond. In the case of two polypeptides, poly-L-methionine and poly-γ-benzyl-L-glutamate, the entire curve for anomalous dispersion can be measured since the absorption band lies near 220 mμ (Fig. 1-33). Close to the absorption band, the optical activity becomes extremely large,

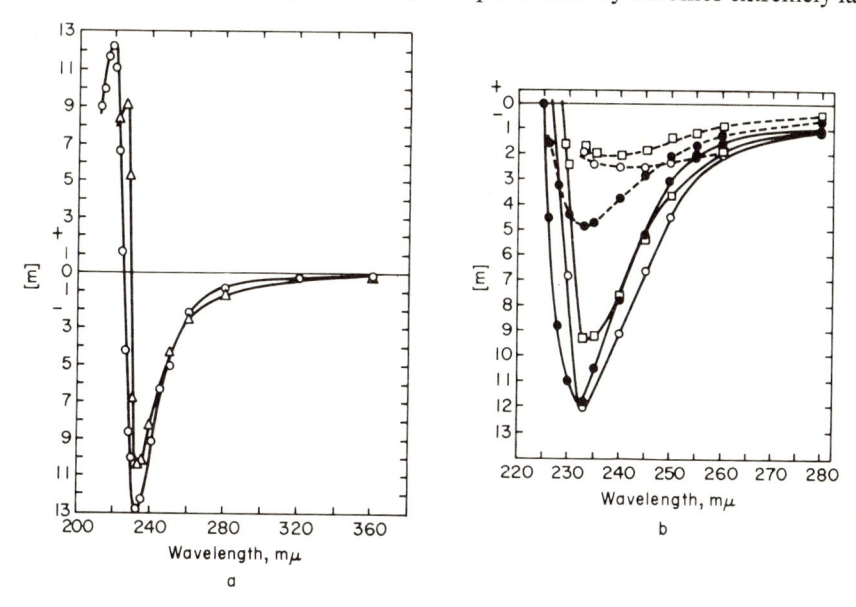

Fig. 1-33. The Cotton effect in polypeptides and proteins. (a) Ultraviolet rotatory dispersion of poly-γ-benzyl-L-glutamate (Δ) in dioxane solution and of poly-L-methionine in methylene dichloride solution (O). (b) Ultraviolet rotatory dispersion of some fibrous α-proteins: paramyosin (●), tropomyosin (O), and myosin (□); solid line represents data for proteins in 0.6 M KCl; dashed line represents data for proteins in 8 M urea. From Simmons et al. (66).

goes through a maximum, drops rapidly, and then, changing sign, passes through a minimum or trough about as great in magnitude as the maximum. Such a change in optical rotation, which is generally accompanied by dichroism, is called the *Cotton effect* (66, 67).

For proteins, whose peptide absorption band lies near 190 mμ, it is difficult to obtain the entire curve. However, a trough at 233 mμ, which can reach an enormous figure (−12,000° for polypeptides and paramyosin), is observed in all proteins and polypeptides with a helical structure. The trough does not appear in the case of protamines which completely lack helical portions. When protein secondary structure is disrupted by 8 M urea, the depth of the trough is reduced at least tenfold. This parameter is roughly proportional to the amount of helix in the molecule. The Cotton effect may therefore be used as a qualitative index of protein secondary structure.

The anomalous dispersion stemming from the β-structure is less dramatic than that displayed by α-helices. Figure 1-34 presents curves for the anomalous dispersion

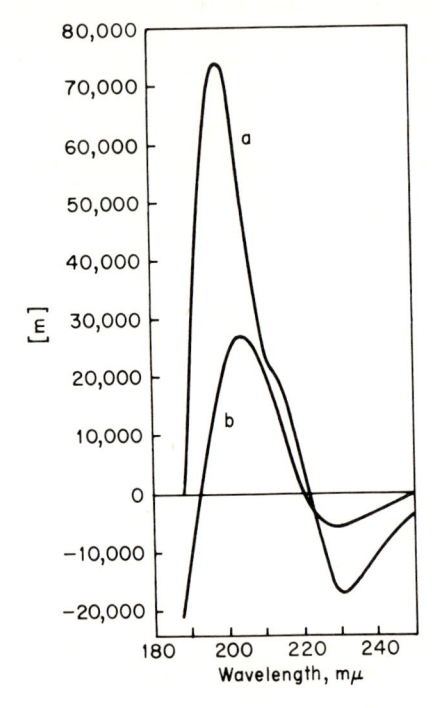

Fig. 1-34. The optical rotatory dispersion of poly-L-lysine in the α- and β-configurations. (a) Polypeptide in the α-configuration at pH 11.06 and 22.5°C. (b) Polypeptide in the β-configuration after heating to 51°C for 15 minutes followed by cooling to 22.5° for measurement of ORD spectrum. From Davidson and Fasman (37).

of poly-L-lysine in both α- and β-forms. The position of the rotation trough in the β-structure is shifted relative to the α-helix from 233 mμ to 230 mμ and its depth is reduced from $-15,700°$ to $-6,000°$. At the same time, the rotation maximum of the β-conformation is 205 mμ compared with 198 mμ for the α-helix, and its value is less by a factor of 2.5.

The ideas and methods developed to investigate the secondary structure of polypeptides can be used to advantage for proteins. The content of ordered regions in the polypeptide chain can be determined by means of isotopic exchange, the hypochromic effect, the specific optical rotation, and optical rotatory dispersion. The proportion of ordered regions can also be calculated on the basis of the effective residue rotation, $[m']_D = 100°$, calculated from data on synthetic polypeptides and paramyosin. The degree of order and its stability are a function of solvent composition and the extent to which the protein is charged. A change in solvent from dichloroethane to dichloroacetic acid, for instance, completely disorganizes α-helical and cross-β-regions in proteins as it does in polypeptides. A comparison of certain optical parameters such as the specific rotation before and after the solvent change provides one approach to evaluating the percent of order in the protein chain.

The content of ordered regions in several proteins, estimated both by optical rotation (based on $[m']_D = 100°$) and optical rotatory dispersion (based on $b_0 = -630°$) is presented in Table 1-5 (compare also Table 1-3, p. 54). The two tables demonstrate that in many cases there is good agreement among the various methods. In particular, where the optical activity and dispersion data are in accord, there is good reason to believe that the protein contains α-helix. The results obtained for ribonuclease by optical rotation, however, differ markedly from figures derived by other methods, suggesting that helix may comprise only part of the ordered structure of the enzyme. This conclusion has been borne out by x-ray diffraction studies (p. 97). A significant discrepancy in the optical data is also found for insulin and ovalbumin. As has been noted, the only wholly satisfactory method available at present for measuring the helix or cross-β-content is x-ray structure analysis. Myoglobin, lysozyme, and ribonuclease, which have been extensively investigated by the latter method, thus provide the best comparison of the different techniques for assessing secondary structure in proteins (Table 1-3).

9. Theory of the Order–Disorder Transition

The theory of helix–coil transitions in proteins and polypeptides was developed independently by Zimm and Bragg and by Gibbs and DiMarzio (68). The exact theory is quite complicated; for purposes of illustration, we shall limit ourselves to a simpler but more graphic exposition of it with particular emphasis on the thermodynamics of the process.

Order–disorder transitions in proteins, whether irreversible or reversible, as in the denaturation–renaturation cycle of ribonuclease and trypsin, are examples of what in statistical mechanics are called cooperative transitions. In a cooperative transition, many similar elements, here the amino acid components of the protein chain, undergo a simultaneous transition from an ordered state to a disordered one. The simultaneity and sharpness of the transition are caused by interaction between the similar elements. These phenomena can be interpreted on the basis of a relatively simple thermodynamic theory. Let us assume that the protein chain consists of n_0 residues, of which n residues are located in disordered, amorphous regions of the macromolecule and $(n_0 - n)$ in regions of helical or cross-β-structure stabilized by hydrogen bonds. The transition of one residue from the ordered to the amorphous state increases the residue energy by an amount, q, the energy necessary to break one hydrogen bond in the Pauling-Corey α-helix or in the β-structure. As the residue goes from the ordered to the disordered state, its entropy also increases since there appear v new rotational or conformational degrees of freedom. The increase in entropy for n residues going to the disordered state is expressed by the formula:

$$\Delta S_1 = (kvn/2)\ln(T/T_0) \qquad (1\text{-}7)$$

where k is the Boltzmann constant, v the number of rotational degrees of freedom per residue, T the absolute temperature, and T_0 the temperature of quantum degeneracy approximately equal to 1°K. The last constant is defined by the expression:

$$T_0 = h^2/8\pi^2 kI \tag{1-8}$$

where I is the moment of rotational inertia per residue, h the Planck constant, and k the Boltzmann constant.

At the outset the cooperative nature of the transition will not be taken into account. In order to find out how n_0 residues are distributed between the two states—hydrogen-bonded and disordered—we must write the expression for the change in free energy ΔF as a function of n, and then find the minimum value of ΔF. The ordered state in which all n_0 elements participate in helix is taken as the standard state. The change in free energy ΔF associated with the transition to disorder is

$$\Delta F = \Delta H - T \Delta S \tag{1-9}$$

where ΔH is the change in enthalpy and ΔS the change in entropy. The entropy actually consists of two parts, ΔS_1 defined by Eq. (1-7) and ΔS_2, the entropy of mixing n disordered residues with $(n_0 - n)$ hydrogen-bonded residues:

$$\Delta S_2 = k[n_0 \ln n_0 - n \ln n - (n_0 - n)\ln(n_0 - n)] \tag{1-10}$$

from which we find

$$\Delta F = nq - \frac{nTvk}{2} \ln \frac{T}{T_0} - kT[n_0 \ln n_0 - n \ln n - (n_0 - n)\ln(n_0 - n)] \tag{1-11}$$

To determine the equilibrium distribution of states, we must minimize ΔF by satisfying the equation $\partial \Delta F/\partial n = 0$. After a simple computation, we obtain:

$$n/(n_0 - n) = (T/T_0)^{v/2} e^{-(q/kT)} \tag{1-12}$$

which is in the form of the familiar Boltzmann equation. It shows that for every temperature there is an equilibrium ratio between disordered and hydrogen-bonded residues and that the ratio changes gradually with increase or decrease in temperature. According to this formulation, in which cooperativity was ignored, a sharp helix–coil transition is not predicted.

Consider now the physical significance of cooperativity. The liberation of a single residue from its hydrogen bonds requires a certain amount of energy, q, but it does not result in an immediate increase of the entropy, S_1. In other words, disruption of the hydrogen bonds of one residue alone does not leave it sufficiently free to rotate. It is necessary to liberate at least two or three neighboring residues in order to realize additional rotational degrees of freedom.

To express cooperativity mathematically, the expression for ΔS_1 must be multiplied by the probability of finding a second free residue next to one already liberated from its hydrogen bonds. Since this probability is simply n/n_0, the corrected value of ΔS_1 is

$$\Delta S_1 = (kvn^2/2n_0)\ln(T/T_0) \tag{1-7a}$$

Putting this into the expression for free energy and differentiating, we have

$$\frac{1}{kT} \frac{\partial \Delta F}{\partial n} = \ln \frac{n}{(n_0 - n)} + \frac{q}{kT} - \frac{nv \ln(T/T_0)}{n_0} = 0 \tag{1-13}$$

9. Theory of the Order–Disorder Transition

In order to solve this equation for n, we must find the intersection of the following two curves:

$$y = \ln n/(n_0 - n) \tag{1-14a}$$

$$y' = -\frac{q}{kT} + \frac{nv \ln(T/T_0)}{n_0} \tag{1-14b}$$

These functions are plotted in Fig. 1-35a. There are actually three intersections and

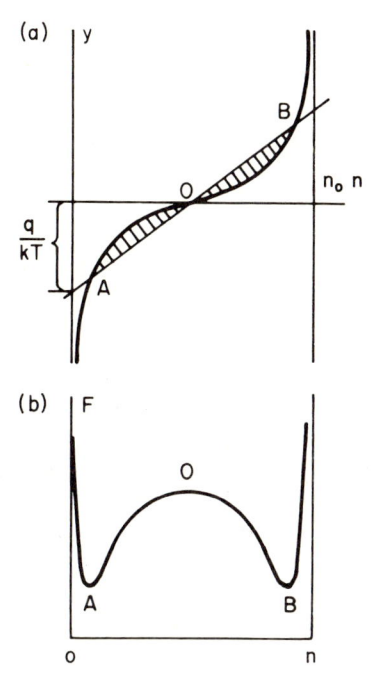

Fig. 1-35. Theory of the cooperative helix–coil transition. (a) Solution of the transcendental equation for location of the free energy minima corresponding to the helical and random coil configurations of the polymer chain. (b) Curve of the free energy, F, as a function of the degree of disorder, n. Points A and B correspond to the coexisting regular and disordered phases, respectively.

hence, three solutions for n. The first, at point A, corresponds to small n, that is, a high degree of order; the second, at B, corresponds to almost complete absence of hydrogen bonding or almost complete disorder. The third solution at O is at a free energy maximum, not a minimum.

So as to be convinced of this interpretation, we shall plot F as a function of n by integrating:

$$dF = kT\left[\ln \frac{n}{n_0 - n} + \frac{q}{kT} - \frac{nv \ln(T/T_0)}{n_0}\right] dn \tag{1-15}$$

The integral $F = \int dF$ is nothing more than the area bounded by the curves y and y' (Eqs. 1-14a and b). It is clear from the plot of $F(n)$ versus n, (Fig. 1-35b) that at a certain temperature both the ordered (A) and disordered (B) states may exist simultaneously and that a cooperative transition between them is possible.

But a cooperative transition can occur only if states A and B have precisely the same free energy, that is, if $F_A = F_B$, in addition to being free energy minima. In other words, $\int_A^B dF$ must be equal to O, or, referring to Fig. 1-35a, the area of the cross-hatch region between A and O must be equal to that between O and B. When this condition is met, the line AB, whose intersections with the function $y(n)$ represent the phase transition, must pass through the point O, where $n = n_0/2$. Mathematically this condition is written:

$$-\frac{q}{kT} + \frac{v \ln(T/T_0)}{2} = 0 \tag{1-16}$$

While another straight line might also intersect with $y(n)$ in three places, it could never fulfill the condition $F_A = F_B$, since the implication would be that either state A or state B was thermodynamically more stable than the other and hence that they could not coexist under any circumstances.

Equation (1-16) can be solved for the temperature of the cooperative order–disorder transition, that is, at which the secondary structure melts or breaks down:

$$T = \frac{\Delta H}{\Delta S} = \frac{q}{k(v/2) \ln(T/T_0)} \tag{1-17}$$

ΔH is the change in enthalpy per residue and ΔS is the change in entropy per residue during the transition. If the value of q, the latent heat of melting, were known, ΔS and finally v, the number of degrees of freedom per residue, could be calculated. Until recently, only a very rough estimate of q was available which suggested a value on the order of 1900 cal/mole.

The latent heat of melting for the helix was recently measured in the laboratory of the author (27). A differential dynamic calorimeter specially built for the purpose was used. This quantity is difficult to assess because the melting interval is quite broad, on the order of 10°. Most of the heat input therefore goes into raising the temperature of the solvent, and very little into the latent heat of melting. The temperature of 100 ml of a 0.2% protein solution in water, for example, must be increased by 10° in order to bring about the transition. This requires about 1000 cal while the latent heat of melting accounts for only about 2 cal per 0.2 gm of protein or 0.2% of the total heat introduced. To measure this tiny difference, two adiabatic calorimeters were immersed in a single thermostatic bath. One contained buffer and the other held the protein solution in the same buffer. Both calorimeters were then heated in an identical fashion for a certain period of time. As a result of the absorption of heat in the melting process, the temperature of the second calorimeter lagged behind that of the first by a few hundredths of a degree. This increment was measured with an accuracy of 10^{-3} degree over several hours until the complete melting curve for the helix was obtained. From this the absolute value of the latent heat of the helix–coil transition was calculated after calibration of the calorimeters. The value of q for horse hemoglobin was found to be 1400 ± 100 cal/mole. This figure is quite precise since the helix content of this protein is known with great accuracy.

Table 1-6 presents calorimetric data on the helix content of several proteins using a latent heat of 1400 cal/mole as a basis for calculation. Results for the same proteins obtained by other methods are included for comparison.

TABLE 1-6

MEASUREMENT OF THE HELIX CONTENT OF SEVERAL PROTEINS[a]

Protein	Percent helix content as measured by		
	Heat of transition	Optical rotation	Optical rotatory dispersion
Horse hemoglobin	75	78	>70
Horse serum albumin	50	48	52
Ovalbumin	48	48	45

[a] Based on Aldoshin et al. (27).

In certain cases the heat of melting can be determined only by conducting the measurements at several different concentrations and extrapolating the observed values of q to zero concentration. This corrects for the effects of aggregation on the latent heat and is particularly important in the case of proteins such as ovalbumin which aggregate readily. Needless to say, experimental conditions should be chosen to minimize aggregation. The pH should not be close to the isoelectric point of the protein and the ionic strength should be kept small.

At issue here is the size of the contribution of electrostatic Coulomb forces between ionic groups in the protein to the total measured value of q. The answer to this can be obtained theoretically. The melting temperature of the secondary structure is given by Eq. (1-17). Since the entropy per residue is practically independent of whether or not a side group is charged, a change of δT in the melting temperature results exclusively from a change δq in the heat of melting. Therefore:

$$\delta T/T = \delta q/q \qquad (1\text{-}18)$$

For ovalbumin a rise in pH from 2 to 9 reduces the melting point by 30° whence $\delta q/q = 30°/340° = 8\%$. The effect of electrostatic forces is therefore inconsequential compared to the forces involved in hydrogen bonding. An 8% change in q is close to the error of measurement and we shall not concern ourselves with it further.

Finally, the entropy change ΔS in a helix–coil transition is a very significant 4 entropy units per mole of residues when $q = 1400$ cal/mole and $T = 340°$K (Eq. 1-17). For a polypeptide composed of 500 residues (molecular weight 50,000), the entropy change amounts to 2000 entropy units per mole upon melting. This figure indicates an average of $v = 0.7$ rotational degrees of freedom per residue, meaning that all the residues never attain complete rotational freedom. This should not be surprising in view of the great complexity of the polypeptide chain.

10. Tertiary Structure of Proteins

Three factors prevent globular proteins from assuming a completely helical conformation in solution. First, intramolecular disulfide bridges unite various parts of the polypeptide chain. Second, the presence of proline and hydroxyproline produces sharp, 130° bends in the polypeptide which disrupt the Pauling–Corey helix. Third, forces of cohesion between amino acid side groups, or to put it another way, differences in the attraction of side groups for the solvent and for each other, cause the chain to fold. Acting together, these phenomena create the tertiary structure of a protein, forcing it to assume a globular configuration.

Alterations in protein tertiary structure in solution can be studied by gradually replacing water with organic solvents such as dioxane and chloroethanol (62, 63). These solvents do not break intramolecular hydrogen bonds, but tertiary bonds are weakened since attraction between hydrophobic side chains and solvent grows as the proportion of water is decreased. Investigation by hydrodynamic and optical methods, in particular, measurement of intrinsic viscosity and optical rotation, indicates that the molecule becomes several times longer and that the proportion of α-helix in it increases.

Ovalbumin, for instance, undergoes a very large elongation when water is replaced by a less polar solvent. Viscosity data can be used to compute the ratio of major to minor axes a/b of an ellipsoid of revolution with the same volume as the molecule. In citrate buffer at pH 1.8, a/b for ovalbumin is equal to 4.0 assuming 20% hydration. The axial ratio in chloroethanol rises to about 35 indicating a large increase in asymmetry. The limiting extension for the ovalbumin α-helix (molecular weight 44,000) corresponds to an axial ratio of 40. Furthrmore, the helix content of the molecule is markedly greater in the organic solvent than in water as judged by its specific optical rotation. In the extended state, ovalbumin loses its tertiary structure and unfolds into one long helix of the type we encountered when discussing synthetic polypeptides.

Of the three factors contributing to the tertiary structure, replacement of a polar solvent by a nonpolar one influences only the interaction between amino acid side groups. Its disruptive effect therefore depends on whether or not the protein is "vulcanized" with disulfide bridges. Ovalbumin contains just one disulfide bridge along its single polypeptide chain, and by all indications it is localized, forming only a small loop in the molecule. The protein thus behaves as if it contained no —S—S— bridges at all. In this regard, note that ovalbumin undergoes a very sharp helix–coil transition (Fig. 1-30), the entire melting process falling within a 12° to 15° interval.

A completely different picture emerges from an examination of human serum albumin. There are seventeen disulfide bridges in this molecule which not only hinder helix formation, but also hold the compact tertiary structure tightly together. If this single-chain polypeptide could be extended into one long helix, its axial ratio a/b would be 70. But when aqueous buffer, pH 3.5, is replaced by 65% dioxane, the axial ratio merely increases from 4 to 14 (Fig. 1-36). This threefold rise in axial ratio appears to be limiting, apparently because the numerous —S—S— bonds prevent further unfolding. A slight increase in helix content is noted when water is replaced by chloroethanol, but the effect is not significant. The change of optical activity with temperature is similarly unspectacular and it does not indicate a phase transition.

Disulfide bonds actually play a rather paradoxical role with respect to helical structure. Although ordered secondary structure breaks down in the neighborhood of these bridges, their presence protects and strengthens the intervening helical regions which do exist, rendering them less susceptible to heat denaturation. To understand this effect, recall that the temperature at which the phase transition occurs is determined by the thermodynamic relationship $\Delta H - T \Delta S = 0$. As stated earlier, the en-

10. Tertiary Structure of Proteins

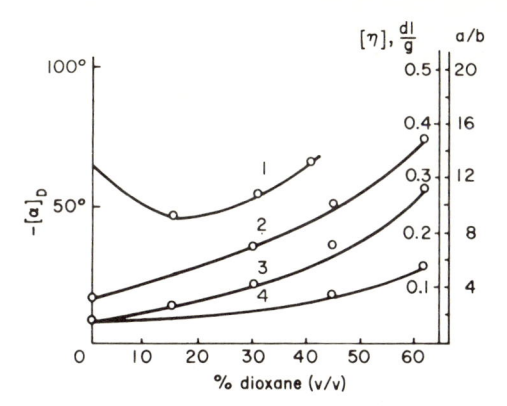

Fig. 1-36. Changes in the secondary and tertiary structures of serum albumin with variation in solvent composition. Values of $[\alpha]_D$, $[\eta]$, and a/b, the axial ratio of the equivalent ellipsoid, are plotted on the ordinate as a function of the volume percent of dioxane in water. (1) Specific optical rotation; (2) axial ratio of the equivalent ellipsoid; (3) intrinsic viscosity in the absence of electrolyte; and (4) intrinsic viscosity in the presence of neutral salt (0.33 M Na acetate). From Bresler (62).

tropy gain per residue in a protein during a helix-coil transition is directly proportional to ν, the number of newly created rotational degrees of freedom (Eq. 1-7). The disulfide bridges which hold the structure rigid greatly limit the freedom of rotation of residues in the chains, or in other words, they reduce the number of effective rotational degrees of freedom. By Eq. (1-17) the melting point of the helix is inversely proportional to ν so that any decrease in the rotational degrees of freedom results in a rise in the transition temperature. This is precisely the situation we meet in serum albumin where moderate heating hardly affects the helical structure at all. An even more striking example is insulin. After the A and B chains of insulin are separated, the A chain turns into a random coil, freely exchanging its —CO—NH— (peptide) hydrogen atoms with deuterium while the B chain retains its helical structure characterized by slow isotopic exchange. The secondary structure of ribonuclease completely breaks down after oxidation of its —S—S— bonds and its peptide hydrogens can then undergo instantaneous exchange with deuterium. But if 8 M LiBr is used as solvent in place of aqueous buffer, the molecule assumes a helical configuration since water molecules are bound by the salt ions and thus are not able to form hydrogen bonds with the —NH—CO— groups (69).

Most of the disulfide bonds in serum albumin can be cleaved by reduction and as a result the tertiary structure of the molecule is partially destroyed. A common method is to use thioglycolic acid in an alkaline medium as reducing agent:

R_1—CH_2—S—S—CH_2—R_2 + 2HS—CH_2—COOH →
   R_1—CH_2—SH + R_2—CH_2—SH + HOOC—CH_2—S—S—CH_2—COOH

Not all of the disulfide bridges are broken under these circumstances, however, since some of them are masked from the reducing agent by other functional groups within the molecular structure, and are not susceptible to attack. In order to achieve complete reduction of the disulfide bridges, it is necessary to carry out the reaction in a concentrated solution of dimethylformamide. Under these conditions, all seventeen of the

—S—S— bonds in human serum albumin are reduced to —SH groups and the tertiary structure of the molecule completely collapses.

The behavior of reduced serum albumin is very different from that of the native protein and its properties resemble those of ovalbumin (70). The helix content of the protein increases somewhat after reduction in dimethylformamide, tending toward 100% in pure organic solvents. In a mixture containing 20% H_2O and 80% dimethylformamide, the axial ratio of reduced serum albumin reaches 60, very close to the theoretical maximum of 70. One further peculiarity of reduced serum albumin is that its secondary structure melts in a narrow concentration interval as a less polar solvent, such as dimethylformamide, is replaced by water, behavior totally unlike that of native serum albumin (Fig. 1-37).

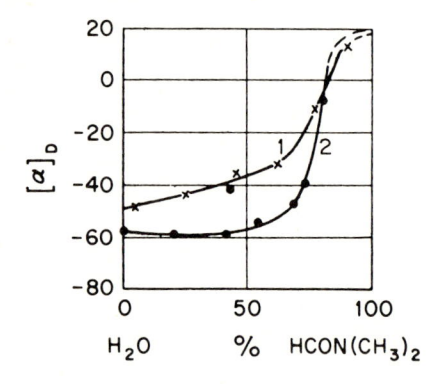

Fig. 1-37. Melting of the secondary structure in human serum albumin as a function of solvent composition. Abcissa shows the solvent composition as the volume percent of dimethylformamide in water. (1) Native protein (noncooperative transition); (2) protein in reduced form with disulfide bonds broken (cooperative transition). From Frenkel' and Horn (70).

The irreversible phenomena which accompany melting of protein secondary structure deserve mention here. If a protein molecule has undergone the transition from intramolecular order to disorder and is then cooled, the disorder becomes "fixed" or "frozen." This is not generally the case with synthetic polypeptides where the helix–coil transition is reversible. In proteins the disordered state becomes fixed owing to the random formation of hydrogen bonds between distant residues along the same polypeptide chain, or between residues belonging to different chains. In place of the α-helix, the protein is capable of forming a new system of intermolecular hydrogen bonds conforming to the peptide β-structure. The denatured protein is said to aggregate under these circumstances. Protein aggregates are generally insoluble and in cases where concentrations are very high the protein forms a gel upon denaturation. Ovalbumin is one example of a protein which strongly aggregates upon denaturation.

Denaturation is reversible in certain proteins, however, especially those well "vulcanized" by disulfide bonds. Enzymes of this type, called thermostable enzymes, are characterized by the complete restoration of catalytic activity even after boiling. Trypsin, ribonuclease, and myokinase fall into this category. From Fig. 1-31, illustrating the order–disorder transition in trypsin, it is evident that the secondary structure lost upon heating to 100° is almost completely restored after the protein solution is

10. Tertiary Structure of Proteins

cooled. Ribonuclease is also completely renatured after undergoing a similar treatment. The foregoing is true only in dilute protein solutions where aggregation is not significant.

One very important question is whether the secondary and tertiary structures of a protein molecule are uniquely determined by its primary structure, that is, by the sequence of amino acids in its polypeptide chain. The problem was answered by Anfinsen and associates through an investigation of the renaturation of ribonuclease and other proteins (71). Disulfide bridges in ribonuclease can be gradually reduced to sulfhydryl groups by β-mercaptoethanol in 8 M urea. As the disulfide bonds are destroyed, the enzyme loses its catalytic activity until it finally becomes completely inactivated. Remarkably enough, reduction of the disulfides is reversible. Anfinsen showed that if urea and β-mercaptoethanol were removed, the —SH groups of ribonuclease could be readily reoxidized by bubbling oxygen through the solution. This procedure restores all of the initial physical and chemical properties of the protein, and, most important of all, 95% of its catalytic activity as well. This means that the denatured molecule folds itself up in exactly the same way as the active enzyme owing solely to the action of intramolecular forces between various residues of the polypeptide chain. Moreover, the disulfide bonds are re-formed in precisely the same places where they occurred initially. The fact that the topology of the protein molecule down to the very positions of the disulfide bridges is predetermined by the amino acid sequence of the polypeptide chain has extremely far-reaching implications for molecular biology.

Many types of molecular forces participate in creating the specific secondary and tertiary structures of a protein molecule. If, for instance, the —SH groups of reduced ribonuclease are oxidized in the presence of 8 M urea, an agent which disrupts protein secondary structure, —S—S— bridges re-form unsystematically and the original structure is restored in only 1% of the cases, judging by enzymatic activity. This is just the figure expected from probability theory if the —SH groups pair at random with each other.

Another interaction vital to the proper packing of ribonuclease is hydrogen-bonding between the —OH radical of tyrosine residues and the side chain carboxyl groups of dicarboxylic amino acids. As noted on p. 51, the presence of this bond in ribonuclease causes a displacement of the tyrosine absorption maximum from 275 mμ to 285 mμ. All methods which inactivate the enzyme apparently destroy this bond, since this anomolous shift in the tyrosine spectrum is not observed in inactivated ribonuclease. We therefore conclude that the topology of active ribonuclease is such that at least some of its tyrosine rings join with carboxyl groups to form supplementary hydrogen bonds. As further confirmation, restoration of the original chain configuration can be inhibited in the absence of urea if —SH groups are oxidized in the presence of phenols. In this case, as with urea, disulfide bridges are seemingly formed at random, resulting in the recovery of less than 1% of the initial enzymatic activity. This experiment shows the important role played by carboxyl-tyrosine interactions in creating the secondary and tertiary structures essential to enzymatic activity.

It has subsequently been possible to reconstitute the structures of many reduced enzymes and proteins by reoxidation of their disulfide bridges. Insulin, lysozyme, takaamylase, alkaline phosphatase, pepsinogen, chymotrypsinogen, and γ-globulin are all of this type. The specific activity of the reconstituted enzymes (and of reconstituted zymogens after activation) generally reaches 50 to 98% of that of the native protein. These results confirm the postulate that protein secondary and tertiary structures are completely determined by their primary structure, even for very complex proteins. γ-Globulin, for instance, consists of four individual polypeptide chains, yet it is almost completely reconstituted by the techniques described above.

Twenty years ago, Bresler and Talmud formulated a hypothesis stating that the secondary and tertiary structures of globular proteins are the result of an equilibrium among three types of molecular forces: *hydrogen bonds* between the peptide groups of each chain which cause the formation of helix; *van der Waals forces* between hydrophobic amino acid side groups; and *electrostatic forces of repulsion* between charges on the surface of the molecule (72). Although the precise configuration of protein secondary and tertiary structures were unknown at that time, the ideas underlying this theory have been wholly confirmed over the intervening years. Nonetheless, estimates of the energies involved in these three types of interaction, originally made in 1949, have had to be corrected on the basis of more recent data. The energy of the hydrogen bond in a number of proteins, for example, has been found to equal 1400 cal/mole of residues participating in the helix (27). In one mole of a hypothetical protein with a molecular weight of 17,000, 60% of whose approximately 150 residues reside in helical segments, the energy of intramolecular hydrogen bonds amounts to $E_1 = 1400 \times 0.6 \times 150 = 130{,}000$ cal.

The most important new principle introduced by the Bresler-Talmud model was that van der Waals interactions among side groups were considered to be responsible for the folding of the polypeptide chain into a compact globule. It is precisely the tendency of nonpolar hydrophobic side groups to form something like hydrocarbon "droplets" within the particle, and of the hydrophilic and charged groups to range themselves on the surface of the molecule, that leads to the formation of a compact molecular structure. In the original formulation it was shown that the minimal dimensions of a globular protein could be predicted from the volumetric proportions of amino acids with hydrophobic and hydrophilic—both polar and ionizable—side groups (72). A similar calculation has been made recently by Fisher on the basis of much more extensive experimental data (73).

Imagine that a globular protein can be approximated by a sphere of volume V_t and radius r, and that it can be further broken down into a hydrophobic core of volume V_1, enclosed by a spherical shell of hydrophilic amino acids of volume V_2. If the thickness of the hydrophilic shell is designated by d, then V_1 and V_2 are given by the following equations:

$$V_1 = \tfrac{4}{3}\pi(r-d)^3 \tag{1-19}$$

$$V_2 = \tfrac{4}{3}\pi[r^3 - (r-d)^3] \tag{1-20}$$

$$V_t = V_1 + V_2$$

V_t can also be written in terms of the ratio:

$$p = V_2/V_1 \tag{1-21}$$

for a given value of d, giving, in effect, the curve of the minimum volume for the protein globule as a function of amino acid composition. This treatment does not, of course,

10. Tertiary Structure of Proteins

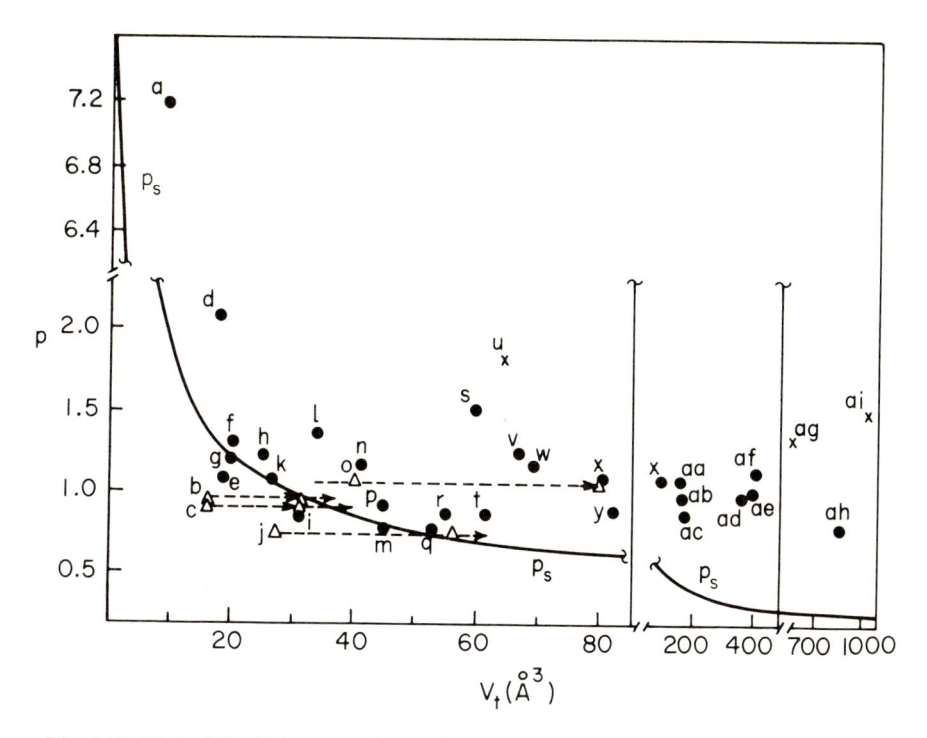

Fig. 1-38. Plot of the Fisher equation and comparison with experimental values. Values of p for various proteins were calculated from their compositions and plotted as a function of V_t. (a) Salmine; (b) and (c) insulin; (d) ribonuclease; (e) lysozyme; (f) myoglobin, whale; (g) myoglobin, horse; (h) papain; (i) chymotrypsinogen; (j) structural protein; (k) corticotropin; (l) ovomucoid; (m) pepsin; (n) carboxypeptidase; (o) prothrombin; (p) β-lactoglobulin; (q) pepsinogen; (r) ovalbumin; (s) edestin; (t) α-amylase; (u) tropomyosin; (v) albumin, human serum; (w) albumin, bovine serum; (x) avidin; (y) hemoglobin, horse; (z) conalbumin; (aa) globulin, human; (ab) aldolase; (ac) triose phosphate dehydrogenase; (ad) leucine aminopeptidase; (ae) phosphorylase; (af) glutamate dehydrogenase; (ag) fibrinogen; (ah) β-galactosidase; and (ai) myosin. P_s denotes the theoretical curve for spherical particles with $d = 4$ Å. From Fisher (73).

take into account the extent to which the shape of a real molecule deviates from a sphere. This curve is plotted in Fig. 1-38 for $d = 4$ Å, along with the empirical values obtained on a great many proteins. For the most part, the experimental points do not diverge greatly from the theoretical curve, though they tend to lie somewhat above and to the right of it, indicating that the respective proteins are not perfectly spherical. Points falling at large distances from the curve all represent large rod-shaped molecules such as myosin and fibrinogen. Experimental values below and to the left of the curve correspond to proteins which tend to aggregate, such as insulin and cell wall structural protein. Dashed lines indicate the displacement of points due to aggregation. The minimum molecular weight of globular proteins for which $p = 2$ is about 10,000, in reasonable agreement with the values given by the Bresler-Talmud theory.

In order to compare "hydrophobic forces" with others acting within the protein molecule, it is necessary to estimate them quantitatively (74). Using data on the temperature dependence of hydrocarbon solubility in water, Scheraga and Némethy evaluated the changes in energy and entropy that accompany the transition of separate hydrocarbon groups, initially surrounded by water, into compact hydrocarbon droplets. Droplet formation under these conditions is considered to be analogous to the formation of the "hydrophobic core" in globular proteins. One interesting result of the calculation was that the enthalpy change in the transition turned out to be positive, its magnitude depending on the dimensions of the hydrocarbon radical. This means that the formation of hydrocarbon droplets requires an expenditure of energy. Table 1-4 presents the thermodynamic data for a number of amino acids,

TABLE 1-7

THERMODYNAMIC PARAMETERS FOR THE TRANSFER OF A NONPOLAR SIDE CHAIN FROM WATER INTO A NONPOLAR MEDIUM[a]

Nature of nonpolar medium	Amino acid side chain	ΔF kcal/mole	ΔH kcal/mole	ΔS entropy units
Aliphatic hydrocarbon	Alanine	−1.3	+1.5	+9.4
	Valine	−1.9	+2.2	+13.7
	Leucine	−1.9	+2.4	+14.3
	Isoleucine	−1.9	+2.4	+14.5
	Methionine	−2.0	+2.7	+16.0
	Cysteine	−1.6	+1.8	+11.4
	Proline	−2.0	+2.2	+14.0
	Phenylalanine	−0.3	+2.7	+10.1
Aromatic hydrocarbon	Phenylalanine	−1.8	+1.0	+9.5

[a] From Némethy and Scheraga (74.)

calculated from their respective heats of solution. There we see that the formation of droplets from separate hydrocarbon molecules always entails a large increase in entropy. The nature of the large entropy changes that occur when hydrocarbons are dissolved in water can be explained very generally as the result of alterations in the short-range order of solvent molecules upon introduction of the hydrophobic substance. In sum, the change in free energy $\Delta F = \Delta H - T \Delta S$ which accompanies the folding of a protein chain into its globular structure will always contain a large negative increment from the hydrocarbon groups. The equilibrium configuration of a globular protein does not depend on whether a particular molecular interaction contributes to the enthalpy or to the entropy, but is determined rather by a minimum in the free energy to which both contribute.

The role of hydrophobic forces in protein tertiary structure has attracted considerable interest in recent years. Some authors, however, incorrectly ascribe to them a greater contribution to the free energy of globular proteins than are made by hydrogen bonds. In fact, the presence of large regions of ordered structure within protein molecules would be impossible if all intramolecular interactions reduced to dispersion forces between the varied hydrocarbon groups of alanine, valine, leucine, phenylalaline, tryptophan, and others. The ordered crystalline regions within the molecule owe their existence to the close packing of identical and repeating elements of structure. In proteins this consistency is provided by the peptide groups which interact with each other through hydrogen bonds. The total free energy of cohesion among hydrophobic side groups E_2 in a protein with a molecular weight of 17,000 can be estimated to fall between 80,000 and 120,000 cal/mole from Table 1-7. This value indicates that the contribution of hydrogen bonds and van der Waals forces to the total configurational energy are of the same order of magnitude. Naturally, the precise energies arising from the different types of bonds varies from protein to protein.

Finally, the energy of electrostatic repulsion between charged residues can be calculated by Coulomb's law:

$$E_3 = (S\beta e^2/2D)\alpha^{3/2} \qquad (1\text{-}22)$$

where S is the surface area of the globular particle, e is the charge borne by an electron, α is the number of ionized groups per unit surface area, β is the coordination number of the surface ions, and D is the dielectric permeability of the solvent. This formula does not account for electrostatic screening of charges and is hence valid only for ionic strengths less than 0.01. Assuming the total number of surface charges per protein molecule to be 20 or 30, a reasonable estimate if the pH is 2 or 3 units away from its isoelectric point, E_3, the energy of interaction between charged groups, is on the order of 50,000 cal/mole.

A peculiarity of proteins is that the contributions of all three kinds of molecular forces are on the same order of magnitude although the energy of the hydrogen bonds somewhat exceeds the other two. Consequently, protein configuration represents a very delicate equilibrium among various forces and structural alterations result when any one of them is acted on individually. It has already been noted that the energy of hydrogen and van der Waals bonds can be either increased or decreased by varying the concentrations of solvent mixtures.

The electrostatic energy can be influenced by changing the pH, which alters the number of charged groups, or by varying the ionic strength which leads to the screening of charges. When a serum albumin solution is acidified to three pH units below its isoelectric point, for instance, the molecule undergoes a 3.5-fold increase in axial ratio a/b. But if the ionic strength of the solution is increased during acidification, the deformation of the molecule may be prevented. Such behavior is typical of polymeric electrolytes or *polyelectrolytes*. These effects are less significant in proteins than in synthetic polymers like polyacrylic acid, since electrostatic repulsion is just one of three major types of intramolecular interaction.

In conclusion, we shall briefly discuss the properties of keratin, one of the most important fibrous proteins. Keratin fibers derived from wool or hair are remarkable in that they can stretch up to 220% of their original length in a humid environment. The crystallographer Astbury first showed that under these conditions the x-ray diffraction pattern of keratin changes, indicating a structural rearrangement. He proposed that native keratin molecules are in the form of a pleated sheet, and that the elimination of pleats in the process of elongation caused the change in diffraction pattern. The original structures suggested by Astbury have since been disproved and are only of historical interest to us at present.

Since the diffraction patterns given by wool keratin were imperfect, porcupine quills, which possess a much higher degree of crystalline order, were used for subsequent studies. In the new material, a strong meridional reflection was observed with a period of 1.5 Å. This value is characteristic of the Pauling-Corey α-helix, since 1.5 Å corresponds to the displacement of one amino acid residue along the axis of the helix. The diffraction patterns of unstretched keratin can be satisfactorily explained if it is imagined that the protein chains exist as α-helices which are intertwined with each other so as to form hexagonally packed, seven-stranded "cables." The regularity of the structure is understandable because there is a specific amino acid sequence in keratin which periodically repeats itself. Furthermore, peptide bond lengths exhibit a slight dependence on the nature of the amino acid side groups. Given a periodic amino acid sequence, this means that the axis of the α-helix will also be deformed at regular intervals. This naturally leads to the formation of a secondary helix consisting of several slightly twisted α-helical strands. This interpretation explains the large, approximately 200-Å, period found in the x-ray diffraction patterns. Stretched keratin yields a diffraction pattern characteristic of the β-structure with intermolecular hydrogen bonds. Stretching evokes the $\alpha \rightarrow \beta$ transition which we discussed earlier.

The crucial test for the proposed keratin structure was provided by infrared spectroscopy, in particular, the measurement of infrared dichroism. By this technique it was demonstrated that the C=O and N—H bonds in the unstretched molecule are parallel to the axis, indicating an α-helical structure with intramolecular hydrogen bonds. The dichroism changes sign upon extension, showing that the bonds become perpendicular to the axis of the molecules, a criterion of the β-structure. The magnitude of the dichroic effect associated with the C=O and N—H stretching frequencies is relatively small if the protein is compared with model polypeptides. This is caused by imperfections in the protein crystal which lead to the formation of amorphous regions. After soaking keratin in D_2O, only the hydrogen atoms within the amorphous sections are exchanged for deuterium. This treatment provides a method of accurately measuring the dichroism of the N—H stretching band in the crystalline regions of keratin because the N—D frequency is shifted by a factor of 1.4 and hence no longer interferes with absorption by hydrogen-bonded N—H groups. The resulting value for the dichroism is a normal one for polypeptides.

In summary, we can state that the Pauling-Corey α-helix and the β-structure provide the basis of secondary structure in both globular and fibrous proteins. When globular proteins are denatured or become aggregated, or when fibrous proteins are stretched, their secondary structure undergoes a transition to the β-form. The tertiary structure of globular proteins results from a topologically complicated packing of the polypeptide chain which itself consists of α-helical and cross-β segments joined together by amorphous regions. Interactions between the secondary and tertiary structures are characteristic. The precise conformation of the tertiary structure can only be discovered through x-ray diffraction analysis, of which more will be said below. Hydrodynamic characteristics such as viscosity, sedimentation constant and diffusion constant, all provide semiquantitative parameters by which to judge protein tertiary structure. They are particularly useful when the protein molecule is subjected to a gross structural alteration. The hydrodynamic properties of proteins will be taken up in Section 12.

11. x-Ray Structure Analysis of Proteins

The most comprehensive technique for investigating protein structure is without doubt *x-ray diffraction analysis*, an approach that has been brought to a high degree of perfection in the last 15 years. In fact, this method is the only one available that permits determination of protein primary, secondary, tertiary, and even quaternary structure at the same time. The first globular proteins extensively studied by x-ray diffraction were sperm whale myoglobin and horse hemoglobin, and in both cases the work was performed at Cambridge University in England. Elucidation of the three-dimensional structures of these two proteins was the crowning achievement of more than a decade of effort during which totally new experimental methods had to be devised and worked out. These pioneering investigations proved that all the necessary information about the structure of a protein might indeed be obtained from the diffraction of x rays and, furthermore, they demonstrated how the mass of data could be processed mathematically. Advances in computer technology contributed greatly to the success of the project since structure analysis requires an enormous volume of numerical calculations. The end result of these measurements and calculations was the determination of the coordinates of all atoms in the two proteins from which exact three-dimensional models of the molecules could be built after appropriate expansion of scale. Of course, x-ray diffraction analysis does not give the positions of hydrogen atoms, but it is not difficult to place them if the coordinates of all carbon, nitrogen, oxygen, and sulfur atoms are known.

In the present discussion, myoglobin will be used to illustrate the application of diffraction techniques to the determination of molecular structure, both because of its important historical role in the high resolution x-ray analysis of proteins and because it exemplifies very well the methods and the problems involved in this kind of study. We shall thus begin with a brief description of the properties and structure of this protein, deferring the consideration of hemoglobin to a later part of this section.

Myoglobin, with a molecular weight of 17,000, consists of a single polypeptide chain containing 153 amino acid residues to which a heme group is tightly bound. The heme group, composed of a ferrous ion and a porphyrin ring, is able to reversibly bind molecular oxygen. The complexity of this protein is evident from the fact that it is made up of more than 1200 atoms, exclusive of hydrogen. Myoglobin is found in the muscle tissue of vertebrates and it is especially plentiful in those animals which spend long periods under water, such as whales, seals, and penguins. Its function is to store up extra oxygen in the muscles to compensate for the loss of a circulatory oxygen supply during prolonged dives.

The diffraction analysis of myoglobin was carried out by Kendrew and associates at Cambridge (75–77). However, before going into the techniques by which the protein was investigated, it will be useful to point out some of the most salient features of its molecular structure. The schematic drawing in Fig. 1-39a, based on x-ray studies at a resolution of 2 Å, shows all three levels of protein structural organization quite unambiguously. The primary structure of the molecule at this resolution is so clear, in fact, that most of the amino acid side groups can be readily identified although they are not included in the diagram. The secondary structure of myoglobin was found to consist of several α-helical segments involving about 75% of the amino acid residues,

80   I. Protein Structure

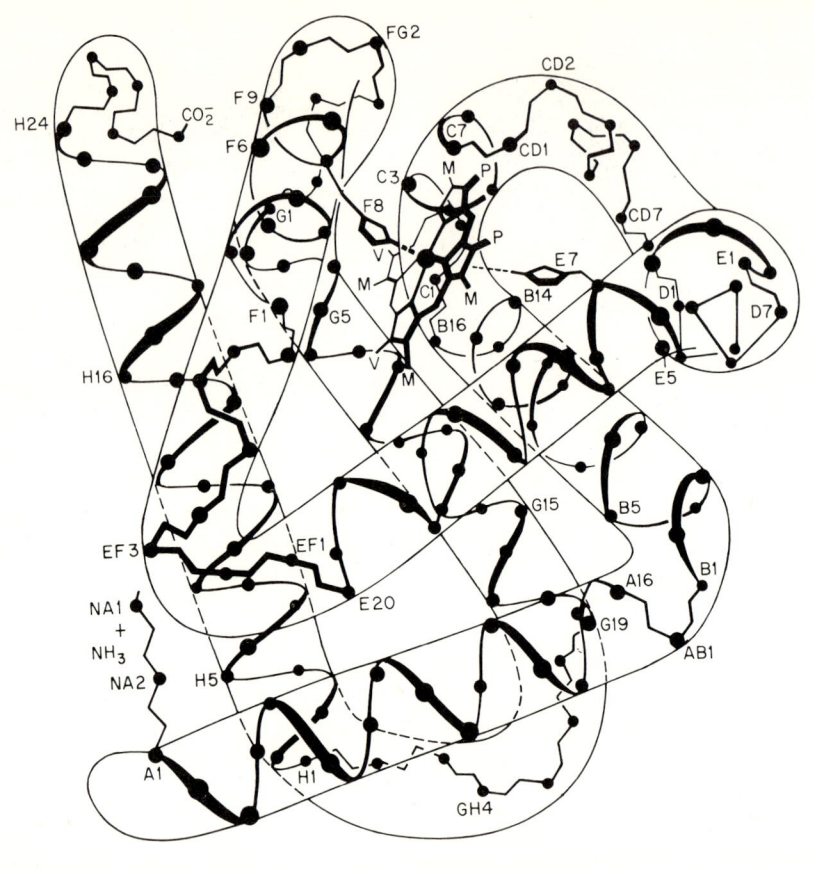

Fig. 1-39. The structure of sperm whale myoglobin. (a) Drawing based on the x-ray structure analysis of myoglobin at a resolution of 2 Å. Continuous tube defines outer contours of the generally helical polypeptide chain. Black line within tube traces polypeptide backbone with positions of α-carbon atoms of the residues marked as black circles. The N-terminus of the chain is at lower left and the C-terminus at upper left. Heme group is sketched in at top center of the diagram. From Dickerson (78).

separated from one another by irregular regions. The overall conformation of the folded and packed chain can be seen from the three-dimensional model illustrated in Fig. 1-39b which, at a resolution of only 6 Å, is sufficient to delineate the serpentine cylinder of the helix. The tortuous structures exhibited by this and other proteins are unprecedented and cannot be compared with those of any other substances in the intricacy of their topology.

Since x-ray diffraction studies require highly specialized methods of experimentation and calculation, only the general principles underlying the analysis of protein structure by this procedure will be discussed here. The reader is referred to any one of a number of standard texts on the subject for a fuller treatment of technical and mathematical details (79).

In order to investigate the structure of a protein by x-ray analysis, the protein must first be obtained in crystalline form. A crystal consists of an orderly three-

11. x-Ray Structure Analysis of Proteins   81

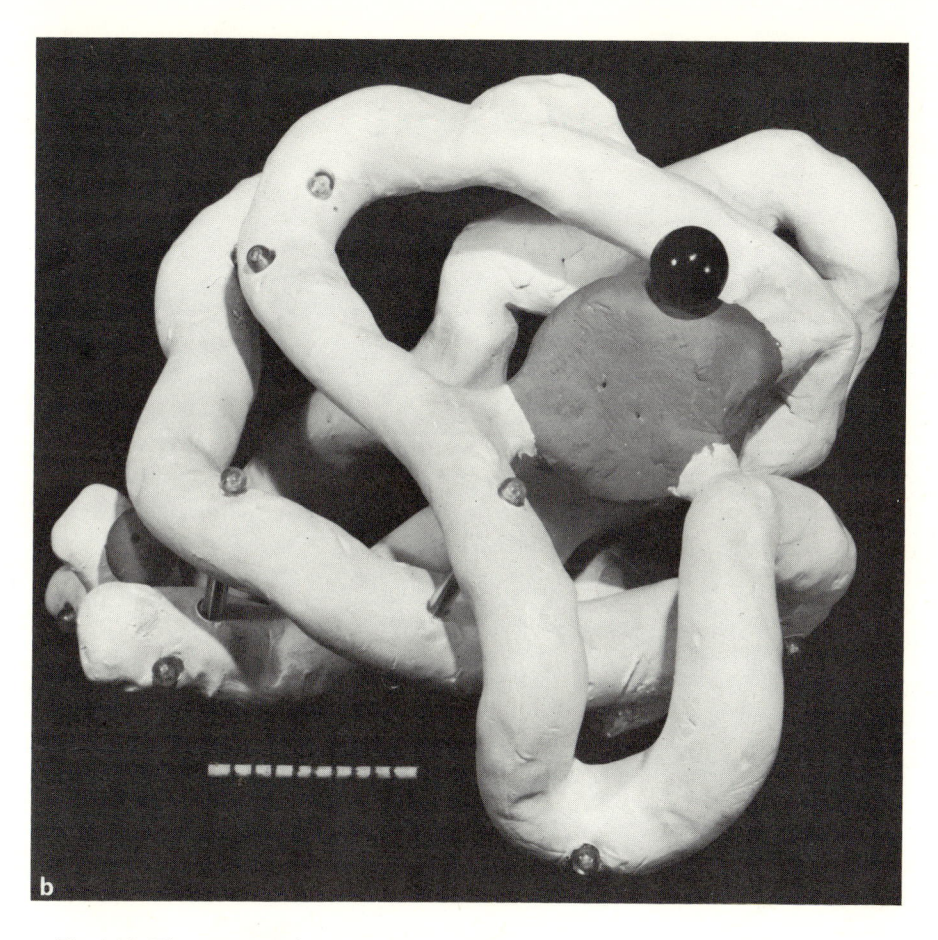

Fig. 1-39. The structure of sperm whale myoglobin. (b) Model of the myoglobin molecule at a resolution of 6 Å. Polypeptide chains are white; the gray disk is the heme group. Dark spheres show sites of attachment of heavy metal atoms. The marks on the scale are 1 Å apart. From Kendrew *et al.* (75).

dimensional distribution of atoms or molecules which is repeated at regular intervals along the axes of the crystal lattice. The fundamental crystallographic element, generally containing one or a small number of molecules, is called the *unit cell* and its dimensions are defined by the smallest translations along each one of the three crystal axes required to bring the lattice into coincidence with itself. An entire crystal can thus be generated, at least in a formal sense, by multiple repetitions of the unit cell in three dimensions.

Information on the structure of a crystal can be obtained by observing the way in which monochromatic x rays interact with atoms in the lattice. x Rays are known to be scattered by electrons and when they are scattered by the electrons of a periodic array of atoms in a regular crystal structure, the scattered radiation will give rise to an interference or diffraction pattern. This pattern, consisting of many separate interference maxima recorded as spots on a photographic film, comprises the primary data

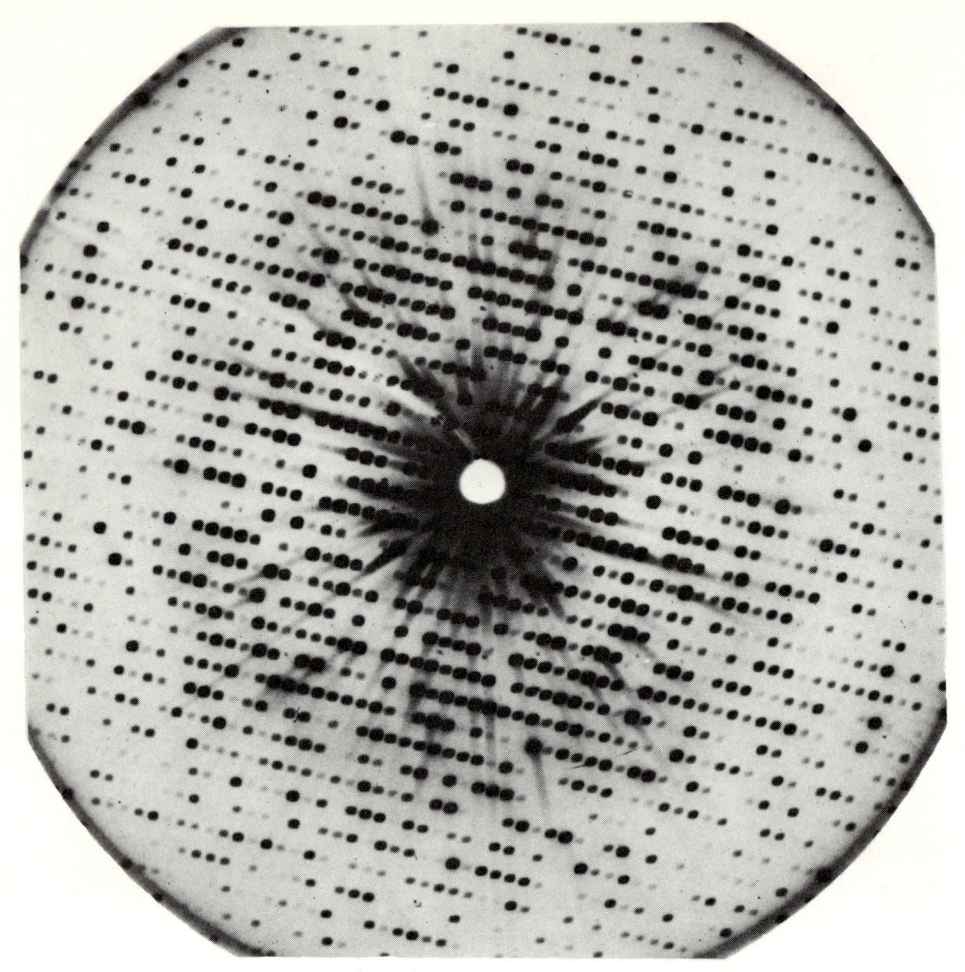

Fig. 1-40. Typical x-ray photograph obtained from a sperm whale myoglobin crystal. Courtesy of Dr. J. C. Kendrew.

of x-ray structure analysis (Fig. 1-40). These maxima can be thought of as reflections of the incident x-ray beam from different sets of imaginary equidistant planes passing through the atoms of the lattice at various angles to the crystal axes. The positions and intensities of the interference maxima depend on the geometrical distribution of atoms in each molecule and on the orientation of the molecules relative to one another in the crystal lattice. Specifically, the distance of a maximum from the center of the pattern is inversely related to the distance between the planes from which it was reflected. In other words, the shorter the periodicities in the lattice, the farther the corresponding maximum will be from the center of the diffraction diagram. In addition, the relative intensities of the maxima provide important information about the density of electrons in the different sets of reflecting planes.

The central problem in x-ray diffraction analysis is how to uniquely determine the structure of the crystal from the observed diffraction maxima. For simple crystals, the answer can be reached by trial and error. Prior knowledge of the symmetry of the

11. x-Ray Structure Analysis of Proteins

crystal and of the chemical structure of the constituent molecules, plus a certain measure of intuition, enables the investigator to avoid serious errors and to pick the correct structure from the relatively small number of acceptable alternatives. Once the distribution of atoms in the lattice has been predicted, the positions and intensities of the corresponding diffraction maxima can be easily calculated. This procedure provides a means of checking proposed structures against actual experimental values. As complex organic compounds gradually came into the sphere of x-ray analysis, however, the techniques that had been suitable for confirming the three-dimensional configurations of simple inorganic and organic molecules were no longer adequate. The problem of structure determination became even more difficult with the advent of the era of proteins. When x-ray diffraction studies on myoglobin were initiated, for instance, the chemical structure, or amino acid sequence, of the protein was as yet unknown. Furthermore, the method of trial and error was in no way applicable to the discovery of its atomic coordinates owing to the very size and complexity of the molecule. Consequently, a general mathematical approach had to be developed that relied solely upon the experimental data, but not on any *a priori* hypothesis about protein structure.

Since x rays are scattered by electrons, it is logical to describe the distribution of atoms within the unit cell by a continuous function, $\rho(x, y, z)$, expressing the density of electrons at all points. The maxima of $\rho(x, y, z)$ thus define the locations of atoms in the lattice. Over the entire crystal, $\rho(x, y, z)$ is a periodic function of all three coordinates and it can therefore be represented by a triple Fourier series. The coefficient of each term in the series is mathematically related to the position and intensity of one of the interference maxima in the diffraction pattern which in turn correspond to the distribution of electrons on a particular set of equidistant planes in the crystal. If all of these coefficients were known, a Fourier synthesis of the electron density could be calculated and the structure of the crystal would be solved. However, a major difficulty is encountered at this point because the complete Fourier coefficients cannot be obtained directly from the data. To understand this, recall that x rays are a form of electromagnetic radiation and that they exhibit the properties of waves. x Rays are consequently characterized by an amplitude and a phase, both of which are required in order to specify the Fourier coefficients of the electron density function. Although the intensities recorded on the diffraction diagram are proportional to the square of the *amplitude* of the scattered radiation and therefore yield part of the necessary information, there is no method for detecting the relative *phases* of the reflected x-ray beams. As a result, neither the electron density nor the locations of the atoms within the crystal can be reconstructed in any simple way from the measured intensities alone. The problem of recovering the phases of all coefficients in the Fourier series for $\rho(x, y, z)$ requires the expenditure of much additional effort, especially in the case of substances as complex as proteins.

The major contribution of the Cambridge University crystallographers Bragg, Perutz, Kendrew, and others to the x-ray analysis of proteins was the development of an ingenious method by which to overcome the phase problem (80). In order to understand the principles underlying their technique for determining phases from interference maxima, we must first consider some of the distinctive properties of protein crystals.

Proteins form a molecular lattice in which the molecules are packed adjacent to one another, as can be seen from the drawing of a myoglobin crystal presented in Fig. 1-41. In this case, the crystal lattice is monoclinic and the unit cell contains two

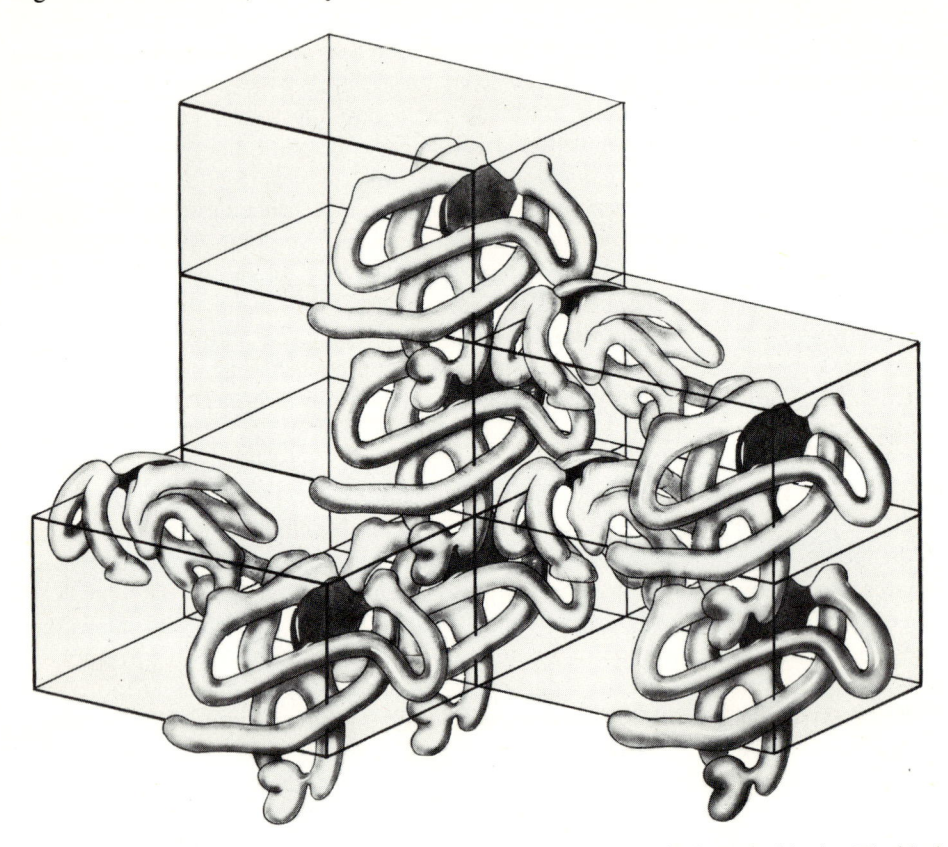

Fig. 1-41. Drawing showing the packing of myoglobin molecules in the crystal lattice. The black disks indicate the positions of heme groups. Note that in each unit cell, the edges of which are marked by lines, there are two protein molecules. From Bodo *et al.* (80).

molecules. The molecular structure of myoglobin itself is extremely complicated, but the way in which the molecules are packed into the lattice is quite simple. Furthermore, when such intricate structures crystallize, spaces remain between individual macromolecules which are occupied only by solvent. The water of crystallization, frequently comprising up to 50% of the total volume, is essential for the integrity of the crystal and dehydration generally destroys its regularity. For this very reason it was long believed that proteins could not form real crystals, since the dehydrated material did not give satisfactory diffraction patterns. The error was finally corrected by Bernal and Hodgkin, who showed that if protein crystals were kept in their mother liquor, it was possible to obtain strikingly detailed diffraction patterns with over 20,000 independent reflections.

Suppose now that a heavy atom can be introduced into every protein molecule at precisely the same location, and that the protein containing the heavy atom crystal-

11. x-Ray Structure Analysis of Proteins

lizes in the same lattice as the original molecule. If these conditions are met, the heavy atoms will alter the intensities of many interference maxima in the diffraction pattern, but they will not change the pattern in any fundamental way. This procedure, known as *isomorphous replacement*, succeeds specifically because the rather loose packing of hydrated protein crystals permits the substitution of bulky groups into the component molecules without significant distortion of their structures. Isomorphous replacement is most effective when the heavy atom can be reacted with only one or a small number of functional groups in the protein. The simplest example of an acceptable heavy atom is mercury. Mercury possesses a specific affinity for many amino acid side chains, particularly those bearing sulfhydryl groups. When the reagent *p*-chloromercuribenzoate (PCMB):

$$ClHg-\langle\bigcirc\rangle-COONa$$

is added to the mother liquor from which hemoglobin is being crystallized, for instance, a derivative is formed in which the four sulfhydryl groups of the molecule are bound to the mercurial in the following way:

$$R-SH + ClHg-\langle\bigcirc\rangle-COONa \longrightarrow R-S-Hg-\langle\bigcirc\rangle-COONa + HCl$$

Thus there will be four atoms of mercury at definite points in every unit cell of the crystal, all of which are able to strongly scatter x rays. By combining the diffraction data from the PCMB derivative with that from the unsubstituted protein in an appropriate manner, a "difference" Fourier synthesis can be calculated and an electron density distribution characteristic of the mercury atoms alone is obtained. Their spatial arrangement is a simple one since there are only four such atoms in the entire unit cell. A related mercurial, *p*-chloromercuribenzene sulfonate (PCMBS), can be introduced into the myoglobin molecule at specific sites. Figure 1-42 illustrates the electron density peaks associated with PCMBS in the myoglobin unit cell, shown in a two-dimensional projection normal to one of the crystal axes. In this case, there are only two mercury atoms per unit cell.

Once the coordinates of the mercury atoms in the unit cell are known, the amplitude and phase of their contribution to the diffraction pattern of the heavy atom derivative of the protein can be computed. Using these values along with the diffraction intensities from both the original protein and the protein plus heavy atom, phases can be found for the interference maxima corresponding to the unsubstituted protein alone. Determination of the electron density function for the protein crystal in two dimensions is generally straightforward and requires only one isomorphous replacement. A two-dimensional projection yields relatively little information, however, since the entire unit cell, which may be some 40 to 50 Å thick in protein crystals, is compressed onto one plane. A three-dimensional Fourier synthesis is much more difficult and data from several isomorphous replacements must be used to obtain consistent and precise phase angles. Consequently, the number of individual calculations that

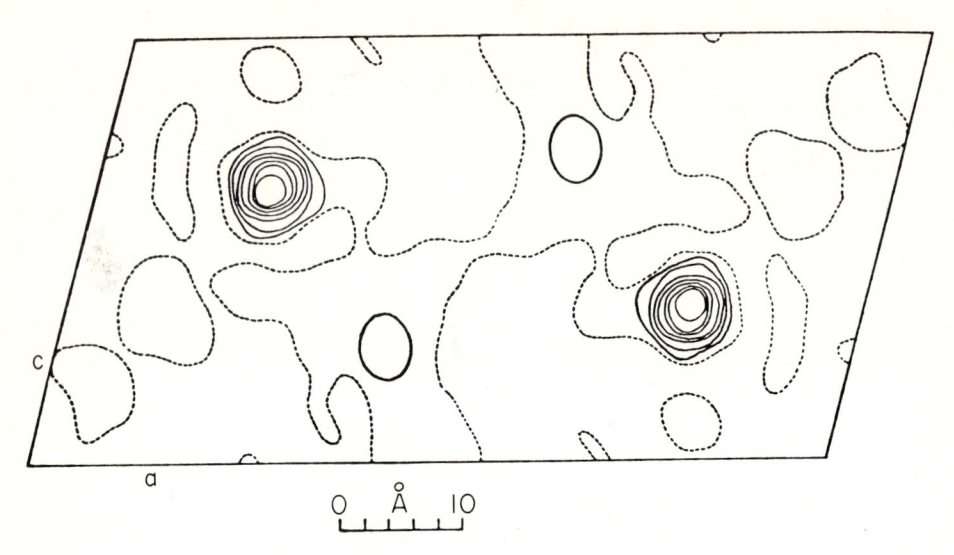

Fig. 1-42. Difference Fourier projection of the complex of PCMBS with myoglobin along the y-axis of the unit cell. Resolution is 6 Å. The continuous lines represent contours of equal electron density. Since there are two peaks of electron density, each molecule in the unit cell must bind one PCMBS group. From Bluhm et al. (81).

must be performed is enormous, considering that several thousand different diffraction maxima must be employed for each derivative of the protein.

The final result of the Fourier synthesis, after all the necessary phases have been determined, is a map of the electron density $\rho(x, y, z)$ at every point within the cell. Each maximum of this function specifies the position of an atom in the structure. To build a three-dimensional model of the molecule from the electron density function, a series of x–y planes are chosen at regular intervals Δz along the z-axis at $z_1, z_2, z_3, \ldots, z_n$. Holding z_i constant, $\rho(x, y, z_i)$ is computed as a function of x and y, and isobars of electron density are constructed in that particular plane. These curves bring to mind the contours on a topographic map which represent equal height above sea level; every "mountain" corresponds to the position of an atom. Figure 1-43 illustrates an electron density map of one such section taken perpendicular to the y-axis of the sperm whale myoglobin unit cell as well as an oblique section selected for the clarity with which the outlines of an α-helical segment can be seen.

By constructing the curves of equal electron density for a set of planes parallel to one of the crystal axes, we arrive at a many-layered, three-dimensional representation of the protein molecule reminiscent of a series of histological sections through a complex tissue (Fig. 1-44). If the successive histological sections are properly oriented relative to one another, a three-dimensional picture of the tissue can be reconstructed. Similarly, a complete three-dimensional model of the protein can be built from a series of parallel planar sections through the crystal unit cell. The myoglobin model at 6 Å resolution shown in Fig. 1-39 was constructed in this way using contours of high electron density to define the external surfaces of the molecule. A more detailed model of myoglobin was based on a Fourier synthesis at 2 Å resolution and required close to 100 parallel electron density sections at a spacing of 2/3 Å for its construction

11. x-Ray Structure Analysis of Proteins

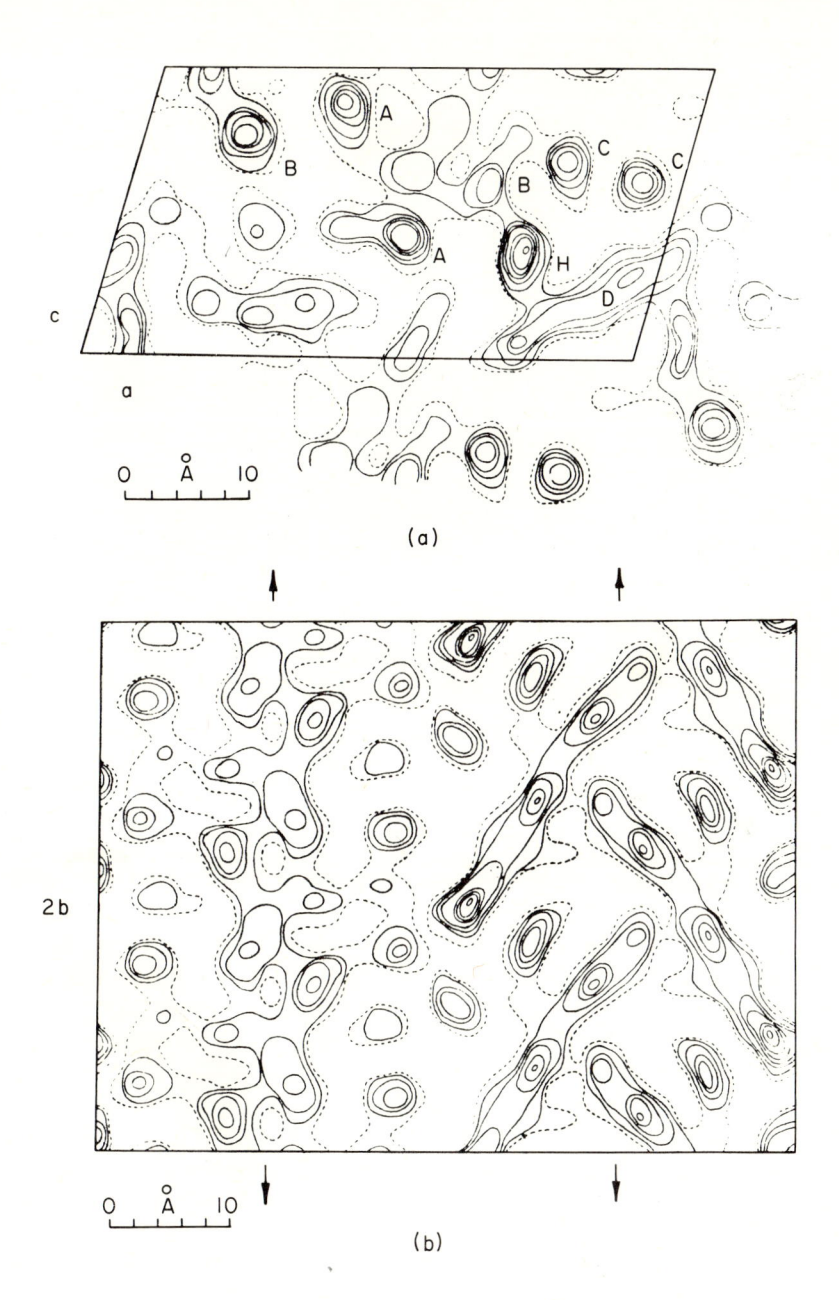

Fig. 1-43. (a) Section of the three-dimensional Fourier synthesis of myoglobin at 6 Å resolution taken perpendicular to the y-axis. Contour lines represent equal electron density. A to D are helical polypeptide chains, three of which (A, B, and C) cross the plane of the section at right angles while the other, D, lies nearly in that plane for a stretch of 20 Å. H indicates the heme group. (b) Oblique section through the 6 Å Fourier synthesis showing polypeptide chain A (on right) as a straight segment running for almost 40 Å in the plane of the section. From Kendrew et al. (75).

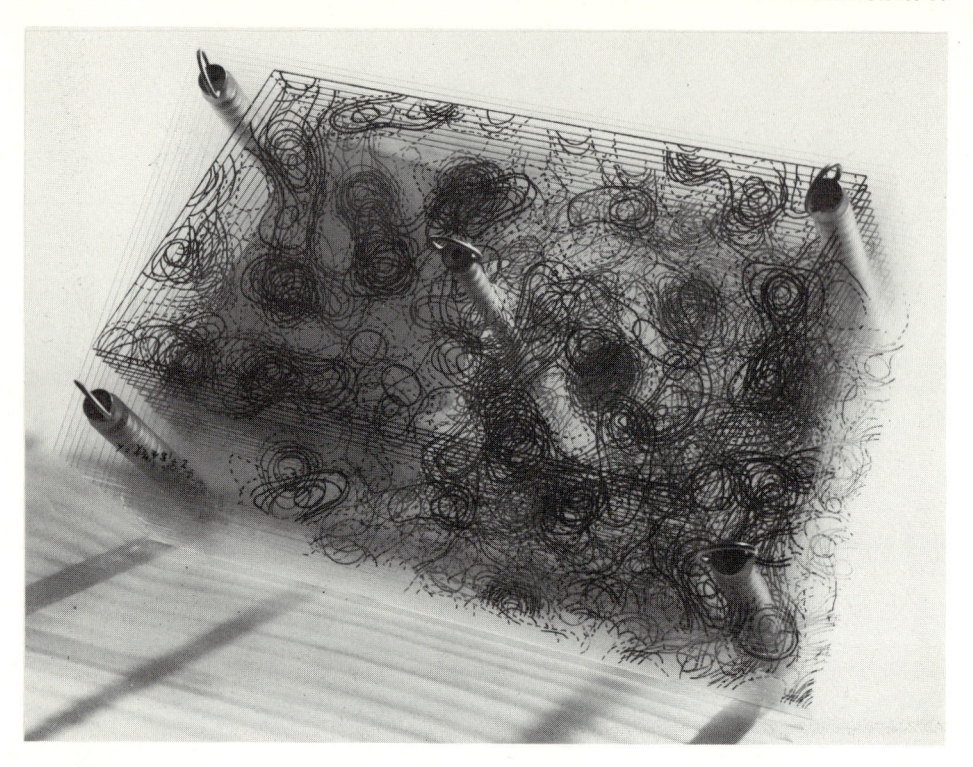

Fig. 1-44. Photograph of the three-dimensional Fourier synthesis of the myoglobin unit cell at a resolution of 6 Å, constructed from sections such as the one illustrated in Fig. 1-43a. Some of the helical polypeptide chains are clearly visible as long rods of high electron density (closely packed contour lines). From Bodo *et al.* (80).

(Fig. 1-45). The contour of the polypeptide chain, the points at which the chains alter their direction, and the amino acid side groups are all clearly visible at this resolution.

Let us now consider some of the intermediate stages in the x-ray analysis of sperm whale myoglobin. The resolution obtained by this method lies to a certain extent in the hands of the investigator. At the outset of the myoglobin work, it was decided to limit the resolution to 6 Å. Specifically, this means that diffraction maxima representing distances of less than 6 Å within the crystal lattice were not taken into account. The first structure was thus expected to be rather crude. Maxima falling within an appropriate distance of the center of the pattern were used for the analysis, while those beyond the limiting value were discarded since the farther a maximum is from the center, the smaller the distance it represents in the crystal lattice. To obtain the myoglobin structure at a resolution of 6 Å it was necessary to analyze some 400 reflections. When the resolution was extended to 2 Å, the volume of data which had to be included in the analysis rose by a factor of $3^3 = 27$, increasing the number of maxima used to 10,000. In the most recent stage, the resolution was increased to 1.4 Å, thereby raising the number of diffraction maxima required to 20,000.

The positions of individual atoms could not be resolved in the 6-Å structure although the α-helix, with a diameter of 10.1 Å, stood out very clearly. This is the structure illustrated by the model in Fig. 1-39. The internal structure of the helix was

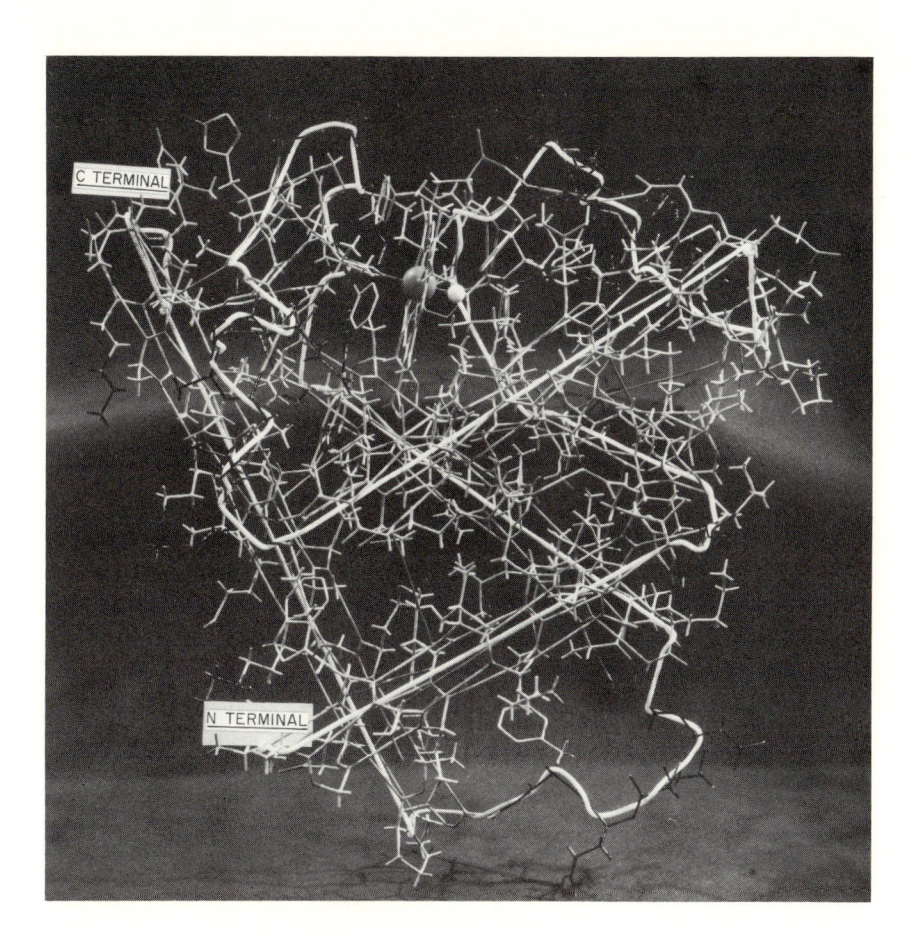

Fig. 1-45. Model of the sperm whale myoglobin molecule at 2 Å resolution, including side chains. The course of the main chain is indicated by a white cord and the N-terminal and C-terminal residues are marked. The iron atom is shown as a gray sphere. From Kendrew et al. (77).

of course not distinguished in this model, just the outlines of the coiled polypeptide chain. Amino acid side groups appear as unstructured, amorphous lumps occupying the space between adjacent segments of the helix although they were not included in the model for purposes of clarity. Since it was not possible to identify individual side groups, no conclusions could be drawn about the sequence of amino acid residues in the polypeptide chain. The secondary and tertiary structure of seal myoglobin was found to be the same as that of whale myoglobin at this resolution even though there are significant differences in the amino acid compositions of the two proteins (82). In sum, the structure of myoglobin which emerges at a resolution of 6 Å confirms the hypothesis that certain proteins, at least, consist of segments of α-helix joined together by short irregular regions.

Because it was necessary to analyze almost 10,000 maxima in increasing the resolution of sperm whale myoglobin to 2 Å, the success of the project in large measure depended on the mechanization of all numerical calculations. Diffraction patterns were made with myoglobin crystals in 22 different orientations. After careful measurement of all intensities, the diffraction patterns were normalized to one another so that a wholly comparable set of 10,000 Fourier coefficients could be obtained. This procedure was then repeated for four isomorphous derivatives of myoglobin containing heavy atoms.

Fortunately, it turned out that the isomorphous derivatives could be formed both easily and in considerable diversity. There are many functional groups in the protein which are capable of binding with heavy metal reagents either covalently or by complex formation. Reagents were selected which would attach their heavy atom to only one or two sites per protein molecule. It was also quite important that the positions of the heavy atoms be definitely fixed in the crystal structure, though their exact coordinates were not known until after the analysis had begun. The final requirement was that the derivatives be strictly isomorphous with the original unsubstituted protein. This last demand is usually not difficult to satisfy since the introduction of one or two heavy metal groups into the protein does not in general alter the lattice packing of the macromolecule to any significant degree. The diffraction patterns of the various heavy atom derivatives themselves serve to determine whether or not the substitution meets all of the above criteria.

Among the best heavy atom reagents for myoglobin is *p*-chloromercuribenzene sulfonate (PCMBS):

$$\text{ClHg}-\!\!\bigcirc\!\!-\text{SO}_3\text{H}$$

This compound binds at only one site in the molecule even when it is present in twentyfold excess. It has not been established which group in the protein fixes the reagent though it is certainly not a sulfhydryl group for there are none in myoglobin. Furthermore, the usual reagent for sulfhydryl groups, *p*-chloromercuribenzoate, is not effective in the case of myoglobin. The second reagent is a complex mercuric–ammonia ion called mercury diammine which results from heating mercuric oxide with aqueous ammonium sulfate. These ions, which have the structure $Hg(NH_3)_2^{2+}$, also attach at a unique site on the molecule, but it is completely different from that interacting with PCMBS. The third derivative suitable for x-ray analysis contained a complex ion of gold, $AuCl_4^-$, added as the salt $KAuCl_4$ to the crystallization solution along with $(NH_4)_2SO_4$. The reaction proceeds very slowly in this case and it requires several months for completion. This ion is once again fixed at a unique site in the myoglobin structure. The fourth derivative used in the 2-Å analysis contained both PCMBS and mercury diammine.

Diffraction patterns for the four isomorphous derivatives were obtained in all 22 positions of the crystal and the 10,000 reflections per derivative were normalized to each other. The difference electron density distributions for the metal atoms were then constructed, thereby defining the coordinates of these atoms in the unit cell. Next, the phases of all diffraction maxima relating to unsubstituted myoglobin were determined using the known contributions of the heavy atoms and a Fourier synthesis of the protein molecule was computed. Finally, contours of equal electron density were plotted in 96 layers and a model of the myoglobin molecule was built from these (Fig. 1-45).

The tertiary structure of myoglobin was quite evident at 6 Å resolution, but the 2-Å model of the protein permitted detailed study of the secondary structure and even of the primary structure. Eight different α-helical segments were found in the sperm whale myoglobin chain and all were of the right-helical configuration. A cylindrical projection of the electron density in one of the helical segments demonstrates how well the observed distribution agrees with the superimposed projection of a Pauling-Corey α-helix (Fig. 1-46). The precise figure for the proportion of residues participating in α-helix turned out to be 77% and was thus consistent with the evaluation of Benson and Linderstrøm-Lang from isotope exchange. Furthermore, the nature of the amino acid side chains could be established in many cases, although an analysis at 1.4 Å resolution was required for the positive identification of all amino acid residues in the molecule. Figure 1-47 shows the appearance of some of the more distinctive side

Fig. 1-46. Cylindrical projection of a helical segment of the myoglobin polypeptide chain with the α-helix structure superposed. A key to the atomic arrangement in the α-helix is shown below. The points marked β and β' are the two alternative projected positions of the β-carbon atom of the amino acid side chain; β is the position in a right helix and β' that in a left helix composed of L-amino acids. Contour lines once again represent an equal number of electrons per unit volume. From Kendrew et al. (76).

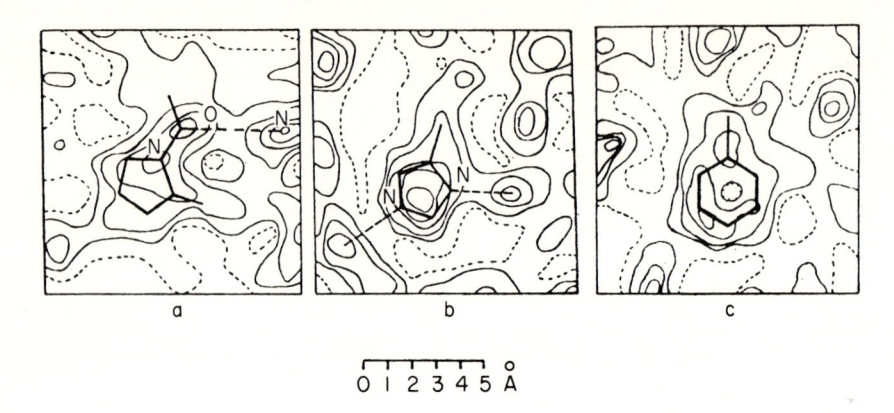

Fig. 1-47. Sections through parts of the 2 Å Fourier synthesis of myoglobin showing the electron density associated with amino acid side chains. Contour interval is 0.5 electrons per Å^3. (a) Proline; the dotted line is a hydrogen bond from the CO group to an NH group along the helix. (b) Histidine; the dotted lines indicate hydrogen bonds to neighboring groups. (c) Phenylalanine. From Kendrew et al. (77).

chains in sections through the 2-Å electron density distribution. There is a general tendency for polar side chains to project outward from the molecule and for nonpolar ones to be buried within it. The iron atom and some of the carbon atoms of the heme group were also resolved in the 2-Å Fourier synthesis; the proposed structure of heme is presented together with the observed electron density in Fig. 1-48. When an electron density distribution based only on the proposed heme configuration was calculated, it was found to coincide with the observed distribution in almost all details (Fig. 1-48c). This example demonstrates how the validity of a model structure can be verified by comparison with the experimental values.

Extensive x-ray diffraction studies of hemoglobin, a protein related to myoglobin in both structure and function, have been performed at Cambridge by Perutz and his co-workers. Hemoglobin is a heme-containing protein from red blood cells or erythrocytes and its physiological function is to reversibly bind molecular oxygen as the erythrocytes pass through the lungs and to deliver it to all tissues of the organism via the circulatory system. Each hemoglobin molecule is made up of four polypeptide subunits, two identical α-chains containing 141 residues each and two identical β-chains containing 146 residues each, with an overall molecular weight of 66,000. The two kinds of subunits can be reversibly dissociated at acidic pH and can be separated from one another by ion exchange chromatography.

Many important analytical techniques, including the method of isomorphous replacement, were initially developed for the investigation of hemoglobin and it is the only case where protein quaternary structure has been investigated in detail. The first analysis of horse hemoglobin in three dimensions was carried out at a resolution of 5.5 Å and required the use of 1200 diffraction maxima from the unsubstituted protein and from each of six isomorphous derivatives (83). A model of the molecular structure, presented in Fig. 1-49, was constructed from 32 parallel two-dimensional electron density maps taken at 2-Å intervals along the y-axis of the unit cell (see also Figs. 1-43 and 1-44). An electron density of 0.54 electrons per Å^3 was used to define

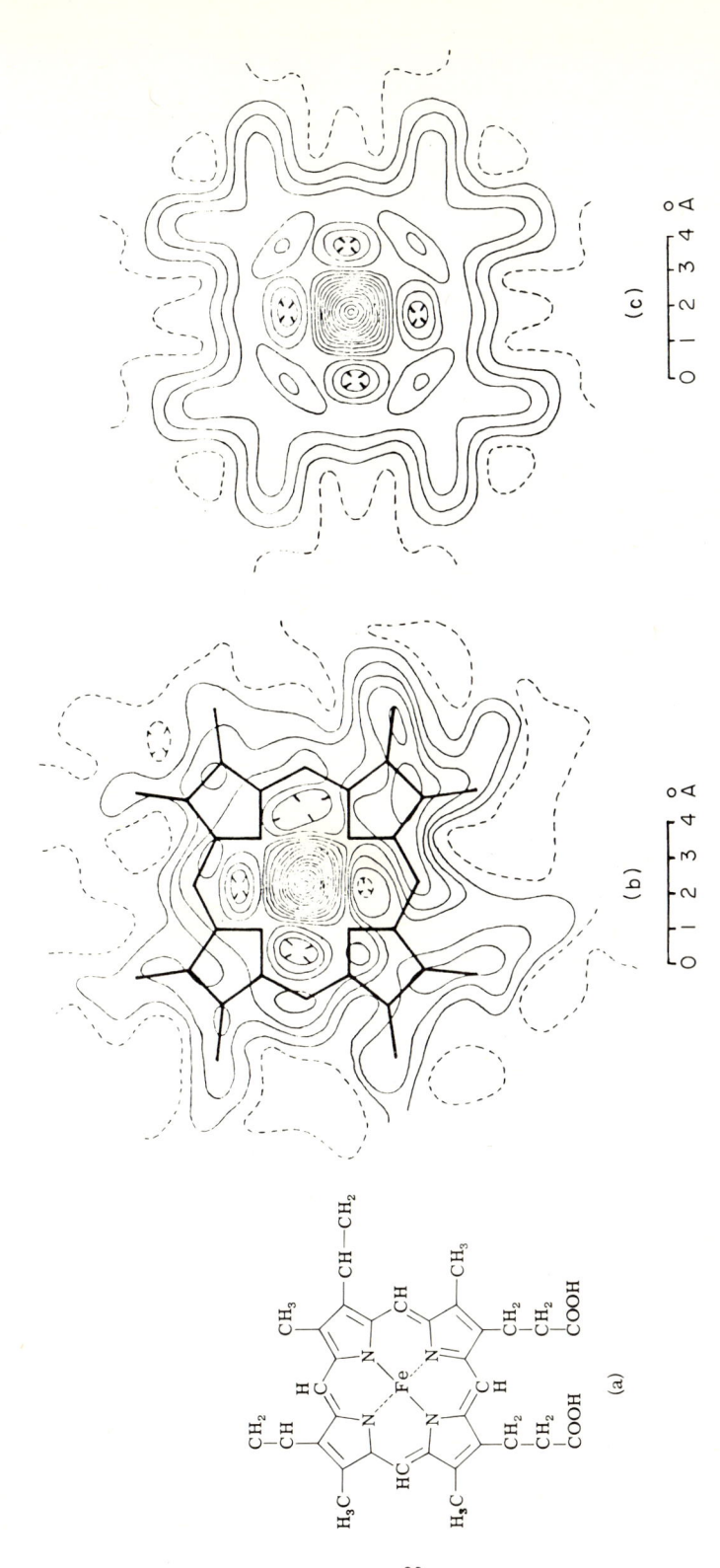

Fig. 1-48. Structure of heme. (a) The chemical formula of heme. (b) Observed electron density distribution in the plane of the heme group at 2 Å resolution. The atomic arrangement in the group is superposed. (c) Calculated electron density distribution computed from the atomic arrangement proposed on the basis of the observed electron density. From Kendrew et al. (76).

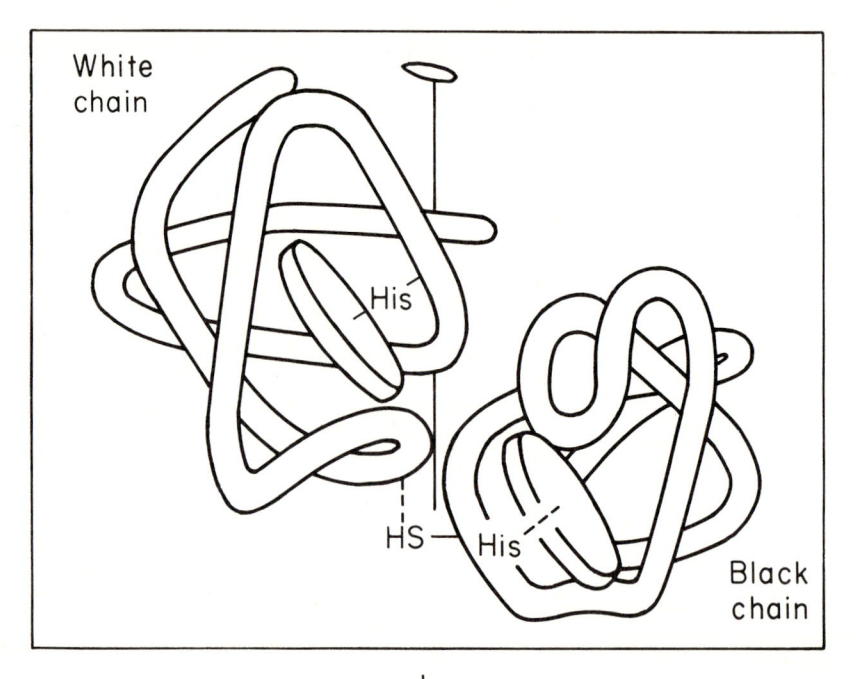

a

b

94

the outer contour of the molecule. In the model, one αβ subunit pair is shown in white and the other in black so as to emphasize the way in which the four polypeptide chains tightly interlock. The resulting structure is nearly tetrahedral and at the center of the molecule there is a cavity filled with solvent. A heme group lies in a "pocket" on the surface of each subunit with its reactive side facing outwards. One of the most striking aspects of the hemoglobin structure is that the α- and β-chains are very similar in conformation to each other and to myoglobin. The resemblance includes the distribution of helical and nonhelical segments, the location of bends in the main chain, the orientation of heme groups, and the tendency of polar groups to be excluded from the interior of each subunit. Due to the uncertainty in atomic positions, the 5.5-Å structure could not be checked by calculation, but the obvious analogies in configuration between myoglobin and the hemoglobin subunits in addition to the self-consistency of the structure as a whole served to establish the validity of the model.

A more recent analysis of horse hemoglobin at 2.8-Å resolution confirmed the original model, although it did point up a number of subtle differences in secondary and tertiary structure which distinguish the α- and β-chains from each other and from myoglobin (84). Nonetheless, the similar environment of the heme group in all three cases is significant. Its porphyrin ring forms a great many contacts with the polypeptide chain and its ferrous ion is coordinately bonded to a specific histidine residue. Furthermore, the pocket into which the heme group fits is lined with nonpolar residues, directly facilitating the reversible binding of molecular oxygen to the heme iron.

The high resolution analysis also permitted investigation of the contacts between subunits in the hemoglobin tetramer. The α- and β-chains are highly complementary to one another and come into close proximity over a large surface area. The α-chain is related to each of its two neighboring β-chains in different ways and to distinguish the two kinds of interactions, the designations $\alpha_1\beta_1$ and $\alpha_1\beta_2$ are employed. In the region of contact between the α_1 and β_1 subunits, 110 atoms belonging to 34 residues lie within 4 Å of each other and several of the side chains are interlocked. Most of the interactions are nonpolar, although five hydrogen bonds are probably formed between the two polypeptides. The $\alpha_1\beta_2$ contact involves about 80 atoms from 19 residues and only one hydrogen bond is formed between this pair. Contacts between like subunits, $\alpha_1\alpha_2$ and $\beta_1\beta_2$ are not extensive and if they do exist, they must be polar in nature.

When comparative diffraction studies on oxygenated and reduced hemoglobins were performed, considerable structural differences between the two forms came to light. Specifically, the protein macromolecule swells slightly when oxygen is bound owing to a realignment of subunits in the quaternary structure. The secondary and tertiary structures of the individual polypeptide chains are not altered to any significant extent, however, when oxygenation occurs. This is consistent with the observation that myoglobin, which contains only a single polypeptide chain, undergoes no significant conformational changes when it binds a molecule of O_2 (85). Furthermore,

Fig. 1-49. The structure of hemoglobin at a resolution of 5.5 Å. (a) Hemoglobin model viewed normal to the x-axis; heme groups are indicated by gray disks. The two white chains represent a half-molecule containing one α- and one β-chain; the black chains represent the other half of the tetramer. (b) Chain configuration in the two subunits facing the observer; the other two chains are omitted for clarity. Note the similarity of these chains with that of myoglobin at about the same resolution depicted in Fig. 1-39. From Perutz et al. (83).

although there is little change in the $\alpha_1\beta_1$ contact of hemoglobin during the transition the relative displacements of interacting atoms in the $\alpha_1\beta_2$ contact may amount to as much as 5.7 Å. Moreover, the polypeptide conformation at the $\alpha_1\beta_2$ interface is such that the two subunits can slide past each other quite smoothly. Since the $\alpha_1\beta_2$ contact lies adjacent to the heme groups, any change in the relative orientation of subunits in this region undoubtedly affects the heme environment and consequently, the binding of oxygen. Thus, the so-called heme–heme interaction in which the binding of O_2 by one subunit increases the affinity of the other subunits for the ligand (see p. 169) may depend largely on changes in the $\alpha_1\beta_2$ region. The $\alpha_1\beta_1$ interface, which lies at a considerable distance from the heme groups, is probably not intimately involved in the cooperative effect. This theory is supported by the observation that all but one of the amino acid residues making up the $\alpha_1\beta_2$ contact in several normal mammalian hemoglobins are invariant while up to one third of the amino acids in the $\alpha_1\beta_1$ contact can be replaced without noticeable changes in function. Furthermore, three mutant forms of human hemoglobin with amino acid substitutions in the $\alpha_1\beta_2$ region exhibit reduced cooperativity in O_2 binding.

The important topic of enzyme conformation and how the relationships between enzyme structure and function can be elucidated by diffraction techniques will occupy the remainder of this section. Hen egg-white lysozyme, with a molecular weight of 14,600, was the first enzyme to be analyzed at a resolution of 2 Å (86). Although derived from an animal tissue, egg-white lysozyme has been found to possess the capacity to hydrolyze polyaminoglucosides in bacterial cell walls. To solve this structure, 9040 reflections from crystals of lysozyme and each of several isomorphous derivatives containing organic mercurials and complexes of the uranyl ions UO_2^{2+} and $UO_2F_5^{3-}$ were analyzed. A three-dimensional model of the enzyme has been constructed, of which a version at 6 Å resolution is illustrated in Fig. 1-50. There are six α-helical regions in the molecule, involving 55 of its 129 amino acid residues. From these numbers we see that the helix content is 42%, confirming the 40% figure based on optical rotatory dispersion. As in the case of myoglobin, the hydrophobic amino acid side chains of lysozyme are concentrated within the molecule while hydrophilic groups are generally located near its surface.

A very clever approach was employed to find the position of the active site of lysozyme by x-ray structure analysis. The protein was first complexed with a specific inhibitor of its enzymatic activity, di-N-acetylchitobiose, and the complex was crystallized. x-Ray diffraction patterns were made and the structure solved for the complex, thereby establishing the topological relation of the active site to the rest of the molecule. Figure 1-50 illustrates the juxtaposition of chitobiose with lysozyme in a three-dimensional model and shows that the inhibitor and, by implication, the active site, lies in a cleft or crevice in the surface of the molecule. From independent experiments on the inhibition of lysozyme, it was known that the active site contained one or more tryptophan residues. In fact, x-ray analysis showed that there were three tryptophans (numbers 62, 63 and 108, Fig. 1-6) in close contact adjacent to the proposed location of the active site.

The 2-Å x-ray analysis of another enzyme, bovine pancreatic ribonuclease, revealed a kidney-shaped molecule with a deep depression in the middle of one side (88). The active site of ribonuclease is believed to be located in this depression since it

11. x-Ray Structure Analysis of Proteins

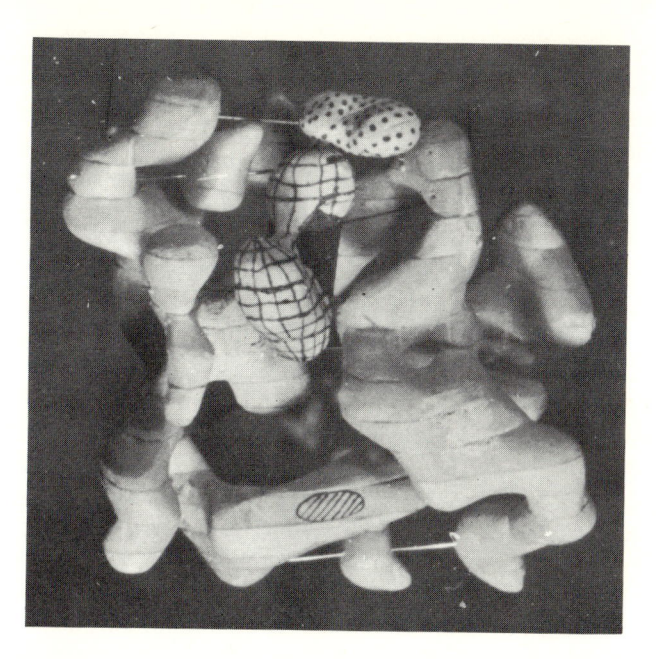

Fig. 1-50. Model of the lysozyme molecule obtained by x-ray analysis at 6 Å resolution. The increase in electron density observed in the presence of the inhibitor di-*N*-acetylchitobiose, marking the active site of the enzyme, is cross-hatched. From Johnson and Phillips (87).

contains the binding site to which both substrate and a substrate analog attach. Of the 124 amino acids in this enzyme, 22 or 16% occur in three helical segments consisting of two turns each. The remainder of the molecule is not totally disordered, however. Some 45 residues are present in three roughly antiparallel stretches that run behind the central depression between two pairs of disulfide bonds at opposite ends of the molecule. Although the partially ordered extended chains cannot be strictly characterized as β-structure, hydrogen bonds are undoubtedly formed between peptide groups in such regions. This finding explains why measurements of hypochromicity and isotopic exchange give values of 40 and 57%, respectively, for the amount of ordered structure in ribonuclease, while optical activity and optical rotatory dispersion yield the much lower estimates of 16 and 17%. The latter figures pertain to amino acid residues actually present in α-helix as demonstrated by x-ray analysis. Different techniques for studying protein secondary structure thus provide different kinds of information about polypeptide chain conformation. Finally, unlike myoglobin and lysozyme, ribonuclease has a rather open structure so that some of its hydrophobic residues lie near the molecular surface. Nonetheless, predominantly polar segments of the polypeptide chain are generally on the exterior and their side chains point toward the solvent.

Our knowledge of the secondary and tertiary structure of enzymes has been substantially increased in the past few years by detailed x-ray studies on four proteases. The structures of both carboxypeptidase and chymotrypsin have been worked out to a resolution of 2 Å, that of bacterial subtilisin to 2.5 Å, and that of papain to 2.8 Å. Through the analysis of enzyme–substrate or enzyme–inhibitor complexes, these investigations have all provided a means for examining the nature of the active site, and the amino acid side groups participating in substrate binding as well as in catalysis have been identified. In addition, the steric relationships among amino acids at the active site have been contrasted with enzymatic mechanisms proposed on the basis of chemical studies alone. The proximity of histidine to a serine hydroxyl or to a cysteine thiol in three of the proteases (chymotrypsin, subtilisin, and papain) supports the view that these side groups act in a concerted fashion during enzymatic hydrolysis of the substrate. The catalytic groups in carboxypeptidase have also been located, but more importantly, the "induced-fit" hypothesis of enzyme catalysis (p. 163) was confirmed in this case by the observation that the protein undergoes a dramatic configurational change when a substrate molecule is bound.

Diffraction studies on bovine carboxypeptidase A revealed three types of conformational organization in substantial amounts (89). About 30% of the 307 amino acid residues in this protein are present in α-helical regions and most of these, including four rather large segments of 16 to 19 residues each, are on the outside of the molecule. Another 20% of the residues comprise a twisted sheet of eight extended chains folded back on themselves in such a way that 42 of the 53 residues involved can form hydrogen bonds characteristic of the cross-β-structure. The sheet lies at the center of the structure and lines one side of a depression or pocket which extends into the interior of the molecule. The pocket, judged to be in the immediate area of the active site by x-ray analysis of the enzyme–substrate complex, is large enough to accommodate the side chain of tryptophan. Finally, a further 20% of the residues are folded and packed into an irregular coil at one end of the molecule and it is this part of the protein which undergoes a large structural rearrangement when substrate is bound.

The analysis of bovine α-chymotrypsin, a protein composed of 241 amino acid residues, required the use of more than 24,000 independent reflections in the native protein and in each of five isomorphous derivatives (90). There are only two turns of α-helix in the entire molecule and the eight participating residues are located at the C-terminus. Otherwise, the chain tends to be fully extended and frequently folds back on itself. In such regions, the distance between antiparallel polypeptide chains makes hydrogen bonding between peptide groups possible. Consequently, much of the molecule is probably characterized by cross-β-structure, although the number of amino acid residues involved is rather difficult to estimate. The majority of the charged groups are found on the surface of the structure and point toward the solvent while hydrophobic side chains, for the most part, are packed in the interior. The active site of the enzyme was identified by binding an inhibitory tosyl (toluene-p-sulfonyl) group to the side chain of a serine residue. There is no pronounced cleft in chymotrypsin as there is in lysozyme, ribonuclease, and carboxypeptidase, but there are open regions in the molecular surface near the catalytically active serine.

Subtilisin BPN', an extracellular protease from the bacterium *Bacillus amyloliquefaciens*, contains 275 amino acids in its polypeptide chain (91). This enzyme is similar to trypsin and chymotrypsin in its catalytic properties, including substrate and inhibitor specificity and the presence of both serine and histidine at the active site, but it bears no relation to either of the pancreatic enzymes in amino acid sequence. Subtilisin is roughly spherical in shape with a core of packed hydrophobic side chains and no cleft or depression at the active site. There are eight segments of right α-helix in the molecule involving 30% of the total residues, and seven of these are approximately parallel to

11. x-Ray Structure Analysis of Proteins

each other with the same N to C orientation. Five segments of extended chain containing five to six residues apiece form a twisted cross-β-sheet running through the molecule. About 80% of the potential hydrogen bonds between segments are sterically possible. Many of these features can be quite clearly seen in Fig. 1-51. Comparison of the amino acid sequence of subtilisin BPN′ with that from another strain of *Bacillus*

Fig. 1-51. Perspective drawing of the subtilisin BPN′ polypeptide chain derived by x-ray analysis at a resolution of 2.5 Å. Vertices represent α-carbon atoms and each length of rod represents a single residue. The active site region lies within the circle containing the side chains of His 64, Met 222, and Ser 221 with its covalently bonded heavy atom group. The side chain orientation of His 64 and Ser 221 in the active enzyme are shown in pale outline. The eight segments of right α-helix can be easily distinguished: they run between residues 5–10, 14–20, 64–73, 103–117, 132–145, 223–238, 242–252, and 269–275. All helical segments except 242–252 are roughly parallel to one another. The twisted, parallel-chain pleated sheet is composed of the five segments of extended polypeptide between residues 148–152, 120–124, 28–32, 89–94, and 45–50, running from top to bottom through the core of the molecule. From Wright *et al.* (91).

showed that the two proteins differed in 84 out of 275 residues, but that the substitutions had occurred in such a way that amino acids whose side groups extended outward from the external surface of the molecule were much more frequently affected than interior residues. This strongly implies that the three-dimensional structure of the molecule is largely determined by the packing of internal side chains.

In papain, the fourth protease studied at high resolution, the ordered part of the macromolecule consists of four short segments of α-helix containing about 40 residues out of 211, and one small region of β-structure consisting of nine residues (92). The helix content of 20% compares favorably with a figure of 17% derived from optical rotatory dispersion, demonstrating good agreement between the two techniques. Much of the molecule is irregular in shape and folded into two main parts lying to either side of a cleft at whose surface the active site is located. Even at a resolution of 2.8 Å, sufficient information about the amino acid side chains was available to permit the correction of a tentative primary structure proposed on the basis of chemical analysis.

The high resolution x-ray analysis of myoglobin, hemoglobin, and six different enzymes demonstrates quite conclusively that ordered secondary structure stabilized by hydrogen bonds plays a major role in defining protein conformation. The importance of α-helix in secondary structure was initially overemphasized, however, since myoglobin, with an unusually large helix content of 75%, remained for many years the only protein whose complete three-dimensional structure was known. Further analysis of protein conformation revealed that polypeptide β-structure also contributes to the internal order of several compact globular proteins, ranging from the highly organized cross-β-sheets of carboxypeptidase and subtilisin to the less regular pairing of extended parallel chains in ribonuclease and chymotrypsin. The joint occurrence of α- and β-configuration in a single molecule appears to be quite common, although depending on the protein, one form or the other may predominate. Of course, the absence of hydrogen-bonded secondary structure in a given region of the protein does not necessarily mean that the polypeptide chain can assume a random configuration at that point. It implies simply that the three-dimensional packing is determined by factors other than secondary forces, such as proline residues, disulfide bonds, or hydrophobic interactions, none of which generate regular structure.

x-Ray diffraction studies on protein structure also confirm the basic postulate of the Bresler-Talmud theory (p. 74) which states that hydrophobic amino acid side groups tend to prevail within the molecule while polar side groups generally dominate the protein surface. Such a distribution undoubtedly stabilizes the molecular structure in an aqueous milieu and it may help to explain why the active site of several enzymes is located in or near a depression caused by the tertiary folding of the polypeptide chain. Besides drawing amino acid residues from many different parts of the protein together at the active site, this conformation provides for the participation of hydrophobic as well as of polar residues in the catalytic process since the depression exposes nonpolar amino acid side chains located in the interior of the molecule. Although the three-dimensional structures of only six enzymes are known, it is nonetheless tempting to speculate that hydrophobic interactions are of material importance in defining the shape of the active site and that hydrophobic groups participate jointly with polar groups in the binding of substrate and in the catalytic attack. Further generalizations about relationships between conformation and function must await the determination of additional enzyme structures. It is safe to predict that within the next few years we shall know the complete primary, secondary, and tertiary structures of scores of proteins and that the problem of interpreting function will therefore be considerably easier. Even now the general features of all three levels of protein structure have been well established and a start has been made toward the detailed explanation of protein function on the scale of atomic dimensions.

12. Physicochemical Techniques for Measuring Molecular Size and Shape

For analytical studies proteins can presently be prepared in *chemically* pure form with contaminating matter amounting to as little as 0.01% by weight. At this level, any impurities that do persist will not substantially affect the measurement of physical constants nor will they distort purely chemical analyses. Indeed, the very fact that a structural formula can be worked out or that a detailed three-dimensional model can be constructed presupposes the availability of a chemically pure and individual molecular type. All preconceptions about the supposed multiplicity or heterogeneity of protein molecules must thus be laid aside in the face of such strong experimental refutation. The structure of each protein is completely determined genetically and, with a few exceptions, only a mutation in the cell can lead to its alteration. In Chapter IV, we shall see how simple point mutations can evoke changes in a unique amino acid residue in the protein chain.

This section will deal with techniques for measuring the most important physical constants relating to protein structure. The constants assist in identifying proteins, in judging their purity, and in understanding the processes which lead to their functional activity.

Molecular weight is one of the most important physicochemical properties of protein molecules. There are many ways of measuring this parameter; in particular, direct chemical analysis frequently makes a very precise determination possible. In the case of zinc-insulin, for instance, a single atom of zinc binds to each molecule of insulin. By measuring the total weight and the zinc content of a certain amount of this material, the molecular weight of insulin may be easily computed. Similarly, myoglobin contains one iron atom per protein molecule, which facilitates molecular weight calculation. Sometimes a particular amino acid is present in a protein in very small quantities, permitting computation of the molecular weight from the content of the infrequent amino acid.

The simplest physicochemical methods for determining the molecular weight of polymers are osmometry and light scattering, both of which have been successfully applied to proteins. Sedimentation in the ultracentrifuge is undoubtedly the best method, however (93). The main virtue of sedimentation analysis is that impurities are separated from the protein during the very process of measurement. Thus, the purity of the preparation, the nature of any contaminants, and the sedimentation constant of the pure protein can all be assessed simultaneously. Proteins were the first polymers that proved to be *monodisperse*, having a unique and well-defined molecular weight. When Svedberg originally made this observation in the 1930's, the reaction among contemporary biochemists was one of astonishment. Linear polymers, synthetic and natural alike, had always appeared to be *polydisperse*, with a very broad molecular weight distribution. Proteins were the first exception to this rule, and nucleic acids the second. Considerable variation was noted in the molecular weight of individual protein species, however, from 6000 for insulin to 6,000,000 for hemocyanin (Table 1-8). There were many attempts by Svedberg and others to find some sort of arithmetic regularity in these values with the hope of identifying a universal subunit from which all proteins were assembled. None of these analyses were rewarded with success, and the accumulated data on protein structure indicated much more diversity among proteins than had been originally supposed.

Ultracentrifugal studies do not yield molecular weight directly. This figure must be calculated from the *sedimentation constant*, s, by definition, the rate of motion of a particle undergoing unit acceleration:

$$s = (dx/dt)/\omega^2 r \qquad (1\text{-}23)$$

where dx/dt is the rate of motion of the particle in the centrifuge cell, ω the angular velocity of the centrifuge rotor (equal to 2π multiplied by the number of revolutions per second), and r the radius of rotation, that is, the distance from the center of the rotor to the moving particle boundary. The sedimentation constant has the dimensions of time and is usually measured in Svedbergs (S), a unit defined as 10^{-13} seconds. If the preparation contains more than one protein, each moves separately and sedimentation constants can be determined for several components simultaneously. The sedimentation constant is related to molecular weight by the Svedberg equation:

$$s = M(1 - V\rho)/fN \qquad (1\text{-}24)$$

where $V = 1/\rho_0$; ρ_0 represents the buoyant density of the particle, M its molecular weight, N the Avogadro number, ρ the density of the medium, V the specific volume of the protein, and f the *frictional coefficient* of the protein. The shape of the macromolecule is reflected in f which is a function of particle dimensions.

In order to evaluate the molecular weight, f must be determined independently of s. Fortunately, there is another technique which simply and precisely yields the value of this coefficient. The frictional coefficient is related to the *diffusion constant D* of the protein. This parameter reflects the net rate at which solute molecules are transported from regions of high concentration to regions of lower concentration when a concentration gradient exists in solution. Although methods for determining D will not be discussed here, it will suffice to remark that small molecules have much greater diffusion constants, that is, they diffuse faster, than large molecules. Thus D is itself related to the mass of the solute even though the functional dependence—D is approximately proportional to the inverse cube root of M—is not a strong one. According to the relationship discovered by Einstein:

$$fD = kT \qquad (1\text{-}25)$$

where k is the Boltzmann constant and T the absolute temperature. Substituting for f and replacing Nk by R, Eq. (1-24) can be solved for M:

$$M = sRT/D(1 - V\rho) \qquad (1\text{-}26)$$

Note that D must be measured in the same solvent used for sedimentation analysis or else this relationship is not valid. The molecular weights of more than 500 proteins have been evaluated in this way. The total number of proteins known is only about 1000 and roughly half of these are enzymes. Table 1-8 presents the data for many of the more important proteins. Sedimentation constants are always corrected to 20°C and a solvent of pure water ($s_{20,w}$), although under experimental conditions, the ionic strength is generally maintained at 0.05 to 0.1 to compensate for undesired electrokinetic effects.

An alternate way to measure molecular weights by ultracentrifugation involves waiting until the average motion of the molecules in the centrifugal field has stopped. This method, called *sedimentation equilibrium*, requires considerably less intense force fields and longer times than does measurement by sedimentation velocity. The spatial distribution of macromolecules in the centrifugal field is governed by the Boltzmann law. If c_1 and c_2 are the concentrations of solute molecules at distances r_1 and r_2 from the rotation axis, and if ΔE is the difference in potential energy between molecules at the two positions, then:

$$\ln(c_1/c_2) = -\Delta E/kT \qquad (1\text{-}27)$$

Since ΔE is equal to the amount of work necessary to transport a molecule from r_2 to r_1 against the centrifugal force $M(1 - V\rho)\omega^2 r$, we find that:

$$\Delta E = -\frac{M(1 - V\rho)\omega^2}{2}(r_2^2 - r_1^2) \qquad (1\text{-}28)$$

Substituting Eq. (1-28) into the Boltzmann distribution, we obtain:

$$\ln\frac{c_1}{c_2} = \frac{M(1 - V\rho)\omega^2}{2kT}(r_2^2 - r_1^2) \qquad (1\text{-}29)$$

This relationship depends neither on the shape of the molecule nor on the extent to which it is solvated, so that M can be computed from the centrifugation data alone. A number of optical methods have been developed to measure the equilibrium concentration of solute molecules as a function of distance from the rotation axis while the centrifuge rotor is in motion. These methods are also applicable to the determination of sedimentation rates.

Another use for the ultracentrifuge is in the study of protein quaternary structure. Quaternary structure refers to the association of two or more individual polypeptide chains into a more complicated structure. Association generally results from the interaction of various functional groups localized on the surface of the molecule. Reversible dissociation of a protein complex was first observed by Svedberg and his associates (95). They used changes in pH to break up hemocyanin, a protein distinguished by an enormous molecular weight running into the millions. Similar types of association–dissociation reaction, involving 12, 24, or even more polypeptide subunits have been noted in ferritin, glycinin (a soybean protein), and other proteins.

Association is usually very specific and distinctive as in the case of hemoglobin. The hemoglobin macromolecule consists of four polypeptide subunits: two α-chains and two β-chains. The abbreviated hemoglobin formula is generally written $\alpha_2\beta_2$. At alkaline pH the protein reversibly dissociates into two identical particles:

$$\alpha_2\beta_2 \rightleftharpoons 2\alpha\beta$$

But in an acid medium, or on the surface of an acidic resin, dissociation proceeds further and results in four separate polypeptide chains:

$$\alpha_2 \beta_2 \rightleftharpoons 2\alpha\beta \rightleftharpoons 2\alpha + 2\beta$$

There are obviously different kinds of groups holding the structure together by what can be roughly classified as acid-labile interactions and base-labile interactions.

Irreversible dimerization as well as other types of protein aggregation are frequently attributable to chemical reactions. Serum albumin contains sulfhydryl groups which very readily react with mercury ions to form dimers of the type: albumin-S-Hg-S-albumin. The product of the dimerization, mercaptalbumin, is like a new protein

TABLE 1-8

Physicochemical Constants of Proteins[a]

Protein	Specific volume	Molecular weight (Daltons)	Sedimentation constant (Svedbergs)	Isoelectric point, pH
Adrenocorticotropic hormone (pig)	—	4,500	—	7.0
Insulin (bovine)	0.71	6,300	1.2	5.6
Ribonuclease (bovine)	0.73	13,700	1.64	9.45
Cytochrome c (bovine)	0.71	12,400	2.5	10.6
Lysozyme (hen egg white)	0.69	14,600	1.9	11.0
Myoglobin (sperm whale)	0.74	17,300	2.04	7.0
Tobacco mosaic virus coat protein	0.75	18,270	2.3	—
Lactalbumin	0.75	17,400	1.9	—
Chymotrypsinogen (bovine)	0.72	24,000	2.54	9.5
α-Chymotrypsin (bovine)	0.72	24,000	2.5	8.1
Trypsinogen (bovine)	0.73	23,700	2.5	9.3
Trypsin (bovine)	0.73	23,800	2.5	10.6
Papain	0.725	20,900	2.4	8.7
Carbonic anhydrase	0.75	31,000	2.8	5.3
Carboxypeptidase	0.75	34,300	3.1	6.0
Pepsin	0.75	36,000	3.3	1.1
Peroxidase	0.7	40,000	3.5	—
β-Lactoglobulin	0.75	35,000	3.2	5.2
Ovalbumin	0.75	44,000	3.6	4.6
Zein	0.75	50,000	1.9	—
Pyrophosphatase	0.75	63,000	4.4	—
Enolase	0.735	67,000	5.6	—
Serum albumin (bovine)	0.74	66,500	4.4	4.7
Serum albumin (horse)	0.75	70,000	4.5	5.0
Serum albumin (human)	0.73	65,000	4.3	5.2
Hemoglobin (bovine)	0.75	66,000	4.6	6.8
Hemoglobin (horse)	—	66,000	4.6	6.6
Hemoglobin (human)	0.75	66,000	4.5	6.8
Actin	—	70,000	4.0	4.8
Diphtheria toxin	0.736	72,000	4.6	4.1
Diphtheria antitoxin	0.75	160,000	7.0	—
Tropomyosin	0.71	93,000	2.55	5.1
Hexokinase	0.74	90,000	—	—
Alkaline phosphatase (E. coli.)	—	80,000	—	4.5

TABLE 1-8—(continued)
Physicochemical Constants of Proteins[a]

Protein	Specific volume	Molecular weight (Daltons)	Sedimentation constant (Svedbergs)	Isoelectric point, pH
Glyceraldehyde phosphate dehydrogenase (rabbit muscle)	0.79	140,000	7.0	—
β-Amylase (sweet potato)	0.75	215,000	8.9	—
γ-Globulin (bovine)	0.725	180,000	7.0	7.1
γ-Globulin (horse)	—	180,000	6.9	7.6
γ-Globulin (human)	0.74	185,000	7.4	7.3
γ-Globulin (rabbit)	0.744	160,000	7.0	6.6
Aldolase (rabbit muscle)	0.74	142,000	8.27	6.1
Fumarase	0.75	194.000	8.5	—
Catalase	0.73	250,000	11.0	—
Phycoerythrin	0.75	290,000	12.0	—
Phosphorylase a (rabbit muscle)	0.74	495,000	13.7	5.8
Fibrinogen (bovine)	0.706	330,000	7.9	—
Fibrinogen (human)	0.72	450,000	9.0	—
Myosin	0.74	500,000	6.1	5.4
β-Galactosidase (*E. coli*)	—	540,000	16.0	—
Ferritin	0.74	747,000	—	—
Urease	0.73	480,000	18.6	—
Thyroglobulin	0.72	630,000	19.2	4.5
Hemocyanin (*Helix pomatia*)	—	9×10^6	103.0	4.7
Actomyosin (rabbit)	0.73	5×10^6	35.0	—

[a] Based on Edsall (94) with additions and corrections.

with twice the molecular weight of the original one and it crystallizes in a completely different manner as well.

An alteration in quaternary structure often affects the functional activity of the protein. The enzyme phosphoglyceraldehyde dehydrogenase from rabbit muscle contains a single sulfhydryl group. Upon oxidation in air the protein dimerizes via a disulfide bridge, losing its enzymatic activity in the process (96). When the disulfide bond is reduced, complete activity is restored. Reactions of the sulfhydryl groups which do not lead to dimerization, however, do not necessarily eliminate the functional capacities of the enzyme. Alkaline phosphatase from *Escherichia coli* illustrates the reciprocal situation. The active form of this protein consists of two identical subunits. If the two chains are separated from each other by either raising or lowering the pH, enzymatic activity drops to zero.

An especially interesting situation arises when proteins are bifunctional, that is, when they each possess two groups capable of interacting with each other. The result is polymerization of the protein molecules into fibers. Two of the best examples of this class are actin and myosin, found in muscle, and fibrinogen. Actin is a small spherical protein with a molecular weight of 70,000 while myosin is a large extended chain of 500,000 molecular weight. At the appropriate pH and in the presence of magnesium and potassium ions, actin can either polymerize with itself, or else copolymerize with myosin to form filaments of actomyosin, the major functional component of muscle tissue. Fibrinogen is found in blood serum and is itself inert. However, under the action of the proteolytic enzyme thrombin, small peptides are removed from the end of the fibrinogen molecule, unmasking groups which have the potential to interact with similar groups on other molecules. The product of this cleavage is called fibrin monomer, and its rapid polymerization leads to the formation of macroscopic fibrous aggregates possessing considerable elasticity. These aggregates are admirably suited to their task of covering small openings in the walls of blood vessels and of preventing the spread of inflammation.

The polymerization of globular proteins plays a curious, but extremely important role in the formation of viruses. One example is tobacco mosaic virus, usually abbreviated TMV, which contains 95% protein and 5% RNA by weight. Some 2300 individual protein subunits, each with a molecular weight of 18,270 ($s = 2.3$ S), are packed into a cylindrical tube (Fig. 1-52). The polypeptide chains can be dissociated in 66%

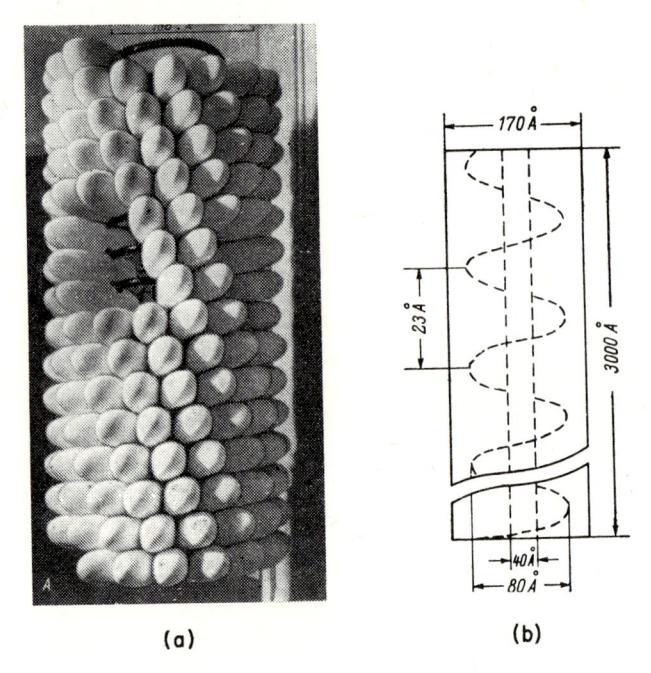

Fig. 1-52. The structure of tobacco mosaic virus (TMV). (a) Model of the virus showing the distribution of polypeptide subunits in the coat. Some of the subunits are removed to reveal the placement of the helical RNA chain. (b) Schematic drawing of TMV in section, showing how the helical RNA chain and protein subunits are distributed about an open core 40 Å in diameter. After Klug and Caspar (97).

acetic acid. After dialysis and neutralization, the subunits repolymerize to give either small closed rings with a sedimentation constant of 25 S or else helical tubes of various sizes whose length sometimes exceeds that of the native virus. The presence of the viral nucleic acid is not material to the process of polymerization, though particles devoid of RNA are not infectious (98). Moreover, the conformation of the nucleic acid strand within its protein sheath determines the length of the virus particle;

12. Physicochemical Techniques for Measuring Molecular Size and Shape

helical protein tubes produced by reassociation in the absence of RNA appear to be of random length. Polymerization of the protein to the disk- or ringlike intermediate occurs only at pH 7.1 and only if the ionic strength is appropriate. The rate of polymerization, strongly dependent on temperature, is practically nil at 0°, but quite rapid at 20°C, indicating the importance of chemical interactions in the process. Consequently, groups in the molecule with a pK near 7.1 were sought and from the titration curve, two carboxyls were discovered which exhibited the proper electrochemical properties. Interactions among subunits might be mediated by bonds between carboxyl and tyrosine phenol groups.

The behavior of TMV is typical of many viruses which contain protein subunits capable of polymerizing into complex three-dimensional structures. In the electron microscope viruses frequently appear to be composed of a large number of identical protein molecules associated with each other in a regular array (Fig. 1-53).

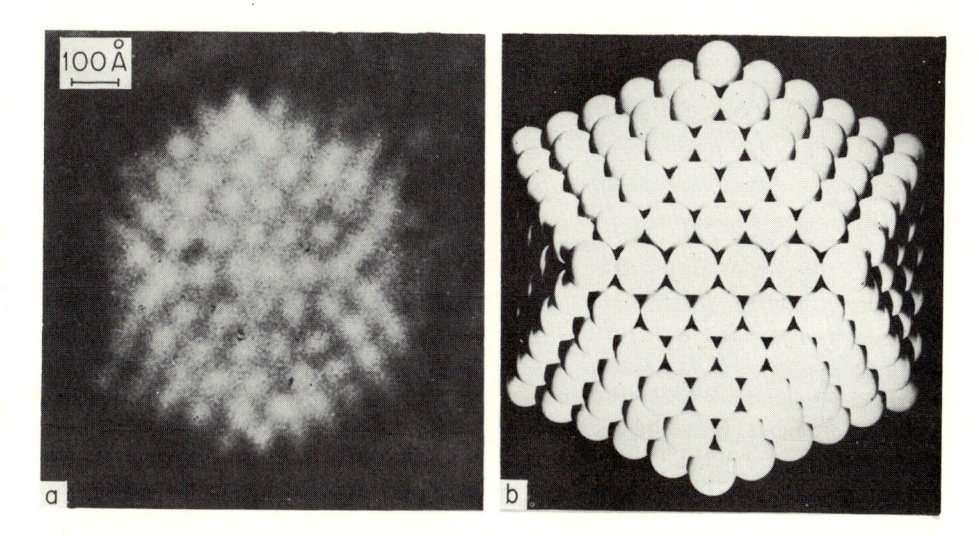

Fig. 1-53. The structure of adenovirus. (a) Electron micrograph of virus particle at high magnification (×700,000). Micrograph shows the three-dimensional packing of the protein molecules in the protein shell. (b) Model of the icosahedral virus constructed from spheres on the basis of the electron micrograph in (a). Orientation is the same in both cases. The distance between centers of the subunits is 70 Å and the distance along the edge of the icosahedron is 420 Å. From Horne *et al.* (99).

There is one aspect of virus structure which demands particular attention. The geometry most commonly assumed by viral protein shells or *capsids* is that of an icosahedron, a solid figure with 20 faces. The remarkable feature of this structure is that the faces are all equilateral triangles. Octahedral viruses are next in order of frequency and it turns out that the faces of an octahedron are also equilateral triangles. These observations undoubtedly point to an overall regularity in the determination of viral form. The protein subunits which constitute the shell are approximately spherical, and the equilateral triangle represents the simplest symmetrical plane figure which can be formed from tightly packed spheres. Furthermore, in order to construct the shell most economically, it is necessary to minimize the ratio of its surface to the volume which it encloses. The icosahedron satisfies this requirement most easily, while the rarer octahedron meets the condition somewhat less perfectly (97).

For the most part, problems of quaternary protein structure have been taken up only recently. It is very likely that the orderly and specific aggregation of protein subunits is extremely important, particularly for understanding of supermolecular structures such as cell organelles and membranes. These components act to compartmentalize the cell interior and play a role in the regulation of the rate of metabolic reactions.

Now it is time to consider some further parameters relating to macromolecular structure, particularly what are called the hydrodynamic constants. These values characterize the behavior of the protein when it is in motion through a fluid and hence depend both on the volume and the shape of the particle. The first of the hydrodynamic constants is the *frictional coefficient*, f, which is calculated from the diffusion constant by Eq. (1-25). If the macromolecule were a sphere, its frictional resistance could be expressed by Stokes' law:

$$f_0 = 6\pi\eta_0 R \qquad (1\text{-}30)$$

where η_0 is the viscosity of the solvent, and R the radius of the sphere. R is related to molecular weight and particle density in the following way:

$$\tfrac{4}{3}\pi R^3 = M/N\rho = MV/N \qquad (1\text{-}31)$$

where N is Avogadro's number, and M, V, and ρ are the molecular weight, specific volume, and density of the particle, respectively. The frictional coefficient thus depends only weakly on M ($f_0 \propto M^{1/3}$). In practice, protein molecules are far from spherical and the measured value of f is usually greater than f_0 calculated on the basis of the protein molecular weight. The ratio f/f_0 characterizes the extent to which the actual shape of the protein diverges from that of a sphere. It is difficult to interpret this ratio when it approaches unity since protein molecules possess an intricate and irregular shape which approximates a sphere only very roughly. But if f/f_0 differs greatly from unity, we may be sure that the protein particles are very asymmetrical. In such cases they can be approximated as elongated or prolate ellipsoids of revolution, for which it is possible to compute f according to the laws of hydrodynamics. If the major semiaxis, or semiaxis of revolution, is designated by a and the minor semiaxis by b, the axial ratio a/b is greater than unity and the following formula applies:

$$\frac{f}{f_0} = \frac{[1 - (b^2/a^2)]^{1/2}}{\left(\dfrac{b}{a}\right)^{2/3} \ln \dfrac{1 + [1 - (b^2/a^2)]^{1/2}}{b/a}} \qquad (1\text{-}32)$$

In practice, this relationship is used to determine the axial ratio from experimental data on the frictional ratio f/f_0.

A second important hydrodynamic constant, derived from measurements on the viscosity of dilute protein solutions, is called the *intrinsic viscosity*:

$$[\eta] = (1/C)(\eta/\eta_0 - 1)_{C \to 0} \qquad (1\text{-}33)$$

where η is the viscosity of the polymer solution, η_0 is the viscosity of the solvent, and C is the polymer concentration in grams per 100 ml. The intrinsic viscosity is obtained by measuring the viscosity of the solution at several different concentrations and then extrapolating $(1/C)(\eta/\eta_0 - 1)$ to infinite dilution. The intrinsic viscosity has a stronger dependence on molecular shape than does the frictional coefficient f.

For perfectly spherical particles, the Einstein relation states that:

$$\eta/\eta_0 = 1 + 2.5 \, nv \qquad (1\text{-}34)$$

where n is the number of particles per unit volume and v is the actual volume of each particle. It is evident that:

$$n = 10^{-2} CN/M \qquad (1\text{-}35)$$

At the same time,

$$v = M/N\rho = MV/N \qquad (1\text{-}31')$$

where V is the *specific volume*, with a value of about 0.75 for most proteins. Consequently, the intrinsic viscosity of spherical proteins will be a constant independent of the molecular weight and size of the particle:

$$[\eta] = 2.5 \times 0.75 \times 10^{-2}$$

The expected intrinsic viscosity for globular proteins is thus on the order of 1/100. But when the molecules are very asymmetrical, the size of $[\eta]$ considerably exceeds the limiting value given by the Einstein relation. In such cases the molecule must once again be approximated as an ellipsoid of revolution, for which a complete theoretical calculation can be made. Equations relating the measured value of $[\eta]$ to the axial ratio of the equivalent elipsoid a/b have been obtained by Simha. Tables were constructed on the basis of these formulas which can be used in making the appropriate computations.

The interpretation of hydrodynamic properties in terms of molecular structure is not a task without pitfalls. For example, many controversies have arisen over the extent to which solvent molecules bind to the protein surface in the form of a solvation shell. Solvation results from the strong attractive forces which a number of different amino acid side groups exert on water molecules. Hydrogen bonds play an important role in this phenomenon, as do interactions between ionic groups and water dipoles. In fact, solvation accounts for the solubility of proteins in water.

In order to measure the extent of solvation or hydration, it is essential to precisely define what is included in this concept. To suppose that the solvation shell is the entire volume of solvent which moves together with the macromolecule in hydrodynamic phenomena would be ridiculous in certain cases. Ordinary linear polymers, for instance, exist as loose coils in solution and of the total volume which they occupy, 97 to 99% is taken up by solvent and only 1 to 3% by the molecule itself. Nonetheless, this heavily impregnated coil behaves hydrodynamically as a single entity, solvent and all. If the interstitial solvent is taken as water of solvation, there will be 30 to 40 gm of water associated with every gram of polymer. This formulation is therefore of slight practical significance in dealing with real substances.

Fortunately, such enormous numbers are not obtained for compact globular proteins. Only water which fills the available cracks and crevices on the molecular surface will move together with the hydrodynamic unit. This is a trivial effect of hydrodynamic flow over an irregular surface. We shall therefore define water of solvation in proteins as that which is bound to the protein molecule by very strong intermolecular interactions. The distinguishing feature of this solvent layer is that it cannot be penetrated by electrolytes. The quantity of solvent so bound in aqueous protein solutions can be assessed by its very impermeability to ions.

The best method for determining the degree of solvation in proteins is based on x-ray diffraction, and it was first applied to hemoglobin. Protein crystals usually contain about 50% water by volume, but very little of this is bound water. Instead, it fills up the interstices between irregularly shaped protein molecules which cannot be packed any more tightly together in the crystal lattice. Salt ions can freely penetrate into these interstices, but there is a certain small volume of water bound by polar groups of the protein molecules which is inaccessible to electrolytes, namely, the water of hydration. This volume was determined by Bragg and Perutz from diffraction patterns of hemoglobin made in the presence of sodium iodide, a salt that strongly scatters x rays. By studying the differences in diffraction patterns made with and without salt, it was apparent that the protein macromolecule is not simply a sponge which indiscriminately soaks up ions. To the contrary, iodide ions remain in the surrounding volume, intensifying scattering from the external contours of the protein macromolecules. If ions were to penetrate within the hemoglobin structure, an enormous change in all diffraction maxima would result. As observed experimentally, only the maxima corresponding to large interplanar distances increased in intensity.

This analysis helps to explain the distribution of hemoglobin molecules in the crystal lattice. In the monoclinic unit cell the macromolecules are arranged in relatively dense layers (*ab*, Fig. 1-54) while the water of crystallization mainly occupies

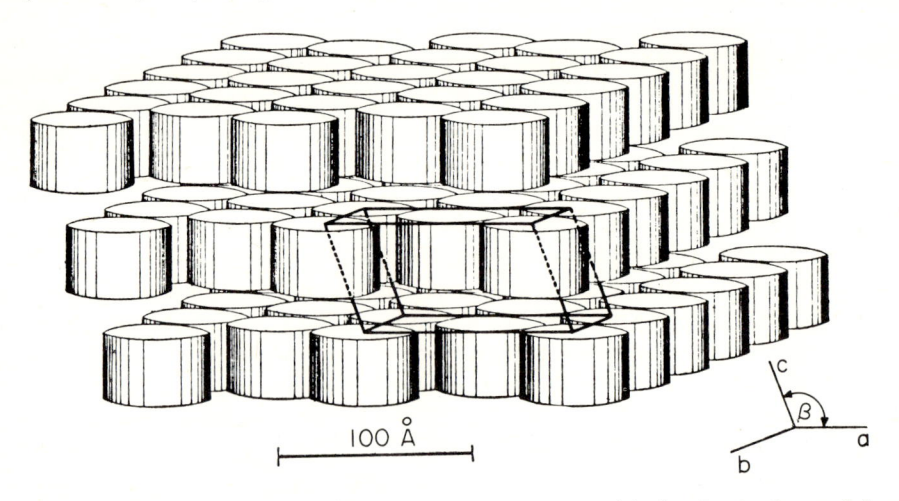

Fig. 1-54. Arrangement of hemoglobin molecules in the crystal lattice showing layers of closely packed molecules separated by liquid. Axes in the right foreground indicate the dimensions of the monoclinic unit cell. From Boyes-Watson et al. (100).

the spaces between successive layers. If the crystal is dehydrated, its volume is reduced mostly as a result of contraction along the *c* axis. From the intensity of the x-ray reflections originating from the *ab* planes, it can be shown quite exactly that 36% of the water of crystallization comprises a tightly bound shell from which salt ions are excluded. The water of hydration thus occupies 18% of the total crystal volume and, expressing this figure in terms of a weight ratio, there are 0.22 gm water per gm protein. Since hemoglobin is a typical example, this number gives us a reliable estimate of the extent to which most globular proteins are hydrated. Similar data have been obtained by Luzzati from low-angle scattering of x rays by proteins in solution (101).

The thickness of the hydration layer covering the protein molecule may be calculated from the data on hemoglobin, though the estimate is rather imprecise in view of the irregular contours of the molecular surface. The average gap between hemoglobin molecules along the c axis is 16.7 Å, of which 6.1 Å (36%) can be attributed to tightly bound water. The aqueous layer enveloping each protein is thus about 3 Å thick, a figure that corresponds to the linear dimensions of a water molecule. x-Ray analysis reveals both that the water of hydration is concentrated on the surface of the protein macromolecule and that it constitutes, on the average, a monomolecular layer. All data published before these precise studies must be regarded with suspicion.

Since solvation amounts to 20% of the protein particle weight, it materially affects the interpretation of the hydrodynamic constants f and $[\eta]$. The axial ratio a/b of the equivalent ellipsoid may be quite strongly influenced by solvation and it is essential that a correction be introduced when this parameter is calculated from experimental measurements. It should be recalled that small changes in a/b are difficult to interpret in terms of structural alterations. When the axial ratio changes by a large factor, however, the corresponding structural alterations may be justifiably assessed using hydrodynamic parameters since there is little danger that small differences in hydration will significantly affect the result.

Consider finally the electrochemical properties of proteins. This class of molecules generally bears a net charge which is the sum of positive charges arising from the side chains of lysine, arginine, histidine, and terminal amino groups, and negative charges contributed by glutamic acid, aspartic acid, tyrosine, and cysteine, as well as by terminal carboxyl groups. Since all of these groups, both positive and negative, are weak acids or bases, their degree of dissociation and hence the net charge of the protein depends on pH. In principle, each type of side group is characterized by a dissociation constant K_i or a pK_i, the pH at which the group is discharged. On titrating a protein a series of inflections in the titration curve is expected at values of the pH corresponding to the dissociation of each weak acid or base. If the situation were really that simple it would not be difficult to determine the quantity and nature of all the ionizable side groups in the protein from a single titration curve.

Unfortunately, the actual state of affairs is very much more complicated. As a concrete example, consider dissociation of the —COOH group. The pK of carboxyl groups in acetic acid is 4.8, but in the polyelectrolyte polyacrylic acid, it is 5.5. Protons dissociate less readily in the latter case because of additional electrostatic interactions with neighboring anionic groups.

If a polar solvent such as water is partially replaced by a less polar organic solvent (ethanol, acetone, dioxane), acidic groups will not dissociate as readily and there will be a rise in their pK. We encounter precisely this situation in the tertiary structure of proteins when ionizable groups inside the macromolecule are surrounded not by water but by other side groups, some of which are nonpolar hydrocarbons.

It is not surprising that the effective pK's of protein carboxyl groups are quite varied, depending on the polarity of their immediate environment, itself a function of position within the macromolecule. The carboxyls in closest contact with the aqueous medium titrate with a pK near 5. Others, more deeply embedded in the structure, require a more basic medium and the pK of certain carboxyl groups is in excess of 7. Finally, there are carboxyl groups so deeply buried that they cannot be titrated at all.

The same reasoning applies to amino groups and other functional groups in the protein as well. Actual titration curves of proteins therefore appear very smooth and often contain no points of inflection whatsoever, since they represent the sum total of a great number of individual titrations. Valuable information on the way in which the net charge of the protein depends on pH may be obtained from these curves, but it is impossible to identify the charge with the ionization of any particular groups without additional data.

The masking of functional groups by the tertiary structure is a well-known effect in the chemistry of globular proteins and it is very characteristic of their native configuration. This phenomenon is not limited to electrolytic dissociation. Any chemical reaction in which the amino acid side chains normally participate may be strongly inhibited when the functional groups in question are buried in the interior of the molecule. The resistance of disulfide bridges to the action of oxidants and reductants may be cited as an example. Ribonuclease contains four disulfide bonds, two of which are very stable to reduction. The inaccessibility of tryptophan and tyrosine residues to iodine provides another example, as do the inability of certain —SH groups to react with specific mercurials, and the stability of side chain amino groups to formaldehyde. In all of these instances it is sufficient merely to denature the protein by destroying its secondary and tertiary structure in order to render the side groups reactive. The titration curve also substantially changes upon denaturation, owing to alterations in the immediate environment of ionizable groups. Of all the methods of denaturation, the use of high concentrations of urea is the one most desirable for unmasking amino acid side chains because it prevents aggregation of the protein molecules, that is, the replacement of one system of hydrogen bonds by another.

From the preceding discussion it is clear that the electrochemical behavior of protein macromolecules is a very complicated affair which cannot be described by simple constants and equations. Nevertheless, electrochemical properties are widely used for the practical purpose of fractionating proteins by *electrophoresis*, the migration of charged molecules in an electric field. The most important constant which can be determined by electrophoresis is the *isoelectric point*, the pH at which the net charge of the protein is equal to zero so that the molecule cannot move in an electric field. The protein is not uncharged at this pH, but simply bears an equal number of positive and negative charges. At pH's above the isoelectric point, the protein is negatively charged and behaves as an anion while below the isoelectric point a net positive charge results in motion toward the cathode. Proteins can bear a maximum net charge of 30 to 40 electrons at extreme pH's.

The isoelectric point of most proteins lies between pH 5 and pH 7 (Table 1-8), but exceptions are sometimes encountered. Although a protein at its isoelectric point is immobilized in an electric field, a change of two or three units in pH permits it to move with a rate of about 10^{-4} cm/V-sec, a value comparable to the mobility of small ions. The mobility of a protein at a given pH is not a constant, for it depends on the nature of the buffer employed and on the ionic strength of the medium. Values for mobility are useful only when the precise conditions of the determination are known.

Complex mixtures of proteins can be fractionated by electrophoresis and the electrophoretic homogeneity of a protein is a valuable criterion of its purity. Nonetheless, in dealing with such intricate substances as proteins, electrochemical criteria of purity are not enough; they must be used in conjunction with other standards such as ultracentrifugation and chromatography.

Blood serum, that is, blood which has been defibrinated and freed of whole cells, serves as a good example of a complex protein mixture which has been fractionated by electrophoresis. Tiselius and Svensson (102) discovered four main electrophoretic components when they first examined blood serum: serum albumin and α-, β-, and γ-globulins. The isoelectric points of serum albumin and γ-globulin are pH 5 and 7, respectively, while those of the remaining components fall at intermediate values. Serum is electrophoresed at pH 8 where all the proteins are negatively charged and move toward the anode; albumin moves most rapidly and γ-globulin is the slowest. As methods were perfected with time, buffers with much greater resolving power came into use. In the best of these (a mixture of trishydroxymethylaminomethane, ethylenediaminetetraacetic acid, and boric acid) normal serum can be separated into nine components.

The electrophoretic study of serum and urine from persons ill with a variety of diseases has been of great practical importance in medicine. A new protein—the so-called Bence-Jones protein—was identified in the urine of individuals suffering from myeloma, a malignancy arising in bone marrow. Quantitative changes in the proportions of serum proteins are characteristic of leukemia, arthritis, and a number of other diseases. Hemoglobin from persons affected by certain diseases of the red blood cells such as sickle-cell anemia differs significantly from the normal protein in both electrophoretic mobility and isoelectric point. These were all important discoveries which linked various illnesses to alterations in the synthesis of specific proteins. The term "molecular diseases" was coined to describe these pathological conditions, but it was not a very accurate choice; a disease affects the organism as a whole, and a change in the structure of one or more proteins may be only one of its symptoms.

In discussing the use of electrophoresis for protein fractionation, it is important to bear in mind the origin of those differences in mobility on which the method is based. The principal difference is of course the charge, ε, which various proteins bear. But the question is not one of charge alone. Electrophoresis is always carried out in buffer containing an electrolyte. According to the Debye-Hückel theory, each charge in such an environment is screened by a diffuse atmosphere of counterions, the density, ρ, of which decreases with distance, r, according to the formula:

$$\rho = \rho_0 e^{-\chi r} \qquad (1\text{-}36)$$

The average thickness of the ionic atmosphere is equal to $1/\chi$ where:

$$\chi = (8\pi e^2 NI/10^3 DkT)^{1/2} \qquad (1\text{-}37)$$

In this equation, e is the charge of an electron, N the Avogadro number, k the Boltzmann constant, T the absolute temperature, D the dielectric permeability of the medium, and I its ionic strength:

$$I = \tfrac{1}{2} \sum_i C_i z_i^2 \tag{1-38}$$

C_i is the molar concentration, and z_i is the charge, of the ith ion. The mobility, U, of a spherical particle moving in an electric field is expressed by the equation of Smoluchowsky which is generally accurate within a factor of 1 to 1.5:

$$U = \Psi D / 4\pi\eta \tag{1-39}$$

D is again the dielectric permeability, η the viscosity of the medium, and Ψ is the difference of potential in the double layer surrounding the particle.

Two possible cases must be considered individually. The first applies to extremely dilute electrolyte solutions with an ionic strength $I << 0.01$. Under these conditions the thickness of the Debye diffuse layer is greater than the dimensions of the particle, and $\Psi = \varepsilon/Da$. Here ε is the charge borne by a protein particle and a is its radius. The mobility is thus:

$$U = \varepsilon / 4\pi\eta a \tag{1-39a}$$

and since the charge of a sphere is approximately proportional to its surface area ($\varepsilon = 4\pi\alpha a^2 e$), where α is the charge density per unit surface area:

$$U = \alpha a e / \eta \tag{1-39b}$$

In this case, the electrophoretic mobility depends on the charge carried by the protein and on its radius which, for a spherical molecule, is proportional to the cube root of its molecular weight.

If electrophoresis takes place in a more concentrated solution in which $I \geqslant 0.01$, the thickness of the double layer becomes smaller than 10 Å, or less than the dimensions of the protein molecule. Here the potential drop Ψ is concentrated within a layer of thickness $1/\chi$. In these circumstances Ψ is calculated as for a flat condenser of the same thickness so that $\Psi = 4\pi\alpha e/D\chi$ and

$$U = \alpha e / \chi \eta \tag{1-39c}$$

where αe is the charge per unit surface area. We see that in this case the mobility of the protein does not depend on molecular weight at all, a situation which more closely conforms to the actual conditions of electrophoresis than the first example. And when it is possible to experimentally ascertain whether or not U depends on the molecular weight of the protein, the answer is generally negative. Such was the case with hemocyanin which can be dissociated into individual subunits. Hence, in buffers of average ionic strength, electrophoresis fractionates proteins by charge and not by differences in molecular weight or dimensions. This is an important fact, for it indicates that proteins are separated on the basis of entirely different properties in electrophoresis and in ultracentrifugation. For that reason, the two methods complement each other very well.

13. Purification and Identification of Proteins

The fractionation of cell juices into individual, highly purified protein species constitutes an entire branch of preparative biochemistry. Just a few decades ago the purification of each new protein was a real event and the techniques by which this was achieved were more akin to art than to science. A large measure of intuition was required to devise new approaches to the problem and many of these proved to be of little avail. But the most characteristic aspect of this work was the enormous expenditure of effort necessary to concentrate, separate, and purify any sort of new enzyme. The extent of purification was usually followed by assaying the increase in catalytic activity per unit weight protein, a parameter called the *specific activity*. Crystallization was then considered the crowning achievement of the purification process, although we know today that it is no absolute guarantee of purity.

Very great changes have come about in protein preparative procedures over the last few years, as they have in all areas of protein chemistry. Methods have been worked out that are both efficient and of general applicability. The job of fractionating proteins has thus lost much of its burdensome nature and the isolation of new enzymes is a routine operation. All this is owing in large measure to developments in the fields of chromatography and preparative electrophoresis. Chromatography is undoubtedly the most selective and general method of isolating closely related substances, but it could not be applied to proteins for a long time because of difficulties relating to denaturation. Chromatography requires that the acts of adsorption and desorption be repeated hundreds of times as the protein passes through the sorbent. The protein must of course undergo these processes reversibly if chromatography is to be effective. However, because of structural peculiarities, proteins are frequently denatured on adsorption to ion-exchange resins; they become insoluble as a result and cannot be chromatographed further.

Only when specialized ion exchange materials based on polymeric carbohydrates, such as cellulose, starch, and dextran, became available was the problem successfully resolved. At the present time the most widely used ion exchangers made from cellulose are the cationic resin carboxymethylcellulose (CMC-cellulose) and the anionic resins diethylaminoethylcellulose (DEAE-cellulose) and ECTEOLA, a sorbent prepared from cellulose, epichlorhydrin, and triethanolamine. All of these are extremely well suited to the chromatographic purification of proteins. A group of resins have recently appeared which are based on a cross-linked polyglucoside called dextran to which various ionic radicals have been attached: DEAE-Sephadex, carboxy-Sephadex, and sulfoxy-Sephadex. These last are distinguished by a specific capacity four to five times greater than the cellulose-based ion exchangers, an important factor in obtaining well-defined chromatograms.

The extent of the revolution in protein fractionation wrought by chromatography can be judged from the following example. If a bacterial extract containing hundreds of different proteins is adsorbed to DEAE-cellulose, and then eluted with buffer in which the salt concentration gradually increases (gradient elution), it is possible to obtain many individual enzymes in almost pure form after only one chromatographic run. In addition, the loss of protein in this process is practically nil. Even a structure as complicated as bacteriophage T2 can be purified on DEAE-cellulose with a yield of 50%. It is important to note that chromatography with gradient elution is a relatively rapid procedure which is suitable even for labile proteins.

An important aspect of protein purification consists in removing low-molecular-weight substances such as electrolytes, the total quantity of which may be quite large. This task used to be accomplished by dialysis through a cellulose membrane which, because of the long time it required, was not appropriate for the preparation of unstable substances. Dialysis has been largely replaced with a technique known as molecular exclusion chromatography, in which protein solutions are passed through specially prepared "molecular sieves." These materials are easily penetrated by small molecules but not by large molecules like proteins. The best of the currently available molecular sieves consist of tiny beads of dextran gel, cross-linked into an insoluble three-dimensional network. The commercial name of this product is Sephadex. Now instead of hours of dialysis, a protein solution can be desalted and freed of low-molecular-weight impurities by a chromatographic run on Sephadex lasting several minutes. Not only does this method represent a large economy in time, but it helps to preserve labile material.

Other methods of purifying proteins which have been developed in recent years occupy a secondary position in comparison to chromatography. It is nonetheless worthwhile to comment on preparative electrophoresis, a method which has been widely used in a number of different ways. In order to fractionate small quantities of proteins into separate zones it is necessary to conduct the electrophoresis in some sort of stabilizing medium which eliminates mixing by convection. Cellulose fiber, and paper in particular, provides a suitable supporting medium for this purpose. Paper electrophoresis of proteins is entirely analogous to paper electrophoresis of amino acids and peptides which was discussed in Section 4. Another widely used supporting medium is starch gel. Upon completion of electrophoresis, the gel can be cut into small pieces and any protein present eluted.

The latest advance in electrophoretic separation of proteins is *zone electrophoresis* on polyacrylamide gels (103). The use of polyacrylamide gel in place of starch as a supporting medium greatly increases the resolving power of this method which is presently one of the most powerful analytical tools in protein chemistry. Polyacrylamide gel electrophoresis is carried out in the following way. A length of glass tubing with an inner diameter of 5 to 6 mm is filled with a buffered solution of acrylamide monomer:

$$CH_2=CH\\ |\\ CONH_2$$

containing a small amount of the cross-linking reagent N, N'-methylenebisacrylamide:

$$CH-CONH-CH_2-NHCO-CH\\ \|\qquad\qquad\qquad\qquad\quad\|\\ CH_2\qquad\qquad\qquad\qquad\quad CH_2$$

and a catalyst to initiate polymerization.

The monomers polymerize to form a homogeneous, cross-linked gel at room temperature in the course of an hour. Next the protein sample solution containing acrylamide monomers is layered on top of the first gel, and it too is polymerized.

The ends of the tubing are placed in two separate reservoirs of buffer to which electrodes are attached. A voltage difference is then applied across the tube via the reservoirs of electrolyte and electrophoresis proceeds until the desired distribution of protein bands has been attained. The gel is then carefully removed from the glass tubing and stained by immersion in a solution of amido black, a specific protein stain. After removing excess stain, the separated proteins are visible as extremely thin dark disks perpendicular to the axis of the long cylindrical gel (Fig. 1-55). The gel can be

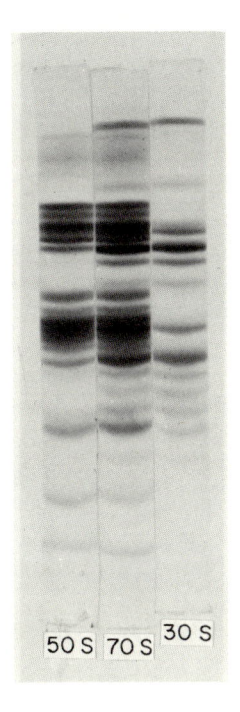

Fig. 1-55. Fractionation of *E. coli* ribosomal proteins by polyacrylamide gel electrophoresis. Ribosomes are ribonucleoprotein particles which play a vital role in cellular protein synthesis. Bacterial ribosomes possess a sedimentation coefficient of 70 S and they can be dissociated into two unequal subunits that sediment at 30 S and 50 S. The protein moiety of these particles and their subunits can be resolved into a large number of distinct components on polyacrylamide gels. Electrophoresis is carried out at pH 4.5 in this case since most of the ribosomal proteins are basic. Only 50 to 100 μg total protein sample, or 2 to 5 μg per protein band, are required for such analyses. From Traub *et al.* (104).

frozen and sectioned with an instrument similar to a microtome to obtain individual fractions.

Zone electrophoresis can also be performed in the absence of a supporting matrix. A density gradient of sucrose or heavy water is established in a long vertical tube. Convection mixing does not occur in the density gradient and the electrophoretic zones are stable. The equalization of the sucrose concentration gradient by diffusion takes place so slowly that it in no way interferes with the electrophoretic separation.

Several kinds of apparatus have been invented for the electrophoretic fractionation of large quantities of protein, but whether they will ever offer advantages over chromatography is questionable.

Zone sedimentation in stabilizing media provides still another means for the separation of protein mixtures (105). A centrifuge tube is filled with a solution of sucrose or other inert, dense material in which a linear gradient of concentration has been established. The protein sample is then layered on top of the gradient and as the tubes are centrifuged, each protein component sediments into the gradient as a separate zone at a rate determined by the size and shape of its constituent molecules. The density gradient stabilizes the sedimenting matter against convection and helps to keep the zones compact, since it greatly reduces hydrodynamic mixing of the macromolecules. After a period of centrifugation sufficient to distribute the components throughout the gradient, the bottom of the tube is punctured with a hollow needle and the contents are divided into individual fractions.

The earliest method of isolating and purifying proteins was by *fractional precipitation* and *coprecipitation*, procedures which relied on differences in solubility among various proteins in a mixture. In cases where differences in solubility were large, precipitation methods were extremely effective and indeed, they are still important today. Suppose we want to isolate a thermostable protein, that is, one which is not denatured at high temperatures, from a mixture containing many other proteins. The mixture can first be heated to denature and precipitate the bulk of the proteins. If the pH is next adjusted to between 5 and 6, near to the isoelectric point of most proteins, any denatured proteins not yet precipitated will become insoluble, leaving only the thermostable protein in solution. Ribonuclease, myokinase, and several other proteins are routinely prepared in this way. With this approach, the mass of unwanted proteins are eliminated in one step, and the desired component is enriched by a very large factor.

Precipitation methods are also sufficiently selective in the case of proteins which form insoluble complexes with other polymers. This technique is called coprecipitation and it can be used to enrich the concentration of an enzyme or other protein. Many acidic proteins, for instance, form insoluble complexes with protamines, basic polypeptides isolated from fish sperm.

Classic methods of salting out proteins in high concentrations of ammonium sulfate are still in widespread use even though they are rather unselective. Rising salt concentration decreases the solubility of the protein until it eventually begins to precipitate. Since the dependence of solubility on the salt concentration varies from protein to protein, a partial fractionation of the mixture is achieved by increasing the amount of ammonium sulfate in a stepwise fashion. At one time this was the only known technique of fractionating protein mixtures, and it was a very laborious task requiring many repetitions. Nonetheless, the first pure protein preparations, which included the blood serum proteins and proteolytic enzymes, were produced by the salting-out procedure. One of the principal drawbacks of the method is the necessity of removing the added salt by means of dialysis or gel filtration. For this reason the precipitation of proteins with organic solvents such as ethanol and acetone was suggested as a substitute for precipitation with salt. The organic solvents may be conveniently eliminated by evaporation, a technique more suited to large-scale production then dialysis. But although ethanol and acetone effectively precipitate proteins,

13. Purification and Identification of Proteins

they also partially denature them so that the procedure must be conducted at very low temperatures ($-20°$ to $-30°C$).

One new method which is only now being developed is an iterative multistep extraction known as *countercurrent distribution* (Fig. 1-56). The method is based on the different ways in which dissolved substances become distributed between two mutually insoluble liquid phases. In order for the extraction to be effective, the distribution coefficient of the soluble compound in the two liquid phases must be close but not equal to unity. Countercurrent distribution has been used with great success to purify a long list of low-molecular-weight substances including coenzymes, antibiotics, vitamins, and alkaloids. It originally appeared impossible to apply this technique to proteins because of the difficulty in selecting a pair of mutually insoluble liquids, each of which would be a good solvent for protein.

The selection of suitable solvents for proteins is not as difficult as it had first seemed, however, if three-component systems are used consisting of water, an organic solvent such as glycerin, glycol, propanol, or butanol, and mineral salts. Such three-component systems divide themselves into two mutually insoluble phases, one of which is richer in organic components, the other in mineral salts. If conditions are chosen so that the composition of the two phases are sufficiently similar to one another, the distribution coefficient of proteins in them approaches unity. In fact, countercurrent distribution has already been successfully used in the case of a few proteins. It is entirely likely that this technique will prove to be practicable in the future for it offers many advantages.

An enumeration of the methods for purifying proteins would be incomplete without mention of *crystallization*. Although the reasons are unclear, it is not possible to crystallize every purified protein. Sometimes the addition of a low-molecular-weight component which can form a complex with one of these proteins leads to success. Outstanding examples of this type of protein are insulin, which crystallizes only after complexing with zinc ions, and human serum albumin whose crystal-forming capacity is considerably increased when a small amount of dodecyl alcohol is added to the medium. Occasionally the proteins of a mixture will crystallize together. Thus crystals of myogen A consist of three different enzymes: aldolase, phosphoglyceraldehyde dehydrogenase, and triosephosphate isomerase.

Although crystallization of proteins is a complex and capricious affair, it is sometimes the best method for the final stage of purification once the protein solution has been enriched and concentrated and most impurities have been removed. Hundreds of proteins have already been obtained in the crystalline state, usually by precipitation from supersaturated solutions. Increasing the protein concentration, adjusting the pH to the isoelectric point, or adding salts which reduce the solubility of the protein are all methods of preparing supersaturated solutions, although extremely gentle conditions must be used to prevent the premature formation of an amorphous precipitate. There is no general recipe for this procedure so that each new protein demands considerable empirical manipulation before proper conditions can be selected. If the preliminary purification of the protein has been adequate, however, these efforts are more often than not rewarded with success, though the resulting crystals are usually of miscroscopic size. If x-ray diffraction studies are contemplated, these are unsuitable and the arduous task of specially growing crystals with linear dimensions on the order of 1 mm must be undertaken.

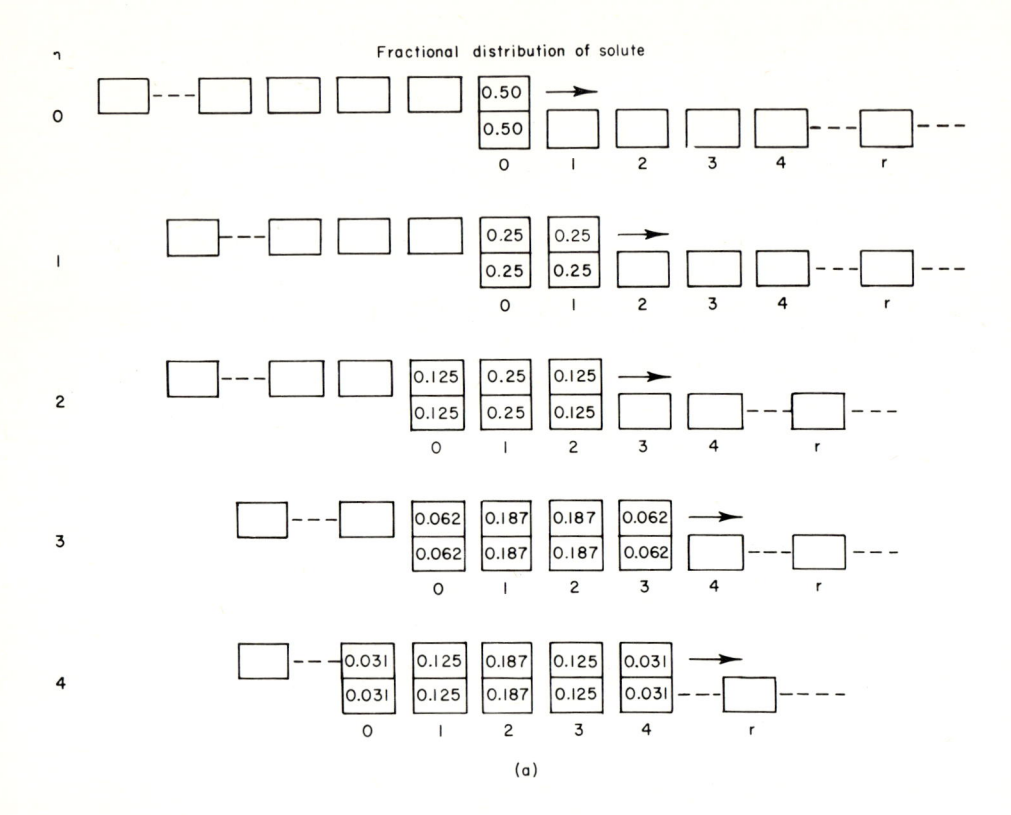

Fig. 1-56. Seperation of macromolecules by the method of countercurrent distribution. (a) A solute, when shaken with a solvent system that will seperate into two phases upon equilibration, will distribute itself between the two liquid phases according to its partition coefficient

$$K = \frac{\text{concentration of solute in upper phase}}{\text{concentration of solute in lower phase}}$$

a parameter dependent on its relative solubility in the solvents employed. In this figure, each square represents an equal volume of the upper or lower phase. The juxtaposition of two squares indicates that the two phases have been shaken together until equilibration has been achieved and the phases permitted to seperate. During countercurrent fractionation, successive volumes of each of the two phases are brought into contact in a serial fashion. If a single solute with a partition coefficient of 1.0 is taken for the purpose of illustration, then at the start of the process (row 0) half the solute will be in each phase. Transfer of the upper phase one tube to the right in order to contact the original upper and lower phases with fresh lower and upper phases, respectively, results in the fractional distribution of the solute shown in row 1. A second shift of the upper phases to the right will give row 2 and so on. The distribution for up to four serial transfers is illustrated. In practice, when a large number of successive contacts must be made to seperate the components of a mixture of solutes, a specially designed apparatus is employed which carries out the process automatically.

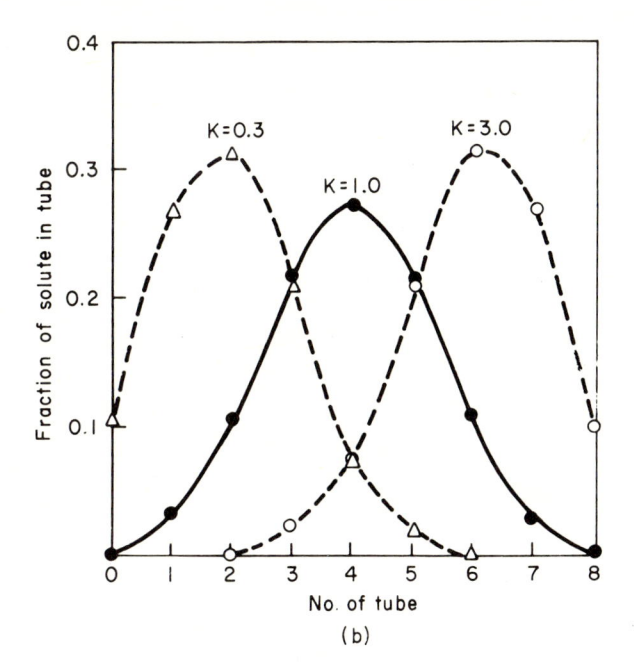

13. Purification and Identification of Proteins

Fig. 1-56. Separation of macromolecules by the method of countercurrent distribution. (b) The distribution of solutes ofter a given number of transfers can be represented as a graph of the fraction in any one tube (upper plus lower phases) vs the serial number of the tube. The figure shows such a plot for substances with $K = 0.3$, 1.0, and 3.0 after eight transfers. The solutes with $K = 0.3$ and 3.0 would be fairly well, but not completely, seperated under these circumstances if they did not interact with each other. The greater the number of transfers, the better the resolution of the procedure, and in the seperation of mixtures of similar substances whose partition coefficients are differentiated by very small increments, hundreds or even thousands of transfers must be made in order to resolve the components into pure or nearly pure fractions. The theoretical distribution curves for a given solute can be calculated as follows. If n is the number of transfers and r is the serial number of the tube, the fraction $T_{n,r}$ of a solute of partition coefficient K in the rth tube after n transfers is given by the binomial term:

$$T_{n,r} = \frac{n!}{r!(n-r)!} \left(\frac{1}{K+1}\right)^n K^r$$

For $n > 20$, the distribution can be more conveniently approximated by a normal curve of error:

$$Y = \frac{1}{[2\pi n K/(K+1)^2]^{1/2}} e^{-\frac{x^2}{2nK/(K+1)^2}}$$

where Y is the solute fraction (ordinate) and x is the number of tubes removed from the maximum of the curve (abcissa). Based on Craig (106).

There are three general methods of checking the purity of protein preparations and identifying the products obtained. The first consists of the purification procedure itself which simultaneously serves as a control. For example, if an individual protein is isolated by chromatography, and behaves in exactly the same way upon rechromatography, it is good evidence that we are dealing with a single component. Second, the important physicochemical constants of the preparation can be measured including sedimentation and diffusion constants, from which the molecular weight can be calculated, electrophoretic mobility, and temperature dependence of solubility. When taken together, all of these properties permit a fairly accurate conclusion to be drawn regarding the purity of the protein. If it sediments as a single component in the ultracentrifuge and moves as a single zone in electrophoresis, and if its solubility is strictly constant at a given temperature independent of the solution volume, then we may be sure that we have a pure preparation. One further control on purity is provided by a determination of amino acid end groups. The N-terminal positions of both α- and β-chains of hemoglobin, for instance, are occupied by valine residues. The chances are good that the terminal residues of contaminating proteins will differ from valine and hence will be identified by end group analysis.

Functional activity, whether it be enzymatic, hormonal, or something else, serves as a third type of control on protein purity. The activity per unit weight of protein or per unit of protein nitrogen, called the specific activity, is an excellent index of purity, particularly if the maximal activity for highly purified standard preparations of the same protein is available for comparison. All measurements of functional activity must of course be conducted under identical conditions.

Immunological techniques may be counted among the most selective for isolation and identification of proteins. In higher organisms, immunological reactions play a vital role in defending the system against invasion by foreign proteins. After repeated introduction of a foreign protein or antigen into the circulatory system, proteins of the γ-globulin class, called antibodies, appear in the bloodstream. They are capable of specifically reacting with the antigen under whose influence they were formed. The chemical nature of the antigen–antibody reaction may be studied *in vitro* (outside the organism) by mixing a solution of the antigen with specific antiserum. A typical result of this procedure is precipitation of an antigen–antibody complex, the so-called precipitin reaction.

One highly perfected method of fractionating protein mixtures, *immunoelectrophoresis*, employs the precipitin reaction in conjunction with electrophoresis in agar gels [Grabar and Williams (107)]. The experimental apparatus is quite simple. Agar is poured into a 5 × 10–15 cm tray so as to form a gel 4 to 5 mm thick. A protein solution containing a small amount of agar is next introduced into a slot at the center of the gel. An electric field of about 4 volts per centimeter is applied across the ends of the gel and the protein mixture is electrophotetically separated into several bands. Proteins may move either toward the anode or toward the cathode from the origin, since a current of fluid owing to endosmosis is superimposed on the electrophoretic motion. The proteins of blood serum, for example, are distributed over an interval of 10 cm after 4 to 5 hours of electrophoresis. While normal serum can be fractionated into nine components under optimal conditions, its composition is actually more complex since most bands represent mixtures of proteins with similar net charges.

13. Purification and Identification of Proteins

After electrophoresis, there are two choices. Either the bands are developed with a stain so as to mark their positions, or else the immunological part of the method is carried out. In the latter case, antiserum to the protein mixture is poured into a channel, running the length of the agar block at a distance of about 1 cm from the end of the slot, in which the sample was originally placed. The entire gel block is then placed for 1 to 2 weeks in a closed vessel whose atmosphere is saturated with water vapor. During this period the antibodies slowly diffuse into the protein bands and wherever the antigen concentration is sufficient to give a precipitin reaction, a thin arc of precipitate forms which is visible to the eye. A schematic drawing and a photograph of these arcs are presented in Fig. 1-57.

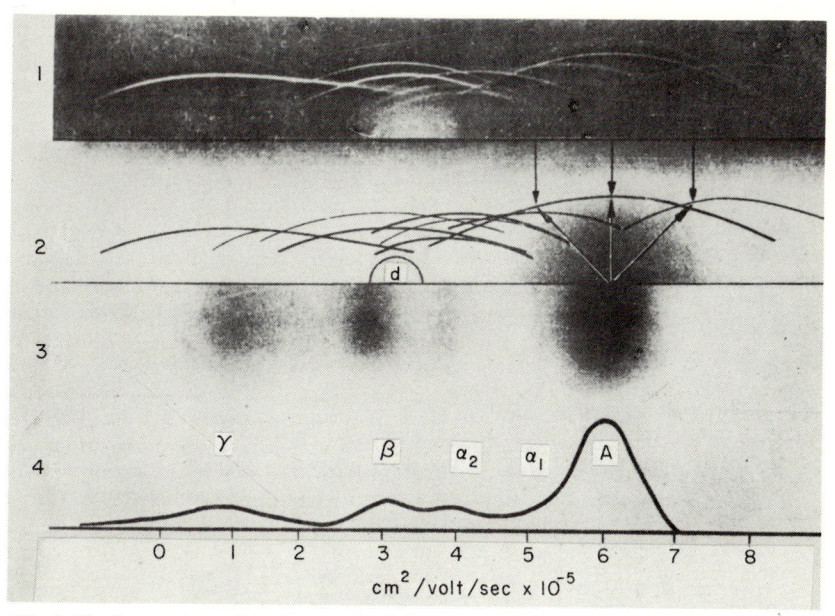

Fig. 1-57. Comparison of various methods for the electrophoresis of serum proteins. (1) Immunoelectrophoresis of normal human serum by the method of Grabar showing the precipitin arcs. (2) Scheme explaining the formation of precipitin arcs. Antibodies diffuse from a channel at the side of the gel in the direction indicated by the upper arrows while the protein antigens diffuse outward from the electrophoretic bands (lower arrows). Where the two meet, an arc of precipitated protein is laid down. (3) Paper electrophoresis and (4) free-boundary electrophoresis of normal human serum in the same buffer. A marks the position of serum albumin, α_1, α_2, β, and γ, the positions of the four major globulin components of serum. From Grabar and Williams (107).

In the apparatus of Grabar, all the electrophoretic bands including very weak ones yield precipitin arcs after the immunological reaction, as in Fig. 1-57. Each arc represents a separate protein component. Nineteen independent components were discovered in normal serum by this method. In the corresponding patterns of sera from persons ill with a variety of diseases, the changes proved to be much more complicated than was supposed earlier from the fractionation of proteins by electrophoresis alone. There are, of course, numerous variations on the methods of immunological analysis, which combine electrophoresis and diffusion with the precipitin reaction.

References

1. G. Natta, *J. Polymer Sci.* **16**, 143 (1955); G. Natta and P. Corradini, *Makromol. Chem.* **16**, 77 (1955).
2. C. B. Anfinsen, "The Molecular Basis of Evolution," Wiley, New York, 1959.
3. "The Proteins" (H. Neurath, ed.), 2nd Ed., Vols. 1–5. Academic Press, New York, 1963–1965.
4. J. Leggett Bailey, "Techniques in Protein Chemistry," 2nd Ed. Elsevier, Amsterdam, 1967.
5. D. H. Spackman, W. H. Stein, and S. Moore, *Anal. Chem.* **30**, 1190 (1958); for a more recent modification, see R. W. Hubbard, *Biochem. Biophys. Res. Commun.* **19**, 679 (1965).
6. For a current listing of known protein sequences, see the annual publication, "Atlas of Protein Sequence and Structure" (compiled by R. V. Eck and M. O. Dayhoff). Nat. Biomed. Res. Found., Silver Spring, Maryland.
7. F. Sanger, *Biochem. J.* **39**, 507 (1945); F. Sanger, *Biochem J.* **45**, 563 (1949).
8. W. R. Gray and B. S. Hartley, *Biochem. J.* **89**, 59P (1963); W. R. Gray and B. S. Hartley, *Biochem. J.* **89**, 379 (1963).
9. P. Edman, *Acta Chem. Scand.* **4**, 283 (1950); P. Edman, *Acta Chem. Scand.* **7**, 700 (1953).
10. F. Sanger, *Biohcem. J.* **44**, 126 (1949); C. H. W. Hirs, *J. Biol. Chem.* **219**, 611 (1956).
11. V. M. Ingram, *Biochim. Biophys. Acta* **28**, 539 (1958).
12. E. Gross and B. Witkop, *J. Biol. Chem.* **237**, 1856 (1962).
13. C. H. W. Hirs, W. H. Stein, and S. Moore, *J. Biol. Chem.* **221**, 151 (1956).
14. D. G. Smyth, W. H. Stein, and S. Moore, *J. Biol. Chem.* **238**, 227 (1963).
15. R. E. Canfield and A. K. Liu, *J. Biol. Chem.* **240**, 1997 (1965).
16. A. P. Ryle, F. Sanger, L. F. Smith, and R. Kitai, *Biochem. J.* **60**, 541 (1955).
17. H. Brown, F. Sanger, and R. Kitai, *Biochem. J.* **60**, 556 (1955); J. I. Harris, F. Sanger, and M. A. Naughton, *Arch. Biochem. Biophys.* **65**, 427 (1956); D. S. H. Nicol and L. F. Smith, *Nature* **187**, 483 (1960).
18. A. Tsugita, D. T. Gish, J. Young, H. Fraenkel-Conrat, C. A. Knight, and W. M. Stanley, *Proc. Natl. Acad. Sci. U.S.* **46**, 1463 (1960).
19. F. A. Anderer, B. Wittmann-Liebold, and H. G. Wittmann, *Z. Naturforsch.* **20b**, 1203 (1965).
20. C. Tanford, "Physical Chemistry of Macromolecules," p. 127. Wiley, New York, 1959.
21a. C. H. Bamford, A. Elliott, and W. E. Hanby, "Synthetic Polypeptides." Academic Press, New York, 1956.
21b. M. Goodman and G. W. Kenner, *Advan. Protein Chem.* **12**, 465 (1957); P. G. Katsoyannis, *J. Polymer Sci.* **49**, 51 (1961).
22. R. B. Woodward and C. H. Schramm, *J. Am. Chem. Soc.* **69**, 1551 (1947).
23. J. C. Sheehan, P. A. Cruickshank, and G. L. Boshart, *J. Org. Chem.* **26**, 2525 (1961).
24a. R. B. Merrifield, *Federation Proc.* **21**, 412 (1962); R. B. Merrifield, *J. Am. Chem. Soc.* **85**, 2149 (1963).
24b. R. B. Merrifield, *J. Am. Chem. Soc.* **86**, 304 (1964); R. B. Merrifield, *Biochemistry* **3**, 1385 (1964).
25. P. G. Katsoyannis, A. Tometsko, and K. Fukuda, *J. Am. Chem. Soc.* **85**, 2863 (1963); J. Meinhofer, E. Schnabel, H. Bremer, O. Brinkhoff, R. Zabel, W. Sroka, H. Klostermeyer, D. Brandenburg, T. Okuda, and H. Zahn, *Z. Naturforsch.* **18b**, 1120 (1963); P. G. Katsoyannis, K. Fukuda, A. Tometsko, K. Suzuki, and M. Tilak, *J. Am. Chem. Soc.* **86**, 930 (1964); P. G. Katsoyannis, A. Tometsko, and C. Zalut, *J. Am. Chem. Soc.* **88**, 166 (1966).
26a. B. Gutte and R. B. Merrifield, *J. Am. Chem. Soc.* **91**, 501 (1969).
26b. R. Hirschmann, R. F. Nutt, D. F. Veber, R. A. Vitali, S. L. Varga, T. A. Jacob, F. W. Holly, and R. G. Denkewalter, *J. Am. Chem. Soc.* **91**, 507 (1969).
26c. R. B. Merrifield, *Science* **150**, 178 (1965); R. B. Merrifield, J. M. Stewart, and N. Jernberg, *Anal. Chem.* **38**, 1905 (1966); R. B. Merrifield, *Advan. Enzymol.* **32**, 221 (1969).
27. V. G. Aldoshin, S. E. Bresler, and E. M. Saminskij, *Vysokomolekul. Soedin.* **4**, 1118 (1962).
28. L. Pauling, R. B. Corey, and H. R. Branson, *Proc. Natl. Acad. Sci. U.S.* **37**, 205 (1951).
29. L. Pauling and R. B. Corey, *Proc. Natl. Acad. Sci. U.S.* **37**, 235 (1951).
30. L. Pauling and R. B. Corey, *Proc. Roy. Soc.* (*London*) **B141**, 21 (1953); B. W. Low and J. T. Edsall, in "Currents in Biochemical Research" (D. E. Green, ed.), p. 378. Wiley (Interscience), New York, 1956.
31. L. Pauling and R. B. Corey, *Proc. Intern. Wool Textile Res. Conf.*, Australia, 1955 Part B, p. 249 (1956).
32. C. H. Bamford, W. E. Hanby, and F. Happey, *Proc. Roy. Soc.* (*London*) **A205**, 30 (1951); C. H

References

Bamford, L. Brown, A. Elliott, W. E. Hanby, and I. F. Trotter, *Proc. Roy. Soc. (London)* **B 141**, 49 (1953)
33. R. H. Karlson, K. S. Norland, G. D. Fasman, and E. R. Blout, *J. Am. Chem. Soc.* **82**, 2268 (1960).
34. L. Pauling and R. B. Corey, *Proc. Natl. Acad. Sci. U.S.* **37**, 251 (1951).
35. L. Pauling and R. B. Corey, *Proc. Natl. Acad. Sci. U.S.* **37**, 729 (1951).
36. R. B. Corey and L. Pauling, *Proc. Roy. Soc. (London)* **B141**, 10 (1953).
37. B. Davidson and G. D. Fasman, *Biochemistry* **6**, 1616 (1967).
38. K. Linderström-Lang, *Chem. Soc. (London) Spec. Publ.* **2**, 1 (1955).
39. A. Berger and K. Linderström-Lang, *Arch. Biochem. Biophys.* **69**, 106 (1957).
40. H. A. Scheraga, *Brookhaven Symp. Biol.* **13**, 71 (1960).
41. J. Hermans, Jr., and H. A. Scheraga, *Biochim. Biophys. Acta* **36**, 534 (1959).
42. P. Urnes and P. Doty, *Advan. Protein Chem.* **16**, 401 (1961).
43. P. Doty, *Collection Czech. Chem. Commun., Spec. Publ.* **22**, 5 (1957).
44. P. Doty, J. H. Bradbury, and A. M. Holtzer, *J. Am. Chem. Soc.* **78**, 947 (1956).
45. A. Wada, *J. Chem. Phys.* **29**, 674 (1958); A. Wada, *J. Chem. Phys.* **31**, 495 (1959).
46. C. E. Hall and P. Doty, *J. Am. Chem. Soc.* **80**, 1269 (1958).
47. P. Doty, A. Wada, J. T. Yang, and E. R. Blout, *J. Polymer Sci.* **23**, 851 (1957).
48. E. R. Blout, P. Doty, and J. T. Yang, *J. Am. Chem. Soc.* **79**, 749 (1957).
49. R. D. Lundberg and P. Doty, *J. Am. Chem. Soc.* **79**, 3961 (1957).
50. P. Doty and R. D. Lundberg, *Proc. Natl. Acad. Sci. U.S.* **43**, 213 (1957).
51. E. K. Rideal and R. Roberts, *Proc. Roy. Soc. (London)* **A205**, 391 (1951).
52. H. A. Scheraga, "Protein Structure," Ch. VII. Academic Press, New York, 1961.
53. S. J. Leach and H. A. Scheraga, *J. Am. Chem. Soc.* **82**, 4790 (1960).
54. C. Tanford and J. Epstein, *J. Am. Chem. Soc.* **76**, 2163 (1954).
55. K. Imahori and J. Tanaka, *J. Mol. Biol.* **1**, 359 (1959).
56. K. Rosenheck and P. Doty, *Proc. Natl. Acad. Sci. U.S.* **47**, 1775. (1961).
57. E. R. Blout, C. de Lozé, and A. Asadourian, *J. Am. Chem. Soc.* **83**, 1895 (1961).
58. P. J. Urnes, K. Imahori, and P. Doty, *Proc. Natl. Acad. Sci. U.S.* **47**, 1635 (1961).
59. T. Miyazawa and E. R. Blout, *J. Am. Chem. Soc.* **83**, 712 (1961).
60. T. Miyazawa, *J. Chem. Phys.* **32**, 1647 (1960).
61. G. D. Fasman and E. R. Blout, *J. Am. Chem. Soc.* **82**, 2262 (1960).
62. S. E. Bresler, *Discussions Faraday Soc.* **25**, 158 (1958).
63. S. E. Bresler, V. P. Kushner, and S. Ya. Frenkel', *Biokhimiya* **24**, 685 (1959).
64. V. P. Kushner and S. Ya. Frenkel', *Dokl. Akad. Nauk SSSR* **141**, 481 (1961).
65. J. G. Kirkwood, *J. Chem. Phys.* **5**, 479 (1937); W. Moffitt, *J. Chem. Phys.* **25**, 467 (1956); W. Moffitt and J. T. Yang, *Proc. Natl. Acad. Sci. U.S.* **42**, 596 (1956); W. Moffitt, D. D. Fitts, and J. G. Kirkwood, *Proc. Natl. Acad. Sci. U.S.* **43**, 723 (1957).
66. N. S. Simmons, C. Cohen, A. G. Szent-Györgyi, D. B. Wetlaufer, and E. R. Blout, *J. Am. Chem. Soc.* **83**, 4766 (1961).
67. N. S. Simmons and E. R. Blout, *Biophys. J.* **1**, 55 (1960).
68. B. H. Zimm and J. K. Bragg, *J. Chem. Phys.* **28**, 1246 (1958); J. H. Gibbs and E. A. DiMarzio, *J. Chem. Phys.* **28**, 1247 (1958); see also T. M. Birshtein and O. B. Ptitsyn, "Conformations of Macromolecules." Wiley (Interscience), New York, 1966.
69. W. F. Harrington and J. A. Schellman, *Compt. Rend. Trav. Lab. Carlsberg, Ser. Chim.* **30**, 167 (1957).
70. S. Ya. Frenkel' and P. Horn, *Vysokomolekul. Soedin.* **3**, 541 (1961).
71. C. B. Anfinsen, *Proc. 5th Intern. Congr. Biochem., Moscow, 1961* **4**, 66 (1963); C. J. Epstein, R. F. Goldberger, and C. B. Anfinsen, *Cold Spring Harbor Symp. Quant. Biol.* **28**, 439 (1963); D. Givol, F. DeLorenzo, R. F. Goldberger, and C. B. Anfinsen, *Proc. Natl. Acad. Sci. U.S.* **53**, 676 (1965); P. G. Katsoyannis and A. Tometsko, *Proc. Natl. Acad. Sci. U.S.* **55**, 1554 (1966).
72. S. E. Bresler and D. L. Talmud, *Dokl. Akad. Nauk SSSR* **43**, 367 (1944); S. E. Bresler, *Biokhimiya* **14**, 180 (1949).
73. H. F. Fisher, *Proc. Natl. Acad. Sci. U.S.* **51**, 1285 (1964).
74. G. Némethy and H. A. Scheraga, *J. Phys. Chem.* **66**, 1773 (1962).
75. J. C. Kendrew, G. Bodo, H. M. Dintzis, R. G. Parrish, H. W. Wyckoff, and D. C. Phillips, *Nature* **181**, 662 (1958).

76. J. C. Kendrew, R. E. Dickerson, B. E. Strandberg, R. G. Hart, D. R. Davies, D. C. Phillips, and V. C. Shore, *Nature* **185**, 422 (1960).
77. J. C. Kendrew, H. C. Watson, B. E. Strandberg, R. E. Dickerson, D. C. Phillips, and V. C. Shore, *Nature* **190**, 666 (1961).
78. R. E. Dickerson, *in* "The Proteins" (H. Neurath, ed.), 2nd Ed., Vol. 2, p. 603. Academic Press, New York, 1964.
79. M. J. Buerger, "X-ray Crystallography." Wiley, New York, 1942; A. A. Kitajgorodskij, "Rentgenostrukturny Analiz Melkokristallicheskich i Amorfnykh Tel." Gosizdat, Moscow, 1952; J. M. Robertson, "Organic Crystals and Molecules." Cornell Univ. Press, Ithaca, New York, 1953; F. H. C. Crick and J. C. Kendrew, *Advan. Protein Chem.* **12**, 133 (1957); H. Lipson and C. A. Taylor, "Fourier Transforms and X-ray Diffraction." Bell, London, 1958; H. Lipson and W. Cochran, "The Determination of Crystal Structures." 2nd Ed., Cornell Univ. Press, Ithaca, New York, 1966; H. R. Wilson, "Diffraction of X-rays by Proteins, Nucleic Acids and Viruses." Arnold, London, 1966.
80. G. Bodo, H. M. Dintzis, J. C. Kendrew, and H. W. Wyckoff, *Proc. Roy. Soc.* (*London*) **A253**, 70 (1959).
81. M. M. Bluhm, G. Bodo, H. M. Dintzis, and J. C. Kendrew, *Proc. Roy. Soc.* (*London*) **A246**, 369 (1958).
82. H. Scouloudi, *Nature* **183**, 374 (1959).
83. M. F. Perutz, M. G. Rossmann, A. F. Cullis, H. Muirhead, G. Will, and A. C. T. North, *Nature* **185**, 416 (1960).
84. M. F. Perutz, H. Muirhead, J. M. Cox, and L. C. G. Goaman, *Nature* **219**, 131 (1968).
85. C. L. Nobbs, H. C. Watson, and J. C. Kendrew, *Nature* **209**, 339 (1966).
86. C. C. F. Blake, D. F. Koenig, G. A. Mair, A. C. T. North, D. C. Phillips, and V. R. Sarma, *Nature* **206**, 757 (1965).
87. L. N. Johnson and D. C. Phillips, *Nature* **206**, 761 (1965).
88. G. Kartha, J. Bello, and D. Harker, *Nature* **213**, 862 (1967).
89. G. N. Reeke, J. A. Hartsuck, M. L. Ludwig, F. A. Quiocho, T. A. Steitz, and W. N. Lipscomb, *Proc. Natl. Acad. Sci. U.S.* **58**, 2220 (1967).
90. B. W. Matthews, P. B. Sigler, R. Henderson, and D. M. Blow, *Nature* **214**, 652 (1967).
91. C. S. Wright, R. A. Alden, and J. Kraut, *Nature* **221**, 235 (1969).
92. J. Drenth, J. N. Jansonius, R. Koekoek, H. W. Swen, and B. G. Wolthers, *Nature* **218**, 929 (1968).
93. H. K. Schachman, "Ultracentrifugation in Biochemistry." Academic Press, New York, 1959.
94. J. T. Edsall, *in* "The Proteins" (H. Neurath and K. Bailey, eds.), 1st Ed., Vol. 1B, p. 549. Academic Press, New York, 1953.
95. I.-B. Eriksson-Quensel and T. Svedberg, *Biol. Bull.* **71**, 498 (1936); T. Svedberg and S. Brohult, *Nature* **142**, 830 (1938).
96. P. Elödi, *Acta Physiol. Acad. Sci. Hung.* **13**, 199 (1958); G. Szabolcsi, E. Biszku, and M. Sajgó, *Acta Physiol. Acad. Sci. Hung.* **17**, 183 (1960).
97. A. Klug and D. L. D. Caspar, *Advan. Virus Res.* **7**, 225 (1960).
98. H. Fraenkel-Conrat, *in* "The Viruses" (F. M. Burnet and W. M. Stanley, eds.), Vol. 1, p. 429. Academic Press, New York, 1959.
99. R. W. Horne, S. Brenner, A. P. Waterson, and P. Wildy, *J. Mol. Biol.* **1**, 84 (1959).
100. J. Boyes-Watson, E. Davidson, and M. F. Perutz, *Proc. Roy. Soc.* (*London*) **A191**, 83 (1947).
101. V. Luzzati, J. Witz, and A. Nicolaieff, *J. Mol. Biol.* **3**, 367 (1961); V. Luzzati, J. Witz, and A. Nicolaieff, *J. Mol. Biol.* **3**, 379 (1961).
102. A. Tiselius, *Biochem. J.* **31**, 1464 (1937); H. Svensson, *J. Biol. Chem.* **139**, 805 (1941).
103. S. Raymond and L. Weintraub, *Science* **130**, 711 (1959); S. Raymond and Y. J. Wang, *Anal. Biochem.* **1**, 391 (1960); L. Ornstein and B. J. Davis, "Disc Electrophoresis." Distillation Products Industries, Division of Eastman Kodak, Rochester, 1961.
104. P. Traub, M. Nomura, and L. Tu, *J. Mol. Biol.* **19**, 215 (1966).
105. R. J. Britten and R. B. Roberts, *Science*, **131**, 32 (1960).
106. L. C. Craig, *in* "A Laboratory Manual of Analytical Methods of Protein Chemistry" (P. Alexander and R. J. Block, eds.), Vol. 1, p. 121. Macmillan (Pergamon), New York, 1960.
107. P. Grabar and C. A. Williams, Jr., *Biochim. Biophys. Acta* **17**, 67 (1955).

Chapter II

Protein Function

1. Introduction

Proteins comprise a large group of biologically active substances whose number includes enzymes, hormones, antibodies, toxins, and many other highly specialized macromolecules. Of all the tasks performed by proteins in the cell, the most important is without doubt enzymatic or catalytic. Enzymes are the material agents upon which cellular metabolism depends, because each one of them catalyzes a specific biochemical reaction. In contrast to known inorganic catalysts, they are distinguished by enormous activity and extraordinary specificity (1).

Starting from the same initial compounds, inorganic catalysts can promote many alternative chemical reactions. Alcohols, for example, undergo partial dehydration to hydrocarbons and dehydrogenation to aldehydes under the influence of certain catalysts; the two kinds of reactions proceed simultaneously on the same catalytic surface, although at different rates. In this respect enzymes are qualitatively different from ordinary catalysts, since they direct the reaction along a unique route in every case. Here their specificity is absolute, although in the choice of substrates for the reaction, they are somewhat less discriminating. Enzymes are frequently capable of acting on a whole gamut of different substrates, all of which nonetheless undergo the same kind of chemical transformation.

Enzymatic catalysis constitutes an extraordinarily powerful means of accelerating chemical reactions. In this regard, too, enzymes differ from inorganic catalysts to an enormous extent, although it is not always easy to draw direct parallels. A typical comparison would be the catalytic degradation of hydrogen peroxide to water and oxygen by the enzyme catalase and by Fe^{3+} ions. The ratio of catalytic activities per gm-mole of catalyst is 10^{10} in favor of the enzyme. This means that 1 mg of catalase

(M.W. = 540,000) is as active a catalyst as 2 kg of ferric ions! The example cited provides some idea of the vast potential inherent in enzymatic catalysis.

Living cells contain a wide variety of enzymes and even the relatively simple bacterial cell possesses 1000 to 2000 different enzymatic activities. The metabolic reactions which are catalyzed by this host of enzymes can be broken down into two classes: *catabolic* reactions in which organic compounds such as sugars, amino acids, and fats are oxidatively degraded with the accrual of energy to the cell; and *anabolic* reactions in which complex compounds, frequently of high molecular weight, are synthesized for cell growth and reproduction.

A systematic study of enzymatic reactions has been one of the principal contributions of biochemistry. In the present text we shall consider enzymes in their general context, stressing the unique aspects of protein catalysis. Particular emphasis will be placed on the relationship between their mechanism of action and the distinctive properties of protein macromolecular structure. For examples we shall have recourse to concrete enzymatic reactions, but questions regarding the significance of these reactions for the organisms will be deferred.

Structural proteins comprise another important class of macromolecules in the cell and some of them take part in active transport. Each cell is like a chemical factory in which thousands of chemical reactions occur at the same time. These reactions must be separated spatially, but coordinated in time. In cells, as in industrial production, there exist mechanisms which automatically regulate the rates of reactions according to the needs of the cell and in conformity with environmental conditions, such as temperature and the external concentration of various metabolites. The cell possesses a complex substructure, specifically for the purpose of separating chemical processes from one another in space.

Subcellular organelles such as nuclei, mitochondria, ribosomes and plastids play the role of reactors in which certain classes of reactions are concentrated. The nucleus produces chemical substances for the transfer of genetic information to the cytoplasm; the ribosomes provide sites where this information is used to direct protein synthesis; the mitochondria contain the enzymes of oxidative phosphorylation, a process which stores up chemical energy for use in other parts of the cell. Regulation of cellular processes is achieved in part by protein-containing membranes which compartmentalize the cytoplasm, surround the organelles, and form the outer layer of the cell. The membranes are selectively permeable, permitting the rapid transport of certain specific compounds, but inhibiting the movement of others.

In living cells, permeability is often an active process having little in common with molecular diffusion or osmotic flow. On the contrary, active transport is generally accomplished against a concentration gradient, for molecules must be moved from regions of low concentration to regions of high concentration. This is a complex phenomenon that demands expenditure of energy since the transport of chemical substances contrary to diffusion results in a decrease in the entropy of those substances in the system. Active transport, whether into the cell from the surrounding medium, or from place to place within cytoplasmic particulates, is the job of special insoluble proteins and protein complexes which make up the outer cell membrane as well as various structures inside the cell. Their functional activity, permeability, is thus closely related to their enzymatic capacities. While active transport answers some of the needs of a regulatory mechanism, the permeability of mitochondrial particles is subject to a further type of control. These particles can contract and relax in much the same way as muscle tissue, due to the presence in them of specialized contractile proteins.

Mechanochemical proteins, a third major group, participate in reactions which directly transform chemical bond energy into mechanical work, circumventing intermediate conversion to thermal energy. The efficiency of this process is very high, approaching 50% in some cases (2). The contractile proteins of muscle therefore differ fundamentally from machines on two counts. In machines, the chemical energy of fuel is also turned into mechanical work, but only after obligatory conversion to heat, and the overall efficiency of the process is much lower, usually between 20 and 30%. Mechanochemical reactions are observed in a wide variety of organisms from mammals to bacteria, and even to viruses.

We shall be concerned only with the three functional classes of proteins indicated above, since they are the most general and the most important. Although other protein groups will not be treated here, they too play extremely important roles, especially in complex differentiated organisms. Among them are protein hormones which regulate specific chemical reactions, proteins of nerve cells which facilitate the conduction of impulses, and antibodies which constitute the organismic defense system against invasion by foreign proteins and cells.

2. Enzymatic Catalysis

Despite the great diversity of enzymatic reactions, there are certain fundamental principles which govern them all. Most enzymatic reactions are exchange processes with the general formula

$$AB + CD \rightleftharpoons AC + BD$$

in which two substrates must obligatorily participate. This is particularly evident in transfer reactions such as transmethylation, transphosphorylation, and transamination where one substrate is the donor of a specific functional group and the second, an acceptor of that group. The enzyme facilitates the transfer of the group from one molecule to the other.

In many reactions, the role of the second substrate is more obscure, especially if it is present in great excess and does not therefore figure in the kinetic equations. Typical examples are the reactions catalyzed by a variety of degradative enzymes, such as proteases which hydrolyze polypeptide molecules according to the formula:

$$R_1-\underset{\underset{O}{\|}}{C}-N(H)-R_2 + H_2O \rightleftharpoons R_1-COOH + H_2N-R_2$$

In a general way, this reaction also fulfills the requirements of *metathesis* or double exchange, but the second substrate is water which is always in excess. The enzyme esterase, which catalyzes the saponification of esters

$$R_1-\underset{\underset{O}{\|}}{C}-O-R_2 + H_2O \rightleftharpoons R_1-COOH + HO-R_2$$

acts according to the same principle.

Synthetic reactions in living tissue are always characterized by condensation. The formation of a polyglucosidic chain, for instance, occurs by the polycondensation of glucose-1-phosphate monomers:

It is apparent that such reactions also exhibit the characteristics of metathesis.

The mechanism of enzymatic action can be formulated as follows. Two substrates, one containing the bond A—B, the other, the bond C—D, attach to certain groups on the enzyme molecule. The proximity and mutual orientation of the two compounds must be appropriate for the ensuing reaction. The role of the enzyme is to promote weakening of the A—B and C—D bonds and to facilitate the formation of the new covalent bonds A—C and B—D. Despite all the services rendered by the enzyme, local increases in thermal energy are nonetheless needed to produce the chemical reaction. The process described by the equation AB + CD → AC + BD occurs within dimensions only slightly greater than the length of a chemical bond, that is, within a few angstroms. This raises the question as to why enzymes consist of comparatively large protein macromolecules, whose dimensions frequently reach 20 to 30 Å or even more. It has been shown that there is a localized site of enzymatic activity on the protein surface which is composed of a small number of closely packed functional groups. These groups may belong to amino acid residues which are quite far from each other along the polypeptide chain, but which are all brought together by the folding of the molecule that creates the secondary and tertiary structures. This explains why enzymatic activity is generally sensitive to denaturation of the protein.

Single protein macromolecules frequently bear more than one active site. While these sites may be identical, as in hemoglobin, there are cases in which active sites of different structure and function are combined in one macromolecule. In yeast, for example, cytochrome b and lactate dehydrogenase activities are both present in a single macromolecular protein complex. The presence of quarternary structure is quite well established in these cases and it appears that each active site resides on a separate polypeptide subunit. As we shall see (p. 168ff), the existence of several active sites in one protein macromolecule is of considerable advantage to the enzyme in fulfilling its catalytic function. There are also enzymes composed of two different kinds of subunits whose active sites are created as a direct consequence of subunit assembly. This type of quaternary structure characterizes both tryptophan synthetase and lactose synthetase. In the process of evolution, selective pressures have apparently favored enzymes with quaternary structure and the complex functional activity which generally accompanies it.

In the tissues and organs of differentiated organisms, enzymes are encountered which have the same function but differ in macromolecular properties, such as molecular weight and isoelectric point. Such related groups of enzymes are called isoenzymes or isozymes by analogy with isotopes. Isozymes consist of two or three types of subunits and differences between them are related to quaternary structure alone; each isozyme is an oligomer containing fixed proportions of the different subunits.

One of the most extensively studied enzymes is ribonuclease (RNase) (3). Since its structural formula is known in entirety, it is likely that its mechanism of action will be completely understood in the near future. RNase from mammalian pancreatic tissue selectively hydrolyzes ribonucleic acid chains wherever pyrimidine residues occur. But a synthetic, low-molecular-weight substrate, the cyclic 2′,3′-ester of cytidylic acid:

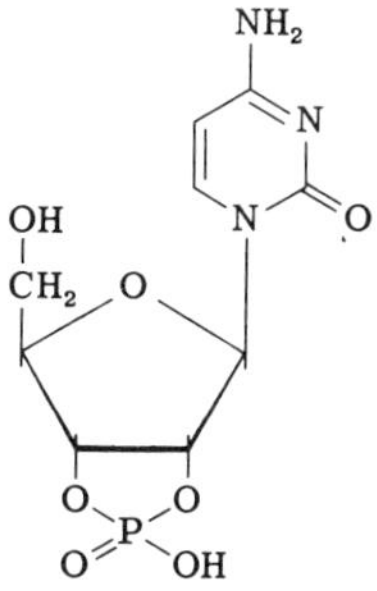

can also be used to assay RNase function. The catalytic activity of the enzyme can be readily estimated from the rate at which the cyclic ester is hydrolyzed.

Much important information about RNase action has been obtained from controlled degradation of the enzyme. With subtilisin, a proteolytic enzyme from *Bacillus subtilis*, Richards succeeded in specifically cleaving a single peptide bond between the twentieth and twenty-first residues (counting from the N-terminal residue of the polypeptide chain) (4). This is a bond joining alanine and serine, as can be seen in Fig. 2-1. The subtilisin-treated derivative is called RNase S to distinguish it from native RNase A. After hydrolysis, the polypeptide "tail" or S-peptide, consisting of twenty amino acids, is not immediately liberated from the remainder of the enzyme molecule, called the S-protein. RNase S furthermore retains full enzymatic activity and it can be separated from RNase A by chromatography on the carboxy cation exchange resin IRC-50. Further studies on purified RNase S showed that S-peptide and S-protein are bound by hydrogen bonds, since the complex can be dissociated in urea. The separation of S-peptide and S-protein can be accomplished by fractional precipitation with trichloroacetic acid which renders the S-protein insoluble while the S-peptide remains in solution. Neither of the fragments possesses enzymatic activity, although the S-protein is capable of binding the 2′,3′ ester of cytidylic acid. Thus the active site of the enzyme is badly damaged, but not completely destroyed by removal of its tail; S-protein can bind the substrate, but it cannot activate it. By mixing S-protein and S-peptide together in equimolar amounts, it is possible to fully restore the catalytic properties of RNase. The activity of the reconstituted complex does not differ from that of the native protein.

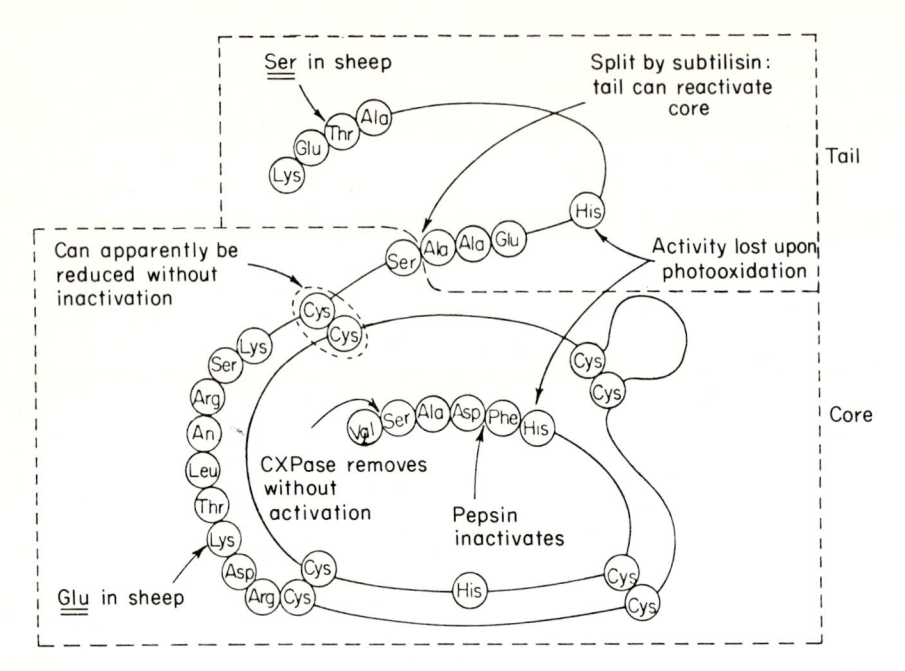

Fig. 2-1. A schematic drawing of the ribonuclease molecule. Various experimental modifications of the native molecule are indicated together with their effects on enzyme activity. The active site of ribonuclease must actually be a complicated three-dimensional structure but may not necessarily involve more than a small fraction of the total protein. From Anfinsen (5, p. 135).

Cleavage of residues from the C-terminal end of the chain leads to quite a different result. The enzyme carboxypeptidase can remove the three C-terminal amino acids—valine, serine, and alanine (residues 124, 123, and 122)—without influencing functional activity. But if RNase is incubated with pepsin for a short time at pH 1.8, the bond between residues 120 and 121 is severed, splitting off the tetrapeptide Asp-Ala-Ser-Val and completely inactivating the enzyme. From these data it was concluded that the aspartic acid residue in position 121 is part of the active site and, in addition, that one or more groups at the N-terminal end are also crucial for enzyme function. Evidently, the macromolecule is tightly packed in such a way that both ends of the polypeptide chain are in direct contact with each other.

The active site of RNase has been investigated by attacking various functional groups of the protein with specific chemical reagents. Using methylene blue as a sensitizer, the enzyme can be totally inactivated by photooxidation. The kinetics of inactivation coincide with the rate at which one of the four histidine residues in RNase is destroyed by oxidation. In other experiments, Barnard and W. D. Stein, using bromoacetic acid under closely controlled conditions, managed to specifically attack the histidine residue in position 119, the sixth amino acid from the C-terminus. Since the bromoacetic acid treatment caused the enzyme to lose its activity, it was inferred that this histidine also constitutes part of the active site in addition to the aspartic acid residue at position 121.

Interesting results were obtained by Vithayathil and Richards as a result of modifying the S-peptide chemically (6). The binding of S-peptide to S-protein is almost independent of electrostatic forces and it was possible to acetylate all three amino side groups of the S-peptide and to convert both its carboxyl groups to methyl esters without destroying its ability to bind with the S-protein to form an active enzyme. This occurs even though there is a net change of five charges on the peptide! The methionine residue at position 13 of the S-peptide is however very important in the binding of S-peptide with S-protein. Oxidation of the methionyl group to methionyl sulfone with performic acid changes a hydrophobic side chain into a charged and hydrophilic one, reducing the affinity between S-protein and S-peptide to a great extent. The initial S-peptide completely reactivates the S-protein at a concentration of 10^{-6} M; after oxidation of methionine at position 13, the peptide concentration must be 1000 times greater in order to form the protein: peptide complex. Nonetheless, even under these circumstances, RNase activity is fully restored. This means that the methyionyl side group is needed to make a firm bond between tail and S-protein, but that it has no direct relationship to the structure of the active site.

Of all the amino acids in the S-peptide, only the histidine at position 12 is vital to enzymatic activity. Evidence for this was obtained by subjecting the S-peptide to photoxidation or to the action of iodo- or bromoacetic acid. Thus, two histidine residues have been placed in the active site. It is a curious fact that RNase is able to dimerize in such a way that the S-peptide of one molecule is bound to the S-protein of the second (7) (Fig. 2-2). When monomeric RNase is treated with iodoacetic acid, it becomes totally inactive upon alkylation of *either* of the histidines at positions 12 and 119. Dimerization of RNase inactivated in this way results in the regeneration of up to 25% of the initial activity (8). It is assumed that the active sites are formed from an undamaged histidine at position 12 of one macromolecule, and an undamaged histidine at position 119 of a second.

The time is now suitable to examine the extremely important role of protein secondary and tertiary structure in enzymatic activity. From what has been said so far, it should be clear that the active site of an enzyme is composed of several functional groups brought together by the distinctive spatial packing of the polypeptide chain. A major stabilizing factor in RNase is the presence of four disulfide bridges which "vulcanize" the macromolecular configuration. For this reason, 8 M urea has little effect on RNase activity, even though physicochemical criteria indicate considerable changes in secondary and tertiary structure. But if thioglycolic acid or β-mercaptoethanol are added to the 8 M urea, all of the disulfide bonds are eventually reduced (Fig. 2-3a) (9). One disulfide bond is readily reduced even in the absence of urea, but its elimination does not decrease enzymatic activity. Cleavage of the three remaining —S—S— bonds is more difficult, requiring preliminary treatment of the protein with urea to partially unfold the molecule. Once again, the first bond broken in urea is apparently not necessary for catalytic activity, but disruption of the two most stable disulfide bridges leads to complete inactivation.

Now these —S—S— bridges possess a remarkable property: they can be totally reformed after removal of urea and reducing agent. It was shown that simple oxidation of the —SH groups by molecular oxygen regenerates all the macromolecular characteristics of the protein and 95% of its catalytic activity (Fig. 2-3b). Specific molecular interactions among the various amino acid residues are thus sufficient to fold the protein back into its native configuration. The accuracy of the process is such that

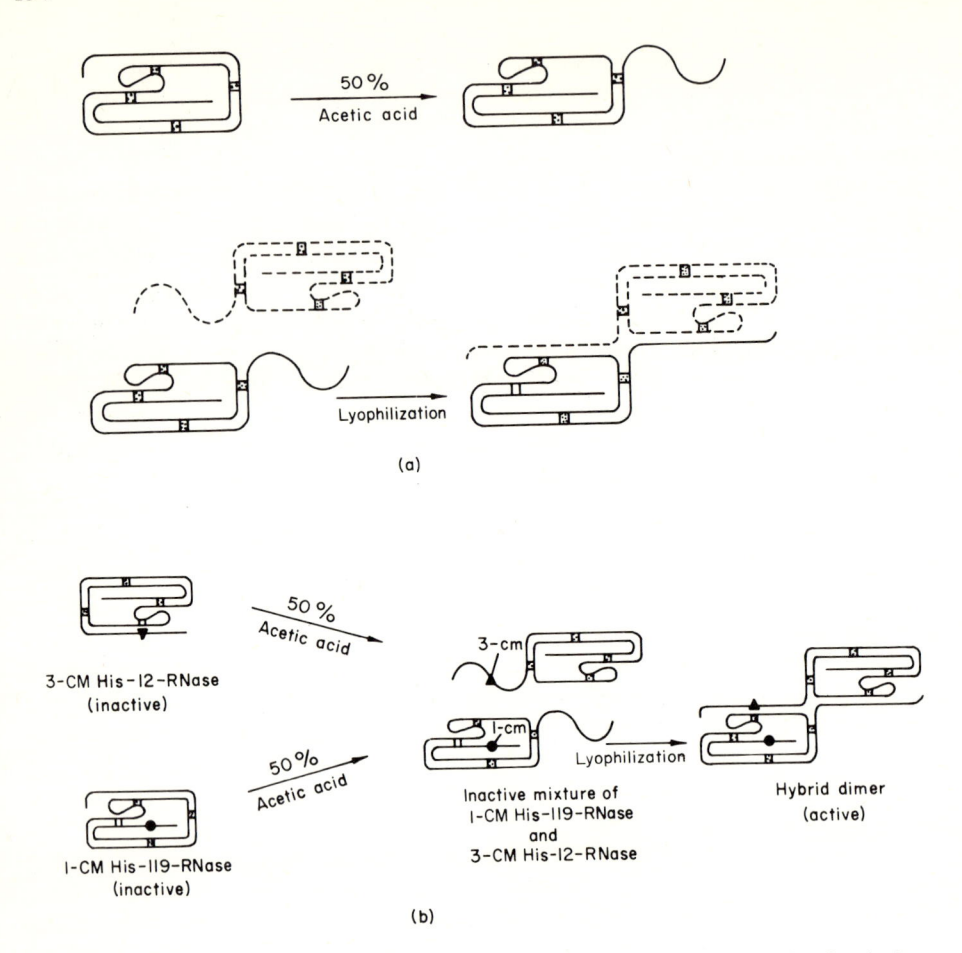

Fig. 2-2. Dimerization of the ribonuclease molecule. (a) Hypothetical mechanism for the formation of enzymatically active dimers of ribonuclease. The line represents the polypeptide chain cross-linked by the four disulfide bonds. In 50% acetic acid the amino-terminal segment of the protein is unfolded; lyophilization then causes aggregation of the two unfolded molecules in a manner such that each dimer has two active sites, one half of each contributed by each monomer unit. This would explain why no loss in enzymatic activity occurs. From Crestfield et al. (7). (b) Formation of active dimers from inactive carboxymethylated monomers. Upon treatment with iodoacetamide, two inactive alkylated derivatives result: 1-carboxymethylhistidine-119-ribonuclease and 3-carboxymethyl-histidine-12-ribonuclease. Alkylation of either histidine-119 or histidine-12 blocks subsequent alkylation of the other. The two derivatives can be separated chromatographically. When treated with 50% acetic acid and lyophilized together, hybrid dimers result with one active site (right) containing non-alkylated histidine residues at positions 12 and 119. The generation of enzymatic activity on hybridization of the two derivatives furnishes impressive support for the idea that histidine-12 and histidine-119 are both at the active site of the enzyme. After Stein (3).

all four disulfide bonds re-form in exactly the same places as in the original structure (11). This proves the important point that the secondary and tertiary structure of a protein is wholly determined by the amino acid sequence of its polypeptide chain.

Enzymes often contain *coenzymes* or *prosthetic groups*, small molecules other than protein which attach to the active site by coordinate, covalent, or ionic bonds,

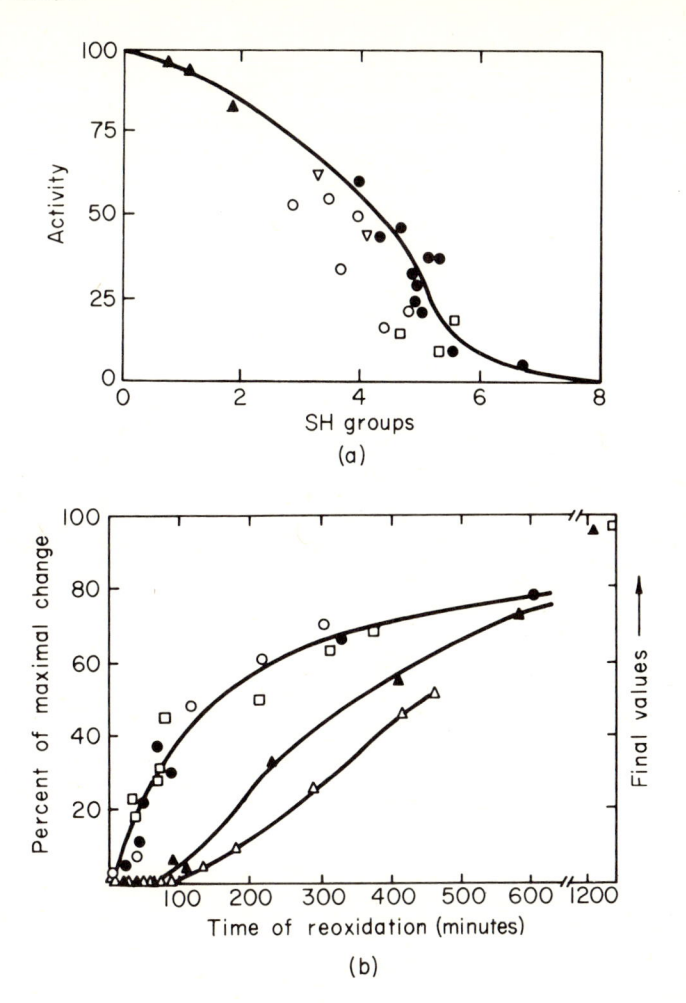

Fig. 2-3. (a) Activity of ribonuclease at various stages of reduction as a function of the number of moles of sulfhydryl per mole of enzyme. Enzymatic activity is expressed as a percentage of the specific activity of native ribonuclease. Reduction in absence of urea (▲) and in presence of 8 M urea (●); reoxidation of fully reduced inactive ribonuclease (□) and partially reduced ribonuclease (○, △). From Anfinsen (5, p. 134). (b) Changes during the oxidation of SH groups in fully reduced ribonuclease as measured by sulfhydryl group content [titration with p-chloromercuribenzoate (○) and reaction with radioactive iodoacetate (●)], by specific optical rotation (□), and by enzymatic activity against two different substrates (△, ▲). The fact that enzymatic activity lags behind the formation of disulfide bonds and restoration of secondary structure suggests that some of the S—S bonds formed during the early stages of oxidation are not identical with those of the native protein, but undergo a later rearrangement to yield the native configuration. From Anfinsen *et al.* (10).

and which are necessary for catalytic activity. Among the oxidative enzymes of the respiratory chain, flavin adenine dinucleotide and nicotinamide adenine dinucleotide serve as coenzymes for flavoproteins and dehydrogenases, respectively. Other enzymes have polyvalent metal ions at their active sites, such as zinc in carbonic anhydrase, cobalt and manganese in peptidases, molybdenum in xanthine oxidase and copper in cytochrome oxidase, to cite just a few examples.

The method most widely employed to investigate the nature of biologically active groups in proteins is inhibition of function with specific chemical substances. Many such poisons irreversibly inactivate protein functional groups. Carbon monoxide, sodium azide, hydrogen cyanide, and sodium sulfide all bind to the iron of heme groups and can therefore be classified as inhibitors of respiratory proteins. Sulfhydryl groups are irreversibly bound by compounds containing cadmium or mercury. They can be quantitatively titrated with such molecules as *p*-chloromercuribenzoate (PCMB):

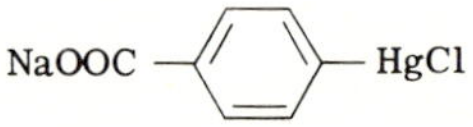

$$NaOOC-C_6H_4-HgCl$$

or alkylated with iodoacetate (ICH_2COOH) or iodoacetamide (ICH_2CONH_2). Arsenate replaces phosphate in many reactions, but since the product formed is usually hydrolyzed quite readily, subsequent reaction stages are prevented. Strong complexing agents like ethylenediaminetetraacetic acid (EDTA) can poison metalloenzymes by binding their divalent metal ions. On the other hand, heavy metal ions of lead, cadmium, mercury, and uranium can themselves act as potent inhibitors when the active site of the enzyme contains acidic —SH and —COOH groups or groups capable of complexing ions by coordination bonds.

Hydrolytic enzymes such as proteases and esterases are very sensitive to the universal, yet very specific inhibitor, diisopropylphosphofluoridate (DFP):

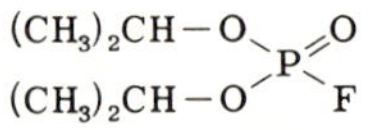

$$(CH_3)_2CH-O\diagdown \diagup O$$
$$ P$$
$$(CH_3)_2CH-O\diagup \diagdown F$$

All enzymes with esterase activity, including cholinesterase, trypsin, chymotrypsin and many others, are irreversibly inactivated by this compound (12), which is one of the most potent of inhibitory substances. Several enzymes of this type contain very similar amino acid sequences at their active sites. Compare, for instance, -Gly-Asp-Ser-Gly- in chymotrypsin with -Glu-Gly-Gly-Ser-Gly- in trypsin. Since serine is the residue in these sequences required for enzymatic activity, we can deduce that it alone of all the amino acids in the protein molecule is attacked by DFP. The serine hydroxyl forms an ester with diisopropylphosphoric acid of the following structure:

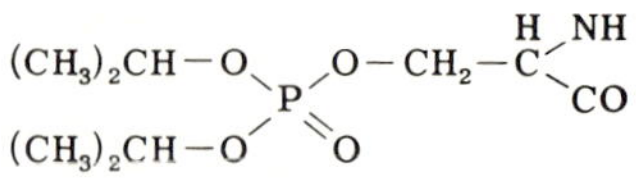

$$(CH_3)_2CH-O\diagdown \diagup O-CH_2-CH(NH)(CO)$$
$$ P$$
$$(CH_3)_2CH-O\diagup \diagdown O$$

All other serine residues in the enzyme remain unchanged. The reaction between serine and DFP most likely reflects the normal catalytic activity of the enzyme, explaining why the serine residue at the active site is the only one affected.

Little more is known about the active site of esterases, although there is reason to believe that histidine as well as serine plays a role. Evidence for this is provided both by photooxidative inactivation of the enzyme, and by the pH dependence of enzymatic activity. Invertase and trypsin activity are plotted as a function of pH in Fig. 2-4. The

2. Enzymatic Catalysis

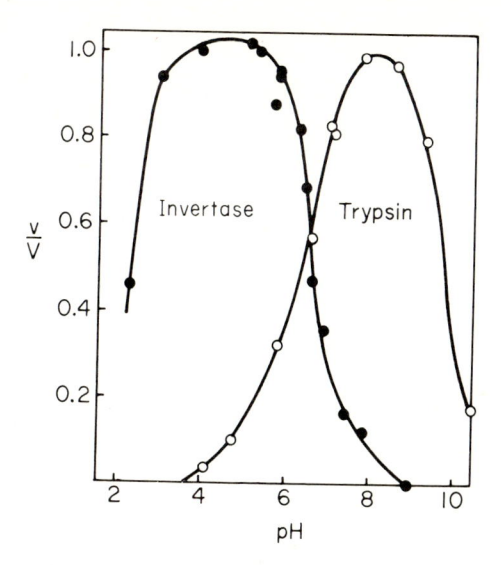

Fig. 2-4. Dependence of the rate of enzymatic reactions on pH. The variation of reaction rate with respect to pH in invertase and trypsin is presented in terms of the ratio of actual rate to the maximum rate measured at the pH optimum. A pH optimum is one of the most characteristic features of enzymatic reactions, differentiating them sharply from other types of catalysis in solution. From Moelwyn-Hughes (13, p. 63).

bell-shaped curves may be interpreted as the superposition of two separate titration curves. In the case of trypsin, for instance, we may assume that ionization of either of two groups, one acidic, with a pK of about 6.5, the other basic, with a pK of 8 to 9, results in loss of enzymatic activity. Chymotrypsin exhibits a similar titration curve in the alkaline region. It is likely that both enzymes have at least one ionizable group in common, probably the imidazole ring of a histidine residue, judging from the position of the pK. Interpretations of this sort must be made with caution, however, since the pK of ionizable groups can be significantly displaced by interactions with neighboring residues of the protein chain. Nonetheless, the role of histidine is open to verification by independent experiments. One such attempt was to determine the catalytic activity of imidazole molecules in the saponification of esters, recalling that trypsin and chymotrypsin have esterase, as well as protease, activity. It turned out that imidazole actually does catalyze the hydrolysis of esters, though it is much less active than the enzymes.

The presence of histidine at the active site of trypsin and chymotrypsin is also strongly indicated by the behavior of certain specially synthesized enzyme poisons called chloromethylketones which selectively attack the imidazole group. These compounds, whose general formula is:

138   II. Protein Function

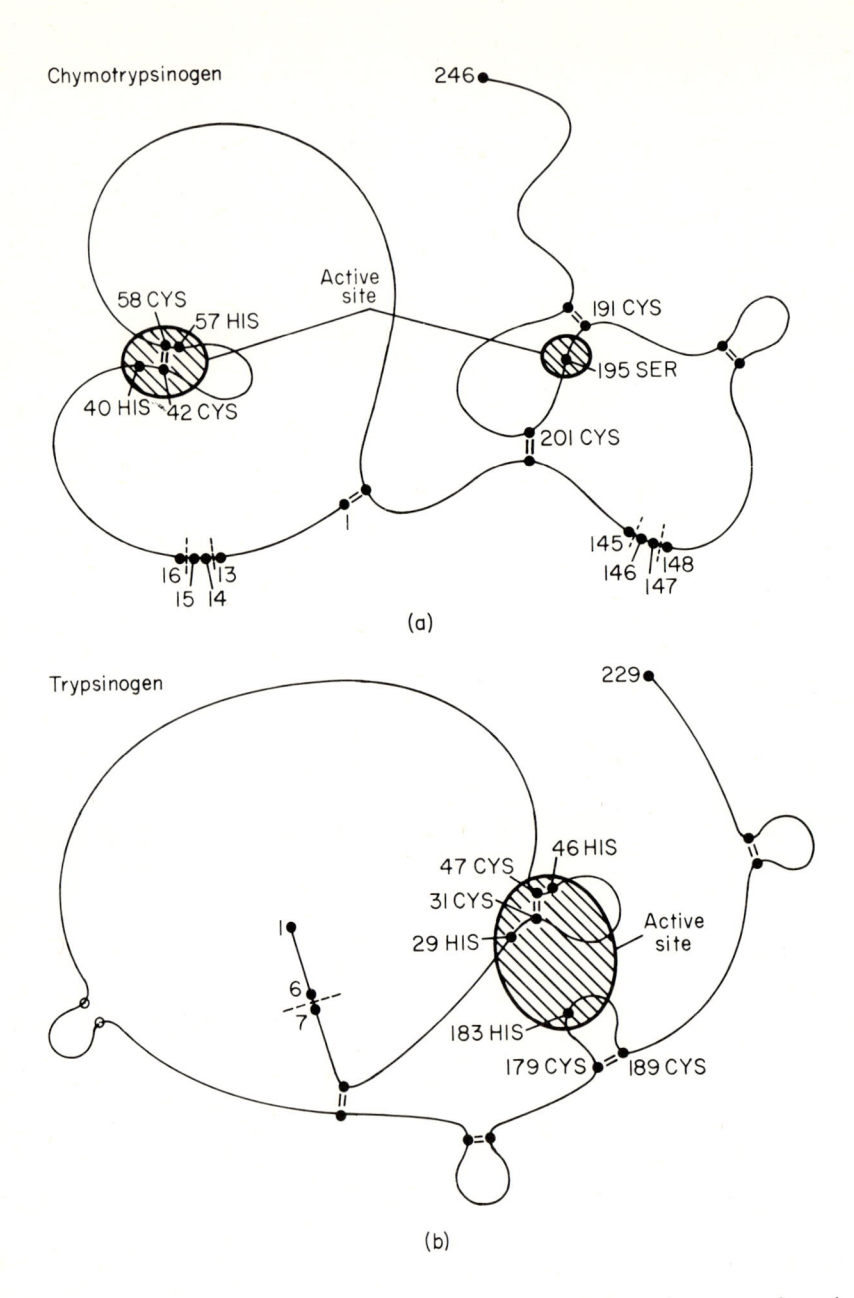

Fig. 2-5. The active sites of chymotrypsinogen and trypsinogen. In both cases the active sites are very similar with two disulfide bridges (indicated as short double bonds) serving to lock the active histidine and serine residues in a fixed position. When cleaved between residues 15 and 16, chymotrypsinogen is converted to the active enzyme chymotrypsin. (Other broken lines in chymotrypsinogen indicate secondary cleavage points.) Cleavage of trypsinogen between residues 6 and 7 yields active enzyme. From Protein-digesting Enzymes, H. Neurath, Copyright © 1964, by Scientific American, Inc. All rights reserved (Reference 14).

are peptide analogs and hence bind very tightly to the active site of these enzymes. The R group must be a phenylalanyl side chain for the compound to be active against chymotrypsin, and a lysyl side chain for activity against trypsin. The chloromethylketone group itself is the part of the inhibitor molecule which attacks the imidazole ring (Shaw and co-workers).

The primary structures of trypsinogen and chymotrypsinogen are illustrated schematically in Fig. 2-5. These are zymogens from which functional enzymes are formed after activation. There are striking similarities in the active sites of the two proteins. In particular, two histidines are brought together by a loop of the polypeptide backbone in each case, and they are joined by a serine from quite a different portion of the chain. The specific serine residue was identified in both cases by reacting it with diisopropylphosphofluoridate, labeled with the radioactive isotope ^{32}P. The peptide bonds which are cleaved upon activation of the zymogens are also shown in Fig. 2-5; these follow the sixth and fifteenth residues in trypsinogen and chymotrypsinogen, respectively. The rupture of these bonds evokes considerable change in the tertiary structure of these proteins, and the ensuing molecular rearrangements in effect assemble the active center by bringing the serine residue together with the pair of histidines.

In higher organisms, digestive enzymes are not the only proteins to be elaborated by the glands as zymogens and then activated by the cleavage of a specific peptide bond. The same method of converting inactive proteins to active ones occurs in coagulation of blood where prothrombin is split to thrombin, plasminogen to plasmin, and fibrinogen to fibrin.

Let us summarize the findings which show that it is not the entire protein, but just a portion of it that participates in the act of catalysis. This will elucidate the relationship between the active site and the overall macromolecular structure of the enzyme molecule.

1. Modification of most amino acid side groups by specific chemical agents does not interfere with enzymatic activity. In addition to the examples cited above, we might mention the attachment of six- to ten-membered polypeptide branches to the ε-amino groups of lysine (Stamann). Here a protein is used as initiator for the polymerization of amino acid carboxyanhydrides (see p. 27ff). The polymerization reaction begins with the ε-amino group of lysine and proceeds outward from the protein until the desired number of residues have been bound. Stamann attached polylysine branches to chymotrypsin in this way, and Anfinsen constructed polyalanine side chains on RNase molecules by a similar procedure. In both cases, 60 to 70 new amino acid residues were linked to the protein molecules, which, in the case of RNase, increased its molecular weight by 50%. Nonetheless, these enormous structural modifications were not reflected in decreased enzymatic capacity in either instance.

2. Functionally identical enzymes from different species may vary from one another in amino acid sequence at almost any point along the polypeptide chain except in the limited region surrounding the active site. Cytochrome c, an important component of the cellular respiratory chain, contains a porphyrin prosthetic group attached to a specific histidine residue. The polypeptide segment -Cys-His-Thr- always occurs in cytochrome c molecules, regardless of the organism from which it was obtained. At greater distances from the crucial histidine residue, there is much more variation in polypeptide sequence from species to species. This is evident from Fig. 2-6

Beef Horse Pig	··· Val · Gln · Lys · Cys · Ala · Gln · Cys · His · Thr · Val · Glu · Lys ···
Salmon	··· Val · Gln · Lys · Cys · Ala · Gln · Cys · His · Thr · Val · Glu ···
Chicken	··· Val · Gln · Lys · Cys · Ser · Gln · Cys · His · Thr · Val · Glu ···
Silkworm	··· Val · Gln · Arg · Cys · Ala · Gln · Cys · His · Thr · Val · Glu ···
Yeast	Phe · Lys · Thr · Arg · Cys · Glu · Leu · Cys · His · Thr · Val · Glu ···
Rhodospirillum rubrum	··· Lys or Arg } · Cys · Leu · Ala · Cys · His · Thr · Phe · Asp · Glu · Gly · Ala · Asn · Lys ···
Common sequence:	Lys or Arg } · Cys · X · Y · Cys · His · Thr ·

Fig. 2-6. Variations in the sequence of the polypeptide chain of cyctochrome c from species to species. Data of Tuppy (15) as presented by Anfinsen (5, p. 157).

which presents the primary structure of polypeptide fragments surrounding the active site of cytochrome c from several diverse organisms. In a number of species, the sequences are identical, but the similarity declines with decreasing evolutionary relatedness until only the tripeptide -Cys-His-Thr- is shared in common. In other regions of the polypeptide chain, which contains a total of 104 residues, the correlation between amino acid sequences is not at all significant. Comparable results have been found with a number of enzymes, including trypsin and chymotrypsin: extensive species differences in amino acid sequence are observed except near the active site.

3. Mutations frequently lead to alterations in the structure of specific proteins which do not affect the active site. Recall the mutant forms of hemoglobin in man, which differ from the normal protein by one or two amino acid replacements. Many important properties of the mutant hemoglobins—solubility and charge, for example—can undergo considerable changes in comparison to normal molecules, but the binding constant for molecular oxygen remains exactly the same. Furthermore, several mutant forms of certain bacterial enzymes are known which differ from one another by a single amino acid residue. The activity of many of such mutant proteins is identical to that produced by the wild-type organism.

It should not be forgotten, however, that there are mutant enzymes totally devoid of activity, as well as those whose catalytic properties have been substantially changed. One of the most interesting mutational changes among enzymes is the acquisition of resistance to specific inhibitors. Bacterial mutants can be selected with this purpose in mind. Inhibitors are usually structural analogs of normal substrates, that is, of normal cellular metabolites. It is precisely because of their structural similarities that they bind to the enzyme; if the association is stable, the enzyme is put out of commission. Owing to their mode of action, enzyme inhibitors of this type are often called *antimetabolites*.

In bacteria, *p*-aminobenzoic acid (PABA) serves as a precursor in the synthesis of folic acid, a vitamin. Structural analogs of PABA are found among the sulfanilamides (including sulfanilamide, streptocide, sulfidine, sulfathiazole, and others), a class of compounds which compete for sites on the enzyme molecules which convert PABA to folic acid (16).

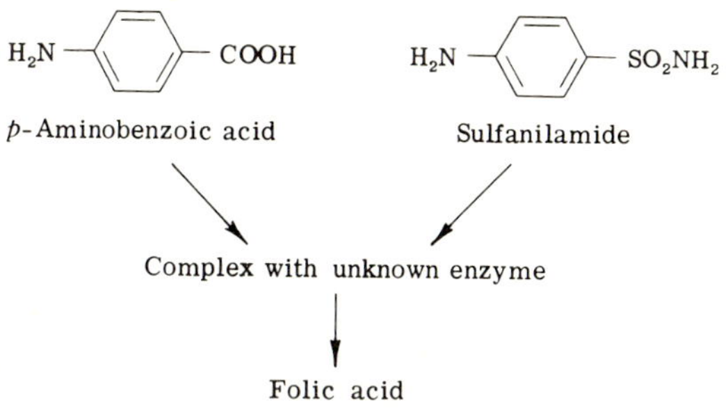

$$H_2N - \langle \rangle - COOH \qquad H_2N - \langle \rangle - SO_2NH_2$$

p-Aminobenzoic acid   Sulfanilamide

Complex with unknown enzyme

Folic acid

A number of bacterial mutants—from *Escherichia coli*, pneumococcus, and *Staphylococcus*—have been isolated with heightened resistance to sulfanilamide (SA). All these mutants differ from one another, but their genetic alterations are always related to the enzyme which acts on PABA. The inhibition of bacterial growth by SA depends on the ratio of SA concentration to PABA concentration in the medium, in accord with the scheme depicted above.

The active site of mutant proteins which no longer respond to inhibition by SA is clearly different from that of the normal protein. The mutant protein becomes more selective with respect to its substrate, indicating that it can now discriminate between the chemical structure of the substrate and that of the analogs. In a comparative study, Hotchkiss found that normal and mutant enzymes differ from one another by amino acid changes at certain points along the polypeptide chain. All of these residues lie close to each other and are therefore deemed to be of significance for the active site of the enzyme. This is a case in which the fine structure of the active site has been mutationally altered.

The sum total of evidence cited confirms the view that enzymatic activity is concentrated in only one portion of the protein molecule. A question naturally arises here. Why must protein macromolecules be so large and complex and why must their configuration be determined with mathematical precision? A complete answer is not possible at present, but we should remember that enzymes are not merely catalysts of specific chemical reactions. Since reactions must be localized in definite regions of the cell, the corresponding enzyme macromolecules must also have the capacity to bind with other substances in order to form the specific particulates and organelles which accomplish this purpose. Furthermore, we have no real basis to believe that the structure of biological polymers is the most economical and efficient one possible. Proteins as we know them have been shaped by an evolutionary process involving the repeated selection of random mutational alterations. As different species have originated, the functional activity of various proteins has changed. Thanks to accidents of evolution, it is likely that present-day protein molecules contain unused or degenerate regions along with those whose function is vital to the organism. A final answer to the question posed above may be possible in the future, once the structure and function of many more proteins have been studied.

Nonetheless, it is certain that the active site of an enzyme, however small it may be, is assembled from the side chains of amino acids quite distant from one another along the polypeptide backbone, yet brought into close contact owing to the specific secondary and tertiary structure of the protein. The work of Anfinsen and co-workers clearly demonstrated that the secondary and tertiary structure of a protein are completely determined by its primary structure. The enzymatic activity of a protein depends on the conformation of the polypeptide chain and the orientation of amino acid side groups and is hence a property of the macromolecule as a whole.

3. Kinetics of Enzymatic Catalysis

Enzymes may be studied quantitatively by determining the kinetics of the specific reactions that they catalyze. In order to obtain clear and reproducible results, it is essential to use highly purified, in particular crystalline, enzyme preparations. The enzymatic reaction itself is studied in a synthetic system, consisting only of buffer, substrate or substrates, and enzyme. The reader should bear in mind that conditions in such a model system are quite unlike those in the cell where the enzyme actually functions, especially since enzymes *in vivo* are often incorporated into supramolecular complexes or organelles. At one time it was thought that by measuring the kinetics of a reaction in the presence of pure, crystalline enzyme, one could realize the maximum catalytic activity of which the given protein was capable. In fact, this may be quite far from the truth. It is entirely possible that the activity of an enzyme is enhanced by virtue of its presence in an organelle since such an association could bring about a favorable change in enzyme configuration.

In this regard, experiments have shown that several important enzymes of glycolysis, such as aldolase and phosphoglyceraldehyde dehydrogenase, exhibit increased activity upon addition of certain nonaqueous solvents to the reaction mixture (P. Elödi). We have already seen that such solvents are capable of specifically altering protein tertiary structure (p. 70). Thus the study of enzymatic catalysis in aqueous solutions may not reveal the maximum potential of the enzyme in promoting a specific reaction. Nonetheless, analysis of enzyme kinetics *in vitro* still remains the most general method for the investigation of catalytic properties.

The theory of enzyme kinetics was developed by Michaelis and a somewhat more general formulation of his original analysis will be presented here (17, 18). The theory is founded on the assumption that enzyme, E, and substrate (or substrates), S, form a complex with each other, ES; the substrate is then either reversibly dissociated or converted to the end product, P, of the reaction with regeneration of free enzyme. We can write the basic relationship for the process in the following way:

$$E + S \underset{k_2}{\overset{k_1}{\rightleftharpoons}} ES \underset{k_4}{\overset{k_3}{\rightleftharpoons}} E + P \qquad (2\text{-}1)$$

where k_1 and k_2 are the rate constants for formation and dissociation of the enzyme-substrate complex, ES, and k_3 and k_4 are the rate constants for the forward and reverse reactions between complex and products. In this scheme, all reactions are considered to be reversible, and both forward and reverse reactions are taken into account simultaneously. It is not difficult to write the kinetic equations for the consumption of substrate, that is, for changes in the quantity of substrate, S, and complex, ES. We shall

3. Kinetics of Enzymatic Catalysis

denote the substrate, enzyme, complex, and product concentrations at any time by (S), (E), (ES), and (P), respectively:

$$d(S)/dt = -k_1(S)(E) + k_2(ES) \tag{2-2}$$

$$d(ES)/dt = k_1(S)(E) - (k_2 + k_3)(ES) + k_4(E)(P) \tag{2-3}$$

Moreover, two equations can be written which express the conservation of mass:

$$(E)_0 = (E) + (ES) \tag{2-4}$$

$$(S)_0 = (S) + (P) + (ES) \tag{2-5}$$

where $(E)_0$ is the total enzyme concentration, (E) is the free enzyme concentration, and $(S)_0$ is the initial concentration of substrate.

The integration of such a system of nonlinear differential equations presents formidable difficulties. Therefore, several simplifications are introduced. Most important is the assumption that the reaction proceeds under steady-state conditions except for an initial transient phase, in other words, that the concentration of the enzyme-substrate complex, (ES), remains constant:

$$(ES) = \frac{k_1(S) + k_4(P)}{k_2 + k_3}(E) = \text{constant}; \quad \frac{d(ES)}{dt} = 0 \tag{2-6}$$

in accord with Eq. (2-3). The assumption embodied in the condition $d(ES)/dt = 0$ means that when enzyme and substrate are mixed, the equilibrium quantity of the complex is formed in a time short compared with the period over which the reaction is observed. Thereafter, the concentration of the complex will change only gradually with changes in (S) and (P). If we should attempt to measure reaction rates in the very first moments after mixing the components, steady-state conditions would not apply, and we would have to return to the original formulation in all its complexity in order to find a solution. Substitution of Eq. (2-6) into Eq. (2-2), with the aid of Eq. (2-4) to eliminate (E), yields the steady-state rate equation in terms of two variables, (S) and (P):

$$-\frac{d(S)}{dt} = \frac{d(P)}{dt} = \frac{(V_S/K_S)(S) + (V_P/K_P)(P)}{1 + (S)/K_S + (P)/K_P} \tag{2-7a}$$

where:

$$V_S = k_3(E)_0 \qquad V_P = k_2(E)_0$$
$$K_S = (k_2 + k_3)/k_1 \qquad K_P = (k_2 + k_3)/k_4$$

Since the concentration of substrate is usually much greater than that of the enzyme, it follows that $(S)_0 \gg (ES)$, and Eq. (2-5) can be reduced to $(S)_0 = (S) + (P)$. This permits us to further simplify the steady-state rate equation to an expression with only one variable:

$$-\frac{d(S)}{dt} = \frac{d(P)}{dt} = \frac{[(V_S/K_S) + (V_P/K_P)](S) - (V_P/K_P)(S)_0}{1 + [(1/K_S) - (1/K_P)](S) + (1/K_P)(S)_0} \tag{2-7b}$$

All four of the parameters V_S, K_S, V_P, and K_P are constants under given conditions of temperature, pH, ionic strength, etc. At the moment of equilibrium $d(S)/dt = 0$, and the ratio between equilibrium concentrations of substrate and product are obtained:

$$K_{eq} = (P)_{eq}/(S)_{eq} = V_S K_P/V_P K_S \tag{2-8}$$

In practice, enzymatic reactions are generally studied under conditions in which very little of the substrate is converted to product, so that $(P) \ll (S)$. This can be accomplished by measuring the reaction velocity in the initial steady-state period, immediately after the transient phase concludes. As long as the reaction is far from equilibrium, (P) will be very small, and the last terms of both numerator and denominator in Eq. (2-7a) can be disregarded if, in addition, $(P)/K_P \ll 1$. The kinetic equation for the forward reaction becomes even simpler as a result:

$$-d(S)/dt = d(P)/dt = k_3(ES) = V_S(S)/[(S) + K_S] \tag{2-9}$$

This formula is called the Michaelis-Menten equation, although similar ideas had been advanced somewhat earlier by Henri. It expresses the dependence of reaction rate on substrate concentration alone.

The two constants in the equation both have straightforward interpretations. From Eq. (2-9), we see that as (S) becomes very large, so that $(S) \gg K_S$, the reaction rate approaches a maximum limiting value of V_S. This is understandable, since all the enzyme is in the form of complex $[(E)_0 = (ES)]$ under these conditions, and $d(P)/dt = k_3(E)_0 = V_S$. Product is formed at a constant rate and the reaction is said to obey zero-order kinetics (Fig. 2-7a). If the substrate concentration $(S) = K_S$, the reaction rate will be equal to half the maximum rate attainable, or $d(P)/dt = V_S/2$. The physical meaning of K_S, the Michaelis constant, becomes clear if we recall that $K_S = (k_2 + k_3)/k_1 \cong k_2/k_1$. If k_3 is small relative to k_2, K_S represents the reciprocal of the equilibrium constant for formation of the enzyme-substrate complex from enzyme and substrate.

When zero-order kinetics apply, that is, when the enzyme is saturated with substrate, a quantitative measure of enzymatic activity can be defined, called the *turnover number*. It is equal to $k_3 = V_S/(E)_0$, where V_S is by convention measured in moles of substrate per minute and $(E)_0$ in moles. The turnover number is thus the number of moles substrate converted by one mole of enzyme in one minute. For enzymes of relatively low activity (e.g., hexokinase), the turnover number is about 10^2 to 10^3, while for more active enzymes (e.g., catalase, peroxidase), it can reach 10^6 to 10^7.

A simple graphical construction is used to determine the two constants, V_S and K_S. While this procedure is usually attributed to Lineweaver and Burk, variants of it had been introduced previously by Haldane and others. By taking the reciprocal of the Michaelis-Menten equation (Eq. 2-9):

$$\frac{1}{-d(S)/dt} = \frac{1}{V_S}\left[1 + \frac{K_S}{(S)}\right] = \frac{1}{V_S} + \frac{K_S}{V_S}\left[\frac{1}{(S)}\right] \tag{2-10}$$

we obtain a linear equation in terms of $1/V_S$ and $1/(S)$. When plotted, the resulting straight line has a slope of K_S/V_S and intersects the ordinate at $1/V_S$ (Fig. 2-7b). Experimental determination of these two constants thus depends on measuring the reaction rate as a function of substrate concentration. It would appear that this could be accomplished in one experiment; after all, if the reaction is carried out over a

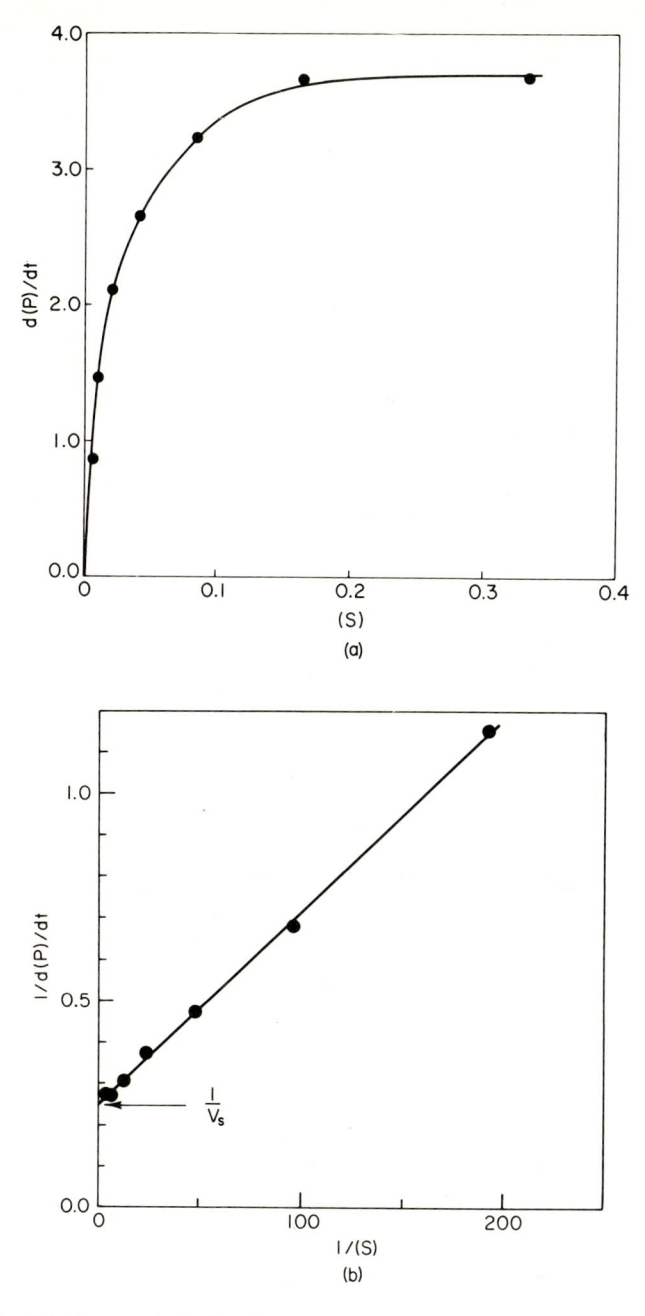

Fig 2-7. Kinetics of sucrose hydrolysis by the enzyme invertase. (a) Dependence of reaction rate $d(P)/dt$ on sucrose concentration (S) according to the Michaelis-Menten equation. Values on the abcissa represent the initial concentrations of sucrose; ordinate gives the initial rate of cleavage of the substrate as measured by decrease in optical rotation per unit time. The concentration of enzyme was the same in all cases. (b) The same dependence plotted according to Lineweaver and Burk; a straight line intersecting the ordinate at $1/V_s$ results. Data of Michaelis and Menten (17).

sufficient period, the substrate concentration will gradually change. By integrating, the Michaelis-Menten equation can be obtained in a form applicable to this case:

$$V_S t = (P) - K_S \ln\left[1 - (P)/(S)_0\right] \tag{2-11}$$

Although V_S and K_S can in principle be calculated by this method, the result may not be reliable owing to a complication that has not yet been considered. The reaction product will naturally have a substantial affinity for the active site of the enzyme since rearrangement of electronic orbitals in the reaction $AB + CD \rightarrow AC + BD$ may not significantly affect those portions of the substrate molecule which contribute to the formation of active complex, ES. Therefore, a certain fraction of the product molecules are bound to the enzyme in an inactive complex: $E + P \rightleftharpoons EP$. The active site of the enzyme is therefore obstructed by product molecules which, because of their competition with substrate for binding sites, effectively reduce the quantity of available enzyme.

Product inhibition is just one example of competitive or reversible inhibition which can be brought about by a variety of substances, particularly structural analogs of the normal substrate or product. Such specific inhibitors of enzymatic reactions, generally called antimetabolites, are widely used in pharmacology and medicine to suppress undesirable metabolic processes. Their mode of action is distinct from that of enzyme poisons like diisopropylphosphofluoridate, which irreversibly attacks the active site, or p-chloromercuribenzoate, which blocks functional sulfhydryl groups. An excess of substrate can even inhibit some enzymatic reactions. In these cases the reaction rate rises with increasing concentration, passes through a maximum, and then declines. The cleavage of sucrose by invertase is one reaction in this category. Substrate inhibition is apparently explained by the formation of an inactive secondary complex consisting of enzyme plus two or more substrate molecules; at high concentrations, the substrate itself behaves as a competitive inhibitor.

Kinetic equations for competitive inhibition can be formulated with the same approximations used to obtain the Michaelis-Menten equation. In addition to formation of active complex, ES, according to Eq. (2-1), an inactive enzyme–inhibitor complex, EI, is also produced:

$$E + I \underset{k_6}{\overset{k_5}{\rightleftharpoons}} EI \tag{2-12}$$

where I represents the inhibitor and EI represents the enzyme-inhibitor complex. Both Eqs. (2-1) and (2-12) must be treated simultaneously, taking into account the conservation of mass:

$$(E)_0 = (E) + (ES) + (EI) \tag{2-13}$$

$$(S)_0 = (S) + (P) \tag{2-14}$$

If steady-state conditions apply to both ES and EI, that is, if $d(ES)/dt = d(EI)/dt = 0$, a simple calculation leads to the kinetic equation:

$$-\frac{d(S)}{dt} = \frac{d(P)}{dt} = \frac{V_S(S)}{K_S[1 + K_i(I)] + (S)} \tag{2-15}$$

3. Kinetics of Enzymatic Catalysis

The constant $K_I = k_5/k_6$ pertains to the equilibrium quantity of enzyme-inhibitor complex formed.

The effect of competitive inhibition is to cause an apparent increase in the Michaelis constant by a factor of $[1 + K_I(I)]$. One consequence is that the inhibitory effect of I can be relieved by increasing the substrate concentration which emphasizes the competition between substrate and inhibitor for active sites on the enzyme molecule. Competitive inhibition is expressed in the Lineweaver-Burk plot by a change in slope, but not in point of intersection with the ordinate (Fig. 2-8a).

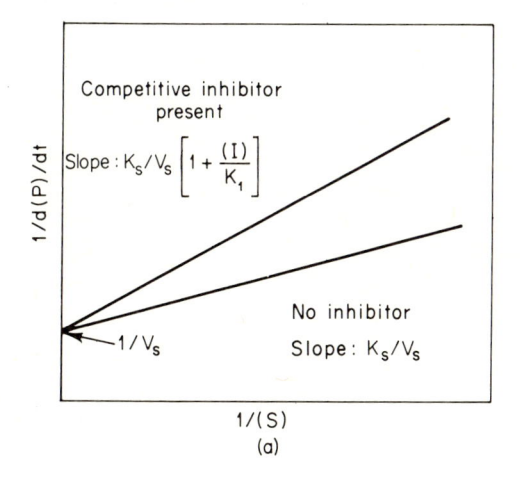

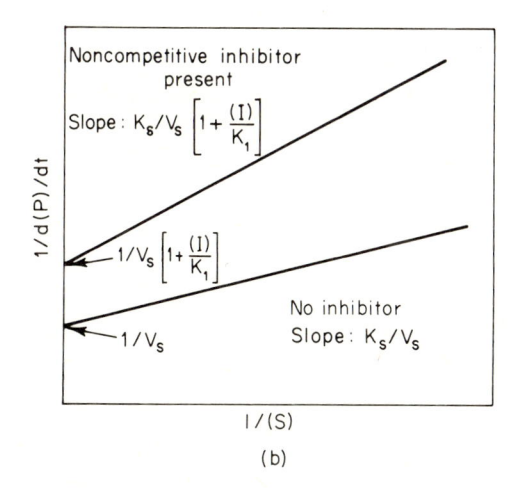

Fig. 2-8. Hypothetical Lineweaver-Burk plots for (a) competitive and (b) noncompetitive inhibition of enzyme action.

Values of K_S, K_I and $k_3 = V_S/(E)_0$ for several substrates and competitive inhibitors of chymotrypsin are presented in Table 2-1. It is clear from the figures given that D-amino acid derivatives have about half the affinity (with one exception) for the active site as the L-isomers which are the normal substrates of this enzyme. This means

TABLE 2-1

The Michaelis Constants K_S and K_I and the Rate Constant k_3 for Substrates and Competitive Inhibitors of the Chymotrypsin-Catalyzed Reaction[a]:

$$R_1-CO-NH-\underset{\underset{R_2}{|}}{CH}-CO-X + H_2O \rightarrow R_1-CO-NH-\underset{\underset{R_2}{|}}{CH}-COO^- + HX + H^+$$

$R_2 =$ p-hydroxybenzyl (Tyrosyl-); $R_2 =$ benzyl (Phenylalanyl-); $R_2 =$ indolyl-methyl (Tryptophanyl-)

Constant	R_1	X		Tyrosyl- L-	Tyrosyl- D-	Phenyl-alanyl- L-	Phenyl-alanyl- D-	Trypto-phanyl- L-	Trypto-phanyl- D-
$K_I \times 10^3\ M$	Acetyl	O⁻	(inhibition)	110	—	—	—	10	5
	Nicotinyl	O⁻	(inhibition)	60	—	—	—	15	—
	Chloroacetyl	O⁻	(inhibition)	150	—	—	—	—	—
$K_S \times 10^3\ M$	Acetyl	NH₂	(hydrolysis)	32	12	31	12	5	2.3
	Nicotinyl	NH₂	(hydrolysis)	12	7	19	10	2.5	1.6
	Chloroacetyl	NH₂	(hydrolysis)	26	9	—	—	—	—
	Acetyl	OC₂H₅	(hydrolysis)	0.7	4.0	—	2.2	—	—
$k_3\ \text{sec}^{-1}$	Acetyl	O⁻		0	—	0	—	0	—
	Nicotinyl	O⁻		0	—	0	—	0	—
	Chloroacetyl	O⁻		0	—	0	—	0	—
	Acetyl	NH₂		0.15	—	0.05	—	0.03	—
	Nicotinyl	NH₂		0.29	—	0.12	—	0.09	—
	Chloroacetyl	NH₂		0.24	—	—	—	—	—
	Acetyl	OC₂H₅		250	—	≈150	—	—	—

[a] From Hearon et al. (19, p. 127).

that the former are quite effective inhibitors of chymotrypsin activity, since they cannot be activated once bound. Reaction products occupy a special place in the table for they also possess a significant affinity for the active site and therefore serve as antimetabolites. They are even more effective than the D-amino acid derivatives since their binding constants with the enzyme are several times greater.

Noncompetitive inhibition of enzyme activity occurs when the agent attacking the active site cannot be displaced by the substrate at any concentration. All irreversible chemical changes at the active site lead to noncompetitive inhibition. Kinetically, the situation reduces to substitution of the initial enzyme concentration (E) by a lower effective concentration $(E)_{\text{eff}}$ in the reaction equations. In essence, an amount of enzyme equal to $(E) - (E)_{\text{eff}}$ has been inactivated. The velocity constant is correspondingly reduced to $V_S = k_3(E)_{\text{eff}}$. In the Lineweaver-Burk plot, both slope and intercept are increased by a factor of $(E)/(E)_{\text{eff}}$, providing a convenient means for determining whether inhibition is competitive or noncompetitive (compare a and b of Fig. 2-8).

All in all, the theory of Michaelis quite satisfactorily fits actual experimental data, thus substantiating the fundamental postulate that a steady-state enzyme-substrate complex is formed during the reaction. Of course, we have considered here only the simplest case where one of the two substrates, water, is present in great excess, where the enzyme-substrate complex has a very simple structure, and where there is only a single reaction product. Kinetic equations for much more complicated situations can be formulated by introducing additional constants, but they are very difficult to verify experimentally owing to the large number of measurements required. In practice, therefore, the kinetics of most enzymatic reactions are described in terms of the simpler Michaelis-Menten equation; it is then up to the skill of the investigator to exclude, insofar as possible, any experimental factor which might cause deviation from this model.

As we have seen, the physical interpretation of both V_S and K_S is quite clear, but their kinetic significance, which we shall presently examine, is also of great importance. In most kinetic measurements the enzyme is completely saturated with substrate and the actual reaction rate is equal to the maximum rate of $V_s = k_3(E)_0$. By determining V_s at various temperatures and inserting the corresponding values into the Arrhenius equation:

$$V_S/(E)_0 = k_3 = Ae^{-E^*/RT} \qquad (2\text{-}16)$$

the activation energy of the reaction, E^*, can be calculated. Other parameters in this equation are T, the absolute temperature; R, the gas constant per mole; and A, a multiplicative factor. A comparison of the activation energies involved in enzymatic and nonenzymatic catalysis of several different reactions is presented in Table 2-2. Although the figures show that enzymes effectively reduce activation energies, the values are not abnormally low for catalytic processes in general.

The comparatively large rates observed in enzymatic reactions are not due solely to lowered energy of activation. The magnitude of the coefficient A in the Arrhenius equation is also of importance. From the theory of absolute reaction rates, the rate constant of any reaction involving an activated intermediate is given by the expression:

$$k = \frac{RT}{Nh} e^{-\Delta F^{\ddagger}/RT} = \frac{RT}{Nh} e^{-(\Delta H^{\ddagger} - T\Delta S^{\ddagger})/RT} = \frac{RT}{Nh} e^{\Delta S^{\ddagger}/R} e^{-\Delta H^{\ddagger}/RT} \qquad (2\text{-}17)$$

where k is the rate constant, h is Planck's constant, R is the gas constant per mole, N is Avogadro's number, T is the absolute temperature, $\Delta F^{\ddagger}$ is the free energy of the activated state through which the (enzyme-substrate) complex must pass prior to degradation, $\Delta H^{\ddagger}$ is the heat of activation, and $\Delta S^{\ddagger}$ is the entropy of activation. All thermodynamic quantities are per mole reactant in the standard state. Now $\Delta H^{\ddagger}$ is

TABLE 2-2

A Comparison of Activation Energies in Enzymatic and Nonenzymatic Catalysis[a]

Reaction	Catalyst	E^*(cal/mole)
Decomposition of hydrogen peroxide	None	18,000
	Colloidal platinum	11,700
	Liver catalase	5,500
Hydrolysis of ethyl butyrate	Hydrogen ion	16,800
	Hydroxyl ion	10,200
	Pancreatic lipase	4,500
Hydrolysis of casein	Hydrogen ion	20,600
	Trypsin	12,000
Hydrolysis of sucrose	Hydrogen ion	25,560
	Invertase (yeast)	11,500
	Invertase (malt)	13,000
Hydrolysis of benzoyl-glycine	Hydrogen ion	22,100
Hydrolysis of benzoyl-L-arginine	Trypsin	15,500

[a] From Moelwyn-Hughes (13, p. 66).

related to the activation energy E^* in the following way: $\Delta H^{\ddagger} = E^* - RT$. The RT term can generally be disregarded, and Eq. (2-17) then becomes equivalent to the Arrhenius equation [Eq. (2-16)] when:

$$A = \frac{RT}{Nh} e^{\Delta S^{\ddagger}/R} \qquad (2\text{-}18)$$

In a vast majority of ordinary reactions, $\Delta S^{\ddagger}$ is a rather large negative quantity. This is so because chemical bonds are broken in the activated state, leading to a reduction in vibrational degrees of freedom. Therefore, $e^{\Delta S^{\ddagger}/R}$ is but a small fraction in most cases, on the order of 10^{-5} to 10^{-8} or even smaller. But the situation is quite different for enzyme-catalyzed reactions. As can be seen from the experimental data for several enzymatic reactions assembled in Table 2-3, the entropy is not greatly reduced in the activated state, and as a result $e^{\Delta S^{\ddagger}/R}$ falls between 10^{-1} and 10^{-2}. This circumstance plays a role no less important than activation energy in enzymatic reactions and helps to explain why enzymes are so efficient in comparison to non-enzymatic catalysts.

The Michaelis constant K_S can also be analyzed in thermodynamic terms. At low substrate concentrations and when $k_2 \gg k_3$, the expression $1/K_S = k_1/k_2$ is the equilibrium constant for formation of the enzyme-substrate complex. In this case:

$$-RT \ln (1/K_S) = \Delta F^{\circ} = \Delta H^{\circ} - T \Delta S^{\circ} \qquad (2\text{-}19)$$

3. Kinetics of Enzymatic Catalysis

TABLE 2-3

Thermodynamic Quantities for Some Enzymatic Reactions[a]

Enzyme	Substrate	$\Delta F^{\ddagger}$ (cal/ mole)	$\Delta H^{\ddagger}$ (cal/ mole)	$\Delta S^{\ddagger}$ (e.u./ mole/	$\Delta F°$ (cal/ mole)	$\Delta H°$ (cal/ mole)	$\Delta S°$ (e.u./ mole)
Chymo-trypsin	Methyl hydrocinnamate	19,700	16,200	−11.3	−1,900	−5,300	−11.4
	Methyl-DL-α-chloro-β-phenylpropionate	18,700	14,800	−13.2	−2,600	−8,500	−19.8
	Methyl-D-β-phenyllactate	18,700	14,500	−14.2	−2,200	−12,000	−33.0
	Methyl-L-β-phenyllactate	17,500	10,500	−23.4	−2,800	−7,300	−15.1
	Benzoyl-L-tryosine ethyl ester	15,000	8,600	−21.4	−3,300	−8,400	−17.1
	Benzoyl-L-tyrosine amide	17,900	14,000	−13.0	900	−9,900	−30.0
Acetyl-choline esterase	Acetylcholine	8,000 (av.)	(14,000–19,000)	(16–34)	−5,500	0	18.5
	Dimethylaminoethyl acetate	9,900 (av.)	(6,700–8,000)	−(6.5–10.5)	−4,400	0	14.6
	Methylaminoethyl acetate	10,700	8,000	−9	−2,900	0	9.7
	Aminoethyl acetate	12,200	9,500	−9	−2,500	0	8.4
Carboxy-peptidase	Carbobenzoxy-L-tryptophan	19,300	16,000	−11	−3,400	5,000	−28
	Carbobenzoxyglycyl-L-phenylalanine	14,400	8,900	−18	−2,500	−400	−7
	Carbobenzoxyglycyl-L-tryptophan	15,300	9,300	−20	−3,400	0	−11.5
Pepsin	Carbobenzoxy-L-glutamyl-L-tyrosine ethyl ester	22,100	20,100	−6.5	−4,700	1,400	20.6
	Carbobenzoxy-L-glyutamyl-L-tyrosine	23,100	16,600	−21.8	−4,300	3,000	24.4
Urease	Urea	11,300	9,100	−7.2	−3,200	−2,900	0.9
Adenosine-triphosphatase	Adenosine triphosphate	14,400	12,400	−8.0	−7,500	8,000	52

[a] From Lumry (20, p. 178).

where $\Delta F°$ is the free energy change, $\Delta H°$ the enthalpy change, and $\Delta S°$ the entropy change, all for the formation of the enzyme-substrate complex. These quantities can be calculated if the Michaelis constant is known as a function of temperature. Table 2-3 presents values of $\Delta F°$, $\Delta S°$, and $\Delta H°$ for a number of enzymatic reactions. We see that the complex is frequently formed at the expense of a large increase in entropy. The relationship between $\Delta F^{\ddagger}$ and $\Delta F°$ are shown for a hypothetical enzymatic reaction in Fig. 2-9.

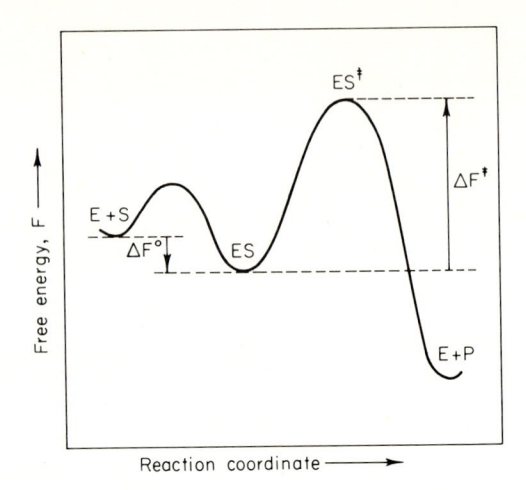

Fig. 2-9. Hypothetical free energy profile for an enzymatic reaction. The reaction coordinate plots enzyme E and substrate S as they are converted to enzyme substrate complex ES and finally to product P and free enzyme. The quantity $\Delta F°$ is the change in free energy upon formation of the enzyme-substrate complex from reactants and is related to the Michaelis constant K_S. The parameter $\Delta F^{\ddagger}$ is the free energy of activation of the enzyme–substrate complex ES prior to release of the product; $\Delta F^{\ddagger}$ can be determined from k_3 at high substrate concentrations when the final reaction step is rate-limiting.

It should be emphasized that when enzyme kinetics are studied as a function of external factors such as inhibitors, temperature, pH, and inorganic ions, both V_S and K_S must be considered since they relate to different aspects of the catalytic process. We have already metntioned the influence of pH on the rate constant. But pH can also produce changes in K_S, the apparent affinity constant. For instance, Fig. 2-10

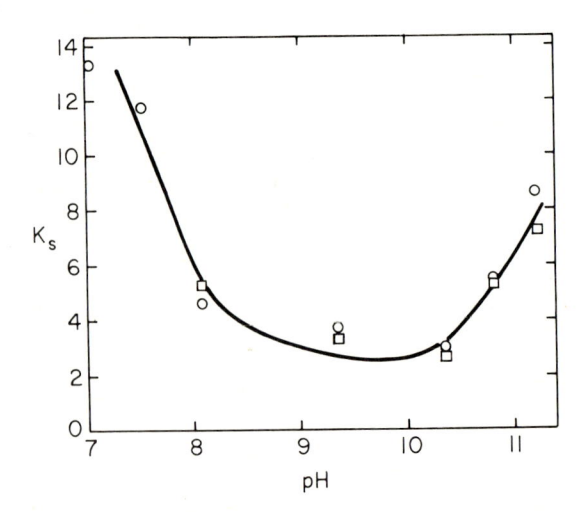

Fig. 2-10. Change in the Michaelis binding constant K_S with pH for the enzyme arginase from beef liver. Measurements were made at 25° (□) and 35° (○). From Roholt and Greenberg (21).

depicts the dependence of K_S on pH for liver arginase, an enzyme which cleaves arginine according to the scheme:

$$\begin{array}{c} \text{HOOC-CH-(CH}_2)_3\text{-NH-C(=NH)-NH}_2 + \text{H}_2\text{O} \\ | \\ \text{NH}_2 \\ \downarrow \\ \text{HOOC-CH-(CH}_2)_3\text{-NH}_2 + \text{H}_2\text{N-C(=O)-NH}_2 \\ | \\ \text{NH}_2 \end{array}$$

The shape of the curve leaves no doubt that arginine must be positively charged in order to bind with the enzyme. The break in the curve at alkaline pH corresponds to the dissociation of the α-amino group of arginine while a similar break in the acid region marks the titration of a group with a pK equal to about 5. This evidently means that there is a charged carboxyl group in the active site of the enzyme which binds positively charged arginine. It is through studies of this type that changes in V_S and K_S can be used to elucidate the role of functional groups in enzyme and substrate, both those responsible for binding and those responsible for activation of the complex.

4. The Mechanism of Enzymatic Reactions

The mechanisms of a number of enzymatic reactions have already been treated in the previous sections. Enzyme action can always be reduced to the rupture of two chemical bonds between A—B and C—D with the formation of new bonds between reciprocal pairs: AB + CD → AC + BD. The difficulty in accomplishing this reaction is that the bonds in the initial molecules, AB and CD, are chemically saturated and the electronic orbitals formed by their valence electrons are oriented along the axes of the A—B and C—D bonds. For reaction to occur, the electron distributions of the two substances must undergo substantial rearrangement and it is for this process that the energy of activation is needed. The electronic structure of the substrate is greatly altered when it becomes bound to the enzyme, and in particular, its electronic orbitals are displaced from their initial unperturbed states. Roughly speaking, the energy of the A—B bond is proportional to the square of the integral of the electron density between points A and B, assuming that these symbols stand for the nuclei of the atoms participating in the bond. If some external force displaces or polarizes the electron distribution, the A—B bond energy will be diminished and the bond itself will be weakened. When this happens, the bond may be more easily cleaved by thermal energy.

Catalysis thus depends on the formation of a bond between substrate and enzyme that is sufficiently strong to produce a significant displacement of electronic orbitals in the region of the A—B bond. In a number of cases, this is effected by the formation of covalent bonds between the substrate molecule and certain functional groups of the enzyme. In other cases, coordination bonds provide the basis for this association.

Coordination bonds are formed by many enzymes which contain metal atoms or ions. These metals—including iron, manganese, copper, cobalt, molybdenum, and others—all belong to the group of transition elements in the periodic system, which means that they possess unoccupied quantum states in their d-orbitals. As a result, transition metal ions are able to form coordinate covalent bonds with atoms which can donate unshared electron pairs. Nitrogen, oxygen, and sulfur atoms may all be counted among potential donors of unshared electron pairs. The coordination number of the metal, or the number of atoms or groups it can complex, is generally greater than its valence.

The formation of complexes between metal ions and molecules, called ligands, is often made dramatically visible by the appearance of brightly colored products from colorless reactants. The displacement of absorption bands from the ultraviolet to the visible region of the spectrum signals a profound rearrangement of electronic orbitals which bind the groups together. A good example is the salt $CuSO_4$: when unhydrated, the compound is colorless, but in the presence of water, ammonia, or other substance capable of forming complexes, $CuSO_4$ takes on a brilliant blue hue.

If the molecule AB is associated with a metal ion through coordination bonds, the A—B bond is likely to be weakened by a displacement of its electronic orbitals, as mentioned above. This is generally believed to be the principal mechanism to which many enzymes owe their activating capacity. In several cases the formation of complexes between enzyme-bound metal ions and substrate molecules has been proved beyond question by spectroscopic methods; Keilin showed this for catalase, Chance for peroxidase, and Smith for various peptidases (22).

A very simple example of interaction between enzyme and substrate occurs with carbonic anhydrase, an enzyme catalyzing the hydration of carbon dioxide:

$$OH^- + CO_2 \longrightarrow O=C\begin{smallmatrix}OH\\O^-\end{smallmatrix}$$

Keilin and Mann showed that carbonic anhydrase contains zinc. At pH 7, zinc is able to form subcarbonates, basic salts in which carbonate and hydroxyl ions are simultaneously bound:

$$Zn\begin{smallmatrix}O-C(=O)-OH\\OH\end{smallmatrix}$$

It is apparent that the zinc ion in carbonic anhydrase can form at least two coordination bonds with the oxygen atoms of the substrates:

$$\begin{smallmatrix}Protein\\Zn^{2+}\\HO^- \quad O=C=O\end{smallmatrix}$$

4. The Mechanism of Enzymatic Reactions

so as to facilitate rearrangement of the molecular groups with formation of carbonic acid: $OH^- + CO_2 \rightarrow HCO_3^-$.

Ions of another metal, cobalt, occur at the active site of some peptidases, enzymes which hydrolyze peptide bonds. Cobalt readily forms coordination bonds with both nitrogen and oxygen, behaving as an acceptor of electrons in both cases. The following is a suggested structure for the enzyme–substrate complex:

$$\text{Protein} \cdots Co^{2+} \begin{array}{c} O=C \\ N-H \\ CHR \\ O=C \\ N-H \end{array}$$

Here cobalt provides one link in a five-membered ring, a characteristic feature of such complexes due to the absence of strained bonds. The Co----O coordination bond draws electrons away from the —C—N— (peptide) bond, which makes hydrolysis of the latter easier. In the schematic diagram given above, arrows indicate the direction in which electrons are displaced by the so-called induction effect during formation of the coordination bond.

All metal-containing enzymes are subject to inhibition by various complex-forming reagents. By cataloging effective inhibitors, it is often possible to judge which metal is present in the enzyme. It should be remembered, however, that the stability constant of a metal–reagent complex in solution may not be quantitatively applicable when the metal ion is bound to a protein.

A favorable reagent for heavy metals is ethylenediaminetetraacetic acid (EDTA):

$$\begin{array}{c} HOOC \\ \diagdown \\ HOOC \end{array} N-CH_2-CH_2-N \begin{array}{c} \diagup COOH \\ \\ \diagdown COOH \end{array}$$

When measured in solution, the stability constant of EDTA with different metal ions is largely a function of the valence of the ion. Some of these binding sites will be occupied, however, when the ion is associated with an enzyme, and its ability to complex with EDTA may be greatly reduced. The extent of reduction depends on how tightly the metal ion is bound to the protein. If the association is weak, the metal may be easily removed by a complex-forming reagent, or even by dialysis alone. The enzyme loses its activity as a result, although it can generally be reactivated by adding back the required ion. In this case, the enzyme is usually referred to as a metalloprotein complex and the metal is known as the activator of the enzyme. If the bond between metal and protein is very stable, it cannot be removed by dialysis, chromatography, or complex-forming substances. In these enzymes, called simply metalloproteins, the metal ion behaves quite differently than when it is free in solution. In reality, all gradations are observed in the stability of metal-containing enzymes, from those in which the association is very stable to those in which the metal ion is very easily dissociated.

A more complicated case of enzymatic catalysis occurs when activity depends on association between the enzyme and a small organic molecule called a coenzyme. A classic example is pyridoxal-5-phosphate which serves as the coenzyme for many enzymes concerned with amino acid metabolism. Braunstein and Shemyakin as well as Snell have presented a unified theory of the way in which such enzymes function (23). Pyridoxal contains an aldehyde group

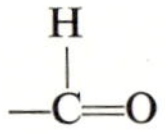

$$\begin{array}{c} H \\ | \\ -C=O \end{array}$$

bound to a pyridine nucleus, which readily reacts with amino groups of amino acids to form aldimines or Schiff bases. In this case the aldimines are of a special type containing conjugated double bonds. The π-electrons are very mobile in such compounds and a variety of polar groups, both within the amino acid structure or external to it, can exert an inductive influence on covalent bonds at any point within the limits of the conjugated chain. Amino acids are rendered extremely labile and reactive under these conditions and several of their bonds may be weakened and then broken. Moreover, the Schiff bases undergo a tautomeric shift from structure II to structure III:

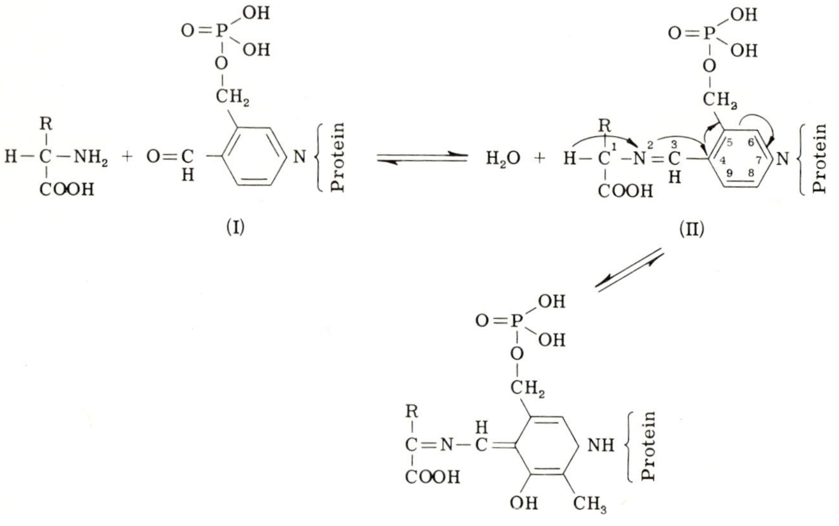

As shown here, the presence of a strongly electrophilic substituent (N) in the pyridine ring exerts an inductive effect, displacing electrons along a whole series of conjugated bonds. As a result, bonds are weakened within that part of the molecule which initially belonged to the amino acid reactant and, in addition, the molecule becomes polarized.

4. The Mechanism of Enzymatic Reactions

The following schematic diagram shows the bonds which could be cleaved in the labile form of a hypothetical amino acid bound to an enzyme through pyridoxal as an aldimine:

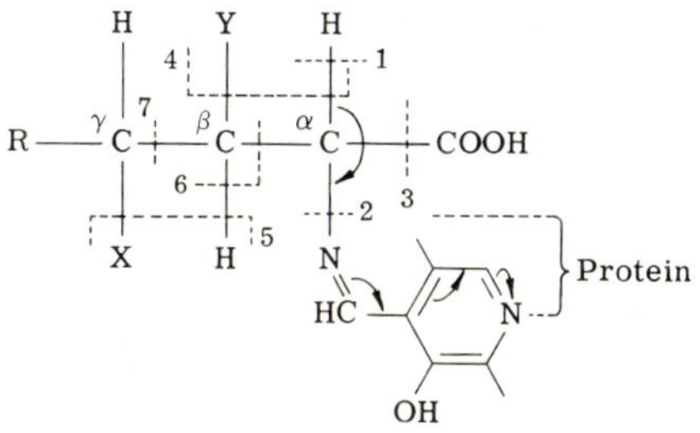

$$R - \overset{H}{\underset{X}{\overset{|}{\underset{|}{C}}}} \overset{\gamma}{\underset{7}{-}} \overset{Y}{\underset{|}{\overset{|}{\underset{|}{C}}}} \overset{4}{\underset{6}{-}} \overset{\beta}{\underset{|}{\overset{|}{\underset{|}{C}}}} \overset{H}{\underset{|}{\overset{|}{\underset{5}{-}}}} \overset{\alpha}{\underset{|}{\overset{|}{\underset{N}{C}}}} \overset{1}{\underset{2}{-}} COOH$$

The amino acid portrayed is much more complex than the compounds of this type usually encountered, but it illustrates the many possibilities for reaction which exist. Let us consider some of the specific reactions which an activated compound might undergo. They are indicated in the scheme by numbers running from 1 to 7. Removal of the hydrogen atom from the amino acid α-carbon (reaction 1) forms the basis for enzymatic tautomerization as well as enzymatic racemization of amino acids. It is clear that the excised hydrogen should be in the form of a proton, H^+, since the shift of electrons in the direction of the ring makes the molecule polar. Another possibility is cleavage of the —C—N— bond (reaction 2). This mechanism is responsible for deamination and transamination, the transfer of amino groups from amino acids to acceptor molecules. Enzymatic decarboxylation results from reaction 3. More complicated processes can also occur. If Y designates a hydroxyl group, for instance, as in serine or threonine, cleavage of the bond at position 4 causes dehydration of the amino acid. If Y represents a sulfhydryl group, then enzymatic cleavage of H_2S can occur. Reactions at positions 5, 6, and 7 yield still further variations of the action of pyridoxal-containing enzymes on amino acids.

An enormous number of possibilities arise from the labilization of amino acids through binding to pyridoxal phosphate. However, we have not yet asked why, in a given situation, the reaction proceeds according to just one of the mechanisms. This question is particularly pertinent in the present example of amino acid reactions where the number of alternatives is so great. But this problem might be appropriately posed in the case of metalloenzymes or esterases too, where the active site is composed in part of amino acid side groups. The fact that certain chemical bonds are weakened does not in itself define the reaction pathway. There must be additional factors which contribute to the remarkable selectivity observed in enzyme-catalyzed transformations. Since the pyridoxal prosthetic group is common to many enzymes, the protein molecule must play a major role in defining enzyme specificity.

Trypsin and chymotrypsin have provided more insight into the significance of protein structure in enzyme catalysis than any other enzymes. These two enzymes catalyze the hydrolysis of various ester and peptide bonds. We have already mentioned that the active sites of both contain one serine and two histidine residues (p. 137ff). Our knowledge is sufficient to permit speculation as to why these residues are necessary, and even as to how the active site functions. We know that an enzyme must activate a saturated chemical bond in each of its two substrates. Now there are two ways in which this might be accomplished. According to the first possibility, the enzyme begins by cleaving a fragment from one molecule and attaching it to its active site, thereby forming an intermediate compound. The second substrate is then activated, and the reaction proceeds to completion. In the second alternative, both bonds are activated simultaneously, resulting in what is called *concerted catalysis*.

Trypsin and chymotrypsin apparently make use of both these mechanisms. The hydrolysis of esters is a two-step reaction in which an alcohol is first split from the ester while the acidic (acyl) remainder is esterified to the serine residue at the active site. This process creates an acyl-enzyme intermediate. During the second stage, a water molecule is activated, after which the acidic group is cleaved from the serine. One of the histidines at the active site plays a key role in initiating this reaction sequence, owing to the ability of its imidazole ring to reversibly accept and dissociate protons. In particular, it can accept a proton from serine, and as a result, serine acquires a negative charge:

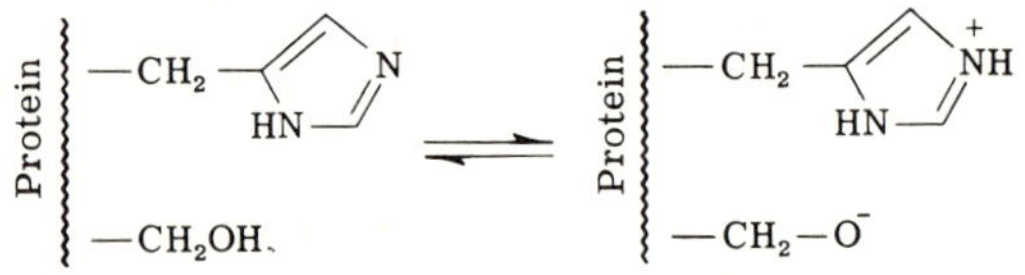

$$\text{Protein} \begin{cases} -CH_2-\langle\text{imidazole}\rangle \\ \quad\quad\quad HN \\ -CH_2OH \end{cases} \rightleftharpoons \text{Protein} \begin{cases} -CH_2-\langle\text{imidazole}\rangle NH^+ \\ \quad\quad\quad HN \\ -CH_2-O^- \end{cases}$$

There is good evidence that nonenzymatic hydrolysis of esters by hydroxyl ions proceeds as follows:

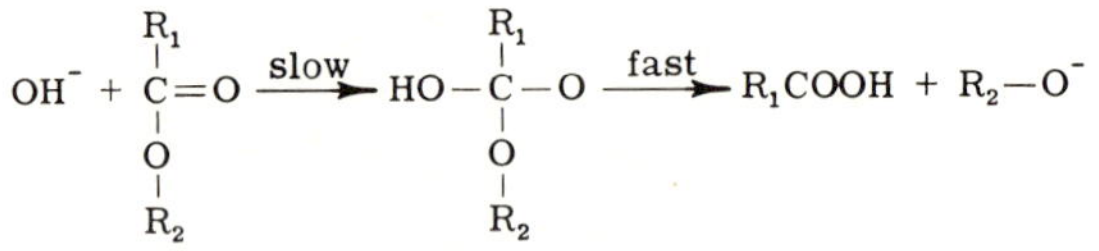

$$OH^- + \underset{\underset{R_2}{\overset{|}{O}}}{\overset{\overset{R_1}{|}}{C}} = O \xrightarrow{\text{slow}} HO-\underset{\underset{R_2}{\overset{|}{O}}}{\overset{\overset{R_1}{|}}{C}}-O \xrightarrow{\text{fast}} R_1COOH + R_2-O^-$$

$$R_2O^- + H^+ \longrightarrow R_2OH$$

The hydroxyl ion donates an unshared electron pair to the ester, forming an unstable intermediate which decomposes into alcohol and acid. This process, termed a *nucleophilic attack*, is a very general mechanism for bond activation. In the enzymatic reaction, it is thought that the negatively charged serine takes the place of hydroxyl ion as nucleophilic agent, attacking the electron-deficient carbonyl group of the ester. This results in formation of a saturated "tetrahedral intermediate" which decomposes into acyl-enzyme and alcohol. A tetrahedral intermediate arises again upon deacylation. A likely mechanism for chymotrypsin and trypsin esterase activity is described in the following scheme:

4. The Mechanism of Enzymatic Reactions

This reaction pathway has been confirmed by a considerable body of evidence. In particular, the presence of an acyl-enzyme intermediate has been convincingly demonstrated by the following observation: when the ester nitrophenylacetate

was used as a substrate for trypsin or chymotrypsin, it was hydrolyzed in spite of its low affinity for these enzymes. As a consequence of its cleavage, the serine residues at the active sites were almost completely acetylated. The resulting acetyl-trpysin and acetyl-chymotrypsin were unable to hydrolyze esters. Deacetylation occurred after several hours at neutral pH and was accompanied by the restoration of esterase activity.

The hydrolysis of peptide bonds occurs by a different route and an acyl-enzyme intermediate is apparently not formed in this case. As shown in the laboratory of the author, acetyl-trypsin, prepared as described previously, retains about half of its native activity when assayed on specific amide substrates. This is good evidence that the serine residue does not participate in the cleavage of amides. It is very probable that amide hydrolysis is an example of concerted acid-base catalysis in which both histidine residues take part. The nonenzymatic hydrolysis of amides proceeds via an ionic complex:

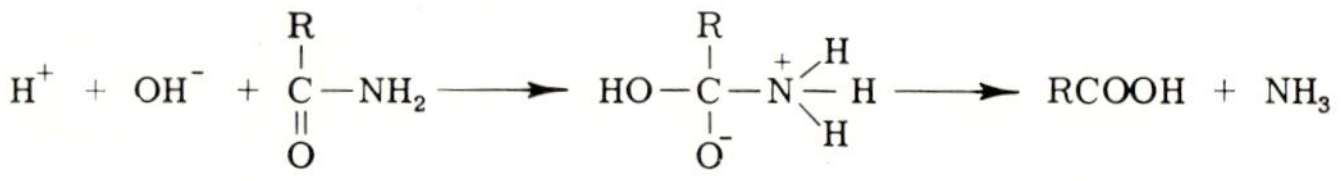

$$H^+ + OH^- + \underset{O}{\overset{R}{C}}-NH_2 \longrightarrow HO-\underset{O^-}{\overset{R}{C}}-\overset{+}{N}\overset{H}{\underset{H}{\diagdown}}H \longrightarrow RCOOH + NH_3$$

From this, the following scheme has been proposed for concerted catalysis by trypsin:

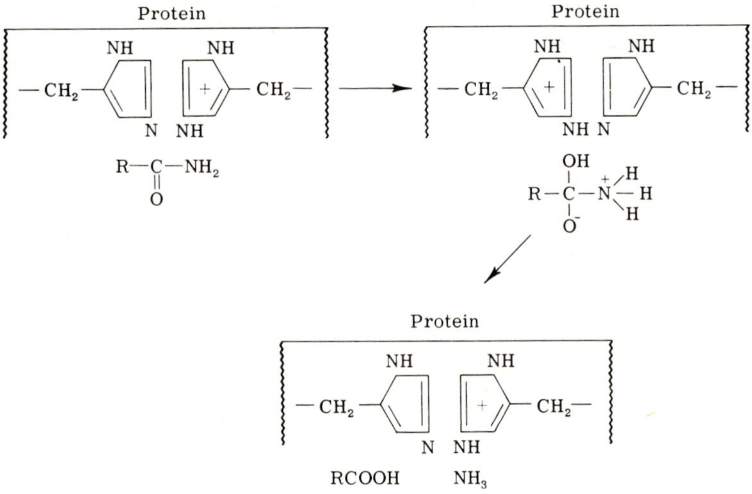

[Reaction scheme showing protein with two histidine imidazole rings (NH) attached via CH₂ groups, reacting with R-C(=O)-NH₂, proceeding through a tetrahedral intermediate R-C(OH)(O⁻)-N⁺H₂H, to give RCOOH + NH₃]

Precisely the same mechanism has been suggested to account for the enzymatic activity of ribonuclease. Concerted acid-base catalysis is brought about by two histidine residues—at positions 12 and 119 of the polypeptide chain—one of which always bears a proton. For this reason, iodoacetic acid can alkylate only one of the histidines, the nonprotonated one, at the active site:

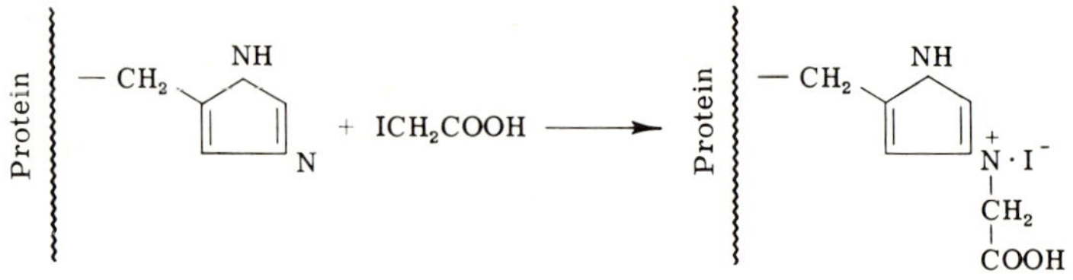

[Reaction: Protein-CH₂-imidazole (NH, N) + ICH₂COOH → Protein-CH₂-imidazole with N⁺·I⁻, CH₂, COOH]

However, if the enzyme is denatured and its active site disorganized, both of the histidines can be alkylated simultaneously. When the RNase dimer illustrated in Fig. 2-2 is reacted with iodoacetic acid, only one of the two histidines at each active site is

4. The Mechanism of Enzymatic Reactions

affected. However, since the residue which reacts can be in either position 12 or 119, 25% of the individual protein molecules will have both histidines alkylated. In fact, this was one of the experiments which supported the hypothesis that concerted acid-base catalysis, mediated by a pair of histidine residues, occurs in this class of enzymes.

Although the interpretation of enzyme mechanisms has met with considerable success, we should take note of some further subtleties. Consider, for instance, the exchange reaction: $B\text{—}X + Y = B\text{—}Y + X$, where X and Y stand for groups of atoms or molecules. If an intermediate compound is formed as in Fig. 2-11, it follows

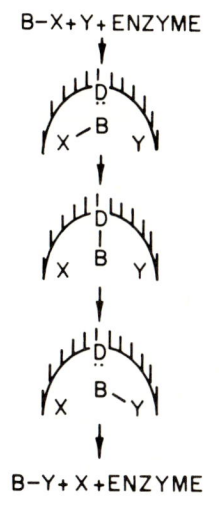

Fig. 2-11. The double-displacement mechanism for the rearrangement of electronic bonds during an enzymatic reaction. In this scheme the enzyme makes the initial attack on the substrate B—X, forming an enzyme—B intermediate; in a subsequent step the acceptor Y makes an attack on the enzyme—B intermediate to form the final product B—Y. From Koshland (24, p. 331).

that a single enzyme must transfer the group B to any acceptor Y which can be bound to the active site in a suitable manner. This situation occurs rather frequently, as in the case of papain and chymotrypsin which can transfer certain amino acids from one peptide to another. This reaction is called transpeptidation. The same enzymes also hydrolyze peptide bonds; in such cases, water serves as the acceptor for the amino acid. At first glance, this appears quite natural. Proteins are highly solvated, and since water molecules are very small, they cannot be displaced from the active site by steric hindrance; in addition, water is in great excess at a concentration of about 55.5 M. We would thus predict that water is a very likely acceptor in transfer reactions. But it has been shown that there are a large number of enzymatic reactions in which water cannot serve as an acceptor. Thus phosphorylases and phosphokinases transfer phosphate to sugars and polysaccharides by nucleophilic substitution at the hydroxyl group. The hydroxyl group of water is not an acceptor here, even though its presence at the active site of the enzyme is beyond doubt. In fact, molecules related to normal substrates, but smaller in size, are often not attacked by the corresponding enzymes. For example, the enzyme amylomaltase can hydrolyze starch by splitting off maltose molecules. But the maltose ester, methylmaltoside, is not hydrolyzed although it can

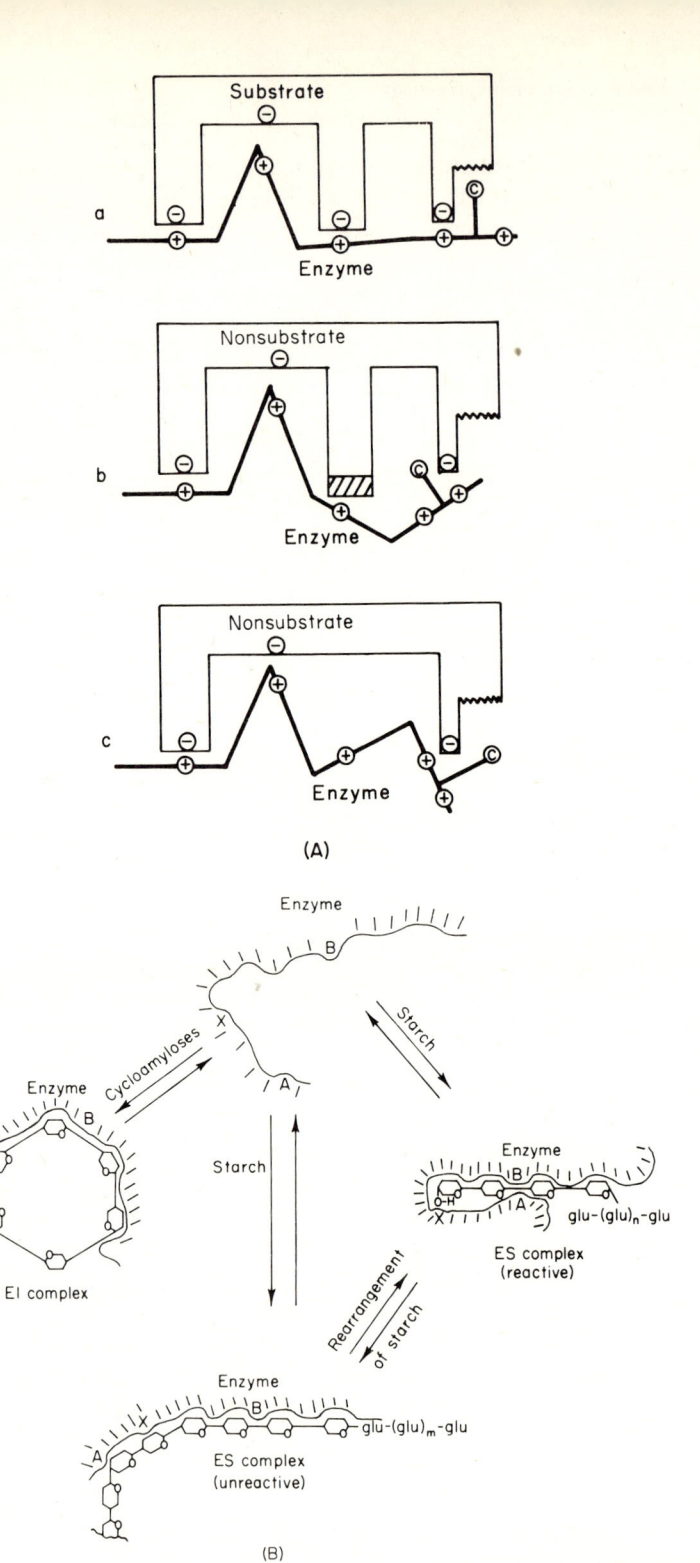

(A)

(B)

4. The Mechanism of Enzymatic Reactions

be bound to the active site. To the contrary, methylmaltoside is an inhibitor of the enzyme.

These examples, as well as many similar ones, can be given a qualitative interpretation according to the "induced-fit" theory of enzyme specificity [Koshland (24, 25); see also Linderström-Lang and Schellman (26)]. Imagine that the active site of the enzyme is flexible and that it can be modified by interaction with small molecules (Fig. 2-12). A precise fit between enzyme and substrate deforms the protein in such a way that its catalytic groups are brought into proper alignment for enzymatic action. Substrates are thought to cause the conformational changes leading to correct alignment of these groups, whereas nonsubstrates do not. The deformed enzyme tends to return to its original configuration and certain bonds in the substrate itself are strained, reducing activation barriers as a consequence.

Since affinity to the enzyme surface is only the first requirement for activity, substrate analogs may be bound without reacting if they fail to provoke the correct realignment of catalytic groups. Very small alterations in conformation may be sufficient and the flexibility of the amino acid side chains at the active site will vary from enzyme to enzyme. It is somewhat surprising that the substrate is bound to the enzyme at all under these conditions since the process necessitates deformation of the protein and thus, an expenditure of energy. The key to the puzzle is that perturbations in the protein structure must produce an increase in entropy. The entropy increase compensates the enthalpy change, so that the overall change in free energy will be negative:

$$\Delta F^\circ = \Delta H^\circ - T\Delta S^\circ < 0 \qquad (2\text{-}20)$$

From the values of enthalpy and entropy differences given in Table 2-3 we note that in most cases an entropy increase is the sole cause of enzyme-substrate formation.

Activation of the substrate is brought about by the strains that are produced in specific bonds. This is a very subtle effect which depends on (1) interactions between the substrate and several groups on the protein, (2) the distribution of these groups on the protein surface, and (3) the extent to which the protein is deformed by the substrate. This explains why small molecules such as water are often not activated by enzymes even though they are bound in significant quantities. Furthermore, the activation of

---

Fig. 2-12. Induced-fit theory of enzyme specificity. (A) In (a) the proper substrate forms a complementary structure with the enzyme so that the catalytic group C is in the appropriate position relative to the bond to be broken (jagged line). The + and − signs indicate attractive and repelling groups, charged or uncharged. When an attractive group in the substrate is too large (b) a disorientation occurs so that the catalytic group is not in juxtaposition with the bond to be broken. When an attractive group in the substrate has been deleted (c) the natural bending of the enzyme may also lead to disorientation in the catalytic groups, even though this third substance is smaller than the natural substrate. From Koshland (24). (B) Schematic illustration of an induced-fit mechanism for β-amylase, an enzyme which specifically cleaves the terminal maltosyl group from polysaccharides called amyloses which are found in starch. It is suggested that groups A and B on the enzyme must be aligned in the reactive complex for cleavage of the 1–4 acetal linkage to occur. The terminal maltosyl unit permits proper folding of the protein chain, whereas substituted substrate analogs such as the interior positions of amylose or cyclic amyloses do not allow the flexible active site to assume the conformation necessary for reaction although they contain the potentially hydrolyzable bond and are bound to the enzyme surface. From Koshland et al. (27).

chemical compounds is clearly affected by certain aspects of protein structure which are not immediately relevant to the binding of the substrate. This kinetic specificity is evident in cases where the conversion of a preexisting enzyme-substrate complex can be studied, as with the acylated trypsin or chymotrypsin considered earlier. There we saw that deacetylation of trypsin is a rather slow reaction. But the cleavage of an esterified acyl group for which this enzyme is specific, such as benzoylarginine, proceeds at a rate 1000 times greater. In the Arrhenius equation (Eq. 2-16), the decrease in activation energy is reflected in the exponential term. Equally important is the coefficient A (Eq. 2-18), however, whose large values in enzymatic reactions indicate that the activated state is characterized by a very high entropy.

Several predictions based on Koshland's hypothesis have already received experimental confirmation. In the first case, Koshland and his co-workers studied the ultraviolet absorption of phosphoglucomutase, an enzyme which transfers the phosphate group of glucose-6-phosphate to the 1-carbon and vice versa (27). When substrate is present, absorption increases in the band corresponding to peptide bonds. This partial disappearance of the hypochromic effect indicates that a certain proportion of the intramolecular hydrogen bonds are broken upon formation of the enzyme-substrate complex.

In a small number of enzyme-substrate reactions, alterations in protein tertiary structure have even been observed. One example is provided by the work of Yagi, Ozawa, and Ooi on D-amino acid oxidase (28). When this enzyme interacts with its coenzyme and substrate, there is a change in the hydrodynamic properties of the protein indicating that the initial prolate ellipsoid gathers up into a compact sphere. More recently, direct x-ray structure analysis was used to demonstrate that the binding of the substrate glycyltyrosine to carboxypeptidase leads to large conformational changes in the protein molecules. One tyrosine residue in the enzyme undergoes a displacement of 14 Å, bringing the tyrosine phenol group together with the peptide bond of the substrate (29). Although hemoglobin is not strictly an enzyme, it is interesting to note that Perutz observed a considerable change in the dimensions of this protein when it binds molecular oxygen (p. 95).

This discussion of the mechanism of enzymatic reactions can be epitomized by considering why the specific catalytic activity of enzymes is so high in comparison to ordinary catalysts. This phenomenon can apparently be explained by the fact that enzymes bear functional groups on their surface, in particular, at their active site, which are topologically and electrostatically complementary to their specific substrate molecules and are hence capable of binding both substrates in the optimal configuration relative to one another. It is precisely this positioning of substrate molecules at proper distances and in favorable spatial orientations which constitutes the primary advantage of enzymes over other kinds of catalysts.

Experiments on simple low-molecular-weight model catalysts illustrates the importance of proper orientation for concerted attack by functional groups. A classic example is the mutarotation of tetramethylglucose. The initial and determining stage of this reaction is the opening of the pyranose ring which can be catalyzed by a mixture of the weak base pyridine and the weak acid phenol. If the two functions are combined in a single molecule such as 2-hydroxypyridine, however, so that acidic and basic groups are adjacent to one another, catalytic activity is increased by as much as

7000-fold in comparison to an equivalent mixture of pyridine and phenol. The following mechanism has been suggested for the reaction (30):

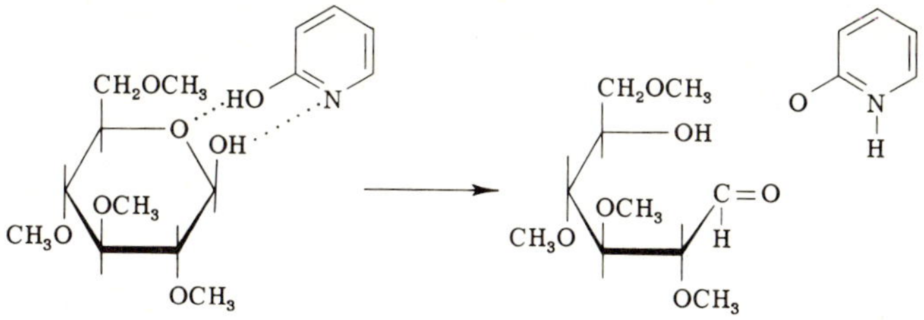

The presence of both necessary functional groups in the catalyst at the required distance from one another accelerates the reaction by four orders of magnitude. Since the binding and activation of *two* substrates occurs at the active site of an enzyme, an increase in reaction rate of $(10^4)^2$ or eight orders of magnitude is expected relative to ordinary catalysts. The mechanism discussed is therefore sufficient to account for the high rates observed in enzymatic reactions.

The following consequences ensue from this model of enzyme action. (1) There will be a high probability of effective collisions between reactants tantamount to a large increase in their concentration. (2) The mutual orientation of reacting molecules will facilitate the choice of the most advantageous reaction pathways. (3) At the points of contact, a concerted nucleophilic and electrophilic attack on the substrate will be carried out by the relevant functional groups of the enzyme. This is called concerted acid-base catalysis. (4) Covalent bonds in the substrates will be weakened owing to the redistribution or polarization of their electronic orbitals through formation of coordinate or covalent bonds with functional groups on the enzyme. (5) The necessary bonds of the substrate molecules will be deformed owing to the induced fit between them and the active site of the enzyme. All of these factors are mutually related and they all enter into the examples cited, though to varying extents.

5. Allosteric Regulation of Enzymatic Activity

A novel aspect of enzymatic function which has only recently received extensive theoretical treatment is the automatic regulation of the rate of metabolic processes. All chemical reactions in the cell must be regulated and their rates must change in accord with alterations in the external medium. Thus, reactions leading to the formation of a given metabolite must be slowed or stopped if the concentration of the metabolite reaches a level sufficient for the needs of the cell, while synthesis must be accelerated if the concentration of the metabolite drops below normal levels. Indeed, chemical mechanisms which react to changes in metabolite concentration have been discovered in a wide variety of instances.

It is a curious fact that biochemical regulatory mechanisms obey the same basic laws as do automatic systems in technology. In the first place, they are characterized by a feedback loop: the end product in a chain of biochemical reactions can exert a regulatory influence on the first link in that chain. In the second place, metabolic regulation proceeds via an oscillatory phase, another property which it shares in common with electronic or mechanical regulatory systems. To illustrate this point, let us take the example of a mechanism that must automatically keep a ship on a given course. It is clear that this task cannot be accomplished with absolute accuracy, and, as a result, the ship undergoes deviations from the desired direction of travel. The regulatory mechanism must sense these deviations and correct them with a compensatory turn of the helm. But due to overcompensation, the ship now deviates from its course to the opposite side, necessitating a new correction. This process continues indefinitely and is evidently oscillatory in nature. As the regulatory mechanism is improved, the oscillations will be damped more rapidly and their amplitude in the stationary phase will be reduced.

Chance and his co-workers developed methods for rapid and inertia-free measurements of enzymatic reaction rates which were used to study rate regulation in response to changes in substrate and inhibitor concentrations (31). From Fig. 2-13, it is apparent

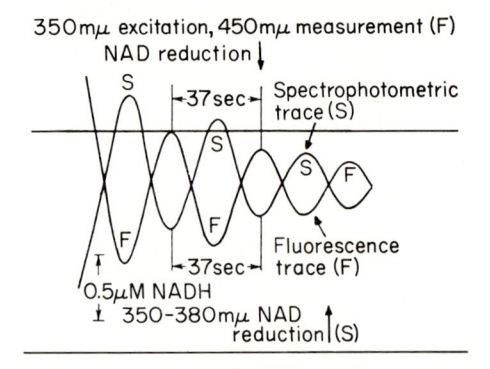

Fig. 2-13. Induction of periodic fluctuations in the level of reduced nicotinamide adenine dinucleotide (NADH) in yeast cells upon transition from aerobiosis to anaerobiosis. Glucose is added to a suspension of starved cells at a concentration of 5.5 mM. Rapid respiration ensues until dissolved oxygen is exhausted, setting the glycolytic system into a damped train of oscillations. Cytochromes remain reduced during these fluctuations in the level of cytoplasmic NADH. The concentration of NADH is measured spectrophotometrically (curve S) (350–380 mμ) and fluorimetrically (curve F) with excitation at 350 mμ and recording at 450 mμ. The spectrophotometric and fluorimetric records are plotted 180° out of phase so that the traces can be more easily compared. From Chance et al. (31).

that such changes give rise to periodic alterations in the reaction rate, the amplitude of which diminishes with time, though it does not go to zero. The analogy between metabolic and physical regulatory systems is thus well supported by this example.

In order for oscillatory regulation to be both sensitive and stable, it is necessary that the regulatory system be nonlinear. This means that changes in the regulated quantity must evoke a response which increases more rapidly than the first power of the stimulus. Such phenomena are commonly encountered in the distinctive biochemical regulatory mechanisms of cellular metabolism.

We shall now see how these basic ideas can be applied to the concrete problems of enzyme kinetics. An extremely important new form of inhibition and activation of enzymatic reactions has been recognized recently by a number of investigators (32, 33). These processes are called *allosteric* interactions to distinguish them from the *isosteric* interactions discussed earlier in which a single active site in an enzyme is affected by competition between substrate and structurally related inhibitors. In allosteric inhibition, by contrast, the inhibitor (or activator) binds to a stereospecific site on the protein molecule which is completely separate from the catalytically active site. There is neither direct competition between inhibitor (or activator) and substrate, nor is any kind of structural homology between the two substances required. This explains why the term allosteric, meaning structurally unlike, has been applied to this phenomenon. One site on the surface of a protein can influence another at some distance only if a "signal" is transmitted by the tertiary and quaternary structures of the molecule. Hence, allosteric inhibition or activation could arise from an overall deformation of the protein molecule, which, by changing the distribution of groups at the catalytic site, alters the enzyme-substrate bond.

Allosteric inhibition was first studied in bacteria where it appears to be one of the most important mechanisms for regulating the synthesis of threonine and lysine, for which aspartic acid serves as a common precursor. The biosynthetic pathways leading to the two amino acids branch soon after aspartic acid and each one individually consists of several enzymatic reactions. The first reaction in each sequence is the same, however, involving the conversion of aspartate to aspartyl-phosphate by the enzyme aspartate kinase. In *E. coli* there are two enzymes which account for most of the aspartate kinase activity, one of which is strongly inhibited by threonine, and the other by lysine. Thus the end products of the two pathways can inhibit their common first step individually and a perfectly flexible control is thereby maintained. Furthermore, there is no other amino acid capable of functioning in this way, nor are any other enzymes in the biosynthetic pathway affected by the two end products. In this case there is no structural affinity between substrate and inhibitor, for threonine and lysine have very little in common with the substrates of aspartate kinase: aspartic acid and ATP. Here we have a clear demonstration of feedback inhibition in which the consumption of a metabolic product determines whether or not the synthesis of that product will be initiated, even when all necessary enzymes and raw materials are present.

Certain allosteric proteins in bacteria have come to light as the result of mutations which alter the site of allosteric inhibition without affecting the catalytic site. Mutant proteins of this type are no longer subject to regulation, although their enzymatic properties remain unchanged. It is sometimes possible to selectively eliminate the allosteric effect *in vitro* by gentle heating or use of specific chemical agents which do not attack the catalytic site (Pardee).

Another example of an allosteric protein is phosphorylase b (Cori), an enzyme of muscle which catalyzes the phosphorolytic cleavage of glycogen to glucose-1-phosphate. There is an allosteric site in the enzyme which is activated by adenosine monophosphate (AMP) and inactivated by adenosine triphosphate (ATP). The mono- and triphosphates compete for binding at this site. The need for such regulation becomes clear when it is realized that the enzyme must cleave glycogen whenever the muscle increases its output of work, since this is a process in which ATP is consumed and AMP accumulates. Monod and his co-workers showed that adsorption of the dye, bromthymol blue, to phosphorylase b is greatly affected by changing the conditions from allosteric activation to allosteric inhibition. This finding indirectly testifies to a rearrangement in the protein macromolecular structure, and provides a basis for the allosteric regulation of phosphorylase b.

An important characteristic of many if not all allosteric proteins is that they possess quaternary structure, and usually consist of either two or four subunits per macromolecule. Indeed, in at least half of the two dozen or more examples of allostery studied, it has been proved that the protein is an oligomer of two or more identical protomer subunits. As Monod, Wyman, and Changeux demonstrated, this structural feature engenders the nonlinearity in enzyme kinetics which is essential for the establishment of a sensitive and stable system of regulation (33). It can be readily shown that for nonlinearity to apply, the enzyme particle must consist of several subunits, each of which contains at least one active site capable of binding substrate and allosteric inhibitor (or activator). Furthermore, there must be interaction between these active or allosteric sites. If the binding were to occur independently at all active sites on the protein, we would obtain simply the Langmuir adsorption isotherm. This relationship reduces to the Michaelis-Menten equation which, as we have seen, can be further simplified to a linear relationship under certain conditions.

On the basis of simple concepts in statistical thermodynamics, we shall show that nonlinear kinetic relationships characteristic of allosteric enzymes can be obtained only by assuming interactions between bound substrate, activator, and inhibitor molecules.

The conditions for equilibrium between free and enzyme-bound substrate demand that the chemical potential of the substrate in both states be equal. If there are n protein molecules per unit volume, each consisting of v subunits, there will be nv sites per unit volume capable of binding substrate, provided that each subunit contains one binding site.

In the *absence* of interactions between subunits, the chemical potential of bound substrate is not difficult to calculate. The number of different ways in which to fill nv free sites with s indistinguishable substrate molecules per unit volume (the number of "microstates" of the system) is the following:

$$W = \frac{(nv)!}{s!(nv-s)!} \tag{2-21}$$

which by means of Stirling's formula:

$$n! \approx (2\pi n)^{1/2} n^n e^{-n} \qquad n \gg 1 \tag{2-22}$$

can be reduced to:

$$W = \frac{(nv)^{1/2}(nv)^n}{(s)^{1/2}(s)^s(nv-s)^{1/2}(nv-s)^{nv-s}} \tag{2-23}$$

Now the bridge between the statistical behavior of microstates and the overall thermodynamic properties of the macroscopic system is provided by the formula relating entropy and probability:

$$S = R \ln W \tag{2-24}$$

Thus the entropy per mole associated with the distribution of states described by Eq. (2-21) is:

$$S = -R\{s \ln (s/nv) + (nv-s) \ln [(nv-s)/nv]\} \tag{2-25}$$

if insignificant terms are disregarded, and the chemical potential is obtained by differentiating the entropy with respect to s, the quantity of bound substrate:

$$\mu = \mu' - T(\partial S/\partial s) \tag{2-26}$$

From Eqs. (2-25) and (2-26) we calculate μ_1, the chemical potential of enzyme-bound substrate:

$$\mu_1 = -RT \ln[(nv - s)/s] - \Delta H \tag{2-27}$$

where ΔH is the heat of association between enzyme and substrate. The particle concentrations of enzyme and bound substrate, n and s, can be conveniently replaced by molar concentrations, since the two parameters are proportional to one another. Accordingly, we insert $v(E)_0$ for nv and (ES) for s. We may therefore rewrite Eq. (2-27):

$$\mu_1 = -RT \ln \frac{v(E)_0 - (ES)}{(ES)} - \Delta H \tag{2-28}$$

The chemical potential of free substrate in solution, μ_2, is:

$$\mu_2 = RT \ln [(S)/C] \tag{2-29}$$

where (S) is the molar concentration of free substrate and C is a constant. At equilibrium $\mu_1 = \mu_2$, and by combining Eqs. (2-28) and (2-29), we obtain:

$$(ES) = \frac{v(E)_0(S)}{(S) + Ce^{-\Delta H/RT}} \tag{2-30}$$

This expression is of the same form as the Langmuir adsorption isotherm which describes the adsorption of identical particles to a surface with independent and non-interacting binding sites. Equation (2-30) reduces to the Michaelis-Menten equation (Eq. 2-9) when we multiply through by k_3, the rate constant, since the reaction rate is proportional to the amount of enzyme-bound substrate.

In order to obtain a result which describes allosteric phenomena, we must introduce the notion of *cooperativity*, or of interactions among bound substrate molecules within a single protein oligomer. Until relatively recently, such interactions were known only in the case of hemoglobin, the protein responsible for the transport of molecular oxygen in the bloodstream. Although hemoglobin behaves as a reversible adsorbent rather than an enzyme, it nonetheless exhibits the same sort of nonlinear isotherm for substrate binding as do allosteric enzymes (Fig. 2-14). x-Ray structure analysis of hemoglobin has greatly contributed to our knowledge of cooperative changes which this protein undergoes when it reversibly binds oxygen. One hemoglobin macromolecule contains four subunits, each of which is capable of binding a molecule of oxygen, O_2. The four heme groups, to which oxygen attaches, are so distant from one another that there is no question of direct interaction between different O_2 molecules. Nonetheless, it may be demonstrated from the x-ray diffraction diagrams that the tertiary structure of hemoglobin is altered when O_2 is bound; the entire macromolecule "breathes" and its dimensions and geometry undergo a substantial change. This strongly implies that the interaction between O_2 molecules is communicated through changes in the tertiary structure of the protein. x-Ray analysis has further demonstrated that there are two stable configurations of hemoglobin distinguished by whether or not oxygen is bound.

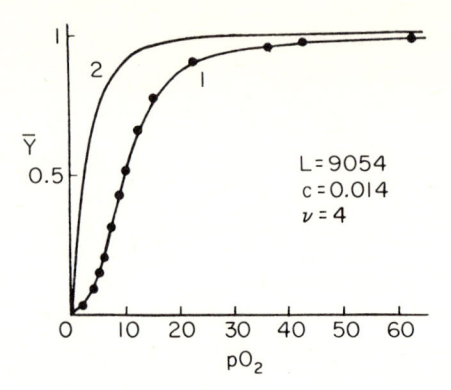

Fig. 2-14. The saturation of hemoglobin and myoglobin with oxygen. Curve 1: experimental points obtained with horse hemoglobin; solid line drawn according to Eq. (2-37) using the values of constants L and c given on the graph. Curve 2: theoretical plot for myoglobin according to Eq. (2-38). pO_2 is the partial pressure of oxygen in mm Hg. From Monod et al. (33).

Simple formulas can be deduced for the binding of substrates to this type of protein in which there are two stable structures or states. It will be assumed that the affinity of a ligand, the substrate or any other molecule that can stereospecifically interact with sites on the protein, differs in the two states. Upon transition from one state to the other, however, it is presumed that the molecular symmetry of the protein is conserved. If the number of subunits in the protein is v, it is necessary to consider separately the ligand-protein complexes containing $0, 1, 2, \ldots, m, \ldots, v$ molecules of the ligand in each of the two possible configurational states. We shall designate the first state by R and the concentration of protein molecules containing $0, 1, 2, \ldots, m$, $\ldots$ and v ligands by $(R_0), (R_1), (R_2), \ldots, (R_m), \ldots$, and (R_v), respectively. In the same manner, the second state will be designated by T and the concentration of protein in this state with $0, 1, 2, \ldots, m, \ldots$, and v ligands by $(T_0), (T_1), (T_2), \ldots, (T_m), \ldots$, and (T_v), respectively. We further assume that there is an equilibrium between T_0 and R_0, $R_0 \rightleftharpoons T_0$ such that $(T_0) = L(R_0)$, where L is an equilibrium constant called the allosteric constant. Equilibrium requires that the chemical potential of the substrate in all its possible complexes be equal to each other. Consider the equilibrium between protein molecules in the R state, which contain (m) and $(m-1)$ ligands, and the substrate, S:

$$R_{m-1} + S \rightleftharpoons R_m$$

Conditions for thermodynamic equilibrium demand that:

$$\ln \frac{(R_m)}{(R_{m-1})(S)} S_0 = -\frac{\Delta H_R - T \Delta S}{RT} \qquad (2\text{-}31)$$

in accordance with Eq. (2-19). S_0 is a constant.

The energy term ΔH_R represents the heat of association between ligand and protein in the R state. This figure is considered to be a constant so that any site in the R state is equally accessible energetically to any ligand, regardless of the number of ligands already bound. The entropy change arises from the fact that the number of permutations of (m) ligands in an oligomer of v subunits and hence of v stereospecific

5. Allosteric Regulation of Enzymatic Activity

sites, is different from the number of permutations of $(m-1)$ ligands. By analogy with Eq. (2-25):

$$\Delta S = R \ln \frac{v!}{m!(v-m)!} - R \ln \frac{v!}{(m-1)!(v-m+1)!} \tag{2-32}$$

$$= R \ln \frac{v-m+1}{m} \tag{2-33}$$

which when substituted into Eq. (2-31) yields:

$$\frac{(R_m)}{(R_{m-1})} = \frac{(v-m+1)}{m}\frac{(S)}{S_0}e^{-\Delta H_R/RT} = \alpha \frac{(v-m+1)}{m} \tag{2-34}$$

where $\alpha = (S)/S_0 \, e^{-\Delta H_R/RT} = (S)/K'_R$, the concentration of the substrate divided by the microscopic equilibrium constant K'_R which depends on the heat of association and temperature. The equilibrium for protein molecules in state T, for which the ligand is assumed to have a different affinity, can be calculated in a similar fashion:

$$(T_m)/(T_{m-1}) = \alpha'[(v-m+1)/m] \tag{2-35}$$

where $\alpha' = (S)/S_0 \, e^{-\Delta H_T/RT} = (S)/K'_T$

The fraction of sites in the protein oligomers occupied by the ligand can now be calculated:

$$\overline{Y}_S = \frac{(R_1 + 2R_2 + \cdots + vR_v) + (T_1 + 2T_2 + \cdots + vT_v)}{v[(R_0 + R_1 + \cdots + R_v) + (T_0 + T_1 + \cdots + T_v)]} \tag{2-36}$$

where the denominator is the total number of sites and the numerator is the sum of the sites occupied subject to the equilibria of Eqs. (2-34) and (2-35). If we define a further constant $c = \alpha/\alpha'$, the fraction $\overline{Y}_S$ assumes a simpler form:

$$\overline{Y}_S = \frac{Lc\alpha(1+c\alpha)^{v-1} + \alpha(1+\alpha)^{v-1}}{L(1+c\alpha)^v + (1+\alpha)^v} \tag{2-37}$$

Equation (2-37) specifies $\overline{Y}_S$, the proportion of sites occupied by substrate in a population of oligomeric proteins, each consisting of v subunits, as a function of α, a quantity proportional to substrate concentration.

The curve of $\overline{Y}_S$ vs α is S-shaped, and by a proper choice of L and c, it can be brought into conformity with the experimental data on the interaction of oxygen with hemoglobin, a tetramer for which $v = 4$ (Fig. 2-14). However, we should not be unduly impressed by the quantitative fit since we have at our disposal three empirical constants, L, c, and K'_R, and it is rather easy to vary them so as to bring about agreement between theory and experiment. The qualitative aspect is considerably more important, particularly the fact that an S-shaped binding (adsorption) curve is generated when we take into account protein quaternary structure and the interaction of individual subunits within a single oligomer.

It is reassuring that Eq. (2-37) can be reduced to the Michaelis-Menten equation under certain conditions. When both states have the same affinity for the substrate, $c = 1$ and we obtain:

$$\overline{Y}_S = \alpha/(1+\alpha) = (S)/[(S) + K'_R] \tag{2-38}$$

The characteristic S-shaped function is reduced to a Michaelian hyperbola and cooperativity disappears. The same simplification can be made when $L \ll 1$, meaning that the R state is greatly favored or, in effect, that there is only one state accessible to the substrate. In this case, all protein subunits behave independently and Eq. (2-38) can be used to describe the binding of oxygen by myoglobin which is known to be a monomer (Fig. 2-14).

The cooperativity of ligand binding is emphasized when L is large, indicating that the equilibrium is strongly in favor of T_0, and when the affinity of the R state for the ligand is large compared with that of the T state ($c \ll 1$). Equation (2-37) simplifies to:

$$\bar{Y}_S = \frac{\alpha(1+\alpha)^{\nu-1}}{L + (1+\alpha)^\nu} \qquad (2\text{-}39)$$

when the second condition is fulfilled.

For allosteric enzymes with stereospecific sites for both activator (A) and inhibitor (I), in addition to those which bind substrate, the model becomes somewhat more complicated. We shall assume, for the sake of simplicity, that substrate and allosteric activator bind only to proteins in the "active" R state and that the inhibitor has affinity only for the T state. This means that the protein must dissociate its substrate and activator before binding inhibitor when all three ligands are present in the medium simultaneously. We shall introduce the following new parameters to describe this situation:

$$\begin{aligned} K'_I &= I_0\, e^{\Delta H_I/RT} & \beta &= (I)/K'_I \\ K'_A &= A_0\, e^{\Delta H_A/RT} & \gamma &= (A)/K'_A \end{aligned} \qquad (2\text{-}40)$$

where (I) = concentration of inhibitor; (A) = concentration of activator. The equilibrium must now reflect these additional interactions, and after a derivation analogous to that for Eq. (2-39), we obtain:

$$\bar{Y}_S = \frac{\alpha(1+\alpha)^{\nu-1}}{L(1+\beta)^\nu/(1+\gamma)^\nu + (1+\alpha)^\nu} \qquad (2\text{-}41)$$

for the fractional saturation of the enzyme with substrate.

This equation gives us the degree to which the active sites of the enzyme are filled, and since $\bar{Y}_S = (ES)/(E)_0$ in the language of enzyme kinetics, it is proportional to the reaction rate according to Eq. (2-9) which states that $d(P)/dt = k_3(ES)$ when the reverse reaction—formation of active complex from product and enzyme—is not significant. According to this model of enzyme action, binding of the substrate, inhibitor, and activator as functions of concentration are often described by S-shaped curves, that is, they are all generally cooperative phenomena.

A classic example of allostery is provided by the work of Umbarger on the enzyme L-threonine deaminase. This enzyme catalyzes the first step in the reaction pathway by which isoleucine is synthesized from threonine. Of all the enzymes in the metabolic chain, only the first one—threonine deaminase—can be inhibited by isoleucine, the end product and allosteric inhibitor. Furthermore, isoleucine alone of all the intermediates is capable of serving as feedback inhibitor. Both theoretical and experimental curves are presented in Fig. 2-15 for the kinetics of L-threonine

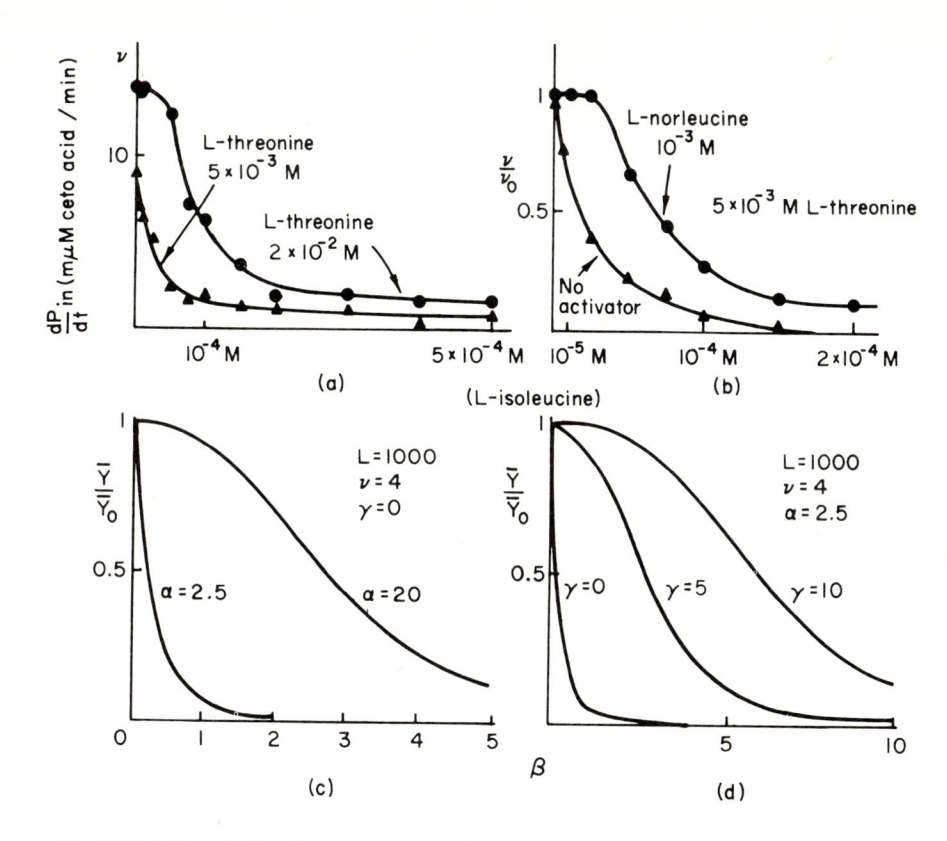

5. Allosteric Regulation of Enzymatic Activity

Fig. 2-15. Effects of the allosteric inhibitor L-isoleucine on the activity of L-threonine deaminase. (a) In the presence of two different concentrations of the substrate (L-threonine). Inhibitor (L-isoleucine) concentration plotted along the abcissa, reaction rate along the ordinate. (b) At low concentration of the substrate in the presence or absence of the allosteric activator (L-norleucine). Inhibitor concentration plotted on the abcissa, relative reaction rate on the ordinate. Compare with theoretical curves (c and d) describing similar situations according to Eq. (2-41). Note that at low substrate concentrations the cooperative effect of the inhibitor is scarcely detectable either in the experimental or in the theoretical curves. An increase in the concentration of substrate, or the addition of an activator, both reveal the cooperative effects of the inhibitor. From Monod et al. (33).

deaminase action as a function of isoleucine concentration. The inhibition curves show that there is a threshold concentration of isoleucine at which the reaction suddenly and cooperatively declines. At low substrate concentrations, cooperativity is hardly noticeable. An increase in substrate or activator (here L-norleucine) makes the allosteric effects of the inhibitor much more evident.

We should bear in mind that only those key enzymes which initiate metabolic chains or which catalyze the first reactions after branch points need be allosteric, since it is through them that entire biochemical pathways are regulated.

An interesting consequence of allosteric behavior is the opposing influence exerted on the quaternary structure of certain enzymes by substrate and inhibitor in the event that the active R state consists of oligomers and T state of inactive protomers (monomers). If the active oligomeric form binds substrate more effectively than the inactive form, then, according to the Le Chatelier-van't Hoff rule, the protein will form oligomers in the presence of substrate while in the presence of the allosteric inhibitor, the oligomer will dissociate into monomers. An example of this behavior is provided by glutamic dehydrogenase isolated from beef liver. The protein has a molecular weight of 10^6 and is capable of reversibly dissociating into four subunits. The substrate, glutamate, causes the subunits to combine into an active enzyme; inhibitors such as leucine, however, cause the protein to dissociate into subunits and enzymatic activity disappears. In all likelihood, inhibitor binding changes the tertiary structure of the subunits so as to render them unable to bind to one another tightly enough.

An unusual quaternary structure occurs in the enzyme aspartate transcarbamylase. The catalytic and allosteric sites are on two separate polypeptide chains. Each oligomer contains two large subunits, on which the catalytic capacity resides, and four small subunits which exclusively bind the feedback inhibitor. It is entirely possible that other enzymes are constructed according to the same principle.

6. Active Transport

The interior of the cell is separated from the external milieu by a membrane layer shown to be about 50 to 70 Å thick by electron microscopy. Membranes also divide the cellular cytoplasm, partitioning organelles such as the nucleus, mitochondria, and ribosomes from one another and from the endoplasmic fluid. The nuclei of higher organisms are themselves surrounded by membranes, while the structural materials of other particulates, including mitchondria, provide barriers to the free diffusion of intracellular solutes. The membranes apparently consist of identical subunits with a molecular weight on the order of 2×10^6 daltons. This observation was first made by electron microscopy (34) and subsequently confirmed by mechanical and chemical fragmentation (35). Structural members of the cell contain a considerable amount of protein and almost always contain lipids, a class of compounds that is insoluble in water. Their apparent purpose is to separate and localize the many chemical reactions which occur simultaneously within the confines of the cell, as well as to provide an osmotic barrier against the surrounding medium. Nonetheless, there is a continuous transport of certain materials across all kinds of membranes, whether internal or external (36).

Cellular transport processes are of two kinds. Diffusion plays an important role, but a biologically much more significant process is *active transport*, the transfer of ions and molecules against a concentration gradient, from regions of low concentration to regions of high concentration. This phenomenon forms the basis of cell nourishment whereby essential metabolites are absorbed from the medium, and of secretion, in which various substances are discharged into the surrounding milieu for use by other cells and tissues. Within the cell, this mechanism directs certain compounds to the nucleus, others to the mitochondria, and still others to the ribosomes.

Active transport differs in principle from normal thermal diffusion which leads to passive transport in the direction of a concentration gradient according to Fick's law. Passive transport plays a very small role in cellular processes by comparison with active transport, although it is still of importance and can be investigated with isotopically labeled compounds. The lipid membrane is distinguished by its low permeability,

6. Active Transport

and sometimes, impermeability, to water-soluble metabolites. These are precisely the substances which participate in metabolic processes for it is known that all enzymatic reactions occur in aqueous solution. In creating obstacles to passive transport or diffusion, the cell, and the organism as a whole, is defended against inhibitors and poisons and is able to exclude these substances even when they are present in the immediate surroundings. Inhibitory compounds cannot be completely excluded, however, since some diffusion occurs anyway.

One of the best known examples of active transport is the "ion pump" which admits potassium to the cell, but excludes sodium. This mechanism is operative in many types of cells, including erythrocytes, nerve cells, and the cells of muscle tissue. In muscle tissue, for instance, the intracellular concentration of potassium is close to 0.1 M while that of sodium is about 0.04 M. The concentration ratio is almost reversed in the extracellular medium where the concentration of potassium is 0.005 M and that of sodium is 0.15 M. Similar figures apply to nerve cells. In erythrocytes, the ratio of concentration is the following: $K_{int} : K_{ext} = 20-30$ and $Na_{int} : Na_{ext} = 0.05-0.1$. These differences in K and Na concentrations within and without the cell are of great importance in cellular function. In nerve cells, for instance, they give rise to potential differences vital for the transmission of impulses. Another example of active transport can be found in bacteria which can concentrate metabolites, such as sugars and amino acids, as well as certain metal ions, by a factor of 10^3 or 10^4 with respect to their surroundings. This enables bacteria to live quite well in media where the concentration of essential metabolites is on the order of 10^{-5} M.

We might suspect that chemical reactions, and in particular, enzymatic reactions, comprise the basis of active transport. In fact, a large body of purely external criteria indicate without a doubt that active transport relies on enzymatic mechanisms. The dependence of transport rates on temperature points to an activation energy on the order of 12,000 to 20,000 cal/mole, figures typical of enzymatic reactions, but too large for normal thermal diffusion. Furthermore, active transport can be inhibited by a number of metabolic poisons such as Hg^{2+}, Cd^{2+}, UO_2^{2+}, CN^-, substances which uncouple oxidative phosphorylation, and others, all known to interfere with enzyme function. Finally, the decisive argument is that active transport requires a source of energy, whether it be oxidative phosphorylation or glycolysis, or simply a store of high-energy phosphates.

Unfortunately, active transport is rather difficult to study since it does not occur in solution but in hydrated lipoprotein structures, where chemical reactions cannot be followed conveniently. For this reason, we still do not know a single case of active transport in which the detailed reaction formulas can be written down. Nonetheless, these processes are of fundamental importance and there is every expectation that with the development of molecular biology, new experimental approaches will be devised to resolve the problem.

Active transport may be described in general as the process of "chemical diffusion" (Fig. 2-16). On the external side of the membrane, the metabolite a is turned into a new compound by means of an enzymatic reaction. The new compound is both soluble and mobile in the lipoprotein which comprises the membrane. We can write the first phase of the transport process as an association reaction:

$$a + C \underset{}{\overset{\text{enzyme}}{\rightleftharpoons}} aC \qquad (2\text{-}42)$$

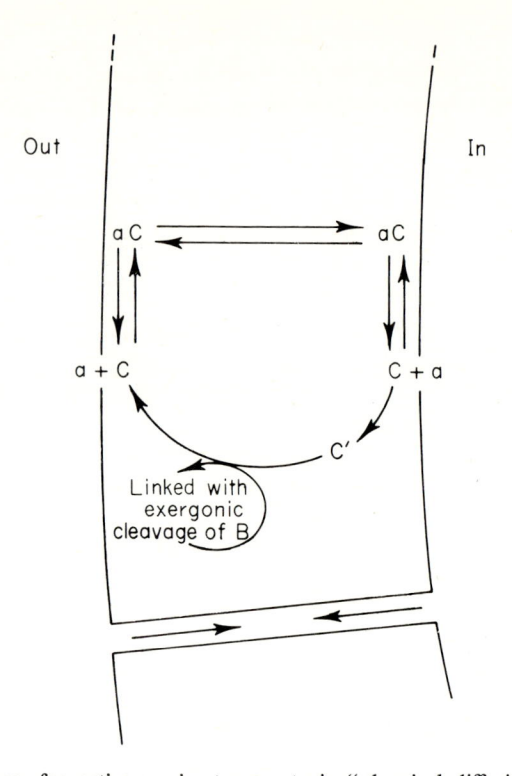

Fig. 2-16. A scheme for active carrier transport via "chemical diffusion". The substrate *a* combines reversibly with the carrier *C* to form a complex which can penetrate the cell. The free reversibility of this reaction converts *C* to *C'* thereby antagonizing the reassociation of *a* and *C*. The reconversion of *C'* to a form which can combine readily with *a* is visualized as being driven by an exergonic reaction linked to the cleavage of compound *B*. Influx of *a* by this pathway does not measure transport, however, because *exchange diffusion* will continue. In addition, some free diffusion in both directions (below) is believed to occur. From H. N. Christensen (1959). *Perspectives Biol. Med.* **2,** 228. With permission of Chicago Univ. Press (Reference 37).

The substance *aC* is considered to be soluble and to diffuse in the medium of the membrane. The molecule *C*, which we shall designate as the carrier, may be either simple or complex, as long as it confers the necessary properties on the product *aC*. On the inner surface of the membrane, *aC* is cleaved and free *a* is regenerated, this time inside the cell. The second reaction, however, cannot be simply the reverse of Eq. (2-42). If the first reaction occurs spontaneously, that is, with a decrease in free energy, cleavage of *aC* via the intermediate *C'* will require an expenditure of chemical energy. Consequently, at least one chemically active compound, *B*, must participate in the second reaction so as to provide a mobile supply of chemical energy. Let us suppose that ATP fulfils this role. Then the reaction occurring on the inside of the membrane can be represented by the following equation:

$$aC + ATP \underset{H_2O}{\overset{enzyme}{\rightleftharpoons}} a + C + ADP + H_3PO_4 \qquad (2\text{-}43)$$

This of course is a much simplified scheme, for there are undoubtedly a number of stages involving activated intermediates. The final result of the chemical processes is the transfer of metabolite *a* across the membrane. This is accompanied by the cleavage of a high-energy phosphate bond in ATP which supplies the chemical energy for work against the diffusion gradient. The transport of the metabolite under these circumstances reduces the entropy and increases the free energy of the substance at the expense of high-energy phosphate bonds. The term chemical diffusion is applied to this phenomenon because a compound has traversed a membrane normally impermeable to it as the result of chemical modification.

The equations for the transport process written above represent only one of several possibilities. Compounds other than ATP may contribute energy to the process, for instance. Moreover, the energy-dependent process need not necessarily occur on the inside of the membrane. In principle, it is quite possible that energy is expended on the outside of the membrane in order to form the complex between metabolite and carrier and that its decomposition within the cell proceeds spontaneously. All of these variants are included, however, in the general scheme of chemical diffusion.

More specifically, we would like to know the reason for the directionality of active transport, why product *a* is concentrated inside the cell even though the concentration of *a* at the internal surface is considerably greater than at the external surface. An obvious answer to this question is that diffusion-mediated flow in the membrane depends on the concentration of *aC* which is maintained at a very low level on the internal surface of the membrane. One reason for this could be the participation of an energy donor *B* which is present inside the cell, but not in the membrane nor on its outer surface.

There are some cases in which chemically similar substances can be concentrated within the cell by a kind of active transport not requiring energy. It was shown in Ehrlich ascites tumor cells that the flow of one amino acid from outside to inside is stimulated by the flow of a second amino acid from within to the external medium. During such an exchange, the flow of the first compound into the cell is approximately equal to the counterflow of the second compound outwards. A similar observation was made on the active transport of several sugars—glucose, xylose, and mannose—in erythrocytes. This phenomenon is consistent with the general scheme of active transport presented above. Insofar as the enzymatic reaction of Eq. (2-42) is reversible, an enzymatic exchange can occur at the inner surface of the membrane if there is within the cell a large concentration of a compound *a'*, similar, but not identical, to *a*:

$$aC + a' \rightleftharpoons a'C + a \qquad (2\text{-}44)$$

The opposite exchange takes place at the outer surface of the membrane:

$$a'C + a \rightleftharpoons aC + a' \qquad (2\text{-}45)$$

In sum, besides inward-directed transport regulated by the energy donor *B* there will be an additional flow of metabolite *a* at the expense of the reverse movement of the analog *a'* toward the external medium. The discovery of the exchange effect has nullified many previous schemes for active transport. The mechanism suggested by Eqs. (2-42 and 2-43) contains very few arbitrary assumptions and is likely to require modification only in detail as more and more examples are studied.

Another question with which we must deal is whether or not specialized active sites play a role in active transport. Since active transport appears to be enzymatic in nature, it is natural to expect that active sites of the type considered in the discussion of water-soluble enzymes will also be encountered here. Experimental evidence for such sites has been provided by studies on transport inhibition. Both competitive and noncompetitive inhibition have been observed, and the second type includes several instances in which specific functional groups have been irreversibly inactivated. Glycerol, for example, readily penetrates human erythrocytes by active transport against a concentration gradient. The compound 1,3-propanediol

$$\begin{array}{c} CH_2-CH_2-CH_2 \\ | \quad\quad\quad\quad | \\ OH \quad\quad\quad OH \end{array}$$

is a competitive inhibitor of glycerol transport, and kinetic studies show that it binds to the active site of the permease—or membrane-bound transfer enzyme—100 times more strongly than glycerol. Several other poisons have been found which inhibit the active transport of glycerol, such as copper ions at 10^{-7} M, mercury ions, and p-chloromercuribenzoate.

At first, it seemed as if metal ions were capable of attacking thiol (—SH) groups in the cysteine residues of the transport protein. However, it was found that the inhibitory effect could be eliminated by the addition of the amino acid histidine. Histidine cannot compete with sulfhydryl groups for mercury ions, but alone it complexes quite effectively with various metals. Therefore it was suggested that histidine must be present at the active site of the glycerol permease and this speculation proved correct. The histidine which occurs in the active site was found to be at the N-terminus of the polypeptide chain. This was established by noting that irreversible inhibition of glycerol transport occurs when the N-terminal groups of the membrane proteins are reacted with specific reagents such as fluorodinitrobenzene and phenylisothiocyanate. When the stroma of red blood cells, outer membranes of erythrocytes prepared by osmotic shock or hemolysis, are used as experimental material, N-terminal histidine residues can be detected by reaction with the above agents. Moreover, when the stroma are previously saturated with 1,3-propanediol (glycol) to block the active sites, N-terminal histidine does not react with phenylisothiocyanate. After the glycol is washed away, the N-terminal histidine residues can once again be attacked by the specific end group reagent. These experiments very convincingly demonstrate that the enzyme which carries out active transport of glycerol contains histidine at its active site and that the active site is located at the N-terminus of the polypeptide chain. In addition, this work calls attention to the experimental difficulties alluded to earlier. All the reactions involved in active transport occur within the membrane into which the transport enzymes are integrated. We still do not know with any certainty a single reaction leading to the chemical diffusion of important metabolites.

Elucidation of the nature of the active site can also be carried out according to a somewhat different approach. Derivatives of a given transportable compound can be synthesized and used to determine which changes in chemical structure affect the capacity of a certain membrane to transport that compound. Although a considerable amount of data has been accumulated in this way, it is not yet possible to draw general conclusions.

Amino acids may be divided into three groups: acidic, basic, and neutral. Within each group there is competition for unidirectional transport across membranes and compensation for counterflow. Consequently, each group appears to have its own type of active sites. While this situation is quite generally observed, there are certain cells in which the grouping of amino acids is more complicated and hence, more specific. The relationship between active transport and amino acid stereoisomers is less regular. Membranes derived from higher organisms have a much higher selectivity for L-amino acids than the membranes of bacteria or ascites tumor cells. In order for active transport to occur, the amino group must be in the α- or β-position with respect to the carboxyl group, but not in the γ-position.

At present there is very little known about the chemical nature of carrier substances. It has been shown that phospholipids participate in the transport of many substances, particularly of ions. During transport, phospholipid exchange intensifies as demonstrated by the penetration of radioactive phosphorous. It is suspected that transport systems generally consist of protein-phospholipid complexes.

As we have mentioned, active transport requires energy and it is possible to interrupt the flow of metabolites by inhibiting cellular glycolysis and respiration. The transport of amino acids and of many sugars can be stopped by specific inhibitors which interfere with oxidative phosphorylation and the formation of energy-rich phosphates such as ATP within the cell. A typical inhibitor of this type is 2,4-dinitrophenol which dramatically halts the transport of amino acids into most cells. For this reason, it is supposed that ATP and other similar compounds serve as energy donors in most but not all cases of active transport. If during active transport ATP is cleaved to ADP and orthophosphate, the membrane must contain enzymes which perform this function. In fact, adenosinetriphosphatase (ATPase) of high activity has been found in the membranes of a variety of cells, including bacteria, erythrocytes, and ascites tumor cells.

The ion pump which operates in the membranes of bacteria and mammalian cells to keep potassium in and sodium out draws its energy from ATP (38). ATP deficiency, inhibition of oxidative phosphorylation, and interruptions in cellular metabolism can all result in an equalization of ion concentrations on both sides of the membrane. When the external concentration of potassium is 5×10^{-5} M, the bacterium $E.$ $coli$ can increase its internal concentration by a factor of 4000 by active transport or ionic pumping. However, this process can occur only during exponential growth. In the stationary phase after the bacteria have stopped growing, the potassium ions exit from the cell until the concentration is equal inside and out. The flow of K^+ ions in one direction must be compensated by an equivalent flow of other cations in the opposite direction. In mammalian cells, the role of counter-ion is played by sodium, Na^+, and under normal circumstances, the inward flow of K^+ ions balances the outward flow of Na^+ ions. This rule is not observed in bacterial cells, where hydrogen ions serve as compensatory cations. *Escherichia coli* cells, for example, are known to gradually and continuously acidify the medium in which they are growing. Various other external factors such as temperature and the presence of fluoride ions exert differing influences on the rate at which Na^+ and K^+ ions are transported through bacterial membranes.

A most remarkable function of certain types of membranes is their *excitability*, or

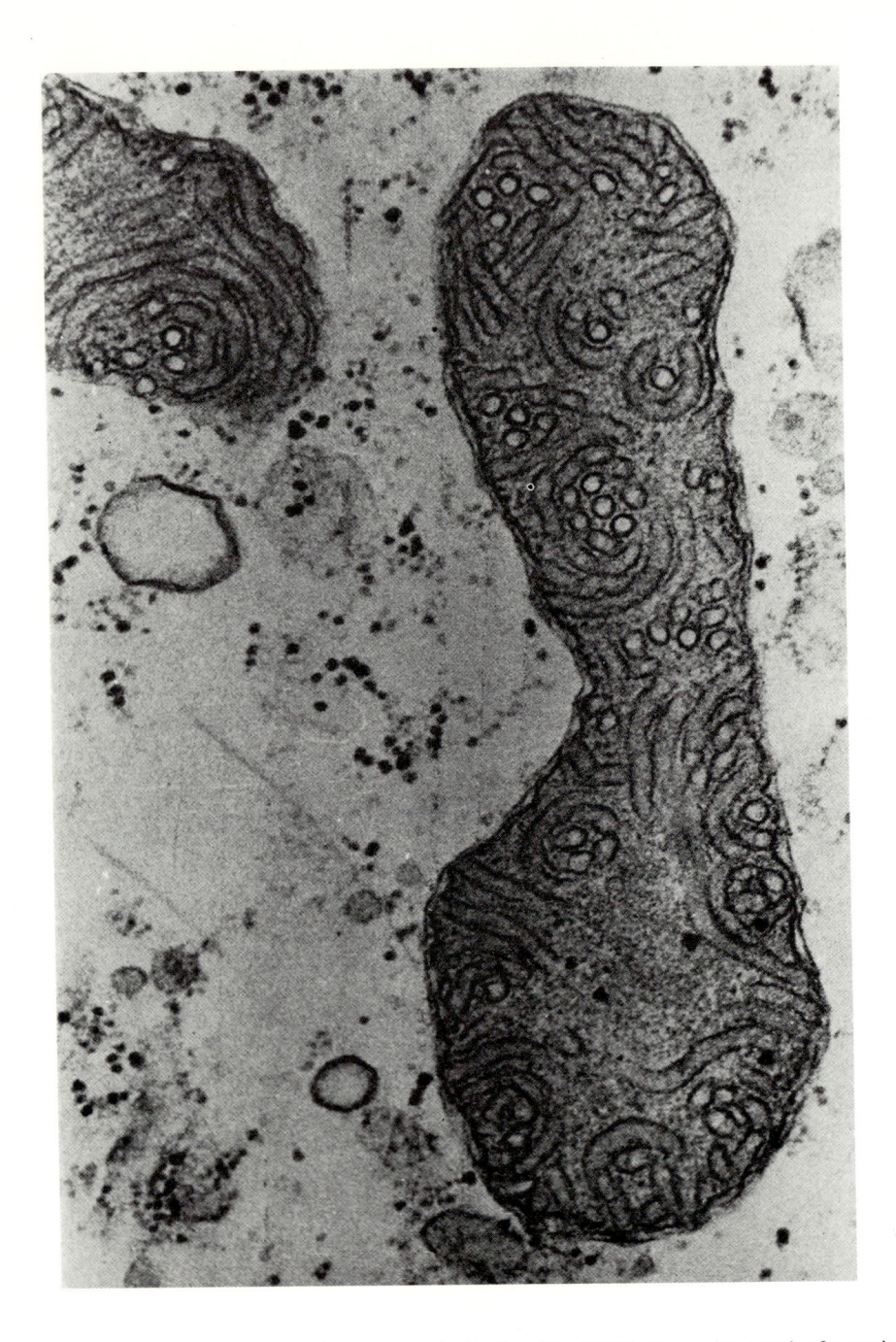

Fig. 2-17. The appearance and structure of mitochondria. (a) Electron micrograph of a section through a mitochondrion in *Paramecium caudatum*, showing arrangement of cristae, a complex system of involutions of the inner mitochondrial membranes. From Lehninger (41, p. 27).

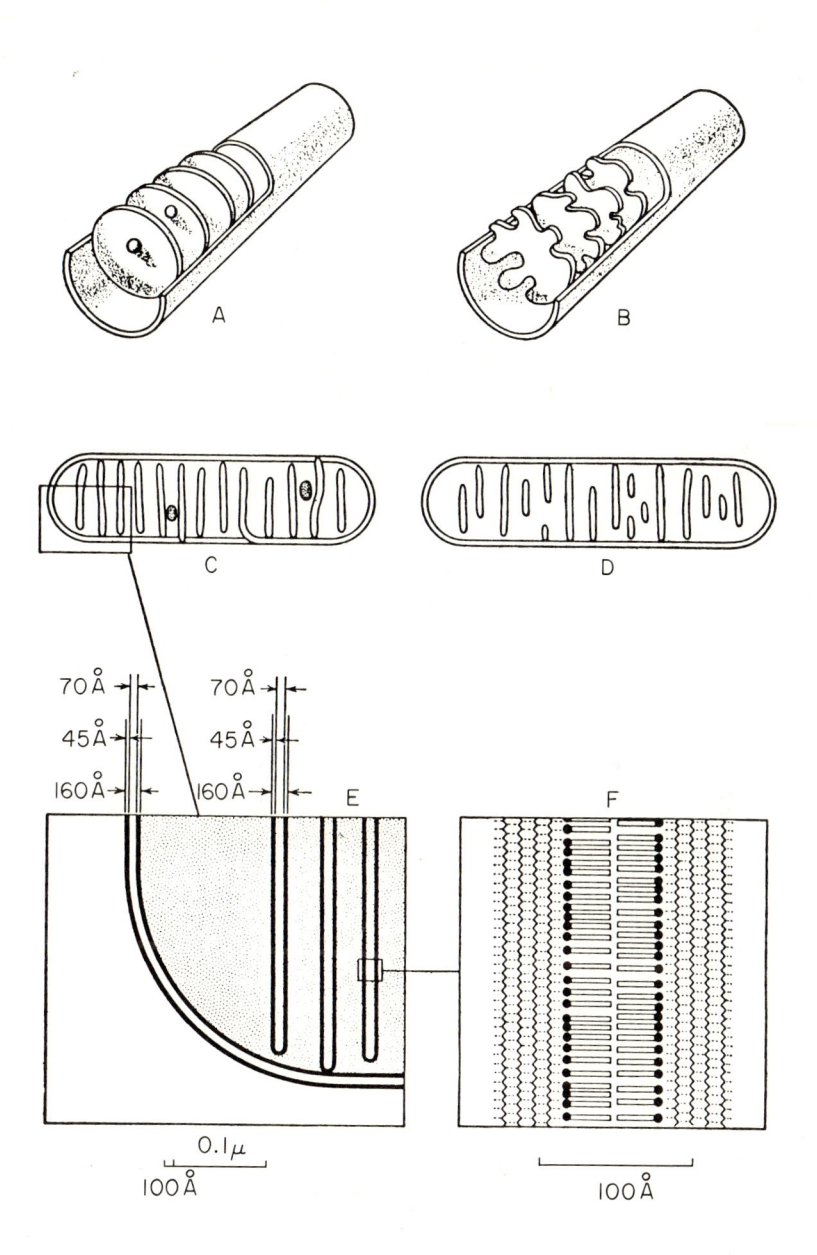

Fig. 2.17. The appearance and structure of mitochondria. (b) Schematic interpretation of the appearance of mitochondria and of the molecular architecture of the mitochondrial membranes. The three-dimensional representations A and B are based on patterns observed in sections such as C and D. E gives some commonly observed dimensions of the mitochondrial membranes and F a scheme of the proposed triple-layered structure of the membrane in which a lipid bilayer is sandwiched between two protein layers. From Sjöstrand (42).

their capacity to generate and conduct electrical impulses. The membranes of both nerve cells (neurons) and muscle cells display electrical activity. In general, this behavior results from the potential difference developed across the membrane by an unequal flow of ions into and out of the cell. In neurons, the membrane potential reaches 0.07 V with the internal surface bearing the negative charge.

The membrane thus behaves as a charged capacitor whose energy is supplied by the sodium-potassium ion pump. In specific regions of the membrane called synapses, special "transmitter" substances such as acetylcholine and noradrenalin can be secreted. When this occurs, the membrane in these regions becomes permeable not only to ions, but to water-soluble substances in general. This evokes a local electrical current and the membrane is depolarized in the vicinity of the synapse. Wherever the potential difference falls below a certain threshold (about 0.015 V), the membrane becomes permeable even in the absence of a "transmitter." An electrical signal is therefore generated and spread along the membrane with a speed of roughly 100 m/sec discharging the membrane as it goes. The molecular mechanism of this phenomenon is not yet known, although it has been suggested that the transmitter substances are allosteric effectors which cause the reaggregation of protein subunits in the neuronal membrane (39). If the membrane-bound lipids normally act as insulators, maintaining the potential difference between the inner and outer surfaces of the membrane, a conformational change in the membrane proteins might bring about a redistribution of the lipids so as to permit a flow of ions. Whatever its ultimate resolution, the problem of membrane excitability remains without doubt one of the most interesting and important in biology.

Up to this point we have considered only the permeation of outer cell membranes. However, active transport is no less important for processes taking place wholly within the cell. It has been suggested recently that transport of substances among intracellular organelles is one of the mechanisms by which metabolic processes are regulated. In fact, the principal mode in which organelles interact with each other is via active transport. Mitochondria, the particulates most extensively studied in this regard, contain the enzymes of respiration and oxidative phosphorylation which convert the chemical energy of carbohydrate metabolites to the high-energy bonds of ATP. Structurally, these particulates consist of a convoluted membrane system forming many internal folds called *cristae* which enclose a certain amount of fluid (Fig. 2-17). Their very structural material, lipoproteins, provides the separatory membranes through which the substrates of respiration, ATP and other substances, must pass by means of active transport. Mitochondria produce ATP at the expense of oxidative phosphorylation to meet the energy requirements of the cell.

Lehninger has shown that certain proteins present in mitochondria possess mechanochemical properties (40, 41). Similar to actomyosin, they contract upon interaction with ATP, and apparently enable the entire mitochondrial structure to change shape under certain conditions. When electrons are transferred from substrate to molecular oxygen along the respiratory chain, the mitochondria swell or inflate (Fig. 2-18). Thus the enzymatic processes which mediate electron transport and phosphorylation are ultimately responsible for the mechanochemical contraction and expansion of the mitochondrial particles.

These conformational changes can be followed by observing the scattering of

6. Active Transport

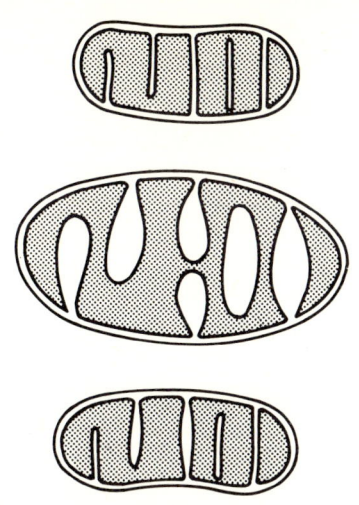

Fig. 2-18. Top drawing shows intact mitochondrion with normal density of matrix. Middle drawing shows inflation of cristae and dilation of matrix during swelling. Bottom drawing shows mitochondrion after contraction by ATP; it retains some inflation of cristae. From Lehninger (41, p. 193).

visible light (520 mμ) from mitochondrial suspensions. In order for the particles to contract, ATP is needed at a concentration of about 10^{-3} M, as are Mg^{2+} ions and a soluble contraction factor which can be washed from mitochondria if they are previously dilated in a solution of glutathione, a tripeptide containing a free thiol group. The soluble factor was separated chromatographically into three separate substances, two of which have been positively identified as the enzymes catalase and glutathione peroxidase. When they are added to a mitochondrial suspension under appropriate conditions, the mitochondria contract in volume and a portion of the internal fluid is squeezed out. This phenomenon is strongly reminiscent of muscle contraction, although mitochondrial contraction does not take place along a single axis, but inward from all points on a spheroidal surface. In this respect, it is more like an actomyosin gel than an oriented fiber. Mitochondrial contraction can be inhibited by a number of different substances. Thus, if ATP is not added to the medium, contraction must occur at the expense of ATP synthesized by the mitochondria themselves. Contraction can be eliminated in this case by cyanide or dinitrophenol which uncouple oxidative phosphorylation. These inhibitors are ineffective, however, when there is an independent supply of ATP present, although azide and various sugars at high concentrations can still inhibit contraction.

When the mitochondria swell, they are penetrated by oxidative substrates and other soluble components of the surrounding aqueous medium called the hyaloplasm. The reverse occurs upon contraction. A study of the rate at which sugars are actively transported into the mitochondria revealed that substrate penetration falls off by a large factor when the particulates are in their contracted state. Mitochondrial contraction can thereby cause a deceleration in oxidative phosphorylation and thus serves to regulate the rate of respiration by regulating permeability. Some possible molecular rearrangements leading to changes in the conformation of mitochondrial membranes are illustrated in Fig. 2-19.

(a) Changes in conformation of coupling enzymes

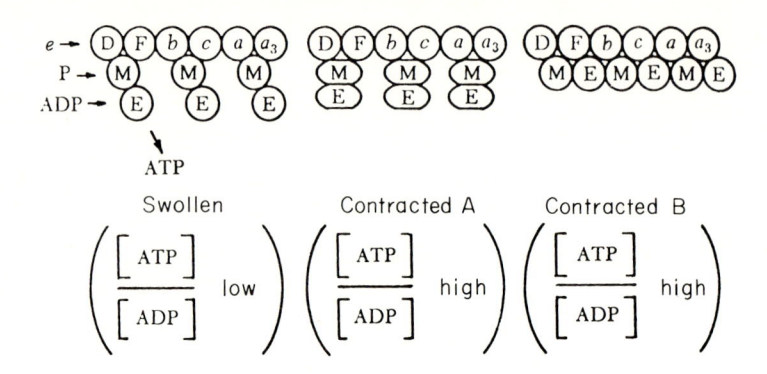

$$\left(\frac{[ATP]}{[ADP]}\text{ low}\right) \quad \left(\frac{[ATP]}{[ADP]}\text{ high}\right) \quad \left(\frac{[ATP]}{[ADP]}\text{ high}\right)$$

Swollen — Contracted A — Contracted B

(b) Changes in conformation of carrier molecules

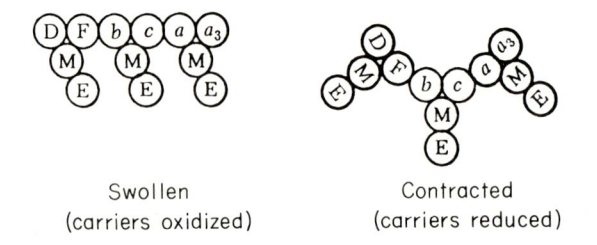

Swollen (carriers oxidized) — Contracted (carriers reduced)

(c) Changes in conformation of structure protein or contractile protein

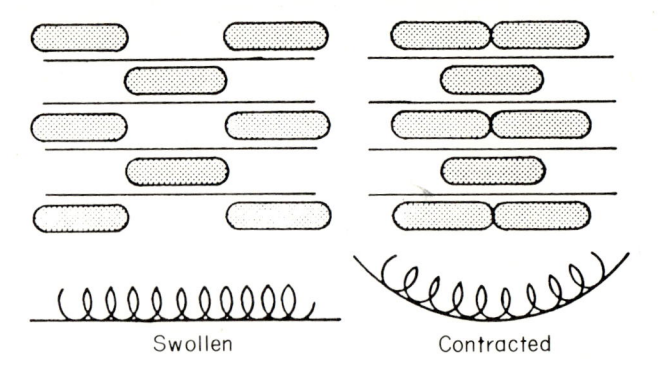

Swollen — Contracted

Fig. 2-19. Schematic representations of some possible molecular mechanisms for the active changes in dimensions or conformation of mitochondrial membranes. In (a) it is suggested that the enzymes (M) and (E) which couple phosphorylation to the electron carrier assembly, consisting of dehydrogenases (D), flavoproteins (F) and cytochromes a, a_3, b, and c, may undergo changes in conformation (A) or packing arrangement (B) as a function of the ATP/ADP ratio. In (b) the carrier molecules may undergo changes in conformation or packing arrangement as a function of oxidation-reduction state. Scheme (c) shows two representations of changes in the packing arrangement of structural or contractile proteins, such as mitochondrial actomyosin, attached laterally to the surface of the membranes. These changes might also be a function of ATP/ADP ratio. From Lehninger (41, p. 230).

Another enzymatic system which generates ATP as the result of glycolytic phosphorylation is concentrated in the hyaloplasm. The glycolytic decomposition of carbohydrates yields a smaller amount of ATP than respiration and is energetically less advantageous than the latter process. Resting cells draw energy only from respiration and the glycolytic pathway is not operative. This is called the Pasteur effect. But when the cell is growing rapidly, or performing some form of work, oxidative phosphorylation does not meet the energy needs and a supplementary generator of chemical energy, glycolysis, is switched on. Both pathways for the synthesis of ATP are therefore regulated. Although many hypotheses have been proposed to explain the regulatory mechanisms involved, they all miss their mark since they fail to account for the behavior of cellular organelles.

Neifakh and his co-workers have recently advanced a new hypothesis for the regulation of glycolysis called the membrane hypothesis (43). They showed that glycolysis as well as respiration is regulated by the reversible structural transitions undergone by mitochondria in response to changes in their own internal ATP concentration. Although all the enzymes of glycolysis are located in the hyaloplasm, they are not functional in the absence of coenzymes such as ATP, ADP, NAD (nicotinamide adenine dinucleotide), and special intensifying factors. Neifakh and colleagues found that two factors, one a nucleotide, the other a protein, are excreted from the mitochondria. Both substances were isolated and studied. It turned out that the nucleotide factor is ATP with small amounts of ADP and AMP and that the protein factor behaves as a lipoprotein. With the excretion of the nucleotide factor from the mitochondria, glycolysis begins in the hyaloplasm. When the lipoprotein appears somewhat later, it appears to have a stimulatory effect on a specific glycolytic enzyme, increasing its rate of action.

The two factors which control glycolysis are excreted into or withheld from the hyaloplasm by the mitochondrial membranes owing to their mechanochemical behavior, that is, owing to their contraction and dilation. The mitochondrial membranes, in turn, obtain information about displacements in energy metabolism from internal ATP levels upon which their conformation depends. When the ATP concentration is large, the membranes are contracted and intensifying factors are not released into the hyaloplasm; when the ATP level drops, the membranes swell, their permeability increases, and both stimulatory substances are transported into the hyaloplasm with a concomitant acceleration of glycolysis.

In this way, a special feedback loop regulates the concentration of ATP in the cell through active transport across mitochondrial membranes. When the demands for energy are great and large amounts of ATP are consumed, oxidative phosphorylation is insufficient, and the mitochondria release stimulatory factors which evoke a burst of glycolytic activity in the hyaloplasm. Additional ATP is formed as a result and its level in the cell is stabilized. Regulation of the rates of metabolic reactions achieved by means of changes in the active transport of substrates, ions, coenzymes, and enzymes is, in all likelihood, a very common mechanism in living cells.

There is considerable evidence that hormones in higher organisms act as chemical regulators of metabolic reactions and that they too sometimes function at the membrane level of organization, perhaps by influencing active transport. Insulin, for instance, is known to substantially affect the permeability of various membranes to sugars and amino acids. This is an influence on active transport, since the transport system remains highly specific and subject to the same competitive phenomena as enzyme-catalyzed reactions. The mechanism by which the hormone changes the properties of the membrane is not known.

7. The Mechanochemical Function of Proteins

One of the most important functions performed by proteins, and one which is encountered in almost all forms of living matter, is the conversion of chemical energy into mechanical work. The most highly perfected proteins of this type make up the cross-striated skeletal muscles of higher animals. But there are examples of mechanochemical activity to be found in much simpler organisms, such as the flagellae of motile bacteria, the vibratory cilia of paramecia, and the contractile tail of bacteriophages which injects the viral DNA into the host cell. Finally, the processes by which mitochondria expand and contract and by which the ribosomes move along the RNA template during protein synthesis undoubtedly reflect the same sort of physical and biochemical function. All biological movement probably has a common origin despite the structural differences between highly specialized muscle tissue and more primitive organelles found in microorganisms.

The study of muscle fiber activity has long been an important field of physiology. A major contribution to this subject was made by the work of Engelhardt and Lyubimova (44), who were the first to show that in the form of a gel, actomyosin, the fundamental contractile protein of muscle, is capable of changing its length, and thereby of performing mechanical work, in the presence of ATP. At the same time, it was demonstrated that purified actomyosin possesses an enzymatic activity capable of cleaving ATP by hydrolyzing the external phosphate group:

$$ATP + H_2O \rightarrow ADP + P_i$$

The muscle protein is therefore said to exhibit adenosinetriphosphatase or ATPase activity.

In the years following the initial investigations, many details of the mechanochemical reaction were clarified, but the fundamental significance of the earlier work endured, for it showed that the most important mechanochemical function is inherent in a specific protein isolated from muscle fibers. Since the mechanochemical act occurs on the molecular level, it is a suitable object for detailed physical and chemical investigation. This discovery also established a close relationship between mechanochemical and enzymatic reactions

In spite of a tremendous amount of data on muscular activity, there are very few unambiguous results; as a consequence, the mechanism of muscle function is still open to speculation. Since both the chemical and physical aspects of this problem remain rather obscure, we shall have to consider mechanochemical activity from a logical point of view, drawing on experimental evidence wherever possible.

The skeletal muscle of animals contains 16 to 20% protein, of which actomyosin constitutes 60%, or 12% of the total muscle weight. Muscle fibers contract when they receive a neuronal impulse and, in doing so, they perform mechanochemical work. The maximal contraction which can be achieved by a striated muscle fiber reduces its

7. The Mechanochemical Function of Proteins

length by a factor of two, although, to a first approximation, the muscle volume does not change. This means that by contracting to half its length, the muscle fiber increases in diameter by a factor of $2^{1/2}$. Work performed by a muscle is proportional to its mass and is equal to 5×10^5 to 10^6 ergs/gm. After the contractile phase, the muscle enters a phase of relaxation, upon completion of which the entire work cycle can be repeated. If the muscle remains in the contracted state for a certain period of time, energy is continuously expended, although additional mechanical work is not produced. Consequently, the relaxed state of skeletal muscle is a state of rest in which work is not performed and energy is not expended. Besides striated muscle, there is another type of tissue called smooth muscle in which the contracted or flexed state may be the state of rest. Although the smooth muscles of mollusks which close the shell and hold it together are under tension for extended periods of time, energy is not expended.

Early attempts to determine the source of chemical energy in muscles provoked a long debate. Considerable evidence supported the view that ATP was the sole supplier of energy to the reactions leading to contraction. But measurement of the decrease in ATP during periods of activity was difficult, since muscle contains several substances which bear phosphate groups characterized by a large free energy of hydrolysis. Furthermore, there are enzymes present which transfer phosphate groups from one compound to another, e.g., from creatine phosphate to ADP. Therefore the consumption of ATP is partially masked by its concurrent resynthesis. Nonetheless, from the specific poisoning of transphosphorylation in muscle fibers by fluorodinitrobenzene, we now know that ATP is in fact the unique energy substrate of muscle protein. During a cycle of contraction and relaxation, the external phosphate group of ATP molecules are hydrolyzed according to the reaction:

$$ATP + H_2O \rightarrow ADP + P_i$$

The free energy of ATP decreases in this process by 10,500 cal/mole. If the efficiency of energy transfer were 100%, the muscle would be able to perform an equivalent amount of mechanical work per mole ATP hydrolyzed. But since this is unlikely thermodynamically, we must actually measure the efficiency by estimating both the mechanical work performed and the heat produced during one cycle of activity.

Extemely delicate measurements of this type were carried out by A. V. Hill (Fig. 2-20), who determined that chemical energy is transformed to mechanical work with an efficiency of 50% in muscle tissue. This figure is not very accurate, since, as we have remarked there are, in addition to resynthesis of ATP, many other processes occurring simultaneously in muscle, including respiration and glycolysis, all of which produce a certain quantity of heat. We shall, however, attempt to isolate the basic mechanochemical reaction from other processes occurring at the same time.

Approximately half of the total heat released is produced during the contractile phase (45). We must consider the heat produced in an entire cycle of activity, however, as for any cyclic machine. A substantial amount of heat is released in the relaxation phase if mechanical work was produced in the contractile phase. If the efficiency of our mechanochemical machine is 50%, the hydrolysis of 1 mole of ATP to ADP will make 5000 cal or 2×10^{11} ergs available for work. As we have seen, the muscle produces 5×10^5 to 10^6 ergs/gm of tissue, or 5×10^6 to 10^7 ergs/gm of actomyosin. Now the molecular weight of myosin is 500,000; therefore, 1 mole of myosin produces

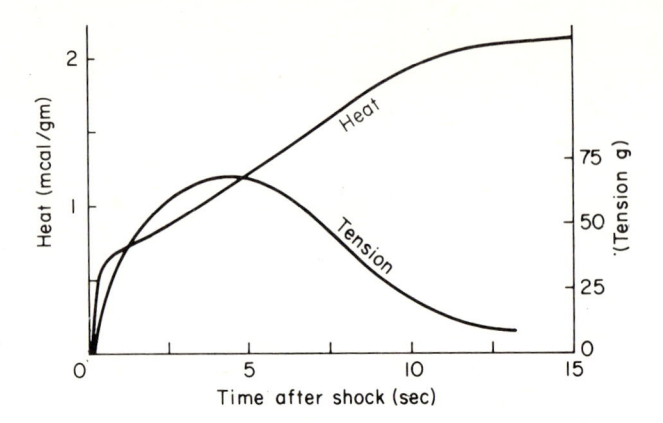

Fig. 2-20. Tension and heat production in tortoise muscle after stimulation under isometric conditions at 0°C. The heat has been corrected for heat loss. Note that heat production begins appreciably before the mechanical response. From Hill (45).

from 3 to 7.5 × 10^{12} ergs of mechanical work. This means that 15 to 30 moles of ATP are expended per mole of protein during one cycle of activity.

By using an inhibitor which prevents the transfer of high-energy phosphate groups from creatine phosphate to ADP, it is possible to directly measure the consumption of ATP in one cycle of muscle contraction. The figures obtained are in good agreement with the estimates based on considerations of efficiency.

Details of the biochemical processes involved in muscle contraction have still not received a complete explanation. It is clear that the equation ATP + H$_2$O → ADP + P$_i$ represents only the overall reaction, providing no insights into the complex events which contribute to the phases of contraction and relaxation. If the process were really described in its entirety by the hydrolysis of ATP, the only result would be the diffusion of energy in the form of heat. This clearly does not occur. In the first phase, ATP participates in a certain reaction with muscle protein, which leads to the performance of mechanical work and a partial conversion of chemical energy to heat. In the second phase of the process, relaxation, a phosphate group is converted to free orthophosphoric acid accompanied by an additional dissipation of thermal energy. The nature of the first reaction is presently unknown, but there is good reason to think that it includes the transfer of a phosphate ester from ATP to the protein, as first suggested by Kalckar in 1941 (46). The hypothetical scheme of the reaction is as follows:

(1)   ATP + protein ⇌ protein-P + ADP   Contractile phase
                                           Work produced
                                           Heat released

(2-45)

(2)   Protein-P + H$_2$O → protein + P$_i$ Relaxation phase
                                               No work produced
                                               Heat released

for which the overall reaction is the familiar ATP + H$_2$O → ADP + P$_i$. This scheme of energy transfer has not yet received reliable confirmation, although it seems logically unavoidable.

7. The Mechanochemical Function of Proteins

Some arguments in favor of the model of muscle function considered above have been deduced from the effects of a number of metabolic poisons on myosin, and by the action of denaturing agents such as urea and formamide. Far from suppressing the ATPase activity of myosin, these substances tend to intensify the cleavage of ATP, but with the following difference: they all apparently uncouple hydrolysis of the phosphate group from phosphorylation of the protein, and thereby prevent the production of mechanical work. The same effect is achieved by substituting Ca^{2+} ions for Mg^{2+} ions at a concentration of about 10^{-3} to 10^{-4} M in the reaction mixture. Effective inhibitors of myosin include *p*-chloromercuribenzoate, which selectively attacks —SH groups, *p*-nitrophenol

$$HO-\langle\bigcirc\rangle-NO_2$$

and *p*-nitrothiophenol

$$HS-\langle\bigcirc\rangle-NO_2$$

The last two compounds are well known in biochemistry as uncouplers of biological oxidation and phosphorylation, and their activity with respect to myosin is not at all surprising.

The physical mechanism of mechanochemical processes are of great interest. In particular, we would like to know how the mechanochemical "machine" turns chemical energy, concentrated in individual covalent bonds, into tensile energy dispersed throughout a large volume at relatively low density. The only type of molecular forces with sufficiently long-range effects to cause the conversion of chemical energy to mechanical work are electrostatic Coulomb forces among charged groups. Proteins, of course, may be regarded as macromolecular electrolytes since many side groups of their constituent amino acids carry charges. In fact, the most logical explanation of actomyosin contraction under the influence of ATP would be to attribute conformational changes in the protein macromolecule to alterations in the balance of electrostatic charges. Polymeric electrolytes of much simpler structure than proteins are easier to deal with in this regard, and by using such substances as model systems it is possible to study and come to an understanding of mechanochemical reactions. The pioneering work in this field was performed by Katchalsky and Kuhn (47).

The behavior of polymeric electrolytes is governed to a large extent by the electrostatic energy of repulsion among ionic groups bound to the polymer chain. Because of mutual electrostatic repulsion, polyelectrolyte molecules assume an extended conformation in solution which greatly increases the viscosity. If individual macromolecules are joined together through chemical bridges, the polymer will form a gel enclosing considerable amounts of solvent. It is possible to choose for this purpose a polyelectrolyte whose net charge can be altered at will; thus, if the polymer contains weakly acidic side chains, the degree of their ionization can be readily changed by raising or lowering the pH. A gel made up of such material will swell when the molecules bear a large net charge and its volume will increase. If the charges are neutralized, the electrostatic forces diminish and as a result of random thermal motion, the macromolecules will tend to contract, displacing solvent from their interstices, and bringing about a reduction in gel volume (syneresis). Since it is considered to be isotropic the gel is expected to expand and contract proportionally in all directions. Even such a simple gel as the one described is capable of giving rise to mechanical forces upon contracting which can be constrained to do work. An extremely anisotropic structure, such as an oriented fiber, can produce work much more effectively than an unoriented one. Kuhn and Katchalsky used these principles to develop a graphic model of muscle fiber consisting of a periodic mechanochemical machine. Chemical energy was supplied to the machine by altering the dissociation of a weakly acidic polyelectrolyte, that is, by promoting the recombination of carboxyl groups with protons according to the reaction: $-COO^- + H^+ \rightarrow -COOH$.

Soon after this model was proposed, many other attempts were made to approximate muscle contraction with systems containing synthetic polymers. The Kuhn-Katchalsky model consists of fine oriented threads of a special copolymer made up of acrylic acid (20 mole %) and vinyl alcohol, which has the following structural formula:

$$-CH_2-\underset{OH}{CH}-CH_2-\underset{COOH}{CH}-CH_2-\underset{OH}{CH}-CH_2-\underset{OH}{CH}-CH_2-\underset{COOH}{CH}-$$

Under tension, the fiber is "vulcanized" with glycerol by means of a short period of heating (30 seconds) at 120°C. During this treatment, occasional ester bonds are formed between neighboring parallel chains. The polymer also loses its ability to dissolve in water because the molecules have been linked together in a three-dimensional network. But the fiber is still permeable to water, a condition necessary for diffusion of the reagent to all carboxyl groups located within the structure. At pH 7, the carboxyl groups are all ionized, but they are completely discharged when the fiber is immersed in dilute HCl (pH 2). After the last treatment, the fiber contracts to roughly one half its original length and performs work amounting to about 2.5×10^6 ergs/gm. This is approximately the same as the output of work by muscle fibers. Upon contraction, the synthetic fiber becomes thicker, preserving its initial volume, which is another property shared in common with muscle. The mechanism of the synthetic model is quite clear. Neutralization of the carboxyl ions leaves the polymer chains uncharged and eliminates the repulsive forces among negative charges along the entire length of the fiber. This in turn disturbs the equilibrium among the various kinds of molecular forces and leads to contraction of the fiber as the oriented chains tend to fold up into a new equilibrium configuration. As is the case with rubber bands, contraction in the synthetic model system is a direct consequence of the entropy difference between the stretched and contracted configurations of the polymer chains. As we shall see, the model is not adequate to explain muscle function in this respect.

Anistropy of the fiber is important because it means that the distance between charged carboxyl groups along the fiber axis is different from their average spacing perpendicular to the axis. Owing to the orientation of macromolecules in the gel, the distances between charges along the fiber are much less and the forces of repulsion act chiefly in a direction parallel to the axis. When the electrostatic forces are liquidated the fiber contracts only along its axis and it becomes thicker, expanding perpendicular

7. The Mechanochemical Function of Proteins

to the axis. If this anisotropy in the distribution of charges did not exist, the fiber could only contract isodimensionally, that is, proportionally in all directions.

Imagine now a cyclic mechanochemical machine that operates on the principle of successive ionization and neutralization of carboxyl groups by base and acid, respectively. The working cycle of the machine is illustrated in Fig. 2-21. Fiber contraction

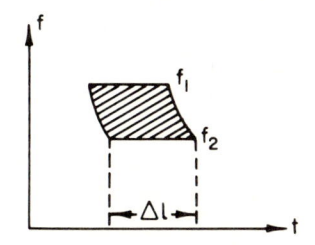

Fig. 2-21. The work cycle of a mechanochemical machine.

occurs under the load f_1, and relaxation takes place under the lesser tension f_2. The work performed in one cycle will be $(f_1 - f_2)\Delta l = f \Delta l$, where Δl represents the change in length of the fiber. At this point we ask where, exactly, does the work come from? At first glance, it would appear that the number of carboxyl groups is fixed and that the process by which they are ionized and neutralized is completely reversible. However, from arguments already introduced, it is clear that in a polyelectrolyte under mechanical tension (the force, f), the dissociation constant of the carboxyl groups will depend on the tension and in the final analysis, it will be determined along with other factors by the mechanical work performed by the fiber. From thermodynamics, we know that the equilibrium dissociation of carboxyl groups occurs when variations in the chemical potentials are equal to zero, $\Delta(H - TS + fl) = 0$, or when $\Delta(H - TS) = -f \Delta l$. The change in dissociation constant K with respect to K_0, the constant applying to the unstretched fiber, follows from the relationship:

$$\ln (K/K_0) = f(\Delta l/RT) \qquad (2\text{-}46)$$

We note that the energy loss in this machine under ideal conditions can take place only as the result of irreversible phenomena. These might arise from the fact that real processes must be carried out not under infinitely small, but at finite, changes in concentration.

The Kuhn-Katchalsky model embodies the basic features of a mechanochemical system, but it does not account for the many complexities that actual muscle proteins exhibit. In this section, we are trying to avoid most complications so as to concentrate attention on the fundamental aspects of mechanochemical phenomena. Hence, actomyosin gels of the type used by Engelhardt and Lyubimova (44); and by Szent-Györgyi (48) are a simpler and more suitable topic for our discussion than native muscle tissue. Actomyosin gels, however, lack the high degree of anisotropy possessed by native fibers. As a result, their contraction in the presence of ATP proceeds proportionally in all directions, thereby displacing some of the interstitial solvent.

The major contractile protein of muscle is actomyosin, or myosin B, actually a copolymer made up of two different kinds of polypeptide chains, myosin and actin. Actomyosin is usually prepared by extracting muscle homogenates with 0.5 to 1.0 M

KCl over a period of 24 hours. At the high ionic strengths used, this protein yields extremely viscous solutions, which testify to the presence of large elongated particles. If the high concentration salt solutions are replaced by water, actomyosin forms a gel, and if a fine jet of the actomyosin solution is introduced into a vessel of water by means of a capillary, strands of actomyosin gel are formed. These strands will contract if the vessel also contains ATP (5×10^{-3} M), KCl (5×10^{-2} M), and Mg^{2+} (10^{-4} M) at pH 7.5. Potassium and magnesium ions are essential to the function of muscle tissue, and they are required also for the contraction of actomyosin gels.

If the extraction of the muscle suspension is interrupted after a short time, on the order of 10 minutes, the solution will contain mainly myosin with a small amount of actomyosin. When the solution is diluted to 0.3 M KCl, actomyosin precipitates and almost pure myosin can be recovered from the soluble phase. In contrast to actomyosin, myosin solutions are much less viscous, consistent with the fact that myosin has a lower molecular weight than the complex. Pure myosin also forms a gel-like substance when introduced into water, but strands of this material are not able to contract in the presence of ATP and Mg^{2+} ions. Consequently, it is evident that the contractile protein of muscle is the actomyosin complex (myosin B) and not myosin, sometimes called myosin A, alone. The weight ratio of myosin to actin in actomyosin is approximately 3:1 and from the molecular weights, it turns out that there are about two molecules of actin for every molecule of myosin. In order to obtain pure actin, most of the myosin is first removed from the muscle homogenate by extracting with 0.6 M KCl at pH 8 to 9. The myosin still remaining bound to the tissue is then denatured with acetone and actin can be extracted with water.

Actin is a globular protein with a sedimentation constant of 4 S and a molecular weight of 70,000. It apparently contains a prosthetic group consisting of a nucleotide, probably adenylic acid. Myosin A is also a globular protein, but its structure is highly asymmetric. It has a sedimentation coefficient of 6.1 S, a molecular weight of 500,000, and an intrinsic viscosity of 2.2. The axial ratio of myosin molecules is $a/b = 60$, which indicates a rod-shaped configuration with an overall length of 1500 to 1700 Å.

Actin alone spontaneously aggregates in 0.6 M KCl and 5×10^{-4} M $MgCl_2$. This process does not appear to disrupt the secondary or tertiary structure of the protein and the long helical chains of actin formed are held together by localized interactions among groups on the molecular surfaces. If actin and myosin are mixed under the above ionic conditions, they jointly aggregate to form even longer chains than actin alone. The resulting product exhibits sedimentation constants up to 40 S and molecular weights reaching well into the millions. Since many properties of the recombined complexes are similar to those of native actomyosin, it is commonly thought that their structures are identical. In water, reconstituted actomyosin forms gel fibers which contract in the presence of ATP and Mg^{2+}, just like native actomyosin. The intrinsic viscosity of reaggregated actomyosin is very large, on the order of 6, which indicates an axial ratio of 130 and a very high degree of asymmetry. Although this figure is only approximate, it shows that when actin and myosin polymerize, the resulting particles grow longer and thicker, simultaneously.

Morales has formulated a set of basic principles to explain muscle action as a polyelectrolytic phenomenon (49). Along with all proteins, actomyosin bears charged groups, but in this case, the isoelectric point strongly depends on the presence of

Mg²⁺ ions. At low ionic strengths, the isoelectric point of actomyosin jumps from pH 5.5 to pH 9 when small amounts of Mg²⁺ (10^{-4} to 10^{-3} M) are added to the solution. This indicates that the actomyosin gel carries a rather strong positive charge. Morales suggested that ATP binds to positively charged proteins until they are completely neutralized, causing them to contract due to thermal motion or entropy changes. In this regard, the theory of Morales is quite similar to that of Kuhn and Katchalsky, though it contains several assumptions that are incorrect and require reconsideration.

Entropy effects play a secondary role in muscle contraction, for the protein fiber is fundamentally different from an elastic fiber. If the work performed by a muscle fiber were to result from entropy changes, the temperature of the system would drop during the contractile phase, and the loss of heat would be equal to the work produced. But as Fig. 2-20 illustrates, muscle fibers heat up slightly when they contract, and the amount of heat released is considerably less than the work done. An additional amount of heat is evolved during relaxation. All of the evidence indicates that entropy is not an important factor in actomyosin function.

Furthermore, we know that the configuration of a protein molecule results from an equilibrium among many different types of forces. Protein secondary structure arises from hydrogen bonds, while tertiary structure depends on attractive forces between hydrocarbon side groups and repulsive forces between electrostatically charged side groups. The conditions for internal equilibrium require that energy be minimized or, stated very crudely, that the forces of attraction and repulsion be equal to one another. Therefore, it is quite natural to suppose that when electrostatic groups on a protein are neutralized, its energy also changes since the repulsive forces which stretch the molecule into a rod disappear. As a consequence, the actomyosin particle contracts in length though its width increases. The mechanical work is done at the expense of the energy of attraction between side groups of the polypeptide chain. In this case, the work will be equal to the drop in potential energy; the evolution of heat is not very substantial during the contractile phase.

Another point in the theory that must be improved is the suggestion that ATP adsorption is responsible for neutralizing the charges and, therefore, for the production of mechanical work. In reality, the work is done at the expense of chemical energy so that there must be a chemical reaction coupled with work production. This reaction is presumed to involve the transfer of the high-energy phosphate group to the protein.

It is possible to estimate the electrostatic forces of repulsion in muscle on the basis of the number of ATP molecules expended in the contraction of one mole of protein. We shall consider a bundle of muscle fibers whose volume is 1 cm³. The tensile force exerted on these fibers as a result of electrostatic interactions can be calculated as follows. In 1 cm³ of muscle, there is present 12% or about 0.12 gm of myosin, or $0.12/(5 \times 10^5) = 2 \times 10^{-7}$ moles. The number of phosphate groups reacting with one molecule of protein during the contractile phase is 30, so for 1 cm³ of muscle, this figure is $30 \times 2 \times 10^{-7} = 6 \times 10^{-6}$ moles or $(6 \times 10^{-6})(6 \times 10^{23}) = 3.6 \times 10^{18}$ phosphate groups. Assuming an equal distribution of charged groups throughout the muscle tissue, an average distance between them will be:

$$(3.6 \times 10^{18})^{-1/3} = (1.5 \times 10^6)^{-1} = 6.5 \times 10^{-7} \text{ cm}$$

But muscle fibers are fundamentally anisotropic, and we must introduce this condition into the distribution of charges. This has to be done somewhat arbitrarily since the true distribution of charged groups within the protein is not definitely known. As an order of magnitude estimate, we will say that the density of charges along the axis is 10 times greater than that perpendicular to it (Fig. 2-22). In order to satisfy this condition,

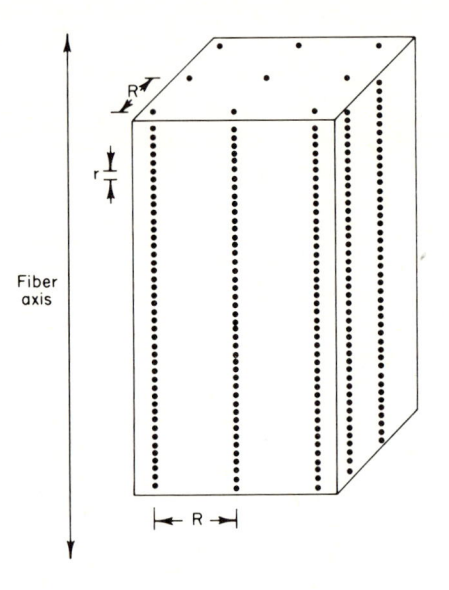

Fig. 2-22. Scheme of the anisotropic distribution of charged groups in an oriented fiber of muscle protein.

the distance along the fiber must be reduced by a factor of $10^{2/3} \approx 5$. Then the intercharge distances will be $r = 1.3 \times 10^{-7}$ cm along the fiber and $R = 1.3 \times 10^{-6}$ cm normal to it.

Now we can evaluate the force of repulsion and the tension that it produces:

$$F = (k\varepsilon^2/Dr^2)v = (2\varepsilon^2/Dr^2) \times (1/R^2) \tag{2-47}$$

where r and R have already been defined, ε is the charge on a phosphate group and equal to 2 electronic charges, D is the dielectric permeability of water, v is the number of

7. The Mechanochemical Function of Proteins

ionized protein chains per cm², and k is a coefficient with a value of about 2 which accounts for the additional interaction of nonadjacent ions. Substituting the appropriate figures, we obtain:

$$F = \frac{(2)(4)(2.5 \times 10^{-19})(10^{-6})}{(80)(1.6 \times 10^{-14})(1.6 \times 10^{-12})} = 1.2 \text{ kg/cm}^2$$

The calculated value is in agreement with the actual figure for the tension arising in muscle fibers and it shows that electrostatic interactions between charged groups in the protein can satisfactorily account for the work performed by muscle fibers.

One further important characteristic is that the heat released upon final hydrolysis of the phosphate groups from the protein during relaxation is considerably less than the heat of hydrolysis of ATP, since mechanical work is accomplished at the expense of this energy source.

All of the theoretical reasoning that we have performed indicates that the contraction of actomyosin gels takes place because of conformational changes in their component protein molecules. However, the native structure of the actomyosin polymer and, in particular, the arrangement of actomyosin strands in the muscle fiber, are still disputed. From electron microscopic studies of muscle fibers, Hanson and Huxley (50) have suggested that myosin and actin are polymerized separately into bundles of strands which mutually penetrate one another (Fig. 2-23). Upon contraction the thicker strands, believed to be myosin, and the finer actin strands slip along one another in the longitudinal direction, but draw apart slightly in the transverse plane. Huxley has attempted to explain the process of contraction in terms of mutual slippage between more or less rigid fibers. While these general traits of muscle structure are widely accepted it is still difficult to escape the notion that conformational changes in proteins must occur at some stage of the contractile process.

A successful attempt to use x-ray structure analysis in the study of contraction was carried out by Astbury, Beighton, and Weibull, who demonstrated that the contractile reaction in bacterial flagellae is accompanied by a conformational transition in the protein. The transition is characteristic of a change from the α-helical structure to the β-pleated sheet, since deep folds develop normal to the direction of contraction (52). It would be important to confirm this observation in the case of other contractile fibers.

In summary, it is apparent that the morphology and dynamics of contractile proteins must be subjected to much additional investigation before the details of mechanochemical processes in muscle and in other biological structures can be understood.

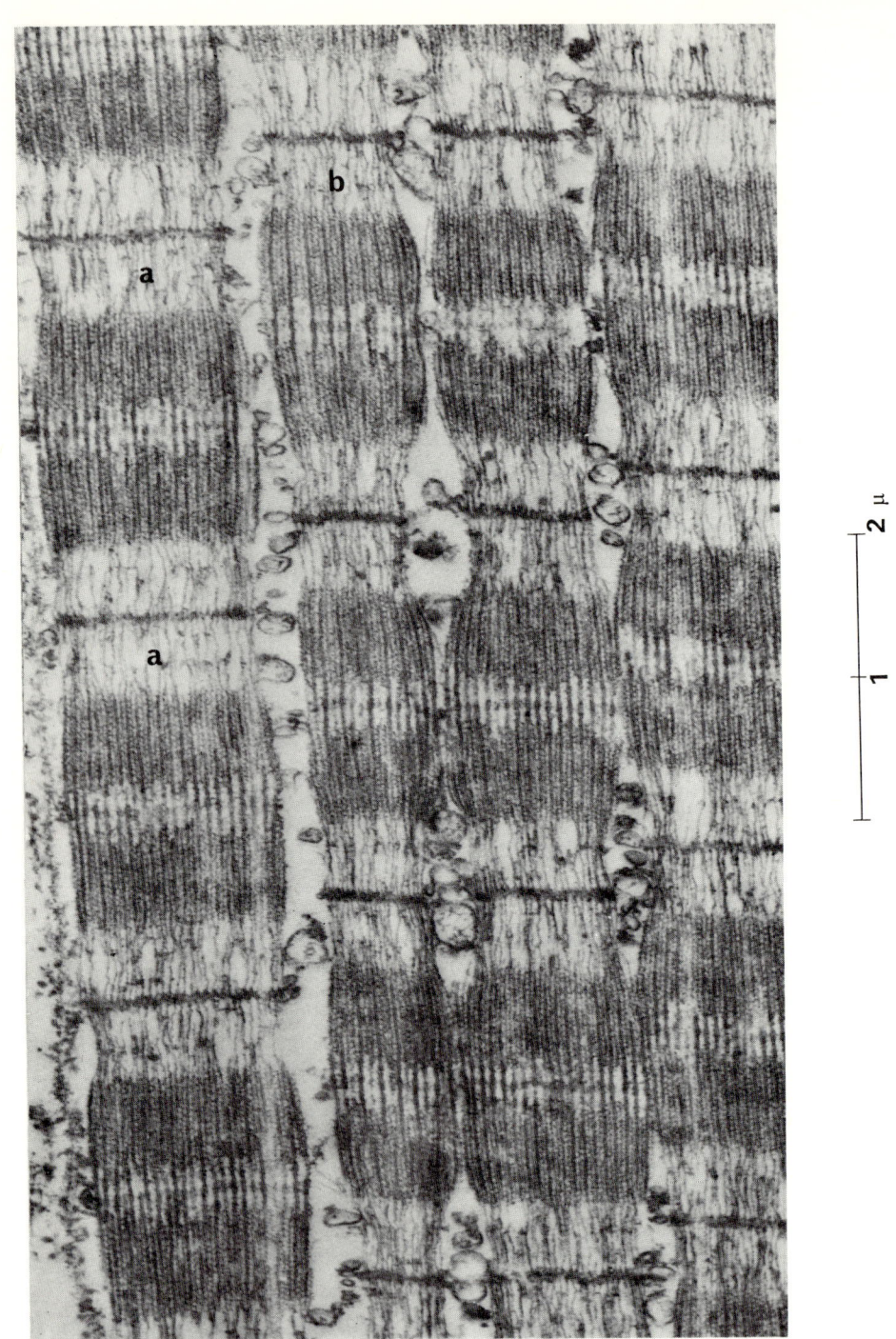

Fig. 2-23. Electron micrographs of muscle fibers fixed in osmium tetroxide. (a) General view of thin section through a number of striated myofibrils from rabbit psoas muscle. The letters a and b denote different orientations of the fibers relative to the plane of sectioning; ×60,000.

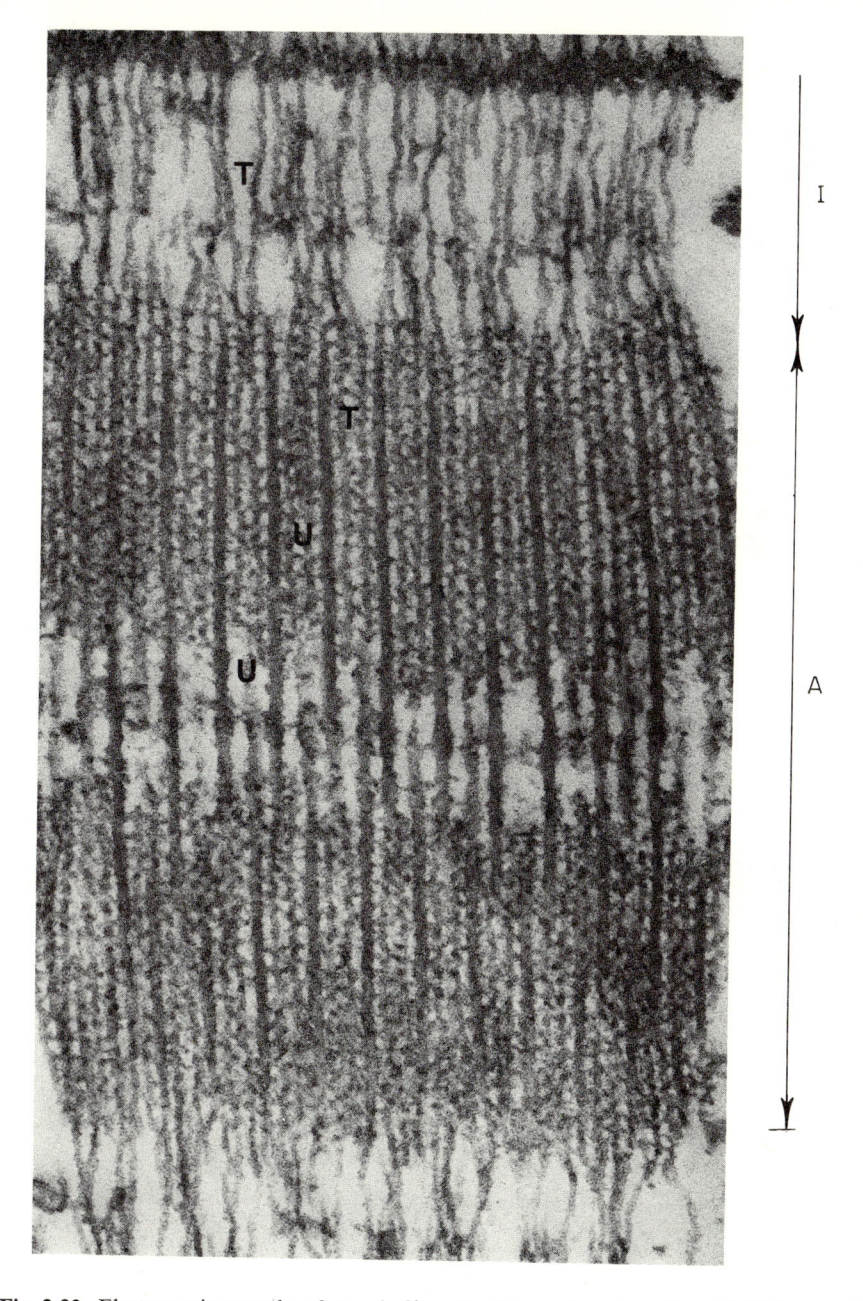

Fig. 2-23. Electron micrographs of muscle fibers fixed in osmium tetroxide. (b) Higher resolution micrograph of section through one sarcomere, the repeating unit of the myofibril, and nearly parallel to the axis of the filaments. In the interpretation of H. E. Huxley, the thick filaments (U) contain the protein myosin and the thin filaments (T) contain the protein actin. During muscle contraction, the interdigitated filaments in the A band are envisioned as interacting through slender transverse bridges, which can be seen to connect the two kinds of filaments, in such a way that the actin and myosin filaments slide along each other, preserving the dimensions of the A band, but causing an apparent shortening of the I band; ×195,000. From Huxley (51).

References

1. J. B. Neilands and P. K. Stumpf, "Outlines of Enzyme Chemistry," 2nd Ed. Wiley, New York, 1958; M. Dixon and E. C. Webb, "Enzymes," 2nd Ed. Academic Press, New York, 1964; "The Enzymes" (P. D. Boyer, H. Lardy, and K. Myrbäck, eds.), 2nd Ed., Vols. 1–8. Academic Press, New York, 1959–1963; "Methods in Enzymology" (S. P. Colowick and N. O. Kaplan, eds.), Vols. 1–12. Academic Press, New York, 1955–1968; "Fermenty" (A. E. Braunshtejn, ed.). Nauka, Moscow, 1964.
2. B. F. Poglazov, "The Structure and Functions of Contractile Proteins." Academic Press, New York, 1966; I. I. Ivanov and V. A. Yuryev, "Biokhimiya i Patokhimiya Myshts." Medgiz, Leningrad, 1961.
3. W. H. Stein, *Federation Proc.* **23**, 599 (1964).
4. F. M. Richards, *Proc. Natl. Acad. Sci. U.S.* **44**, 162 (1958); F. M. Richards and P. J. Vithayathil, *Brookhaven Symp. Biol.* **13**, 115 (1960).
5. C. B. Anfinsen, "The Molecular Basis of Evolution." Wiley, New York, 1959.
6. P. J. Vithayathil and F. M. Richards, *J. Biol. Chem.* **235**, 1029 (1960); P. J. Vithayathil and F. M. Richards, *J. Biol. Chem.* **235**, 2343 (1960).
7. A. M. Crestfield, W. H. Stein, and S. Moore, *Arch. Biochem. Biophys. Suppl.* **1**, 217 (1962).
8. A. M. Crestfield, W. H. Stein, and S. Moore, *J. Biol. Chem.* **238**, 2413 (1963); A. M. Crestfield, W. H. Stein, and S. Moore, *J. Biol. Chem.* **238**, 2421 (1963).
9. F. H. White, Jr., *J. Biol. Chem.* **235**, 383 (1960).
10. C. B. Anfinsen, E. Haber, M. Sela, and F. H. White, Jr., *Proc. Natl. Acad. Sci. U.S.* **47**, 1309 (1961).
11. E. Haber and C. B. Anfinsen, *J. Biol. Chem.* **236**, 422 (1961); F. H. White, Jr., *J. Biol. Chem.* **236**, 1353 (1961).
12. H. S. Jansz, R. A. Oosterbaan, F. Berends, and J. A. Cohen, *Proc. 5th Intern. Congr. Biochem. Moscow*, 1961 **4**, 45 (1963).
13. E. A. Moelwyn-Hughes, *in* "The Enzymes" (J. B. Sumner and K. Myrbäck, eds.), 1st Ed., Vol. 1, Pt. 1, p. 28. Academic Press, New York, 1950.
14. H. Neurath, *Sci. Am.* **211**, No. 6, 68 (1964).
15. H. Tuppy, *in* "Symposium on Protein Structure" (A. Neuberger, ed.), p. 66. Wiley, New York, 1958.
16. R. D. Hotchkiss and A. H. Evans, *Federation Proc.* **19**, 912 (1960).
17. L. Michaelis and M. L. Menten, *Biochem. Z.* **49**, 333 (1913).
18. R. A. Alberty, *Advan. Enzymol.* **17**, 1 (1956).
19. J. Z. Hearon, S. A. Bernhard, S. L. Friess, D. J. Botts, and M. F. Morales, *in* "The Enzymes" (P. D. Boyer, H. Lardy, and K. Myrbäck, eds.), 2nd Ed., Vol. 1, p. 49. Academic Press, New York, 1959.
20. R. Lumry, *in* "The Enzymes" (P. D. Boyer, H. Lardy, and K. Myrbäck, eds.), 2nd Ed., Vol. 1, p. 157. Academic Press, New York, 1959.
21. O. A. Roholt and D. M. Greenberg, *Arch. Biochem. Biophys.* **62**, 454 (1956).
22. B. Chance, *Advan. Enzymol.* **12**, 153 (1951).
23. A. E. Braunshtejn and M. M. Shemyakin, *Biokhimiya* **18**, 393 (1953); A. E. Braunstein, *in* "The Enzymes" (P. D. Boyer, H. Lardy, and K. Myrbäck, eds.), 2nd Ed., Vol. 2, Pt. A, p. 113. Academic Press, New York, 1960; E. E. Snell, *Brookhaven Symp. Biol.* **15**, 32 (1962).
24. D. E. Koshland, Jr., *in* "The Enzymes" (P. D. Boyer, H. Lardy, and K. Myrbäck, eds.), 2nd Ed., Vol. 1, p. 305. Academic Press, New York, 1959.
25. D. E. Koshland, Jr., *Proc. Natl. Acad. Sci. U.S.* **44**, 98 (1958).
26. K. U. Linderström-Lang and J. A. Schellman, *in* "The Enzymes" (P. D. Boyer, H. Lardy, and K. Myrbäck, ed.), 2nd Ed., Vol. 1, p. 443. Academic Press, New York, 1959.
27. D. E. Koshland, Jr., J. A. Yankeelov, Jr., and J. A. Thoma, *Federation Proc.* **21**, 1031 (1962).
28. K. Yagi, T. Ozawa, and T. Ooi, *Biochim. Biophys. Acta* **77**, 20 (1963).
29. G. N. Reeke, J. A. Hartsuck, M. L. Ludwig, F. A. Quiocho, T. A. Steitz, and W. N. Lipscomb, *Proc. Natl. Acad. Sci. U.S.* **58**, 2220 (1967).

7. References

30. C. G. Swain and J. F. Brown, Jr., *J. Am. Chem. Soc.* **74**, 2538 (1952).
31. B. Chance, R. W. Estabrook, and A. Ghosh, *Proc. Natl. Acad. Sci. U.S.* **51**, 1244 (1964).
32. J. Monod, J.-P. Changeux, and F. Jacob, *J. Mol. Biol.* **6**, 306 (1963); M. Freundlich and H. E. Umbarger, *Cold Spring Harbor Symp. Quant. Biol.* **28**, 505 (1963).
33. J. Monod, J. Wyman, and J.-P. Changeux, *J. Mol. Biol.* **12**, 88 (1965).
34. H. Fernández-Morán, *Circulation* **26**, 1039 (1962).
35. D. G. McConnell, A. Tzagoloff, D. H. MacLennan, and D. E. Green, *J. Biol. Chem.* **241**, 2373 (1966).
36. H. N. Christensen, *Advan. Protein Chem.* **15**, 239 (1960).
37. H. N. Christensen, *Perspectives Biol. Med.* **2**, 228 (1959).
38. A. K. Solomon, *Biophys. J.* **2**, Suppl., 79 (1962).
39. J.-P. Changeux, J. Thiéry, Y. Tung, and C. Kittel, *Proc. Natl. Acad. Sci. U.S.* **57**, 335 (1967); J.-P. Changeux and T. R. Podleski, *Proc. Natl. Acad. Sci. U.S.* **59**, 944 (1968).
40. A. L. Lehninger, *Federation Proc.* **19**, 952 (1960).
41. A. L. Lehninger, "The Mitochondrion." Benjamin, New York, 1964.
42. F. S. Sjöstrand, *Fine Struct. Cells, Symp. 8th Congr. Cell Biol., Leiden, 1954*, p. 16, (1955).
43. S. A. Nejfakh, T. B. Kazakova, M. P. Melnikova, and V. S. Turovskij, *Dokl. Akad. Nauk SSSR* **138**, 227 (1961); S. A. Nejfakh, V. S. Gajtskhoki, T. B. Kazakova, M. P. Melnikova, and V. S. Turovskij, *Dokl. Akad. Nauk SSSR* **144**, 449 (1962).
44. V. A. Engelgardt and M. N. Lyubimova, *Biokhimiya* **7**, 205 (1942).
45. A. V. Hill, *Proc. Roy. Soc. (London)* **B137**, 268 (1950).
46. H. M. Kalckar, *Chem. Rev.* **28**, 71 (1941).
47. W. Kuhn, *Experientia* **5**, 318 (1949); A. Katchalsky, *Experientia* **5**, 319 (1949); W. Kuhn and B. Hargitay, *Experientia*, **7**, 1 (1951); A. Katchalsky, S. Lifson, I. Michaeli, and M. Zwick, *in* "Size and Shape of Contractile Polymers" (A. Wasserman, ed.), p. 1. Macmillan (Pergamon), New York, 1960.
48. A. Szent-Györgyi, "Chemistry of Muscular Contraction." Academic Press, New York, 1947; F. B. Straub and G. Feuer, *Biochim. Biophys. Acta* **4**, 455 (1950).
49. M. F. Morales, *in* "Enzymes: Units of Biological Structure and Function" (O. H. Gaebler, ed.), p. 325. Academic Press, New York, 1956.
50. J. Hanson and H. E. Huxley, *Symp. Soc. Exptl. Biol.* **9**, 228 (1955); H. E. Huxley, *Proc. Roy. Soc. (London)* **B160**, 442 (1964).
51. H. E. Huxley, *J. Biophys. Biochem. Cytol.* **3**, 631 (1957).
52. W. T. Astbury, E. Beighton, and C. Weibull, *Symp. Soc. Exptl. Biol.* **9**, 282 (1955).

Chapter III

Nucleic Acid Structure

1. Introduction

Nucleic acids, as their name indicates, are found as the major chemical constituents of the cell nucleus. Two classes of nucleic acids are present in that subcellular organelle: deoxyribonucleic acid (DNA) and ribonucleic acid (RNA), but they are both found elsewhere in the cell as well. RNA is widely distributed throughout the cytoplasm, while DNA is localized in subcellular particulates such as the chloroplasts of green plants and the mitochondria of yeast and higher organisms. Extranuclear DNA is responsible for the phenomenon of cytoplasmic inheritance. Both kinds of nucleic acids consist of linear polymers, each residue of which contains a phosphate moiety, a five-carbon sugar—deoxyribose in DNA and ribose in RNA—and a heterocyclic purine or pyrimidine base as a side group (1).

Deoxyribonucleic acid is insoluble under normal cellular conditions, that is, at pH 7 and a salt concentration of about 0.1 M. DNA molecules are usually associated with proteins in complexes known as nucleoproteins which impart to the nucleus the properties of a water-saturated gel. Extraction of the nucleoprotein complex from a tissue can be effected with a 1 to 2 M salt solution. Moreover, removal of protein from the nucleoprotein complex renders the DNA soluble. The action of detergents, such as sodium dodecyl sulfate, or of phenol results in deproteinization, after which the molecular characteristics of the dissolved DNA can be studied by normal physical and chemical techniques. One of the most striking features of individual deoxyribonucleic acid chains is their tremendous length. Molecular weights of 10^8 and higher are not uncommon; since the average residue weight is about 330 daltons, the number of residues per molecule frequently exceeds 3×10^5.

RNA, the second nucleic acid component of the cell, is found in several different

1. Introduction

forms. One class is of relatively low molecular weight and is soluble in the cellular milieu. For reasons to be discussed in Chapters IV and V, this type of RNA is called transfer RNA or tRNA. Its molecular weight is about 25,000, indicating a chain length of approximately 80 residues. Other types of RNA are characterized by relatively high molecular weight (5×10^5 to 10^6) and are found associated with proteins in subcellular patriculates called ribosomes. Ribosomes or microsomes, small spherical ribonucleoprotein structures with a diameter of from 150 to 200 Å, are the sites of protein synthesis in the cell. Although ribosomes are not visible in the ordinary light microscope, electron microscopy has revealed their extremely wide distribution throughout the cytoplasm of almost all cells. In animal and plant cells, these ribonucleoprotein granules are attached to the surfaces of lipid-rich membranes which comprise the endoplasmic reticulum. In this form the RNA can be solubilized only after deproteinization with high concentrations of phenol or with surface-active substances such as detergents which destroy the nucleoprotein complex. There is yet another kind of high-molecular-weight RNA which is frequently associated with ribosomes, although its properties are quite different from those of ribosomal structural RNA (rRNA). In bacterial cells, this RNA species appears to turn over very rapidly owing to constant degradation and resynthesis. Without further explanation at this time, we shall designate this kind of RNA as messenger RNA or mRNA.

To better understand the structure of nucleic acids, it is necessary to say a few words about their biological function. Life processes require an unbroken flow of energy, matter, and information. The flow of energy and matter is completely determined by the function of various proteins, discussed in detail in the preceding chapters. Information flow is another matter, however. As pointed out in the Introduction, there is imprinted in each cell and each virus particle information specifying the compounds which must be synthesized and the way in which they must be organized in order to ensure that the major task of the organism, self-reproduction, will be accomplished. Such information must be continuously transmitted from the nucleus, where it is stored, to the synthetic centers of the cell, where it is turned into new components. Furthermore, when a cell divides, a complete set of "blueprints" must be passed on to the daughter cells; this type of information flow must also remain uninterrupted as long as life continues.

Nucleic acids serve as the material agents of both kinds of information flow in living organisms. DNA stores information in the nucleus and passes it on to the mRNA under appropriate circumstances. The mRNA carries the information to ribosomes, where the cellular proteins are synthesized. The information contained in mRNA is there directly impressed on new protein molecules as they are being synthesized. Since each mRNA molecule bears the complete structural information for one or more proteins, it is often called template RNA. Besides this flow of information, which can be characterized by the scheme,

$$\text{DNA} \rightarrow \text{mRNA} \rightarrow \text{protein}$$

there is also a flow of information from the DNA of the maternal cell to that of the daughter cell. This is accomplished by a mechanism which replicates or copies the maternal DNA so that each daughter cell receives all the information originally present in the nucleus of the maternal cell.

Insofar as the function of nucleic acids is to transmit information, it is logical to imagine that the information is fixed in the form of a code. Since the nucleic acids are linear copolymers containing four different sorts of residues, the only possible code must arise from the sequence in which these residues are arranged. The code must be able to specify the primary structure of all proteins required by the cell. Since the enzymes which catalyze the synthesis of low-molecular-weight metabolites are prominent among these, all cellular components are ultimately specified by the nucleic acid code. In effect, the code links the chemical structure of the polynucleotide chains of RNA and DNA with the polypeptide chain of the protein.

Understanding the nature of the genetic code is one of the most important problems in molecular biology. Let us enumerate the most significant discoveries providing direct proof that the function of nucleic acids is to store and transmit genetic information.

1. In several species of bacteria, the hereditary constitution of a particular strain can be altered by introduction into the cell of purified DNA isolated from strains with different genetic properties. This method of transmitting new hereditary characteristics to a given strain is called bacterial transformation.

2. The infectivity of RNA-containing plant viruses is attributable to their nucleic acid component. Administration of a purified viral RNA solution to the plant is sufficient for infection and the production of new virus particles.

3. Similarly, infectivity of DNA-containing viruses from animals and bacteria is a property of purified viral DNA molecules.

4. When a bacterial cell is infected with an intact bacteriophage (bacterial virus), the phage injects its protein-free nucleic acid into the host.

Besides these four direct observations implicating nucleic acids in genetic phenomena, there is a large body of indirect evidence which confirms the basic conclusion.

Before considering concrete problems in nucleic acid structure, it is important to note some of the differences between RNA and DNA. The proportion of nucleic acids by weight varies from organism to organism, as does the protein content. In tobacco mosaic virus, RNA comprises 5% and protein 95% of the total particle weight. In yeast, the RNA content reaches 10% while that of DNA is only about 1%. Proteins comprise 60 to 70% of the dry weight of bacterial cells, and the nucleic acids range from 10 to 25%. The contribution of DNA to this figure is almost constant for any given strain, and usually falls between 5 and 7% of the dry weight, equal to about 10^{-14} gm per cell. The RNA content, however, varies considerably with conditions of growth.

In cells of higher organisms the amount of DNA is remarkably constant in a given species of plant or animal; moreover, this quantity does not significantly vary from one tissue to another. Table 3-1 gives the average DNA and RNA contents per cell for several different tissues of the rat. The figures for nuclear DNA are all similar to one another with the exception of liver which is known to contain a certain proportion of polyploidic cells. This consistency is not reflected in the figures for RNA, however. RNA content is extremely variable from tissue to tissue and the large range in certain of the determinations is undoubtedly associated with different conditions of growth and nutrition. Secretory glands evidence particularly large variations in RNA

1. Introduction

TABLE 3-1

AMOUNTS OF DNA AND RNA IN VARIOUS CELLS OF THE RAT[a]

Tissue	Amount of DNA per cell (gm × 10¹²)	RNA/DNA ratio
Liver	9.13	4.38
Pancreas	7.12	4.1
Small intestine	7.38	0.57
Lung	6.51	0.57
Kidney	6.52	0.9–2.46
Heart	6.27	1.03
Thymus	7.18	0.19
Spleen	6.33	0.36–1.03
Bone marrow	6.70	0.57–0.97

[a] From Leslie (2).

concentration; for example, the ratio of RNA to DNA in pancreas can be as large as 15. Embryonic tissues exhibit a similar preponderance of RNA, the RNA:DNA ratio reaching 17 in larvae of *Drosophila*.

The content of DNA in the nucleus of every young cell in a given species of higher organism is almost invariant. It depends neither on the nutrition nor on the growth rate of the cells, nor on any other external conditions. As cell division approaches, the quantity of DNA precisely doubles, so that after division, the level of DNA characteristic of the particular organism is restored. Moreover, the nuclei of most cells in a complex differentiated organism contain exactly the same quantity of DNA. This regularity serves to emphasize the role played by nuclear DNA in storing genetic information, since the information content remains constant. It is convenient to express the amount of DNA per cell in terms of the number of monomeric units of mononucleotides. In ΦX174, one of the smallest of DNA-containing bacteriophages, the total number of nucleotide residues is about 5000; in bacteriophage T2 DNA, the comparable figure is approximately 4×10^5. In each cell of the bacterium *Escherichia coli*, there are 10^7 mononucleotide residues. The number of residues per cell of most higher organisms is on the order of 5×10^9 to 10^{10}. These figures are characteristic of the particular species, changing only before cell division, when nuclear material is elaborated for the new daughter cells.

RNA presents quite a different pattern. It was noted some time ago that intracellular RNA concentration depends on the rate of cell growth and on the rate of protein synthesis in them (Kedrovski, Brachet). Particularly clear quantitative results were obtained by Magasanik who showed that within certain limits, the RNA content of a bacterial cell is inversely proportional to the generation time of the culture. Thus the RNA concentration can be made to vary by altering either the nourishment provided the cells or the temperature of growth. This peculiarity strongly implicates RNA as a direct participant in the synthesis of cellular proteins.

2. Chemical Structure of the Nucleic Acids

The chemical structure of the monomeric constituents of nucleic acid chains, called mononucleotides, was worked out by the usual methods of organic chemistry: the molecules were split up into many fragments and the structure of the fragments was analyzed. On the basis of such investigations, a structural formula was proposed for the complete mononucleotide and then confirmed by synthesis (3). The structural formulas of the mononucleotides found in DNA and RNA are presented in Fig. 3-1. Each mononucleotide consists of a nitrogenous purine or pyrimidine side group, a five-carbon sugar, and a phosphate group. In the illustration, the monomeric units are joined together by phosphodiester linkages characteristic of the polynucleotide chain. Although only a tetranucleotide is depicted, naturally occurring nucleic acids may consist of tens of thousands of residues linked together in this way. The sequence of residues along the chain is quite varied with no suggestion of a short repeating polynucleotide subunit as was supposed at one time.

The backbone of the polynucleotide chain is formed by alternating phosphate and sugar groups—ribose in RNA and deoxyribose in DNA—bound by ester linkages between phosphoric acid and the 3'- and 5'-hydroxyl groups of successive furanose rings. Deoxyribose is distinguished from ribose by the absence of a hydroxyl group on the 2'-carbon atom of the sugar. A purine or pyrimidine base is attached to the cyclic sugar through the 1'-carbon atom to complete the nucleotide. It is apparent that the polynucleotide chain is asymmetric. The direction of the chain is conventionally designated as 5' to 3', that is, running from the 5'-phosphate of the first sugar of the backbone to the 3'-hydroxyl of the last sugar in the sequence.

The purine bases adenine and guanine occur in both RNA and DNA, but the two nucleic acids differ somewhat in their pyrimidine composition. While cytosine is a constituent of both, uracil occurs only in RNA and thymine only in DNA. The nucleotides derived from these bases are called adenylic acid, guanylic acid, cytidylic acid, uridylic acid, and thymidylic acid, respectively, although the nucleotide residues are often referred to by the name of their purine or pyrimidine side chains alone. If the phosphate group is removed from a nucleotide, the resulting compound is called a nucleoside and contains a nitrogenous base joined to ribose or deoxyribose. The nucleosides derived from adenine, guanine, cytosine, uracil, and thymine are called adenosine, guanosine, cytidine, uridine and thymidine. Nucleotides can also be designated as phosphate derivatives of nucleosides, as adenosine triphosphate (abbreviated ATP) or uridine monophosphate (UMP).

Nucleic acid chains frequently contain minor bases which differ in structure from the five major ones illustrated in Fig. 3-1. Thus, in the DNA of higher plants and animals there is usually a certain amount of 5-methylcytidine:

2. Chemical Structure of the Nucleic Acids

in addition to cytidine. In animals and man, the proportion of this base is about 1.5%, while in plants it sometimes accounts for as much as 5 to 7% of the total residues. In T-even bacteriophages (T2, T4, T6), cytidine is completely replaced by 5-hydroxymethylcytidine. Furthermore, a significant fraction of the hydroxyl groups at the 5-position are bound to a molecule of glucose via a glucosidic linkage:

The proportion of glucose residues so bound depends on the species of phage.

Transfer RNA molecules contain a wide variety of minor base components, though their concentration is relatively small. Pseudouridine, in which the sugar is linked to the 5-carbon atom of uracil rather than the 1-nitrogen atom, constitutes 5 to 9% of the total residues in unfractionated *E. coli* tRNA:

Some minor bases differ in the substitution of additional methyl groups, such as 1-methylguanosine:

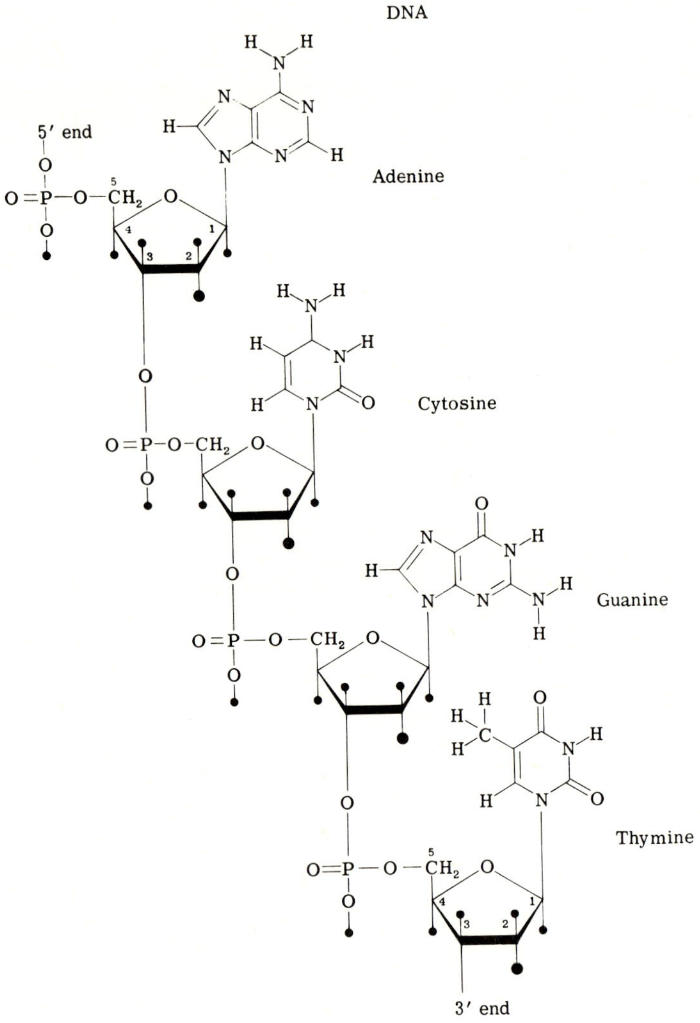

Fig. 3-1. Structure of the nucleic acids. (a) Deoxyribonucleic acid (DNA), showing constituent nucleotide bases and the 3′ → 5′ phosphodiester bonds that link the monomer units.

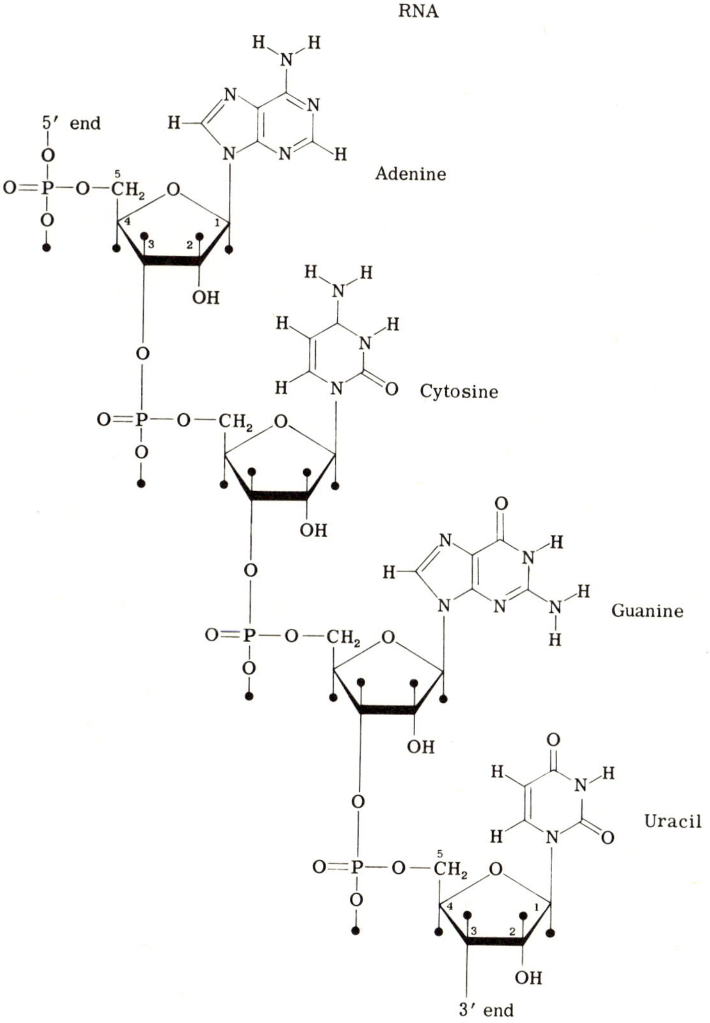

2. Chemical Structure of the Nucleic Acids

Fig. 3-1. Structure of the nucleic acids. (b) Ribonucleic acid (RNA), showing nucleotide bases and 3′→5′ phosphodiester bonds. The RNA base uracil is analogous to thymine in DNA.

and N^2-dimethylguanosine:

Among the minor bases, inosine, a product of guanosine deamination:

1-methylinosine:

and several other derivatives are also encountered.

The principal chemical differences between RNA and DNA arise from the behavior of their sugar moieties. Compounds containing ribose are chemically more labile than those with deoxyribose, owing to the presence of the hydroxyl group on the 2'-carbon atom. Alkaline hydrolysis of the polynucleotide is facilitated by the 2'-hydroxyl groups, for example. Moreover, ribose containing hydroxyl groups at both 2' and 3' positions is readily attacked by various oxidants:

Deoxyribose is devoid of this chemically vulnerable site.

DNA and RNA are both long, unbranched, linear polymers with an alternating sugar–phosphate backbone. The nucleotide composition of the chain may be determined

2. Chemical Structure of the Nucleic Acids

by complete hydrolysis of the backbone into mononucleotides and subsequent fractionation and identification of the mononucleotides by chromatography. The hydrolysis of DNA is accomplished by extended heating in the presence of 6 N HCl. The chromatographic separation of mononucleotides is usually performed with ion exchange resins, particularly strongly anionic resins of the type illustrated on p. 12. Paper electrophoresis and paper chromatography can also be used for this purpose.

In order to discover the structural formula, or nucleotide sequence, of a nucleic acid chain, the approach is essentially the same as that used for sequencing proteins. The first stage is to obtain the nucleic acid in pure form. Next, the polynucleotide must be cleaved into short, well-defined fragments, or oligonucleotides, which are then separated from one another. The third step involves determination of the nucleotide sequence within each of the oligonucleotides. This phase of the work requires a method by which to sequentially remove individual mononucleotides from the ends of the oligonucleotide segments. Finally, by cleaving the polynucleotide in alternative ways, the sequences overlapping the original set of oligonucleotides must be obtained in order to permit reconstruction of the structural formula for the entire chain. Although this procedure appears very simple in theory, it is quite difficult to apply to polynucleotides in practice.

Let us consider first the purification of individual DNA or RNA molecules. Nucleic acids can be freed from protein relatively easily by treating the cells or cell lysate with a mixture of concentrated phenol and sodium dodecylsulfate. Proteins are completely denatured by these agents and can be removed, whereas the nucleic acids remain soluble in the phenol-water medium. Phenol is removed from the preparation by precipitating the nucleic acids with ethanol. DNA can now be freed from RNA, or RNA from DNA, by using the purified pancreatic enzymes ribonuclease or deoxyribonuclease, respectively. After enzymatic digestion of the unwanted polynucleotide, the remaining nucleic acid molecules are once again deproteinized with phenol and precipitated with ethanol. While RNA and protein contaminants in DNA preparations may be reduced to less than several tenths of a percent by this method, the DNA is not necessarily homogeneous in a chemical sense. Such a preparation will consist of identical molecules only when the DNA is extracted from viruses. Unless great care is exercised to avoid degradation of the giant molecules, DNA obtained from bacteria will consist of a varied set of molecular segments. The heterogeneity of DNA prepared from higher organisms is much greater still. Heterogeneity in DNA is not at all surprising, since it must store the coded information necessary to specify all the protein and RNA components of the cell. The more complex the cell, the more the information required, so that a greater quantity of DNA is necessary to specify different cellular components. It is easier to obtain homogeneous DNA from viruses precisely because the information they require is relatively limited, and their DNA molecules are accordingly smaller. The DNA molecules of the very smallest bacteriophages have a molecular weight of about 1.5×10^6 daltons, and contain about 5000 nucleotides. Although the preparation of homogeneous DNA populations from such organisms is possible, the problem of establishing the sequence of nucleotides within the polynucleotide chain is nonetheless beyond our ability at the present time.

For sequence work, a much smaller and simpler nucleic acid molecule was required. For this reason, Holley and his co-workers turned their attention to a transfer

RNA, one of the smallest of naturally occurring polynucleotides, and they were able to completely solve the chemical structure (4). There are several circumstances that made tRNA a particularly apt subject for this kind of analysis. First of all, the molecule is quite small, consisting, in the case of alanine tRNA, of 77 nucleotide residues. The molecular weight of its sodium salt is 26,600. Second, the presence of minor bases greatly facilitated the analysis since they render the oligonucleotide fragments less similar and serve as a kind of natural label. Finally, the studies were carried out with ribonucleases of remarkable specificity which permitted hydrolysis of the chain into a relatively small number of rather large fragments.

This work is worthy of further attention here since it represents one of the most significant achievements in nucleic acid chemistry in recent years. There are from 30 to 40 different kinds of tRNA present in every cell, each variety serving as a specific transport agent for one kind of amino acid. Since there are only 20 different amino acids, there is frequently more than one variety of tRNA specific for a given amino acid. All of these tRNA varieties are almost identical in molecular weight, and quite close in nucleotide composition although certain differences are noted in the content of minor bases. The initial problem was thus the isolation of one chemical species from a mixture of 40 very similar substances. The extraordinary difficulties involved may be likened to those encountered in the separation of different isotopes of the same element. As was true in the case of isotopes, tRNA's were successfully fractionated by using a series of several hundred separations, each of which partially enriches for the desired component. For tRNA's, countercurrent distribution has proved to be the most effective means of accomplishing this and two solvent systems are in general use. In one system, tRNA components are separated in phosphate buffer, pH 6, isopropyl alcohol, and formamide while in a second, tRNA salts with quaternary amines (such as trimethylcetylammonium chloride) are fractionated between water and butanol. It is necessary to pass the tRNA's through a 250-tube apparatus four times, that is, to carry out 1000 serial extractions, in order to obtain a given tRNA in a form sufficiently pure for sequence analysis. The following figures will give some idea of the work involved. From 10 kg of yeast cells, it is possible to extract about 10 gm of mixed tRNA's by usual preparative procedures. After countercurrent distribution the final yield of any one individual tRNA species is on the order of 0.1 gm.

Holley selected alanine tRNA for structure analysis. Various nucleases were employed to cleave the polynucleotide chain of alanine tRNA into fragments, and to further divide the fragments into even shorter oligonucleotides. Ribonucleases are essentially phosphodiesterases, capable of splitting one of the bonds of a double ester of phosphoric acid at selected points along the RNA chain. Of the many ribonucleases known, several are *exonucleases*, which can hydrolyze only the terminal bond at one or the other end of the chain. They then work down the chain one residue at a time until the entire chain has been hydrolyzed to mononucleotides. Such enzymes are extremely useful for investigating the sequence of nucleotides within short oligonucleotide fragments. The fragments themselves are produced by *endonucleases* which, depending on their specificity, can attack certain internal phosphodiester bonds.

Of the exonucleases, the most important is snake venom phosphodiesterase. This enzyme cleaves one residue at a time from the 3'-terminus of the polynucleotide chain, by breaking the bond between the phosphate group and the 3'-carbon atom of the next ribose group. Thus the products of this hydrolysis are nucleoside-5'-phosphates. Another widely used exonuclease, spleen phosphodiesterase, cleaves residues stepwise from the 5'-terminus of the molecule, severing the bond between phosphate and the ribose 5'-carbon atom, thereby yielding nucleoside-3'-phosphates.

The best known endonuclease is pancreatic ribonuclease. Its specificity is such that it breaks every bond following a pyrimidine nucleotide so that the phosphate group remains attached to the 3'-carbon atom. A limiting hydrolyzate prepared with pancreatic ribonuclease contains mostly di- and trinucleotides with a smaller amount of tetra- and pentanucleotides. Takaribonuclease (T_1-ribonuclease), isolated from the mold, *Aspergillus oryzae*, displays an even greater selectivity. This endonuclease attacks the chain only at guanine residues, liberating the 5'-hydroxyl group of the ribose. The way in which the various hydrolytic enzymes act on the RNA chain is schematized in Fig. 3-2.

Holley showed that if hydrolysis is carried out at 0°C, T_1-ribonuclease reproducibly yields fragments 30 to 40 residues in length. This circumstance made it possible to order the smaller oligonucleotides produced under normal conditions of hydrolysis, for it provided the necessary overlapping sequences.

The various oligonucleotides obtained after enzymatic hydrolysis, even the very large ones, are usually isolated by ion-exchange chromatography. Particularly useful in this respect are long, narrow columns of DEAE-cellulose or DEAE-Sephadex, approximately 3 mm in diameter and 2 to 3 m long. In order for the chromatographic peaks to be sharp, it is essential that the column work solely on the principle of ion exchange. The number of charges borne by any polynucleotide is roughly equal to the number of phosphate groups, at least when the pH is between 5 and 8. The oligonucleotides can be eluted from the column by gradually increasing the salt concentration. Ammonium bicarbonate is frequently used for this purpose because it can be easily removed at a later time by evaporation. The main obstacle to obtaining good chromatographic separations arises from intermolecular hydrogen bonds which polynucleotides form very readily. To circumvent this difficulty, the chromatography of oligonucleotides, especially large ones, is usually carried out in concentrated urea (6 to 8 M), a substance which disrupts hydrogen bonds.

The methodology described above led directly to the complete primary structure of alanine-specific tRNA. Before discussing particular aspects of this structure, it should be mentioned that there is an abbreviated transcription by which the structural formulas of nucleic acids are conventionally presented. Two of the most common systems are shown below for the sequence adenine, guanine, uracil, cytosine, guanine (see Fig. 3-1 for the complete structural formula):

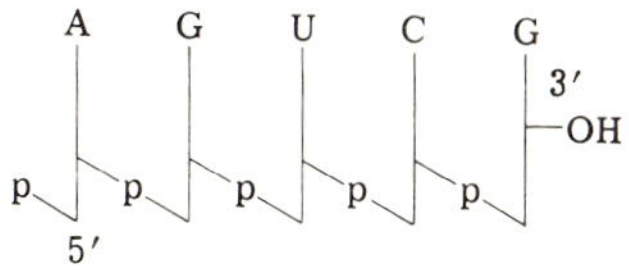

pApGpUpCpG

The left end of the sequence is generally taken as the beginning of the chain and is known as the 5'-end. The letter p at the left indicates that the terminal residue bears a phosphate group attached to the 5'-carbon atom of ribose.

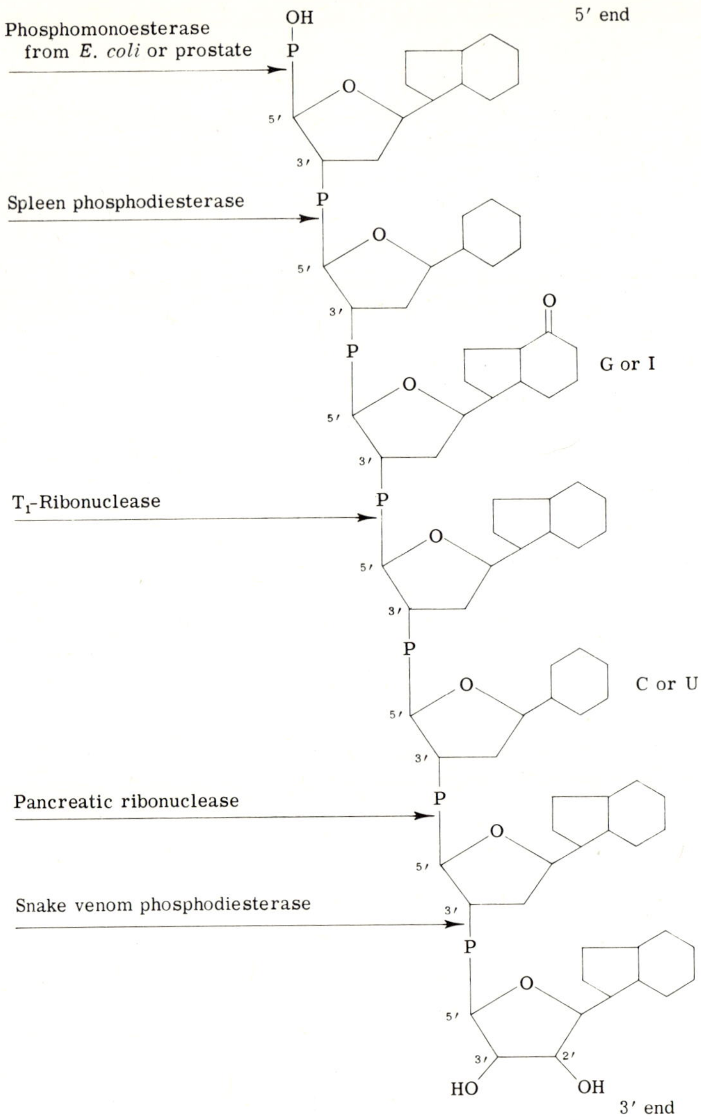

Fig. 3-2. Scheme showing the specificity of various enzymes that cleave the RNA chain. Pancreatic and T_1-ribonucleases are *endonucleases* that attack *internal* bonds in RNA with the specificity noted in the diagram. Spleen and snake venom phosphodiesterases are both *exonucleases*, that is, their attack starts at one of the *termini* of the molecule after terminal phosphate groups have first been removed by *E. coli* or prostatic phosphomonoesterase and continues sequentially along the chain. The spleen and venom exonucleases are particularly useful for identifying the nature of 3'- and 5'-residues, respectively, since, as they proceed down the RNA molecule, the final residue liberated is a nucleoside that can easily be separated from the nucleotides that result from cleavage of all other bonds.

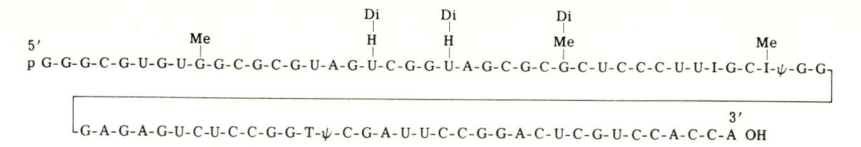

```
                        Di        Di            Di
         Me             H         H             Me                    Me
5'       |              |         |             |                     |
p G-G-G-C-G-U-G-U-G-G-C-G-C-G-U-A-G-U-C-G-G-U-A-G-C-G-C-G-C-U-C-C-U-U-I-G-C-I-ψ-G-G┐
                                                                                    │
                                                                                  3'│
└G-A-G-A-G-U-C-U-C-C-G-G-T-ψ-C-G-A-U-U-C-C-G-G-A-C-U-C-G-U-C-C-A-C-C-A OH
```

Fig. 3-3. The primary structure of alanine tRNA from yeast. In addition to the four usual bases, this RNA also contains 1-methylguanine (Me-G), 5,6-dihydrouracil (Di-H-U), N^2-dimethylguanine (Di-Me-G), inosine (I), 1-methylinosine (Me-I), pseudouracil (Ψ), and ribothymidylic acid (T). From reference 4, R. W. Holley, J. Apgar, G. A. Everett, J. T. Madison, M. Marquisse, S. H. Merrill, J. R. Penswick, and A. Zamir (1965). *Science* **147**, 1462–1465, 19 March 1965. Copyright 1965 by the American Association for the Advancement of Science.

The structural formula of alanine tRNA, as worked out by Holley and his colleagues, is presented in Fig. 3-3. In addition to the usual nucleotides adenine (A), guanine (G), cytosine (C), uracil (U), and thymine (T), there are six minor bases in the molecule, the key for which is included in the figure legend. Since completion of the work on alanine tRNA, the structures of many other tRNA's have been worked out: tyrosine tRNA, two kinds of serinet RNA and phenylalanine tRNA from yeast, several tRNA's from *E. coli*, and at least one tRNA from an animal tissue. There is a large measure of similarity in the sequence of all tRNA's investigated so far. Knowledge of the nucleotide sequence permits us to draw many important conclusions about the secondary structure of tRNA molecules and we shall return to this question in a subsequent section.

3. Macromolecular Structure of DNA

In a study of the nucleotide composition of DNA molecules, Chargaff discovered some surprising quantitative regularities (5). Regardless of the source, the molar ratios of adenine to thymine and of guanine to cytosine are always equal to one. This may be summarized as $A/T = G/C = 1$. It follows from these simple experimental facts that (1) the sum of the purine bases in any DNA molecule is equal to the sum of the pyrimidine bases $(A + G)/(T + C) = 1$, and (2) the number of 6-amino groups in the nucleotide bases is equal to the number of 6-keto groups, $(G + T)/(A + C) = 1$. These rules hold quite well for all kinds of DNA studied if 5-methylcytosine and 5-hydroxymethylcytosine are included in the figure for cytosine.

Important work on the chemical composition of microbial DNA's was done by Belozersky and Spirin, who measured the $(G + C)/(A + T)$ ratio in several different varieties of microorganisms (6). This factor varies from 0.45 to 2.8, depending on the species; in many bacteriophages, for example, $(G + C)/(A + T) = 0.5$. The limits of variation are much narrower in higher plants and animals, for which the ratio generally lies between 0.55 and 0.93. A similar ratio $(G + C)/(A + U)$ can be defined for RNA; in bacteria, it is found that this ratio falls within a very narrow range, from 1.05 to 1.35, although it does exhibit a slight correlation with the DNA base ratio. The fact that the nucleotide composition of DNA varies within such wide limits is a significant factor in the problem of the genetic code, as we shall see in Chapter V (p. 437).

The macromolecular structure of DNA has been much more extensively studied than that of RNA mainly because of its higher degree of internal order. This property was made apparent by the very detailed x-ray diffraction patterns obtained from oriented fibers of purified high-molecular-weight DNA (Wilkins and co-workers, Rosalind Franklin). The diffraction patterns, which contained up to 100 independent reflections, permitted an evaluation of the general properties of DNA structure, particularly its helical nature. The diameter of the DNA helix is about 20 Å and the radius of the cylindrical surface on which the phosphorus atoms of the backbone fall is 9 Å. One complete turn of the helix encompasses ten mononucleotide residues, and represents an axial displacement of 34 Å, or 3.4 Å per residue. This distance is also called the pitch of the helix and is the translational distance along the helix axis necessary to bring the structure into coincidence with itself. The unit cell of the macromolecule is monoclinic with the dimensions $a = 22.2$ Å, $b = 40$ Å, $c = 28.1$ Å, $\beta = 97°$ and it contains two separate chains, apparently in the form of a double-stranded helix. The purine and pyrimidine side groups are packed in the center of the helix like a stack of coins (7–9).

Such is the information that can be gleaned from the diffraction patterns without any sort of special hypotheses. However, it is impossible to produce a structural analysis of DNA as complete as that made in the case of proteins. Such an effort would require a hundred times more diffraction information than is obtainable from the relatively imperfect DNA fibers. Accordingly, any reasonable hypothesis for the complete structure of a DNA macromolecule must be advanced only on the basis of many different kinds of data.

It will be recalled that Pauling and Corey postulated the structure of the polypeptide chain in protein molecules on the basis of general concepts of molecular physics; only afterward was the structure confirmed with synthetic model polypeptides. A similar approach led Watson and Crick to their celebrated model of DNA structure. This brilliant conjecture was confirmed time after time in many different ways until it became a fundamental principle of molecular biology. It is nevertheless unfortunate that no direct proof of the structure by x-ray diffraction techniques is possible.

The Watson-Crick model is based on known atomic dimensions and bond angles. Moreover, the following rule served as a helpful guideline: DNA is a double helix with the maximum possible number of hydrogen bonds joining the purine and pyrimidine side chains which are packed into the area between the two helical strands (Figs. 3-4 and 3-5). The steric relationships, especially the similar overall dimensions of the side groups, require that a pyrimidine on one chain be opposite a purine on the other chain. Combining this with the maximization of hydrogen bonding, we find that these conditions are satisfied only by the pairs adenine:thymine (two H-bonds) and guanine:cytosine (three H-bonds) as illustrated in Fig. 3-4. An interesting consequence of this is that the nucleotide sequences of the two helices are not independent of each other in the double-stranded model. Specification of a nucleotide base in one chain defines the one opposite to it in the second chain: adenine must lie opposite thymine, and guanine opposite cytosine. Both A:T and G:C base pairs have exactly the same dimensions perpendicular to the axis, a necessary condition for regularity of the double helix. Moreover, the two molecular strands are antiparallel with respect to one another, defining the sense of the chain according to the orientation of the deoxyribose moiety along the polynucleotide backbone.

The regularities observed by Chargaff are fully understandable on the basis of the Watson-Crick model, the essence of which is the interdependence of nucleotide

3. Macromolecular Structure of DNA 215

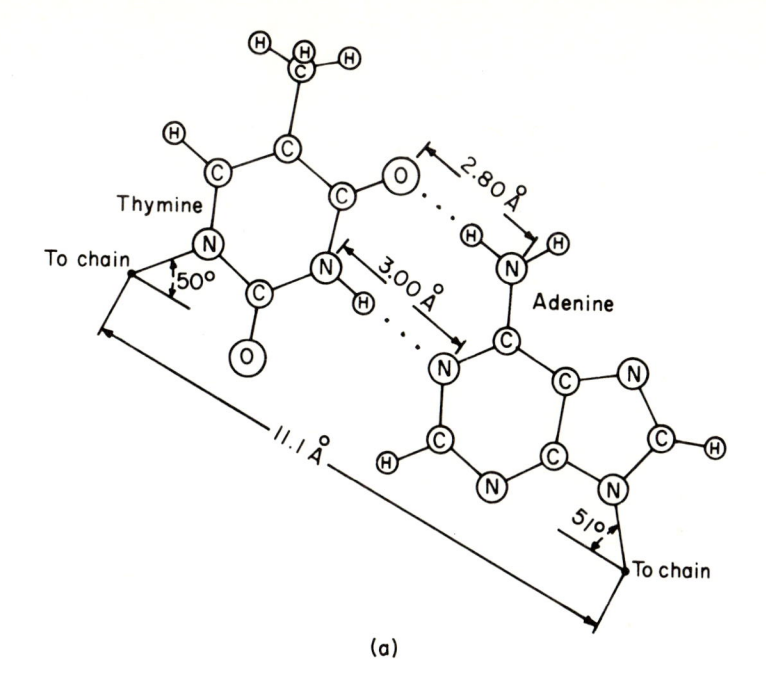

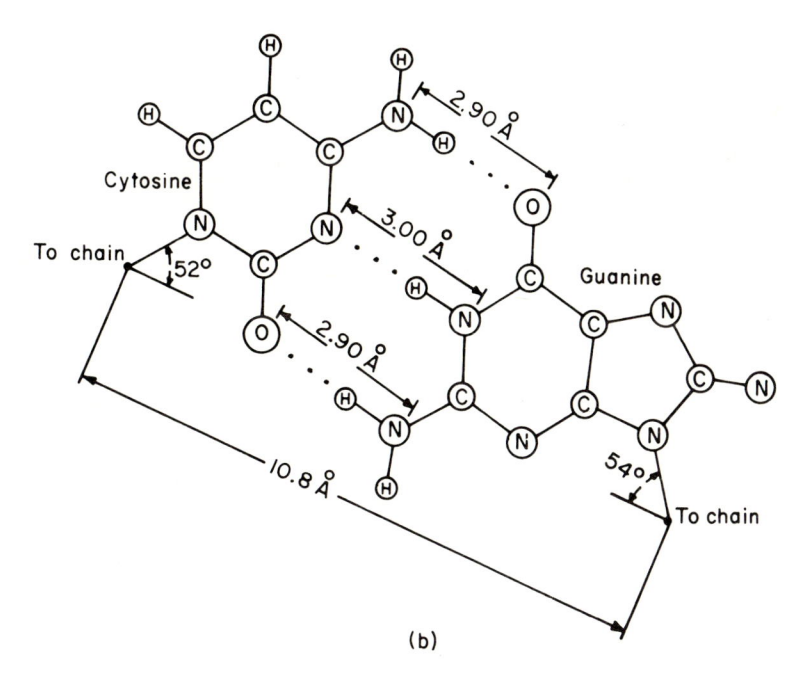

Fig. 3-4. Orientation of complementary base pairs in double-stranded DNA, drawn in a section perpendicular to the helix axis. Adenine and thymine (a) are held together by two hydrogen bonds, while cytosine and guanine (b) are held together by three hydrogen bonds. From Pauling and Corey (10).

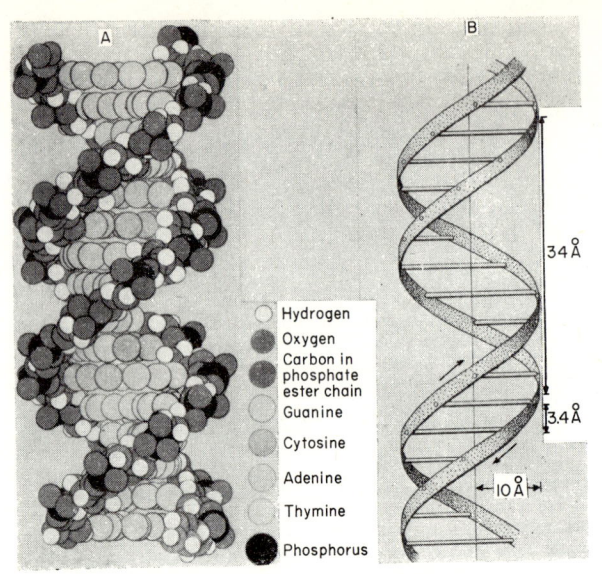

Fig. 3-5. Structure of the DNA molecule according to Watson and Crick. (Left) Drawing of the DNA duplex showing two helical grooves of unequal size on the outside of the molecule. After Feughelman *et al.* (11). (Right) Schematic drawing of double-stranded DNA indicating the dimensions of the helix. From Watson and Crick (7).

sequences in the two chains. Each of the two strands is like a replica of the other, and both thereby contain exactly the same information, even though it may not be equally available to the cell in the two cases. Thus the principle of complementarity and the mechanism of replication proposed by Watson and Crick are both anchored in the very structure of the DNA molecule.

Let us imagine that a DNA double helix is in solution along with adenine, guanine, cytosine, and thymine mononucleotides. If the ends of the two strands pull apart, then free mononucleotides should adsorb to each exposed half via hydrogen bonds in a complementary fashion (thymine with adenine, cytosine with guanine, etc.). If the unwinding of the helix now proceeds further and if the adsorbed bases are joined into a single chain, we will eventually obtain two identical double-helical DNA molecules. This process is known as DNA *replication*.

Another important feature of the Watson-Crick model is that it permits one to envision the possibility of errors in replication caused by the formation of incorrect base pairs. Such errors represent heritable, replicable alterations in genetic information and hence provide a molecular model for mutation. In particular, changes in the amino and keto groups, which play a crucial role in defining the base-pairing capacities of a given nucleotide, will ultimately result in alterations of base sequence upon replication. It is known, for instance, that mutagenic agents such as nitrous acid and formaldehyde attack the amino groups of adenine, cytosine, and guanine. Another source of mutability in both purine and pyrimidine bases arises from possible tautomeric transformations between keto–enol and amino–imino forms. Thymine, shown as the keto form in Fig. 3-4, can also assume the enol form in which it could hydrogen bond with the normal keto form of guanine (Fig. 3-6a). Similarly, adenine can be transformed from its normal amino form to the imino form, which is capable of

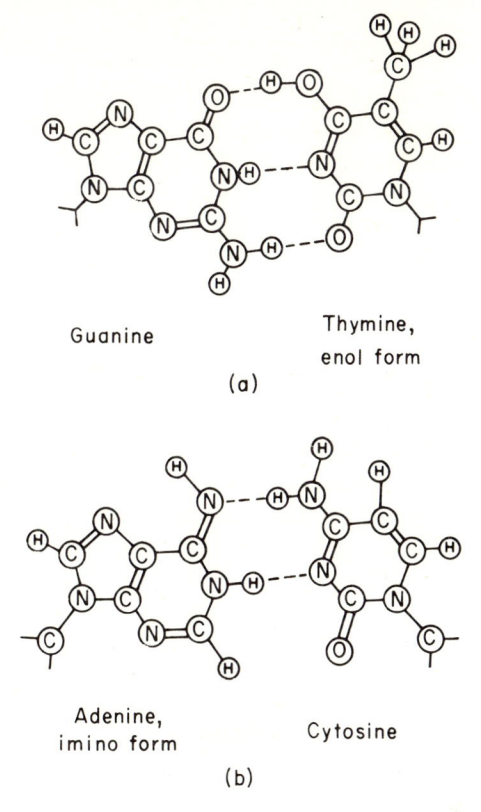

Fig. 3-6. Possibilities for "mispairing" between bases during DNA replication. (a) Guanine and the rare enol form of thymine can form three hydrogen bonds. (b) Cytosine can form two hydrogen bonds with the rare imino form of adenine.

hydrogen bonding with cytosine (Fig. 3-6b). The enol and imino forms illustrated occur only rarely under normal conditions and they are quite unstable. Nonetheless, there is always a definite probability of tautomerization, dependent in part on pH and temperature, and it may be responsible for coding errors to which spontaneous mutations can be attributed.

The equilibrium between tautomers of the nucleotide bases is displaced very strongly towards the forms which are found in the Watson-Crick structure (Fig. 3-4). For many of the bases, the measured equilibrium constant for tautomerization is on the order of 10^{-4} to 10^{-5}, but for the abnormal base, 5-bromouracil, the equilibrium constant is about 10^{-2}. Accordingly, substitution of 5-bromouracil for thymine in the DNA of bacteria, accomplished simply by growing a thymine-deficient strain on medium containing the abnormal nucleotide, leads to cells of unusually high mutability.

The fidelity with which DNA is replicated *in vivo* must be extremely great, since spontaneous mutation rates, at least for microorganisms, are quite low. Although a full explanation of how base substitution errors are avoided during replication is not yet possible, the problem is of sufficient interest to warrant some speculations based on known mutation rates in bacteria. The frequency of spontaneous mutations in any given bacterial gene has been found to lie between 10^{-5} and 10^{-8} per cell generation. Since in an average gene there are about 10^3 nucleotide base pairs, the probability that an incorrect residue will be spontaneously introduced into the DNA is on the order of 10^{-8} to 10^{-11} per base pair per generation (12). This figure can be considered as the ratio of rates between two competing reactions taking place during replication. In the normal reaction, a base is inserted into the new DNA strand according to the rules of complementarity postulated by Watson and Crick, while in the second, these rules are violated and a noncomplementary base is introduced. If the rate of noncomplementary pairing is, on the average, 10^{-9} that of complementary pairing, then the difference in free energy between the two corresponding activated complexes is:

$$\Delta F = RT \ln K = RT \ln 10^{-9} = -13,000 \text{ cal/mole}$$

Such a large free energy barrier ensures that DNA replication will be virtually free from the incorporation of noncomplementary bases.

One may validly ask at this point where such a large free energy difference comes from. Calculation of binding energies between bases in the Watson-Crick helix by the methods of quantum chemistry reveals that hydrogen bonds between complementary base pairs account for only about 25% of the total energy of association. The principal contribution is made by the energy of dispersion forces between planar purine and pyrimidine bases stacked one on top of the other at regular intervals within the twofold helix (13). From this result, it is sometimes argued that the principle of complementarity is without physical basis insofar as hydrogen bonds do not contribute the fundamental stabilizing forces to the duplex DNA molecules. In fact, however, this conclusion is too hasty. It is essential in such calculations to take into account the effects of substituting noncomplementary bases for complementary ones not merely among isolated base pairs, but under the conditions of a regular and highly ordered Watson-Crick helix. The importance of this derives from the following considerations. Owing to the strict geometrical regularity determined by the DNA helix, noncomplementary base pairs experience a mutual repulsion which will tend to destabilize the structure over limited regions. Thus, although we cannot at present calculate the free energy of association with sufficient reliability by theoretical means, it is nonetheless clear that this energy results from cooperative interactions inherent in the Watson-Crick helix and that it is not simply the sum of hydrogen bond energies determined for isolated base pairs. Moreover, an altered DNA polymerase from a temperature-sensitive strain of phage T4 has been found to increase the mutation frequency throughout the T4 genome *in vivo*, and to cause a significant rise in the occurrence of base substitution errors *in vitro* (14). This experiment implies a direct link between spontaneous mutation and fidelity to correct base pairing during DNA replication. In sum, the evidence accumulated since the Watson-Crick model was proposed in 1953 demonstrates that the principle of complementarity between A:T and G:C base pairs is the cornerstone of " molecular recognition " between nucleic acids, and that in DNA replication, this mechanism is characterized by almost error-free operation.

We shall now devote our attention to the purely macromolecular properties of DNA. The three general approaches to the physical chemistry of DNA are: (1) measurement of optical properties in solution, both absorption of light in the near ultraviolet and optical activity; (2) investigation of hydrodynamic characteristics such as sedimentation constant and intrinsic viscosity; and (3) study of light scattering which permits the determination of molecular weight and dimensions.

Nucleotides and nucleosides possess characteristic absorption bands in the near ultraviolet due to the structural properties of the purine and pyrimidine rings. The absorption maxima of all nucleotide bases except cytosine fall close to 260 mμ; the maximum absorption of cytosine is near 270 mμ. As a result of superposition, the absorption bands of both high-molecular-weight DNA and RNA have their maxima near 260 mμ (Fig. 3-7). The optical density (O.D.) of highly polymerized nucleic acids

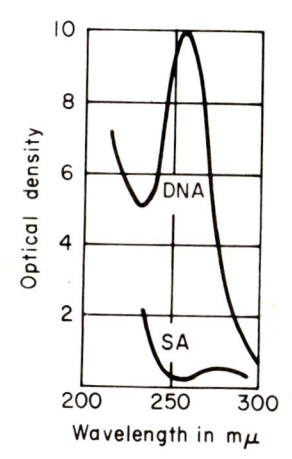

Fig. 3-7. Ultraviolet absorption spectrum of calf thymus DNA. The optical density of a 0.05% solution of the sodium salt of the DNA is plotted versus wavelength. For comparison, the absorption spectrum of a 0.05% solution of bovine serum albumin (SA) is presented on the same scale.

at 260 mμ is approximately 40% less than the total O.D. of an equal quantity of mononucleotides. This phenomenon is called the *hypochromic effect* and as we saw in the case of proteins, it is determined by the spatial relationship between constituents of the polymer chain. In nucleic acids it is a result of the parallel alignment of heterocyclic bases perpendicular to the axis of the double helix. Hypochromicity is not limited to high-molecular-weight polynucleotides; though the effect is much smaller, it is observed even in di- and trinucleotides.

A quantitative theory of the hypochromic effect was elaborated by Tinoco (15). The absorption of light by chromophoric groups, nucleotide bases in the case at hand, depends in part on the interaction of their dipole transition moments. This dependence is particularly noticeable when the chromophores are both close to one another and correlated in spatial orientation, conditions characteristic of the DNA double helix. If the dipole moments are distributed randomly in space, as in a solution of mononucleotides, their interactions neutralize one another and there is no net effect. But when the dipole transition moments are oriented in a parallel fashion, it may be readily shown that there will be a decrease in optical density or absorption. A theoretical evaluation which accounts for the electrostatic interaction of dipole transition moments leads to an answer of the same order of magnitude as the observed effect. Hence the hypochromic effect in DNA is attributed to the specific mutual orientation of its heterocyclic side groups; the hydrogen bonds between bases on opposing chains are relevant only insofar as they stabilize the regularity of the structure. By measuring the size of the hypochromic effect in terms of optical density at 260 mμ, it is possible to calculate the degree of regularity of any DNA sample.

Further important information can be obtained from other optical measurements. If a thin film of polynucleotide solution is deposited on quartz, concentrated by evaporation, and then carefully oriented by tension, the DNA molecules will become preferentially aligned in the direction of stretching. Such a preparation is highly dichroic, absorbing different amounts of radiation when the light source is polarized perpendicular and parallel to the direction of molecular orientation (16). The broad region of absorption extending from 220 to 300 mμ is actually a superposition of a major band with a maximum at 270 mμ (in the case of polycytidylic acid) and a second, much weaker one with a maximum at 280 mμ. The first absorption band, which is usually used in nucleic acid studies, is characterized by a large dichroism. The component of light polarized perpendicular to the molecular axes is the one chiefly absorbed (Fig. 3-8), signifying that the dipole transition moments are oriented per-

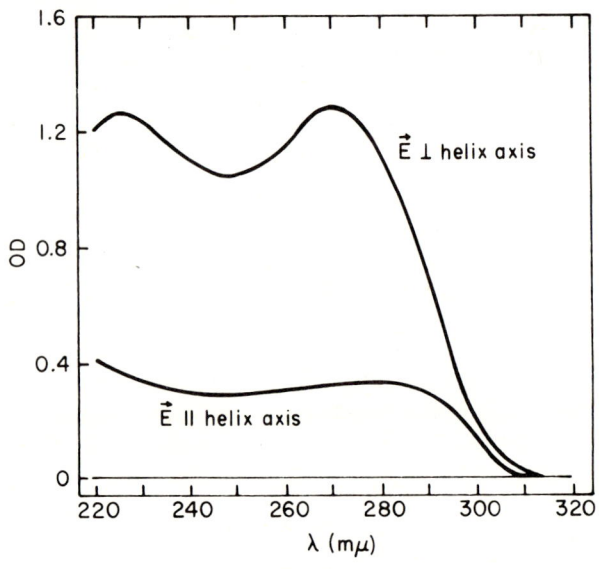

Fig. 3-8. Ultraviolet dichroic absorption curves for an oriented film of polycytidylic acid. Electric component of the radiation is (1) perpendicular and (2) parallel to the axis of the helix. From Rich and Kasha (16).

pendicular to the helix axis in the plane of the heterocyclic bases. In polynucleotides, incident light energy is adsorbed by the delocalized π-electrons of the heterocyclic rings and the transition of one of these electrons to an excited state is referred to as a $\pi \to \pi^*$-transition. As is known from quantum chemistry, the dipole moment of such a transition lies in the plane of the purine or pyrimidine ring. The dichroism of oriented DNA molecules leads to the conclusion that the planes of the purine and pyrimidine rings must be perpendicular to the helix axis. This observation is a direct confirmation of the Watson-Crick model of DNA structure.

The origins of the second, weaker absorption maximum of polycytidylic acid at 280 mμ are different, however. There are groups bound to the heterocyclic ring, such as the keto group of cytosine, in which the electrons are localized and do not participate in the system of conjugated π-bonds. The electron localized in the C=O bond, for instance, normally occupies what is called the n state. It may be excited to a π^*-state, however, in which it becomes part of the system of conjugated electrons in the heterocyclic ring. The dipole moment of this transition is oriented perpendicular to the plane of the purine and pyrimidine rings. Thus it is natural that the dichroism of the absorption band associated with the $n \to \pi^*$-transition will be the reverse of that associated with the $\pi \to \pi^*$-band. In other words, light polarized parallel to the molecular axes will be absorbed to a greater extent than that polarized perpendicular to it. There is still another interesting detail related to this phenomenon. Dipole transition moments perpendicular to the plane of the nucleotide bases will be aligned linearly and hence would be expected to interact electrostatically. Nonetheless, the theory of Tinoco predicts that a *hyperchromic effect* should be observed in this orientation, that is, there should be an increase in absorption upon formation of a highly regular helix. Such an effect is actually observed on the shoulder of the absorption spectrum near 280 mμ. The orientation of the dipole moments for both $\pi \to \pi^*$- and $n \to \pi^*$-transitions are

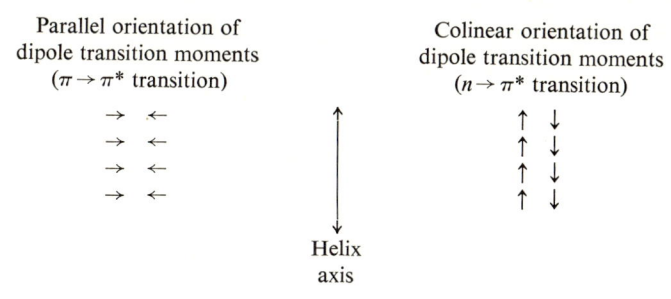

Parallel orientation of dipole transition moments ($\pi \to \pi^*$ transition)	Colinear orientation of dipole transition moments ($n \to \pi^*$ transition)
→ ← → ← → ← → ← ↓ Helix axis	↑ ↓ ↑ ↓ ↑ ↓ ↑ ↓

if the helix axis is assumed to run from top to bottom on the page.

Another important property of helical DNA is its ability to rotate polarized light. Its specific rotation is rather large and is in the positive or righthand sense: $[\alpha]_D \cong 150°$. The specific rotation of mononucleotides at the same wavelength is almost six times less, even though the ribose moieties contain asymmetric carbon atoms. Helical macromolecules strongly rotate polarized light since they contain no symmetry plane.

When DNA is heated to a temperature of about 80°C, the double helix melts within a rather narrow temperature range and assumes the random coil configuration (17). The steeply rising portion of the melting curve (Fig. 3-9) is indicative of the helix–coil transition and the midpoint temperature is conventionally taken as the melting temperature, T_m, of the helix. The melting temperature is a function of ionic strength (Fig. 3-10), thereby demonstrating the contribution of ionic interactions to the stability of the helix (18). After such heat denaturation, the hypochromic effect is reduced by 80% and the optical activity becomes almost identical to that of free mononucleotides (Fig. 3-9). It is thus clear that both these optical parameters provide information about internal order in the DNA molecules.

Optical activity strongly depends on the wavelength of light employed. It becomes larger as the wavelength decreases, as stated in the Kirkwood-Moffitt formula for helices or the Drude equation for random coils (see p. 61). In an absorption band, the rotation becomes especially large, a phenomenon called the anomolous dispersion of

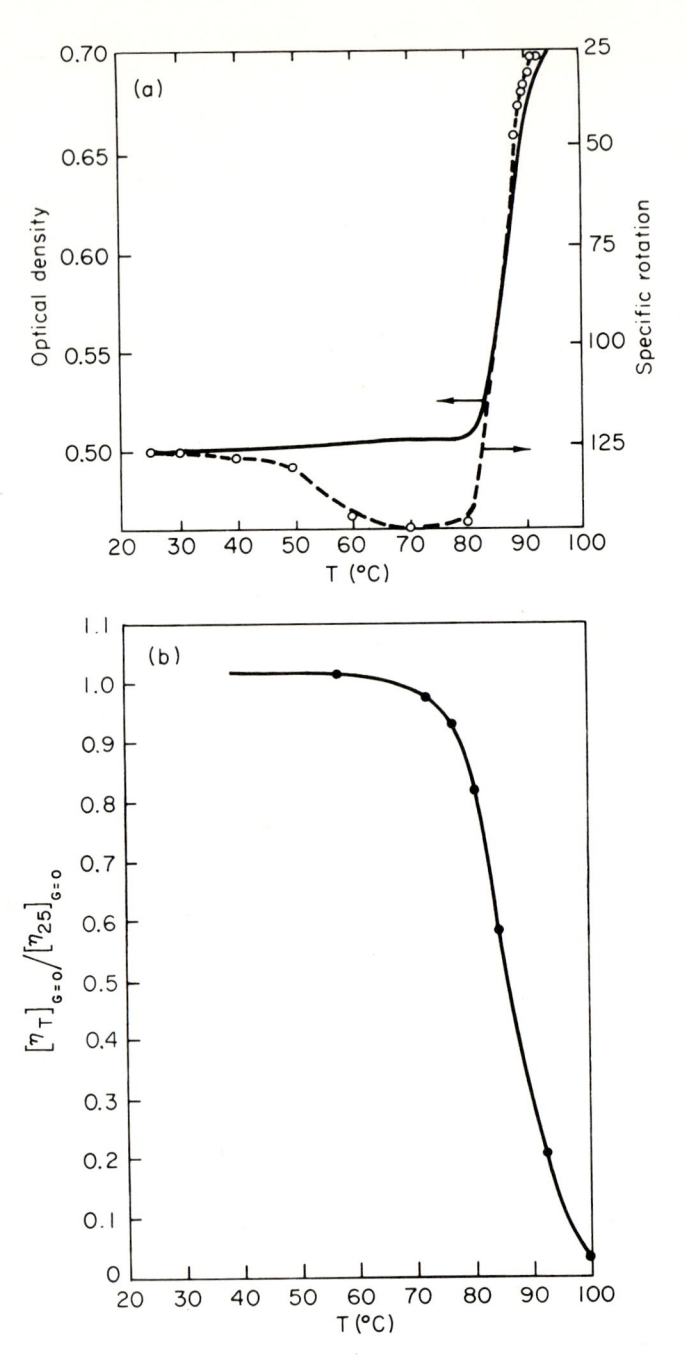

Fig. 3-9. Thermal denaturation of calf thymus DNA. (a) Changes in optical rotation (dashed line) and disappearance of the hypochromic effect (solid line) with increasing temperature. From Doty *et al.* (18). (b) Decline in the relative intrinsic viscosity of the DNA solution with temperature. Measurements were carried out at an ionic strength of 0.15. From Rice and Doty (17).

3. Macromolecular Structure of DNA

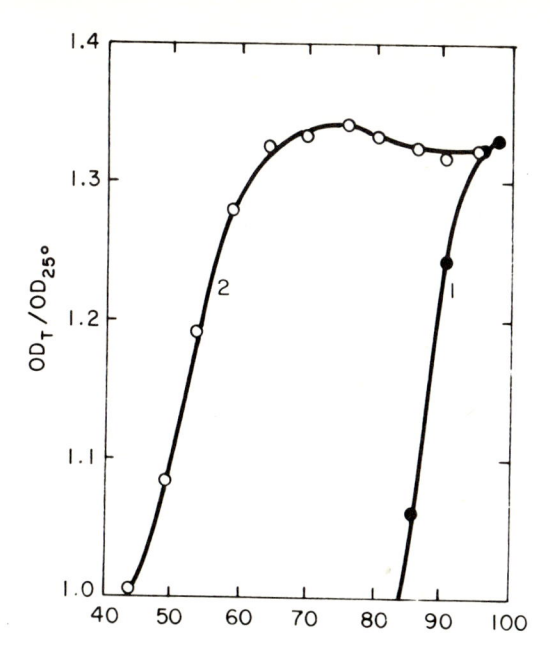

Fig. 3-10. The influence of ionic strength on the melting of helical DNA. Variations in the optical density (259 mμ) of a calf thymus DNA solution were plotted as a function of temperature at ionic strengths of (1) 0.15 and (2) 0.001. Adapted from Doty et al. (18).

optical activity. Figure 3-11 presents curves for the anomolous dispersion of DNA in both native and partly denatured states.

DNA molecules are of very high molecular weight and are quite rigid owing to the nature of the double-stranded helix. If molecules weighing 10^7 daltons were to form individual rigid rods, their linear extent would be 5 μ, based on the dimensions provided by x-ray structure analysis. However, light scattering measurements indicate that the maximum dimension of molecules of this size is about 0.5 μ, ten times less than the figure cited above. We must conclude that even twofold helices can bend and fold into a sort of loose coil when they are of very great length. The radius of gyration of DNA molecules in solution, determined from the angular dependence of light scattering, may be used to calculate what is called the *statistical segment*. In essence, this parameter is a measure of the flexibility of a polymer chain and is related to the least distance between two links of the chain which may assume random orientations with respect to each other. In the case of DNA, the length of this segment is about 700 Å, while for ordinary flexible polymers it is on the order of 10 Å. It is only because the DNA molecules are so large that we can speak of a statistical random coil even in spite of the enormous segment length. The following expression for the radius of gyration, R_G, of DNA coils as a function of molecular weight was obtained by Doty, McGill, and Rice (20): $R_G = 8.3 \times 10^{-9} M_w^{0.58}$ where M_w is the weight average molecular weight. If DNA molecules behaved as ideal random coils, the above relationship would be of the form $R_G = AM^{0.5}$; if they were completely rigid rods, the dependence would be linear, $R_G = AM$. The experimentally determined relationship demonstrates that very high-molecular-weight DNA behaves essentially as a coil.

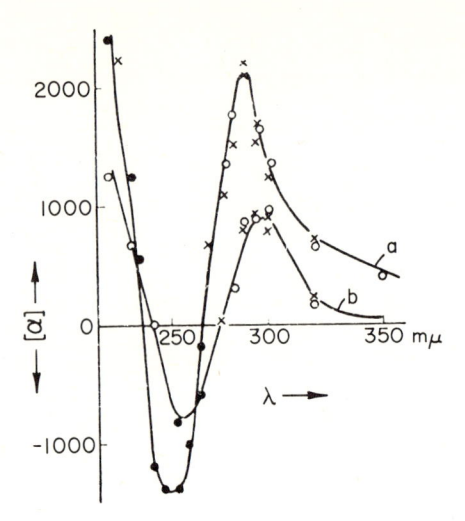

Fig. 3-11. The optical rotatory dispersion of calf thymus DNA showing anomalous dispersion near absorption maximum. (a) Native DNA. (b) DNA denaturated by heating to 100°C. From Dvorkin *et al.* (19).

The macromolecular structure of DNA in solution has been studied by means of hydrodynamic properties and various measurements have confirmed the "stiff" random coil configuration. Doty showed that sonication and other forms of hydrodynamic shearing forces can disrupt DNA molecules, breaking them into short pieces without, however, disturbing the double helix. It became apparent that all hydrodynamic methods in which a gradient of flow rate is established in the solution lead to the disruption of this polymer, and into fragments of rather well-defined dimensions at that.

The most reliable technique for evaluating the molecular weight of polymers is to measure the angular dependence of light scattered from them in solution. The main difficulty with this method is that it requires extrapolation of the data to zero angle. As shown by Zimm and his colleagues (21), most instruments permit light scattering measurements only in the range of angles between 30° and 65°, restricting their precision to polymers with molecular weights of 2×10^6 or less. Use of a specially designed apparatus with a range of scattering angles from 10° to 25°, however, yields accurate data on molecular weights up to 30×10^6 daltons. The molecular weight of still larger DNA molecules can be estimated directly from autoradiographs (p. 267f) or from electron micrographs. Application of these methods to the DNA of a number of different bacterial viruses led to the discovery that the molecules are quite homogeneous as to molecular weight. The following values were obtained for several phages that infect the bacterium *E. coli*: 31×10^6 daltons for λ, 30×10^6 for T1, and 130×10^6 for T4.

By measuring the sedimentation constant and the intrinsic viscosity (extrapolated to zero flow rates) for DNA, the dependence of these hydrodynamic parameters on molecular weight can be determined empirically (Fig. 3-12). The curves in Fig. 3-12a can be approximated by several intersecting straight lines, each of which gives the dependence quite accurately over a defined range of molecular weights. The relevant relationships are presented in Table 3-2. The dependence of s and $[\eta]$ on the molecular weight

TABLE 3-2

Empirical Formulas Relating Sedimentation Constant and Intrinsic Viscosity to Molecular Weight of Native DNA[a]

Experimental parameter	Formula	Applicable molecular weight range (daltons)
Sedimentation constant, $s^\circ_{20, w}$	$s^\circ_{20, w} = 0.116 \times M_w^{0.325}$	0.3–4×10^6
	$s^\circ_{20, w} = 0.034 \times M_w^{0.405}$	4–130×10^6
Intrinsic viscosity, $[\eta]$	$[\eta] = 1.05 \times 10^{-7} M_w^{1.32}$	0.3–2×10^6
	$[\eta] = 6.9 \times 10^{-4} M_w^{0.70}$	2–130×10^6

[a] From Eigner and Doty (22).

of highly polymerized DNA is characteristic of molecules in the random coil configuration. For ordinary random coils, the intrinsic viscosity falls within the following range: $[\eta] = KM^{0.5-0.7}$. For completely rigid rods, both experiment and theory give $[\eta] = KM^{1.8}$. As noted above, the behavior of DNA in solution is between these two extremes. When the molecular weight is relatively small, less than 2×10^6 for instance, they are more similar to rigid rods, while above that value, their molecular characteristics become more and more like those of a random coil. This is not surprising since the Watson-Crick model of the double helix permits a certain degree of flexibility. When the chain length is short, the molecule can be imagined to consist of a small number of very large statistical segments. The statistical segment of DNA lies between 600 and 700 Å, equivalent to 200 nucleotide pairs or 1.3×10^5 daltons. Thus a DNA chain with molecular weight 1.3×10^6 will contain only ten flexible segments, insufficient to constitute a random coil. When the DNA molecular weight is very high, however, the random coil model of macromolecular configuration is quite suitable.

In macromolecular random coils there is a unique relationship between molecular weight, sedimentation constant, and intrinsic viscosity. This is the so-called Flory-Mandelkern equation (24):

$$M_w^{2/3} = \frac{S[\eta]^{1/3} \eta_0 N \times 10^{-13}}{\beta(1 - V\rho)} \tag{3-1}$$

where M_w is the weight-average molecular weight of the polymer, S the sedimentation coefficient in svedbergs, $[\eta]$ the intrinsic viscosity, η_0 the viscosity of the solvent, N the Avogadro number, ρ the solvent density, and β an empirical coefficient which for most polymers falls between 2 and 2.5×10^6. DNA obeys the Flory-Mandelkern equation when $\beta = 2.5 \times 10^6$. It may thus be reasoned that there exist empirical formulas relating M to the measured values of η and S which hold for any molecular weight. In fact, such relationships were found by Crothers and Zimm (23) and they have proved extremely useful for measuring the molecular weight of DNA:

$$0.665 \log_{10} M = 2.863 + \log_{10}([\eta] + 5)$$
$$0.445 \log_{10} M = 1.819 + \log_{10}(s - 2.7)$$

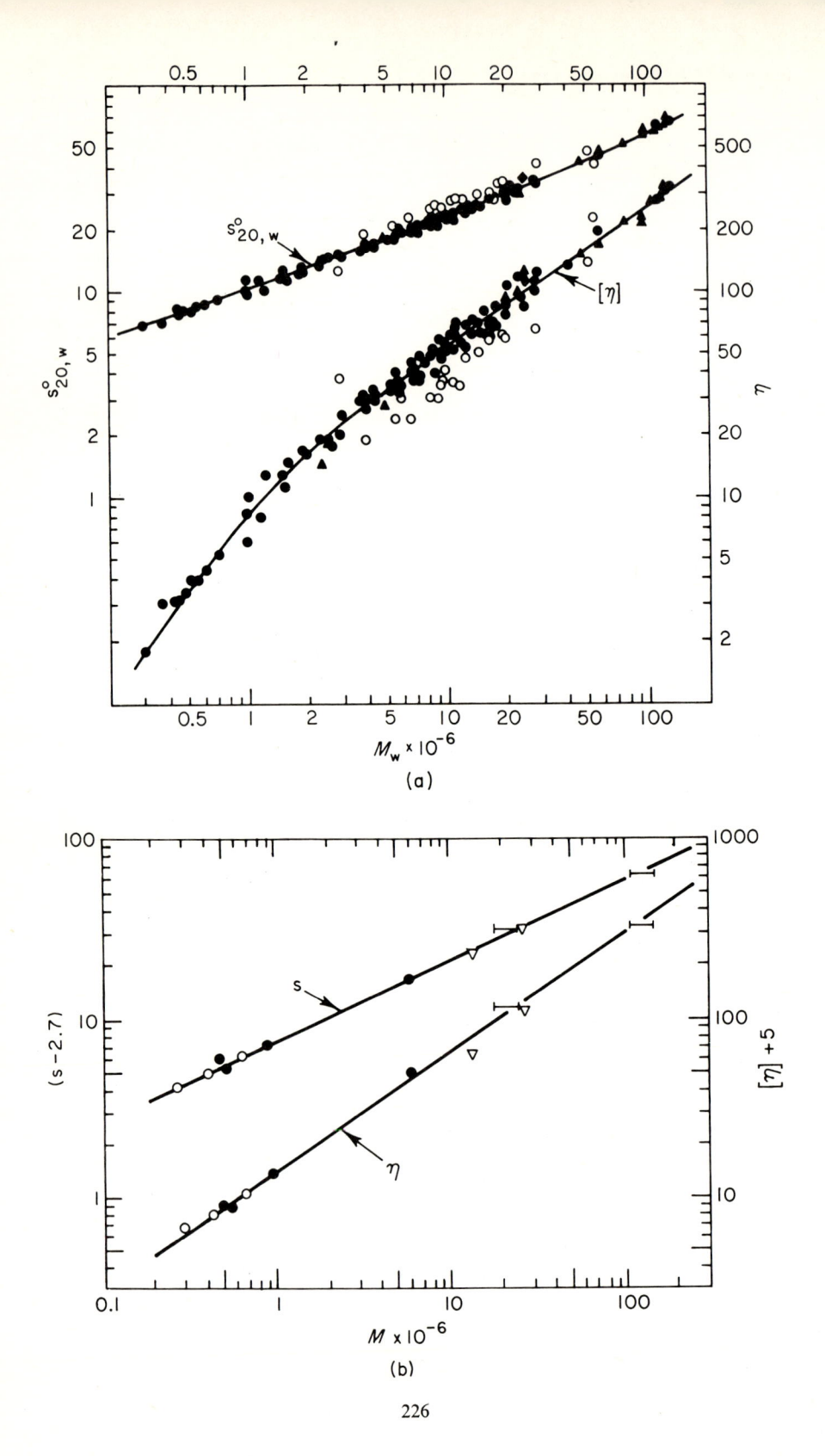

(a)

(b)

226

3. Macromolecular Structure of DNA

The dependence of s and $[\eta]$ on molecular weight is given by the general expression $s = KM^{\alpha_s}$ and $[\eta] = K'M^{\alpha_\eta}$. Values of the coefficients, α, are presented in Table 3-3 for DNA, typical random coil polymers, collagen (a fully rigid cylinder), and cellulose trinitrate (a linear polymer with a relatively rigid chain).

TABLE 3-3

COEFFICIENTS RELATING MOLECULAR WEIGHT TO PHYSICOCHEMICAL
PARAMETERS FOR MACROMOLECULES IN VARIOUS CONFIGURATIONS[a]

Polymer	Sedimentation constant $s = KM_w^{\alpha_s}$	Intrinsic viscosity $[\eta] = K'M_w^{\alpha_\eta}$
	α_s	α_η
Collagen	0.20	1.80
DNA, low-molecular-weight range	0.33	1.32
Cellulose trinitrate	0.29	1.00
DNA high-molecular-weight range	0.41	0.70
Typical random coil polymers	0.5–0.4	0.5–0.8

[a] From Eigner and Doty (22).

There is still another aspect of DNA which is important for an understanding of its macromolecular behavior. Nucleic acid molecules bear a significant charge and therefore, it would seem, they should have the properties of polyelectrolytes. However, in the pH interval where the native DNA structure is stable (from pH 4 to 11) there are practically no indications of polyelectrolytic behavior. In fact, as long as the double helix is intact, the macromolecular characteristics of DNA are not significantly affected by the ionic strength of the solution. This is particularly true of the hypochromic effect, which is almost completely independent of the ionic strength. Even the hydrodynamic properties of the macromolecules are influenced only slightly by ionic screening. The insensitivity of the hypochromic effect is easily understood if we consider that it depends on short-range interactions between neighboring residues. Hydrodynamic properties are somewhat more sensitive to electrostatic interactions since they depend on the configuration of the molecules as a whole.

At pH values from 3 to 4.5 or above 11, the secondary structure of DNA molecules is destroyed and the double helix collapses (25–27). This phenomenon is illustrated in Fig. 3-13. At acid pH the phase transition is brought about by the association of

Fig. 3-12. Sedimentation and viscosity data for native DNA as a function of molecular weight. (a) $s^\circ_{20,w}$ and $[\eta]$ were measured for each of a series of DNA samples and the molecular weight was calculated according to Eq. (3-1). From Eigner and Doty (22). (b) The hydrodynamic parameters were plotted as a function of molecular weight according to the empirical relationships of Crothers and Zimm (23) (p. 225).

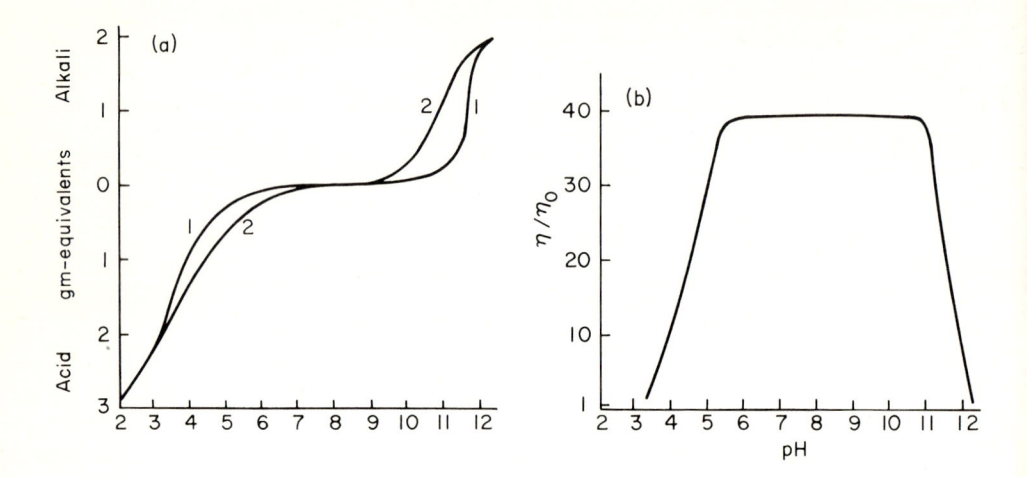

Fig. 3-13. Melting of DNA secondary structure at extreme pH values. (a) Curves for the (1) forward and (2) back titration of calf thymus DNA in gram-equivalents of acid or base per nucleotide residue. (b) The relative viscosity of calf thymus DNA as a function of pH. From Gulland and Jordan (25).

protons with the amino groups of adenine, guanine, and cytosine according to the reaction $-NH_2 + H^+ \rightarrow -NH_3^+$, which disrupts the system of internal hydrogen bonds. Denatured DNA behaves as a genuine polyelectrolyte and is very sensitive to the ionic strength of the solution. In order to study the properties of DNA under these conditions, it is necessary to screen the charges on the molecule with an ionic atmosphere, usually provided by 0.3 M NaCl. The macromolecular structure of DNA is also destroyed at alkaline pH's, apparently owing to the dissociation of a proton from the hydroxyl groups formed in guanine, cytosine and thymine as a result of keto–enol tautomerization. Thus, as is shown in Fig. 3-13, either acid or base titration of the DNA molecule leads to the breakdown of the double helix. It is curious to note that the titration curve of denatured DNA is displaced relative to that of the native DNA. This is explained by the fact that in the native polymer, the energy of the helix-coil transition is added to the energy of ionization. In the pH interval where DNA is stable, only the phosphate groups of the molecule are charged. The pK of this dissociation

$$O=P-OH \rightleftharpoons O=P-O^- + H^+$$

is equal to 1.5 and hence, the phosphate groups are always charged. They are generally screened by the presence of monovalent or divalent cations in solution. Electrostatic repulsion between phosphate groups becomes significant only at very low ionic strength ($I < 10^{-3}$) and when this occurs, the DNA structure loses it stability. For this reason, DNA is always kept in a 0.1 to 1.0 M salt solution.

The macromolecular characteristics of denatured DNA are rather different from those of native, double-helical molecules The denatured strand folds up into a random coil, a state manifested by a considerable decrease in viscosity. The form and structure of denatured DNA in solution can be studied quantitatively only if aggregation is avoided. Denatured DNA strands readily form disordered and unstable aggregates which defy consistent physicochemical characterization. There are two effective methods of combating aggregation. The first is to use very low concentrations, on the order of 0.002% or 20 µg/ml, in experiments with denatured DNA. Even at such great dilution, sedimentation constants can be accurately measured with the aid of ultraviolet absorption. An alternative method is to maintain the DNA in a charged state by using buffers of low ionic strength—on the order of 0.01 M. When these conditions are maintained, denatured DNA behaves as an almost ideal random coil and obeys all the theoretical equations relating sedimentation, viscosity and molecular weight.

The Flory-Mandelkern equation applies quite satisfactorily to denatured DNA when β is taken as 2.2×10^6. Under these conditions, the denatured DNA markedly exhibits the characteristics of a polyelectrolyte. If aggregation is completely avoided, it is possible to measure the dependence of s and $[\eta]$, and also of the radius R calculated from them, on molecular weight as a function of ionic strength. At low ionic strengths the polymer is highly charged and extended as the result of electrostatic repulsion while at high ionic strengths the charges are well screened by counterions and the molecules behave as ordinary linear polymers. Typical relationships among the hydrodynamic parameters, for both low and high ionic strengths, are given in Table 3-4. That denatured DNA departs from a theoretical random coil at low ionic strength

TABLE 3-4

Empirical Formulas Relating Physicochemical Parameters to the Molecular Weight of Denatured DNA[a]

Experimental parameter	Formula for $[Na^+] = 0.015$	Formula for $[Na^+] = 0.195$
Sedimentation constant, s_0	$s_0 = 0.060 M_w^{0.35}$	$s_0 = 0.022 M_w^{0.48}$
Intrinsic viscosity, $[\eta]$	$[\eta] = 2.6 \times 10^{-5} M_w^{0.94}$	$[\eta] = 4.9 \times 10^{-4} M_w^{0.55}$
Molecular radius, R	$R = A M_w^{0.64}$	$R = A' M_w^{0.51}$

[a] Based on data of Eigner and Doty. (22).

is reflected most obviously in the relationship between radius of gyration and weight average molecular weight; empirically, R depends on $M_w^{0.64}$ rather than on $M_w^{0.5}$ as expected for ideal random coils. Interestingly enough, in spite of this deviation, charged DNA molecules strictly adhere to the Flory-Mandelkern equation with the same value of β that is applicable to uncharged chains.

Although the native structure of DNA from most living tissue is that of a double-stranded helix, there are some notable exceptions. In 1959 Sinsheimer initiated studies on ΦX174, one of the smallest of the bacteriophages (28). Its overall molecular weight was found to be 6.2×10^6 by light scattering, and the radius of the rodlike phage particles was estimated to be 24 mµ from electron micrographs. This may be compared with the relatively large phage T2, whose molecular weight is 280×10^6. Chemical analysis revealed that each ΦX174 particle contains 1.6×10^6 daltons of DNA, equivalent to 25.5% of the total viral weight. When viral DNA was deproteinized and purified by phenol extraction, light scattering measurements showed that its molecular weight was 1.7×10^6, proving that the total DNA complement of each particle was contained in a single molecule. This DNA did not behave at all like a Watson-Crick double helix, however, and in many ways it resembled denatured DNA.

From the melting curve of normal, double-stranded calf thymus DNA depicted in Fig. 3-10, it is apparent that the optical density remains constant until about 85°C and then rises very steeply. The optical density of a solution of denatured DNA, however, rises continuously over the entire temperature interval from 0° to 90°C. Solutions of ΦX174 DNA behave in exactly the same way. Furthermore, the optical density of DNA double helices remains constant at all ionic strengths. By contrast, both denatured DNA and that from phage ΦX174 exhibit a rather high optical density at low ionic strengths, and there is practically no hypochromic effect. As the ionic strength increases, however, some hypochromicity appears as the optical density drops owing to the screening of charged phosphate groups. A likely explanation of the behavior of denatured and ΦX174 DNA's is that there are many small regions of ordered structure spread along the entire length of the single-stranded chain. It is possible that such regions result from the formation of small loops in the molecule. There is yet another peculiarity of ΦX174 DNA. Its extinction coefficient is somewhat greater than that of DNA's from other organisms, directly signaling a breach of Chargaff's rule and indicating that the total quantities of purine and pyrimidine bases are not equal to each other. In fact, chemical analysis revealed that the molar ratio of nucleotide bases in ΦX174 DNA is 1 adenine: 1.32 thymine: 0.98 guanine: 0.80 cytosine. Such a strong divergence from Chargaff's rule leaves no doubt as to the absence of double helix in this DNA.

Final confirmation of the fact that ΦX174 DNA is not double-stranded comes from its reaction with formaldehyde. Formaldehyde characteristically interacts with primary amino groups and in native DNA, this reaction proceeds slowly. In denatured DNA, however, formaldehyde is bound quite rapidly, leading to hydroxymethylated amino groups ($-NH_2 + CH_2O \rightarrow -NH-CH_2OH$) and other, more complex derivatives. In solutions of either viral DNA or of intact virus particles, this reaction can be followed by measuring the absorption of light at 265 mµ. In both cases the reaction rate is approximately the same for ΦX174 DNA as for denatured DNA from any of a number of different sources.

Its macromolecular structure notwithstanding, ΦX174 DNA is replicated in just the same way as any other DNA. From cells infected with this phage, it is possible to isolate a DNA component having all the properties of a Watson-Crick double helix, which is the intermediate or *replicative form* of the phage DNA. The chief criteria used to judge double-strandedness are density, nucleotide composition, and the ability to undergo the helix–coil transition. It follows from the data on the nucleotide composition, however, that when the phage particle is assembled, only one specific strand of the two is actually included. The second strand is apparently synthesized only to permit replication of the first; each original strand serves as a template for formation of a second, complementary strand, which in turn provides a template for the synthesis of molecules identical to the original.

ΦX174 is not the only phage that contains single-stranded DNA. Another small DNA phage of *E. coli*, called S13, was found to be of the same sort (Tessman). In practically all of its properties—DNA, dimensions, and host range—S13 is very similar to ΦX174.

When gently extracted from infected cells, single- and double-stranded phage and viral DNA, as well as the duplex replicative form of ΦX174 DNA, characteristically possesses a closed circular structure. This is also true of DNA from mitochondria. In the case of viruses, there is a special enzyme in the host cell called polynucleotide ligase which transforms the linear DNA molecules typical of mature virus particles into closed circular structures. The circular configuration of such DNA's was identified by a marked increase in sedimentation coefficient relative to linear molecules of the same type, and by electron microscopy. An interesting peculiarity of circular DNA is that it assumes a *supercoil* structure in which the duplex helix is twisted back on itself. If a single chemical bond is broken in one of the DNA strands, the supercoiled circles unwind to form a flat loop. A double-strand break leads to the formation of linear molecules.

4. Thermodynamics and the Mechanism of Phase Transition in Polynucleotides

Proteins and nucleic acids are both characterized by enormous intramolecular forces which lead to the formation of regular helical configurations. In both cases, a distinctive kind of intramolecular crystallization occurs which imposes an intimate, as well as an overall, order on each individual macromolecule. At a definite temperature, or, more precisely, within a narrow temperature range, the intramolecular structure melts, giving rise to the helix–coil transition, a transition from order to disorder. Although we are most interested in these effects as they relate to complicated natural copolymers, the very same phenomena are observed in simple polypeptides and polynucleotides, often in a much more exaggerated form.

Intramolecular melting and crystallization are examples of cooperative transitions and there are a number of statistical theories which correctly describe these phenomena. Their common feature is that at the average melting temperature, T_m, the following thermodynamic relationship obtains:

$$\Delta H - T_m \Delta S = 0 \tag{3-2}$$

where ΔH is the change in enthalpy or heat of melting per monomeric residue and ΔS is the increase in entropy per residue. The magnitude of the enthalpy, ΔH, has been measured calorimetrically on a great many samples of DNA in recent years (29). Denaturation was performed by acidifying the medium, causing the amino groups of adenine, guanine, and cytosine to acquire a proton. Additional repulsive forces between neighboring stacked bases within the molecule arise as a result, and the double helix becomes disorganized. To the heat of transition measured experimentally, it is necessary to add the heat of ionization of the amino groups which can be determined separately on mononucleotides. The heat of melting of DNA secondary structure is found to be 8000 cal/mole of nucleotide pairs.

As the molar proportion of G:C pairs in DNA increases from 37 to 64%, there is no noticeable alteration in ΔH. According to the precise measurements of Marmur and Doty (p. 270), the melting temperature of the DNA helix changes by 0.42°C when the proportion of G:C pairs increases by 1 mole%. Since the relation $T_m = \Delta H/\Delta S$ applies at the transition temperature, we obtain for the variation of temperature with nucleotide composition,

$$\delta T_m/T_m = \delta(\Delta H)/\Delta H \pm \delta(\Delta S)/\Delta S \qquad (3\text{-}3)$$

where δT_m, $\delta(\Delta H)$, $\delta(\Delta S)$ are the variations in transition temperature, enthalpy, and entropy, respectively. If the helix–coil transition is characterized by a certain increment of entropy, independent of composition, then $\delta T_m/T_m = \delta(\Delta H)/\Delta H$. It is thus evident that a change in the content of G:C pairs from 37 to 64% should result in a change in the heat of melting equal to $\delta(\Delta H)/\Delta H = 13/340 = 4\%$. Consequently, the variation in the heat of transition will be within experimental error.

In an independent determination, the heat of helix formation in the case of the simple polyribonucleotide, poly (A + U), produced by mixing solutions of poly A and poly U together in a calorimeter, was estimated to be 7000 ± 170 cal/mole. This figure is in satisfactory agreement with the heats of transition measured for the helix–coil transition in DNA.

In their consideration of the mechanism of helix–coil transitions in DNA, Doty and Marmur suggested that all hydrogen bonds between purines and pyrimidines are broken cooperatively, resulting in the appearance of single-stranded molecules which fold up into random coils. This view of the helix–coil transition in DNA was accepted without dispute at the time. Accordingly, the principal contribution to the heat of melting, ΔH, should be made by the hydrogen bonds between base pairs. At the present time, this point of view must be reconsidered in the light of new data. Tinoco first questioned the role of hydrogen bonds in this phenomenon on the basis of the energy involved in van der Waals bonds between hydrophobic groups of the nucleotide bases. The contribution of the van der Waals or dispersion forces turned out to be quite large, about 5000 cal/mole of nucleotide pairs. On the basis of this calculation, Tinoco questioned the significance of hydrogen bonds in stabilizing the Watson-Crick structure. This argument was not convincing, however, since the principle of complementarity between A:T and G:C pairs is strictly adhered to and cannot be accounted for except in terms of specific hydrogen bonding.

In addition to the calorimetric data, there were two studies of isotopic phenomena which proved extremely important in the understanding of intramolecular cooperative transitions. In a number of laboratories, it was shown that the melting temperature of DNA does not change by more than a tenth of a degree when water is replaced by deuterium oxide (30). DNA basically differs from protein in this regard, and the finding was taken as proof that the contribution of hydrogen bonding to the heat of melting, ΔH, is insignificant. This work seemed to confirm completely the calculations of Tinoco, since the only attractive forces which could explain the large value of ΔH were

4. Thermodynamics and the Mechanism of Phase Transition in Polynucleotides

the van der Waals interactions between the hydrophobic heterocyclic bases. Thus, the paradox referred to above was even further from resolution. At about the same time, studies were being carried out on the kinetics of isotope exchange between DNA and water labeled with tritium. They showed that all the hydrogen atoms of the —NH₂ and —NH groups which participate in hydrogen bond formation in the double helix exchange not instantaneously, but at a measurable rate. The time required for one half these atoms to exchange with water is about 5 minutes (31). This means that hydrogen bonds actually do exist and that they inhibit isotopic exchange between the solvent and the atoms so occupied. As a control, the amount of hydrogen bonding was measured in tRNA (32). As we shall see (p. 285), there are several double-helical regions in tRNA which result from intramolecular hydrogen bonding. The maximum possible proportion of helix in tRNA can be calculated from its known primary structure, while the actual figure can be determined experimentally by means of its hypochromic effect or structural optical activity. All methods yield a helix content between 60 and 75%. Investigation of the kinetics of isotopic exchange showed that in tRNA about 20% of the bases exchange —NH₂ and —NH hydrogen atoms instantaneously while about 80% exchange at the same rate found for DNA. This served to justify the applicability of isotopic exchange to the study of hydrogen bonds in polynucleotides.

Additional information as to the forces conferring stability on the double helix was provided by a study of DNA which had been heated to 100°C and then cooled very rapidly. After rapid cooling, the melted state of the molecules was in part preserved or "frozen," as judged from measurements of optical density and optical activity. However, roughly half the bases in the irreversibly denatured DNA still undergoes slow isotopic exchange just as in native DNA. In other words, the number of amino and imino groups participating in hydrogen bonds is diminished by a factor of two. We must thus conclude that in the helix–coil transition, hydrogen bonds within the DNA molecules are only partially broken and that the separation of the two antiparallel strands from one another is far from complete. Melting of the helical structure must therefore be due to disruption of van der Waals interactions between adjacent bases. The helix apparently unwinds and stretches out, turning into a relatively flexible double strand. The stacked purine and pyrimidine bases pull apart from one another and become surrounded with water molecules as if fully dissolved; as a result, their parallel alignment is upset wherever the chain flexes. Figure 3-14 depicts this type of helix–coil transition in which there is neither rupture of hydrogen bonds nor separation of the strands. The hypochromic effect disappears under these circumstances, however, as does structural optical activity. The heat of melting ΔH will thus be determined by van der Waals interactions between nucleotide bases and the heat of solution of hydrophobic groups, explaining its insensitivity to the replacement of H_2O by D_2O.

If DNA is heated at high enough temperatures, the intrachain hydrogen bonds will eventually break and the two strands of the double helix will dissociate. This process should be subject to a considerable isotope effect. Indeed, Maslova and Varshavsky (33) found a substantial isotope effect, on the order of 2°, when they measured the temperature dependence of elution of DNA from a Bolton-McCarthy column (p. 283), a process that requires strand separation. The isotope effect was absent, though, when the transition curve was followed by hypochromicity.

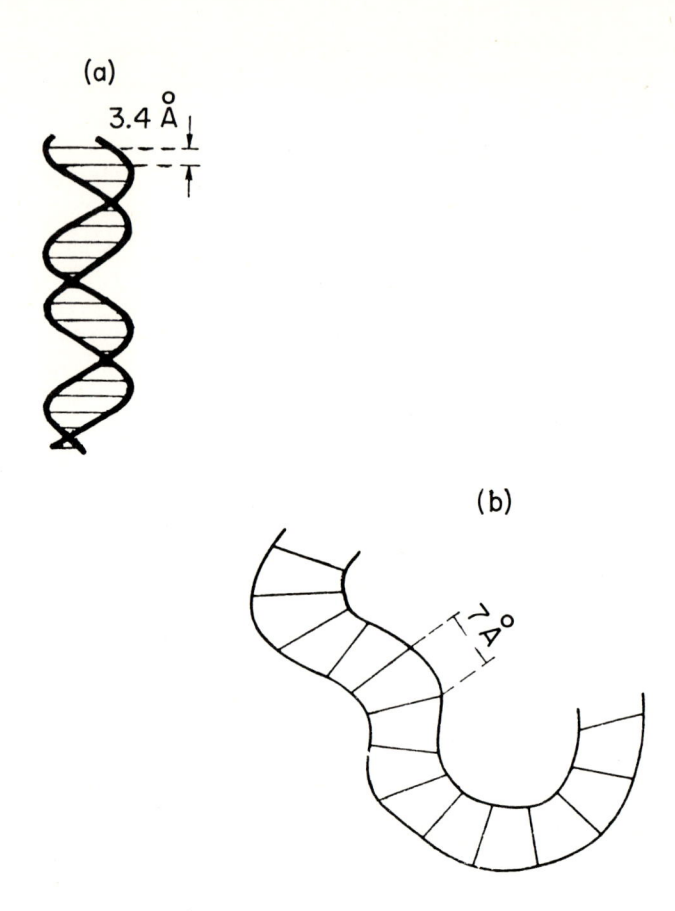

Fig. 3-14. Depiction of DNA secondary structure (a) before and (b) after melting of the double-stranded helix.

The mechanism of the helix–coil transition in DNA discussed here would not be convincing were it not for more direct evidence of the structural changes suggested. Such data were obtained by Luzzati and co-workers, who studied low-angle x-ray scattering from DNA in solution (34). The scattering produced by a solution of native DNA coincides perfectly with that predicted for rigid rods. Measurement of the absolute value of the scattering provides a means of determining the number of electrons per unit length of the rod. In the case of native DNA, this figure came out to 105 electrons/Å, in precise agreement with the Watson-Crick model. After melting of the helix, as judged by the disappearance of hypochromic effect, the scattering pattern, contrary to expectation, continues to correspond to that obtained from rods, though with a somewhat smaller effective diameter (14 Å instead of 18 Å). Moreover, the number of electrons per unit length is almost exactly halved to 47 electrons/Å. The structural data are consistent with an extended double strand in which hydrogen bonding is maintained. First, a twofold DNA strand retains considerable resistance to bending and scatters x rays like a rod rather than a random coil. Second, if a single turn of the Watson-Crick helix, 34.6 Å in length, is fully extended, its length increases to about 70 Å, explaining a drop in the linear electron density by a factor of exactly 2.

A study of the hydrodynamic properties of DNA molecules shows that the viscosity begins to decrease precipitously near the melting temperature T_m. The decrease is not limited to the interval of the helix–coil transition, though, and continues right up to 100°C. This means that hydrogen bonds between bases continue to break and that the flexibility of the DNA rods continues to grow with temperature from the melting point of the helix to the boiling point of water (35). Figure 3-15 presents two

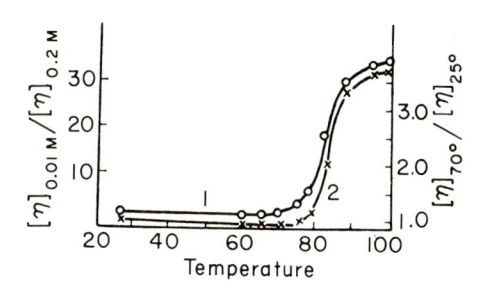

Fig. 3-15. Increase in the flexibility of calf thymus DNA molecules after thermal denaturation in 0.001 M NaCl. Flexibility was measured (1) by the ratio of intrinsic viscosities at 70° and 25°C and (2) by the polyelectrolyte effect expressed as the ratio of intrinsic viscosities in 0.01 M and 0.2 M NaCl. From Ryabchenko et al. (35).

curves which testify to the gradual increase in chain flexibility above the transition temperature. Curve 1 corresponds to values of $[\eta]_{70°}/[\eta]_{25°}$ for calf thymus DNA undergoing denaturation at the temperatures indicated along the abcissa. Curve 2 shows the polyelectrolytic behavior of denatured DNA whose structure was "frozen" by rapid cooling. In the second case, values along the ordinate represent $[\eta]_{c=0.01 M}/[\eta]_{c=0.2 M}$, the ratio of intrinsic viscosity measured at low ionic strength to that measured at high ionic strength. We see the gradual increase in chain flexibility, in particular, a polyelectrolyte effect, with temperature of denaturation. The denaturation was carried out at an ionic strength of 0.001 and at temperatures much greater than the T_m. The T_m of calf thymus DNA at this ionic strength is 64°C, but the sharpest change in flexibility is observed at 85°C, 20° above the temperature at which the hypochromic effect vanishes. To ascertain whether the DNA is single- or double-stranded at the higher temperature, it is sufficient to study the kinetics with which the molecules are degraded by DNase or x-rays. In order to sever double-stranded DNA there must be simultaneous "hits" or breaks in both chains and the decrease in molecular weight will be proportional to the square of the number of incident electromagnetic quanta; in single-stranded DNA, there will be a direct proportionality between depolymerization and radiation dose. Sinsheimer verified this rule by comparing the rate of degradation of ΦX174 DNA with that of T2 DNA as a function of x-ray dose. Using this criterion, it was shown that when DNA was denatured at temperatures lower than 85°–90°C and then rapidly cooled, the molecules behaved as if double-stranded.

Only at temperatures above the sharp rise in Fig. 3-15 did the molecules become predominantly single-stranded.

If analogies were sought between the behavior of DNA and other known phenomena of physical chemistry, the cooperative transition occurring at the T_m could be compared to the melting of a crystal, while the complete elimination of hydrogen bonds and the dissociation of the two strands might be likened to vaporization of a liquid. The increase in equilibrium vapor tension, or in our case, the increase in number of dissociated DNA strands with rising temperature is greatest above the melting temperature. Below this temperature, the heat of melting of the helix contributes to the heat of dissociation. Therefore, separation of the two strands would be expected to take place only after the T_m is exceeded. We shall return to this topic later in the present chapter (p. 273ff).

As a result of the dynamic equilibrium between double-stranded and single-stranded DNA, an exchange of strands in the double helix can occur even when the concentration of single strands is quite small. The equilibrium of the dissociation and the kinetics of exchange are not independent, but the relation between the two phenomena is indirect.

5. Synthesis and Properties of Model Polynucleotides

Just as simple synthetic polypeptides played an important role in the study of proteins, simple synthetic polynucleotides are of fundamental significance in the investigation of nucleic acids. The synthesis of polyribonucleotides became possible for the first time in 1955, when Grunberg-Managó and Ochoa isolated a new enzyme from the bacterium *Azotobacter vinelandii* (36). This enzyme was capable of catalyzing the polycondensation of ribonucleoside diphosphates into high-molecular-weight polynucleotides:

This reaction is reversible and in the presence of excess phosphate the same enzyme degrades polynucleotides by phosphorolysis, the introduction of a molecule of phosphoric acid between the residues and cleavage of the chain into nucleoside diphosphates. Because of the last-mentioned function, the enzyme was called *polynucleotide phosphorylase*. Similar enzymes have been isolated from a variety of bacterial species (37).

Condensation reactions between nucleotides exhibit certain typical peculiarities. Thus, an ester bond cannot be formed directly between two nucleoside monophosphates in aqueous solution, since the reaction would result in the exclusion of a molecule of water. Such an event is thermodynamically impossible because reaction equilibrium strongly favors hydrolysis under these conditions. In order to displace the equilibrium toward polycondensation, it is essential that the initial reactants have a sufficiently high free energy and that a small molecule other than water be eliminated in every act. If this condition is met, the formation of ester bonds will proceed with a reduction in free energy and the equilibrium will consequently favor polymerization. Nucleoside diphosphates fill this role quite well because they yield a molecule of orthophosphate on reaction rather than water. The free energy released in this reaction is sufficient to ensure that condensation will occur.

Polynucleotide phosphorylase was used to synthesize a large number of polyribonucleotides from various purine and pyrimidine bases, including those which are not normally found in natural RNA's. Among the homopolymers obtained were polyriboadenylic acid (poly A), polyuridylic acid (poly U), polyriboguanylic acid (poly G), polyribothymidylic acid (poly T), polyribocytidylic acid (poly C), and polyriboinosinic acid (poly I) which contains the base hypoxanthine. In addition, many copolymers were synthesized that contained combinations of two or three different types of mononucleotides.

The enzyme polynucleotide phosphorylase is widely distributed among bacteria and it has also been found in a variety of tissues from higher organisms. There was initially some question as to whether this enzyme bore the main responsibility for RNA synthesis within the cell. It has become clear, however, that this is not its principal physiological role. In living matter there exists a special chemical mechanism for producing specific RNA's of defined nucleotide sequences. Numerous experiments have shown that nucleotide sequences in such RNA's are invariably transcribed from complementary sequences in specific DNA templates. This mechanism is not reflected in the reaction discovered by Grunberg-Manago and Ochoa. Although priming with short oligonucleotides accelerates the polymerization reaction and helps to overcome the lag preceding initiation of synthesis in certain cases—with poly G, for instance—this phenomenon in no way approximates the transcription of RNA in living cells. The base sequence in a polynucleotide formed from a mixture of the four naturally occurring monomers is as random as that in any other synthetic copolymer. In addition, oligonucleotide primers are always chemically bound at the beginning of the chains and hence serve as initiators rather than templates for polynucleotide synthesis. If the synthetic reaction catalyzed by polynucleotide phosphorylase has any physiological significance, it must be concerned with the formation of nonspecific RNA which does not participate in the transmission of genetic information. It is more likely that in the conditions of the cell, polynucleotide phosphorylase functions as a nonspecific degradative enzyme much the same way in which cathepsins are needed for the degradation of proteins, although they play no role in protein synthesis. Our present interest in the polynucleotide phosphorylase reaction will be confined to the means it provides for obtaining various model polynucleotides.

In addition to polyribonucleotides, the study of polydeoxyribonucleotides was also of considerable importance. Early attempts to synthesize the latter compounds did not meet with much success and the problem was finally resolved with the techniques of organic chemistry. Khorana developed a stepwise condensation of mononucleotides which requires the presence of a special reagent called dicyclohexylcarbodiimide to take up the molecule of excluded water (cf., p. 28):

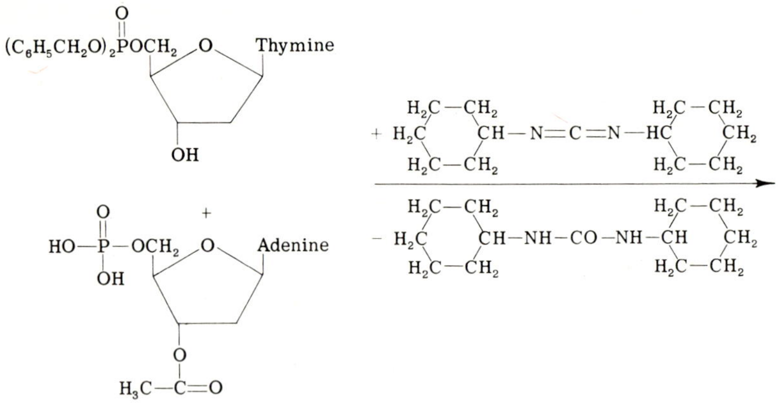

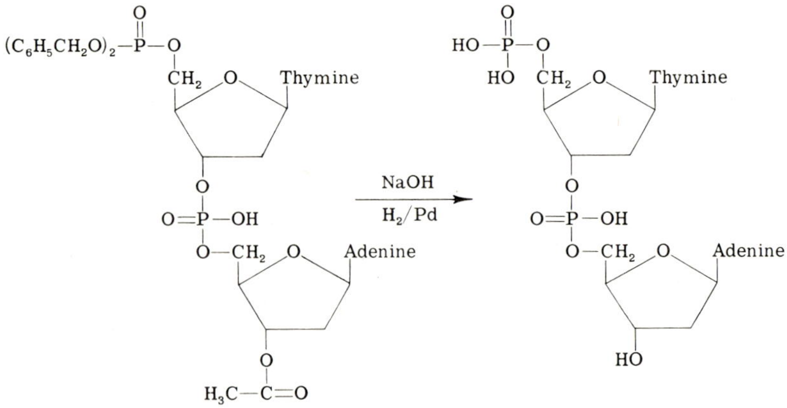

As usual, it was necessary to mask unwanted reactive groups during the reaction and to regenerate them again afterwards. The 3'-OH groups were rendered inactive by acetylation and the 5'-phosphate groups were masked as benzyl esters. A number of different oligonucleotides containing up to fifteen individual residues were obtained by this reaction. The products were readily fractionated by ion exchange chromatography on DEAE-celloluse. Khorana and his associates have recently performed the synthesis of duplex polydeoxyribonucleotides containing up to 60 residues in defined sequence by using the enzyme polynucleotide ligase to catalyze formation of phosphodiester bonds between shorter segments synthesized by chemical methods (38).

Soon after some of the simplest polyribonucleotides had been synthesized—homopolymers such as poly A and poly U—it was shown that they can interact with each other under appropriate conditions. In particular, pairs of complementary homopolymers form stable bimolecular complexes which sediment and electrophorese as single substances. By mixing polynucleotide solutions and then concentrating them under vacuum, Rich and Davies observed the formation of gelatinous precipitates from which it was possible to draw fine filaments. These fibers were further oriented by additional concentration and tension and they were finally subjected to x-ray structure analysis (39, 40). The x-ray patterns, which contained up to twenty independent reflections, were almost as detailed as those obtained from DNA. On the basis of these data, it was inferred that the synthetic complexes were double-helical in nature and that the disposition of polynucleotide strands within them was consistent with the Watson-Crick model. The pitch of the poly (U + A) helix is approximately 34 Å, and the axial distance between adjacent nucleotides is indicated by a strong meridional reflection at about 3.4 Å; a single turn is thus composed of ten individual residues. These dimensions are all nearly identical to those of DNA. The adenine and uracil side groups are, of course, situated within the helix and the pattern of hydrogen bonding between uracil and adenine in this case is most likely analogous to that of the T : A pair [Fig. 3-4(a)] in DNA (39, 41). Synthetic poly (U + A) even crystallizes into the same lattice as DNA, although the dimensions of the unit cell are somewhat different. Consequently, the only dissimilarity between native DNA and the synthetic RNA is the distance between the centers of neighboring helical molecules in the crystal lattice. For the double-stranded polyribonucleotide, this distance is equal to 28.8 Å, while in DNA it is 6 Å less. This difference arises from the two kinds of sugar residues found in RNA and DNA. Ribose contains one more oxygen atom than deoxyribose and as a result it requires more space, leading to an increase in the effective diameter of the helix.

Complex formation between poly A and poly U in solution, depicted schematically in Fig. 3-16, can be conveniently studied by the hypochromic effect at 260 mμ. Figure 3-17 shows how the optical density varies with the ratio between components in the mixture. All the solutions in this experiment contained 0.1 M NaCl and were at neutral pH. Under these conditions, all the nucleotide residues are charged due to the dissociation of the phosphate groups:

$$-\text{O}-\overset{\overset{\text{O}}{\|}}{\underset{\underset{\text{ONa}}{|}}{\text{P}}}-\text{O}- \longrightarrow -\text{O}-\overset{\overset{\text{O}}{\|}}{\underset{\underset{\text{O}^-}{|}}{\text{P}}}-\text{O}- + \text{Na}^+$$

The ions of a neutral electrolyte such as NaCl provide an atmosphere of counterions which electrostatically screens the charges borne by the polynucleotide chain in

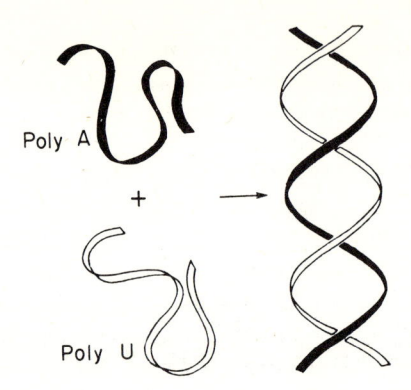

Fig. 3-16. Diagram illustrating the formation of a helical complex between polyadenylic acid (poly A) and polyuridylic acid (poly U) from randomly coiled polymers in solution. From Rich (42).

accordance with the Debye-Huckel theory. The average statistical thickness of the Debye-Huckel atmosphere is:

$$R = (10^3 DkT/8\pi e^2 NI)^{1/2} \tag{3-4}$$

where D is the dielectric permeability, k the Boltzmann constant, e the electronic charge, N the Avogadro number, and I the ionic strength. At ionic strengths near 0.1, R has a value of about 10 Å, sufficient to neutralize the Coulomb repulsion between individual chains so that they may form helical complexes with one another. It is apparent from Fig. 3-17 that when the molar concentrations of poly U and poly A are equal, the optical density of the mixture is at a minimum, corresponding to the greatest hypochromic effect. From the magnitude of the latter we find that more than 95% of the bases are associated with one another in a 1:1 ratio.

The possibilities for complex formation are not exhausted by the above example, however. If divalent magnesium ions are introduced into the solution at a concentration of about $10^{-2} M$, the charges on the polynucleotide chain are neutralized more

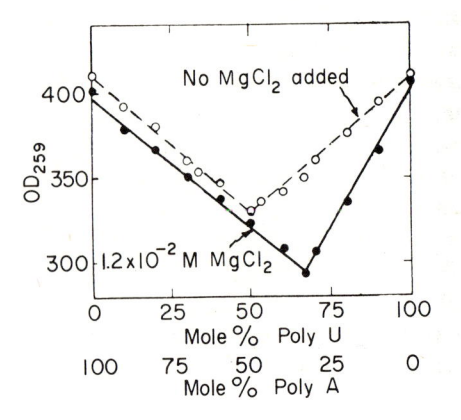

Fig. 3-17. Complex formation between poly A and poly U as measured by the dependence of optical density at 259 mμ on the proportion of the two polynucleotides in the mixture. From Felsenfeld et al. (43).

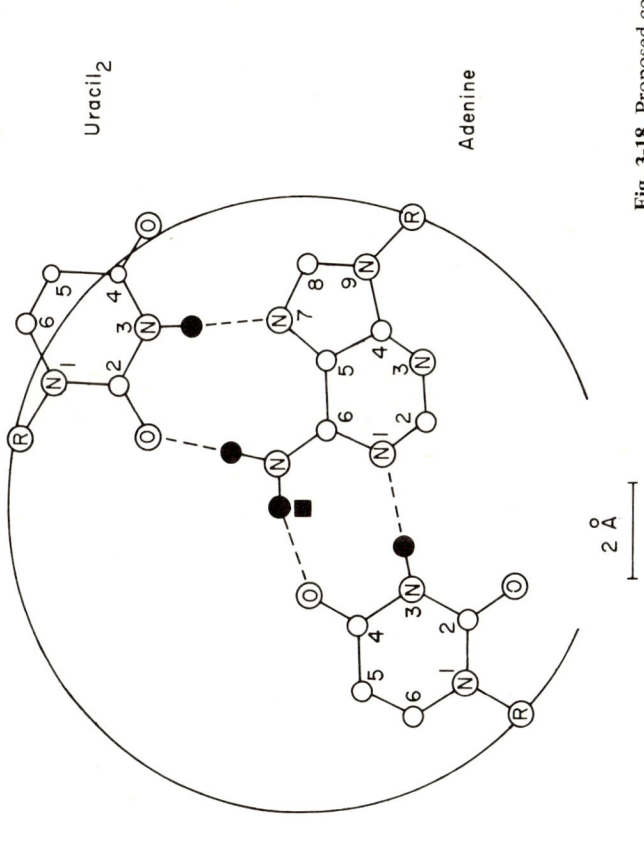

5. Synthesis and Properties of Model Polynucleotides   241

Fig. 3-18. Proposed configuration of bases in the triple-stranded poly (U + U + A) complex perpendicular to the helix axis. In this model, based primarily on infrared spectral data, uracil₁ and adenine are hydrogen bonded in the classic Watson-Crick fashion, while uracil₂ is hydrogen-bonded to the 6-amino group and N-7 of adenine through its C-2 oxygen and N-3, respectively. There are no bonds between the uracil residues, however. The two poly U strands in this scheme are antiparallel to poly A. Open circles represent carbon atoms, and solid circles represent hydrogen atoms. Adapted from Miles (44).

fully than by monovalent cations. Bonds of the type —P—O—Mg—O—P— are formed which are more than ionic interactions, since Mg^{2+} is capable of establishing coordination bonds with phosphate ions. As a result, the polynucleotide chains are quite effectively discharged and associate so as to form triple helices containing one strand of poly A and two strands of poly U, designated poly (A + U + U). Magnesium ions are not essential for this interaction, since the same result is produced by $MnCl_2$ and even by NaCl when its concentration is increased to 0.7 M. The formation of triple-stranded helices was confirmed both by crystallographic and ultracentrifugal analyses. It is of interest that the Watson-Crick model permits the third strand to position itself in the so-called deep groove of the double-stranded helix. When this happens, three bases, one from each strand, are found to be associated by hydrogen bonds in the plane perpendicular to the helix axis (Fig. 3-18). Precisely the same kinds of structures are formed between polyriboadenylic acid and polyribothymidylic acid when the latter replaces poly U.

Because polyguanylic acid (poly G) and polycytidylic acid (poly C) are complementary to each other with respect to hydrogen bonding, they interact in much the same way as poly A and poly U, forming the complexes poly (G + C) and poly (G + C + C) (45a). In some studies, it was expedient to replace poly G in the model system by polyinosinic acid (poly I), a polymer whose nitrogenous side chain consists of hypoxanthine (45b). Hypoxanthine is a purine distinguished from guanine only by the absence of a primary amino group on the heterocyclic ring. Poly I and poly C complex to form double-stranded poly (I + C) which possesses a typical DNA-like structure (Fig. 3-19). A triple helix is not possible in this case since base pairing in the double helix exhausts all opportunities for hydrogen bonding.

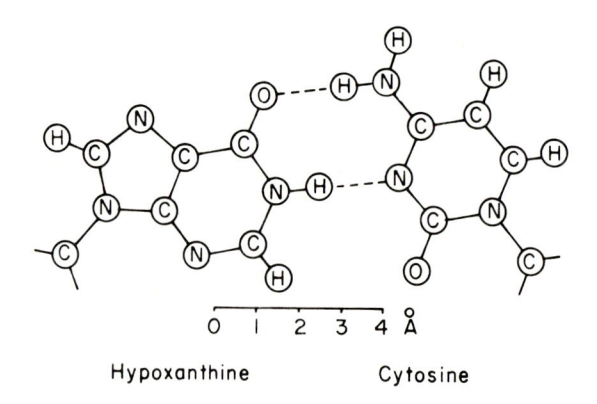

Fig. 3-19. The arrangement of bases in the double-stranded complex poly (I + C). From Davies (45b).

In the course of studies on complex formation, it became apparent that simple polynucleotides have the capacity to form much more diverse helical complexes than naturally occurring, heterogeneous copolymers. For example, chains composed of a single type of pyrimidine nucleotide such as poly U are able to complex with themselves.

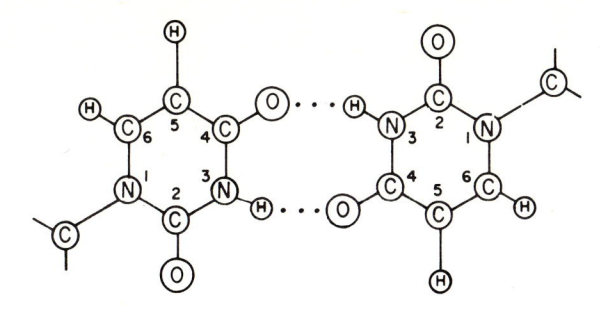

5. Synthesis and Properties of Model Polynucleotides 243

Fig. 3-20. A parallel chain model for helical poly (U + U) showing possible hydrogen bonds between uracil residues.

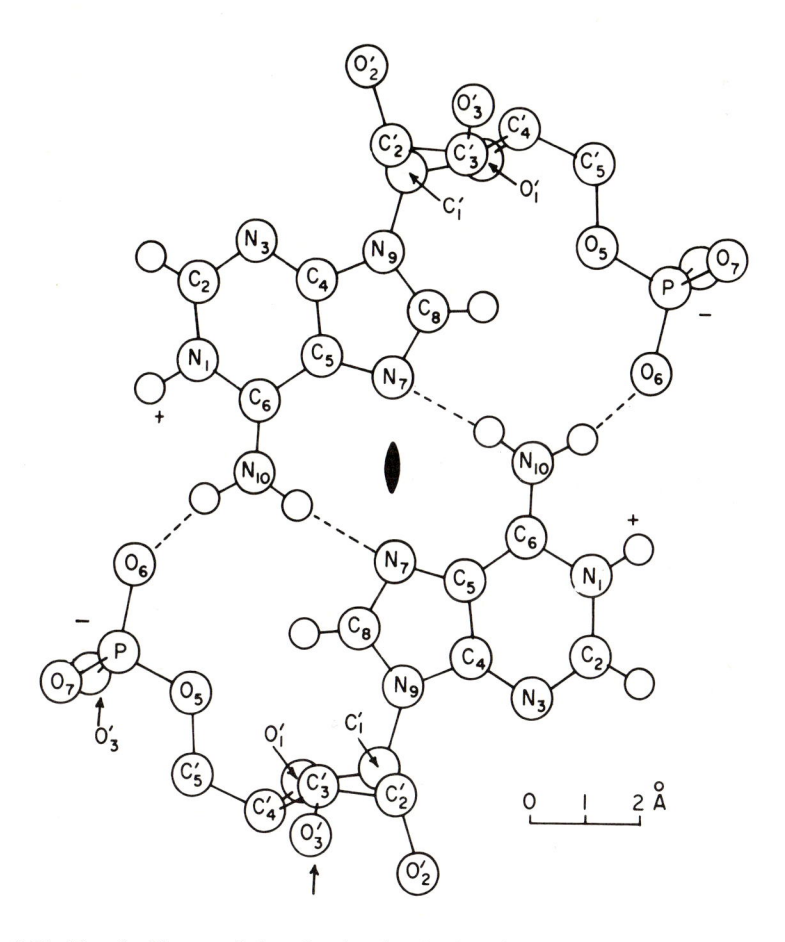

Fig 3-21. The double-stranded molecule of polyadenylic acid as viewed down the fiber axis. The two polynucleotide chains are parallel to one another rather than antiparallel as in DNA. There are two hydrogen bonds between each pair of adenine residues and an additional hydrogen bond between the adenine amino group (N_{10}) and the phosphate oxygen atom (O_6) of the opposite chain. An extra O'_3 atom is shown to indicate the course of the backbone chain. From Rich et al. (48).

Thus poly U in solution has been found to possess considerable secondary structure below 5°C as indicated by characteristic changes in ultraviolet absorption and Cotton effects between 0° and 20°C (46). The precise configuration of poly U at low temperatures has not yet been confirmed by x-ray diffraction, but if poly U forms a twofold helix under these conditions, it is likely that the strands run parallel to each other as shown in Fig. 3-20, and not antiparallel as in DNA. This finding in no way contradicts theory, since all requisites for hydrogen bonding are met by the U:U base pair. Poly C can also complex with itself to form a double helix, but only under acidic conditions in which its amino groups are charged (47). Ionic forces apparently help to stabilize such complexes.

Complexes composed of two poly A strands exhibit some unusual properties (Fig. 3-21). The diameter of poly (A + A) double helices is somewhat greater than in the case of DNA. Furthermore, the double helices form only below pH 5 and are unstable at neutral pH. It was shown that at low pH, the N-1 atom of the purine ring binds a proton giving rise to a system of ionic bonds within the twofold helix. The bond is formed between negative phosphate groups in one chain and oppositely positioned positive groups in the second chain localized at the N1 atom of the adenine ring. This phenomenon lends great stability to the poly (A + A) complex and could be termed intramolecular salt formation. At pH 4.25 its melting temperature (T_m) is 90°C. The distinctive structure of this complex was worked out by x-ray diffraction

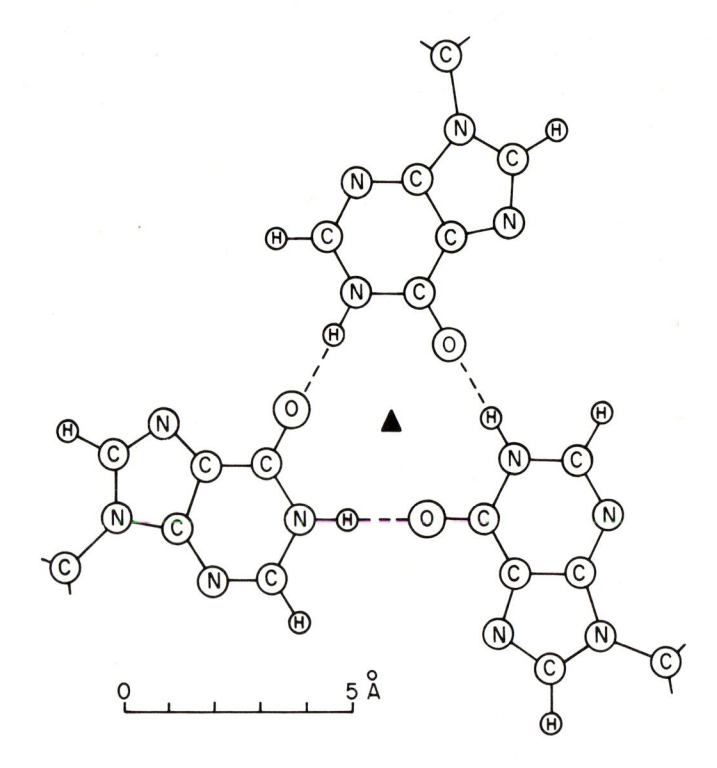

Fig. 3-22. The hydrogen-bonding system for three-stranded polyinosinic acid. Each of the three hypoxanthine bases forms one hydrogen bond with each of its partners. From Rich (49).

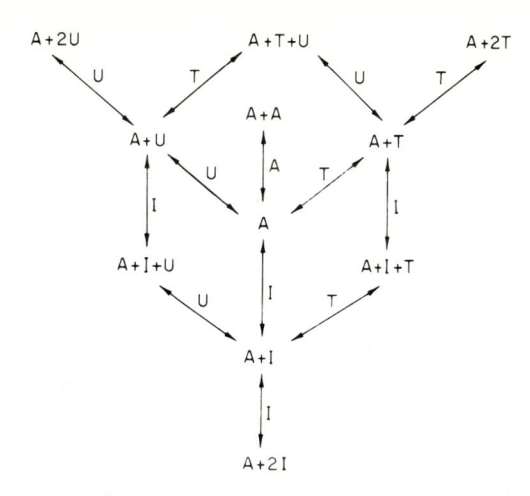

5. Synthesis and Properties of Model Polynucleotides

```
   A+2U           A+T+U          A+2T
      ↘ U    T ↙      ↘ U    T ↙
            A+A
      A+U    ↑↓      A+T
       ↑↓  U↘  ↗T    ↑↓
       I        A     I
            ↑↓
             A
      A+I+U   ↑↓     A+I+T
       ↘ U   ↗ T ↙
            A+I
            ↑↓ I
            A+2I
```

Fig. 3-23. Diagram illustrating the formation of complexes between polyadenylic acid (A), polyuridylic acid (U), polyinosinic acid (I), and polyribothymidylic acid (T). The pairs or triplets of letters represent two-stranded or three-stranded helical complexes, respectively, and the double arrows indicate that the reactions can go in either direction under suitable conditions. From Rich (40).

analysis (Rich, Davies, Crick, and Watson (48)). It was found that the specificity of complex formation resided mainly in hydrogen bonding between bases, but that there also occur hydrogen bonds between phosphate groups in the molecular backbone and amino groups attached to adenine.

Purine polynucleotides can also form triple helices at high ionic strengths or in the presence of Mg^{2+} ions. The structure of poly $(I + I + I)$, illustrated in Fig. 3-22, was solved by x-ray analysis (49). The three interacting side groups lie in a plane perpendicular to the helix axis as was the case with poly $(A + U + U)$, but in this case there are identical hydrogen bonds linking each pair of bases.

A scheme showing permitted complexes between sets of purine and pyrimidine homopolymers is presented in Fig. 3-23. Complex formation is reversible and from the scheme it is apparent that some of the more complicated structures can be obtained directly from simpler ones. In order to obtain double- and triple-stranded complexes, it is generally necessary to start with polynucleotide solutions in which the components have not already reacted with each other. In other words, it is desirable that the initial material be in the random coil configuration. This is achieved through the proper choice of pH, ionic strength, and temperature. We have seen, for instance, that acid pH leads to the ionization of amino groups with significant consequences for polynucleotide conformation. High pH also induces instability in polynucleotide complexes. When the pH exceeds 10, poly $(I + I + I)$ dissociates into individual chains, apparently because the keto-enol tautomerization is shifted in favor of the enol form and the resulting hydroxyl groups become ionized. Finally, a most important factor in the stability of twofold and threefold helices is temperature. Such complexes melt in a rather narrow temperature interval (Fig. 3-24). The cooperative transition which occurs is very similar to that observed in DNA and proteins.

Let us consider the more complicated case of association between the copolymer

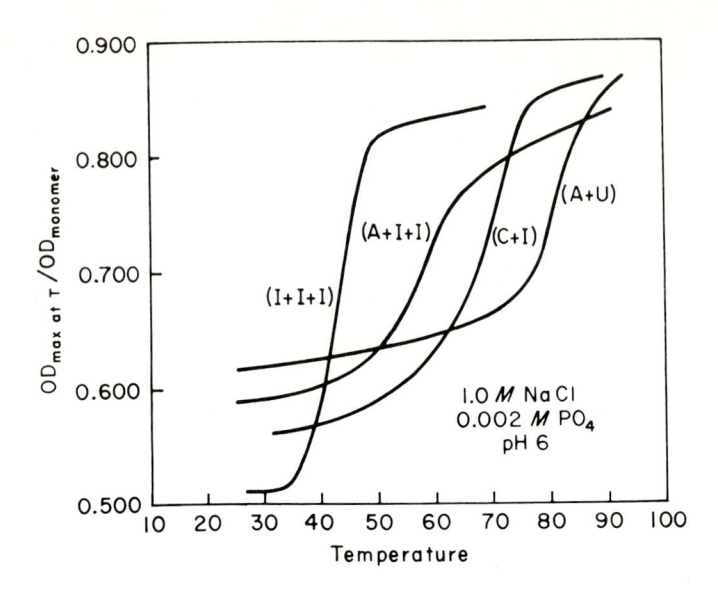

Fig. 3-24. The melting of secondary structure in two- and three-stranded helical polyribonucleotide complexes. The transition from order to disorder is measured by the variation in optical density with temperature in a solution containing 1.0 M NaCl in 0.002 M phosphate buffer, pH 6. A stands for polyadenylic acid, I for polyinosinic acid, C for polycytidylic acid, and U for polyuridylic acid. From Doty et al. (18).

poly AU and the homopolymers poly U and poly A. Despite the imperfect complementarity between the two kinds of polymers they nonetheless form double-helical complexes. By varying the ratio of A:U in a series of copolymers, it was observed that the greatest hypochromic effect, corresponding to the maximum amount of complex formation, occurs when the number of residues in the homopolymer is exactly equal to the number of complementary residues in the copolymer. It was concluded that complex formation does not take place as

$$\begin{array}{c}-A-U-U-A-A-U-A-A-U-A-\\ \vdots\quad\vdots\quad\vdots\quad\vdots\quad\vdots\quad\vdots\\ -U-U-U-U-U-U-U-U-U-U-\end{array}$$

which would leave a large fraction of the uracil residues in the homopolymer unbonded, but rather as

$$\begin{array}{c}\text{U—U}\quad\text{U}\quad\text{U}\\ |\quad|\quad\wedge\quad\wedge\\ -A\quad A-A\quad A-A\quad A-\\ \vdots\quad\vdots\quad\vdots\quad\vdots\quad\vdots\quad\vdots\\ -U-U-U-U-U-U-\end{array}$$

where the unpaired uracil residues of the copolymer are squeezed out of the chain into loops. The latter structure provides the greatest possible opportunities for hydrogen bond formation.

In order to avoid anomalous effects at the ends of the helical regions, both components were chosen with high molecular weights: about 1,000,000 for the homopolymer and about 50,000 for the copolymer. Each molecule of the homopolymer binds on the order of twenty copolymer chains. Under these conditions, most of the residues of the homopolymer will be hydrogen bonded to complementary bases in the copolymer. These considerations indicate that while the overall effect of double-helical structure is to confer considerable rigidity on the polynucleotide complex, the individual strands nonetheless retain significant flexibility. This is reflected in the ease with which unpaired bases can form loops that do not contribute to the helix (50).

It is possible that loop formation might serve as a model for the generation of errors in DNA replication, that is, for the production of spontaneous mutations. A point mutation, defined as an alteration in a single base pair of a DNA sequence, could be explained by the occurrence of a loop at the appropriate position in a replicating gene. That particular part of the gene would be miscopied with a resultant change in coded information. The manner in which loop formation might produce known types of point mutations is illustrated in Fig. 3-25.

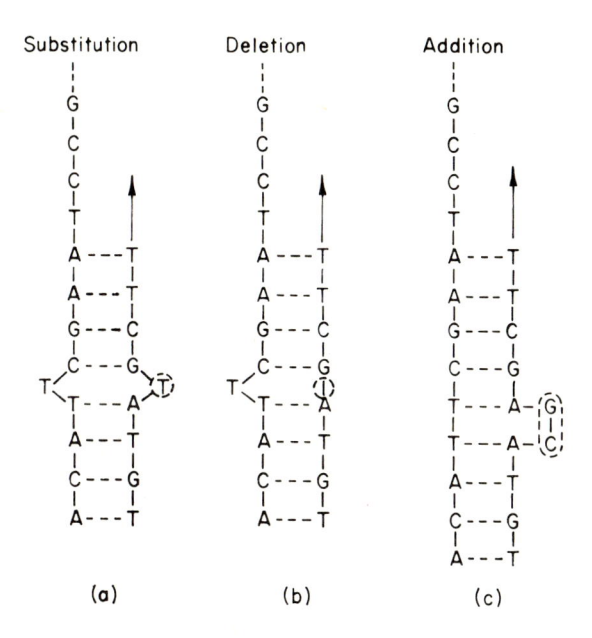

Fig. 3-25. Hypothetical models for the occurrence of point mutations during DNA replication. In each case, the template strand is on left and errors in the growing chain on right are enclosed by dashed circles. (a) Both the incorrect base and its noncomplementary partner have looped out of the growing helix, enabling succeeding regions to pair normally. (b) The incorrect base has paired with a succeeding base on the template, causing the intervening template residue to loop out of the helix. (c) The incorrect bases inserted into the growing strand have formed a loop, permitting chain growth to proceed normally. From Fresco and Alberts (50).

6. Enzymatic Synthesis of Nucleic Acids

The enzymatic synthesis of DNA *in vitro* was first achieved by Kornberg and his co-workers, marking one of the most significant advances in modern nucleic acid biochemistry. The synthetic reaction, as may be seen from the accompanying scheme, consists of an enzyme-catalyzed condensation of nucleoside triphosphates into a polymer chain with the liberation of pyrophosphate:

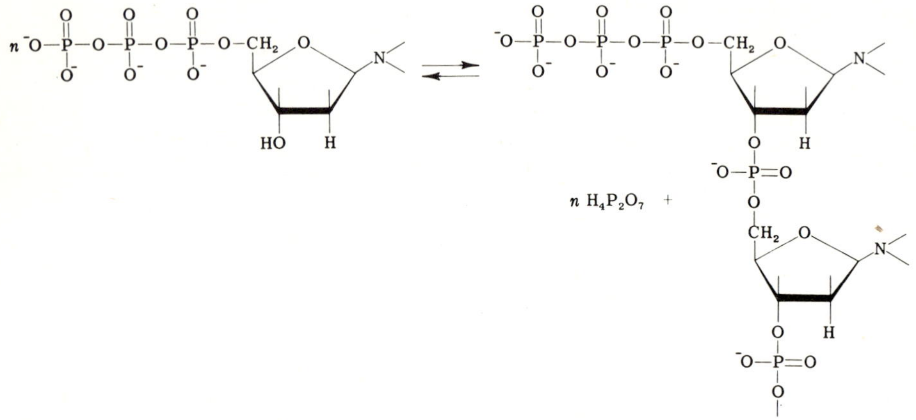

The necessary enzyme, *DNA Polymerase*, was first isolated from *E. coli*. A kilogram (dry weight) of bacteria yields a few hundred milligrams of the protein. At present, similar enzymes have been obtained from many varieties of bacteria and from numerous enzymes have been obtained from many varieties of bacteria and from numerous tissues of higher organisms (51–53). *E. coli* DNA polymerase consists of a single polypeptide chain with a molecular weight of 109,000 (54).

The Kornberg reaction exhibits two remarkable properties. First, in order that the synthetic reaction proceed at a rapid rate, it is necessary for the mixture of monomers to contain all four nucleoside triphosphates which normally occur in the DNA chain. Second, the reaction requires priming by a DNA template. If these conditions are met, the newly synthesized nucleic acid frequently exceeds the quantity of primer by a factor of twenty or more. The primer DNA must possess the specificity necessary for self-replication in its own structure since the enzyme itself serves as a universal agent, capable of catalyzing the duplication of DNA from any source whether derived from bacteria, viruses, or from the tissues of higher organisms. By most criteria the synthetic DNA molecules are copies of the priming substance. In such properties as $(G + C)/(A + T)$ base ratio and melting temperature, T_m, for instance, synthetic DNA reflects the characteristics of native DNA. It may thus be concluded that the priming DNA serves as a template for new DNA and that it is copied in compliance with the principle of complementarity inherent in the Watson-Crick model.

The principle of complementarity may be subjected to even more rigorous and detailed proof with the help of DNA polymerase (55, 56). One of the monomers in the enzymatic incubation can be labeled with radioactive phosphorus (^{32}P) in its

6. Enzymatic Synthesis of Nucleic Acids

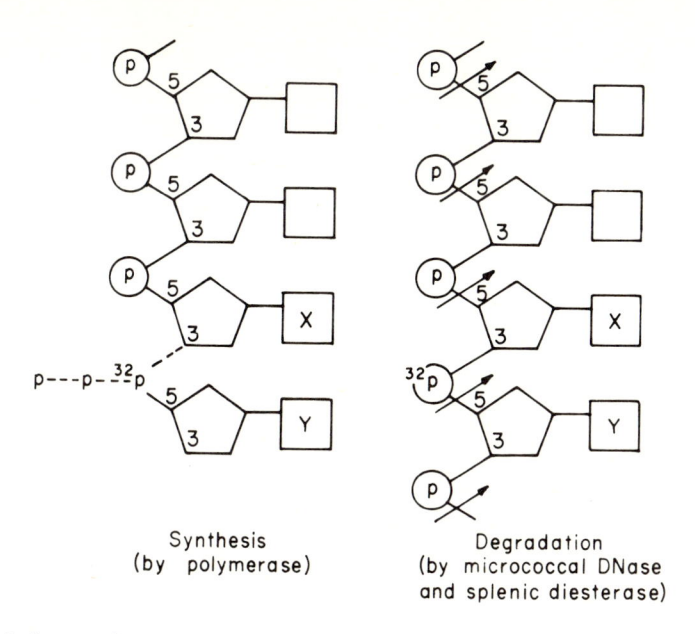

Synthesis
(by polymerase)   Degradation
(by micrococcal DNase
and splenic diesterase)

Fig. 3-26. An experiment for verifying the principle of complementarity during *in vitro* DNA synthesis. New DNA strands are synthesized from ^{32}P-labeled 5′-deoxyribonucleotides using DNA polymerase. The DNA is then degraded enzymatically to 3′-deoxyribonucleotides by the combined action of micrococcal DNase and calf spleen phosphodiesterase. This results in the shift of the labeled P atom from the original 5′-nucleotide to its neighbor. From Josse *et al.* (56).

innermost phosphate group. Let us take labeled adenosine triphosphate as an example; the radioactive phosphorus atom is bound through an ester linkage to the 5′-carbon atom of deoxyribose (Fig. 3-26). After polymerization, the labeled DNA is isolated from the incubation mixture and then completely hydrolyzed with two enzymes, spleen phosphodiesterase and micrococcal DNase from *Staphylococcus aureus*. Both enzymes are of identical specificity, cleaving the bond between the 5′-carbon atom of deoxyribose and phosphorus. The hydrolyzate thus consists of a mixture of nucleoside-3′-monophosphates. After cleavage, each labeled posporus atom which was originally attached to the 5′-carbon of adenylic acid is now attached to the 3′-carbon of the neighboring residue in the chain. By separating the four nucleoside-3′-phosphates by paper electrophoresis and measuring the quantity of radioactive phosphorus in each fraction, the distribution of residues adjoining adenine, specifically, before adenine in the direction of synthesis, can be determined. If the same sort of analysis is performed successively with ^{32}P-labeled guanylic, thymidylic, and cytidylic acids, the entire distribution of the nearest nucleotide neighbors along the DNA chain can bedetermined (Fig. 3-27). Using this method of investigation, it is possible to verify some of the details of the Watson-Crick model, particularly the antiparallel nature of the two strands in the double helix. If the chains are antiparallel, the proportion of a particular dinucleotide sequence in one strand should be equal to the proportion of the complementary sequence in the second chain. For example, the number of TpCp pairs in one chain should correspond to the number of GpAp pairs in the complementary chain, the number of ApGp pairs should be equal to the number of CpTp pairs, and so on.

(a)

Labeled triphosphate	Isolated 3'-deoxyribonucleotide			
	Tp	Ap	Cp	Gp
dAT³²P	a TpA 0.053 I	b ApA 0.089	c CpA 0.080 II	d GpA 0.064 III
dTT³²P	b TpT 0.087 I	a ApT 0.073	d CpT 0.067 IV	c GpT 0.056 V
dGT³²P	e TpG 0.076 II	f ApG 0.072 IV	g CpG 0.016	h GpG 0.050 VI
dCT³²P	f TpC 0.067 III	e ApC 0.052 V	h CpC 0.054 VI	g GpC 0.044
Sums	0.283	0.286	0.217	0.214

(a)

(b)

Labeled triphosphate	Isolated 3'-deoxyribonucleotide			
	Tp	Ap	Cp	Gp
dAT³²P	a TpA 0.059	b ApA 0.088 I	c CpA 0.078 II	d GpA 0.063 III
dTT³²P	b TpT 0.083 I	a ApT 0.075	d CpT 0.068 IV	c GpT 0.056 V
dGT³²P	e TpG 0.076 II	f ApG 0.074 IV	g CpG 0.011	h GpG 0.057 VI
dCT³²P	f TpC 0.064 III	e ApC 0.051 V	h CpC 0.055 VI	g GpC 0.042
Sums	0.282	0.288	0.212	0.218

(b)

Fig. 3-27. Nearest neighbor analysis of native and enzymatically synthesized calf thymus DNA. (a) Native DNA. (b) DNA synthesized by the Kornberg enzyme with calf thymus DNA as template. The added template comprised only 5% of the total DNA after incubation. Identical Roman numerals designate those sequences which should be equivalent in a Watson-Crick DNA model with strands of opposite polarity; identical lower case letters designate sequence frequencies which should be equivalent in a model with strands of similar polarity. Data from Josse et al. (56).

All of these predictions were admirably borne out by the experimental evidence. That the dinucleotides are not randomly distributed is demonstrated by the fact that in calf thymus DNA, the number of CpGp pairs differs from the number of GpCp pairs by a factor of three. Both doublets contain the very same nucleotides but differ in sequence. Nearest-neighbor analysis of fifteen natural DNA's, all of which came from different sources varying greatly in their nucleotide compositions, confirmed the equivalence of complementary dinucleotides. The evidence lent strong support to the Watson-Crick model in which the double helix consists of two complementary strands with opposite polarity.

The Kornberg reaction is carried out at pH 7.5 in the presence of 6×10^{-3} M Mg^{2+}. Its equilibrium is strongly displaced toward synthesis, and even at a very low concentration (10^{-5} M) the monomers are almost completely polymerized in the presence of a primer. However, addition of a hundredfold excess of pyrophosphate (2×10^{-3} M), the other reaction product, inhibits the synthesis by about 50%. Furthermore, it has been ascertained that both condensation and cleavage (pyrophosphorolysis) occur simultaneously. As the concentration of pyrophosphate increases, the rate of the degradative reaction approaches that of the synthetic reaction. Studies on the mechanism of the Kornberg reaction indicate that DNA polymerase molecules form a complex with the substrate at the very beginning of the reaction. Chain growth then gradually proceeds with continuous renewal of the enzyme-substrate complex until the supply of monomers is all used up.

The synthetic reaction places certain demands on the structure of the primer DNA. Bacterial DNA polymerases generally replicate native and denatured DNA equally well. But there are other, more fastidious DNA polymerases which require denatured or single-stranded DNA, such as that from phage ΦX174, as a template. Such enzymes have been extracted from *E. coli* infected with phage T2 (Kornberg), and from calf thymus (Bollum).

An interesting peculiarity of DNA polymerase action is that the priming DNA can serve both as template and initiator at the same time. If one of the strands of a DNA molecule is partially digested with a nuclease, repairs can be mediated by DNA polymerase; mononucleotides are added to the end of the shorter chain while the complementary strand serves as a template (Fig. 3-28). This aspect of the process can be established experimentally by isotopically labeling the newly synthesized DNA so as to differentiate it from the primer.

Detailed investigations have shown that DNA polymerase joins nucleoside-5'-phosphates to the 3'-carbon atom of deoxyribose. If the 3'-position is occupied by a phosphoryl ester instead of the normal hydroxyl group, further addition of nucleotides to the chain is blocked. In this case, the 3'-ends can be activated by a special phosphoesterase which cleaves the phosphate groups, leaving the 3'-hydroxyls available for further chain growth.

In recent years, Khorana and his colleagues have chemically synthesized low-molecular-weight oligonucleotides of defined sequence which they have used as primers for polymerase-mediated synthesis of DNA (57). The newly synthesized DNA was found to consist of very large molecules in which the sequence complementary to the short template was repeated several thousand times. Thus, if a mixture of the two decanucleotides $(dTC)_5$ or $(dAG)_5$ is taken as primer, and all necessary deoxy-

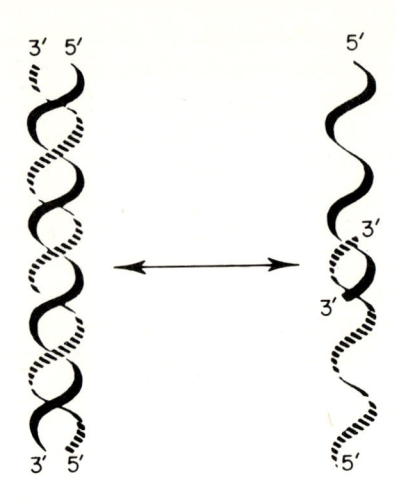

Fig. 3-28. Diagram of the hydrolysis of the 3′-hydroxyl ends of each DNA strand by an exonuclease and the subsequent resynthesis of the complementary strand with DNA polymerase. From Richardson et al. (52).

nucleoside triphosphates are added to the system, a 3-hour incubation yields double-stranded (dAG)$_n$ and (dTC)$_n$ polymers with molecular weights in the millions. One such preparation was characterized by a sedimentation constant of 16 S and a melting temperature (T_m) of 90° at an ionic strength of 0.02. After hydrolysis of these polymers it was shown that the distribution of neighboring nucleotides was precisely complementary to that of the original template. The amount of newly synthesized DNA exceeded the amount of primer by a factor of more than 200. This remarkable phenomenon indicates that one DNA chain must be able to slide along the other whenever a nucleotide doublet or triplet is periodically repeated in the chain. If such sliding did not occur, it would be very difficult to imagine how the growing chain could continue beyond the limits of the template while at the same time preserving the same strictly determined sequence of nucleotide residues.

When native double-stranded DNA is employed as a template for synthesis, the reaction proceeds somewhat irregularly *in vitro*. Starting from either end of the duplex, only one of the two strands can serve as a template since the chain grows in the direction 5′ → 3′. The complementary strand must be replicated from the other end. This process need not necessarily result in the complete copying of the corresponding strand, for a change in direction usually occurs much earlier than that. The growing chain makes a 180° turn and starts to copy either itself or the second strand of the original template. As a result, the newly synthesized DNA differs greatly from the natural product in possessing many branching points and a generally irregular structure. The synthesis and branching of DNA *in vitro* has been studied by electron microscopy (58). When DNA from bacteriophage λ (molecular weight, 3 × 10^7) was used as a primer, it was noted that after a 30% increase in DNA mass about half of the newly synthesized strands were already branched.

The problem of obtaining the synthesis of regular double-stranded DNA molecules *in vitro* was finally resolved by the use of *single-stranded, circular* DNA as a

6. Enzymatic Synthesis of Nucleic Acids

template or primer. Such templates are replicated by DNA polymerase without production of the physical abnormalities that had been noted in earlier studies with double-stranded DNA primers. Using this approach, Goulian, Kornberg, and Sinsheimer demonstrated that completely synthetic DNA with full biological activity could be formed *in vitro* (59, 60). In these experiments, native, single-stranded, circular DNA molecules from bacteriophage ΦX174, designated (+) strands, were converted to a circular duplex form, called the replicative form or RF, by the combined action of DNA polymerase and polynucleotide ligase. Ligase is an enzyme capable of catalyzing closure of the newly synthesized complementary (−) strand at the point where its 3'- and 5'-termini abut, thereby producing a completely covalent circular duplex molecule (Fig. 3-29a). Furthermore, the (−) strand was labeled with ^{32}P to aid in its

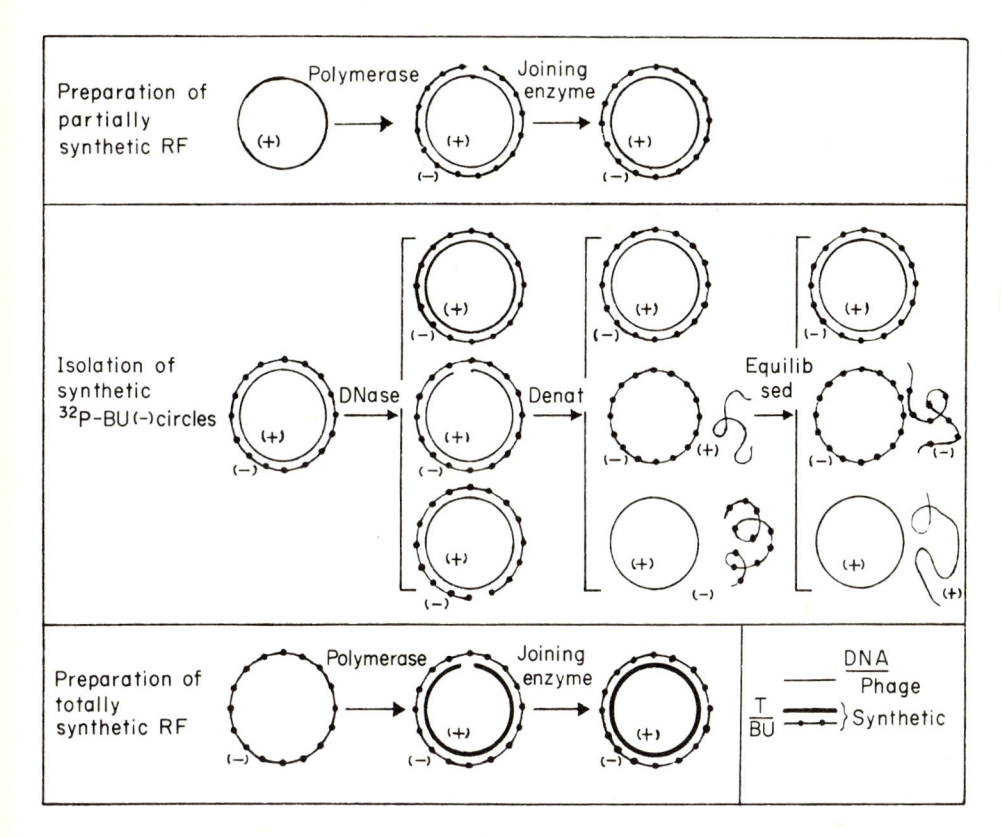

Fig. 3-29. The preparation of synthetic (−) strands and RF (replicative form) from phage ΦX174 DNA. See text for explanation. From Goulian *et al.* (60).

identification and thymine was replaced by 5-bromouracil to facilitate its separation from the native (+) phage DNA by density gradient centrifugation. In short, the first stage of the experiment yielded partially synthetic duplexes of ΦX174 DNA in which the synthetic (−) strand, labeled with ^{32}P and bromouracil, was obtained by using a native (+) strand from the phage as primer.

In order to separate the native (+) and synthetic (−) strands from each other, the duplex circles were exposed to low concentrations of pancreatic DNase for a time sufficient to rupture a single phosphodiester bond in one of the strands of about half of the molecules (Fig. 3-29b). The introduction of single-strand nicks was necessary to permit the two strands of the duplex to draw apart from one another upon thermal denaturation. Since the nicks occurred randomly in both (+) and (−) strands, the denatured material included, in addition to intact duplex RF, both circular and linear (+) single strands and circular and linear (−) single strands. The (+) and (−) strands were then fractionated by equilibrium centrifugation in CsCl (p. 262ff) in which the densities of the bromouracil- and thymine-containing molecules differed by 0.08 to 0.09 gm/cm^3; the intact hybrid duplexes that were not nicked by DNase occupied a position in the gradient intermediate to the two peaks of single-stranded DNA. The circular single strands in each peak could be separated from the linear molecules by velocity centrifugation in sucrose gradients because the closed circles sediment more rapidly than the linear forms. Isolated (−) circles were next used as templates for a second round of replication which produced a fully synthetic duplex RF (Fig. 3-29c). Finally, synthetic (+) circles, corresponding to the original phage DNA, were isolated from the fully synthetic duplex by limited DNase digestion and gradient centrifugation, much as in the isolation of synthetic (−) circles from partially synthetic RF.

The biological activity of synthetic DNA was assayed by its ability to infect *E. coli* spheroplasts. Spheroplasts are prepared by solubilizing the external bacterial cell wall with the enzyme, lysozyme. Devoid of their rigid wall, the bacteria assume a spherical shape, but they remain viable and will not rupture if they are suspended in a slightly hypertonic medium, that is, if the osmotic pressure of the medium is somewhat higher than that within the cell. The cell membrane which surrounds the spheroplast is permeable to DNA and consequently, spheroplasts can be infected with purified bacteriophage DNA. As a result, complete phage particles containing both protein and DNA are formed. Although their specific infectivities differ, (+) strands, (−) strands and duplex RF all exhibit biological activity in this assay so that each one of the intermediates in the production of synthetic (+) strands could be tested. It was found that partially and fully synthetic duplex DNA as well as synthetic (+) and (−) strands were all active in the spheroplast assay and that they all possessed roughly the same specific infectivity as the corresponding *in vivo* forms. An interesting aspect of the spheroplast assay is that only circular molecules are infective. This observation underscores the importance of polynucleotide ligase in the synthetic reaction, for without it, the production of circular synthetic DNA would have been impossible. Since the infectivity assay also requires that all of the genetic information necessary to reproduce a phage particle be present intact in the infecting DNA molecule, the infectivity of the synthetic DNAs testifies to the high degree of fidelity with which the original native material was copied *in vitro*. Few, if any, mistakes could have been made in the replication of the initial DNA sequence which consisted of several thousand nucleotides.

Since the enzyme polynucleotide ligase appears to be of prime importance in cellular DNA synthesis and in genetic recombination, it is worth taking note of its properties in greater detail. The ligases characterized so far are enzymes which

6. Enzymatic Synthesis of Nucleic Acids

catalyze the covalent closure of single-strand interruptions in double-stranded DNA, linking appropriately positioned 3'- and 5'-termini together into a 3' → 5' phosphodiester bond. Specifically, the reaction effected by polynucleotide joining enzyme, or ligase, from *E. coli* requires that a 5'-phosphoryl terminus and a 3'-hydroxyl terminus in one strand be juxtaposed by hydrogen bonding to the second and intact DNA strand before closure can occur. The *E. coli* enzyme further requires the cofactor nicotinamide adenine dinucleotide (NAD) which is cleaved to 5'-AMP and nicotinamide mononucleotide (NMN) in the course of the reaction (61, 62). The virus-specified ligase obtained from *E. coli* cells infected with phage T4 acts in much the same way, except that it requires ATP rather than NAD as cofactor (63, 64). Neither of these enzymes has activity toward single-stranded DNA. Further functions of ligase will be discussed at appropriate places in the text.

In addition to DNA synthesized from all four types of nucleotides, it is possible to use the Kornberg enzyme to produce the copolymer poly dAT from ATP and TTP. When primer is absent, initiation of the reaction is extremely slow; an induction period of several hours is usually required before the first polynucleotides appear. After that the polymer already synthesized begins to serve as a template and the reaction becomes autocatalytic. The polymer product obtained is of very high molecular weight and is characterized by a strictly alternating base sequence, ATATAT... This sequence was confirmed by using an enzyme which cleaves the chain into dimers. The only type of dinucleotides recovered from such preparations had the sequence ApTp.

If the Kornberg enzyme acts on a mixture of monomers containing only G and C, the result is quite different; the homopolymers poly dG and poly dC are obtained in equal amounts. Furthermore, they form twofold helical complexes with one another. Thus with DNA polymerase it is possible to obtain two kinds of simple Watson-Crick helices of the form:

$$\begin{array}{c} -A-T-A-T-A-T-A- \\ \vdots\;\;\vdots\;\;\vdots\;\;\vdots\;\;\vdots\;\;\vdots\;\;\vdots \\ -T-A-T-A-T-A-T- \end{array} \quad \text{and} \quad \begin{array}{c} -G-G-G-G-G-G-G- \\ \vdots\;\;\vdots\;\;\vdots\;\;\vdots\;\;\vdots\;\;\vdots\;\;\vdots \\ -C-C-C-C-C-C-C- \end{array}$$

Cellular RNA synthesis is mediated by a similar enzymatic reaction, first described by Weiss (65). It shares the following characteristics with DNA synthesis mediated by DNA polymerase: (1) the presence of all four *ribo*nucleoside triphosphates is necessary for polymer formation; (2) the reaction requires a DNA primer or template; and (3) Mg^{2+} ions are required, while pyrophosphate is inhibitory. This process is termed the *transcription* of RNA from DNA. *DNA-dependent RNA polymerase*, the enzyme which performs this remarkable reaction, was prepared from the nuclei of rat liver cells. The origin of the enzyme is in itself quite significant, for it is precisely in the nucleus that cellular RNA synthesis occurs, even though highly polymerized RNA is subsequently concentrated in the ribosomes.

Weiss used ^{32}P-labeled ribonucleotides to detect newly synthesized RNA. Evidence was adduced to show that all four radioactive bases participate in the reaction simultaneously and that they are all polymerized into new chains. The reaction product was investigated by akaline hydrolysis which cleaves polynucleotides to nucleoside-3'-monophosphates, though some nucleoside-2'-monophosphates form as a result of isomerization. In the original nucleotides, the phosphate groups were attached to the 5'-carbon of the ribose moiety. Transfer of phosphate from the 5'- to the 3'-carbon can occur only as the result of polymerization followed by hydrolysis of the chain. Through enzymatic cleavage of the new RNA polymers and measurement of the kinetics with which the labeled residues appear in solution, it was established that the radioactive nucleotides were uniformly distributed along the RNA chain. Dependence of RNA synthesis on DNA was proved by showing that the reaction is completely inhibited in the presence of deoxyribonuclease (DNase), an enzyme which specifically hydrolyzes DNA without affecting RNA.

Hurwitz and Berg isolated a similar enzyme from *E. coli*, while Ochoa and colleagues extracted an RNA polymerase from *Azotobacter vinelandii* (66–70). The *E. coli* enzyme consists of a large number of subunits with an aggregate molecular weight of about 720,000 daltons. There are many sites along the DNA duplex to which RNA polymerase molecules can bind, indicating that RNA chains are simultaneously transcribed from numerous regions of the template. The association of RNA polymerase with DNA is visualized in the electron micrograph presented in Fig. 3-30. The various RNA polymerases all catalyze a condensation reaction between ribonucleoside triphosphates and the growing end of the polymer chains. Native and denatured DNA from phages ΦX174 and T2, *E. coli*, and calf thymus have all proved effective as templates for enzyme-mediated RNA synthesis. By using radioactive monomers in the reaction, it is possible to determine the relative proportions of all four nucleotides in the product RNA without extensive purification. Studies have shown that the composition of newly synthesized RNA corresponds to that of the priming DNA, if thymine is replaced by uracil. Furthermore, both Weiss and Nakamoto and Hurwitz demonstrated by an analysis of nearest-neighbor frequencies that RNA synthesized *in vitro* is also complementary to its DNA template (69, 70).

DNA-dependent RNA synthesis thus provides a second example of polynucleotide synthesis in which the principle of complementarity is observed, consistent with the fundamental postulate of the Watson-Crick model. RNA synthesis on a DNA template is called transcription in order to differentiate it from DNA-primed synthesis of DNA for which the term *replication* is used.

There is an interesting relationship between the DNA template and the strand of RNA whose synthesis it is directing. If single-stranded ΦX174 DNA is used as a template, the first phase of the reaction consists of the formation of a "molecular hybrid"; this macromolecule contains one strand of primer DNA and a complementary strand of newly made RNA, and has the conformation of a Watson-Crick double helix. The DNA-RNA hybrid differs only slightly from the DNA-DNA double helix as judged by its hypochromic effect, its optical activity, and the sharpness of its melting point. The hybrid melts somewhat more easily, however, undergoing the transition at a lower temperature (72). After all the single-stranded DNA has been converted into hybrid double helices, the hybrid is further copied in a *semiconservative* fashion. This means that the ends of the hybrid duplex unravel, and that new complementary strands are synthesized along both the RNA and DNA strands. This process results in the formation of another RNA-DNA hybrid and two complementary strands of RNA which can be converted into a RNA-RNA double helix after a period of heating followed by slow cooling (73). The last procedure is known as *annealing*.

If double-stranded DNA is taken as a template, RNA synthesis *in vitro* proceeds

6. Enzymatic Synthesis of Nucleic Acids 257

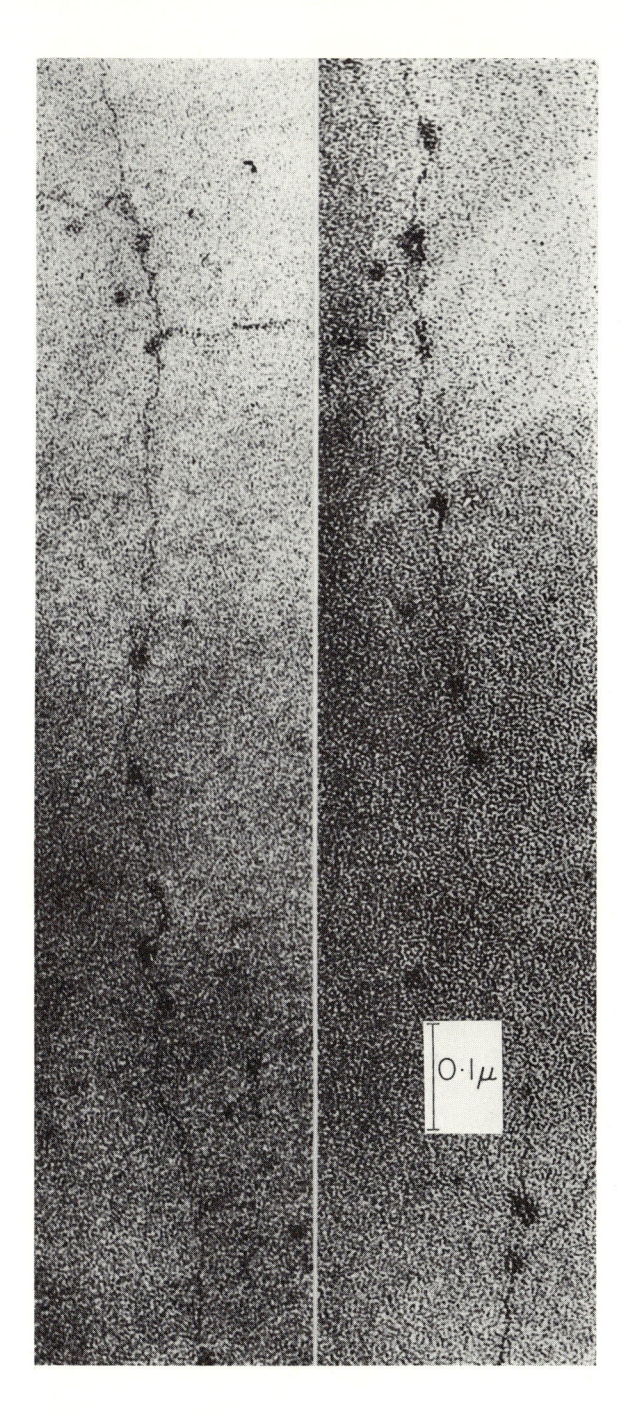

Fig. 3-30. Electron micrographs of segments of phage T7 DNA (continuous strands) with attached RNA polymerase molecules (dark globules negatively stained for contrast). From Slayter and Hall (71).

in a similar fashion. Transcription takes place simultaneously on both DNA strands, although hybrid DNA-RNA helices do not occur under these circumstances. Upon completion of synthesis, the RNA chain is displaced from the template because the DNA-DNA interaction within the double helix is stronger than that between RNA and DNA. This restores the template to its original form and produces free strands of RNA. The intermediate DNA-RNA-enzyme complex can be observed if the reaction mixture is analyzed in the ultracentrifuge without prior deproteinization.

Short oligodeoxynucleotides such as those synthesized by Khorana can also be used as templates for transcription. Thus $d(TTC)_3$ is transcribed into long RNA polymers of the form $(AAG)_n$ (74). The total amount of polyribonucleotide synthesized, however, is almost exactly equal to the initial quantity of template, indicating that the short oligodeoxyribonucleotide segments bind to newly synthesized RNA so as to form a double-helical complex. Since the length of the RNA chains can reach 200 nucleotides, it may be assumed that the chain being synthesized slides along the template. Insofar as the template structure is repeated every three residues, it is sufficient that the mutual "slippage" be only three nucleotides in length in order for chain complementarity to be reestablished. Such slippage probably occurs as the result of thermal motion of the short oligodeoxynucleotide along the longer RNA chain. The possibility of obtaining long polyribonucleotide chains containing repeating triplets of defined sequence is extremely useful for investigating the genetic code (p. 437).

There is at least one fundamental difference between the synthesis catalyzed by RNA polymerase *in vitro* and the process of transcription *in vivo*. *In vitro* synthesis is symmetrical in the sense that both DNA strands are copied, resulting in the formation of two kinds of RNA strands complementary to one another. *In vivo*, however, all species of RNA found in the cell are single-stranded and it has been shown that RNA synthesis is asymmetric. Cellular transcription apparently makes use of only one strand of the DNA double helix. Depending on what portion of the DNA is being transcribed, either one strand or the other is copied, but both strands are never copied simultaneously. RNA polymerase joins nucleoside 5'-phosphates to the hydroxyl group of the terminal 3'-carbon of the growing polymer. Transcription of both strands, which we recall are antiparallel with respect to one another, requires that synthesis start at opposite ends of any given region in the DNA. It is not difficult to imagine that in the cell there exist a series of points on the chromosome from which transcription can be initiated in one direction, but that transcription along the antiparallel strand in the opposite direction is blocked.

A number of cases have been described in which *in vitro* synthesis mediated by RNA polymerase results in the copying of only one of the two DNA strands. The replicative form of bacteriophage ΦX174, for instance, is a closed circle of double-stranded DNA. If this intermediate is used as a template for transcription, a copy complementary to only one of the two strands is obtained, precisely the result found *in vivo*. But if the closed ring is disrupted, both strands are transcribed simultaneously [Hayashi, Hayashi, and Spiegelman (75)]. Similarly, Green noted that *E. coli* RNA polymerase transcribes phage T4 DNA assymetrically *in vitro*, even when the DNA is broken into fragments with molecular weights on the order of 5 to 6 × 10^6 (compared to 130 × 10^6 daltons for the undegraded DNA strand) (76). If the DNA is further fragmented to a molecular weight of 8×10^5, or if it is denatured by heating, both of its strands are transcribed *in vitro*. An asymmetry in the transcription of DNA from phages T2 and T4 *in vitro* has been confirmed by Geiduschek and his associates (77).

The antibiotic actinomycin D is a specific inhibitor of RNA polymerase and is active at very low concentrations. *In vivo*, this inhibitor effectively halts the synthesis of all kinds of RNA. From such data it was concluded that all cellular RNA is transcribed from DNA by RNA polymerase.

In addition to the DNA-dependent RNA polymerases discussed above, there are other RNA synthetases which catalyze the polymerization of ribonucleoside triphosphates under the direction of RNA templates. These unusual polymerases are elaborated in cells infected with RNA viruses. They produce replicas of the viral RNA which are incorporated into new virus particles. Although RNA viruses generally contain single-stranded RNA, a double-stranded replicative form is found in infected cells which has the structural characteristics of a Watson-Crick double helix (78–80). This discovery indicates that the mechanism of RNA replication is essentially the same as that for DNA, although only one of the strands is replicated in significant quantities. In other words, RNA synthesis is asymmetrical under the conditions of infection with RNA viruses. Furthermore, only one of the two types of RNA is selected for incorporation into virus particles; the second and complementary chain is apparently hydrolyzed to mononucleotides by a phosphodiesterase present in the cell.

Spiegelman and his co-workers have investigated the properties of the RNA-containing virus called Qβ (81). Using RNA-polymerase from Qβ-infected *E. coli*, they succeeded in showing that the replication of Qβ RNA *in vitro* results in a product identical to the original phage RNA. The infectivity of the newly synthesized RNA coincides with that of native phage RNA when tested against *E. coli* protoplasts. This experiment provides good evidence that RNA synthesis *in vitro* proceeds with complete fidelity to the initial RNA template, since it leads to the formation of a product with full biological activity.

7. The Synthesis of Unusual Nucleic Acids

One interesting way to change the properties of cells is to replace the natural bases found in DNA and RNA with synthetic derivatives into which various atoms and groups have been introduced. The Watson-Crick model permits us to predict the replacements which are allowable on the basis of steric considerations and the possibilities for hydrogen bonding. A scheme of possible replacements with their hydrogen-bonding capacities is presented in Fig. 3-31.

The efficacy of substituting unusual bases for natural ones was demonstrated in experiments based on the Kornberg system for *in vitro* DNA synthesis. Normal DNA polymerase from *E. coli* and the usual spectrum of natural DNA primers continue to stimulate the production of DNA when certain unusual bases are introduced into the reaction mixture. Specifically, uracil and bromouracil can replace thymine; bromocytosine and methylcytosine can substitute for cytosine; and hypoxanthine can be incorporated in place of guanine. In many cases, such as the substitution of hypoxanthine for guanine, the reaction rate is reduced severalfold. While this indicates that the enzyme is more poorly suited to such unusual substrates, the reaction does not

Hydrogen bonding of adenine to thymine

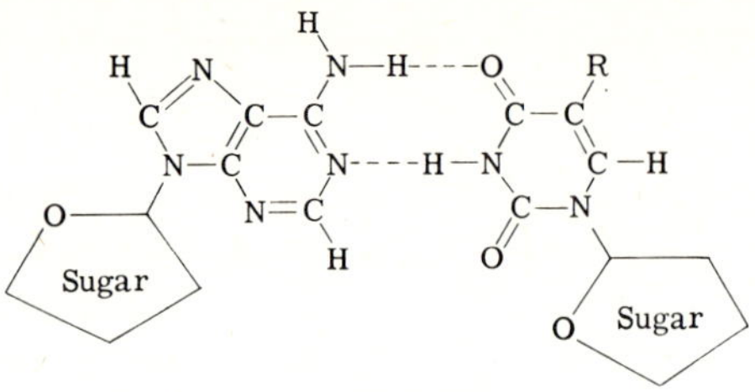

R =
- −CH₃ Thymine
- −H Uracil
- −Br Bromouracil
- −I Iodouracil
- −F Fluorouracil
- −Cl Chlorouracil

Hydrogen bonding of guanine to cytosine

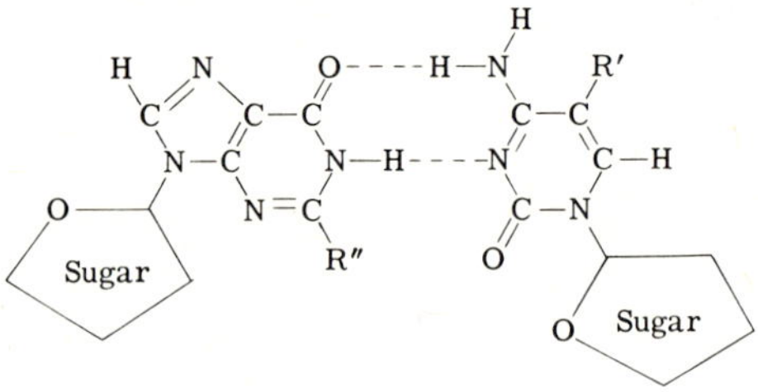

R′ =
- −H Cytosine
- −CH₃ 5-Methylcytosine
- −Br 5-Bromocytosine
- −CH₂OH Hydroxymethylcytosine

R″ =
- −NH₂ Guanine
- −H Hypoxanthine

Fig. 3-31. Base pairing between deoxyribonucleotide bases showing permissible substitutions on the heterocyclic rings consistent with normal hydrogen bonding. After Kornberg (82).

differ in principle from the normal one. We conclude that bases may be replaced by derivatives as long as there is no breach in their ability to form hydrogen bonds with complementary purines and pyrimidines.

The replacement of nucleotide bases can also occur in living organisms. If a thymine-requiring mutant of *E. coli* (that is, one which has lost the ability to synthesize thymine endogenously) is grown on a medium containing 5-bromouracil in addition to thymine, it is possible to obtain a population of cells in which roughly half of the thymine has been replaced by bromouracil. A smaller, but nonetheless substantial, proportion of thymine residues can be exchanged for chlorouracil and iodouracil, though not by uracil or fluorouracil (83). To a more limited extent, bromouracil can be introduced into the DNA of wild type *E. coli* if the medium contains very high concentrations of this base. It is not surprising that cell growth is slower under these conditions and that certain enzymatic reactions are inhibited. Much more amazing is the fact that bacteria retain their viability when 50% of their normal thymine residues have been replaced by bromouracil. A similar experiment was carried out with tobacco mosaic virus grown in medium containing 5-fluorouracil.

Although the rate of virus production was diminished, fully infectious particles were formed from RNA in which up to 40% of the uracil residues had been replaced by the halogenated derivative. Significant quantities of 5-fluorouracil can also be introduced into the RNA of bacteria.

We can draw two main conclusions from the incorporation of unusual bases into nucleic acids. First, it is clear that such replacements do not fundamentally alter the translation of genetic information into cellular components, since the organisms continue to live and reproduce. Second, the substitution of bromouracil for a large proportion of the thymine in *E. coli* is accompanied by an enormous increase in mutability, suggesting that such DNA is unstable. The probability of spontaneous mutations, such as those preventing the synthesis of essential cellular metabolites, is enhanced to the almost unbelievable value of 3% per cell generation. Under ordinary conditions, the mutation frequency per generation is on the order of 1 in 10^5 to 10^7 cells. *Escherichia coli* mutants obtained in this way are themselves quite unstable and the probability of back mutations or reversions is extremely large in comparison to normal cells. Moreover, the presence of bromouracil in *E. coli* DNA brings about immediate changes in cellular phenotype, in particular, an increase in length and a greater sensitivity to ultraviolet light.

The replacement of up to 30 to 40% of the uracil residues in *E. coli* RNA by fluorouracil is accomplished by growing a uracil auxotroph on medium supplemented with 5-fluorouracil. As a result, newly synthesized proteins are considerably altered with respect to amino acid composition, physicochemical properties, and enzymatic capacities. Another unusual base, azaguanine, can replace guanine in cellular RNA:

Guanine   8-Azaguanine

Azaguanine is a potent inhibitor of many synthetic reactions in bacterial cells, especially of the production of proteins and ribosomal RNA. However, the specific RNA which serves as a template for protein synthesis continues to be made with azaguanine taking the place of certain guanine residues in the polymer chains. If the cells are transferred to normal medium after a nonlethal exposure to azaguanine, protein synthesis and growth resume, but, as in the case of 5-fluorouracil, altered proteins are formed.

8. The Use of Isotopes in Studies on DNA Replication

The postulate that DNA reduplicates itself by means of a chain-copying mechanism leads to an important logical consequence. We expect that the DNA of a cell or virus should not be dispersed in successive generations, but that it should be transmitted either intact or in the form of a half-molecule. We can imagine that in replication the DNA double helix separates into two chains, each of which can serve as a template for the synthesis of a complementary strand. Complementary nucleoside triphosphates are first adsorbed to the free side groups of the polymer where they are next condensed into a new DNA strand with the elimination of pyrophosphate (Fig. 3-32).

If this scheme for the replication process is correct, the original DNA double helix must divide into two strands before or upon cell division so that each daughter cell will receive half of the maternal DNA. In order to test this hypothesis, Meselson and Stahl designed an experiment that has become a classic (84). They differentially labeled old and new DNA molecules in a growing culture of *E. coli* with ^{15}N and ^{14}N, the heavy and normal isotopes of nitrogen. The DNA was then extracted and fractionated, and the distribution of isotopes in it was analyzed.

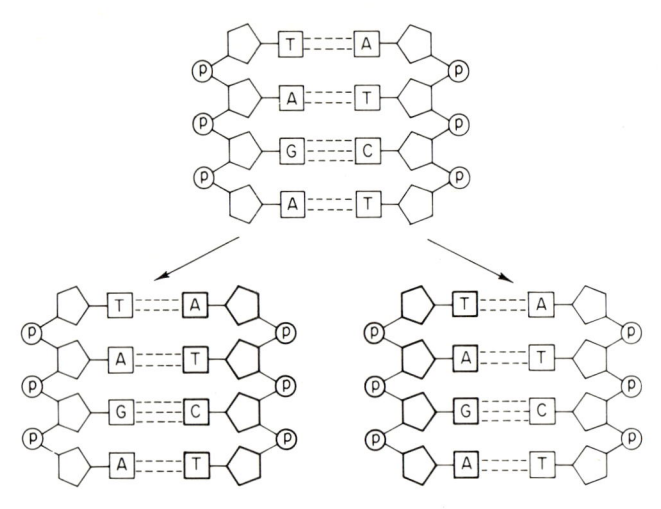

Fig. 3-32. Proposed scheme for the replication of a DNA molecule with the structure postulated by Watson and Crick. Bold-lined polynucleotide chains represent newly synthesized strands. From Josse *et al.* (56).

For isotopic analysis of the DNA, a clever and extremely sensitive method was developed called *density gradient centrifugation*. This method, developed by Meselson, Stahl, and Vinograd, consists of an ultracentrifugal separation of macromolecules according to density rather than molecular weight (85). To establish the gradient an electrolyte of high density, such as CsCl ($\rho = 3.7 \text{ gm/cm}^3$ in the solid state), is introduced into the solution. CsCl distributes itself according to the Boltzmann equation in a centrifugal force field; after the equilibrium distribution is established in the cell of the ultracentrifuge, the salt concentration will be a function of distance from the rotor axis with approximately the following form:

$$C/C_0 = e^{\frac{M[1-(\rho_0/\rho)]\omega^2 \bar{r}x}{RT}} \tag{3-5}$$

where ρ_0 is the density of water, ρ the buoyant density of the salt, M the molecular weight of the salt, ω the angular velocity of the ultracentrifuge, $\bar{r}$ the radius of the rotor measured from the axis to the center of the cell or centrifuge tube, x the distance from the axis to any point in the cell or tube, R the gas constant, and T the absolute temperature.

The equilibrium salt distribution is established relatively rapidly, on the order of the time necessary for the electrolyte to diffuse throughout the cell. Thus, if the height of the cell is designated by L, the equilibrium distribution will be established within a period of about $\tau = L^2/D_0$, where D_0 is the diffusion constant of the salt. In practical terms, τ is equal to about 4 to 6 hours when $L = 1$ cm.

The initial concentration of salt is chosen in such a way that at equilibrium, the density at the midpoint of the cell is approximately equal to the density of the polymer under investigation. In the case of DNA, this figure is roughly equal to 1.7 gm/cm³. Under these conditions, the Boltzmann function can be reduced to a series in which only the linear term (and rarely, the quadratic term) is significant:

$$\frac{C}{C_0} = 1 + M \frac{[1 - (\rho_0/\rho)]\omega^2 \bar{r}}{RT} x \tag{3-6}$$

In other words, the concentration gradient and thus the density gradient will be constant, designated by α.

When homogeneous macromolecules are sedimented in such a gradient, all of them will eventually be concentrated in a narrow band centered at the point where their density is equal to the density of the medium. The molecules will approach this point from both sides, stopping only when the forces acting on them are in equilibrium. Some molecules will sediment from the top of the cell (nearest the axis) in the direction of the centrifugal field while those from the bottom will float upward against this field as they are displaced from the denser medium. If there are several polymers with different densities in the cell, they will separate into a number of discrete bands. Each polymer component seeks its own equilibrium position in the density gradient and forms its own distribution about the corresponding point.

We shall now calculate the equilibrium distribution of macromolecules in a density gradient by first considering the flow of material through a surface perpendicular to the radius of rotation at a distance x' from the point of equilibrium density (Fig. 3-33). The flow of particles consists of a sedimentation term $-vC$ and a diffusion term $-D(dC/dx')$, where v is the rate of sedimentation, and C and D are the concentration and diffusion constant of the polymer, respectively. At equilibrium the net flow, F_{net}, is equal to zero:

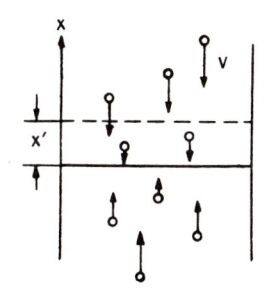

$$F_{net} = -vC - D(dC/dx') = 0 \quad \text{or} \quad d(\ln C)/dx' = -v/D \tag{3-7}$$

Fig. 3-33. Diagram depicting the flotation of particles from a region of higher density and the sedimentation of particles from a region of lower density toward equilibrium in a density gradient. See text for explanation of symbols.

The rate of sedimentation v is equal to $\omega^2 \bar{r} s$, where s is the sedimentation constant:

$$s = (MD/RT)[1 - (\rho'/\rho^*)] \qquad (3\text{-}8)$$

Here ρ' is the density of the medium and ρ^* the density of the polymer. In a CsCl gradient at $x' = 0$, $\rho' = \rho^*$ and $s = 0$. Since the density gradient is approximately linear, $\rho' = \rho^* + \alpha x'$, from which we obtain:

$$s = (MD/RT)(\alpha/\rho^*)x' \qquad (3\text{-}9)$$

Finally, for the equilibrium distribution, the following relation can be calculated:

$$d(\ln C)/dx' = -(\omega^2 \bar{r} \alpha/\rho^*)/(M/RT)x' \qquad (3\text{-}10)$$

which after integration yields:

$$C/C_0 = e^{-[(\omega^2 \bar{r} \alpha/2\rho^*)(M/RT)]x'^2} \qquad (3\text{-}11)$$

Thus we see that the polymer band in the cell conforms to a Gaussian function. The width, or dispersion, of the distribution is

$$\sigma_T = (2\rho^* RT/\alpha \omega^2 \bar{r} M)^{1/2} = (K/M)^{1/2} \qquad (3\text{-}12)$$

and is determined by the square root of the molecular weight of the polymer as well as a number of experimental conditions such as the steepness of the density gradient α, the centrifugal acceleration $\omega^2 \bar{r}$, and the temperature T.

In the experiments of Meselson and Stahl (84), a culture of *E. Coli* was grown for several generations on a medium containing the heavy isotope of nitrogen, ^{15}N. The cells were then transferred to a "light" medium that contained the usual isotope, ^{14}N, in which further cell growth and division took place (Fig. 3-34). At specific times after the transfer, aliquots of the culture were removed. After collecting and lysing the cells, their DNA was extracted under gentle conditions so as to avoid degradation of the polymer and analyzed by the density gradient technique. Nucleic acids fully labeled with the heavy nitrogen isotope ^{15}N are more dense than those containing pure ^{14}N by 0.04 gm/cm^3. If after transfer to light medium the ^{15}N-labeled nucleic acids are dispersed among daughter molecules, we would expect a population of DNA molecules with densities running from pure ^{14}N-DNA to pure ^{15}N-DNA following the first cell division. In fact, as can be concluded from the actual sedimentation diagrams in Fig. 3-35, Meselson and Stahl observed three types of molecules: those containing only ^{15}N, those containing only ^{14}N, and those with equal amounts of ^{15}N and ^{14}N. The result of this experiment provides strong evidence for a scheme of *in vivo* DNA replication in which the double helix is separated into two strands without degradation or further division of any kind, accompanied or followed by *de novo* synthesis of a complementary strand. This mechanism of replication is called *semiconservative*. It has been found that the DNA of the alga *Chlamidomonas* also replicates semiconservatively. Finally, this mechanism has been confirmed by autoradiography of DNA from bacteria (Cairns, p. 328ff) and from plant cells (Taylor). The use of tritium label permits the visualization of newly synthesized DNA strands when cell extracts are spread on a photographic emulsion at various times after the introduction of radioactive nucleotides.

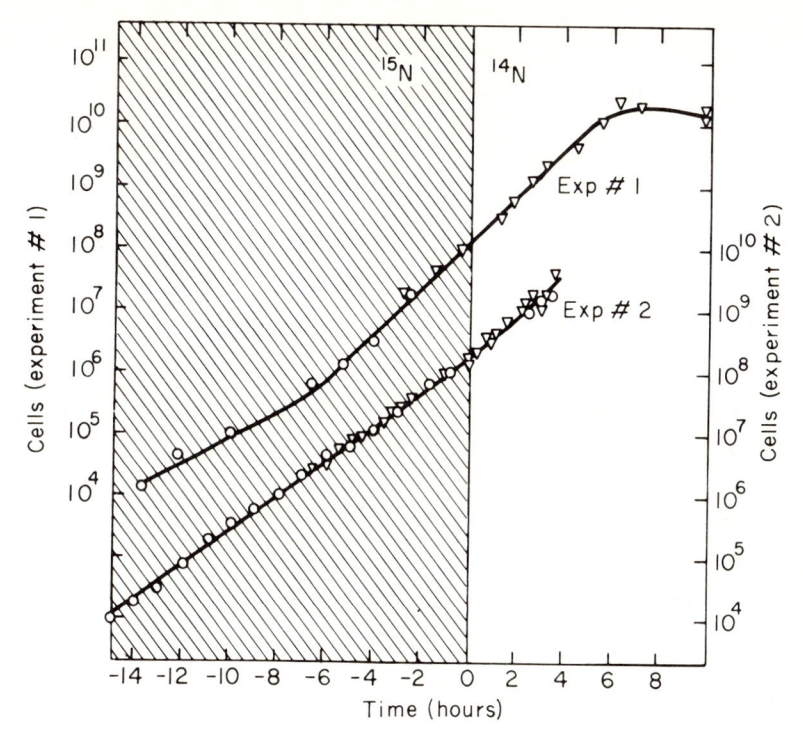

8. The Use of Isotopes in Studies on DNA Replication

Fig. 3-34. Growth of *E. coli* in ^{15}N medium followed by the addition of excess ^{14}N substrates; O = colony count; ∇ = microscopic cell count. Cell titer is plotted along the ordinate; generation time was 0.8 to 0.85 hours. Samples taken at various times after the addition of ^{14}N were analyzed by CsCl density gradient centrifugation (Fig. 3-35). From Meselson and Stahl (84).

It was of considerable interest to ascertain whether or not the semiconservative mechanism of DNA replication was observed *in vitro*. For this experiment, the repeating polymer ATATAT... was used as a primer to test the incorporation of adenosine and bromouridine triphosphates with the Kornberg enzyme. Bromouridine was used as a heavy-atom label replacing thymidine, since both compounds form the same hydrogen bonds with adenine. The resulting reaction may be symbolically represented as follows:

$$\begin{array}{c} T\cdots A \\ A\cdots T \\ T\cdots A \\ A\cdots T \\ T\cdots A \end{array} + n\,ppA + n\,ppBr \rightarrow \begin{array}{cc} T\cdots A & Br\cdots A \\ A\cdots Br & A\cdots T \\ T\cdots A & Br\cdots A \\ A\cdots Br & A\cdots T \\ T\cdots A & Br\cdots A \end{array} + 2n\,pp$$

The Kornberg enzyme causes bromouridine triphosphate to polymerize with adenosine triphosphate to produce a mixed alternating chain. It is not difficult to confirm this result by using density gradient centrifugation since pure poly dAT has a density of

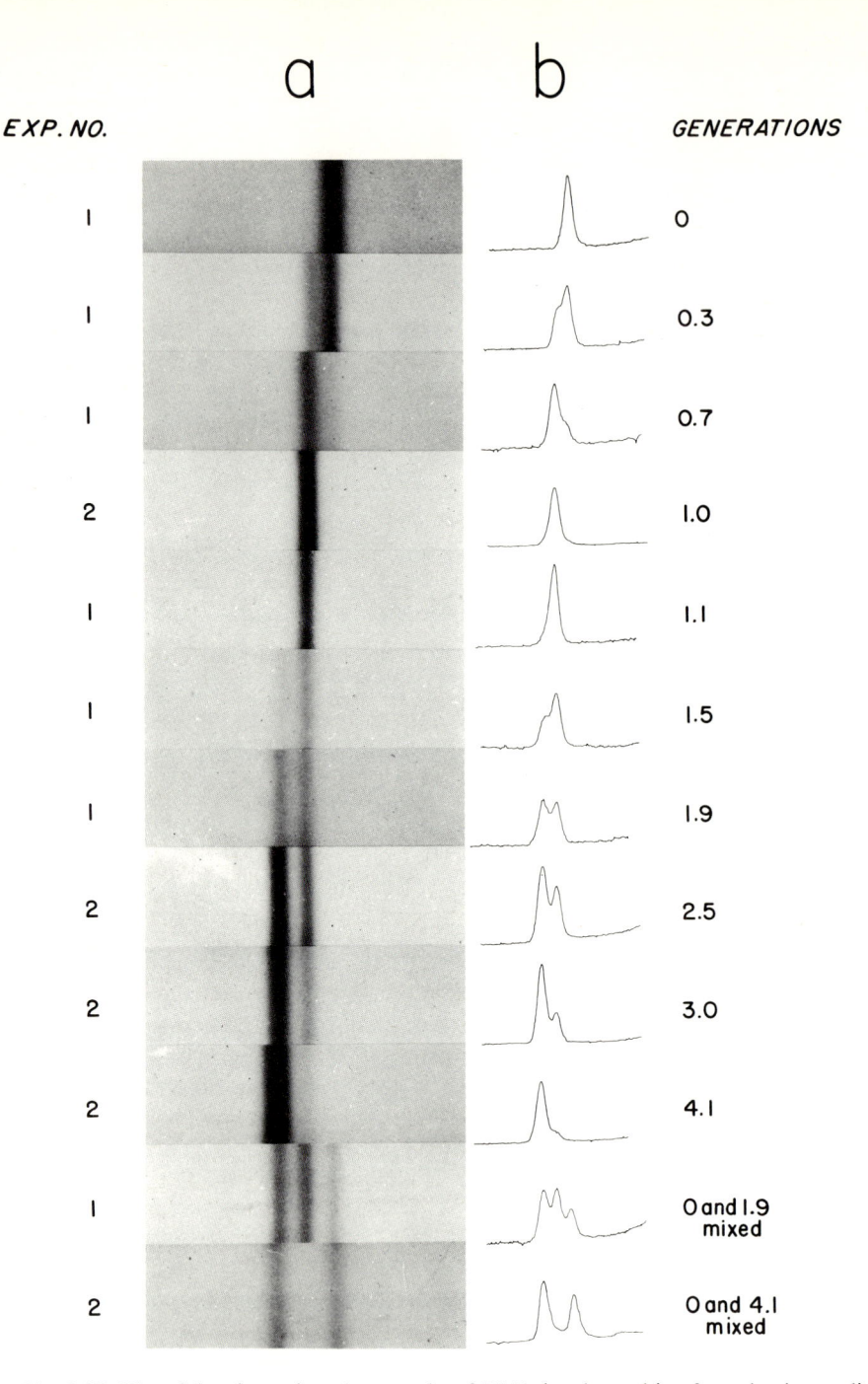

Fig. 3-35. Ultraviolet absorption photographs of DNA bands resulting from density gradient centrifugation of *E. coli* lysates following the addition of ^{14}N substrates to a culture growing in ^{15}N medium. Density of the CsCl solution increases toward the right. The time of sampling is measured from the time of addition of ^{14}N in units of generation time. (a) The photographs and (b) microdensitometer tracings of the DNA bands shown in the photographs. The lowermost frame shows bands corresponding to a mixture of fully labeled and unlabeled DNA. The density of the intermediate band present from 0.3 to 3.0 generations is exactly halfway between the reference bands and presumably contains duplex DNA with one preexisting ^{15}N strand and one newly synthesized ^{14}N strand. From Meselson and Stahl (84).

8. The Use of Isotopes in Studies on DNA Replication

1.69 gm/cm^3 while that of pure poly dABr is 1.87 gm/cm^3. The differences in density are so great that it is very easy to apply the method of Meselson, Stahl, and Vinograd to this problem. When the products of the AT-primed reaction were subjected to analysis, the synthetic scheme shown above was fully corroborated, proving that *in vitro* DNA synthesis adheres to the semiconservative mechanism. The two strands of the double helix separate in the Kornberg reaction just as in cellular DNA synthesis.

Another experiment relating to the structural integrity of DNA was carried out by Levinthal and Thomas with bacteriophage T2 (86). A culture of *E. coli* cells, pregrown in medium containing highly radioactive phosphorus (specific activity: 30 to 50 mcuries/mg), were infected with T2. The new bacteriophages synthesized from *E. coli* components therefore contain radioactive phosphorus (^{32}P) in their DNA at the same specific activity. The phage particles were next carefully separated from contaminating bacterial fragments and purified by chromatography on a DEAE-cellulose column. A drop of the phage suspension was next imbedded in a thick electron-sensitive photographic emulsion with a high concentration of silver bromide. The emulsion was exposed for about a week and then developed. In the course of a week one quarter of the original ^{32}P nuclei had disintegrated, leaving 10 or 15 tracks associated with each radioactive phage particle in the pattern of a "star" (Fig. 3-36). From

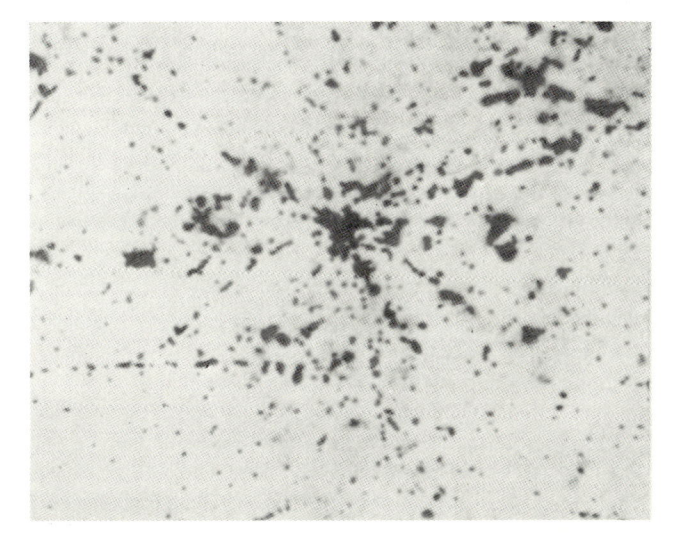

Fig. 3-36. Photomicrograph of a "star" on the surface of a photographic emulsion, formed by a phage particle containing ^{32}P-labeled DNA. Because of the small depth of field, only a few of the many tracks from this particle are visible in the photograph. From Levinthal and Thomas (86).

a count of the average number of disintegrations per star the number of ^{32}P atoms in every particle can be determined. Since the specific activity of the ^{32}P is known, it is possible to calculate the molecular weight of DNA per bacteriophage. This method of analysis is known as *track autoradiography*. Furthermore, all of the DNA in a phage particle is contained in a single molecule. This was ascertained by comparing stars from intact phages with those from phage DNA prepared by phenol deproteinization or by osmotic shock. Phage particles are subjected to osmotic shock by placing them in a salt-free solution: sufficient water is absorbed under these conditions to make the protein coat burst open, liberating the phage DNA. From all of these methods it was estimated that the molecular weight of T2 DNA is 1.3×10^8.

The hereditary transmission of radioactive phage DNA was also studied by Levinthal. The experiment was designed so that radioactive phage particles multiplied in a culture of nonradioactive *E. coli*. Progeny phages were taken from each of several successive generations and their star patterns were examined by track autoradiography. It was found that over many generations the phages either contained no radioactive molecules or else that they harbored large radioactive fragments with molecular weights on the order of 15×10^6. For the most part, the phage DNA is split into fragments in the first generation, and is afterwards transmitted intact. The explanation of this phenomenon has to do with genetic recombination during phage multiplication, the details of which will be given later (p. 380).

Long DNA chains are extremely labile in solution and are particularly sensitive to gradients in flow rate [Levinthal and Davison (87)]. Very small gradients, such as those arising from blowing the solution out of a pipette, are sufficient to initiate the mechanical degradation of the DNA molecules. Interestingly enough, the molecules are degraded in a very regular fashion: they first break in half, then into four roughly equal pieces, and so on. The regularity of this effect gave the impression that degraded DNA possesses a very strict molecular weight distribution, even when the preparations really consisted of fragments. It is quite likely that until the experiments of Levinthal and Davison most measurements of DNA molecular weight were made on mechanically fragmented molecules and not on native material.

The mechanical degradation of DNA chains has now been studied both theoretically and experimentally. The first theory concerned with the disruption of polymers in flow gradients was developed by Frenkel (88). However, his theory was not applicable to relatively rigid DNA molecules. An estimate of the disrupting forces can nonetheless be made from Buerger's formula for the frictional force experienced by an elongated ellipsoid or rod in fluid streamlines. The force per unit length and velocity is

$$P_1 = \frac{2\pi\eta}{[\ln(l/\rho) - 0.72]} \tag{3-13}$$

if the flow is parallel to the axis of the rod, and

$$P_2 = \frac{4\pi\eta}{[\ln(l/\rho) + 0.5]} \tag{3-14}$$

if the flow is perpendicular to the axis. Here η is the viscosity of the fluid, l is the length of the rod, and ρ its radius. By interpolation, an approximate formula is obtained:

$$P = \frac{3\pi\eta}{\ln(l/\rho)} \tag{3-15}$$

If the rod is in a flow velocity gradient and if its axis is inclined at an angle θ to the direction of flow, it will be subject to a longitudinal disruptive force, F. This force will be maximal at the middle of the rod and can be expressed in the following form:

$$F = \int_0^{l/2} PGl \sin\theta \cos\theta \, dl \tag{3-16}$$

where G is the magnitude of the velocity gradient. The maximal force is developed when the rod is inclined at an angle of 45° to the flow. In this case:

$$F_{\max} = PGl^2/16 \qquad (3\text{-}17)$$

By measuring the critical gradient at which the polymer chain begins to break, it is possible to find the value of $F_{\max}$, the tension just necessary to disrupt the chemical bond. Thus it was determined that in a capillary of 0.12 mm radius, the critical flow rate for phage DNA was 0.05 ml/sec, and $F_{\max}$ was equal to 10^{-3} dynes. This is on the same order as the force necessary to disrupt a chemical bond—the P—O bond, for instance—according to the Morse potential curve from which the maximal force for bond stretching can be calculated. DNA labeled with radioactive phosphorus is more sensitive to mechanical degradation than unlabeled DNA, since the disintegration of incorporated ^{32}P atoms frequently introduces breaks into one strand of the duplex polymer.

9. Molecular Heterogeneity, Reversible Denaturation, and the Formation of Molecular Hybrids

One of the complex problems of nucleic acid chemistry is how to isolate chemically individual substances. Chromatography on serum albumin was used by Mandell and Hershey to fractionate DNA from phage T2 (89). Serum albumin is turned into a basic protein by esterifying its —COOH groups with methanol and the product is then spread on kieselguhr; this sorbent is generally referred to by the initials MAK for methylated albumin-kieselguhr. Contrary to his predecessors, Hershey always obtained a single sharp DNA peak when very gentle conditions were used for the isolation of DNA. Ten minutes of stirring with an ordinary mixer were sufficient to partially depolymerize the DNA, which in turn led to the production of several chromatographic peaks. The homogeneous and undegraded DNA obtained by Hershey was known to have a molecular weight of 1.3×10^8 daltons according to autoradiographic measurements on the same phage by Levinthal. But with the exception of viral RNA and DNA it is still difficult to separate nucleic acids into chemically individual species. We can, however, measure the degree of molecular inhomogeneity in nucleic acid preparations, a parameter of considerable importance for certain purposes. The best methods for assaying this property were worked out in the laboratories of Doty and Meselson (see, for instance, ref. 90).

One method for determining the inhomogeneity of nucleic acids is to measure the temperature at which the cooperative helix–coil transition occurs. The transition temperature linearly increases with the proportion of G : C pairs in the macromolecule (Fig. 3-37). Interestingly enough, the melting points of DNA from the most varied organisms—from phage T4 to calf thymus—fall on a straight line when plotted versus (G + C) content. Extrapolation to zero (G + C), corresponding to pure poly dAT, gives a value of 69°C. By comparison, the melting point of synthetic poly dAT obtained from the Kornberg system, is 65°C. The 4° difference is not surprising since the synthetic poly dAT has a regularly alternating base sequence while natural DNA is quite irregular in sequence. The complete melting curves of a number of helical DNA preparations are presented in Fig. 3-38. The melting temperature is taken as the midpoint of the transition curve. Synthetic poly dAT and phage DNA both undergo

270　　　　　　　　　　　　　　　　　　　　　　　　III. Nucleic Acid Structure

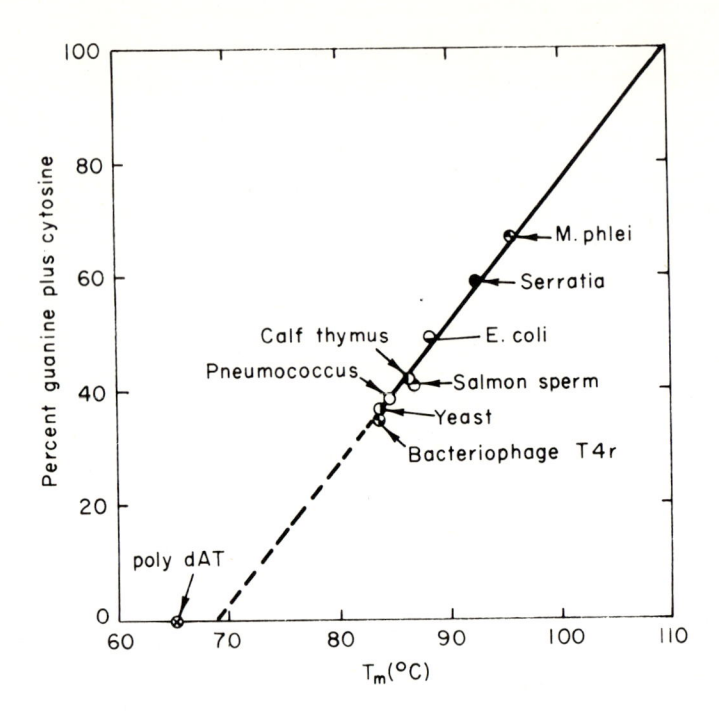

Fig. 3-37. Dependence of the denaturation temperature T_m on the content of guanosine + cytosine pairs in various samples of DNA. From Marmur and Doty (91).

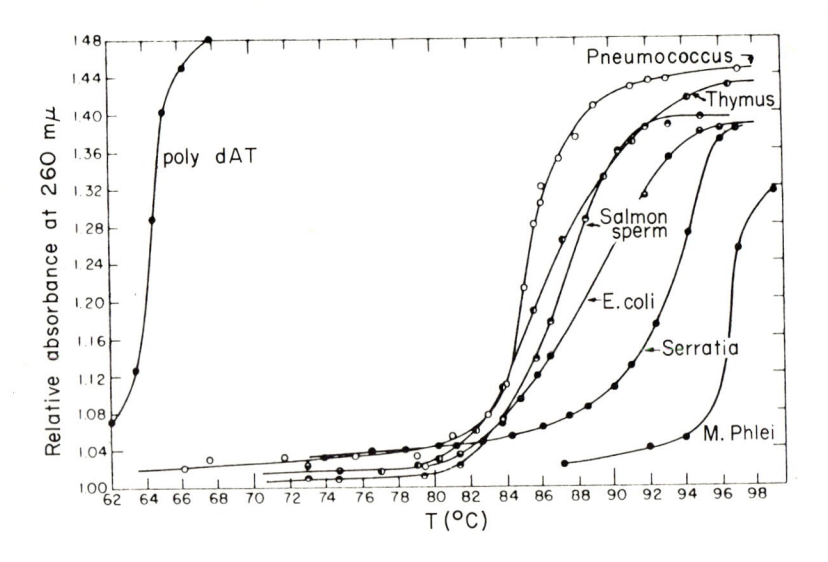

Fig. 3-38. Melting curves for the secondary structure of double-stranded DNA from several different organisms. Variations in the optical density of the DNA solution are plotted as a function of temperature. From Marmur and Doty (91).

extremely sharp transitions, and that of bacterial DNA is comparatively sharp. The transition curves for DNA form animal tissues, however, are much more gradual. This pattern is consistent with the fact that phages contain only one kind of DNA molecule. Bacteria also contain one or a small number of molecules, but because of their great size they are usually fragmented during extraction, presumably into heterogeneous chunks. The number of DNA molecules from animal cells, however, may run into the hundreds or thousands, depending on the integrity of the preparation. It may be supposed that the more gradual the melting curves, the more heterogeneous are the constituent molecules in their $(G + C)/(A + T)$ ratios.

If we assume that the base composition in every statistical population of DNA molecules obeys a Gaussian distribution, we can calculate its standard deviation, σ, from the width of the melting curve. Recall that melting temperature is linearly related to base composition, the temperature rising about 1°C for an increase of 2.5 mole% in $(G + C)$ content. To find the standard deviation of the distribution, we first locate the midpoint on the melting curve where the hypochromic effect is 50% of the maximum. Next we find the temperature at which the hypochromicity is $1/e \times 50\%$ or 18% and the temperature corresponding to the symmetrical point where the hypochromicity is 82% of the maximum. The temperature difference between these two outer points can be converted into percent $(G + C)$ content. The standard deviation for poly dAT (3°C) is taken as that of an ideal helix; even for identical molecules the helix–coil transition will not be infinitely sharp and its width is determined by bond energies.

Table 3-5 presents the figures for variation in nucleotide composition at one

TABLE 3-5
STANDARD DEVIATION OF THE NUCLEOTIDE COMPOSITION DISTRIBUTION
IN SEVERAL KINDS OF DNA[a]

Sample	ΔT	$\Delta T - 3°$	Standard deviation (mole % G + C) $2.5(\Delta T - 3°) = 2\sigma$	$2\sigma_B$
Poly dAT	3.0	0	0	—
Bacteriophage T4	2.6	0	0	—
Diplococcus pneumoniae	4.4	1.4	3.5	3.4
Shigella dysenteriae	4.0	1.0	2.5	—
Escherichia coli K12	5.2	2.2	5.5	—
Calf thymus	7.4	4.4	11.0	9.5
Salmon sperm	6.0	3.0	7.5	—

[a] Based on Doty *et al.* (90).

standard deviation to either side of the midpoint, as calculated from the width of the melting curves. In the last column are similar figures computed by an independent method described below.

The density gradient method of Meselson, Stahl, and Vinograd has also been used to measure the heterogeneity of DNA. DNA of varying $(G + C)/(A + T)$ ratios

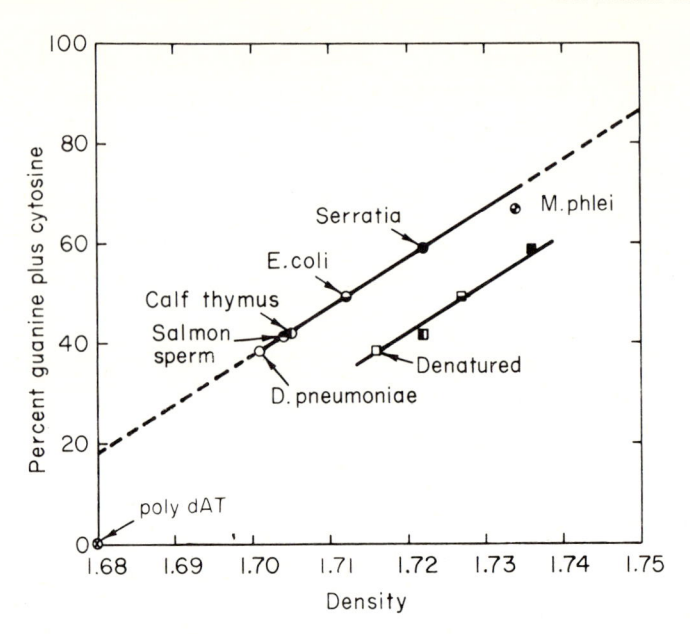

Fig. 3-39. Relationship of buoyant density to the guanosine + cytosine content of various samples of native and denatured DNA. From Sueoka et al. (92).

was extracted from a number of different organisms; Fig. 3-39 shows how the DNA density depends on (G + C) content. There is a difference of 0.015 gm/cm³ between the two lines representing native and denatured DNA because double-helical macromolecules are rather loosely packed and contain considerably greater quantities of bound water into which the CsCl cannot penetrate. From Fig. 3-39 it is clear that a 1 mole% change in the content of guanine and cytosine increases or decreases the density by 0.001 gm/cm³.

The width of the bands formed by DNA in CsCl density gradients is related to the homogeneity of the sample. Measurement of the standard deviation of each band permits us to determine the statistical variation in nucleotide composition exhibited by that DNA preparation. The width of the density gradient distribution depends on two factors: the inhomogeneity in nucleotide composition already mentioned, and the diffusion caused by thermal motion of the macromolecules. It can be shown theoretically that the experimentally measurable deviation, σ^2, is simply the sum of the two effects: $\sigma^2 = \sigma_T^2 + \sigma_B^2$, where σ_T^2 is the deviation due to thermal motion (Eq. 3-12, p. 264). Furthermore, $\sigma_T^2 = K/M$, where M is the molecular weight and K is a known constant relating to experimental conditions. This value can be calculated since the molecular weight of the DNA can be determined independently. By subtracting σ_T^2 from the experimental value of σ^2, we find σ_B^2, the deviation due to inhomogeneity in nucleotide composition. Two values obtained in this manner are presented in Table 3-5 for comparison with similar data from melting curves. The agreement between these two methods of measurement is quite satisfactory.

The heterogeneity in composition observed in preparations of bacterial DNA's probably results from random hydrodynamic degradation of the polymers during preparation since the bacterial chromosome is known to consist of a single enormous

DNA molecule [Cairns (93)]. Furthermore, it has been shown in experiments with artificially degraded DNA from bacteriophage λ, which also contains a single DNA molecule, that the fragments are not homogeneous in base composition (94). In fact, the inhomogeneity in composition along the DNA chain exceeds by a factor of four or more the fluctuations which would occur if the bases were randomly distributed. This explains how degradation can lead to increases in the apparent polydispersity of a DNA sample.

The polydispersity of DNA from higher organisms undoubtedly arises both from degradation and the presence in the native state of many chemically distinct DNA molecules. Attempts to partially fractionate polydisperse DNA's from higher organisms have met with limited success. The standard deviations of the new fractions are always somewhat less than that of the starting material, but none of the methods have proved to be wholly satisfactory. One fractionation procedure consists of partially denaturing the DNA sample and then precipitating most of the denatured material with ethanol. The melting point of the DNA remaining in solution is appreciably increased (2° to 3°) and its melting curve is steeper.

An important group of phenomena related to the renaturation of DNA have been studied in the laboratory of Doty (95, 96). We have already seen that the formation of bi- and trimolecular complexes from simple synthetic polynucleotides is a reversible process. Slow cooling (annealing) restores the complexes to their original state after the polymers have been separated by heat denaturation. The reversible melting of poly (dAT + dAT), a very regular copolymer synthesized by Kornberg, is an example of this type of transition. Restoration of the double-helical properties of heat denatured DNA depends on three factors: the extent of molecular heterogeneity, the temperature to which it is heated, and the rate at which it is cooled.

As a rule, the more homogeneous the DNA, the more completely it renatures since the kinetics are most favorable under these conditions. Thus, bacteriophage DNA, which consists of identical molecules, renatures both fully and quickly after heat denaturation. The greater heterogeneity of bacterial DNA reduces the efficiency of renaturation to about 40 to 50% if the cooling period lasts for about 1 hour. Thymus DNA is much more heterogeneous still and as a result, there is no significant reformation of double-helical molecules under conditions of renaturation. It is not difficult to explain these differences if we bear in mind that a collision between two complementary DNA strands is necessary before a Watson-Crick duplex can be reestablished. For bacteriophage, the probability of collision between complementary strands is 50%. In preparations of *E. coli* DNA where the large native molecules have been degraded to 100 or more fragments, the probability of a collision between two complementary strands is about 1 in 10^5; renaturation is still possible, but the process requires considerable time. The frequency of effective collisions among single strands from fragmented thymus DNA may be as little as 1 in 10^{10}, so that the probability of renaturation becomes extremely small in this case.

The kinetics of renaturation are strongly dependent on the temperature to which the DNA solution has been heated. In our earlier treatment of heat denaturation (p. 232ff), it was pointed out that melting of the DNA helix need not necessarily be accompanied by complete rupture of all hydrogen bonds nor by the total separation of strands. If a DNA solution is heated to some temperature within the limits of its melting curve and then rapidly cooled, renaturation occurs immediately. This is clearly a monomolecular process which, since it does not require collisions, proceeds both quickly and completely. Upon heating the solution to 5° or 10° above the temperature at which the hypochromic effect disappears, the DNA enters a new state in which it undergoes slow renaturation (97). At the higher temperature, renaturation is very sensitive to the homogeneity of the DNA preparation; the very heterogeneous DNA of higher organisms ceases to be renatured although at lower temperatures, close to the melting point, renaturation proceeds reasonably well.

From all of these observations, the following picture emerges. Within the melting range, a fraction of the double helices begin to dissociate into single strands. But this is a dynamic process and at any given temperature within this range, hydrogen bonds re-form as well as dissociate and as a consequence, double-stranded molecules continually separate and reassociate. The equilibrium between the double- and single-stranded forms is determined by the temperature and there is a constant exchange between the two states. This description has been confirmed by experiments in which the equilibrium was artificially displaced in favor of single-stranded chains by the chemical attachment of formaldehyde to the amino groups of the nucleotide bases (98). In such cases it was possible to verify the separation of helical duplexes by using molecules with ^{15}N in one strand and ^{14}N in the other. Separate strands could subsequently be resolved by density gradient centrifugation.

Even under conditions of irreversible denaturation, in which the amino groups have been blocked, it is necessary to heat the DNA for 10 minutes at a temperature 5° to 6° above that at which the hypochromic effect completely disappears in order to effect chain separation. Unfortunately, the temperature dependence of reversible strand separation has not been studied quantitatively. However, the bulk of the data on DNA denaturation and renaturation leads to the conclusion that there is a mobile association-dissociation equilibrium which is spread over a considerable temperature interval above the melting point, T_m.

Curves showing the kinetics of renaturation of several types of DNA are presented in Fig. 3-40. The double helices were melted apart and then "frozen" in their denatured state by rapid cooling. Only in the case of bacteriophage DNA is it impossible to completely freeze the amorphous form, since renaturation occurs even during very rapid cooling. The temperature was then quickly raised to 67°C, slightly below the normal transition point. At this temperature, the single strands are gradually and reversibly converted to their native form as the result of annealing.

The chains which are joined with one another during renaturation must be complementary, but it is statistically unlikely that they will be the same two halves which were separated upon heat denaturation of the original molecules. If two preparations of homologous DNA, each labeled with a different nitrogen isotope, are mixed, denatured, and then annealed, a portion of the newly formed duplexes will be ^{14}N/^{15}N hybrids (100). When analyzed by density gradient centrifugation, three peaks are expected: two of them corresponding to the native forms ^{14}N/^{14}N and ^{15}N/^{15}N, one representing the ^{14}N/^{15}N molecular hybrids. If the initial mixture contains equal parts of the two isotopically pure DNA's, the hybrid peak should for statistical reasons be twice as great as either of the native peaks after annealing. There should also be two other bands corresponding to denatured single strands of each isotopic composition.

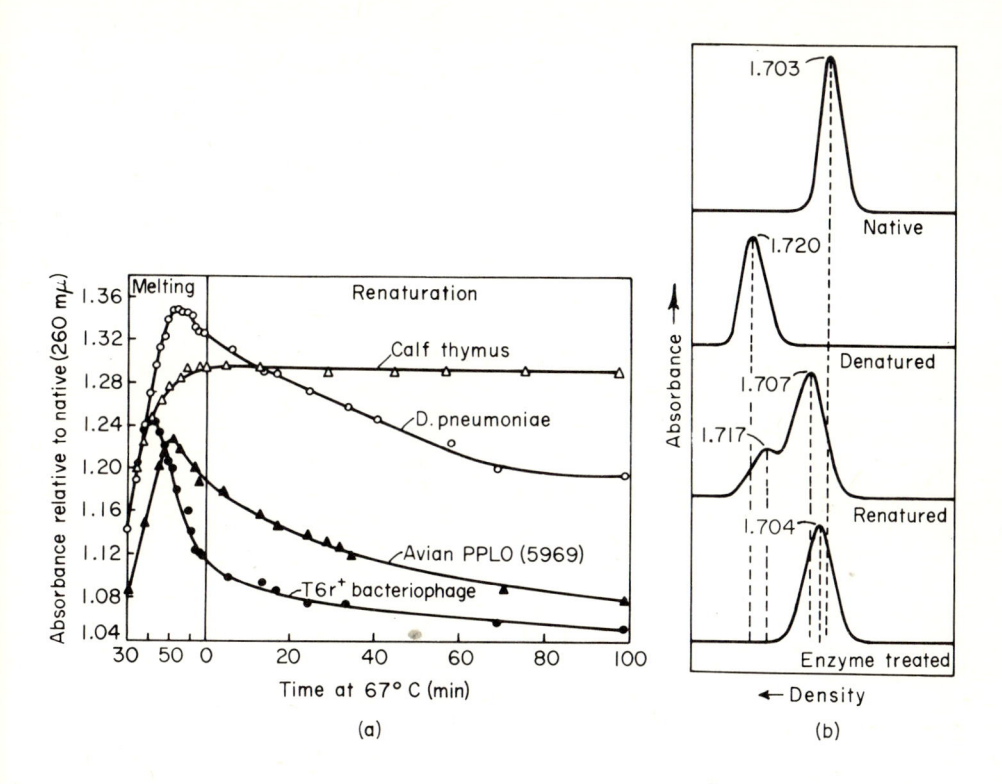

Fig. 3-40. The melting and renaturation of DNA molecules in solution. (a) Kinetics of renaturation of DNA from several organisms. After the DNA was heated to 100°C for 10 minutes and quickly cooled in order to separate the strands, samples were reheated to 67°C for annealing and changes in the optical density were measured as a function of time. There is an initial rise in the O.D. as the solution of denatured DNA is heated to the annealing temperature. The subsequent decrease in the relative optical density at 67° marks re-formation of double-helical molecules. From Marmur and Doty (99). (b) Density gradient investigation of the renaturation of *Bacillus subtilis* DNA. When native *B. subtilis* DNA, which has a buoyant density of 1.703 (first frame), is denatured by heating for 10 minutes at 100°C and quickly cooled, its density increases to 1.720 (second frame), characteristic of the transition to single-stranded DNA. Most of this DNA renatures when annealed (frame 3). The remaining denatured material disappears (frame 4) when the solution is treated with *E. coli* phosphodiesterase, an enzyme that digests single-stranded DNA. From Schildkraut *et al.* (100).

It can be seen from the microdensitometer tracings in Fig. 3-41 that this is precisely the result obtained experimentally. For this experiment, DNA was extracted from two cultures of *Bacillus subtilis*, one grown in normal medium and the other labeled with the ^{15}N isotope. Equal quantities of solution, each containing 10 µg/ml DNA were mixed, heated to 100°C, and slowly cooled. The formation of hybrid can be clearly seen in frame C, but is even more evident when the two kinds of denatured DNA are removed (frame D). Denatured DNA is degraded by *E. coli* phosphodiesterase an enzyme specific for cleaving single-stranded polynucleotides (101). Double helices are not affected by its action, so that it can be used directly on the renatured mixture to rid it of denatured DNA. Three bands remain after phosphodiesterase digestion:

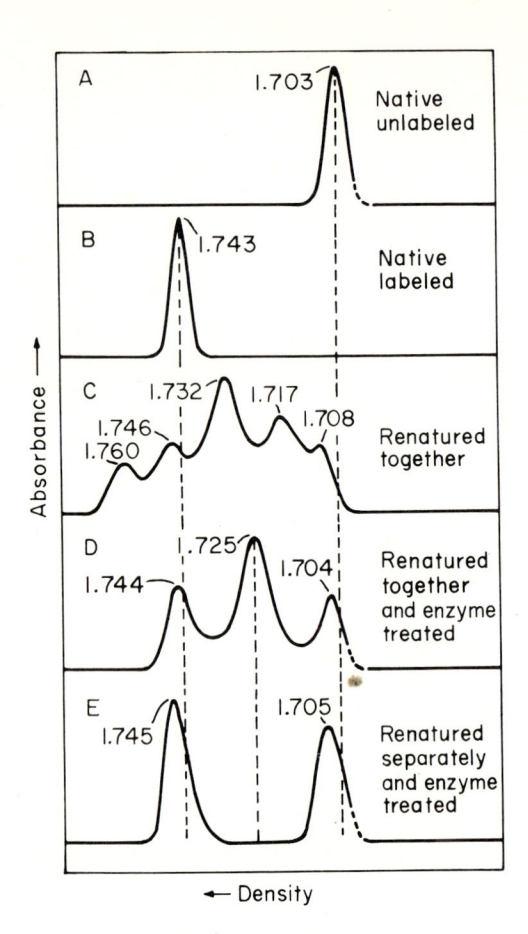

Fig. 3-41. The hybridization of unlabeled and density-labeled DNA from *B. subtilis* by denaturation and annealing. (A) Density gradient profile of native DNA containing ^{14}N. (B) Density gradient profile of native ^{15}N-labeled DNA. (C) A mixture of ^{14}N- and ^{15}N-containing DNA after strand separation and annealing. (D) Same sample as (C) after treatment with *E. coli* phosphodiesterase. Note that removal of denatured material reveals the presence of a hybrid ^{14}N : ^{15}N duplex. (E) ^{14}N- and ^{15}N-containing samples heated and annealed separately, treated with phosphodiesterase, and then mixed. From Schildkraut *et al.* (100).

the two original DNA's and the hybrid. The pattern resulting when the two DNA's are separately denatured and annealed (frame E) provides a control which illustrates the high resolving power of the density gradient technique.

There are still other details of the renaturation process that are worthy of mention. It was of particular interest to establish whether or not renaturation of DNA could be measured by biological activity, specifically its ability to transform bacteria. The phenomenon of bacterial transformation will be considered in detail in the next chapter but, in short, it is a method for transferring specific hereditary characteristics, such as resistance to a bacteriocidal agent, with DNA extracted from a strain of bacteria resistant to that agent. Transforming activity, inherent only in native DNA, can be used as a quantitative index for evaluating the extent of denaturation and renaturation.

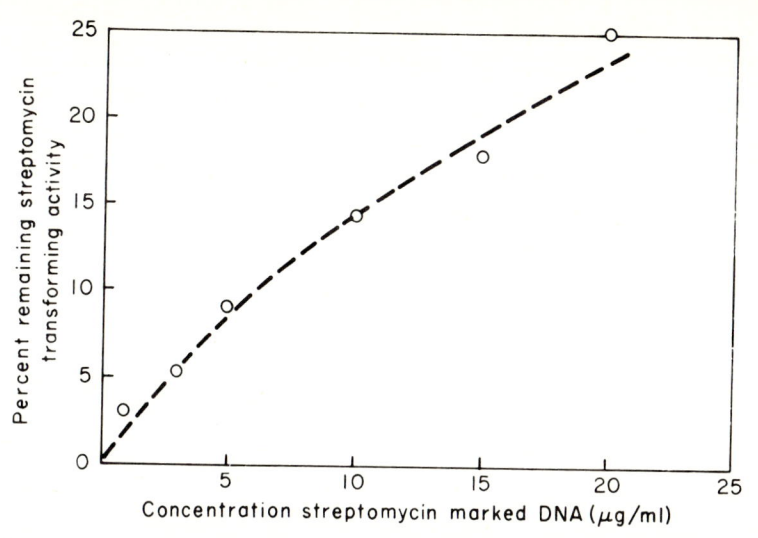

Fig. 3-42. The extent of renaturation as a function of DNA concentration. Samples of pneumococcal DNA were denatured by heating at 100°C for 10 minutes, diluted to the concentrations indicated, and then slowly cooled to permit annealing. Renaturation of the DNA, which carried the determinant of resistance to streptomycin, was measured by its capacity to transform recipient cells for the genetic trait. Denatured DNA does not possess transforming activity. From Marmur and Lane (102).

In the course of kinetic studies on DNA renaturation, it became clear that both the extent and rate of renaturation increase with DNA concentration (Fig. 3-42). In this experiment, renaturation was achieved by slowly cooling a DNA solution previously heated to a temperature above the T_m. The dependence of reaction rate on concentration is to be expected since the annealing of separate single-stranded molecules is a bimolecular process requiring collisions between complementary strands. However, since the separation of double-stranded molecules is far from complete, the order of the renaturation reaction should fall between 1 and 2. Moreover, the phenomenon is further complicated by disordered aggregation which increases with concentration and effectively removes a portion of the molecules from the renaturable population. Therefore, even though an increase in concentration facilitates renaturation, the effect is not nearly so great as would be expected in the absence of undesired interactions. So as to avoid nonspecific aggregation, the DNA concentration is generally kept quite low, on the order of 10 μg/ml.

Low DNA concentrations were used to produce the curve of Fig. 3-43 in which the rate of renaturation, or more accurately the extent of renaturation reached after 1 hour of annealing, is plotted as a function of temperature. These experiments were performed exactly as the ones illustrated in Fig. 3-40: the DNA was heated, rapidly cooled to fix the molecules in their denatured state, and then annealed at a constant temperature for one hour. The curve of Fig. 3-43 is remarkably similar to curves representing the dependence of rate of crystallization on temperature. Crystallization is influenced by two factors. First, a crystal nucleus must be formed, an event whose probability increases with supercooling. Second, however, the molecules must be mobile so that the nucleus will grow, and molecular motion is diminished by decreases in temperature. The renaturation of DNA evidently reflects the same two antagonistic tendencies.

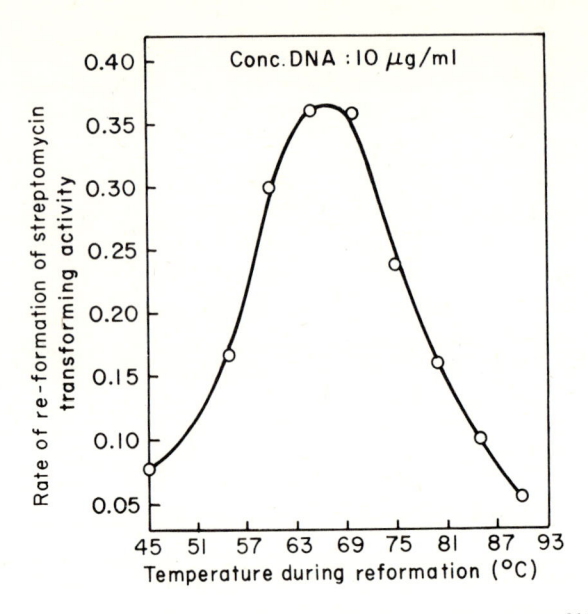

Fig. 3-43. The effect of renaturation temperature on the rate of recovery of biological activity by denatured pneumococcal DNA. The capacity of the renatured DNA to transform recipient bacteria to streptomycin resistance was assayed as a function of the temperature at which the DNA solution was annealed following strand separation at 100°C. From Marmur and Doty (99).

We may pursue this analogy still further in our attempt to better understand the detailed mechanism of renaturation. While renaturation of course requires an effective collision between two complementary DNA strands, this event ends in duplex formation only if certain other interactions occur. Specifically, a nucleus of regular structure must be formed in which hydrogen bonds are established between complementary bases. Furthermore, the nucleus must grow to such an extent that there is a low probability of strand dissociation and separation under the influence of thermal fluctuations. The appearance of a nucleus for helix formation and its growth is dependent on a special kind of Brownian motion in which one chain slides along the other as if to "read" or "study" it. Of course, this process is absolutely devoid of any kind of orderliness or direction, but if a short sequence of nucleotides on one chain happens to meet a complementary sequence on the other, a nucleus of regular structure will be formed. Such a nucleus will either grow, if the two chains are really complementary, or else it will break apart, if there is only a limited and local complementarity between the chains. Fluctuations in thermal energy will cause the mispaired chains to dissociate and Brownian motion will be resumed. This process will continue until two chains find a mutual orientation in which hydrogen bonds can form along their entire length, starting from the initial center of "recognition." The two activities characteristic of any crystallization process can therefore be identified in the renaturation of DNA. The formation of a nucleus for the growth of hydrogen bonding requires a reduction in temperature so that the intermolecular bonds are not immediately broken. On the other hand, the proper positioning of the chains relies on Brownian motion which decreases when the temperature is reduced. This state of affairs explains the occurrence of an optimum temperature for DNA renaturation as illustrated in Fig. 3-43.

9. Molecular Heterogeneity, Reversible Denaturation, and the Formation of Molecular Hybrids

Systematic experiments have been performed on the hybridization of DNA molecules isolated from a variety of organisms (103). Denaturation and annealing of two types of DNA together in one mixture provides a sensitive physicochemical test of genetic compatibility and the results of this analysis are in general agreement with purely genetic experiments. Within any particular species, hybridization occurs without difficulty. For instance, *E. coli* strains B, C, K12, and others are all compatible by the hybridization test. That is, their DNA molecules are for the most part complementary to one another in nucleotide sequence. Similarly, *Bacillus subtilis* DNA is compatible with that of *B. natto* (Fig. 3-44). It was previously known that the DNA

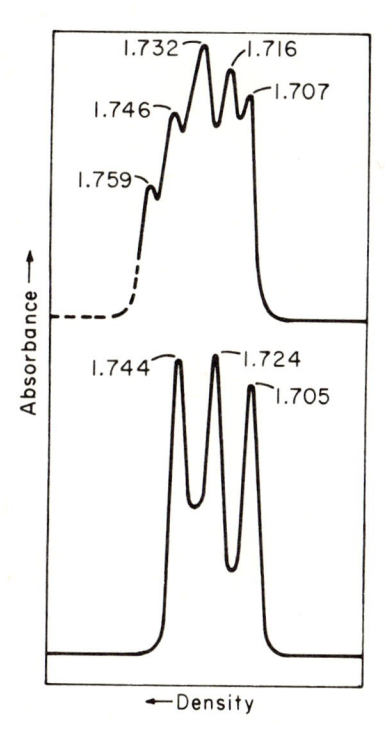

Fig. 3-44. Hybridization of DNA from two different species of bacteria. Density-labeled *Bacillus subtilis* DNA and normal *Bacillus natto* DNA were mixed, heated, and annealed. The frames show the density distribution of the renatured mixture before (upper) and after (lower) treatment with *E. coli* phosphodiesterase to digest denatured DNA. In the lower frame, renatured *B. subtilis* and *B. natto* DNA's are on the left and right, respectively. The presence of the intermediate band, consisting of hybrid double-stranded DNA indicates that considerable sequence homology exists between the two DNA's. From Schildkraut *et al.* (100).

from either one of the strains is capable of transforming the other. In the case of more distantly related organisms, the minimal requirement for hybridization is that the nucleotide compositions of the two DNA's be identical within 1 mole %. When DNA from *E. coli* and *Shigella dysenteriae* are annealed together, a certain amount of hybridization is observed, although decidedly less than between two strains of *E. coli*. Attempts to obtain DNA hybrids between *E. coli* and several strains of *Salmonella* have not been successful. A slight difference in the nucleotide compositions of the two

DNA's apparently reflects a much more significant difference in nucleotide sequence.

Molecular hybridization has proved to be a powerful tool for the investigation of other fundamental problems in molecular biology, especially those of transcription. The use of DNA : RNA hybridization experiments to establish complementarity of nucleotide sequences in the two kinds of nucleic acids is illustrated in Fig. 3-45. The

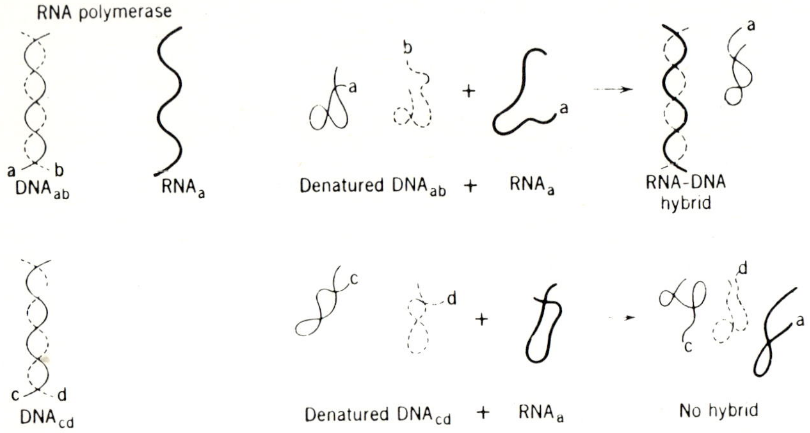

Fig. 3-45. Diagram of the hybridization test for identifying complementary sequences in DNA and RNA. RNA is copied as a complementary molecule from one of the two DNA strands by RNA polymerase. After the DNA strands have been separated by denaturation, the RNA can form a hybrid RNA : DNA duplex molecule with its complementary DNA strand, but not with DNA strands from another source. After Luria and Darnell (104).

development of new experimental approaches to supplement density gradient centrifugation have been essential in this work. The discovery that DNA : RNA hybrids are resistant to RNase concentrations on the order of 20 μg/ml at 20°C, for instance, was an important advance. Using this method, it is quite easy to distinguish complementary double-stranded molecules from various nonspecific and random aggregates which are rapidly hydrolyzed by RNase. Another method consists of using two types of radioactive isotopes, such as ^{3}H (tritium) and ^{32}P, to label RNA and DNA separately. In light of the enormous energy difference between the β-particles emitted by tritium and phosphorus, both isotopes can be assayed together in the same sample. Since these isotopes can be obtained at very high specific activities, CsCl gradient centrifugation can be carried out with concentrations of RNA and DNA too small for optical detection. Furthermore, DNA: RNA complexes need not necessarily be studied by density gradient centrifugation in CsCl. It is possible to recover them directly by any of the following methods: sedimentation from solution after RNase treatment, gel filtration on Sephadex, deposition on a membrane filter, or purification on a column made from kieselguhr coated with methylated albumin (MAK, p. 269.)

Hybridization techniques were applied to the question of ribosomal and transfer RNA synthesis: were these molecules transcribed from specific regions of the DNA or were they produced by some other mechanism? The answer was found in the laboratory of Spiegelman (105). They showed that by hybridizing bacterial DNA with an excess of ribosomal RNA, it is possible to obtain a DNA : RNA complex which is insensitive to

RNase. Moreover, in a CsCl gradient the complex sediments in roughly the same place as double-stranded DNA. If a certain amount of DNA is saturated with ribosomal RNA, it is possible to find what percent of the chromosome serves as a template for the transcription of the particular molecular species. Now *E. coli* contains two kinds of ribosomal subunits, one sedimenting at 50 S and the other at 30 S, and each kind possesses its own variety of structural RNA: 23 S RNA with a molecular weight of 1.1×10^6 daltons in 50 S particles and 16 S RNA with a molecular weight of 6×10^5 in 30 S particles. The regions which hybridize with 23 S ribosomal RNA comprise some 0.2% of the total cellular DNA. Since the molecular weight of the chromosome is 3×10^9 daltons, the regions complementary to 23 S RNA amount to 6×10^6 daltons, sufficient for five or six copies of the template for 23 S RNA synthesis. Spiegelman also found that 16 S ribosomal RNA is transcribed from a region of the DNA distinct from that for 23 S RNA and which occupies approximately 0.1% of the chromosome. Thus, there are also about five or six templates for 16 S ribosomal RNA. It was possible to demonstrate that saturation of each template with its particular RNA proceeds quite independently of the other. In other words, there is no competition between the two types of ribosomal RNA for homologous regions in the DNA.

An entirely analogous experiment was performed independently by Spiegelman and Rich and their associates to find the template for the transcription of transfer RNA (106). *Escherichia coli* DNA was hybridized with labeled tRNA from the same strain. All unhybridized RNA was removed with RNase and the hybridized product, unharmed by the RNase treatment, was purified in a CsCl gradient. The amount of labeled tRNA complexed with the DNA could then be accurately evaluated. It turned out that about 0.025% of the cellular DNA was responsible for tRNA synthesis. If the *E. coli* cell contains 30 different types of tRNA (taking account of degeneracy in the genetic code—see p. 437), each with a molecular weight of 25,000, we can calculate that a DNA complement of 3×10^9 daltons contains on the order of one template for the transcription of each type of tRNA.

A large part of the cellular DNA participates in the synthesis of messenger RNA, used by the cell in protein synthesis. Hybridization experiments have permitted the investigator to follow transcription of a specific messenger RNA in a number of cases. Phage infection, for instance, introduces genetic information into the bacterial cell which is essential for the synthesis of new proteins, such as the T2-specific enzyme for hydroxymethylating cytosine (see p. 205). For this purpose, new messenger RNA's are produced in the infected cells which can be selectively extracted by annealing with phage DNA (107). Control experiments show that this specific mRNA does not exist in the cell prior to infection.

A very subtle question which was investigated by the method of hybridization consists of whether one or both strands of DNA are copied during transcription. This problem was resolved owing to the discovery of bacterial viruses containing DNA whose two strands differed widely in nucleotide composition. Both phage sp8 of *B. subtilis* and phage α of *B. megaterium* are of this type. DNA from these phages was denatured so as to completely separate the double helices into single-stranded molecules. The two complementary strands were then fractionated on MAK columns and checked for homogeneity in a CsCl gradient. The two fractions yield peaks which differ in density by 0.006 gm/cm³. Messenger RNA specific for phage α was extracted from virus-infected cells, and was annealed with separate "heavy" and "light" strands of phage DNA. After RNase treatment, it was found that only the "heavy," purine-rich strand formed a significant number of molecular hybrids with α-specific RNA. The "light," or pyrimidine-rich strand complexed 10 to 20 times less RNA than its complementary strand (108).

A similar experiment was performed with phage ΦX174, exploiting the unique properties of this virus. Phage DNA was prepared in two forms: single-stranded DNA from the phage particles themselves, and double-stranded replicative form from infected *E. coli* cells. Phage-specific mRNA, also from infected cells, was hybridized with both forms of ΦX174 DNA. The results established that the phage mRNA is complementary not to the phage molecule, but to the newly synthesized strand of the duplex replicative form (109).

Several new techniques have recently been developed for separating the two complementary strands of duplex DNA molecules from one another. In one approach, DNA strands are fractionated by centrifugation to equilibrium in an alkaline CsCl gradient at pH 12. At this pH, the Watson-Crick helix is completely dissociated, but single-stranded DNA, unlike RNA, is not degraded. Although the resolving power of the alkaline CsCl gradient is limited, the two strands of λ-phage DNA, for instance, exhibit a density difference of 0.004 gm/cm^3 under these conditions (110). No strand separation is observed after density gradient centrifugation at neutral pH's. In a second method, DNA is denatured at 90°C, frozen in the single-stranded state by rapid cooling, and then mixed with a solution of synthetic polyribonucleotides such as poly G or poly IG, a copolymer of inosinic and guanylic acids (111). If one of the DNA strands contains an excess of pyrimidines, and of deoxycytidine residues in particular, it will form a complex with poly G or poly IG that can be separated from the purine-rich strand, which does not interact with the polyribonucleotide, by density gradient centrifugation in neutral CsCl. Specifically, the pyrimidine-rich DNA, being at least partially double-stranded owing to the presence of the complexed polyribonucleotide, will possess a greater buoyant density than the complementary strand and the two will be well resolved from each other after equilibrium centrifugation in CsCl. This technique has been successfully applied to DNA from viruses, bacteria, and the cells of higher organisms. Finally, a method for the preparation of individual DNA strands from the bacterium *Bacillus subtilis* by chromotography on MAK columns has recently been reported (112). In this procedure, denatured DNA is eluted from the column by a combination of stepwise and continuous changes in ionic strength, a technique called intermittent gradient elution.

Current methods for the preparative isolation of complementary DNA strands have been used to reexamine the transcription of mRNA from DNA templates. As a result, it has been clearly established that both DNA strands can in fact be transcribed, but that each specific type of mRNA is copied from only one of the two strands. In other words, the transcription mechanism can operate on either of the two DNA strands, but both strands are never copied simultaneously in the same functional region of the genetic material. A particularly elegant demonstration of this was performed by hybridizing mRNA formed at various times after infection of an *E. coli* cell by λ-phage with isolated single strands of λ-DNA (113). In the first minutes after infection, mRNA is preferentially transcribed from a limited region of one of the two DNA strands. After a short time, transcriptions of new mRNA's begins from regions of both the first and second strands whose protein products are necessary for the expression of later functions in phage development. Messenger RNA's specific for the late proteins themselves are transcribed exclusively from the second strand, and their formation starts 10 to 20 minutes after infection. The time at which transcription switches from copying primarily the first strand to copying primarily the second strand can be determined with considerable accuracy.

Before concluding, mention should be made of two new and extremely effective methods of molecular hybridization that do not require any complicated technical procedures. In the first, DNA is denatured in a solution of agar and the mixture is cooled, producing a gel in which the individual single strands are immobilized and interspersed through the highly viscous medium. The gel is then passed through a fine-mesh screen which divides it into small particles, and the latter are packed into a chromatographic column [Bolton and McCarthy (114)]. A solution containing the DNA or RNA undergoing the test for hybridization is then passed through the column at a very low rate and the temperature is maintained at a value favorable for annealing. The molecular weight of the nucleic acids in solution is kept beneath 10^6 so as to facilitate the diffusion of molecules throughout the gel. The hybridizing potential of two polynucleotides can be readily assessed by this method, at least qualitatively. But even more important, it is suitable for the hybridization of DNA molecules from higher organisms, a feat which could never be performed with the earlier techniques of annealing in solution.

Many interesting experiments have been carried out with the Bolton-McCarthy column. It was shown, for example, that DNA from organisms close to one another in the evolutionary scale, such as men and monkeys, can hybridize, while no such interaction occurs between the DNA's of unrelated species (115). DNA–agar gels were also used to study changes in mRNA which take place during embryological development and it was concluded that these molecules are highly specific to the stage of differentiation. The mRNA isolated from different tissues of the same organism could also be distinguished (116).

The most effective approach to the formation of hybrid nucleic acid molecules consists in prior fixation of denatured single-stranded DNA to nitrocellulose membrane filters [Gillespie and Spiegelman (117)]. To prepare the sorbent, a solution of denatured DNA is first passed through a membrane filter to trap the single-stranded molecules. The filter is then thoroughly dried at 80°C in order to irreversibly fix the DNA strands. Hybridization is carried out by immersing the filter in a solution of the RNA whose complementarity with the immobilized DNA is to be tested. Annealing generally requires incubation at an appropriate temperature (60° to 65°C) and ionic strength (0.3 M NaCl or more) for periods up to 24 hours. Since DNA–RNA hybrids remain tightly attached to the filter, vigorous washing procedures can be employed to remove unhybridized RNA and the filter can be treated with RNase to further reduce the "noise" level resulting from nonspecifically bound RNA. After inactivation of the RNase with iodoacetic acid, RNA can be eluted from the filter by washing with a hot buffer of low ionic strength so that the charges on the nucleic acid molecules will no longer be effectively screened.

Using this simple procedure, it is possible to carry out hybridization experiments with great quantitative accuracy, particularly if radioactively labeled RNA is employed, because the background of nonspecifically adsorbed RNA can be depressed to as little as 0.001 % of the input. A further advantage of this technique is that it circumvents renaturation of the DNA which reduces the amount available for the formation of complexes with RNA. In order to establish whether two RNA preparations are identical, that is, whether they are complementary to the same region or regions of the DNA, a competition experiment can be performed. The ability of a fixed amount of labeled RNA from one preparation to hybridize with homologous DNA is assayed in the presence of increasing amounts of unlabeled RNA from the second preparation. By measuring the quantity of radioactivity fixed in each case, it is possible to ascertain whether the two RNA's compete for binding sites in the DNA, and if so, to what extent.

10. Macromolecular Structure of RNA

As was already indicated, there are several kinds of RNA, each with its own specialized function. About 10% of the total RNA of a bacterial cell consists of relatively low-molecular-weight transfer RNA (tRNA). Every cell contains a set of at least 30 to 40 different tRNA's, each of which transports a specific amino acid. Sedimentation, diffusion, hypochromic effect, and electron microscopy have all been used to characterize transfer RNA.

The sedimentation constant of tRNA was found to be 4.3 S. It was shown, moreover, that there is no discernible difference in the hydrodynamic properties of tRNA's specific for different amino acids when they are isolated from a single species. Using the sedimentation constant and the viscosity in conjunction with the Flory-Mandelkern equation, the molecular weight of tRNA is estimated to be approximately 25,000.

The macromolecular properties of *E. coli* tRNA are evident from its hypochromic effect (curve 1, Fig. 3-46) (118). For purposes of comparison, the curve obtained for

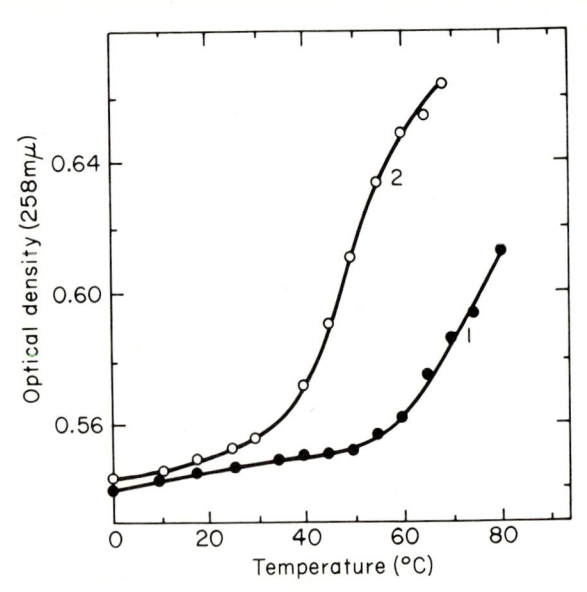

Fig. 3-46. The thermal denaturation of RNA secondary structure. The two curves correspond to (1) low-molecular-weight tRNA and (2) high-molecular-weight ribosomal RNA from bacteria in 0.1 M NaCl. From Cox and Littauer (118).

10. Macromolecular Structure of RNA

high-molecular-weight ribosomal RNA, also from *E. coli*, is reproduced in the same figure (curve 2). From these profiles, it is apparent that there is no sharp melting temperature for either tRNA or ribosomal RNA. In both cases, a gradual change in properties is observed upon increasing the temperature.

The secondary structure of tRNA can be judged from the magnitude of the hypochromic effect, the amount of optical activity, the number of hydrogen atoms in purine and pyrimidine bases which exchange slowly with deuterium, and finally, by the primary structure which is already known in several instances (p. 213). By all criteria, the helical portions of tRNA constitute between 60 and 70% of the total. Three alternative structures (Fig. 3-47) can be generated by folding the tRNA chain

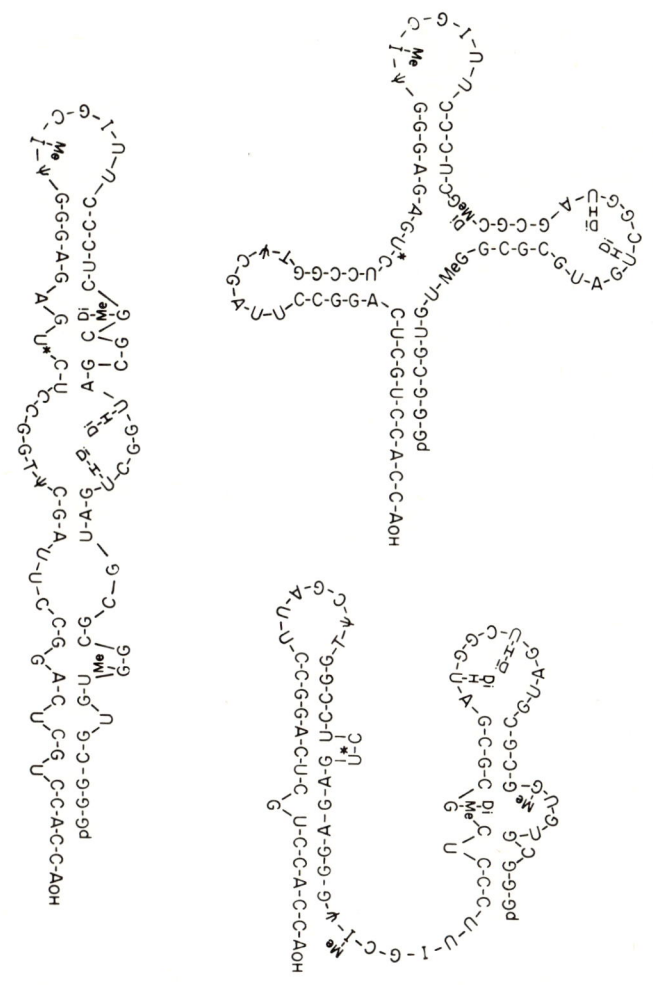

Fig. 3-47. Possible secondary structures of yeast alanine tRNA. The three configurations were chosen so as to maximize pairing between complementary bases. The evidence favors the cloverleaf structure at upper right, particularly since each of the ten to fifteen different kinds of tRNA whose sequences are known can be folded into this pattern with only minor differences. From Reference 4, R. W. Holley, J. Apgar, G. A. Everett, J. T. Madison, M. Marquisee, S. H. Merrill, J. R. Penswick, and A. Zamir (1965). *Science* **147**, 1462-1465, 19 March 1965. Copyright 1965 by the American Association for the Advancement of Science.

so as to form the maximum number of possible complementary base pairs (4). According to current notions, the most likely conformation is the branched one, dubbed the "cloverleaf." In all tRNA's whose primary structure are known, almost identical configurations are permitted by the criterion of maximum base pairing, and in all cases the anticodon (p. 435) which is the nucleotide triplet responsible for positioning the tRNA on the messenger RNA template during protein synthesis, is found in the central petal and specifically, in a region where there is no base pairing.

Ribosomal RNA (rRNA), the nucleic acid constituent of ribonucleoprotein particulates called *ribosomes*, comprises about 85% of the total cellular RNA. Ribosomes generally consist of two unequal subunits which in the case of *E. coli* possess sedimentation coefficients of 30 S and 50 S. The isolation of rRNA in its native state presented great difficulties at one time. Only by freezing the cells and then rapidly deproteinizing them with 50% phenol was it possible to circumvent the degradative action of RNase (ribonuclease) to which rRNA is very sensitive. After such procedures were applied, it was found that ribosomal RNA gives three sharp peaks in the ultracentrifuge with sedimentation constants of 5 S, 16 S, and 23 S if isolated from bacteria, and of 5 S, 18 S, and 28 S if obtained from animal tissues. The 16 S to 18 S rRNA's derive from the smaller subunit of the ribosome, while both 5 S and 23 S to 28 S rRNA are associated with the larger subunit. The complete nucleotide sequence of 5 S RNA from *E. coli* has been worked out by Sanger and associates (119) who developed an elegant two-dimensional electrophoretic procedure for separating the large number of oligonucleotides resulting from partial enzymatic hydrolysis of the RNA molecule. Efforts are presently under way to sequence portions of the larger ribosomal RNA's, a problem of great complexity considering the size of these molecules. By measuring the light scattering of rRNA preparations from *E. coli*, it was possible to find the relationship between the average sedimentation constant, s_w, and the average molecular weight, M_w:

$$s_w = 0.98 \times 10^{-2} \times M_w^{0.56}$$

A similar dependence, $s = AM^{0.5}$, was found by Doty and Boedtker for viral RNA. From this relationship, the molecular weights of the three ribosomal RNA's from *E. coli* are found to be 3.5×10^4, 6×10^5, and 1.1×10^6, respectively. Application of the Flory-Mandelkern equation to the sedimentation and viscosity data leads to similar values for the molecular weight when the constant, β, is equal to 2.2×10^6 (see p. 225 for details).

The behavior of RNA in solution differs from that of DNA in a number of respects, reflecting a difference in secondary structure. Polyelectrolyte effects are very weak in DNA since its native configuration is that of an extremely rigid and stable twofold helix. Consequently, native DNA in solution has a very high viscosity; when the double helix is broken down by heat denaturation, however, the viscosity drops significantly. The RNA molecule is rather flexible since its ordered portions are irregularly distributed, interspersed with disordered segments. The main difference between the two nucleic acids thus lies in the rigidity of their chains, or in quantitative terms, in the magnitudes of their statistical segments, a measure of molecular flexibility (p. 223). The statistical segment of DNA is on the order of 500 to 700 Å while that of RNA is

10. Macromolecular Structure of RNA

about 100 Å. Moreover, RNA is significantly polyelectrolytic. When heated, both its viscosity and the sensitivity of its charges to electrostatic screening increase.

Ribosomal RNA is characterized by a helical configuration even though it is not made up of duplex chains. We may assume that the structure of RNA is such that the molecule is folded back on itself at several points so as to form short double-stranded portions owing to hydrogen bonding between complementary nucleotide sequences. These helical portions are linked by disordered, amorphous regions, as illustrated in Fig. 3-48. Such a model of secondary structure elucidates several basic properties of

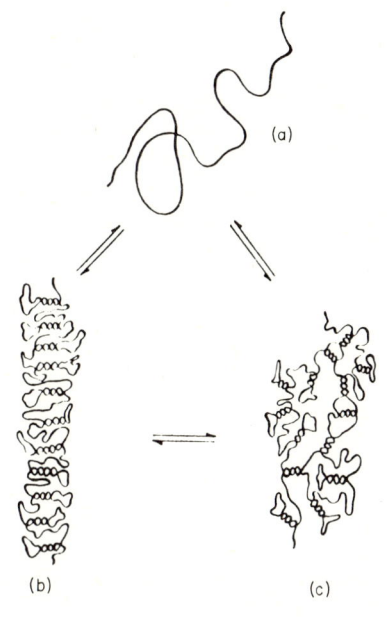

Fig. 3-48. Schematic representations of the possible conformations and configurational transitions of high-molecular-weight RNA molecules in solution, depending on ionic strength, temperature, and pH. (a) The single-stranded polymer exists as a random coil in the absence of salts or at high temperatures. (b) A compact rod with regularly arranged helical regions is favored by low ionic strengths and low temperatures. (c) A compact coil with helical regions is thought to be characteristic at high ionic strengths and low temperatures. From Spirin (120).

RNA molecules. The gradual disappearance of helical regions and of hydrogen bonds with rising temperature is explained by the irregularity of the helix and by its semi-amorphous nature. When such a chain is electrically charged, a direct result is the straightening out and extension of the macromolecule, although hydrogen bonds between neighboring residues are not necessarily disturbed when this happens.

Another RNA which has been studied in detail, high-molecular-weight RNA from tobacco mosaic virus (TMV), behaves in much the same way. In this case, the RNA is also clearly single-stranded and its molecular weight, as measured by light scattering, is 2.1×10^6. That TMV RNA consists of a single molecular chain can be ascertained in a number of ways. In the first place, Franklin and Klug studied the x-ray diffraction patterns of intact TMV and of its protein coat; the latter was obtained by separating the RNA from viral protein and then reaggregating the protein in the absence of

RNA. The difference in the two patterns constitutes the diffraction pattern of the viral RNA. A schematic drawing of the virus particle is given in Fig. 1-52 (p. 106). Viral protein comprises 95% of the total viral mass, and in the intact virus it is aggregated in the shape of a cylinder with an outer diameter of 170 Å. The RNA (5% of the mass) is enclosed in the cylinder in the form of a single-stranded helix. From its diameter (80 Å) and pitch (23 Å), the molecular weight of the RNA is calculated to be 2×10^6.

A single break at any point on the RNA chain is sufficient to eliminate biological activity, even when the helical structure of the RNA is supported by the protein coat. If the viruses are deproteinized, the resulting RNA is monodisperse and gives a single extremely sharp peak in the ultracentrifuge. The secondary structure of the extracted RNA is completely different from that within the virus particle. In the isolated state, intramolecular hydrogen bonds are formed between nucleotide bases of the RNA chains while within the protein coat the only hydrogen bonds which exist are those between nucleic acid and protein.

TMV RNA behaves in an interesting way when it is heated. At low temperatures, its sedimentation constant is 27 S; but when the RNA solution is heated to 80°C, hydrogen bonds are disrupted and the helical structure melts out (Fig. 3-49). The same thing happens at 55°C in 6 M urea because urea facilitates the disruption of hydrogen bonds. Light scattering measurements at high temperature prove that heating of short

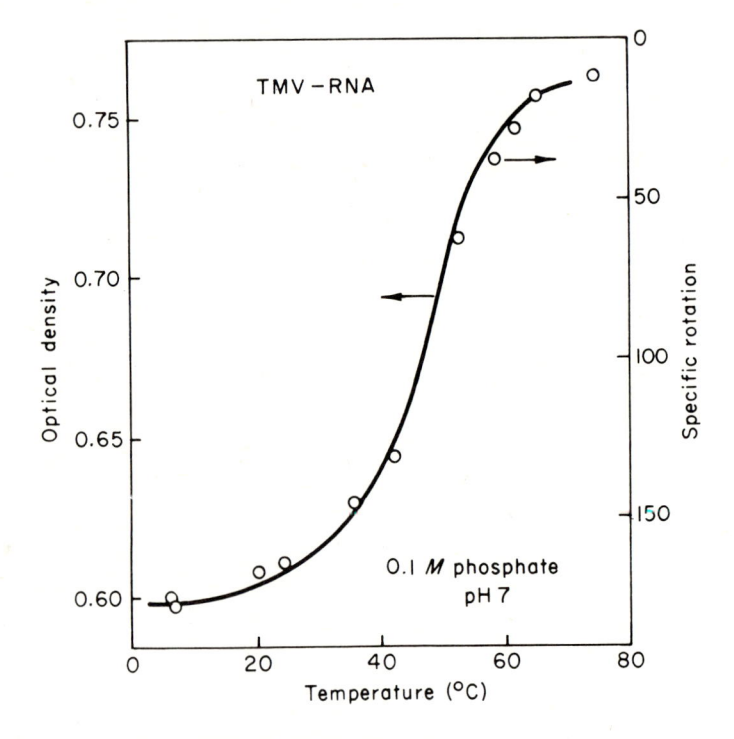

Fig. 3-49. Thermal denaturation of RNA from tobacco mosaic virus (TMV). Melting of the RNA secondary structure was measured by changes in specific rotation (open circles) and optical density at 260 mμ (solid line) of the solution as a function of temperature. From Doty *et al.* (18).

duration does not change the molecular weight of TMV RNA, although its asymmetry is increased and as the helical sections unwind, the linear extension of the molecule becomes greater. The sedimentation constant of TMV RNA measured at 55°C in urea is 10 S. This is in good agreement with the model presented in Fig. 3-48, since elongation of the molecule leads to an increase in hydrodynamic resistance without a change in mass. Furthermore, at 55°C TMV RNA exhibits a broad peak in the ultracentrifuge, indicating that the macromolecules assume a variety of different configurations—a polydispersity of form. When the viral RNA is cooled, all its physicochemical and biological characteristics are restored. A sharp 27 S peak is once again observed in the ultracentrifuge and the ultraviolet absorption and optical activity return to their normal values. The specific infectivity of RNA which has been heated and cooled is just the same as that of freshly extracted RNA. The melting of the helical portions of TMV RNA is thus shown to be reversible.

High-molecular-weight viral RNA behaves as a polymeric electrolyte and its properties strongly depend on the ionic strength of the solution (Fig. 3-50). When

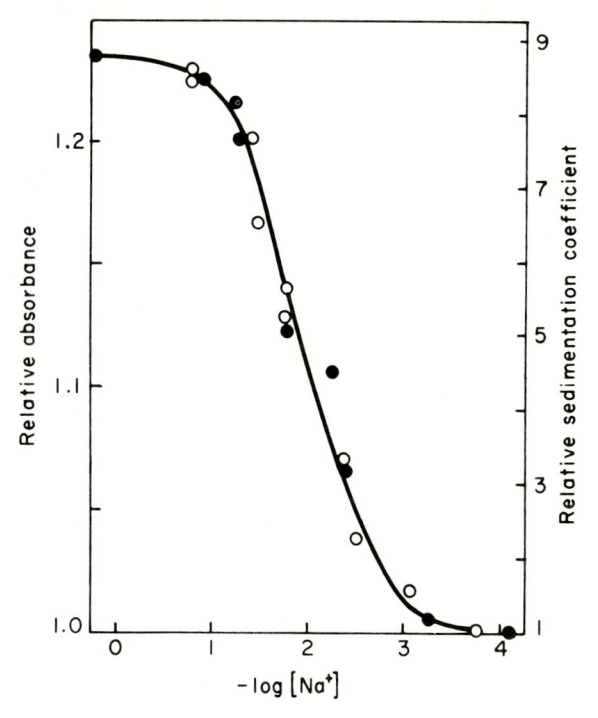

Fig. 3-50. The dependence of the secondary and tertiary structures of tobacco mosaic virus RNA on the ionic strength of the solution. Conformational differences were measured by the decline in relative optical density (hypochromic effect) at 258 mμ (●) and the relative rise in sedimentation coefficient (○) with increasing concentrations of Na$^+$ ions. From Boedtker (121).

helical regions are eliminated by using buffers of low ionic strength, hypochromicity disappears and an eightfold drop in sedimentation coefficient is observed [Boedtker (121)]. However, this procedure does not break all hydrogen bonds; at room temperature, about 20% remain intact. Another aspect of the polyelectrolytic behavior of

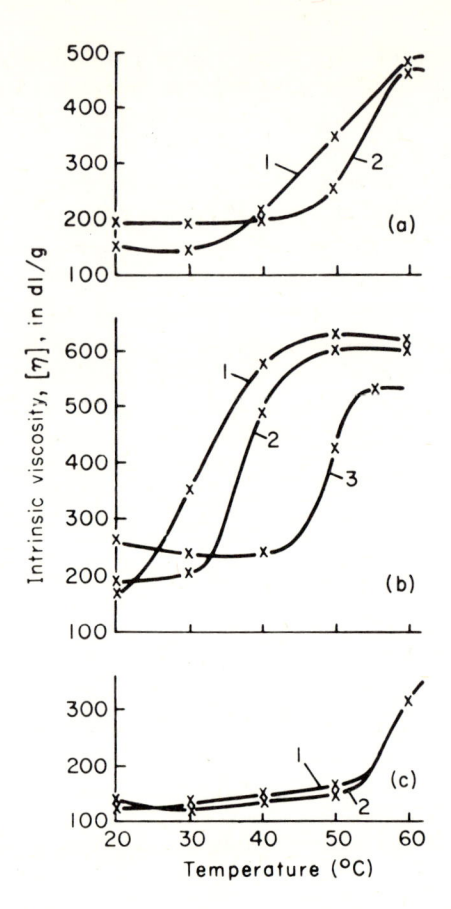

Fig. 3-51. Variations in the intrinsic viscosity of tobacco mosaic virus RNA solutions as a function of temperature. RNA concentrations are 0.02% (1), 0.1% (2) and 0.25% (3). (a) RNA solution in phosphate buffer, pH 7.2; ionic strength, 0.1. (b) Solution in 6 M urea, pH 7.2, ionic strength, 0.1. (c) Solution in 6 M urea, pH 7.2; ionic strength, 0.6. From Spirin *et al.* (122).

RNA is illustrated in Fig. 3-51, where changes in specific viscosity are plotted as a function of temperature in both buffer and 6 M urea (122). Concentrated urea is used to reduce the temperature at which the helix-coil transition occurs. From frames (a) and (b) of Fig. 3-51, it is apparent that the transition interval depends on the concentration of RNA employed. This phenomenon results from the influence of ionic strength, both of the buffer, which has an ionic strength of 0.1 in this case, and of the polymer itself which bears a Na$^+$ counterion on every phosphate group. The contribution of RNA to the ionic strength at this level is significant. If the ionic strength of the buffer is increased to 0.6, however, as in Fig. 3-51c, all curves are superimposable regardless of the RNA concentration and they are displaced toward higher temperature.

The influence of divalent and polyvalent ions on the macromolecular structure of RNA is considerable. As has already been indicated, the effect of ions such as Mg^{2+} and Mn^{2+} cannot be satisfactorily explained on the basis of simple electrostatic interactions. For example, Fig. 3-52 presents average helix-coil transition temperatures

(T_m) as a function of the logarithm of Mg^{2+} and Na^+ concentration. From a comparison of the two curves, it is evident that Mg^{2+} ions are 25,000 times more effective in stabilizing helical structure in RNA than Na^+ ions. This is certainly not an electrostatic effect. Rather, the stabilizing capacity of Mg^{2+} ions is due to the coordination bonds which they form with phosphate groups, bonds characterized by much greater energy than electrostatic interactions. This type of slightly polar bond is responsible for the low solubility of salts such as magnesium and calcium phosphates.

Coordination bonds can be reversibly dissociated, however, as is illustrated in the competition between Mg^{2+} and Na^+ ions for binding sites in RNA molecules. If Na^+ ions are added to a solution of RNA already containing magnesium ions, the helix–coil transition will occur at a lower temperature owing to the replacement of Mg^{2+} with Na^+. Here the melting point decreases even though the total ionic strength of the solution is raised. From these observations we draw the conclusion that the interaction between RNA and inorganic ions is not determined solely by ionic strength. In Fig. 3-52, the T_m approaches a limit of about 70°C when the Mg^{2+} concentration reaches 10^{-4} M. The RNA concentration in this experiment was 15 $\mu g/ml$, or 5×10^{-5} M with respect to phosphate groups, that is, of roughly the same order of magnitude as the limiting Mg^{2+} concentration. A significant proportion of the Mg^{2+} ions is apparently associated with RNA molecules under these circumstances. It is also worthy of note that the width of the transition interval is sharply reduced by the presence of Mg^{2+}, indicating the formation of more extended helical regions in the RNA. Several other divalent cations, such as Ni^{2+}, act in the same way as Mg^{2+}. Thermal denaturation of RNA in the presence of polyvalent ions is in general irreversible. It has been shown, for instance, that tRNA loses its functional activity under these conditions. In order to exclude polyvalent ions from RNA solutions, a strong complexing agent such as ethylenediaminetetraacetic acid (EDTA) can be used. EDTA forms stable complexes with Mg^{2+} as well as other di- and trivalent ions, effectively removing them from solution.

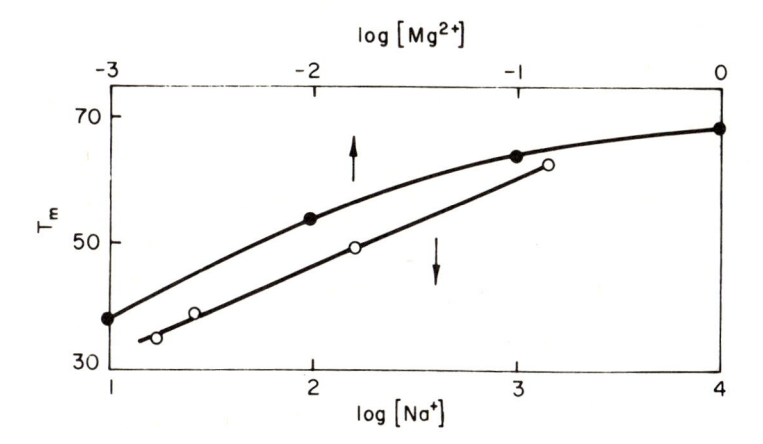

Fig. 3-52. Displacement of the mean melting temperature of tobacco mosaic virus RNA as a function of the concentration of Na^+ (○) and of Mg^{2+} (●) in mM. From Boedtker (121).

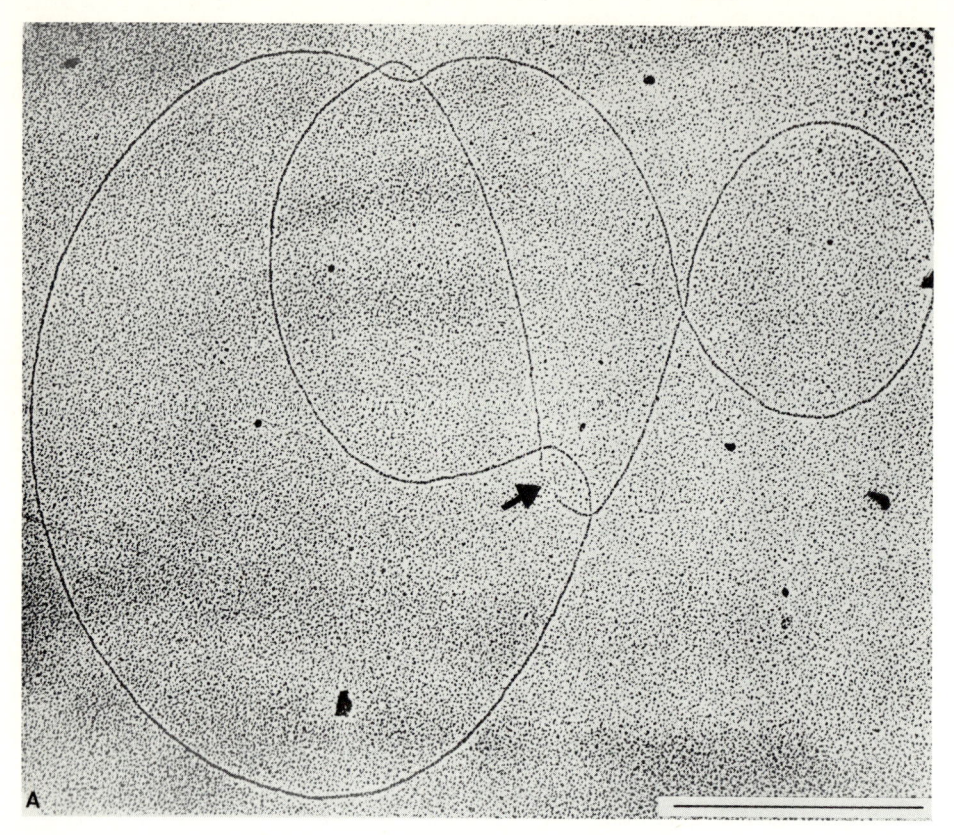

Fig. 3-53(A).

All of the evidence cited indicates that RNA possesses a secondary structure consisting of helical regions maintained by hydrogen bonds. The presence of some tertiary structure is also suggested, mainly by the considerable elongation that the macromolecules undergo at low ionic strengths which permit electrostatic repulsion among residues to become significant.

Finally, mention should be made of electron microscopy and x-ray structure analysis in the investigation of nucleic acid structure. These techniques have been used to study tRNA, ribosomal RNA, and viral RNA from TMV and many other sources. Certain viruses, such as reoviruses, have been found to contain double-stranded RNA with a secondary structure almost identical to that of DNA (123).

Electron micrographs of DNA molecules are presented in Fig. 3-53. The width of the double helices estimated by this procedure is 22 Å, in complete agreement with the x-ray data. Figure 3-54 presents electron photomicrographs of ribosomal RNA and tRNA. Here the strands are about 30 Å in diameter, characteristic of the RNA double helix. In the RNA photographs, the molecular strands appear to be much thicker than they really are, since samples for electron microscopy are often sprayed obliquely with heavy metal atoms. This process leaves an electron-transparent "shadow" alongside the molecules where the beam of metal has been blocked and, as a consequence, the contrast of the nucleic acid chains is increased relative to the background.

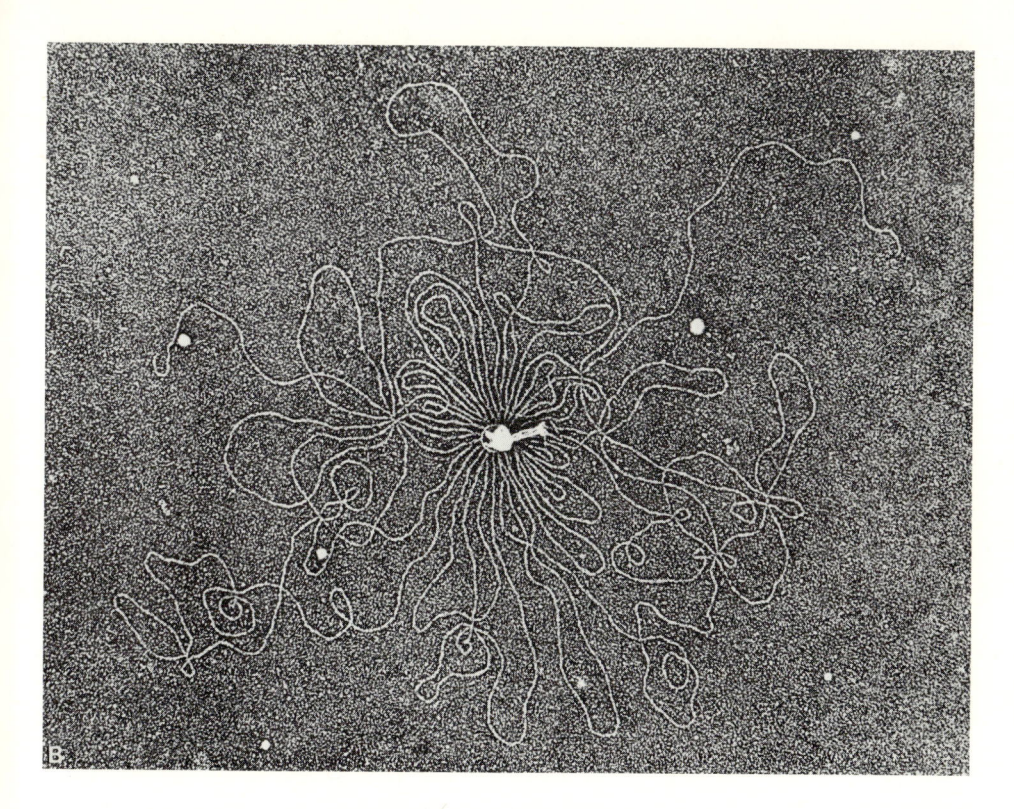

Fig. 3-53. Electron micrographs of DNA. (A) λ Phage DNA molecule; arrow marks discontinuity in the double-stranded chain; ×44,000. From Ris and Chandler (124). (B) DNA released from an osmotically shocked T2 bacteriophage particle; ×90,000. From Kleinschmidt et al. (125).

X-ray analysis has also been used in investigations of RNA structure. One important technique consists of measuring the low angle x-ray scattering from nucleic acid molecules in aqueous solutions [Luzzati and co-workers (127)]. The advantage of this method is that the macromolecules are studied in solution, although it is necessary that the concentration of nucleic acids be rather high, on the order of 1 % or more. Applications of the diffraction formulas to low angle x-ray scattering data shows that the regular regions responsible for scattering consist of relatively short rods, on the order of 100 Å in length. By comparing the intensity of the scattered beams with that of the incident x-ray beam, it is possible to calculate the number of electrons per unit length along the rods. With no further assumptions, it follows from this estimate that the scattering regions are double helices packed according to the Watson-Crick structure. Apparently, the most probable tertiary structure for ribosomal RNA in solutions is a coil in which the segments are about 100 Å in length. Every such segment is a rigid double-helical rod linked to other similar segments by amorphous regions in which hydrogen bonds are not formed owing to a lack of base complementarity (see Fig. 3-48, p. 287). The amorphous regions serve as "hinges" which impart flexibility to the RNA chain.

The x-ray patterns of partially dehydrated ribosomal RNA strands are not rich in reflections, but they nonetheless provide additional evidence for the presence of

Fig. 3-54. Electron micrographs of RNA. (a) ribosomal RNA. (b) tRNA. From Bogdanova et. al. (126).

double-helical structure. A prominent meridional reflection corresponds to a translational displacement of 3.3 Å per nucleotide residue within helical regions of the RNA molecules.

By far the most promising application of x-ray diffraction techniques to RNA, however, lies in the realm of transfer RNA structure. The most important prerequisite for the determination of detailed molecular structure by x-ray analysis is the availability of crystalline samples. This requirement has now been met in the case of tRNA and well-oriented crystals have recently been obtained from a variety of purified *E. coli* and yeast tRNA's (128–132), as well as from a mixture of unfractionated yeast tRNA's (133). Single crystals were grown in a number of different ways and at several temperatures from 5°C to room temperature. Preparative methods have included crystallization from aqueous ethanol in the presence of high salt concentrations, depletion of H_2O by equilibration with dioxane, and slow evaporation of solvent from a two-phase water–chloroform system.

The tRNA crystals produced thus far have been small, on the order of 1–2 mm in their maximum dimension, but the detailed x-ray patterns obtained from them have already yielded considerable information. Some of the best data for purified tRNA afford an overall resolution of 10 Å with individual diffraction maxima to 3 Å resolution. The lattice symmetry is either hexagonal or orthorhombic and in most cases the unit cell is quite large, ranging from 100 to 250 Å on a side and containing between

20 and 40 individual molecules, depending on the particular tRNA. Moreover, the sharp diffraction patterns given by single crystals of unfractionated tRNA's indicate that at a resolution of 20 Å all the molecules have very similar structural features, although it is not known whether they are present in a mixture of crystals of pure tRNA species, or in co-crystals of several tRNA species.

Isomorphous derivatives will of course be necessary before the complete three-dimensional configuration of any one purified tRNA can be solved. With this aim, several purified tRNA's have been crystallized in the presence of heavy metal ions. While the isomorphism of these derivatives and, indeed, the presence of the metal ions in the crystals remains to be proven conclusively, the results at present are quite encouraging. By utilizing isomorphous replacement together with other structural peculiarities of tRNA, such as the relatively limited number of electron-dense phosphate groups and the occurrence of extensive helical regions, it is likely that we shall know the complete spatial conformation of a number of tRNA's at high resolution in the near future.

References

1. For specialized reviews on these topics, see "The Nucleic Acids" (E. Chargaff and J. N. Davidson, eds.), Vols. 1–3. Academic Press, New York, 1955–1960.
2. I. Leslie, in "The Nucleic Acids" (E. Chargaff and J. N. Davidson, eds.), Vol. 2, p. 1. Academic Press, New York, 1955.
3. D. O. Jordan, "The Chemistry of Nucleic Acids," Butterworth, London and Washingron, D.C., 1960; H. G. Khorana, "Some Recent Developments in the Chemistry of Phosphate Esters of Biological Interest," Wiley, New York, 1961; "Khimiya Prirodnykh Soyedinenij" (M. M. Shemyakin, ed.). Nauka, Moscow, 1961; A. M. Michelson, "The Chemistry of Nucleosides and Nucleotides," Academic Press, New York, 1963.
4. R. W. Holley, J. Apgar, G. A. Everett, J. T. Madison, M. Marquisee, S. H. Merrill, J. R. Penswick, and A. Zamir, *Science* 147, 1462 (1965).
5. E. Chargaff, *Experientia* 6, 201 (1950); E. Chargaff, in "The Chemical Basis of Heredity" (W. D. McElroy and B. Glass, eds.), p. 521. Johns Hopkins Press, Baltimore, Maryland, 1957.
6. A. S. Spirin, A. N. Belozerskij, N. V. Shugayeva, and B. F. Vanyushin, *Biokhimiya* 22, 744 (1957); A. N. Belozersky and A. S. Spirin, *Nature* 182, 111 (1958); A. N. Belozersky and A. S. Spirin, in "The Nucleic Acids" (E. Chargaff and J. N. Davidson, eds.), Vol. 3, p. 147. Academic Press, New York, 1960.
7. J. D. Watson and F. H. C. Crick, *Nature* 171, 737 (1953).
8. J. D. Watson and F. H. C. Crick, *Nature* 171, 964 (1953).
9. R. Langridge, H. R. Wilson, C. W. Hooper, M. H. F. Wilkins, and L. D. Hamilton, *J. Mol. Biol.* 2, 19 (1960).
10. L. Pauling and R. B. Corey, *Arch. Biochem. Biophys.* 65, 164 (1956).
11. M. Feughelman, R. Langridge, W. E. Seeds, A. R. Stokes, H. R. Wilson, C. W. Hooper, M. H. F. Wilkins, R. K. Barclay, and L. D. Hamilton, *Nature* 175, 834 (1955).
12. E. Freese, *J. Theoret. Biol.* 3, 82 (1962).
13. B. Pullman, in "Molecular Associations in Biology" (B. Pullman, ed.), p. 1. Academic Press, New York, 1968.
14. Z. W. Hall and I. R. Lehman, *J. Mol. Biol.* 36, 321 (1968).
15. I. Tinoco, Jr., *J. Am. Chem. Soc.* 82, 4785 (1960).
16. A. Rich and M. Kasha, *J. Am. Chem. Soc.* 82, 6197 (1960).
17. S. A. Rice and P. Doty, *J. Am. Chem. Soc.* 79, 3937 (1957).

18. P. Doty, H. Boedtker, J. R. Fresco, R. Haselkorn, and M. Litt, *Proc. Natl. Acad. Sci. U.S.* **45**, 482 (1959).
19. G. A. Dvorkin, E. I. Golub, L. P. Gorbachev, L. G. Koreneva, and M. I. Mekshenkov, *Dokl. Akad. Nauk SSSR* **151**, 1211 (1963).
20. P. Doty, B. B. McGill, and S. A. Rice, *Proc. Natl. Acad. Sci. U.S.* **44**, 432 (1958).
21. J. A. Harpst, A. I. Krasna, and B. H. Zimm, *Biopolymers* **6**, 585 (1968).
22. J. Eigner and P. Doty, *J. Mol. Biol.* **12**, 549 (1965).
23. D. M. Crothers and B. H. Zimm, *J. Mol. Biol.* **12**, 525 (1965).
24. L. Mandelkern and P. Flory, *J. Chem. Phys.* **20**, 212 (1952); H. A. Scheraga and L. Mandelkern, *J. Am. Chem. Soc.* **75**, 179 (1953).
25. J. M. Gulland and D. O. Jordan, *Symp. Soc. Exptl. Biol.* **1**, 56 (1947).
26. J. M. Gulland, D. O. Jordan, and H. F. W. Taylor, *J. Chem. Soc.* p. 1131 (1947).
27. L. G. Bunville and E. P. Geiduschek, *Biochem. Biophys. Res. Commun.* **2**, 287 (1960).
28. R. L. Sinsheimer, *J. Mol. Biol.* **1**, 37 (1959).
29. L. G. Bunville, E. P. Geiduschek, M. A. Rawitscher, and J. M. Sturtevant, *Biopolymers* **3**, 213 (1965).
30. H. R. Mahler, G. Dutton, and B. D. Mehrotra, *Biochim. Biophys. Acta* **68**, 199 (1963).
31. M. P. Printz and P. H. von Hippel, *Proc. Natl. Acad. Sci. U.S.* **53**, 363 (1965).
32. S. W. Englander and J. J. Englander, *Proc. Natl. Acad. Sci. U.S.* **53**, 370 (1965).
33. R. N. Maslova and Ya. M. Varshavsky, *Biochim. Biophys. Acta* **119**, 633 (1966).
34. V. Luzzati, A. Mathis, F. Masson, and J. Witz, *J. Mol. Biol.* **10**, 28 (1964).
35. N. I. Ryabchenko, D. M. Spitkovskij, and P. I. Tsejtlin, *Biofizika* **8**, 19 (1963).
36. M. Grunberg-Manago and S. Ochoa, *J. Am. Chem. Soc.* **77**, 3165 (1955); M. Grunberg-Manago, P. J. Ortiz, and S. Ochoa, *Biochim. Biophys. Acta* **20**, 269 (1956).
37. M. Grunberg-Manago, *Progr. Nucleic Acid Res.* **1**, 93 (1963).
38. N. K. Gupta, E. Ohtsuka, V. Sgaramella, H. Buchi, A. Kumar, H. Weber, and H. G. Khorana, *Proc. Natl. Acad. Sci. U.S.* **60**, 1338 (1968).
39. A. Rich and D. R. Davies, *J. Am. Chem. Soc.* **78**, 3548 (1956).
40. A. Rich, *Brookhaven Symp. Biol.* **12**, 17 (1959).
41. V. Sasisekharan and P. B. Sigler, *J. Mol. Biol.* **12**, 296 (1965).
42. A. Rich, in "Biophysical Science" (J. L. Oncley, F. O. Schmitt, R. C. Williams, M. D. Rosenberg, and R. H. Bolt, eds.), p. 191. Wiley, New York, 1959.
43. G. Felsenfeld, D. R. Davies, and A. Rich, *J. Am. Chem. Soc.* **79**, 2023 (1957).
44. H. T. Miles, *Proc. Natl. Acad. Sci. U.S.* **51**, 1104 (1964).
45a. F. Pochon and A. M. Michelson, *Proc. Natl. Acad. Sci. U.S.* **53**, 1425 (1965).
45b. D. R. Davies, *Nature* **186**, 1030 (1960).
46. M. N. Lipsett, *Proc. Natl. Acad. Sci. U.S.* **46**, 445 (1960); E. G. Richards, C. P. Flessel, and J. R. Fresco, *Biopolymers* **1**, 431 (1963); P. K. Sarkar and J. T. Yang, *J. Biol. Chem.* **240**, 2088 (1965).
47. E. O. Akinrimis, C. Sander, and P. O. P. Ts'o, *Biochemistry* **2**, 340 (1963); R. Langridge and A. Rich, *Nature* **198**, 725 (1963); G. D. Fasman, C. Lindblow, and L. Grossman, *Biochemistry* **3**, 1015 (1964).
48. A. Rich, D. R. Davies, F. H. C. Crick, and J. D. Watson, *J. Mol. Biol.* **3**, 71 (1961).
49. A. Rich, *Biochim. Biophys. Acta* **29**, 502 (1958).
50. J. R. Fresco and B. M. Alberts, *Proc. Natl. Acad. Sci. U.S.* **46**, 311 (1960).
51. A. Kornberg, *Advan. Enzymol.* **18**, 191 (1957); A. Kornberg, "Enzymatic Synthesis of DNA". *Ciba Lectures Microbial Biochem.* Wiley, New York, 1961.
52. C. C. Richardson, C. L. Schildkraut, and A. Kornberg, *Cold Spring Harbor Symp. Quant. Biol.* **28**, 9 (1963).
53. F. J. Bollum, *Cold Spring Harbor Symp. Quant. Biol.* **28**, 21 (1963).
54. P. T. Englund, M. P. Deutscher, T. M. Jovin, R. B. Kelly, N. R. Cozzarelli, and A. Kornberg, *Cold Spring Harbor Symp. Quant. Biol.* **33**, 1 (1968).
55. J. Josse, *Proc. 5th Intern. Congr. Biochem., Moscow, 1961* **1**, 79 (1963).
56. J. Josse, A. D. Kaiser, and A. Kornberg, *J. Biol. Chem.* **236**, 864 (1961).
57. C. Byrd, E. Ohtsuka, M. W. Moon, and H. G. Khorana, *Proc. Natl. Acad. Sci. U.S.* **53**, 79 (1965).

References

58. R. B. Inman, C. L. Schildkraut, and A. Kornberg, *J. Mol. Biol.* **11**, 285 (1965).
59. M. Goulian and A. Kornberg, *Proc. Natl. Acad. Sci. U.S.* **58**, 1723 (1967).
60. M. Goulian, A. Kornberg, and R. L. Sinsheimer, *Proc. Natl. Acad. Sci. U.S.* **58**, 2321 (1967).
61. B. M. Olivera and I. R. Lehman, *Proc. Natl. Acad. Sci. U.S.* **58**, 1700 (1967).
62. S. B. Zimmerman, J. W. Little, C. K. Oshinsky, and M. Gellert, *Proc. Natl. Acad. Sci. U.S.* **57**, 1841 (1967).
63. B. Weiss and C. C. Richardson, *Proc. Natl. Acad. Sci. U.S.* **57**, 1021 (1967).
64. A. Becker, G. Lyn, M. Gefter, and J. Hurwitz, *Proc. Natl. Acad. Sci. U.S.* **57**, 1841 (1967).
65. S. B. Weiss, *Proc. Natl. Acad. Sci. U.S.* **46**, 1020 (1960).
66. J. J. Furth, J. Hurwitz, and M. Goldmann, *Biochem. Biophys. Res. Commun.* **4**, 362 (1961).
67. M. Chamberlin and P. Berg, *Proc. Natl. Acad. Sci. U.S.* **48**, 81 (1962).
68. S. B. Weiss and T. Nakamoto, *Proc. Natl. Acad. Sci. U.S.* **47**, 1400 (1961).
69. J. Hurwitz, J. J. Furth, M. Anders, and A. Evans, *J. Biol. Chem.* **237**, 3752 (1962).
70. D. P. Burma, H. Kröger, S. Ochoa, R. C. Warner, and J. D. Weill, *Proc. Natl. Acad. Sci. U.S.* **47**, 749 (1961).
71. H. S. Slayter and C. E. Hall, *J. Mol. Biol.* **21**, 113 (1966).
72. M. Chamberlin and P. Berg, *J. Mol. Biol.* **8**, 297 (1964).
73. E. P. Geiduschek, J. W. Moohr, and S. B. Weiss, *Proc. Natl. Acad. Sci. U.S.* **48**, 1078 (1962).
74. S. Nishimura, T. M. Jacob, and H. G. Khorana, *Proc. Natl. Acad. Sci. U.S.* **52**, 1494 (1964).
75. M. Hayashi, M. N. Hayashi, and S. Spiegelman, *Proc. Natl. Acad. Sci. U.S.* **51**, 351 (1964).
76. M. H. Green, *Proc. Natl. Acad. Sci. U.S.* **52**, 1388 (1964).
77. E. P. Geiduschek, L. Snyder, A. J. E. Colvill, and M. Sarnat, *J. Mol. Biol.* **19**, 541 (1966); L. Snyder and E. P. Geuiduschek, *Proc. Natl. Acad. Sci. U.S.* **59**, 459 (1968).
78. C. Weissmann, P. Borst, R. H. Burdon, M. A. Billeter, and S. Ochoa, *Proc. Natl. Acad. Sci. U.S.* **51**, 682 (1964); C. Weissmann, M. A. Billeter, M. C. Schneider, C. A. Knight, and S. Ochoa, *Proc. Natl. Acad. Sci. U.S.* **53**, 653 (1965).
79. W. Shipp and R. Haselkorn, *Proc. Natl. Acad. Sci. U.S.* **52**, 401 (1964).
80. D. Baltimore, Y. Becker, and J. Darnell, *Science* **143**, 1034 (1964).
81. S. Spiegelman, I. Haruna, I. B. Holland, G. Beaudreau, and D. Mills, *Proc. Natl. Acad. Sci. U.S.* **54**, 919 (1965).
82. A. Kornberg, in "Biophysical Science" (J. L. Oncley, F. O. Schmitt, R. C. Williams, M. D. Rosenberg, and R. H. Bolt, eds.), p. 200. Wiley, New York, 1959.
83. W. Szybalski, in "The Molecular Basis of Neoplasia," M. D. Anderson Hosp. and Tumor Inst., p. 147. Univ. of Texas Press, Austin, Texas, 1962.
84. M. Meselson and F. W. Stahl, *Proc. Natl. Acad. Sci. U.S.* **44**, 671 (1958).
85. M. Meselson, F. W. Stahl, and J. Vinograd, *Proc. Natl. Acad. Sci. U.S.* **43**, 581 (1957); J. Vinograd and J. E. Hearst, *Fortschr. Chem. Org. Naturstoffe* **20**, 372 (1962).
86. C. Levinthal and C. A. Thomas, Jr., *Biochim. Biophys. Acta* **23**, 453 (1953).
87. C. Levinthal and P. F. Davison, *J. Mol. Biol.* **3**, 674 (1961).
88. J. Frenkel, *Acta Physiocochim. URSS* **19**, 51 (1944).
89. J. D. Mandell and A. D. Hershey, *Anal. Biochem.* **1**, 66 (1960).
90. P. Doty, J. Marmur, and N. Sueoka, *Brookhaven Symp. Biol.* **12**, 1 (1959).
91. J. Marmur and P. Doty, *Nature* **183**, 1427 (1959).
92. N. Sueoka, J. Marmur, and P. Doty, *Nature* **183**, 1429 (1959).
93. J. Cairns, *Cold Spring Harbor Symp. Quant. Biol.* **28**, 43 (1963).
94. J. Miyazawa and C. A. Thomas, Jr., *J. Mol. Biol.* **11**, 223 (1965).
95. P. Doty, J. Marmur, J. Eigner, and C. Schildkraut, *Proc. Natl. Acad. Sci. U.S.* **46**, 461 (1960).
96. J. Marmur, R. Rownd, and C. L. Schildkraut, *Progr. Nucleic Acid Res.* **1**, 231 (1963).
97. E. P. Geiduschek, *J. Mol. Biol.* **4**, 467 (1962).
98. D. Freifelder and P. F. Davison, *Biophys. J.* **2**, 249 (1962).
99. J. Marmur and P. Doty, *J. Mol. Biol.* **3**, 585 (1961).
100. C. L. Schildkraut, J. Marmur, and P. Doty, *J. Mol. Biol.* **3**, 595 (1961).
101. I. R. Lehman, *J. Biol. Chem.* **235**, 1479 (1960).
102. J. Marmur and D. Lane, *Proc. Natl. Acad. Sci. U.S.* **46**, 453 (1960).
103. J. Marmur, S. Falkow, and M. Mandel, *Ann. Rev. Microbiol.* **17**, 329 (1963).
104. S. E. Luria and J. E. Darnell, "General Virology," 2nd Ed., p. 108. Wiley, New York, 1967.

105. S. A. Yankofsky and S. Spiegelman, *Proc. Natl. Acad. Sci. U.S.* **48**, 106 (1962); S. A. Yankofsky and S. Spiegelman, *Proc. Natl. Acad. Sci. U.S.* **48**, 146 (1962); S. A. Yankofsky and S. Spiegelman, *Proc. Natl. Acad. Sci. U.S.* **49**, 538 (1963).
106. D. Giacomoni and S. Spiegelman, *Science* **138**, 1328 (1962); H. M. Goodman and A. Rich, *Proc. Natl. Acad. Sci. U.S.* **48**, 2101 (1962).
107. B. P. Sagik, M. H. Green, M. Hayashi, and S. Spiegelman, *Biophys. J.* 2, 409 (1962).
108. G. P. Tochini-Valentini, M. Stodolsky, A. Aurisicchio, M. Sarnat, F. Graziosi, S. B. Weiss, and E. P. Geiduschek, *Proc. Natl. Acad. Sci. U.S.* **50**, 935 (1963).
109. M. Hayashi, M. N. Hayashi, and S. Spiegelman, *Proc. Natl. Acad. Sci. U.S.* **50**, 664 (1963).
110. W. Doerfler and D. S. Hogness, *J. Mol. Biol.* **33**, 635 (1967).
111. Z. Opara-Kubinska, H. Kubinski, and W. Szybalski, *Proc. Natl. Acad. Sci. U.S.* **52**, 923 (1964).
112. R. Rudner, J. D. Karkas, and E. Chargaff, *Proc. Natl. Acad. Sci. U.S.* **60**, 630 (1968).
113. K. Taylor, Z. Hradecna, and W. Szybalski, *Proc. Natl. Acad. Sci. U.S.* **57**, 1618 (1967).
114. E. T. Bolton and B. J. McCarthy, *Proc. Natl. Acad. Sci. U.S.* **48**, 1390 (1962).
115. B. H. Hoyer, B. J. McCarthy, and E. T. Bolton, *Science* **144**, 959 (1964).
116. B. J. McCarthy and B. H. Hoyer, *Proc. Natl. Acad. Sci. U.S.* **52**, 915 (1964).
117. D. Gillespie and S. Spiegelman, *J. Mol. Biol.* **12**, 829 (1965).
118. R. A. Cox and U. Z. Littauer, *J. Mol. Biol.* **2**, 166 (1960).
119. G. G. Brownlee, F. Sanger, and B. G. Barrell, *J. Mol. Biol.* **34**, 379 (1968).
120. A. S. Spirin, *J. Mol. Biol.* **2**, 436 (1960).
121. H. Boedtker, *J. Mol. Biol.* **2**, 171 (1960).
122. A. S. Spirin, L. P. Gavrilova, S. E. Bresler, and M. I. Mosevitskij, *Biokhimiya* **24**, 938 (1959).
123. R. Langridge and P. J. Gomatos, *Science* **141**, 694 (1963).
124. H. Ris and B. L. Chandler, *Cold Spring Harbor Symp. Quant. Biol.* **28**, 1 (1963).
125. A. K. Kleinschmidt, D. Lang, D. Jacherts, and R. K. Zahn, *Biochim. Biophys. Acta* **61**, 857 (1962).
126. E. Bogdanova, L. Gavrilova, G. Dvorkin, N. Kiselev and A. Spirin, *Brokhimiya* **27**, 387 (1962).
127. S. N. Timasheff, J. Witz, and V. Luzzati, *Biophys. J.* **1**, 525 (1961); J. Witz, L. Hirth, and V. Luzzati, *J. Mol. Biol.* **11**, 613 (1965).
128. B. F. C. Clark, B. P. Doctor, K. C. Holmes, A. Klug, K. A. Marcker, S. J. Morris, and H. H. Paradies, *Nature* **219**, 1222 (1968).
129. H. H. Paradies, *FEBS Letters* **2**, 112 (1968).
130. F. Cramer, F. von der Haar, W. Saenger, and E. Schlimme, *Angew. Chem.* **80**, 969 (1968).
131. A. Hampel, M. Labanauskas, P. G. Connors, L. Kierkegard, U. L. Raj Bhandary, P. B. Sigler, and R. M. Bock, *Science* **162**, 1384 (1968).
132. S. -H. Kim and A. Rich, *Science* **162**, 1381 (1968).
133. J. R. Fresco, R. D. Bleake, and R. Langridge, *Nature* **220**, 1285 (1968).

Chapter IV

DNA Function

1. Introduction

The essential functions of deoxyribonucleic acid (DNA) are to store and transmit genetic information in the living organism. By means of a special chemical code, DNA molecules specify all the hereditary characteristics of an organism which together comprise its phenotype: metabolic capacities, macromolecular composition, and morphological peculiarities.

Investigation of DNA function on the molecular level has been closely associated with the development of the genetics of bacteria and of the viruses which infect them. The microbial cell is particularly well suited to genetic studies for a number of reasons.

In the first place, an enormous population of bacteria can be grown in liquid medium to densities of 10^8 to 10^9 cells per milliliter in less than a day. Owing to the availability of such large numbers of individuals, many types of statistical experiments can be carried out on bacteria with great precision and rapidity. On a solid nutrient medium, a single bacterium can give rise to a colony containing between 10^7 and 10^9 cells, which greatly simplifies the task of selecting desired phenotypes. A large number of technical approaches to selection have been worked out and we shall consider some of them later in this chapter.

A second advantage of working with microorganisms is that constant and well-defined growth conditions can be established quite easily. Temperature, pH, ionic composition, and concentrations of both metabolites and antimetabolites can all be closely controlled.

A third important property of bacteria and viruses is their comparatively high accessibility to mutagenic agents. Hereditary changes in microorganisms can occur either spontaneously or by induction with chemical mutagens or radiation. There is no doubt that all mutations ultimately derive from chemical changes in the DNA molecule.

The fourth characteristic of microorganisms that is of value in genetic investigations is the relative simplicity of their biological organization. The number of nucleotides in the bacterial chromosome is on the order of 10^7 and the overall length of an extended molecule would be about 1 mm. By comparison, the total length of DNA in the chromosomes of an average human cell is 2 m.

The functional unit of genetic material, called the *cistron*, is that which governs the synthesis of a specific protein. In contrast to the older term *gene*, the concept of the cistron is very strictly defined and has a simple molecular meaning. The possibility of introducing a functional unit of the chromosome arose from the development of biochemical genetics by Beadle and Tatum (1). Their experiments on the inheritance and variability of essential metabolic processes led them to state a general principle:

$$\text{one gene} \rightarrow \text{one enzyme}$$

As defined by this formula, the gene is a portion of the chromosome specifying the structure of one enzyme. However, since many enzymes possess more than one polypeptide chain, it became evident that the most elementary unit of hereditary material is a region of the DNA which governs the synthesis of a single polypeptide chain. This newly developed concept of the functional unit can be formulated as follows:

$$\text{one cistron} \rightarrow \text{one polypeptide chain}$$

A cistron is much larger in mass than the protein for which it codes. For example a protein with a molecular weight of 50,000 daltons contains on the order of 500 amino acid residues, since the average molecular weight per amino acid residue is approximately 100. The coding ratio, defined as the number of nucleotides necessary to specify the nature of one amino acid in the polypeptide chain, has been found to equal three. Therefore, the DNA segment corresponding to the protein in our example must consist of at least 1500 nucleotide residues, each with a molecular weight of about 330. Moreover, the DNA molecule is a twofold helix, so it is necessary to double the number of nucleotides involved. It follows that the molecular weight of a cistron should be on the order of 10^6, or twenty times the weight of the protein it determines.

From the dimensions of individual cistrons and from their relative distribution along the chromosome, it is possible to construct a *genetic map*. Assuming that all the DNA of a bacterial cell carries information, its genetic map would have to account for about 2000 to 3000 distinct functional units. In *Escherichia coli*, the most extensively studied species of bacteria, only about 400 cistrons have been identified thus far. Nonetheless, these *genetic markers* have provided a firm basis for the study of genetic processes in bacteria. The genetics of other organisms has been investigated to a much smaller extent.

Experiments in bacterial genetics are possible only because there exist mechanisms for the exchange of genetic material between different cell lines. When such exchanges succeed, the daughter cells will possess a mixed heredity, exhibiting characteristics of both parent cells. Three types of genetic exchange have been discovered in bacterial systems.

In *bacterial transformation*, chemically purified DNA extracted from one strain is mixed with cells of a second strain carrying different genetic markers than the first.

The DNA molecules must penetrate from the external medium into the recipient cells in order for genetic markers from the two strains to recombine. Penetration can occur only when the recipients are in a special physiological state. Among the relatively few microorganisms which can be transformed with homologous DNA are pneumococcus, *Bacillus subtilis*, *Hemophilus influenzae*, *Streptococcus*, and *Pseudomonas*.

A second method of genetic transfer is called *transduction*. In this process, a bacteriophage transmits bacterial DNA fragments from one cell to another within its own protein coat instead of its own DNA. In both transduction and transformation, DNA is the only substance transferred. By infecting bacteria with radioactively labeled phages, it was demonstrated that the DNA injected into the host cell is accompanied by less than a few percent of viral protein. Moreover, bacterial cells can be infected with purified phage DNA under certain conditions. Deproteinized RNA from plant viruses also retains its infectivity. Transduction has been observed in *E. coli*, *Salmonella*, *Shigella*, *B. subtilis*, *Proteus*, and *Staphylococcus*, as well as several other species and it appears to be the most general method of genetic exchange to occur in bacteria.

The third approach to bacterial genetics is *conjugation*, a direct sexual process in which one of the cells, the "male," serves as a donor of genetic material and another, the "female," as recipient. During conjugation, the DNA passes through a protoplasmic bridge which connects the two participating cells. Here as in the other modes of genetic transfer, DNA is the main substance transmitted from one bacterium to another. Conjugation has been demonstrated in *E. coli*, *Salmonella*, *Shigella*, *Serratia marcescens*, and *Pseudomonas aeruginosa*. Using conjugation, the genetic map of strain K12 of *E. coli* has been worked out in great detail. The map was verified by independent transduction studies in the same bacterial strain.

2. Bacterial Genetics

The first task that confronts the bacterial geneticist is the creation and selection of bacterial mutants which provide a source of genetic markers. The usual starting material for this procedure is the *wild-type strain*, which may be thought of as a strain evolved through natural selection, possessing the necessary physiological and biochemical functions in order to survive in its normal environment. Mutations are sought as identifiable changes in these capacities. Further, conditions must be found under which an exchange of genetic material occurs between cells carrying different genetic markers. Reliable methods of crossing were developed only after an enormous amount of work had been done on the hereditary mechanisms of bacteria. A third requirement is for methods of selection adequate to distinguish the progeny resulting from genetic exchange. Experimental data on the exchange of genetic markers are usually expressed in terms of *recombination frequency*, from which information on the structure of the genetic material can be obtained. In particular, one can ascertain the relative positions of markers or *loci* along the chromosome. The reader may wish to consult one of the many excellent reviews of this topic for supplementary information (2–5).

Five major groups of genetic markers used in genetic studies on *E. coli* correspond to (1) the ability to metabolize sugars other than glucose, such as galactose, lactose,

Fig. 4-1. Linkage map of *Escherichia coli*. The inner circle with its associated time scale depicts the whole linkage group. The reason for measuring map distances in minutes will be explained later. Portions of the map are displayed on arcs of the outer circle expanded 4 × in order to accommodate markers in crowded regions. The position of markers in parentheses is known only approximately From Taylor and Trotter (6). Meanings of some of the more important gene symbols are as follows

Marker	Ability to synthesize	Marker	Ability to synthesize
thr	threonine	gua	guanine
leu	leucine	tyr	tyrosine
pro	proline	phe	phenylalanine
pur	purines	asp	aspartic acid
bio	biotin	ser	serine
pyr	pyrimidines	cys	cysteine
trp	tryptophan	met	methionine
his	histidine	ilv	isoleucine-valine
arg	arginine	thi	thiamine
lys	lysine	aro	aromatic amino acids
thy	thymine		

Marker	Ability to utilize	Marker	Ability to utilize
ara	arabinose	rha	rhamnose
lac	lactose	xyl	xylose
gal	galactose	mtl	mannitol
mal	maltose	rbs	ribose

Symbol	Definition
pho	alkaline phosphatase; ability to utilize organic esters of phosphoric acid
rns	ribonuclease I
arg R	regulation of arginine synthesis
trp R	regulation of tryptophan synthesis
pho R	regulation of alkaline phosphatase synthesis
rel	regulation of RNA synthesis
phe S	phenylalanyl-tRNA synthetase
val S	valyl-tRNA synthetase
arg P	arginine permease
ura P	uracil permease
azi	resistance or sensitivity to azide
str	resistance, dependence, or sensitivity to streptomycin
spc	resistance or sensitivity to spectinomycin
amp	resistance or sensitivity to penicillin
ton	resistance or sensitivity to T1 phage
tsx	resistance or sensitivity to T6 phage
tfr	resistance or sensitivity to phages T4, T3, T7, and λ
att λ	integration site for prophage λ
attΦ80	integration site for prophage Φ80
pil	presence or absence of pili
uvr	repair of ultraviolet radiation damage
rec	competence for genetic recombination
mut	generalized high mutability
sup D	suppressor of amber mutations, also called su_I
sup E	suppressor of amber mutations, also called su_{II}
sup F	suppressor of amber mutations, also called su_{III}
sup B	suppressor of ochre mutations
sup C	suppressor of ochre mutations, also called *su*-4
sup G	suppressor of ochre mutations, also called *su*-5

2. Bacterial Genetics

xylose, and maltose, as sole carbon sources; (2) the ability o the bacterium to synthesize various amino acids, nucleotide bases, and vitamins *de novo* from inorganic nitrogen and carbon derived from sugars; (3) the ability to use phosphate esters in addition to free phosphate groups; (4) sensitivity or resistance to bacteriocidal agents such as streptomycin, penicillin, sodium azide, and sulfanilamides; and (5) sensitivity or resistance to various kinds of bacteriophages. Other genetic markers which have been used to a lesser extent include the presence or absence of specific antigens in the cell wall or of hairlike *pili* that protrude from the cell surface.

The genetic map of *E. coli* K12, presented in Fig. 4-1, shows the relative chromosomal placement of about 200 genetic loci (6). The chromosome appears to be a closed circle. Later in this chapter, we shall see why the experimental data necessarily lead

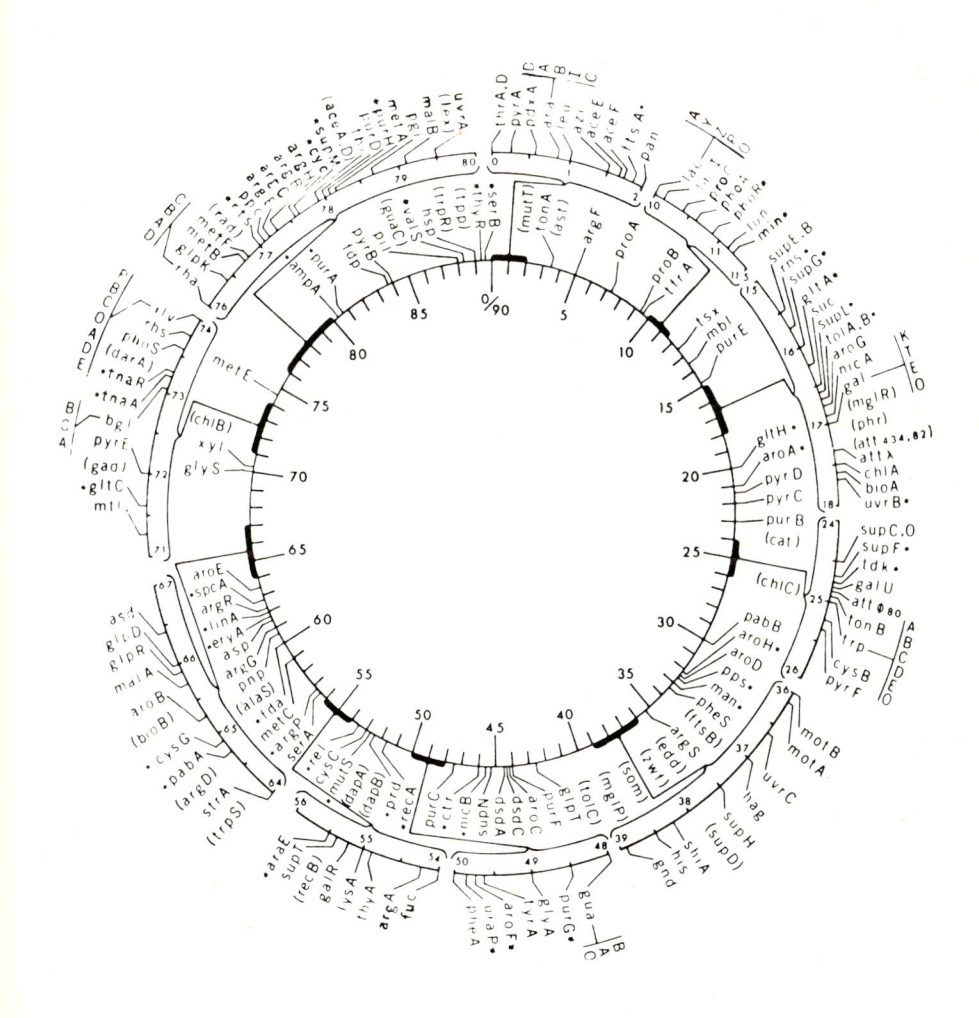

Fig. 4-1

to the construction of a circular map. The numbers inside the circle represent distances measured in minutes required for the transfer of a male chromosome into the female cell during conjugation (p. 342). In a number of instances, there are several loci associated with a single marker, designated by the letters, A, B, C, ... This is due to the fact that several different enzymes participate consecutively in the synthesis of most intracellular metabolites. For example, there are five separate steps in tryptophan biosynthesis, and eight in the case of arginine. Each of the loci in Fig. 4-1 therefore represents a distinct functional unit. Although all details of the genetic map are not yet known, separate regions, comprising from one quarter to one third of the total, have been subjected to a more intensive kind of investigation called fine structure analysis. Since this sort of study demands great effort, data on the genetic map of microorganisms other than *E. coli* are still quite limited. Nonetheless, it has been demonstrated that the genetic map of *Salmonella typhimurium* is also a closed circle, resembling that of *E. coli* in many details (7).

3. Mutation and Mutagenesis

As the first step in our discussion of how the genetic map of *E. coli* was established, we shall take up the production of bacterial mutants. Although there exist a certain number of spontaneous mutants in any wild-type population of bacteria and viruses, mutations can also be induced by ultraviolet light or x rays, or by the action of special chemical compounds called *mutagens*. Mutagenesis generally entails the death of a majority of the treated cells since all mutagenic agents do harm to functions necessary for life in one way or another. But some cells usually survive which harbor one or several nonlethal lesions at specific loci in the chromosome. Each detectable mutation leads either to the loss or alteration of a particular enzyme or RNA molecule, such as transfer RNA. All mutants carrying lesions within a single cistron, the functional unit of chromosomal DNA, are considered to be allelomorphic or homologous, since they differ from the wild type and from each other with respect to just one character.

Because all mutational events require chemical changes in DNA molecules, we shall denote the mutational process in the manner of a chemical reaction. For instance, $pho^+ \rightarrow pho^-$ will mean a heritable alteration at the locus which directs the synthesis of the enzyme, alkaline phosphatase, in which the capacity to form this enzyme is lost. Bacteria carrying this mutation will be unable to grow in a medium containing organic phosphate esters in place of inorganic phosphates. But there are other mutations possible at this locus of the type $pho^+ \rightarrow pho^+_\alpha, pho^+_\beta, pho^+_\gamma, \ldots$ in which the bacterium retains its ability to synthesize phosphatase, although the properties of the enzyme are not identical to those of the wild type owing to an alteration in protein structure. If the kinetic constants of the enzyme change as a result of such a mutation, it is indicated that the alteration has affected the active site of the enzyme macromolecule.

There are also cases in which the catalytic properties of the enzyme remain the same while one or more of its physicochemical characteristics is different. In studies on tryptophan synthetase, an enzyme in the tryptophan biosynthetic pathway of

E. coli, Yanofsky detected mutant proteins that sharply differed in their sensitivity to thermal and acid denaturation (8). Mutations in the alkaline phosphatase cistron of the same organism were found by Garen, Levinthal, and Rothman to result in the synthesis of enzymes differing in electrophoretic mobility from the wild-type protein, although they retained enzymatic activity (9). Thus mutation frequently leads to a change in the protein specified by the altered cistron, but by no means is the functional activity of the protein always impaired.

The mutagenic effect of radiant energy is in all likelihood connected with the formation of free radicals or of excited atoms and molecular groups in the cell. Free radicals can attack the DNA directly by eliminating hydrogen atoms. Mutagenesis with ultraviolet light is usually performed in the following manner. A bacterial culture is centrifuged and resuspended at high density in a medium having a low absorbance for ultraviolet light. The suspension is then subjected to a large dose of radiation which kills most of the cells. The dose is generally chosen so that the number of surviving cells is about 10^{-3} to 10^{-4} of those present initially. The bacteria are next transferred to a nutrient medium and permitted to grow out to a concentration of 10^9 cells per milliliter. Mutants can be selected at this stage or else the entire procedure may be repeated several times to further enrich the culture for mutants. The mutagenic effect of radiation is additive and the probability of inducing a mutation is directly proportional to the total time of irradiation. However, the dose of radiation administered to the cells each time is critical and it is necessary to choose carefully the factor by which the number of survivors is to be reduced.

A kinetic theory of this effect can be formulated as follows. The number of surviving cells, n_s, decreases with time as

$$dn_s/dt = -KIn_s \qquad (4\text{-}1)$$

where I is the intensity of radiation and K is a coefficient related to the rate at which cells are killed. By integrating, we obtain for the number of survivors

$$n_s = n_0 e^{-KIt} \qquad (4\text{-}2)$$

The number of mutant cells, n_m, changes at the following rate

$$dn_m/dt = K'In_s - KIn_m \qquad (4\text{-}3)$$

The probability of mutation, K', is much less than K, since the number of mutants among the surviving cells is relatively small. By integrating Eq. (4-3), we obtain

$$n_m = K'In_0 t e^{-KIt} \qquad (4\text{-}4)$$

The concentration of mutant cells among the survivors increases linearly with time of irradiation, since

$$n_m/n_s = K'It \qquad (4\text{-}5)$$

If we define the survival as $\alpha = n_s/n_0$, then

$$n_m/n_s = (K'/K)\ln(1/\alpha) \qquad (4\text{-}6)$$

After N-fold repetition, this equation becomes

$$n_m/n_s = (NK'/K) \ln(1/\alpha) \qquad (4\text{-}7)$$

From Eq. (4-7), it is apparent that the efficiency of mutagenesis increases as a weak logarithmic function of the decrease in survivors. Therefore, there is only a small advantage realized by setting the number of survivors at 10^{-4} rather than at 10^{-3}, amounting to a 30% enrichment in the relative concentration of mutant cells. If the ratio n_m/n_s is not increased by any additional selective procedures during subsequent outgrowth of the culture, the practical resolution of this method is about one mutant in 10^4 surviving cells. This means that if the treated culture is spread on 100 Petri dishes so that each dish contains 100 colonies, only one of the colonies on one of the plates will result from a mutant bacterium. Hence, the experimental parameters, N and α, are always chosen so that n_m/n_s is greater than 10^{-4}.

As mentioned above, mutagenesis undoubtedly requires a chemical reaction. The discovery of chemical mutagenic agents by Lobashov and Sakharov in 1936 was thus of fundamental importance. At the present time there are a great variety of chemical mutagens known, including base analogs, which cause mistakes in base pairing during DNA replication, and other compounds such as alkylating and deaminating agents, which directly alter nucleotides in the DNA chain. Heating and acid treatment also induce mutations, presumably by denaturing the DNA double helix and thereby rendering individual nucleotide residues more susceptible to chemical reaction. An example of this is depurination, the scission of purine side groups from the DNA molecule. We shall return to a detailed discussion of chemical mutagenesis in a later section.

For the most part, we shall be concerned with what are referred to as *point mutations*, that is, mutations that affect only a small region of the DNA macromolecule. In some cases, it has been ascertained that the "region" in the chromosome altered by mutation is a single nucleotide base pair! The distinguishing feature of point mutants is that they can change back or *revert* to the original wild-type phenotype after a second mutation at or near the site of the forward mutation. The chemical change resulting from a given point mutation can often be deduced by studying its reversion under the influence of mutagens whose effects on DNA are known. Other kinds of mutations cause damage to fairly large and extended regions of the chromosome. These include *translocations* of segments from one place to another on the chromosome, *inversions* of gene sequences, and *deletions* whereby certain segments of the chromosome are lost altogether. For obvious reasons, none of these large-scale alterations can be corrected by single-site reversion.

The detection of bacterial mutants according to the traditional "all-or-nothing" principle requires that the mutation eliminate a biochemical function or enzyme activity in the cell which can be compensated for by the addition of a growth factor to the medium. But large numbers of mutations would not be recognizable by this criterion if they led to the loss of such essential gene products as DNA polymerase that could not be replaced by any added metabolic supplement. Mutational changes in many genes are thus expected to be lethal and, consequently, undetectable by usual methods. The same sort of argument applies to bacteriophage systems, where the classic mutants are identified mainly by differences in host range and plaque morphology, that is, by changes which are compatible with the continued viability of the virus. The loss of certain essential functions, however, would prevent the formation of any phage particles at all. This situation is reflected in the failure to recover mutations at many points on the genetic maps of both bacteria and viruses, even though it is evident that chemical alterations can be induced at almost any residue of bacterial or viral DNA with ultraviolet light or with chemical mutagens.

To circumvent this problem and to obtain a more complete genetic map, a new approach to mutagenesis and selection was developed which depends on the isolation of *conditional lethal mutants*, that is, mutants that can grow under certain conditions of temperature, pH, and so on, but not under others. One class of conditional lethals can be explained by an increase in the thermolability of an essential protein or enzyme. This technique was first applied to the study of bacteriophages (10) and was subsequently extended to bacteria. In the case of bacteria, a typical experiment is performed as follows. Wild-type cells are mutagenized by exposure to ultraviolet radiation and then spread on petri dishes containing minimal medium in agar. The petri dishes are incubated at 25°C until small, scarcely visible colonies appear on the agar surface, whereupon the incubation temperature is raised to 42°C. Most of the colonies quickly grow to normal size, but those formed by *temperature-sensitive* (*ts*) *mutants*, in which an essential gene product has become thermolabile, cannot grow at the higher temperature and hence remain very small. If these colonies are reincubated at 25°C, however, normal growth resumes. It is clear that the conditions for selection on the basis of thermolability are much less demanding than those of the all-or-nothing principle. Furthermore, *ts* mutants are easy to select since it is not necessary to know which function is altered; any one of many essential gene functions requiring protein synthesis may be involved. This method enables us to readily obtain a large number of mutants, equally distributed along the entire genetic map. In quantitative terms, this means that the ratio n_m/n_s can be greatly increased because the many lesions that formerly would not have been recognized as mutations will be classified as such by the method described.

Another important and distinctive class of conditional lethal mutations has been discovered and characterized in recent years. As with *ts* mutations, the new mutations can arise in any bacterial or bacteriophage gene and, consequently, they are distributed throughout the entire genome. All mutants carrying such lesions fail to synthesize some vital protein, although in its place, incomplete polypeptide fragments are formed. It appears that synthesis of the protein or enzyme specified by the altered gene begins normally but then breaks off abruptly before the polypeptide chain is completed. Mutations of this kind are called suppressor-sensitive or *suppressible* because the lost function can be restored by an independent *suppressor mutation* at a distinct locus in the bacterial genome. Surprisingly, the suppressor mutation usually maps at a considerable distance from the mutationally impaired structural gene. Moreover, suppressor genes in the bacterial chromosome can also restore function to bacteriophages carrying otherwise disabling suppressor-sensitive mutations. Therefore, the suppressor gene is not specialized for the function it restores, but for a certain class of mutations that can occur in any gene.

At present, there are about ten distinct suppressor genes known in *E. coli*, each of which corrects a certain group of mutations with varying efficiency (11). Suppressor genes are denoted by the symbol *su* plus an identifying letter or number. The first group of suppressible mutations to be characterized, called *amber mutations*, are suppressed by su_I, su_{II}, and su_{III}. When an amber mutation is introduced into an *E. coli* cell carrying any of these suppressor genes, the product of the deficient gene is synthesized and the impaired function may be restored. However, the protein is usually not identical in all respects to that of the wild-type cell. If it is an enzyme, for instance, its specific activity may be less than that of the wild-type enzyme. Thus, the suppressor can permit synthesis of the protein controlled by a gene containing an amber mutation, but it may not completely restore the original structure of the protein. A second group of suppressible mutations, called *ochre mutations*, are suppressed in *E. coli* by *su*-4, *su*-5, *su*-C, *su*-F, and several others. The ochre suppressors also suppress amber mutations, though at much lower efficiency than the specific amber suppressors. The mechanism of genetic suppression will be discussed in a Section 17.

4. Selection of Bacterial Mutants

We shall define the wild-type bacterium as one that is able to grow on the most primitive medium, containing a simple sugar as sole carbon source, and inorganic salts as sources of nitrogen and sulfur. Such an organism is called a *prototroph*, and the medium on which it grows is referred to as a *minimal* or *synthetic* medium. Prototrophic bacteria can synthesize all amino acids, nucleotides, and vitamins which they require for growth. Through mutation, prototrophs give rise to cells with metabolic deficiencies which are incapable of synthesizing some vital cell component, or of assimilating a certain sugar as sole carbon source. Mutants unable to form enzymes necessary to catalyze the synthesis of essential metabolites are called *auxotrophs*. If we do not know which function has been impaired in a given mutant, we can grow it on an *enriched* or *maximal* medium, usually containing peptone, a protein hydrolyzate, yeast extract, and inorganic salts. In an enriched medium, amino acids, nucleotides, and vitamins are all present in an accessible form. Partially enriched media can also be compounded which contain, in addition to salts and a carbon source, only those substances that cannot be synthesized by the mutant in question.

Indirect selection of auxotrophic mutants can be carried out in the following manner. The test culture is spread on a series of petri dishes containing enriched medium in an agar gel, which will be called enriched agar. Of the hundreds of colonies that grow out, a few will result from auxotrophic mutants. The colonies of each plate are next imprinted in fresh petri dishes containing minimal medium in agar (minimal agar). An ingenious technique for imprinting, called *replica-plating*, was developed by Lederberg and Lederberg (12). The surface of a petri dish covered with individual, well-separated colonies is pressed lightly against a piece of sterile velvet. A fresh plate with minimal medium is then pressed against the same piece of velvet. The pile fibers of the velvet act as small inoculating needles, picking up cells from the enriched agar and depositing them on the surface of the minimal plates, and at the same time, preserving the original orientation of the colonies relative to one another. After the minimal plates are incubated, it becomes apparent that a few of the transferred colonies do not grow. These are the auxotrophic mutants which can now be identified on the original enriched plate and spread on a series of fresh plates, each containing a different metabolic supplement. By observing which metabolites support growth of a given mutant, we can determine the function that has been impaired in it. In many cases, it is possible to pinpoint the precise enzyme which the mutant lacks. As should be evident from what has been said, this method is based on the all-or-nothing principle of functional activity.

A faster and more elegant method of direct selection can be employed when the mutant under study causes the nutrient medium to change color. There are several systems for which specific color tests have been developed. Suppose that we are seeking *E. coli* mutants which have lost the ability to metabolize lactose, that is, which have undergone the mutation $lac^+ \to lac^-$. The culture is spread on agar containing lactose and protein hydrolyzate as carbon sources. Lac^+ cells can obviously grow on this medium by assimilating lactose, but lac^- cells can grow on it too despite their inability to metabolize the sugar, for they can utilize the protein hydrolyzate as a carbon and energy source. In order to identify lac^- mutants, a pH indicator consisting of eosin and methylene blue (EMB) is introduced into the medium. When the pH drops below 6, the indicator acquires a violet color. If the bacteria can metabolize lactose, they will excrete lactic acid and the colony will turn violet. Lac^- colonies remain pale red, for they do not excrete lactic acid. It is therefore possible to identify the mutants without transferring them from one medium to another.

A color test can also be used in the selection of *E. coli* mutants unable to synthesize the enzyme alkaline phosphatase, a mutation of the type $pho^+ \to pho^-$. Both pho^+ and pho^- cells are grown on a minimal medium containing inorganic phosphate at a very low concentration (2 µg/ml). When the colonies are large enough, they are sprayed with a specific substrate of alkaline phosphatase, nitrophenylphosphate. This substance is colorless, but if phosphatase is present in the cells, it is cleaved to nitrophenol, a bright yellow compound:

$$O_2N\text{-}C_6H_4\text{-}O\text{-}P(=O)(OH)_2 \xrightarrow[H_2O]{\text{phosphatase}} O_2N\text{-}C_6H_4\text{-}OH + H_3PO_4$$

In this manner, all of the pho^+ wild-type colonies will be "developed" as in a photograph, while all the colonies composed of pho^- mutants will remain colorless.

A third kind of color test is employed to select *Bacillus subtilis* cells unable to utilize starch as a sole carbon source, that is, which cannot synthesize the enzyme, amylase. The wild-type strain is denoted amy^+ (capable of synthesizing amylase) and we shall be looking for amy^- derivatives. The test culture is grown on a minimal plate containing glucose as well as starch. Once again, both wild type and mutant cells are viable. Next, an iodine solution is sprayed over the plate to "develop" the colonies. Amy^+ colonies appear as light regions on a dark background since they have metabolized the starch in their surroundings, but amy^- colonies will be colored.

These examples all demonstrate how direct selection of a number of different markers can be accomplished. However, without prior enrichment for the mutants sought, none of these methods is very effective, because the rare mutational events occur amidst an enormous background of unaltered cells.

In some instances it is desirable to select for back mutations, that is, for cells which have regained some biochemical function lost in an earlier mutational step. The phenomenon of back mutation is called *reversion*, and the resulting bacteria, *revertants*. It is generally much easier to select revertants than mutants with metabolic deficiencies. Inasmuch as an enzymatic activity reappears that had formerly been absent, the revertants can be grown on minimal medium, while the parent cells require the presence of an essential metabolite. Selection of revertants affords very high sensitivity since millions of cells may be spread on a single petri dish and only those few will grow which have mutated back to functional activity. In the selection of metabolically deficient mutants, by contrast, it is necessary to grow colonies of the unaltered cells along with the mutants, thereby severely limiting the number of individuals which it is practical to screen.

As will be demonstrated later (p. 428ff), revertants frequently synthesize an enzyme which, although active, is not identical to the protein formed by the initial wild-type organism. Consequently, a back mutation may restore a functional enzyme, but not the initial primary structure of the unmutated polypeptide chain.

Another very characteristic feature of back mutations is that they occur much less frequently than forward mutations. Consider the process of forward and back mutation in the alkaline phosphatase cistron. When $pho^+ \to pho^-$ mutations are produced by ultraviolet light at a survival rate of $n_s/n_0 = 10^{-3}$, the number of mutants among the survivors, n_m/n_s, is about 2×10^{-4} after a fourfold exposure, or 5×10^{-5} per exposure (see p. 305f). These pho^- cells are now used to obtain revertants to pho^+. After a fourfold irradiation, with n_s/n_0 once again set at 10^{-3}, the frequency of revertants among the survivors is only about 2×10^{-7} per exposure, indicating that this event is 250 times less probable than the $pho^+ \to pho^-$ mutation. It is concluded that there are many more ways to damage a cistron than there are to restore its functional activity once it has been altered at a specific point.

Davis and Lederberg (13) independently worked out a method for the selection of auxotrophic mutants based on eliminating the prototrophs from the culture. Although it might appear impossible to sustain metabolically defective mutants under conditions lethal for prototrophs, the technique succeeds owing to the distinctive physiological action of the antibacterial drug, penicillin. In the presence of penicillin, cells continue to synthesize proteins and increase in size, but cell wall formation is strongly inhibited. As the cell outgrows its cell wall, the cell membrane, left unsupported, finally ruptures, and the cell dies. The important point here is that the cells must be growing in order for penicillin to kill them; cells that cannot grow in the presence of the antibiotic will not be killed.

With this background in mind, let us follow a typical penicillin selection experiment. After irradiation, the bacteria are grown out on an enriched medium and then transferred to minimal medium with about 200 μg/ml penicillin. The optimal concentration must be ascertained in each case. Most of the prototrophic cells die under these circumstances, but the mutants, which cannot grow on minimal medium, survive. The penicillin is now washed from the cells and they are spread on supplemented agar plates. The deficiencies in the auxotrophs can be analyzed in the usual way by spreading or replica-plating the survivors onto agar plates containing single metabolites. Owing to the enrichment of mutants by the penicillin technique it is possible to detect mutational events occurring with a frequency of less than one in 10^8.

Mutations conferring resistance to inhibitors and metabolic poisons possess distinctive characteristics. Wild-type bacteria are not resistant to most inhibitors, apparently because such substances are not commonly encountered in natural environments. Assume, for example, that a mutation from sensitivity (s) to resistance (r) has taken place at the streptomycin (*str*) locus: $str^s \to str^r$. The selection of resistant mutants is accomplished simply by introducing streptomycin into the growth medium. Sensitive cells will not give rise to colonies under these conditions.

The maximal concentration of inhibitor to which such mutants are resistant may vary from individual to individual, and this property does not therefore adhere to the all-or-nothing principle. This can be most easily determined by streaking the mutants across the surface of an agar plate in which there is a gradient of inhibitor concentration. The gradient is formed in a petri dish by pouring two agar wedges on top of one another, with inhibitor present in the bottom wedge (Fig. 4-2). Vertical diffusion rapidly produces a linear concentration gradient of the inhibitor along the surface of the plate.

Fig. 4-2. Petri dish with a streptomycin concentration gradient. Agar medium containing streptomycin (1) is first poured into a tilted petri dish and allowed to harden. Petri dish is then placed on a horizontal surface and agar medium without streptomycin is layered on top (2). Streptomycin diffuses upwards from wedge (1) to form a linear concentration gradient of the antibiotic along agar surface

The action of many inhibitors depends on their ability to bind and block specific enzymes within the cell. This generally occurs when the inhibitor is a structural analog of the substrate and can substitute for it at the active site of the enzyme molecule. One classic example of such a substance is sulfanilamide, a structural analog of *p*-aminobenzoic acid which is an intermediate in the synthesis of the vitamin, folic acid:

$H_2N-\langle\bigcirc\rangle-SO_2NH_2$ $H_2N-\langle\bigcirc\rangle-COOH$

   Sulfanilamide   Paraamino-
                          benzoic acid

The wild-type enzyme, inhibited by sulfanilamide, lacks sufficient specificity, since its active site apparently does not distinguish between substrate and inhibitor. A mutation to sulfanilamide resistance indicates that the active site of the protein has changed in the direction of greater structural specificity. The mutant enzyme will bind the substrate, but not the inhibitory analog.

Mutations conferring resistance to bacteriophages also have a well-defined structural basis. We shall consider two mechanisms by which phage resistance or immunity is acquired, the first of which relates to a large group of coliphages, including T1, T2, T4, T6, T7, and others. Before injecting their DNA into the host cell, all of these bacteriophages must first become attached to the surface of the cell wall. The phages adsorb to specific structures in the wall, and wild-type bacteria usually possess these binding sites or *receptors*. However, the cell may lose the relevant sites on its surface by mutation, and phages will no longer be able to become adsorbed. Such mutants may be viewed as deficient in a capacity to synthesize a cell wall component which is normally present in the wild-type organism.

The resistance of lysogenic strains of *E. coli* K12 to phage λ, as well as several others, is mediated by a completely different mechanism. This group of so-called temperate phages can exist in two forms. In the vegetative form, the infecting particles can multiply at the expense of the host and eventually lyse the cell wall; in the form of prophage, the DNA of the infecting particles becomes integrated into the host chromosome and is temporarily inactive (see p. 391ff). The association seems to be quite stable and the phage chromosome is replicated along with that of the bacterium during cell division. Bacteria carrying prophage are said to be lysogenic, and in this state they are immune to superinfection by the same kind of phage. But a lysogenic culture is not completely stable; there is a small probability that the cells will be lysed if the phage can spontaneously enter the vegetative phase. Many wild-type bacterial strains are lysogenic, with various prophages integrated into their chromosomes. Sometimes this is rather difficult to demonstrate, however, because the probability of spontaneous lysis can be extremely small.

In some, though not all, cases, the prophage can be induced by an external agent such as nonlethal doses of radiation, certain chemical substances, or moderate increases in pressure. We shall return to the nature of these phenomena in section 12 of this chapter. It is sufficient at present to know that the lysogenic cell is resistant to homologous phages and that this property is transmitted to both daughter cells every division.

5. The Luria-Delbrück Fluctuation Experiment

The occurrence of mutants resistant to phages and metabolic inhibitors poses one of the most general questions in bacterial genetics. Do such changes arise in a population independently and at random, or do they result from the adaptation of certain individuals to the phage or inhibiting substance itself? The answer is not obvious, since the selection of resistant mutants requires that the agent in question be present in the medium. It is therefore difficult to determine whether or not the resistant organisms existed prior to their first contact with the inhibitory substance or to their first exposure to the phage. This dilemma reflects the fundamental and age-old problem of whether variation among individuals is inherited or acquired under environmental influence, as suggested by Lamarck. An ingenious experiment devised by Luria and Delbrück settled the question conclusively (14).

Suppose that we inoculate C separate portions of medium with very small amounts of the same bacterial strain, placing approximately 100 cells in each tube. We will obtain C different cultures growing under what appear to be exactly equivalent conditions. After the cultures have grown out, we can take a series of samples from every tube and determine the average number of resistant cells, $\bar{r}$, by spreading the samples on agar plates with a large excess of bacteriophage particles. The majority of cells will be infected and die, but a certain number will be resistant and will form colonies.

Consider first the probability distribution of the number of resistant bacteria according to the adaptation hypothesis. If resistance is acquired as the result of adaptation, then every cell will have an equal probability of becoming resistant, regardless of the specific culture tube from which it was taken. Since all acts of adaptation are considered to be independent events, the number of resistant cells will fluctuate

5. The Luria-Delbrück Fluctuation Experiment

somewhat from culture to culture according to the Poisson distribution. The width of any statistical distribution is characterized by the variance (var):

$$\text{var}(r) = \left[\sum_1^C (r_i - \bar{r})^2\right]/C \qquad (4\text{-}8)$$

where C is the number of individual cultures tested. If the adaptive events are independent of each other, and if the deviation of r_i, the number of resistant cells in the i-th culture, from $\bar{r}$, the average number cells over all of the cultures, obeys Poisson's law, then the variance is equal to the average: $\text{var}(r) = \bar{r}$. The standard deviation in a statistical distribution of this sort is $[\text{var}(r)]^{\frac{1}{2}} = \bar{r}^{\frac{1}{2}}$, and the relative deviation is

$$[\text{var}(r)]^{\frac{1}{2}}/\bar{r} = 1/\bar{r}^{\frac{1}{2}} \qquad (4\text{-}9)$$

This will be a very small quantity if the average number of altered cells, $\bar{r}$, is sufficiently large. In this case, good agreement is expected between all the values r_i for individually grown samples of the culture.

One further observation must be made. In calculating the number of resistant cells, the entire population is not generally sampled, only an aliquot of volume V. Therefore, the number of resistant cells on each petri dish will be much smaller than $\bar{r}$. But by taking S identical samples, each of volume V, from one of the cultures, we should be able to find the average number of resistant cells per culture, $\bar{r}$, from the average number of resistant cells, $\bar{n}$, in each of the samples:

$$\bar{n} = \left(\sum_1^S n_i\right)/S \qquad (4\text{-}10)$$

$$\bar{r} = (\bar{n}/V)V_0 \qquad (4\text{-}11)$$

where V_0 is the total volume of one culture. However, the variance will be greater in this case than in the true deviation calculated above, because the samples are smaller:

$$\text{var}(n) = \left[\sum_1^S (n_i - \bar{n})^2\right]/S \qquad (4\text{-}12)$$

$$[\text{var}(n)]^{\frac{1}{2}}/\bar{n} = 1/\bar{n}^{\frac{1}{2}} \qquad (4\text{-}13)$$

Thus we see that the method of measurement itself increases the statistical error. However, this error may readily be made as small as desired; in particular, if the calculation is based on 1000 resistant colonies, the relative standard error will be only 3%. Table 4-1 presents results from an experiment in which resistant cells were assayed in ten separate aliquots from a *single* culture. The deviation of the distribution of individual measurements is in good agreement with the Poisson law. If the acquisition of phage resistance were adaptive, we should obtain the same result by comparing samples taken from *separate* parallel cultures.

But the result would be entirely different if bacterial variability results from random mutation. First, mutations are expected to occur throughout the entire period of growth of each culture. Second, since the probability of mutation is assumed to be small, there should be very few mutations at the beginning of the growth period, and the statistical fluctuations from culture to culture should be relatively large. Both mutant and wild-type cells multiply with time of growth in the absence of the selective

TABLE 4-1

THE NUMBER OF BACTERIOPHAGE-RESISTANT BACTERIA IN DIFFERENT SAMPLES OF THE SAME CULTURE[a]

Sample number	Number of colonies resistant to bacteriophage	
	First culture	Second culture
1	14	4
2	15	2
3	13	2
4	21	1
5	15	5
6	19	2
7	26	4
8	16	2
9	20	4
10	13	7
Mean, $\bar{n}$	16.7	3.3
Variance, var(n)	15	3.8

[a] From Luria and Delbrück (14).

agent and the incidence of new mutations constantly adds to the number of resistant cells. As this happens, fluctuations in the number of mutants per culture increase enormously. Qualitatively, this is quite easy to understand. Assume that in a given time interval, Δt, there arise a certain number of mutants whose average is $\overline{\Delta m}$. Until such time as growth of the culture is terminated, each of the mutants formed in Δt will have multiplied λ times. In this case, it is clear that var(Δm) = $\overline{\Delta m}$, since individual mutations are independent of one another and their distribution is therefore subject to the Poisson law. In an actual experiment, however, we do not measure the distribution of mutants, but that of resistant cells descended from the original mutants that have already multiplied λ times, that is, $\Delta r = \lambda \Delta m$. The variance in C independent cultures is thus:

$$\text{var}(\Delta r) = \left[\sum_{1}^{C} \lambda^2 (\Delta m_i - \overline{\Delta m})^2\right] / C = \lambda^2 \overline{\Delta m} = \lambda \overline{\Delta r} \quad (4\text{-}14)$$

Under these circumstances, the variance of the distribution is λ times greater than in the case of a simple Poisson distribution, because the resistant cells do not all arise from independent events; rather, they are the descendents of the original mutants. Since the value of λ is large, we expect enormous fluctuations in the number of resistant cells in each of the parallel cultures.

For mutants arising during any time interval between 0 and t, λ will vary so that the average number of resistant cells is:

$$\bar{r} = \sum \lambda_i \overline{\Delta m_i} \quad (4\text{-}15)$$

5. The Luria-Delbrück Fluctuation Experiment

and the variance of r is:

$$\operatorname{var}(r) = \sum_i \operatorname{var}(\Delta r_i) = \sum_i \lambda_i^2 \overline{\Delta m_i} \qquad (4\text{-}16)$$

or, by making the time intervals infinitely small, we obtain:

$$\bar{r} = \int_0^t \lambda \, dm \qquad (4\text{-}17)$$

and

$$\operatorname{var}(r) = \int_0^t \lambda^2 \, dm \qquad (4\text{-}18)$$

It is now necessary only to express the average number of mutants, m, and the factor by which the cells have multiplied, λ, as functions of time, and we will obtain the desired solution.

We assume that the cells grow exponentially according to the first-order rate law:

$$n_t = n_0 e^{T/\tau} = n_0 2^{T/\theta} \qquad (4\text{-}19)$$

where T is the time in minutes. The generation time of the culture, τ, is somewhat larger than the doubling time, θ, and the two are related as $\tau = \theta/\ln(2) = \theta/0.69$. In the following, τ will be taken as the unit of time, and Eq. (4-19) will be used in the form:

$$n_t = n_0 e^t \qquad (4\text{-}19')$$

In addition to the mathematical simplifications which this permits, we can define a mutation rate independent of precise physiological conditions and of differences in susceptibility to mutation that might occur during the bacterial life cycle by selecting a time unit proportional to the division time. Thus the mutation rate, a, is the probability of mutation per bacterium during τ.

The number of mutations, dm, which occurs during the time interval, dt, is the product of a, dt, and the number of cells in the culture at that moment, n_t:

$$dm = a n_t \, dt \qquad (4\text{-}20)$$

By substituting from Eq. (4-19') and integrating, we find the number of mutations, m, which occurs between 0 and t:

$$m = a(n_t - n_0) \qquad (4\text{-}21)$$

But since the number of mutations cannot be directly observed, we must now find the number of resistant cells in the culture, which result from the original mutants.

Suppose that mutations arise in a certain group of cells at a time t' after the initiation of growth. Assuming that the mutants grow at the same rate as the original cells, they will multiply throughout the interval $(t - t')$, and the factor by which they increase is $\lambda = e^{(t-t')}$. Consequently, the contribution of these mutants to the total number of resistant cells will be:

$$dr = \lambda \, dm = e^{(t-t')} a n_0 e^{t'} \, dt' = a n_0 e^t \, dt' \qquad (4\text{-}22)$$

From Eq. (4-22) we discover that all time intervals contribute equally to the final number of resistant cells, since the exponential increase in the culture is compensated by an exponential decrease in the number of progeny to which each mutant will give rise. The initial number of mutants is small, but they will undergo many divisions; mutants formed at later times, though more numerous, will not produce as many progeny. Because the number of mutants is very small in the first moments of growth, the statistical differences among various individual cultures are expected to be quite large. By integrating dr, we obtain the average number of resistant cells:

$$\bar{r} = \int_0^t dr = an_0 e^t t = an_t t \qquad (4\text{-}23)$$

It should be apparent, however, that Eq. (4-23) yields too large a value of $\bar{r}$. By integrating from zero time, we have included the initial periods of growth in which the average number of mutants per culture is less than unity and this introduces a substantial error into the final result. Use of this equation is permissible only if the number of cultures is infinitely large. For a finite number of samples, it is necessary to integrate not from zero time, but from a later time which we shall designate t_0. We may reasonably choose t_0 so that in the interval $(t_0 - 0)$ just one mutation, on the average, occur in a group of C similar cultures. According to Eq. (4-21), this condition is written $1 = aC(n_{t_0} - n_0)$. The corrected average number of resistant cells is:

$$\bar{r} = an_t(t - t_0) \qquad (4\text{-}24)$$

But it is easy to express n_{t_0} in terms of n_t, since $n_{t_0} = n_t e^{-(t-t_0)}$ or $(t - t_0) = \ln(aCn_t)$ and Eq. (4-24) can be rewritten as:

$$\bar{r} = an_t \ln(aCn_t) \qquad (4\text{-}25)$$

While this equation is more accurate than Eq. (4-23), one should bear in mind that the correction is only approximate.

The variance of r can now be calculated from Eqs. (4-18) and (4-22):

$$\text{var}(r) = \int_{t_0}^t \lambda^2 dm = \int_{t_0}^t an_t e^{(t'-t_0)} dt' = an_t[e^{(t-t_0)} - 1] \qquad (4\text{-}26)$$

But the 1 in the brackets can be neglected since $e^{(t-t_0)} \gg 1$ and $(t - t_0)$ can be replaced by $\ln(aCn_t)$ as before so that:

$$\text{var}(r) = a^2 Cn_t^2 \qquad (4\text{-}27)$$

$$\frac{\text{var}(r)}{\bar{r}} = \frac{C\bar{r}}{[\ln(aCn_t)]^2} \qquad (4\text{-}28)$$

It is easy to see that this value is much greater than unity, whereas if the Poisson distribution had been applicable $\text{var}(r)/\bar{r}$ would have been equal to 1. The relative standard deviation of r in a limited number of cultures is:

$$\frac{[\text{var}(r)]^{\frac{1}{2}}}{\bar{r}} = \frac{C^{\frac{1}{2}}}{\ln(aCn_t)} \qquad (4\text{-}29)$$

In distinction to adaptation, the mutational hypothesis predicts that there will be a great variance in the number of resistant cells in several independent parallel cultures. Such large statistical fluctuations in similar cell populations were at one time taken for poor reproducibility. But Luria and Delbrück showed that variations of this type

5. The Luria-Delbrück Fluctuation Experiment

follow necessarily from the low probability of spontaneous mutation and the exponential growth which characterize bacterial cultures.

Table 4-2 presents some of the data obtained by Luria and Delbrück for variations in the number of *E. coli* mutants resistant to bacteriophage T1. The experimental results are on the same order of magnitude as the theoretical prediction, and a better fit could not have been expected. The fluctuations themselves are enormous and the values of $\text{var}(r)/\bar{r}$ are indeed much greater than unity. Consequently, the hypothesis of adaptation was completely ruled out.

TABLE 4-2

DISTRIBUTION OF THE NUMBER OF BACTERIOPHAGE-RESISTANT
BACTERIA IN SERIES OF SIMILAR CULTURES[a]

	Experiment 1 ($c = 100$)		Experiment 2 ($c = 87$)	
	Resistant bacteria	Number of cultures	Resistant bacteria	Number of cultures
	0	57	0	29
	1	20	1	17
	2	5	2	4
	3	2	3	3
	4	3	4	3
	5	1	5	2
	6–10	7	6–10	5
	11–20	2	11–20	6
	21–50	2	21–50	7
	51–100	0	51–100	5
	101–200	0	101–200	2
	201–500	0	201–500	4
	501–1000	1	501–1000	0
Average per culture, $\bar{r}$:	10.12		28.6	
Variance, $\text{var}(r)$:	6270		6431	
Bacteria per culture, n_t:	2.7×10^8		2.4×10^8	
Mutation rate, a:	$2.3 + 10^{-8}$		2.37×10^{-8}	
$[\text{var}(r)]^{\frac{1}{2}}/\bar{r}$ experimental:	7.8		2.8	
calculated:	1.5		1.5	
$\text{var}(r)/\bar{r}$:	627		225	

[a] From Luria and Delbrück (14).

Besides its theoretical significance, this experiment is of considerable methodological importance, because it provides an example of the statistical approach to the resolution of biological problems. After the initial successes of this method, fluctuation experiments were repeated for a wide variety of mutations and the results consistently supported the theory of random mutational events.

The Luria-Delbrück fluctuation test is not the only means of determining whether variability arises from mutation or adaptation. The replica-plating method of Lederberg, which was described on page 308, also gives a very clear answer to this question. We can grow a large number of colonies on nonselective media and then imprint them onto plates containing an antimetabolite such as streptomycin. Some resistant colonies will be found, even though the cells were originally grown on medium free of the inhibitor. In these cells, the property of resistance must have arisen independently, regardless of whether or not they had ever been in contact with the antibacterial agent.

6. Sex in Bacteria; Conjugation

In higher organisms, new individuals are formed when the genetic materials of maternal and paternal sex cells, or gametes, fuse together in one cell, the *zygote*, which thereby acquires two sets of homologous chromosomes. Evidence that bacteria also undergo genetic exchange by a mechanism akin to the sexual processes of higher organisms was obtained in 1946 by Lederberg and Tatum (15), who observed that cells of mixed heredity are formed when two strains of *E. coli*, differing from one another in a number of genetic markers, are incubated together under appropriate conditions. The transfer of genetic traits between sexually differentiated bacteria, called *conjugation*, has been investigated very thoroughly in the years since its discovery. Although the events leading to genetic exchange by conjugation have proved to be both numerous and complex, most of them can now be given molecular interpretations.

For convenience, bacterial conjugation can be broken down into four successive stages.

(1) When two bacterial cells of different mating type, or "sex," one destined to serve as the donor of genetic material and the other as recipient, collide in a suitable medium, they can form a stable union. Successful pair formation apparently demands complementarity between the cell surfaces of the two partners. Furthermore, conjugal union has been shown to require the presence in the donor or male bacterium of long fibers called pili which protrude from the cell exterior. Once the union is securely established, a tubelike structure, or bridge, is formed between conjugating pairs, as can be seen from the electron micrograph presented in Fig. 4-3. Donor and recipient are distinguished in electron microscopic studies by choosing two strains which differ from one another in some outward characteristic.

(2) The second stage of conjugation consists of a one-way transfer of genetic material from the donor or male to the recipient or female through the intercellular bridge connecting the cytoplasms of the paired bacteria. The principal chemical substance transferred from donor to recipient during conjugation was shown to be DNA by the following experiments (17). Donor cells resistant to phage T6 were grown on media containing radioactive precursors to one of the following cellular components: DNA, RNA, or protein. The different males were then mated with phage-sensitive females and, after conjugation, the sensitive female cells were lysed from without with phage T6. The phage-resistant donors were next removed by filtration, leaving in solution the molecular components of the lysed recipients, including any radioactive material transferred to them by the donors. Analysis of the soluble radioactivity in each case revealed that substantial amounts of DNA penetrated the female cells, but that the quantity of RNA and protein transmitted was not significant. We shall assume that this DNA comprises the bacterial chromosome, and that it is

6. Sex in Bacteria; Conjugation   319

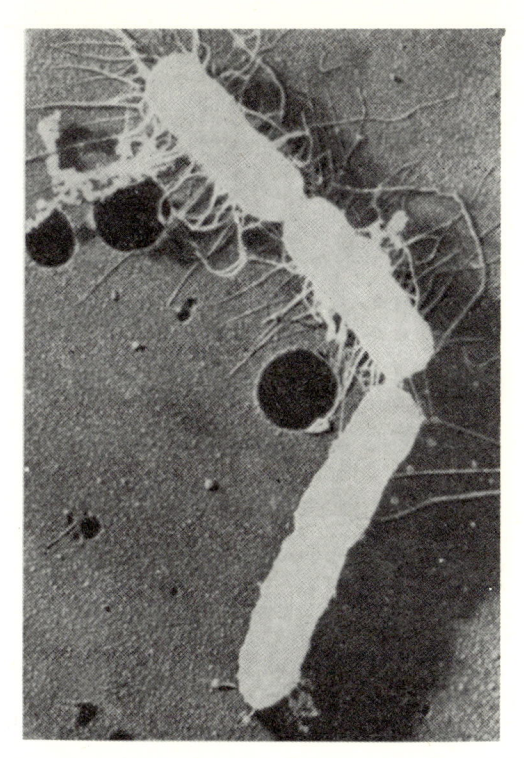

Fig. 4-3. Electron micrograph of conjugating bacteria. Cross between HfrH λ^{sens} and F$^-$ λ^{res}. Inactivated λ bacteriophages were adsorbed on the Hfr cells to identify them. The F$^-$ cells failed to adsorb phages, but can be recognized by their numerous "bristles." In this view, a bridge 1500 Å wide can be seen connecting the end of a dividing F$^-$ bacterium with the end of a dividing Hfr to which bacteriophage particles are attached; $\times$ 21,000. From Anderson *et al.* (16).

specifically the chromosome that is transferred in conjugation. Since no radioactive material from females was found in males when the former were labeled, these experiments also confirmed the unidirectional nature of DNA transfer.

(3) A period of about 100 minutes is required for the transfer of the entire donor chromosome to the recipient. Complete transfer is rare, however, and it more frequently transpires that only a portion of the male chromosome penetrates the recipient. Regardless of whether transfer is partial or complete, the recipient becomes a zygote, diploid for at least some, if not all, of the bacterial genome in contrast to the normal bacterium which is haploid. If there is extensive homology between the two different parental chromosomes, pairing or *synapsis* occurs in such a way that functionally identical regions in them lie side-by-side. After this happens, an exchange may take place between the two synapsed chromosomes, leading to the formation of a new chromosome which contains genetic information from both parents, *recombined* in a new arrangement. Recombinant cells are detected by using parental strains that carry distinguishable alleles at several different genetic loci. The mechanism of recombination will be discussed in Section 8.

(4) The zygotes formed after conjugation may be stable for a few generations, and even if chromosomal exchange occurs, the new assortment of maternal and paternal

characteristics inherited by the recombinants may not be immediately recognizable. In particular, there is a *phenotypic lag* in the expression of recessive alleles if the zygote also contains a dominant allele for that character. This situation is quite a likely one because each bacterium usually possesses more than one chromosome, that is, it is "multinucleate" even though every "nucleus" is identical to all the others and the cell as a whole is thus haploid. In conjugation, only one of the recipient chromosomes will become recombinant and, consequently, a recessive character may remain unexpressed until the second or third division cycle, when recombinant and nonrecombinant chromosomes finally segregate from one another. After segregation, a haploid cell clone that contains chromosomes of the recombinant type only can be recovered by the use of proper selective techniques.

There are two fundamental mating types in *E. coli* which are distinguished by their relationship to the process of genetic exchange. About 90% of the wild-type strains can act as females, or recipients of genetic material, during conjugation, while roughly 10% possess the capacity to serve as donors or males. Since donor cells can easily be turned into recipients by exposure to ultraviolet light or by treatment with a class of chemical compounds known as acridine dyes, there must exist some material agent that confers the property of "maleness" on the donors. The agent responsible has been termed the F (fertility) or *sex factor* and the donor cells which carry it are designated by the symbol F^+. In the female or recipient, the sex factor is either absent or else it is masked or inactive, and such bacteria are consequently referred to as F^-. When F^+ cells conjugate with F^- cells, the male chromosome is transferred only in rare cases with a probability of 10^{-5} to 10^{-6}, but the F factor enters the females with an efficiency of almost 100%, converting them to F^+ donors. Thus, in $F^+ \times F^-$ crosses, F is transmitted to recipient cells independent of the bacterial chromosome. This process takes only about 10 minutes to complete, and even after transferring an F factor to the female cell, the original donor remains F^+.

All of these observations indicate that F is a genetic determinant which, in F^+ cells, is both structurally independent of the bacterial chromosome and capable of autonomous replication. As will be seen in a later section, there is good evidence that the sex factor, like the chromosome, is composed of DNA, although it is much smaller in size. Its DNA content amounts to only a few percent of that in the chromosome.

A breakthrough in the study of conjugation occurred in 1953, when Cavalli-Sforza and Hayes independently discovered the existence of a second class of donor bacteria that are capable of transmitting their chromosome to recipient cells with high probability (18). Since penetration of the donor chromosome into a recipient generally leads to recombination, these new strains were given the symbol Hfr, which stands for *high frequency of recombination*. The original Hfr's, as well as a number of others that have been discovered subsequently, were all isolated from F^+ strains. In fact, the recovery of recombinants for chromosomal genes at low frequency in $F^+ \times F^-$ crosses can be attributed in large measure to the presence of spontaneous Hfr "mutants" in the F^+ cultures. When Cavalli, Hayes, and others used purified Hfr strains in mating experiments, the probability of obtaining recombinants rose by a factor of 10^4 or 10^5 to a level such that 10 to 50 recombinants were formed for every 100 donors. A further consequence of the $F^+ \to$ Hfr transition is that the Hfr's lose their capacity to transmit the sex factor to F⁻ cells at high frequency. But since the Hfr's retain many other characteristics of F⁺ bacteria, one might suppose that the sex factor is still present in them, although in a modified form.

Hfr's from various independent isolates are not necessarily identical to one another. For instance, the Hfr strain found by Hayes (HfrH) transfers its genetic markers in the linear sequence *thr, leu, azi, ton, lac, tsx, gal,* and so on (see Fig. 4-1, p. 302). By contrast, the Hfr isolated by Cavalli (HfrC) transmits its markers in the order *lac, ton, azi, leu, thr, thi,* ..., that is, from a different starting point and in the opposite direction. Thus, every Hfr strain comprises a homogeneous population of donor bacteria, all of whose members transfer their chromosomes with the same orientation and from the same origin, *O*, so that the recipient bacteria are penetrated by donor genes in a fixed, linear order. Despite the difference of polarity and origin in HfrH and HfrC, the chromosomal segments which they transfer to recipients at high frequency partially overlap one another and it is evident that the relative orientation of the genes is the same in both cases. This was also found to be true for a number of other Hfr's with quite different origins, making it impossible to define any unique extremities on the chromosome map. Since all of these Hfr's originated from similar, and in some cases the same, F⁺ strains, it was concluded that the chromosome in F⁺ bacteria is continuous, or circular, and that Hfr strains arise, at least in a formal sense, by an opening up of the circle at a point characteristic of each Hfr type, to yield a linear, transferable structure.

During conjugation, the genetic markers of the Hfr donor do not all have the same probability of appearing among the progeny of the F⁻ recipients. At first it appeared as if there were two groups of markers, one transmitted with high frequency, the other with low frequency. Among the characteristics transmitted with very low frequency is the capacity to transform the recipient into an Hfr; as a consequence, most of the recipient cells which acquire donor markers remain F⁻ with regard to mating type. It was later demonstrated, however, that the genetic markers could not be placed in two, sharply differentiated groups. Instead, the probability with which a character is transmitted to recipient cells decreases continuously with distance from the origin of the Hfr chromosome, presumably due to random breakage of the chromosome during transfer. In HfrH, the markers *thr* and *leu* are transmitted with highest frequency, followed by *azi, ton, lac,* and *gal,* in descending order. This order exactly corresponds to the distribution of these markers on the genetic map as determined by other methods, assuming that the closed circular chromosome is broken open near the *thr* locus. Markers at the far or distal end of the chromosome are transferred with extremely low frequency. The terminal position of the locus determining the donor state can be identified by the fact that the few recombinants for distal markers, such as *cys* in the case of HfrH, also inherit the Hfr donor property as well as other male characteristics which we shall discuss presently. In other Hfr's, too, regardless of their polarity and origin, the Hfr character is always found to be closely linked with the distal end of the chromosome, that is, with the last marker transferred.

The manner in which the determinant of the Hfr donor state is inherited suggests that it is in fact the sex factor which has in some way become associated with the chromosome of the Hfr bacterium so as to define its terminus during conjugal transfer. The persistence of the sex factor in Hfr cells is indicated by several lines of evidence.

First, Hfr's can spontaneously undergo a transition to the F^+ state, showing that the F factor was never lost in these strains while in the Hfr state. Second, presence of the sex factor in a cell is reflected by certain physiological peculiarities, most of which are common to both Hfr and F^+ cells. Fertility, or the ability to serve as a donor of genetic material, is of course the most obvious similarity. Furthermore, cytoplasmic contact via the conjugal bridge can be established between Hfr and F^+ bacteria, as well as between either one of them and F^- bacteria, although in the former case, genetic exchange is limited. Such contact is never established between pairs of F^- strains, however. This indicates that the sex factor directly affects the properties of the cell surface and endows its bearer with the capacity to form an intercellular protoplasmic tube. Loeb and Zinder observed another and perhaps related difference between donor and recipient cells (19). A special kind of bacteriophage, f2, which contains RNA as its genetic material rather than DNA, can adsorb to and infect only F^+ and Hfr cells. The exteriors of bacteria inhabited by the sex factor, regardless of its condition in the cell, must therefore differ in some structural feature. This was confirmed by discovery that up to four long fibers, called F-pili, extend from the surfaces of F^+ and Hfr bacteria. It is specifically to the F-pili that f2 phages adsorb, and it has also been shown that possession of these fibers by the donor cell is required for conjugation. Although their precise role remains to be elucidated, F-pili may participate in formation of the conjugal bridge between mating bacteria.

Dissimilarities between the two types of donors also provide insight into the relationship between sex factor and "maleness." In F^+ strains, where it is clear that presence of the sex factor determines male properties, the factor itself is autonomous with respect to replication and transfer, whereas in Hfr's, the determinant of maleness behaves as a chromosomal marker. Moreover, the autonomous sex factor can be eliminated from F^+ bacteria by exposure to ultraviolet light, after which the cells become F^- recipients. Ultraviolet radiation has no such influence on Hfr bacteria, however. Finally, acridine dyes, a class of compounds that interact with nucleic acids (20), are especially effective in curing F^+ cells of their cytoplasmic sex factors, which are known to consist of DNA. Both proflavin and acridine orange (see p. 411 for structural formulas) act very readily on exponentially growing F^+ cultures at a concentration of 20 to 30 μg/ml. At these concentrations, the viability of the cells is not impaired, but the F factor is eliminated with almost 100% efficiency. By contrast, the male characteristics of Hfr bacteria are wholly insensitive to treatment with acridine dyes.

A unified hypothesis explaining the differences between F^+ and Hfr bacteria, and, in particular, the way that these differences relate to the process of genetic transfer, was advanced by Jacob and Wollman. Bear in mind that linkage among the many known genes of *E. coli* define not a linear genetic map, but a circular one (see Fig. 4-1). The inference from the genetic data was that the *E. coli* chromosome also comprises a circular structure, a supposition since directly confirmed. How then is the circular chromosome, without a "head" or "tail," converted into a linear structure of defined origin and polarity during high frequency transfer of the bacterial genome to F^- recipients? Jacob and Wollman suggested that in the spontaneous $F^+ \rightarrow$ Hfr transition, the sex factor becomes attached to a homologous region on the chromosome by a process analogous to recombination (see Section 8). Integration of the sex factor

6. Sex in Bacteria; Conjugation

would define for the circular chromosome both an origin and, depending on its orientation, a polarity, if during conjugation the circle opens up at the integration site so as to permit sequential transfer of the chromosome from that point onward (Fig. 4-4a). Since the origins of different Hfr's are distributed throughout the entire *E. coli* genome (Fig. 4-4b), there must be numerous sites of homology for the sex factor on the bacterial chromosome where integration can take place.

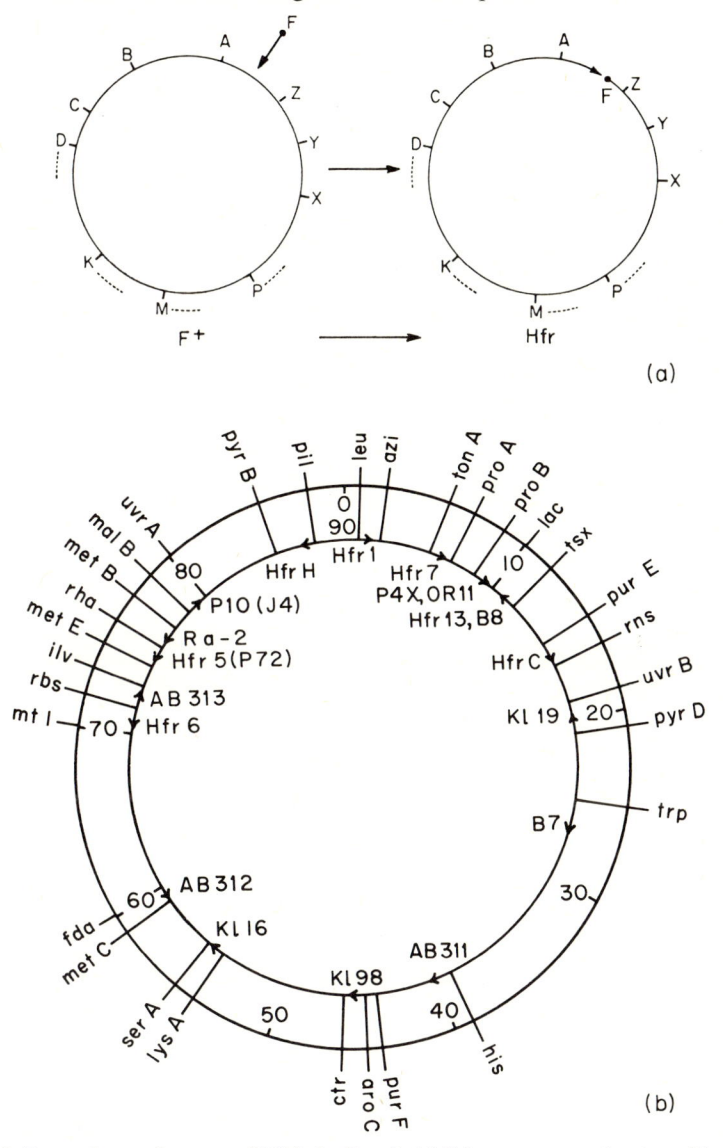

Fig. 4-4. Formation and nature of Hfr's in *E. coli*. (a) Scheme representing transition from F⁺ mating type to Hfr. Modified from Jacob and Wollman (2). (b) Abbreviated linkage map showing the points of origin of chromosome transfer for several Hfr strains. Arrowheads on the inner circle indicate the direction of transfer. The first and last markers known to be transferred by each Hfr are displayed on the outer circle. From Taylor and Trotter (6).

We shall see in Section 7 that the sex factor actually fixes the point on the circular chromosome from which a special round of replication is initiated at the moment of conjugation. As the free end of the newly replicated chromosome increases in length, its leading extremity, designated O, is squeezed through the conjugal bridge into the recipient cell. The other markers follow in an ordered, linear fashion and since 100 minutes or more are required for the entire chromosome to enter the female cell, the kinetics of this process can be analyzed quite easily. The sex factor itself must be attached to the tail of the linear structure because the male properties are the last to penetrate the recipient. When transfer is complete, the F^- recipient may become an Hfr if recombinants for distal markers of the donor are formed. Even after complete chromosomal transfer, the original donor remains viable and is still an Hfr, supporting the idea that chromosome transfer results from a supplementary round of DNA synthesis.

Convincing evidence that Hfr's are created by insertion of the sex factor into the male chromosome by genetic recombination, that is, by integration into the bacterial genome, is provided by the following experiment. A mutant F^- strain, in which one of the enzymes essential for recombination is impaired or absent, was used for this purpose. Although such strains can engage in conjugation and can accept the chromosome from an Hfr, they are unable to produce recombinants and are hence called rec^- (see p. 340). It was then found that F^+ strains constructed from rec^- females cannot give rise to Hfr's, proving that the capacity for recombination must play a vital role in the $F^+ \rightarrow$ Hfr transition (21). The sex factor thus exhibits the properties of an *episome*, that is, of a genetic element supplementary to the normal genome of the bacterium which can be replicated vegetatively and transferred from cell to cell either autonomously or as an integral part of the chromosome. Integration of the sex factor undoubtedly accounts for the insensitivity of Hfr's to acridine dyes.

In accordance with the Jacob-Wollman model, the sex factor should have the capacity to become integrated at any one of several points on the bacterial chromosome. Therefore, it should be possible to isolate, from a single population of F^+ cells, different types of Hfr's with a variety of origins. The replica-plating technique of Lederberg can be used to determine whether this is so. A prototrophic strain with the genotype F^+ ($thr^+ leu^+ str^s$) is spread on minimal agar without streptomycin so that there are between 200 and 300 colonies per plate. The colonies are next replica-plated onto petri dishes containing minimal medium with streptomycin, on whose surfaces an excess of female cells with the genotype F^- ($thr^- leu^- str^r$) had been previously spread. After a suitable period of incubation, prototrophic streptomycin-resistant recombinants are scored. Such recombinants will have resulted from events of the following type:

$$\text{Hfr}(thr^+ leu^+ str^s) \times F^-(thr^- leu^- str^r) \rightarrow F^-(thr^+ leu^+ str^r)$$

The original colony on the master plate from which the new Hfr derived can be traced back from the location of the recombinants. By this test, Hfr's are identified whose chromosomal origins, O, lie close to *thr* and *leu*. By employing other auxotrophs as recipients in analogous experiments, we can select from the same population of F^+ cells, a set of Hfr's with chromosomal origins near *met, trp, rha, his,* and so on. This method is limited only by the number and position of sex factor integration sites in the bacterial genome, which, as can be seen from Fig. 4-4b, must be relatively numerous. Of course, the various kinds of Hfr's are probably not present in equal concentrations in any one F^+ population, and a considerable heterogeneity in "Hfr spectrum" is to be expected when various wild-type F^+ strains are compared. Mating experiments with purified Hfr strains, however, yield highly reproducible results. Although its significance is not yet clear, there is also some evidence that F^+ strains are capable of oriented transfer of their chromosome into F^- cells at very low frequency even when the sex factor is not integrated into the chromosome (21).

The *E. coli* mating types considered so far, F^-, F^+, and Hfr, do not exhaust all of the possibilities. New varieties of *E. coli*, discovered in Hfr populations by Jacob and Adelberg (22), are able to transfer a small segment of the bacterial chromosome to F^- cells with a frequency exceeding 50%. These strains appear to contain cytoplasmic episome consisting of both sex factor and several genetic loci normally found on the chromosome. The transfer of these loci thus proceeds with the same efficiency as that of the F factor alone in $F^+ \times F^-$ crosses. Strains of this type are said to carry a substituted sex factor, denoted F', and transfer of bacterial markers by them has been termed *sexduction* or *F-duction*.

Several Hfr strains carrying F' episomes have been isolated. Of these, the one best studied is F'-*lac*, whose episome contains the markers *lac*, *tsx*, *pho A* (structural gene for alkaline phosphatase), and *pho R* (one of two phosphatase regulatory genes). When the F'-*lac* strain is mated with normal F^- cells, transmission of the markers *lac*, *tsx*, *pho A*, and *pho R* occurs with very high frequency. All other cistrons are transferred in the usual Hfr sequence, but with only about 10% the frequency characteristic of Hfr's. In particular, the chromosomal *pho A* and *lac* regions enter the zygote last and with low efficiency when the classic F'-*lac* strain is used as a donor, for these markers lie at the distal end of the chromosome near the sex factor. It appears that F' cells arise from Hfr's by genetic exchange between the sex factor and chromosome, each of which contains a region of homology for the other. When the F' episome is integrated into the chromosome, the bacteria display normal Hfr behavior. In the cytoplasmic state, however, the episome is replicated autonomously and can be independently transferred to F^- cells at high frequency. A female which receives the F' episome becomes stably diploid for those genetic loci borne by the episome and itself becomes a F' donor. There are also physiological consequences for the partially diploid cells. For instance, the transfer of F'-*lac* means that there will be two copies of the *lac* region present in the recipient, one in the chromosome, and one in the episome, both of which are capable of directing the synthesis of the enzyme β-galactosidase. Such cells are able to produce enormous quantities of the enzyme, which sometimes reaches 8% of the total cellular protein by weight.

Among other episomal strains that have been isolated are F'-*pro*, which contains the proline locus, F'-*gal*, with loci for both galactose fermentation and λ prophage, and F'-*ara*, which bears the genes for arabinose utilization.

The selection of bacterial strains with substituted sex factors is carried out in two steps. By the method of replica-plating, Hfr cells are selected from F^+ cultures which contain the desired locus near the distal end of the chromosome. If episomes carrying *lac* are sought, it is necessary to choose Hfr's that transfer markers lying adjacent to *lac*, such as *thr* and *leu*, with high frequency at early times after the start of conjugation.

In such cases, it is likely that *lac* will be transferred late in conjugation and with low probability, therefore indicating its location at the distal end of the chromosome.

In the second stage of selection, cells are sought among the Hfr population which transmit the *lac* marker to female cells in a short time and at high frequencies, characteristics not typical of the particular Hfr strain chosen. Replica-plating is once again utilized to find these cells, which contain the F'-*lac* episome.

Factors determining sex in bacteria, as well as conjugation itself, are not limited to *E. coli*. Sexual phenomena occur in other enterobacteria such as *Salmonella* and *Shigella*. Moreover, it is possible to cross an *E. coli* Hfr with a *Salmonella* F$^-$ strain and obtain recombinants, although at low frequency. Even more surprising is the fact that sexduction proceeds with high efficiency not only from *E. coli* to *Salmonella* and *Shigella*, but also from *E. coli* to *Serratia marcescens*, a bacterium belonging to another family (23).

The DNA's of *E. coli* and *Serratia* can be easily separated from each other by density gradient centrifugation, since they differ substantially in nucleotide base composition. Guanine and cytosine residues comprise 58 mole % in *Serratia* and only 50 mole % in *E. coli*. When DNA is isolated from a strain of *Serratia* containing an *E. coli* episome and centrifuged in a density gradient, two distinct and independent bands are formed, each corresponding to the DNA composition of the two parental strains. Since this situation is observed even after many generations of growth, it is concluded that the episome is never integrated into the *Serratia* chromosome; rather, the episome is capable of continuous self-replication independent of, but in synchrony with, replication of the host chromosome.

Let us now consider the genetic definition of the often used term, *cistron*, which we have employed to identify the chromosomal unit of functional activity. Each cistron governs the synthesis of one specific polypeptide chain, to be sure, but the concept of the cistron can also be defined in purely genetic terms. Although the original formulation of this idea by Benzer (24) pertained specifically to bacteriophage systems, it can be generalized to include phenomena that are observed in bacterial zygotes.

Imagine a zygote or diploid cell in which there are two mutational lesions, *A* and *B*. If the two are very close together, they may both fall within a single elementary genetic region. In this case, the zygote remains viable if the two mutations are in the *cis* orientation relative to one another, that is, if they are on the same chromosome, for the unimpaired homologous region of the second chromosome can serve as a source of "good" genetic information (see Fig. 4-5). But if mutations *A* and *B* lie *trans* to one another, the zygote is not viable since the products specified by the two chromosomes do not *complement* each other. The boundaries of an elementary genetic region, or cistron, are defined by such relationships. If *A* and *B* occur in different cistrons, then the zygote will be viable regardless of whether the mutations are *cis* or *trans*, because there will always be an undamaged allele present, and the synthesis of the corresponding proteins can proceed unhindered. In this way, the concept of the cistron fully coincides with an elementary unit of functional activity, even though it may be inferred from an experiment relating to the viability of a diploid cell. The necessity for a strict definition of the cistron becomes evident in the relatively frequent cases where the synthesis of a single metabolite requires the participation of many enzymes, each catalyzing one step in a sequential process. Inactivation of any of the enzymes could

6. Sex in Bacteria; Conjugation

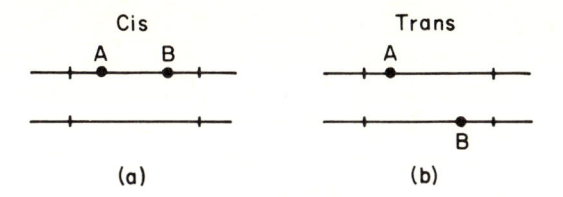

Fig. 4-5. Genetic definition of the term cistron. The letters *A* and *B* refer to two mutations, either of which renders the cell inviable. If both the mutations lie in the same chromosome, that is, if they are in the *cis* configuration with respect to one another (a), the zygote is still viable since the second chromosome is unimpaired and bears all the necessary genetic information. If the mutations lie in different chromosomes, *trans* with respect to one another as in (b), the zygote will be viable only if the mutations affect different functional units; in this case, the mutations are said to complement each other. When *A* and *B* affect the same functional unit or cistron, however, the zygote is inviable, since there is no intact copy of the genetic information present and the two mutations therefore do not complement each other.

lead to the same phenotype, although the particular cistron affected might be different in different mutants.

Complementation tests in bacteria require the production of strains that are stably diploid for the genetic loci to be studied. There are two general approaches to the insolation of diploids. In one procedure, a substituted sex factor which includes the relevant loci is introduced into a recipient carrying different markers at the same loci to form what is called a *heterogenote*. Heterogenotic F′ strains therefore provide a source of stable diploids, but only for limited portions of the genome. Bacteria diploid for much larger regions can be constructed by conjugation. In normal recipients, the donor chromosome will be fragmented as it undergoes recombination and the zygote will eventually segregate into pure haploid clones. But if a recipient is used in which recombination is suppressed, the chromosomal segment introduced by the donor can be replicated in synchrony with the recipient chromosome under certain conditions and the resulting merodiploids will be stable indefinitely.

As a hypothetical example of how a complementation test might be employed in bacteria, consider its application to the arginine pathway of *E. coli* which consists of eight separate stages, catalyzed by eight different enzymes. Mutations in any one of the cistrons governing the synthesis of these enzymes makes the cell an arginine auxotroph, unable to synthesize arginine from simple precursors. Suppose now that there are two mutants that have lost the ability to synthesize arginine and that we wish to know whether the mutations are in the same or different cistrons. The question may be resolved by a complementation test, provided the necessary diploids can be obtained. The two mutations will be assigned to different cistrons if the merodiploids are viable on arginine-free medium, because growth on that medium requires the cell to contain at least one undamaged allele corresponding to each enzymatic step. Complementation is in no way related to recombination, because two mutants with lesions in the same cistron can recombine so as to form viable progeny. The two phenomena are easy to distinguish, however, since complementation affects all the cells of a population and recombination only a small fraction.

Complementation tests in heterodiploid bacteria can be exploited for the study of still another important genetic concept: *character dominance*. In the diploid cells of higher organisms, phenotypic characteristics frequently reflect the properties of only one of the two homologous alleles, the donimant one. Although dominance is not completely understood in the genetics of higher organisms, it has a clear significance in the biochemical genetics of bacteria. If a bacterial heterogenote or merodiploid contains two alleles of the A cistron, A^+ and A^-, signifying ability and inability to synthesize a specific enzyme, the cell will always be able to synthesize the enzyme, because synthetic capacity is dominant over lack of that capacity, In other words, if the cell contains at least one cistron with the genetic information required to specify a given protein, its synthesis will not be affected by the presence in the same cell of a homologous, though impaired, allele.

Generally speaking, resistance to inhibitors is also a dominant characteristic in bacteria. If the target of inhibition is an enzyme, resistance may be manifested by a structural alteration in that enzyme which renders it insensitive to the inhibitor. In other cases, resistance is mediated by some new cellular component which either prevents entry of the inhibitor into the cell, or else destroys it before it has any harmful effects.

There are certain exceptions to this rule, however. Streptomycin resistance is not a dominant characteristic, but a *recessive* one. Streptomycin apparently acts at the ribosomal level by interfering with protein synthesis. The ribosomes of streptomycin-resistant (str^r) strains are not affected by the antibiotic. But in str^r/str^s merodiploids, both types of ribosomes are synthesized and the cell contains a sufficient number of sensitive ribosomes to block normal cell growth in the presence of the drug.

Cellular resistance to bacteriophage infection is almost always a recessive character, too, since resistance results from the loss of special receptor regions in the bacterial cell wall to which the phages are adsorbed. Recessiveness in this case is consistent with the loss of ability to synthesize certain proteins.

7. Replication of the Bacterial Chromosome during Vegetative Growth and during Conjugation

When bacteria divide, each daughter cell must receive at least one complete chromosome, or in other words, one complete set of instructions for synthesizing its essential components. Sexually differentiated bacteria transmit their mating properties to progeny cells as well, and in F^+ and F' strains, the episome must be autonomously replicated along with the chromosome. A sensitive regulatory mechanism is therefore needed to keep chromosome and episome replication apace of cell division. Moreover, the transfer of genetic material from donor to recipient during conjugation also requires replication. The nature of replication in vegetative growth and in conjugation, as well as the relationships between these two processes, will be the subject of the present section.

Replication of the bacterial chromosome during normal growth has been studied in a very graphic manner by Cairns (25, 26). A strain of *E. coli* unable to synthesize its own thymine was grown on a medium containing tritium-labeled thymine, which was incorporated into the bacterial DNA. After the cells were broken open, their contents were smeared onto a photographic emulsion, and the radioactive chromosome was observed by autoradiography. Since each tritium decay usually yields one grain of

7. Replication of the Bacterial Chromosome during Vegetative Growth and during Conjugation

silver upon development of the film, the chromosome appears as a fine black line, consisting of many separate granules (Fig. 4-6). During growth, radioactive thymine was administered for various time intervals, none of which exceeded the generation time of the culture. In this way, it was possible to follow continuously the increase in chromosomal radioactivity as a function of time until the completed molecule formed a closed loop with a total contour length of 1.2 mm, equivalent to a molecular weight of 3×10^9 daltons. Under these conditions, both the F^- and Hfr chromosomes appear to be continuous closed circles. This implies that the opening of the circle into a linear molecule in Hfr strains must occur at the moment of conjugation.

Figure 4-6 shows the autoradiogram of a DNA molecule whose replication is not yet complete (see diagram in upper right corner). The picture is particularly clear because the maternal chromosome XAYC contains only half as much tritium per unit length as the incomplete daughter chromosome YBX owing to the fact that only one chain of the maternal DNA was synthesized in the presence of radioactive thymine. However, over a small region of the maternal chromosome, lying between X and C, both DNA chains are labeled. This means that when the radioactive precursor was introduced into the medium, synthesis of the maternal chromosome had been completed only to the point marked C. In the presence of radioactive thymine, synthesis was continued to the point X and a new cycle of replication was initiated in which the second strand of the maternal chromosome XAYC was formed. Finally, another cycle was begun which resulted in the partially completed daughter molecule whose strands were both labeled. The point designated by X is identified as the place at which replication begins and ends. The new synthesis was proceeding in a counterclockwise direction and Y marks the growing point at the time the cells were broken open.

The location of X, the point on the bacterial chromosome where replication is initiated, has been tentatively established in an experiment carried out as follows. It is possible to label the origin of the chromosome, that is, the region immediately following X, by exposing a synchronized culture of Hfr cells to a short pulse of the heavy pyrimidine, bromouracil, which quantitatively replaces thymine in the DNA strands. Since bromouracil makes the DNA into which it is incorporated more dense, the fragment at the origin of the chromosome can be distinguished by CsCl density gradient centrifugation. But in order to orient the chromosomal segment first replicated relative to the genetic map, it is necessary to determine what genetic markers are borne by the labeled DNA fragment. Generalized transduction with phage P1 can be used for this purpose (see p. 397f). Transducing phages grown on the pulse-labeled Hfr culture will carry segments of bacterial DNA from all regions of the chromosome, but they can be fractionated according to density by centrifugation in CsCl. This operation simultaneously fractionates the phages according to bacterial genetic markers since only those phages carrying initial segments of the chromosome will be distinguished by greater density. Infection of appropriately marked bacteria with the dense transducing phages and subsequent selection of recombinants demonstrated that the origin of replication X lies between *lys* and *xyl* in both Hfr and F^- cells and that the direction of replication is clockwise relative to the genetic map illustrated in Fig. 4-1 (27).

A second intriguing question concerns the structure assumed by double-stranded DNA molecules in the vicinity of Y where replication of the chromosome is taking place. Evidently, the helical secondary structure of the DNA cannot be maintained at

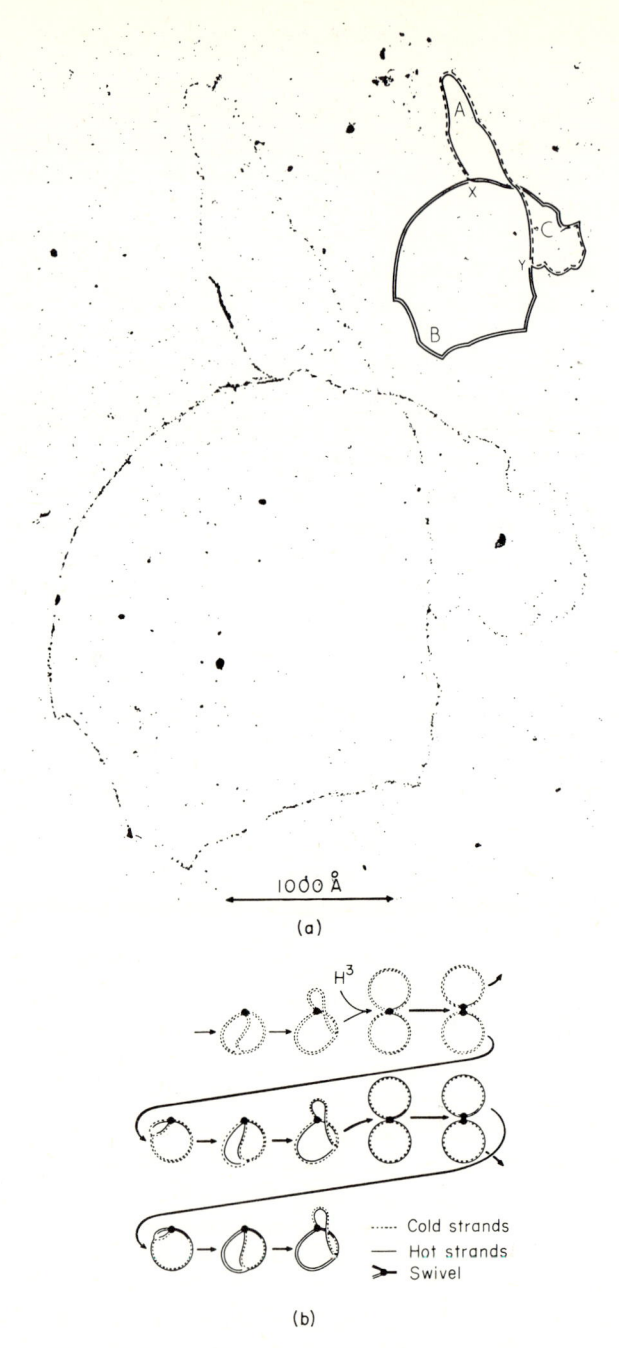

Fig. 4-6. Replication of the *E. coli* chromosome. (a) Autoradiograph of the chromosome of *E. coli* K12 Hfr labeled with ^{3}H-thymine for two generations before extraction. Inset: The same structure is shown diagramatically and divided into three sections (A, B, and C) that arise from the two forks (X and Y). (b) A model showing two rounds in the replication of a circular double-stranded chromosome. Labeled thymine is added as indicated at some arbitrary point in the cycle. It is assumed that replication starts at the same place each time and proceeds in the same direction (here counterclockwise). The final frame in the series corresponds to the stage at which the autoradiograph was made. From Cairns (26).

this point. It has been suggested that a loop of denatured DNA forms in the region of Y and that the DNA polymerase molecule fits into the segment over which hydrogen bonding is disrupted. The loop then gradually travels along the chromosome with the point of replication. Experimental support has been adduced for this proposal by briefly treating a synchronized culture of *E. coli* cells with the strong mutagen, N-methyl-N'-nitro-N-nitrosoguanidine. When exposure to the mutagen was carried out at different times in the division cycle, the occurrence of mutants with lesions in any given locus was found to be a periodic function of time. Furthermore, the period was equal to the division time. A natural explanation of these observations is that the mutagen acts most strongly in the region of DNA replication, which advances along the entire chromosome within one cycle of cell division (28). The origin and direction of replication could thus be confirmed in the same experiment by relating the most frequent mutants produced during different time intervals with their position on the genetic map. This mutagen was independently shown to evoke mutational alterations in isolated bacterial DNA only if it is first heated almost to its denaturation temperature (29). From these two lines of evidence, we may justifiably conjecture that the region of the chromosome undergoing replication has the characteristics of denatured DNA.

A third problem surrounding replication of the chromosome concerns the fact that both antiparallel strands of DNA are copied simultaneously. However, the enzyme DNA polymerase is known to catalyze synthesis of deoxyribopolynucleotides only in the 5'- to 3'-direction, that is, by adding nucleotide residues to the 3'-end of the DNA chain. No other DNA polymerase has been found in the cell to account for synthesis of the complementary strand in the 3'- to 5'-direction. We must therefore find out how both strands can be synthesized at the same time, but in opposite directions. A preliminary answer to this question posits that both DNA strands are copied toward one another discontinuously over small segments of the chromosome, perhaps within the limits of the denatured replicative loop alluded to in the preceding section, and that the newly synthesized segments are then joined together. This explanation was suggested by the following experiments (30). A short pulse of phosphate, labeled with the radioactive isotope ^{32}P, was administered to a culture of growing cells, actively synthesizing DNA. The pulse was terminated by addition of ice and KCN to a final concentration of 0.02 M. DNA was next extracted and purified in the usual fashion, and its sedimentation properties were studied at alkaline pH (0.1 N NaOH). Under these conditions, DNA is chemically stable, but the two strands separate because hydrogen bonds are broken. After pulses of 5 to 30 seconds, the radioactive label was found to be concentrated in comparatively short segments of DNA with molecular weights between 300,000 and 600,000 daltons, corresponding to 1000 to 2000 nucleotides. If exposure to radioactive phosphate was carried out for 40 seconds or longer, only a portion of the radioactivity incorporated into DNA was observed in these relatively small polynucleotide segments.

Synthesis of the entire *E. coli* chromosome according to the above scheme would require the linking together of from 5000 to 10,000 individual segments during the course of replication. This is possible owing to the presence of *polynucleotide ligase*, or joining enzyme, in bacterial cells which can join double-stranded DNA segments together via phosphodiester bonds (see p. 254f, 337). Thus the synthesis of DNA strands in the form of short segments would not cause any particular difficulty for the cell.

Temperature-sensitive mutants with a lesion in the ligase gene provide a good control for this hypothesis. Such mutations are conditionally lethal in that they do not affect growth at 30°C, but cause cell death at 42°C. An examination of DNA synthesis in these mutants at 42° demonstrated that significant quantities of DNA fragments with molecular weights of 300,000 to 600,000 are accumulated. This indicates that proper replication of the chromosome is prevented when ligase is inactivated by heat. These last experiments were performed very recently and still await independent confirmation. It is nonetheless useful to introduce such preliminary results because of their importance for an understanding of chromosomal replication.

DNA replication is fundamental to the process of conjugation as well as to normal cell division. But while vegetative DNA synthesis begins at a definite point on the chromosome, independent of the sex of the bacterium, replication in the Hfr during conjugation is initiated at the site of sex factor integration. The events taking place during conjugation may be envisioned in the following manner. The bacterial chromosome, or episome, receives a signal when the conjugal bridge is formed with another cell. If the donor is an Hfr, a new cycle of replication is begun near the sex factor locus at that moment, and synthesis then proceeds in a unique direction around the bacterial chromosome. One pre-existing DNA strand enters the protoplasmic bridge and is pushed into the recipient where the complementary strand is produced. In F^+ cells, the DNA of the sex factor is replicated autonomously and forced through the conjugal bridge in a similar fashion. One virtue of this model is that it accounts for the retention of donor properties by both F^+ and Hfr bacteria after conjugation is completed. Since the DNA of the male must be replicated in order for transfer to occur, the donor always retains a copy of any genetic information passed on to the recipient cell, including the sex factor, whether it is in free or integrated form.

The idea that replication of the chromosome or episome is necessary for transfer of genetic material from donor to recipient has one very evident consequence. The process of conjugation should be accompanied by DNA synthesis. In fact, when DNA synthesis is inhibited with the specific poisons phenethyl alcohol or nalidixic acid, conjugation is interrupted and the entry of distal markers is affected in particular.

A direct experiment showing that DNA synthesized during conjugation is observed in the recipient cell was performed in the laboratory of the author (31). Hfr and F^- cells were grown up individually on normal medium and then were permitted to conjugate in the presence of radioactive phosphate. After 20 minutes, the radioactivity was diluted by a factor of 1000 with nonradioactive phosphate, and conjugation was continued. Any newly synthesized DNA would thus be pulse-labeled with radioactive phosphate. When conjugation was complete, the culture was frozen and stored for an amount of time comparable to the half-life of ^{32}P. Under these conditions, the merozygotes formed during conjugation undergo "radioactive suicide," that is, any segments of the DNA which contained radioactive phosphorus would be partially destroyed owing to the disintegration of ^{32}P nuclei. The results of this experiment are shown in Fig. 4-7. It is apparent that the yield of thr^+leu^+ recombinants is severely curtailed by radioactive suicide, while recombination at the *trp* locus, which enters the zygote much later than *thr* or *leu*, is affected to a significantly smaller extent.

The explanation of these results is quite simple. The *thr* and *leu* loci penetrate

7. Replication of the Bacterial Chromosome during Vegetative Growth and during Conjugation

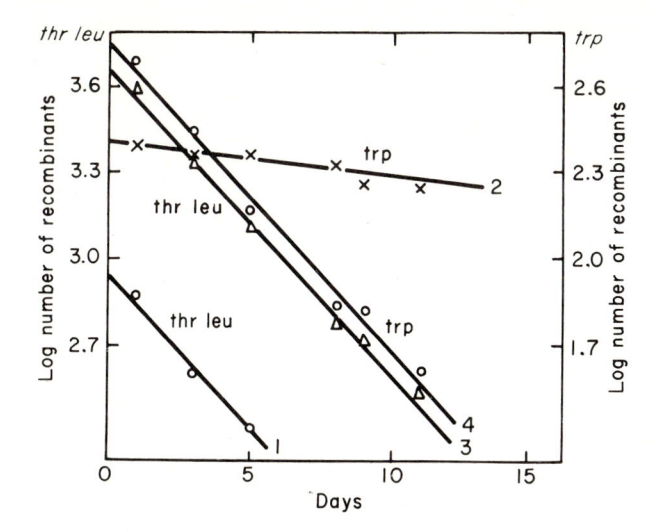

Fig. 4-7. Kinetics of radioactive suicide in recombinants subsequent to conjugation in the presence of radioactive phosphate. (1, 2) Kinetics of decay of thr^+leu^+ and trp^+ recombinants after a 20-minute pulse of $^{32}PO_4^{2-}$. (3, 4) Kinetics of decay of thr^+leu^+ and trp^+ recombinants after a 55-minute pulse of $^{32}PO_4^{2-}$. From Blinkova et al. (31).

the F⁻ cells 9 minutes after the start of conjugation, whereas the *trp* locus is transferred after 33 minutes. Consequently, the *thr* and *leu* loci were transmitted by radioactive DNA, but the *trp* locus was transmitted after the radioactivity of the medium had been diluted. Therefore, the *trp* locus remained stable to ^{32}P decay. If the time of incubation in the presence of radioactive phosphate is extended to 55 minutes, the *trp* locus is also radioactively labeled and the kinetics of its inactivation by "suicide" are identical to those of *thr* and *leu* (Fig. 4-7).

A scheme for the regulation of chromosomal replication has been proposed by Jacob and Brenner (32). The basic new concept which they introduced is that of the *replicon*, a genetic element consisting of DNA that carries the determinants necessary to initiate its own replication. According to this definition, sex factors, episomes, and chromosomes would all be classified as replicons. A DNA molecule that enters a cell during transformation would not be considered a replicon, however, since it cannot be independently replicated, but must first be integrated into the host chromosome.

The replicon hypothesis requires that DNA replication be initiated by a positive chemical stimulus, specific for each kind of replicon. According to this theory, each replicon carries two regulatory loci. One of them, the *initiator*, produces a specific cytoplasmic substance which may be thought of as an initiating enzyme. This enzyme recognizes a second locus called a *replicator*, which presumably consists of a specific sequence of bases in the DNA chain. The initiator might act by rupturing hydrogen bonds between particular sets of bases at the replicator site so as to separate the two DNA strands from one another. The resulting single strands can then serve as primers for new DNA synthesis. After initiation, synthesis continues until the entire replicon is copied, but reinitiation cannot occur without another stimulus from the initiator.

Existence of the initiator enzyme is suggested by the occurrence of temperature-sensitive F' mutants in which the episome cannot be autonomously replicated at 42°C. This function is not impaired at 30°C, however. These mutants may form a thermosensitive enzyme that is unable to initiate replication at the higher temperature. When such a temperature-sensitive F' factor is integrated into the chromosome, its thermolability disappears since the episome becomes part of the chromosomal replicon and its replication is mediated by the thermostable enzyme produced by the chromosome.

In order to explain the equipartition of genetic material between daughter cells, Jacob suggested that bacterial chromosome and unintegrated episome alike are attached to the inside of the cell membrane at the place where the two new cells draw apart. This would provide the bacterium with a mechanism for segregating daughter chromosomes into its progeny. Attachment of DNA to such a site in the bacterial cell was subsequently confirmed by electron microscopy of individual cell sections (33). Such a configuration could also explain the synchrony between replication and cell division, if the act of division itself triggers a new round of DNA synthesis. During conjugation, the site of attachment of episome or chromosome in the donor could define the place where contact with the recipient is established. Formation of the conjugal bridge might then supply the necessary signal for initiation of replication and transfer of either the F^+ episome or the Hfr chromosome, depending on the nature of the donor.

The replicon hypothesis of Brenner and Jacob adequately explains all of the fundamental observations relating to chromosome replication in bacteria, although some geneticists still object to certain of its features. Universal acceptance of the replicon hypothesis is impeded most frequently by ambiguity in the interpretation of experiments on conjugation. Since, as has been emphasized, the process of conjugation consists of four successive stages, it is often difficult to define the precise stage affected in studies employing metabolic poisons, temperature-sensitive mutants, or cells starved for an essential metabolite. While these uncertainties must be taken into account, the evidence presently available tends to favor the replicon hypothesis so far as the transfer of one DNA strand into the recipient is concerned.

8. Mechanism of Recombination

The physical events underlying recombinaton do not directly bear on the construction of the genetic map, but they are of great interest insofar as concerns an understanding of DNA function in relation to its structure. Let us focus our attention on recombination in the chromosomal segment which contains the loci *leu* and *str*. During conjugation, the male introduces its chromosome into the female cell, and a partially diploid zygote, or merozygote, is usually formed. Relevant fragments of the parental chromosomes are shown schematically in Fig. 4-8 as they might exist in the merozygote. A recombinant of the genotype $leu^+ str^r$ makes use of the upper, paternal chromosome to obtain the leu^+ character, and the lower, maternal chromosome as the source of its str^r property.

One mechanism by which recombination might occur, called *reciprocal crossing-over*, requires that two complete chromosomes actually exchange strands with one

8. Mechanism of Recombination

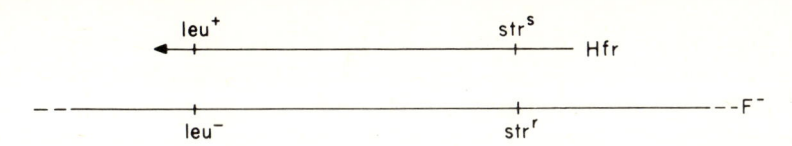

Fig. 4-8. Donor and recipient chromosomal segments in a merozygote. The symbols indicate that the donor (Hfr, above) is sensitive to streptomycin (str^s) and capable of synthesizing the amino acid leucine (leu^+); the recipient (F^-, below) is resistant to streptomycin (str^r) and unable to synthesize leucine (leu^-).

another at a certain point, x, lying between the *str* and *leu* loci (Fig. 4-9a). For this to take place, both chromosomes must break at x and the ends must rejoin so that the upper segment of the left-hand chromosome is attached to the lower segment of the right-hand segment, and vice versa. The zygote would then divide into two haploid cells, one of which would have the desired $leu^+ str^r$ genotype, and the other, $leu^- str^s$.

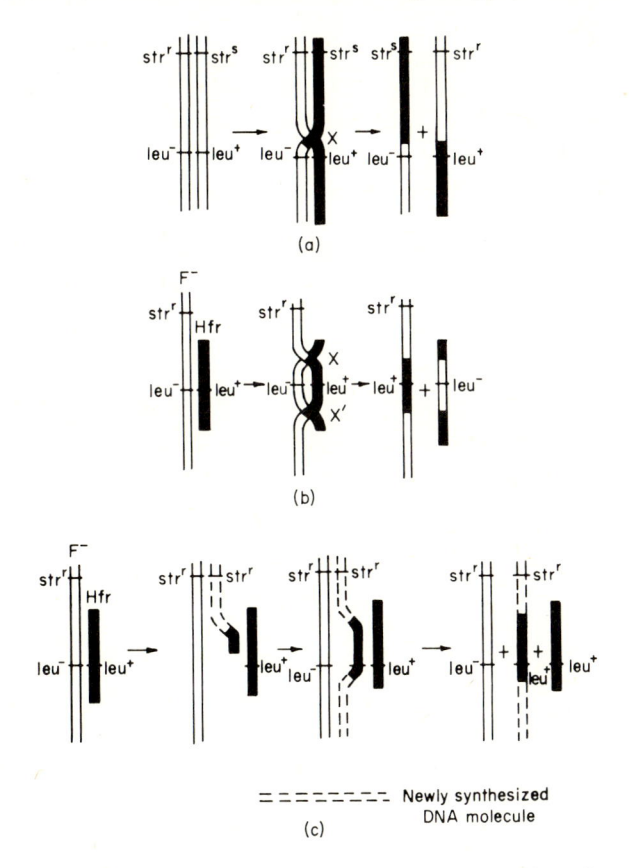

Fig. 4-9. Three models of genetic recombination. (a) Breakage and reunion: Single crossover involving a reciprocal exchange of homologous segments between homologous chromosomes. Replication of DNA is not required. (b) Breakage and reunion: Double crossover in which a segment of the Hfr chromosome replaces a homologous segment in the F^- chromosome. DNA replication is not required. (c) Copy-choice: during chromosome replication, copying mechanism alternates between Hfr and F^- parental chromosomes as templates. Adapted from Jacob and Wollman (2).

Reciprocal crossovers of this type have been shown to occur in higher organisms that reproduce sexually.

If single crossing-over were operative in bacteria, then both reciprocal types of recombinants should be found in equal quantities after the original diploid segregates into two haploid cells. This is not observed in bacteria, primarily because the entire male chromosome is very rarely transferred in conjugation. Instead of complete zygotes, merozygotes are formed in which maternal and paternal genomes are represented unequally. Furthermore, reciprocal recombinants may be difficult to identify since it is often true that only one allele at a given locus can be readily selected for in terms of its physiological characteristics.

In another possible mechanism, a segment of the male chromosome is inserted into the female chromosome in place of the homologous female segment by *breakage and reunion* (Fig. 4-9b). This can be viewed as the result of two crossover events which occur simultaneously within the limits of the diploid region. The reciprocal product is lost and eventually degraded.

A third mechanism of recombination was suggested by Lederberg (34). When chromosome replication occurs in the merozygote prior to cell division, copying alternates between maternal and paternal chromosomes. This mechanism, called *copy-choice*, is depicted in Fig. 4-9c, where the dashed line indicates the newly synthesized chromosome which will be passed onto the recombinant daughter cell. No parental DNA should be found in the recombinant chromosome according to the copy-choice hypothesis. Since alternations between the two chromosomes are presumed to be random, the probability that a copy-choice transition will take place in the region between *leu* and *str* is proportional to the distance between the two loci. However, the probability of recombination by actual physical exchange of genetic material is also proportional to the distance between loci if it is assumed that the frequency of crossing-over per unit length of the chromosome is a constant.

The results of investigations into the mechanism of recombination in bacteria favor the double crossover model involving breakage of both maternal and paternal chromosomes and subsequent reunion of their fragments in new combinations. In one experiment, Siddiqi analyzed the appearance of radioactively labeled maternal DNA in recombinant *E. coli* cells (35). The Hfr's used in this work were resistant to bacteriophage T6 (tsx^r) and sensitive to streptomycin (str^s). The F$^-$ cells, conversely, were sensitive to T6 infection (tsx^s), resistant to streptomycin (str^r), and were also thy^-, that is, they lacked the capacity to synthesize thymine, an essential component of DNA. Since they were unable to elaborate thymine endogenously, the thy^- cells were dependent on an external supply of this precursor. The DNA of these cells could thus be readily labeled prior to mating by growth in the presence of radioactive thymine. Hfr cells were then mated with females containing radioactive DNA, and time was allowed for the formation of recombinants. The total number of F$^-$ cells was determined by plating the exconjugants on agar with streptomycin. At the same time, $tsx^r str^r$ recombinants in the F$^-$ population were assayed on the same medium after lysing sensitive cells with phage T6. The $tsx^r str^r$ cells represented genuine recombinants and not merodiploids, for the later would have been lysed by phage owing to the dominance of sensitivity. Such recombinants comprised 6% of all the females, and accounted for 5% of the total DNA radioactivity initially present in the F⁻ population. This experiment demonstrates that most, if not all, of the maternal chromosome is transmitted to its recombinant progeny, and therefore argues against the copy-choice mechanism. Furthermore, it was shown in the laboratory of the author that a fragment of the paternal chromosome actually does become established in the DNA of the recombinant as a result of genetic exchange (36).

We conclude that recombination occurs by insertion of a portion of the male chromosome into that of the female. A prime requirement for this mechanism is that homologous regions of the maternal and paternal DNA be able to recognize one another so as to bring about synapsis. The only type of molecular forces capable of causing the two DNA chains to pair over regions of homology are hydrogen bonds between complementary bases. Since both recombining chromosomes consist of double-stranded DNA helices, synapsis must require that single-stranded regions in the DNA must be enzymatically produced. A detailed scheme describing synapsis and recombination in presented in Fig. 4-10.

The first step in recombination occurs when an endonuclease introduces some double-stranded breaks along the duplex macromolecule. This reaction may well limit the overall rate of recombination. An exonuclease then begins to hydrolyze one of the strands in the 3'- to 5'-direction from the point of the initial break and DNA fragments with single-stranded ends result. Single-stranded segments from different molecules can recognize each other and synapse if they are homologous, that is, if they contain complementary base sequences. A DNA polymerase next begins its work, adding nucleotide residues to the remaining single-stranded regions of the chain with the second strand as template. Finally, the single-stranded nicks between neighboring residues at the ends of the newly synthesized segments are joined so that the recombinant macromolecules are covalently bonded throughout. The enzyme catalyzing formation of the concluding phosphodiester bonds, called *polynucleotide ligase*, has been discovered in normal *E. coli* cells as well as in those infected with bacteriophage T4 (38). This enzyme is the last in a series of at least four which together accomplish recombination.

An interesting special case of recombination is the integration of the F'-*lac* episome into the chromosome. Scaife and Gross (39, 40) suggested that the episome is a closed circular structure similar in form to the chromosome itself. Indeed, the circularity of DNA from F' episomes has recently been confirmed by sedimentation in alkaline medium (41) and by electron microscopy (42). During synapsis, the episome and chromosome interact through their region of homology which in this case happens to be the *lac* region. Recombination occurs by crossing-over near this locus as illustrated in Fig. 4-11, and the first circle is simply inserted into the second. Note that this model requires but a *single* crossover. The suggested mechanism may be tested experimentally since it implies a very definite result. There will be two *lac* regions present in the newly formed Hfr chromosome, one of which will be the first (proximal) marker to enter the female during conjugation, while the other will be the last (distal) marker transferred. The two loci can be distinguished if the chromosome initially carries the *lac*⁺ allele, and the episome, the *lac*⁻ allele. It can be proved that two types of Hfr are formed upon integration of the episome, corresponding to variants

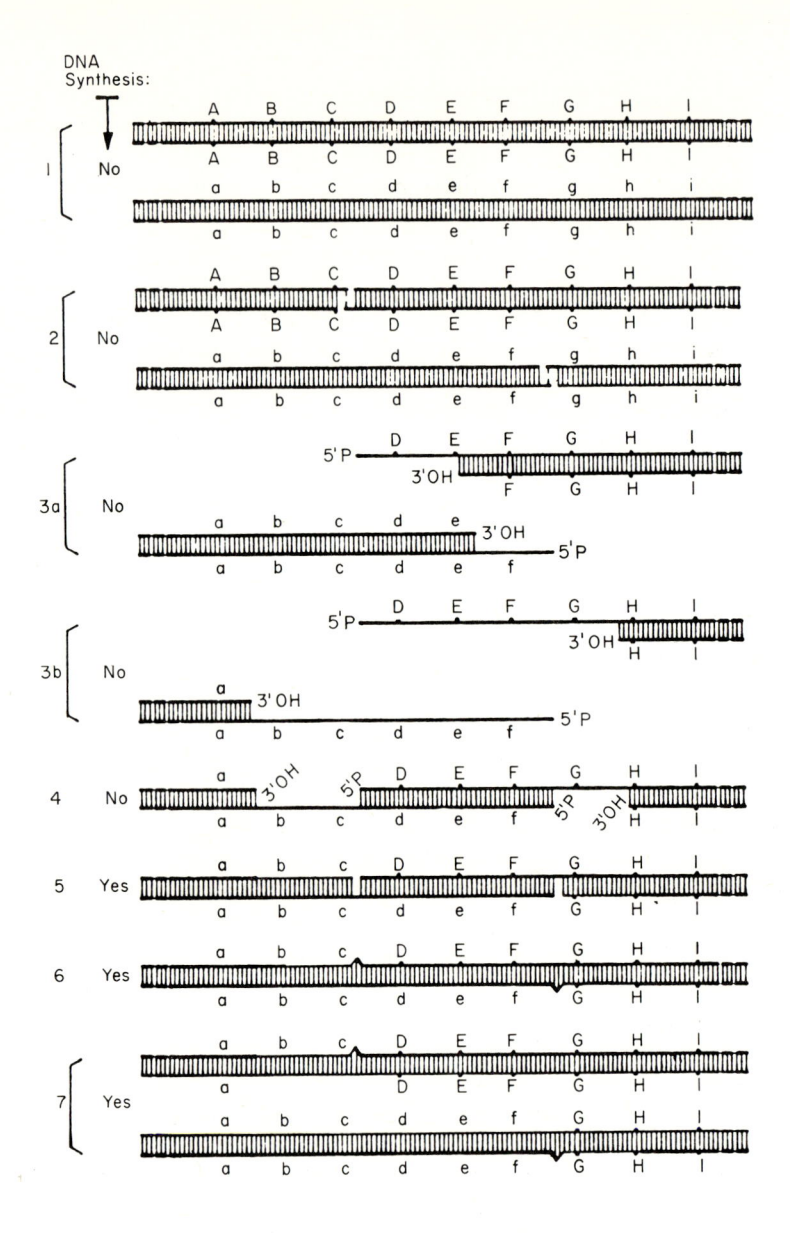

Fig. 4-10. Schematic representation of genetic recombination at the molecular level. (1) Two parental double-stranded DNA molecules, represented by heavy double lines, one with genetic markers indicated by upper case letters, and the other with markers in lower case letters. (2) Double-strand breaks are introduced into the DNA duplexes by an endonuclease. (3) The 3′-ends of single strands are degraded by an exonuclease. (4) Single strands with complementary 5′-ends overlap and pair. (5) DNA repair polymerase inserts nucleotides complementary to those of the opposite strand, starting from the 3′-ends and filling the wide gaps; newly synthesized strands indicated by thin lines. (6) Closing of final 3′-OH–5′-P link (kinked line) by polynucleotide ligase or joining enzyme. (7) Replication of partially "heterozygous" DNA molecule with production of two "homozygous" recombinant duplexes. From Szybalski (37).

338

8. Mechanism of Recombination

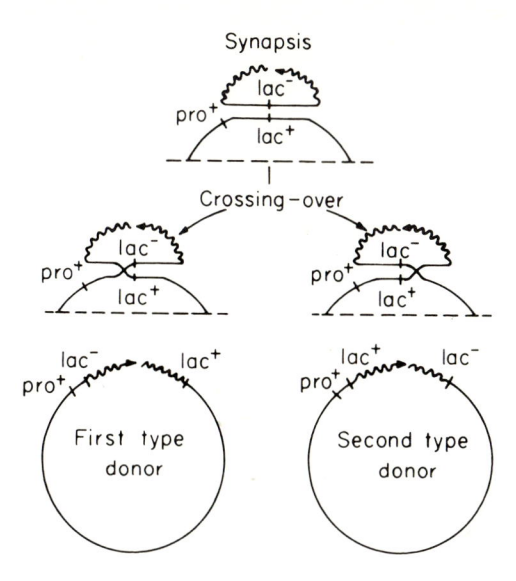

Fig. 4-11. Alternative models for the integration of the F'-*lac* episome into the Hfr chromosome. The genetic material of the sex factor is indicated by a wavy line, and that of the chromosome by a straight line. Arrow represents leading extremity of structure transferred during conjugation. From Scaife and Gross (40).

1 and 2 in Fig. 4-11. In one of the variants, the *lac*⁺ allele enters the female first, and in the second, *lac*⁻; in both cases, the proximal *lac* allele will be closely linked with *pro* (proline), as shown in the scheme. This result has been independently confirmed by Jacob.

The experiments described above provide weighty support for the suggestion that the F' episome is integrated into the chromosome by crossing-over. There is also evidence that the F factor itself becomes integrated into the chromosome of an Hfr by the same process. This conclusion was inferred from the finding that F'-*lac* can be inserted at many alternative sites over the entire *E. coli* genome (43). The region of homology in most cases is apparently defined by the DNA sequence of the F factor. The resulting bacteria bear a second *lac* region derived from the episome in one of several positions, each of which represents a *translocation* of genetic material relative to the normal *lac* region. The F factor alone is sometimes integrated within a genetic locus; this can alter the locus, thereby providing a means of specifically selecting for such events. This mechanism of integration may well be applicable to other episomes as well, such as prophages. In a later section (p. 395f), additional arguments will be adduced in favor of this hypothesis. In principle, the integration of an episome in this manner should be fully reversible. In the case of F'-*lac*, a loop would be formed due to homology between two regions of the chromosome, and then excised by crossing-over.

An aspect of limited DNA synthesis in the cell distinct from normal replication of the chromosome is the repair of DNA segments damaged by exposure to radiant energy or treatment with specific chemical agents. For instance, the primary effect of ultraviolet light on DNA is to cause dimerization of adjacent thymine bases via a cyclobutane ring:

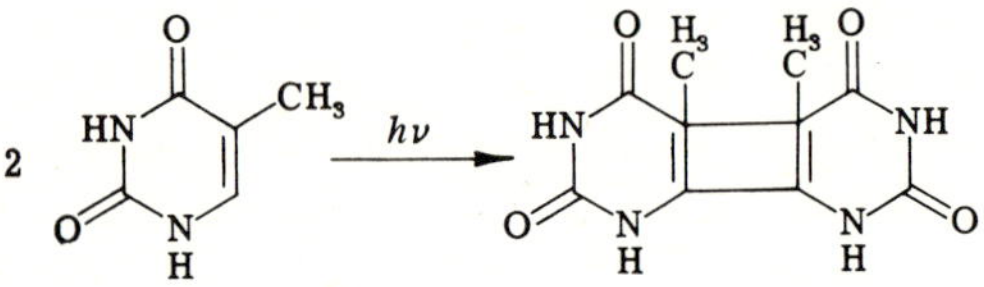

In the wild-type strain there is a special enzyme that excises a single-stranded portion of the chain containing the dimer, whereupon the undamaged complementary strand is converted to a duplex by the combined action of DNA polymerase and ligase.

A number of mutants have been isolated which can no longer carry out recombination or repair of damaged DNA. Certain of these strains have sustained lesions in cistrons specifying enzymes common to both recombination and repair, while others affect only the repair process. Mutants of the first class, independently discovered by Howard-Flanders and Theriot and by Clark and Margulies (44) are designated *rec⁻*. Besides their inability to perform recombination, the *rec⁻* mutants also exhibit heightened sensitivity to ultraviolet light and x rays. These strains apparently lack one of the enzymes essential for recombination. The second class of mutants is unable to repair damage caused by ultraviolet light and alkylating agents, although its capacity for recombination is not lost. Such *uvr* (ultraviolet reactivation) negative mutants are deficient in the enzyme that excises the damaged segment from the DNA duplex. Although there is still relatively little known about these processes they are nonetheless of very great interest.

9. Genetic Mapping by Conjugation

In order to construct a genetic map such as the one illustrated in Fig. 4-1, it is necessary to amass a very large amount of experimental data on linkage among various markers and groups of markers. There are three general ways in which conjugation and the analysis of recombinants resulting from bacterial crosses can be used to map the relative order of genes in the *E. coli* chromosome and to estimate the relative distances between them. Two methods are based on the particular circumstances surrounding chromosomal transfer during conjugation: mapping by the *time of entry* of markers into the zygote and mapping by the *gradient of transmission* of markers to the zygote. These two techniques are useful in establishing linkage between makers separated by considerable distances on the chromosome, but they are not sufficiently accurate for fine structure mapping. Furthermore, both of these approaches are independent of recombination frequency, although they employ the recovery of recombinants to demonstrate that a given marker has actually been transferred from the donor bacterium to the recipient. A third method involves mapping by *frequency of recombination* between two or more loci and it is most sensitive when applied to a relatively limited region of the chromosome. This last topic will be treated in detail in the following section.

All types of genetic analysis utilizing conjugation require an Hfr strain with one easily selectable dominant gene for reference in addition to the unselected markers to be mapped. These markers should all be located on the proximal half of the Hfr chromosome so that they will be transferred at high frequency. Hfr's with different origins can then be used to relate the positions of genes in other regions of the chromosome. The recipient strain must of course carry distinguishable alleles at the loci to be studied or else it would be impossible to determine whether or not recombination has occurred between the two parental chromosomes after conjugation.

Genetic mapping studies are greatly facilitated by the use of sensitive techniques for the selection of genetic traits among recombinant progeny; to this end, Lederberg and others proposed several fundamental guidelines. The most efficient arrangement is to select prototrophic recombinants formed in crosses of prototrophic males with auxotrophic females. In this case, as in the selection of revertants, large numbers of cells can be tested on a few minimal agar plates because the metabolically deficient recipients will not grow. But there must also be a way to get rid of the original Hfr donors which can grow on minimal medium as well as the prototrophic recombinants. This is generally accomplished by choosing an Hfr that is sensitive to an inhibitor or bacteriophage to which the recipient is resistant. If, after mating, the cells are spread on minimal agar containing the inhibitor or phage, only those recombinants will grow that have inherited the resistance of their female parents and the intact biosynthetic capacity of their male parents. Neither of the original parental strains will survive, however.

Proper selection thus requires the elimination of parental bacteria as well as those which derive from them by simple cell division, so that only the recombinants in which the genetic markers of both strains have been mixed will survive for further analysis. The recombinants must simultaneously receive at least one character from the donor bacterium and another from the recipient bacterium. Selection is therefore related to at least two different loci on the chromosome, and recombination has reference to events occurring in the chromosomal region lying between these two points.

To illustrate, consider two markers, capacity to synthesize leucine and resistance to streptomycin. The constitution of the male cells will be taken as Hfr $(leu^+ str^s)$, which in genetic terminology means a streptomycin-sensitive Hfr with the ability to synthesize leucine, and that of the females will be $F^-(leu^- str^r)$, that is, resistant to streptomycin, but unable to synthesize leucine. Selection will be carried out for the genotype $leu^+ str^r$:

$$\text{Hfr}(leu^+ str^s) \times F^-(leu^- str^r) \to F^-(leu^+ str^r)$$

In such a cross, most of the daughter cells will be females. The recombinants can be selected under extremely favorable conditions since the background of parental bacteria will be completely eliminated on a medium containing streptomycin but no leucine. Even if the probability of recombination between the two loci is very small, on the order of 0.01%, for instance, the occurrence of recombination can be readily detected and, if necessary, the actual recombination frequency can be measured with great accuracy. Under suitable conditions the efficiency of conjugation is about 10%, so that the demands placed upon the selective procedure require the identification of one cell in 10^5. Thus, 100 recombinant colonies can be obtained on a single petri dish out of a total of 10^7 cells plated. These are certainly very favorable circumstances!

Since Hfr bacteria transfer their chromosomes to recipient cells in a linear fashion, it should be possible to map the sequence of genes on the donor chromosome by measuring the times at which each one enters the F$^-$ cell and thereby becomes available for genetic exchange or recombination with the recipient chromosome. This problem was investigated by Wollman and Jacob in a series of classical experiments employing the method of *interrupted mating* between streptomycin-sensitive donors and streptomycin-resistant recipients carrying sets of allelic markers (45). In order to study the kinetics of chromosome transfer, mating bacteria were separated at various times during conjugation by subjecting them to intense agitation in a blender. This treatment was sufficient to break apart conjugating cells so that, on the average, only a certain proximal segment of the donor chromosome penetrated the recipients, but it did no damage to the zygotes already formed. By selecting for each of the donor characteristics individually after specific periods of conjugation, it was possible to determine when each of them was transferred to the recipient bacteria. If HfrH was used as a donor, the females were found to receive the *thr* and *leu* loci first, followed by a further sequence of donor markers whose extent depends on the time at which mating was interrupted. Figure 4-12 presents kinetic curves for the formation of

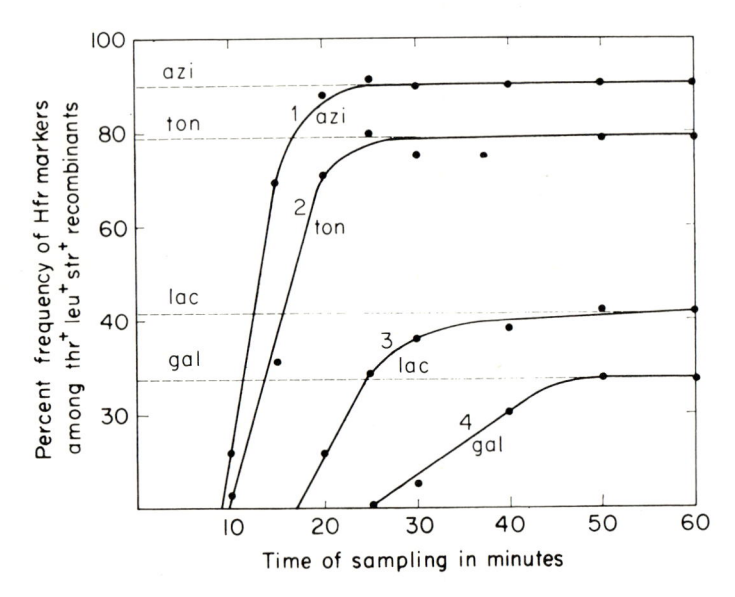

Fig. 4-12. Time sequence of transfer of different characters during conjugation. The distribution among *thr*$^+$*leu*$^+$*str*^{r} recombinants of the characters *azi*, *ton*, *lac* and *gal* from the Hfr donor is plotted versus time at which the mating pairs were separated by agitation in a blender (solid curves). The dashed lines represent the maximum proportion of recombinants recovered for the given marker. The ratio of Hfr to F$^-$ cells was set at 1:20 so as to prevent infertile conjugation between Hfr's. From Wollman *et al.* (46).

recombinants at the *azi*, *ton*, *lac*, and *gal* loci. Although the recovery of these markers is expressed as a percentage of recombinants for the most proximal markers, this procedure is not essential for the measurement of transfer kinetics. The entry of the *thr* and *leu* markers into the zygote from HfrH begins after an interval of 8 minutes, and the number of recombinants per Hfr reaches 20% after 20 minutes. Transfer of the *gal* marker begins after 25 minutes and the yield of *gal*$^+$ recombinants after an hour is about 4% per Hfr bacterium.

Interrupted mating can thus be used to define the relative orientation of genes on the chromosome, and even more information can be obtained from such experiments under the following circumstances. If the chromosome is transferred from the donor to the recipient at a constant rate, the time interval between the initial appearance of each pair of markers should represent the time taken for the corresponding segment of the chromosome to travel through the conjugal bridge. This would mean that the transfer time can be translated directly into distances between markers. Such a mapping technique would depend only on the transfer process and would be quite independent of chromosome breakage or recombination efficiency. The time scale of transfer was indeed found to be truly representative of physical distances on the chromosome by using a series of Hfr's that transfer the same chromosomal segment in different directions and from different origins. For markers transferred early in conjugation, the time interval between appearance of two given markers is always the same. The experiments of Wollman and Jacob therefore demonstrate that the time elapsed before a given marker enters the merozygote is a constant proportional to the actual distance between the chromosomal origin, *O*, and the corresponding locus. Since the segment *O-thr leu-gal*, constituting about one fourth of the total chromosome, is transferred in 25 minutes, it would require about 100 minutes for transmission of all the markers if the rate of transfer were constant throughout conjugation.

Mapping in time units is quite sensitive to environmental conditions. By reducing the temperature from 37° to 32°C, for instance, the time of transfer is more than doubled. Thus, interrupted mating experiments must be done under closely controlled conditions, and even then, it is not accurate for measurement of map distances on the order of 1 minute of transfer or less. On the other hand, this technique is extremely useful in determining linkage over large segments of the chromosome.

An independent method of mapping relatively long distances on the chromosome is made possible by the monotonic decrease or gradient in the frequency with which genetic markers are transmitted from Hfr to F$^-$ cells during chromosome transfer. The necessary experiment is performed by mating bacteria for a long period and then selecting recombinants which have inherited a certain *proximal* marker which is transferred by the donor early in conjugation. The frequencies with which various unselected markers *distal* to the selected marker occur among the resulting recombinants are then determined. As can be seen from the heights of the plateaus for each marker in the kinetic curves of Fig. 4-12, the relative proportions of the unselected markers are inversely related to their distances from the proximal reference marker selected. The gradient in marker transfer can be attributed to random breakage of the chromosome as it traverses the conjugal bridge. During conjugation, mating pairs are broken apart spontaneously and incomplete merozygotes are formed. The farther a given marker is from *O*, the greater the probability that the chromosome will be disrupted before that marker can be transferred. Even when all possible precautions are taken to ensure that the culture is not mechanically perturbed and that its temperature remains constant, the frequency of chromosomal breakage is only slightly reduced. Thus, mapping by the gradient of transmission, as by the kinetics of marker entry, is somewhat sensitive to environmental changes and must be carried out under standardized conditions.

The order of genes on the *E. coli* chromosome derived from interrupted mating experiments and from the gradient in frequency of marker transmission for HfrH are compared in Fig. 4-13. They are in good agreement with one another and with

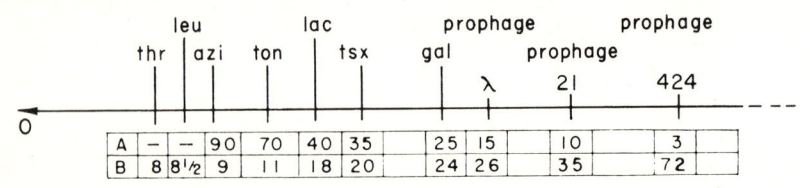

	thr	azi	ton		tsx		gal	λ	21	424
A	–	–	90	70	40	35	25	15	10	3
B	8	8½	9	11	18	20	24	26	35	72

leu / lac / prophage / prophage / prophage

Fig. 4-13. The proximal segment of the chromosome of HfrH bacteria. (A) The frequencies with which various characters of HfrH donors are found per hundred *thr⁺leu⁺str^r* recombinants formed in crosses of the type HfrH × F⁻. (B) The times at which the genetic determinants of these characters begin entering recipient bacteria in experiments of interrupted matings. O is the origin of the chromosome which enters the F⁻ cell first. From Jacob and Wollman (2).

map positions found by other techniques. While the figures on time of entry can be directly translated into distances on the chromosome, data on the gradient of transmission yields relative distances between markers only if chromosomal breakage is random. If this condition obtains, that is, if it is equally likely that the chromosome will be disrupted at any point during its transfer through the conjugal bridge, then the probability that a given marker will enter the zygote, $w(l)$, can be calculated as a function of its distance from the origin of the donor chromosome.

Consider a population of N conjugating cells, and suppose that n of them are still transferring unbroken chromosomes after a length equal to l has already penetrated the female cells. When l increases by dl, the number of males engaged in chromosome transfer will decrease by dn due to additional breaks. Since there is an equal probability of chromosomal breakage per unit length, χ, it is clear that:

$$dn = -\chi n dl \text{ and } n = Ne^{-\chi l}$$

whence: (4-30)

$$w(l) = n/N = e^{-\chi l}$$

This relationship has been verified experimentally, and the value of χ was determined to be 0.06 min⁻¹ under the most gentle mating conditions. Here genetic distance is expressed in units of time, which as we have seen, can be used to measure the length of the chromosomal segment transferred to recipient cells. Using the known value of χ, Eq. (4-30) can be used to determine map distances by the gradient of transmission method.

10. Genetic Mapping by Recombination Frequencies

The analysis of recombination frequencies constitutes a very precise tool for genetic mapping in haploid bacteria as well as in higher organisms. According to this method, the distance separating any two markers on the map can be represented in terms of the measured frequency of recombination between them, that is, the fraction of individuals resulting from a cross that acquires one of those markers from the female parent and the other from the male parent. By convention, one unit of map distance corresponds to a recombination frequency of 1%. Recombination frequencies can be directly converted into genetic map distances only when the markers lie within fairly limited regions of the chromosome. In studies on bacteria, the techniques discussed in the previous section must be used to properly position more distant markers relative to one another.

The basic principle of recombination analysis is independent of the physical mechanism by which the genetic markers are actually redistributed. This means that the map could be constructed in much the same way, regardless of whether recombination takes place by crossing-over or by copy-choice and, conversely, that mapping experiments do not provide information as to the precise mechanism of genetic exchange. In its simplest form, the "law of genetic recombination" states that W_{AB}, the probability of recombination between loci A and B, is proportional to the distance between them, Δl_{AB}, or that $W_{AB} = \alpha \Delta l_{AB}$, where α is a constant. To understand this important concept, consider not two, but three different genetic markers distributed along the chromosome in the order A-B-C. If the recombination frequency between A and B is denoted by W_{AB}, and that between B and C by W_{BC}, then the recombination frequency for the two outside markers, A and C, should be $W_{AC} = W_{AB} + W_{BC}$. In theory, this should apply to any three markers situated on one chromosome, but owing to certain aspects of recombination that will be discussed later in this section, its practical applicability is limited to rather short regions of the chromosome for which $W_{AC} < 1$.

One consequence that stems from the relationship between map distances and recombination frequencies is that the markers must be arranged linearly on the chromosome. This is an essential condition for additivity of recombination frequencies. Furthermore, different forms of the same cistron, such as *leu*⁺ and *leu*⁻—called *alleles* of the *leu* cistron—are considered to be equivalent in the recombination process. Thus one allele possesses no inherent advantages over another in its ability to be copied or integrated. If this were not so, the proportionality between map distance and recombination frequency would not be maintained and mapping would be considerably more difficult. Finally, it must be assumed that the factors in the cell which effect recombination act with equal probability at any point on the chromosome.

Several technical difficulties must be overcome before construction of the genetic map in bacteria by this method is possible. The main problem lies in the normalization of recombination frequencies. When higher organisms are mated, the number of zygotes is always known and the frequency at which recombinants with a specific genotype are recovered can be readily calculated from the ratio of such recombinants to the total number of zygotes. During bacterial conjugation, however, the number of zygotes formed is not precisely known so that special measures must be taken to normalize recombination data. This can be done by scoring multiple events such as quadruple crossovers that require the use of at least three genetic markers in each mating experiment. Although the number of zygotes is still unknown, the analysis of multiple crossovers provides a means of circumventing the problem. The difficulty of obtaining a sufficient amount of data of course remains, since the probabilities of multiple recombination events are usually very small.

A group of closely linked bacterial genes can be mapped by the analysis of recombination frequencies as follows. Conjugation between Hfr and F⁻ cells is carried out under strictly standardized conditions and recombinants *for the most distal locus* of the region of the genome under study are selected. This procedure automatically yields those merozygotes which are homogeneously diploid for the relevant region of the chromosome, that is, which have been penetrated by a segment of donor DNA carrying the markers to be mapped *proximal to the selected marker*. Unselected markers are then classified by replicating the viable clones into a series of plates containing appropriate nutrients, inhibitors, or carbon sources. Analysis of the unselected markers in each recombinant class permits identification of the regions of the chromosome in which crossovers occurred.

Small portions of the chromosome are used for genetic mapping or else recombination frequencies will not be linearly related to map distances. In fact, when the chromosomal segment under investigation is longer than two to three time units, the frequency of recombination approaches a limiting value of 50% and the resulting figure is not at all related to the distance between loci. The entire map can be constructed nonetheless by fitting together small overlapping segments in each of which the linkage among genetic markers has been determined individually.

A small segment of the *E. coli* chromosome surrounding the *lac* region is illustrated in Fig. 4-14, with the *pur E* locus (purine biosynthesis) at the right or distal end and

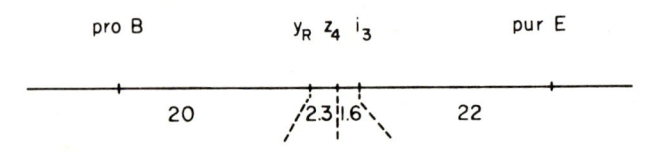

```
      pro B            y_R z_4 i_3              pur E
  ────────────────────┼─┼─┼─────────────────────
         20           2.3 1.6        22
```

Fig. 4-14. Linkage in the *lac* region of the *E. coli* genetic map. Distances between genetic loci represented in recombination units. From Jacob and Wollman (2).

the *pro B* locus (proline biosynthesis) at the left or proximal end. In between these two markers, which represent the limits of the segment under investigation, there are several other cistrons. Determination of the recombination frequency requires that there be an "internal standard" for normalizing the data, and this is precisely why selection is arranged so as to register two crossover events. Figure 4-15 provides a more detailed scheme of the genetic sytem employed in the analysis. Hfr cells bore mutations in the *lac z*, *y*, or *i* cistrons, governing the synthesis of β-galactosidase, galactoside permease, and *lac* repressor, respectively, but were otherwise prototrophic.

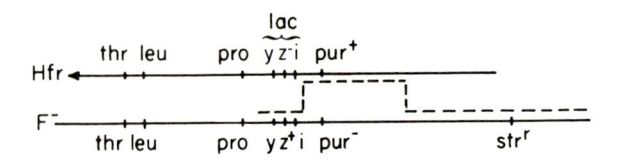

```
                            lac
           thr  leu   pro  y z⁻i  pur⁺
  Hfr ◄────┼────┼─────┼────┼┼┼────────────────
                              ┘     └──────────
  F⁻ ──────┼────┼─────┼────┼┼┼──────────────────
           thr  leu   pro  y z⁺i  pur⁻        str^r
```

Fig. 4-15. Measurement of recombination in the *lac* region. Diagram shows how data on recombination in the *lac* region are normalized to recombination between *pur E* and *str* loci. Dashed line traces the recombinational events (crossovers) which are demanded by the selective procedures (see also Table 4-3).

The F⁻ cells contained an unimpaired z cistron and in addition, were deficient in purine biosynthesis (pur^-) and resistant to streptomycin (str^r). The primary selection was performed for purine while the *lac z, y,* and *i* cistrons comprised the unselected markers which were subsequently studied by replica-plating. A typical mating and the recombinants selected from it can be represented as follows:

$$\text{Hfr}(z^- pur^+ str^s) \times F^-(z^+ pur^- str^r) \to F^-(z^+ pur^+ str^r)$$

It should be noted that three markers were used in the selection process: ability to metabolize lactose, ability to synthesize purines, and resistance to streptomycin. The probability of this complicated rearrangement is equal to the product of the recombination frequencies between z and pur over a small portion of the chromosome, and between pur and str over a relatively large distance:

$$W_{z\text{-}pur\text{-}str} = W_{z\text{-}pur} \times W_{pur\text{-}str}$$

The yield of recombinants is:

$$w_{z\text{-}pur\text{-}str} = k \times W_{z\text{-}pur\text{-}str}$$

where k is a coefficient which depends on the number of zygotes formed, that is, the probability of entry of a marker into the recipient cell, and on other factors relating to the mechanics of recombination. The number of $pur^+ str^r$ recombinants:

$$w_{pur\text{-}str} = k \times W_{pur\text{-}str}$$

is measured independently in the same experiment. Normalization is performed by dividing $w_{z\text{-}pur\text{-}str}$ by $w_{pur\text{-}str}$, which gives us the probability of recombination between *pur* and *z*. In this way, parameters such as the number of zygotes formed does not have to be separately determined.

Strictly speaking, the probability of recombination in the segment *O-pur*, that is, from the beginning of the Hfr chromosome to the *pur* marker, should enter into the number of $pur^+ str^r$ recombinants as a multiplicative factor. The formation of recombinant progeny with characteristics from the Hfr always requires at least a double crossover, insofar as the practical result is an insertion of a small portion of the Hfr chromosome into the complete F⁻ chromosome. Actually, multiple crossovers are likely, and if a segment comprises more than 1/20 of the entire *E. coli* chromosome, there will be an equal probability that even and odd numbers of multiple crossovers will occur within it. Since recovery of $pur^+ str^r$ recombinants requires that an odd number of crossovers take place in the *O-pur* region, the probability of such recombinants must be multiplied by 0.5. But a factor of 0.5 will also stand in front of the number of $z^+ pur^+ str^r$ recombinants for the same reason: the choice of one or the other of the two z alleles is equally likely if we consider all the recombinant progeny and not just z^+ individuals. Thus, the probability of no or an even number of crossovers in the *O-z* segment, required for incorporation of the z^+ character into the recombinant, is 0.5. Since the quotient of the two factors is unity, recombination between *O* and the region under investigation can be ignored in practice if the former is very large in comparison to the latter.

Figure 4-14 shows the distances in recombination units between various markers of the *pro-pur* region, and Table 4-3 gives the data on matings from which these distances were calculated.

TABLE 4-3

Mapping of the lac Segment[a,b]

			Crosses							
Hfr	y_R^-	z_4^+	i_3^+	pur^+	str^s		y_R^+	z_4^-	i_3^-	pur^+ str^s
F$^-$	y_R^+	z_4^-	i_3^-	pur^-	str^r		y_R^-	z_4^+	i_3^+	pur^- str^r
Region	1	2	3	4			1	2	3	4

Number of cross-overs	Percent recombinants	Genotype of pur^+str^r recombinant	Region of cross-over	Number of cross-overs	Percent recombinants	Genotype of pur^+str^r recombinant	Region of cross-over
1	22	z_4^-	2 or 3	1	26	z^+	2 or 3
1	2.3	$z^+ y^+$	1	2	1.9	$z^+ y^+$	2 or 3, 1
2	0.25	$z^+ i_3^-$	2, 3	1	1.6	$+ i_3^-$	2
3	0.05	$z^+ i_3^- y^+$	2, 3, 1	2	0.22	$z^+ i_3^- y^+$	2, 1

[a] Pur^+str^r recombinants were selected in the two reciprocal crosses between Hfr and F$^-$ cells indicated in top lines of table. The genetic constitution of more than 10,000 recombinants was analyzed by replica platings on suitable media and assay for enzyme activity. See also Figs. 4-14 and 4-15.
[b] Data from Jacob and Wollman (2).

One of the great successes of genetic mapping has been achieved in *fine structure analysis*, the determination of the relative positions of a series of mutations *within* a single cistron. The fine structure of the cistron directing the synthesis of the enzyme alkaline phosphatase (*pho*) in *E. coli* was studied by Garen and Levinthal (47). More than 100 mutations were obtained within the *pho* cistron, and the relative position of each altered site or *muton* was determined with great precision by means of a very large number of two-point crosses (Fig. 4-16).

To illustrate the analysis of genetic fine structure, let us suppose that we wish to determine the map position of *U24*, a new mutant in the *pho* cistron induced by ultraviolet light. Mutagenesis is usually performed on males because they can transfer their altered character to female cells during conjugation, while the reverse is not possible. The new mutant thus has the genetic Hfr(*pho$^-$ str^s*). In order to study its location on the map, the mutant is first crossed with the female strain F$^-$ (*pho$^+$str^r*), and recombinants are sought which carry the paternal *pho$^-$* marker. We have already discussed how these mutants are selected (p. 309).

Other *pho$^-$* males, whose mutations have been previously located on the genetic map, are now crossed with the *pho$^-$* female cells carrying *U24*: Hfr(*pho$_m^-$*) × F$^-$(*pho$_{U24}^-$*). For *pho$_m^-$*, it is natural to choose two mutations which lie near opposite ends of the cistron, such as *U3* and *U18*. If the positions of *pho$_m^-$* and *pho$_{U24}^-$* are indicated by j and k (Fig. 4-17), the probability of recovering *pho$^+$* recombinants, W_{jk}, will be proportional to the length of the *jk* segment. Now the selection of recombinants is performed on a medium devoid of inorganic phosphate and in principle, all maternal, paternal,

10. Genetic Mapping by Recombination Frequencies

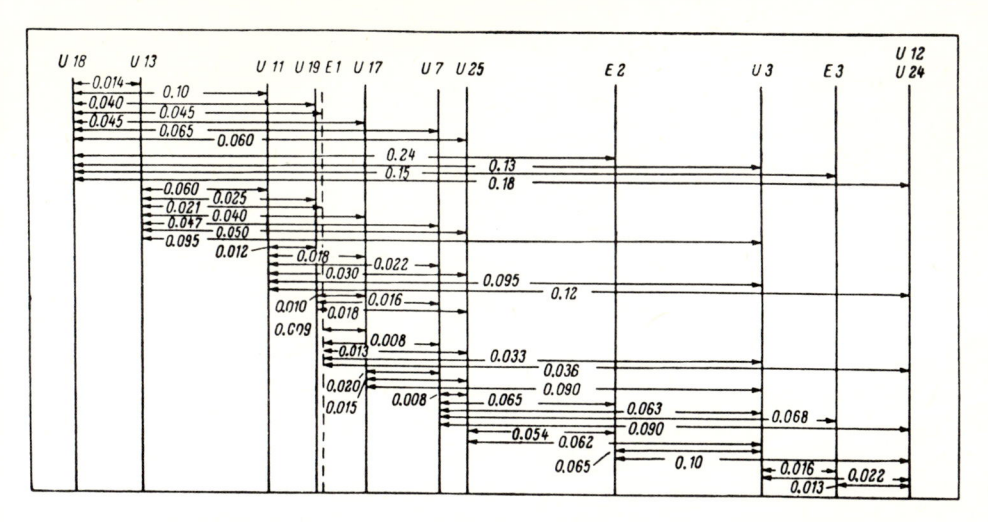

Fig. 4-16. Fine structure map of the alkaline phosphatase gene. Recombination frequencies are calculated as the percent pho^+ per thr^+leu^+ recombinants, both values being determined in the same cross (see fig. 4-17). Each recombination frequency represents the average of reciprocal crosses after correction for the "noise level" of spontaneous revertants. In general, the results from reciprocal crosses agreed to ±20%. Assuming the pho^- mutants to represent a random sample of mutable sites, the total size of the gene may be estimated to be about 0.3 map units. From Garen (47).

and daughter cells will be inviable except those in which recombination has taken place between j and k. But because W_{jk} is quite small, it is essential to exclude background events, such as spontaneous $pho^- \rightarrow pho^+$ reversion, which would obscure the scoring of recombinants.

In order to reduce the background or "noise level," a male is chosen with markers thr^+, leu^+, and str^s, and a female with thr^-, leu^-, and str^r. Recombinants are thus selected for four characteristics:

$$\text{Hfr}(pho_m^- leu^+ thr^+ str^s) \times \text{F}^-(pho_{U24}^- leu^- thr^- str^r) \rightarrow \text{F}^-(pho^+ leu^+ thr^+ str^r)$$

that is, they must be able to grow on streptomycin plates not containing threonine, leucine, or inorganic phosphate. Spontaneous pho^+ revertants among maternal cells will not grow because of threonine and leucine deficiencies, and any pho^+ cells among the males will be killed by streptomycin. Therefore, it is possible to obtain a very

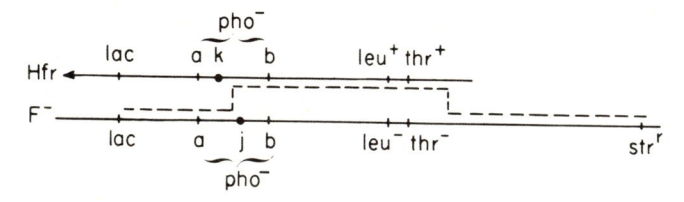

Fig. 4-17. Recombination at the phosphatase locus. Scheme showing how recombination frequencies between the pho^- mutations cataloged in Fig. 4-16 were normalized to the number of recombinants acquiring the thr^+leu^+ markers from the Hfr chromosome. Positions of crossovers are indicated by dashed lines.

precise figure for the product of W_{jk} and $W_{thr\ leu-str}$, the probability of recombination between the *thr leu* and *str* loci, because the recombinational events being scored are very specific. At least one, or any odd number, of crossovers must occur within the relatively large *thr leu–str* region of the chromosome, and another crossover must take place within the *pho* cistron. In a parallel experiment, the same culture is tested on minimal medium containing streptomycin and inorganic phosphate. Under these conditions, recombinants between *thr leu* and *str* are selected, and $W_{thr\ leu-str}$ is thereby determined. Dividing the first number by the second, W_{jk} is obtained. In using this approach, we are necessarily comparing two probabilities, one relating to a complicated rearrangement of chromosomal markers, $W_{jk} \times W_{thr\ leu-str}$, and the other relating to the simpler event, $W_{thr\ leu-str}$. By normalizing the former to the latter, we obtain W_{jk} directly in recombination units and need perform no further corrections. Furthermore, this technique significantly reduces the "noise level" which results from events unrelated to recombination.

The position of mutant *U24* is shown in Fig. 4-16 to be outside the limits of the *pho* cistron defined by *U3* and *U18*. The recombination frequencies between *U24* and various mutations located previously are in excellent agreement with the principle of additivity of recombination frequencies.

In bacterial genetics, however, fundamental difficulties are encountered when the additivity principle is applied to loci more than a few cistrons apart. Recombination frequencies are very large even for neighboring loci; the total length of the *pro-pur* region, for instance, is about 50 recomination units (Fig. 4-14). Furthermore, from studies of other similar small regions, we find that each of them covers a segment of comparable length. By adding up the short-range recombination frequencies along the entire chromosome, we obtain for W_{AZ} the paradoxical value of several thousand recombination units. Nothing of this sort was ever observed in the genetics of higher organisms where the recombination frequency between two loci is always a rather small number, not exceeding a few percent. Therefore, the additivity principle, $W_{AC} = W_{AB} + W_{BC}$ or $W_{AB} = \alpha \Delta l_{AB}$, cannot be rigorously applied to bacteria. This discrepancy is even more prominent in the genetics of bacteriophages.

A more general formulation of the additivity principle must be derived to account for such phenomena. To begin, we shall assume that the simple additivity principle is still valid for small regions of the bacterial chromosome so that the increment in recombination frequency over dl will be αdl. Consider now a region of the chromosome bounded by the loci A and B (Fig. 4-18). We wish to find dW, the probability of recombination within a given segment of the A–B region lying between l and $l + dl$. We shall represent the probability of recombination between A and l by W. Now the two conditions imposed on the required event are (1) that no crossover shall occur between

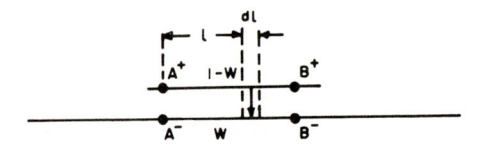

Fig. 4-18. A scheme for the deduction of Haldane's formula. See text for derivation and meaning of symbols.

A and l, which has a probability of $(1-W)$, and (2) that a crossover shall occur within the small segment dl, for which the probability is αdl. Consequently, the overall probability of this event may be written as the product of the two individual probabilities:

$$dW_1 = (1 - W)\alpha dl$$

But in a fraction of the synapsed chromosome pairs equal to W, a crossover does occur between A and l, and there is a probability, αdl, that these recombinants will undergo a second crossover in the opposite sense within dl. Hence, the overall probability of this class of events is

$$dW_2 = -W\alpha dl$$

The total change in W over the chromosomal segment dl will thus be:

$$dW = dW_1 + dW_2 = (1-W)\alpha dl - W\alpha dl = \alpha(1 - 2W)dl \qquad (4\text{-}31)$$

By integrating between points A and B, we obtain:

$$-\tfrac{1}{2}\ln(1 - 2W_{AB}) = \alpha \Delta l_{AB} \qquad (4\text{-}32)$$

or:

$$W_{AB} = \tfrac{1}{2}(1 - e^{-2\alpha \Delta l_{AB}}) \qquad (4\text{-}33)$$

This is a generalized form of the relationship between recombination frequency and map distance, first derived by Haldane in 1919 (48). It takes into account both single and multiple crossovers. Genetic map distance is not expressed by recombination frequency, but by its logarithm. This is understandable, since from general statistical principles, it follows that the probability of independent events is multiplicative, which implies additivity of their logarithms.

At small recombination frequencies, $-(1/2)\ln(1 - 2W_{AB})$ can be replaced by W_{AB}, and we obtain the simple form of the additivity principle as an asymptotic approximation valid for $W_{AB} \ll 1$. Within small regions of the chromosome, the approximate formula can still be used to determine distances on the genetic map directly from recombination frequencies. By summing the separate small distances, we shall always correctly estimate the distance between any two loci, no matter how far apart they are. It is necessary to recall only that large map distances represent not true recombination frequencies, but more complex logarithmic functions of these parameters. In studies of genetic recombination over a region of the *E. coli* chromosome large enough to exhibit pronounced nonlinearity, Verhoef and de Haan verified that map distances are related to recombination frequencies by the logarithmic relationship derived above (49).

As Δl_{AB} increases, W_{AB} approaches a limiting value of 50%. This means that it is equally likely that either B allele will be recombined with a given A allele. Furthermore, it is evident from this that the recombination frequency between a given genetic locus and any other can never exceed 0.5 if measured in a single experiment. From Eq. (4-33), we see that W_{AB} is almost equal to its maximal value when the length of the chromosomal segment between A and B is $1/\alpha$. We can find α from the experimental observation that in *E. coli* 1 minute of chromosome transfer during conjugation is equivalent to twenty recombination units. Thus, from the relationship $W = \alpha \Delta l$ which is valid over this relatively short segment, $\alpha = 0.2$ min^{-1}. Note that the length of the chromosome is measured in minutes here. Since the entire *E. coli* chromosome has a length of 90 minutes, and $1/\alpha = 5$ minutes, we conclude that the recombination frequency between two loci separated from each other by a distance equal to 1/18 of the chromosome or more will be approximately equal to the maximum value of 0.5. To construct the genetic map by recombination frequencies, it is therefore essential to study segments of the chromosome not in excess of a few percent of the total length. This can be accomplished in *E. coli* by using a variety of Hfr strains with different origins so as to cover all portions of the chromosome.

In addition to the value of $1/\alpha$, experiments on genetic recombination confirm the limiting value of 0.5 for the probability of integration of paternal markers, a figure in precise agreement with theory. Specifically, this occurs when both the donor and recipient cells are derived from *E. coli* strain K12. In cases where there is less genetic homology between strains, the maximum recombination frequency is much lower. For instance, when K12 Hfr's are mated with F^- cells from *E. coli* strain B, the maximum frequency with which a paternal marker is integrated into the maternal chromosome approaches 0.02. This result testifies to the nonequivalence of pairs of alleles in heterologous chromosomes (49).

11. The Physical Scale of the Genetic Map; Experiments with Radioactive Cells

Bacterial genetics represents one of the most powerful methods for investigating DNA function. Furthermore, a fundamental link has been forged between genetic measurements and the molecular structure of DNA by using radioactive isotopes in the study of conjugation and recombination. The experimental techniques employed in such efforts typify the approach of contemporary molecular biology to problems of structure and function.

An important use of radioactivity in genetic experiments is based on the principle of "radioactive suicide." When radioactive phosphorus is introduced into DNA at the very high specific activity of 100 mCi/mg, 1 phosphorus atom in 3000 will be labeled with ^{32}P. The total number of phosphorus atoms in the *E. coli* chromosome is approximately 10^7, so that each chromosome will contain roughly 3000 atoms of ^{32}P if the cells are grown on a very "hot" medium of this sort. Radioactive suicide occurs when these atoms disintegrate. Owing either to direct rupture of chemical bonds or to energy of recoil of the resulting ^{32}S nuclei, breaks are produced in the DNA strands. As a consequence, part of the chromosome is inactivated and the accumulation of such events is lethal. Hershey, Kamen, Kennedy, and Gest first used the method of radioactive suicide in 1951 to measure the kinetics of bacteriophage inactivation (50). It was adapted to bacterial cells by Stent soon after.

In bacteriophage, about ten ^{32}P disintegrations are sufficient to inactivate one particle. It may be assumed that most of the ^{32}P disintegrations break only one of the two DNA strands. Stent postulated that phage inactivation requires the simultaneous rupture of both strands, an event which occurs much less frequently. The inactivation ratio of ten disintegrations per particle has been found valid for a great many viruses. However, in phages ΦX174 and S13, whose DNA is single-stranded, the efficiency of each disintegration is equal to unity.

The method of suicide is also applicable to bacteria (51). The chromosomes of

11. The Physical Scale of the Genetic Map; Experiments with Radioactive Cells 353

Hfr cells grown on radioactive phosphate are inactivated with regard to genetic transfer if the bacteria are first stored at low temperature for several days to allow for partial decay of ^{32}P nuclei. This is demonstrated by mating the radioactive Hfr's with nonradioactive F$^-$ cells and measuring the quantity of recombinants as a function of the fraction of ^{32}P atoms disintegrated. The number of recombinants declines exponentially, producing straight lines on a semilogarithmic plot, as illustrated in Fig. 4-19. These kinetics indicate that the length of the Hfr chromosome transferred to the female during conjugation depends on the number of breaks sustained.

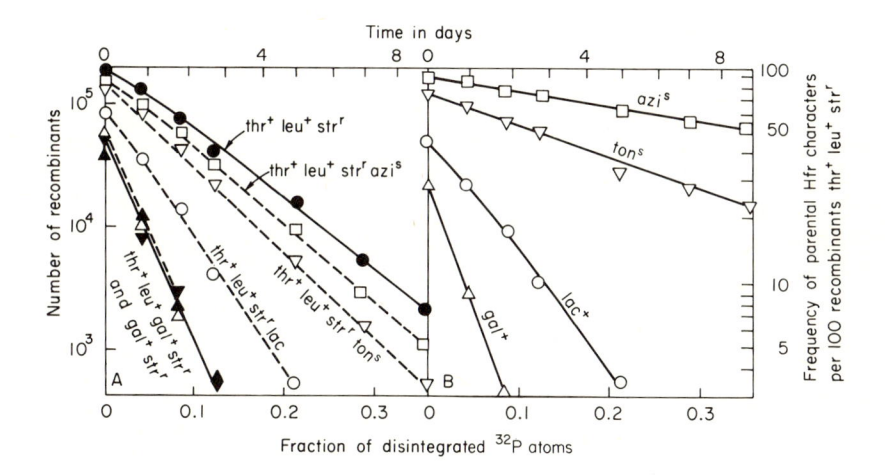

Fig. 4-19. Effect of ^{32}P decay occurring in the Hfr before mating. HfrH donor bacteria grown in medium with 70 mCi/mg ^{32}P are centrifuged, washed, and resuspended in fresh medium. Samples are then frozen in liquid nitrogen. Every day a sample was thawed, resuspended in buffer and mixed with an excess of F$^-$ recipient cells. After 1 hour at 37°C, samples are diluted and aliquots are plated on selective media. The *number* of recombinants having the markers *thr$^+$leu$^+$str^r*, *gal$^+$str^r*, and *thr$^+$leu$^+$gal$^+$str^r* (solid lines on left) and the *proportion* of *thr$^+$leu$^+$str^r* recombinants having one of the Hfr characters *azi^s*, *ton^s*, *lac$^+$*, and *gal$^+$* (graph on right) are plotted on a logarithmic scale versus the time in days and the fraction of disintegrated ^{32}P atoms. On the left figure are also plotted in dotted lines the *numbers* of *thr$^+$leu$^+$str^r* recombinants having one of the Hfr characters *azi^s*, *ton^s*, or *lac$^+$* as calculated from the curves on right. From Jacob and Wollman (52).

This method should provide a new and independent means of constructing the *E. coli* genetic map. The slopes of the lines in Fig. 4-19 are proportional to the probability of chromosomal breakage between a given locus and *O*, the origin. Assuming that the radioactive phosphorus is equally distributed along the chromosome, the slopes of these lines are also proportional to the distance between the locus and *O*. For the markers *thr*, *leu*, *azi*, *ton*, *lac*, and *gal*, the inactivation rates are related as 0.35:0.39:0.43:0.87:1.0, respectively. Corresponding distances on the genetic map constructed from the time sequence of marker penetration during conjugation are in the ratio 0.34:0.36:0.44:0.72:1.0. The agreement between these two sets of data is quite satisfactory.

Suicide experiments can also be used to determine the actual molecular scale of the genetic map. The segment between the chromosomal origin and the *gal* locus of

HfrH, $\Delta l_{O\text{-}gal}$, is transferred to the recipient during a 25-minute mating period. The number of *gal* recombinants decreases with radioactive decay of ^{32}P according to the equation:

$$n = n_o e^{-kavN_{O\text{-}gal}} \text{ or } \ln(n) = \ln(n_o) - kavN_{O\text{-}gal} \tag{4-34}$$

where $N_{O\text{-}gal}$ is the number of nucleotide pairs in the *O-gal* segment, v is the fraction of ^{32}P atoms which have disintegrated up to the time of mating, a is the relative concentration of ^{32}P atoms, and k is the efficiency, or the number of chromosomal breaks per disintegration. Assuming k equal to 0.1 for double-stranded DNA as in the phage experiments, we can calculate $N_{O\text{-}gal}$ from the slope of the curve for *gal* recombinants in Fig. 4-19. In this case, $N_{O\text{-}gal}$ is equal to 25×10^5 nucleotide pairs.

The dimensions of the sex factor and of the F'-*lac* episome have also been estimated by techniques utilizing the disruptive effects of ^{32}P decay in *E. coli* (53). A culture carrying one of these factors was grown on a medium containing radioactive phosphate, and the radioactive episome was transmitted to F$^-$ recipient cells grown on a normal medium. The donor cells were then eliminated with phage T6 to which the female cells were resistant. The F$^-$ cells which had become F$^+$ after conjugation were frozen for a period corresponding to the half-life of ^{32}P. During this interval, the episome underwent partial inactivation, and its dimensions were evaluated by a method similar to that described for the bacterial chromosome. In particular, the efficiency, k, was taken as 0.1, as for any double-stranded DNA. The molecular weight of the sex factor was judged to be 5×10^7 daltons, or about 2% of the total *E. coli* chromosome. A similar result was obtained when the amount of radioactive DNA transferred in F$^+ \times$ F$^-$ crosses was measured (54). Finally, the size of the F'-*lac* episome was measured very accurately by hybridizing labeled episomes with nonradioactive DNA from F$^-$ cells on a Bolton-McCarthy column (55). So as to reduce the background, the labeled F'-*lac* episome was isolated from *Serratia marcescens*, which had originally received the episome from an *E. coli* donor (p. 424f). Since *E. coli* DNA served as an adsorbent on the column, the background from nonhomologous *Serratia* DNA was minimal. In this particular experiment, the size of the F'-*lac* factor was estimated to equal about 2.5% of the bacterial chromosome, and to contain about 0.4 to 0.7% bacterial genes in addition to the sex factor. Other episomes contain even more chromosomal DNA.

The relationship between genetic map distances measured by recombination and the physical scale of the chromosome which it represents varies widely from organism to organism. In *E. coli*, 20 recombination units are equal to 1 minute of transfer or to 10^5 nucleotide pairs, and one recombination unit is thus equal to 5×10^3 nucleotide pairs. According to Benzer, one recombination unit in bacteriophage T4 corresponds to 250 nucleotide pairs, while in *Drosophila* the recombination unit is equal to 3×10^5 nucleotide pairs, and in mouse to 3×10^6 nucleotide pairs (Pontecorvo).

Knowing the scale of the genetic map in bacteria, we can estimate the physical dimensions of a single cistron and thereby deduce the coding ratio. The genetic map distance between the most extreme mutants of alkaline phosphatase studied is approximately 0.3 recombination units (56). This corresponds to 1500 nucleotide pairs with a molecular weight of $1500 \times 330 \times 2 = 10^6$ daltons. The phosphatase protein has a molecular weight of 80,000, and since it consists of two identical subunits, the phosphatase cistron must code for a sequence of about 400 amino acid residues. If 400 amino acids are coded by 1500 nucleotide residues, the coding ratio must be approximately 3. Benzer obtained similar numbers from his studies of bacteriophage genetics.

12. Bacterial Transformation

The phenomenon of bacterial transformation was discovered in 1928 by Griffith, who studied the nature of induced alterations in the polysaccharide capsule of pneumococcus (*Diplococcus pneumoniae*). Some pneumococcal mutants lack the ability to synthesize capsular materials, but the capacity can be transmitted to them by heat-killed cells from a capsule-forming variety (57). Moreover, the type of capsule synthesized is characteristic of the added heat-killed cells and not that possessed by the nonencapsulated strain before its mutation. Capsular differences among various strains of pneumococcus were detected by immunological procedures. When inactivated pneumococci of a given strain are introduced into the bloodstream of a rabbit, a specific antiserum can be obtained after several days. The antiserum contains protein antibodies which strongly adsorb to cells of the type injected, causing them to stick together or agglutinate. Since the antibodies are specifically fitted to the surface of the original cell strain, they are incapable of causing other strains to agglutinate. This provides a simple but very sensitive technique for distinguishing the strain specificity of transformed pneumococcal cells. A few years after Griffith's discovery, it was shown that cell extracts from the encapsulated donor strain also possessed transforming ability. These experiments were the first cases in which directed transfer of a heritable property was mediated not by living cells, but by a chemical substance contained within them.

In 1944, Avery, Macleod, and McCarty demonstrated that the transforming factor is DNA (58). It was found that this substance could be extracted from the transforming cell lysate, and purified from RNA and protein, without causing any losses in transforming activity. This activity, however, is completely liquidated by the enzyme deoxyribonuclease (DNase). In later work, the residual protein in preparations of transforming DNA was reduced to less than 0.2% without affecting its transforming capacity. Using this figure together with measurements of the quantity of DNA taken up by each cell during transformation, it can be calculated that, on the average, less than one molecule of protein penetrates each transformed cell. Consequently, one may be sure that the specific transforming factor is DNA and not protein, as was originally suspected. Proof that purified DNA from one cell strain could penetrate and alter the hereditary characteristics of a strain with different properties was a most important stage in the elucidation of DNA function.

Bacterial transformation is not a very widely distributed phenomenon, and its efficiency reaches 1 to 5% of the total cells only in such bacteria as *Diplococcus pneumoniae, Bacillus subtilis, Haemophilus influenzae*, and *Streptococcus salivarius* (59). This is because molecules of transforming DNA, which usually have a molecular weight of about 10^7 daltons, cannot as a rule penetrate cell walls. Only in certain species and at certain specific periods of their life cycle do pathways become available for the penetration of DNA, owing to a mechanism that has not yet been fully explained. Cells in a physiological state appropiarte for transformation are said to be *competent*. Depending on the bacterial species employed, there are definite procedures for the preparation of competent cultures. The bacteria are usually grown to the end of exponential phase, diluted in fresh medium, once again grown to the end of exponential phase, and again diluted. This procedure leads to the partial synchronization of the cells, that is, they pass through all phases of their life cycle and divide at approximately the same time.

A culture which has reached stationary phase and has stopped growing is not likely to contain any competent cells. Even in a competent culture, periodic oscillations may be observed in the number of transformable cells, and the probability of transformation may vary by a factor of 1000 depending on the conditions. A bacterial population can be enriched for competent cells by differential and density gradient centrifugation. The competent cells appear to be the young ones which have recently divided for they are generally lighter or less dense than the bulk of the cell population, probably due to a lower content of nucleic acids. In order to reduce experiments on the kinetics of tranformation to standard conditions, aliquots of a competent culture are quick-frozen and stored at $-70°C$. Ten percent glycerol is often added to hinder the formation of large ice crystals which would damage the cells. The frozen cultures can then be thawed as needed for individual transformation experiments.

In the bacterium *Streptococcus salivarius*, competence has been shown to depend on the presence of a special enzyme which renders the cell wall permeable to DNA at certain periods during its development (Pakula and Walczak, 60). It was demonstrated that addition of this enzyme caused nontransformable cells to become competent. This enzyme also permits radioactively labeled DNA to penetrate treated cells. Similar factors have been found in both pneumococcus and *Bacillus subtillis*.

Detailed investigation of transformation has revealed that it is a complex process consisting of several successive stages: (1) reversible binding of DNA molecules by the cell; (2) irreversible penetration of DNA molecules into the cell by a slower and more specific mechanism; (3) integration of the foreign DNA molecule into the chromosome of the host; and (4) replication and segregation of the recombinant chromosome among progeny cells. Each of these stages can be isolated from the others and studied individually (Hotchkiss, Fox, Spizizen, 61, 62).

Quantitative analysis of transformation has been greatly facilitated by the use of suitable genetic markers. While the classic experiments of Griffith on pneumococcus demonstrated that the capacity to synthesize a capsule could be transmitted by an extract from nonliving cells, capsule formation was a very difficult character to measure quantitatively. More convenient markers were sought as a consequence. These included resistance to various inhibitors such as streptomycin, erythromycin, cathomycin (novobiocin), and sulfanilamide, the ability to synthesize amino acids such as tryptophan, histidine, and tyrosine, and the capacity to utilize various carbon sources such as maltose, lactose, and mannitol. The transfer of antibiotic resistance is especially easy to test, for the transformed cells can simply be plated on medium containing the inhibitor after a certain amount of time, necessary for the phenotypic expression of the character, has elapsed. Unaltered recipient cells die, but the transformed cells grow and multiply. Hence, selection is carried out against a zero background and is thus very sensitive.

When transforming DNA is extracted from a bacterial cell, the chromosome is generally broken down into 100 to 200 different fragments, each with a molecular weight of about 1 to 2×10^7 daltons, depending on the method of extraction. If selection is carried out for any one marker, then only one type of fragment in the heterogeneous population of DNA molecules will be effective; others are inert with respect to the given character and can even serve as a source of competitive inhibition to the transformation process, since there appears to be an equal probability that any given DNA molecule will be adsorbed to and taken up by the cell.

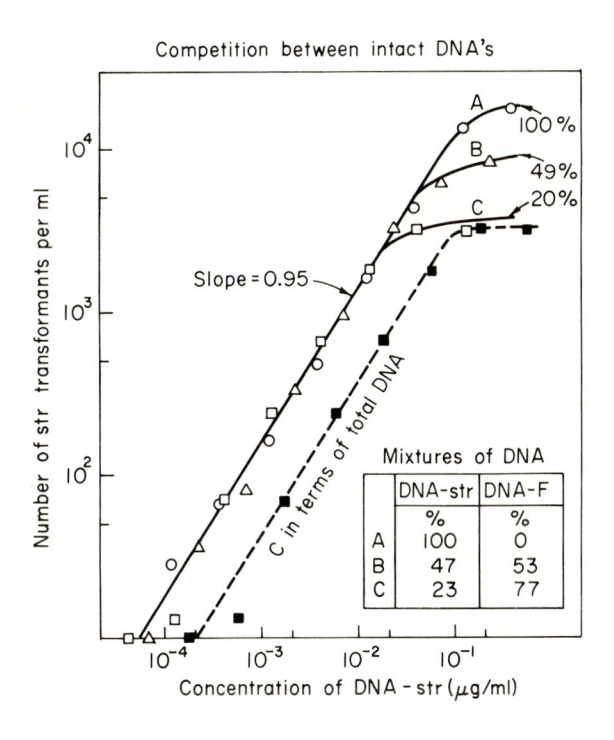

Fig. 4-20. Concentration dependence curve of streptomycin-resistant transformants plotted against concentration of high-molecular-weight DNA bearing the str^r marker. Both scales are logarithmic. The indicated mixtures with pneumococcus DNA bearing the F marker (sulfanilamide resistance) but not the str^r marker were made and added, in increasing quantities, to aliquots of the same culture. There is no detectable competition between different DNA's in the linear, low-concentration range; at saturation, however, the yield of transformants is proportional to the fraction of DNA carrying the str^r marker in the mixture. From Hotchkiss (61).

The way in which the number of transformed pneumococcal cells depends on the DNA concentration in the medium is illustrated in Fig. 4-20. The genetic marker in this case is streptomycin resistance. Reliable and reproducible results were obtained by the use of frozen cell cultures and by restriction of contact between cells and DNA to intervals on the order of 5 minutes. The reaction was interrupted by the addition of DNase which rapidly degrades extracellular DNA. The number of cells transformed is proportional to the amount of DNA present over a concentration range of 10^3, showing that the uptake of a single molecule of DNA can result in transformation. However, at a DNA concentration of about 0.1 µg/ml, the cells become saturated with transforming DNA and the number of transformants reached a plateau.

The influence of genetically inactive DNA was ascertained by mixing transforming DNA prepared from a str^r strain with DNA from another strain which carried the str^s marker. Within the linear portion of the curve (Fig. 4-20), the second and inactive DNA was wholly ineffective as a competitive inhibitor, indicating that the various DNA molecules are bound and taken up by recipient cells independently of each other. However, the height of the plateau was reduced in the presence of inactive DNA, demonstrating that active and inert DNA molecules compete for certain sites on or in the cell when their concentrations are sufficiently great. Such behavior resembles phenomena described by the Langmuir adsorption isotherm for cases in which the number of binding sites is limited. The probability that homologous molecules, one genetically active, the other inactive, will be taken up is precisely the same, thus accounting for competitive inhibition. Nonhomologous DNA, isolated from other strains of bacteria or from higher organisms, may be irreversibly fixed by competent cells but it is genetically inert and its markers are never expressed. One exception however, is DNA from phages to which the bacteria are normally sensitive.

The quantity of DNA irreversibly fixed by recipient cells during transformation was measured with ^{32}P-labeled transforming molecules. After all reversibly adsorbed DNA had been degraded by DNase and washed from the cells, it was found that the amount of bound radioactive DNA is proportional to the number of transformed cells. From the maximum amount of ^{32}P fixed in this way it should be possible to estimate how many radioactive molecules must be taken up in order to produce one transformed cell. According to measurements in two different laboratories, the least amount of DNA fixed by the entire culture per transformed cell is 2×10^{-15} gm for pneumococcus and 3×10^{-15} gm form *Haemophilus*. It is of interest to note that the weight of the pneumococcal genome is 1.5×10^{-15} gm. If the molecular weight of transforming DNA from pneumococcus is 8×10^6 (Doty) and that of *Haemophilus* is 15×10^6 (Herriot), a simple calculation reveals that in both cases the number of molecules fixed by the population as a whole is 150 to 200 per bacterium transformed, an amount comparable to the total DNA complement of a single cell. But since the total number of transformants is small, it is not clear from this result whether transforming DNA enters all the cells equally, or whether it enters only the small number of cells destined to be transformed.

While all varieties of transformable bacteria may not be the same in this regard, the question of DNA uptake in *B. subtilis* was resolved by testing individual cells of a population for the penetration of ^{32}P-labeled transforming molecules (Bresler *et al.*, 63). After transformation, cells were spread on a photographic emulsion which was developed after a suitable period. When bacterial cell images were compared with autoradiographic tracks, it was evident that the bacterial population was extremely inhomogeneous. Only a fraction of a percent of the cells took up polymeric DNA from the medium, although the quantity of DNA entering each cell was large, equivalent to the cellular genome when the external DNA concentration was high.

A decided inhomogeneity in the competence of bacterial populations helps to explain a paradoxical result relating to the simultaneous transmission of two or more genetic markers when the total number of transformants is a small proportion of cells in the culture. If both genetic markers occupy neighboring loci on a single molecule of transforming DNA, then it is quite likely that both characters will be transmitted together. Suppose that in *B. subtilis* the average probability of transmission of one marker, such as trp^+ (tryptophan) and his^+ (histidine), is 0.1% over the entire cell population. If both markers were transferred independently, then double transformation would be expected with a probability of $(0.001)^2$ or 10^{-4}%. In fact, for two linked markers, located next to each other on a single DNA molecule, the probability of double transformation reaches 70% that for the transfer of either marker singly. But even more surprising, the simultaneous transmission of two *unlinked* markers can also occur very frequently, with a probability on the order of 10% that for single transformation. This apparently happens because each of the few competent cells takes up not just one or a small number of DNA molecules, but a quantity of DNA equivalent to that contained in the bacterial genome.

The proportion of competent cells in a population can also be determined genetically. A strain of *B. subtilis*, doubly auxotrophic for unlinked tryptophan and histidine markers ($trp^- his_1^-$) was chosen as a recipient. Although at least one histidine marker is linked to tryptophan, the majority of histidine loci are located in a different linkage group. Among the latter is the his_1 cistron which was used in these experiments. Figure 4-21 presents frequency curves for both single and double transformation of these

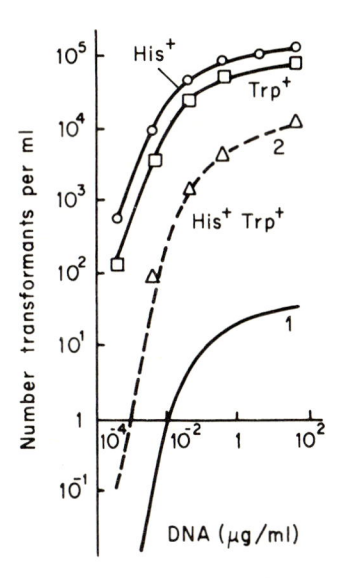

Fig. 4-21. The number of single and double transformants as a function of DNA concentration. After transformation, the recipient cells were tested for ability to grow in the absence of histidine (his^+), in the absence of tryptophan (trp^+), and in the absence of both supplements ($his^+ trp^+$). When the theoretical value of double transformants is calculated from the number of single transformants (Eq. 4-37) and α is chosen equal to 100% (curve 1), the fit between calculated and experimental values is poor. A value of α equal to 0.3% must be selected in order to obtain a good fit between theory and experiment. From Bresler *et al.* (63).

markers. If all the cells were competent, the probability of double transformation would be:

$$W_2 = n_1 n_2 / N^2 \qquad (4\text{-}35)$$

where n_1 and n_2 are the number of single transformants, and N is the total number of recipient cells. The number of double transformants should thus be:

$$n_{12} = W_2 N = n_1 n_2 / N \qquad (4\text{-}36)$$

In fact, the theoretical curve based on this equation (curve 1, Fig. 4-21) lies very far from the experimental points. In order to obtain a fit between experiment and theory (curve 2, Fig. 4-21), it is necessary to introduce a coefficient, α, equal to the fraction of competent *B. subtilis* cells:

$$n'_{12} = n_1 n_2 / \alpha N \qquad (4\text{-}37)$$

For the strains investigated, $\alpha = 0.3\%$, showing that the cell population is extremely heterogeneous with respect to competence. These results are in good agreement with the figure derived by autoradiography. In pneumococcus, by contrast, the cell population is almost uniformly competent. Thus for double transformation in pneumococcus, Eq. (4-36) is obeyed with satisfactory accuracy (Hotchkiss and Marmur, 64).

Integration of transforming DNA into the genome of the recipient and formation of recombinants constitute the final stages of transformation. The question of whether integration is accomplished by breakage and reunion or copy-choice was resolved by the experiments of Fox on pneumococcus (65) and of Voll and Goodgal on *Haemophilus influenzae* (66). In both cases, the kinetics of recombination were studied in the following way. Recipient cells were lysed at various times after the addition of transforming DNA. DNA from the primary recipients was then isolated, deproteinized, and its transforming capacities were tested in a second experiment.

Particularly clear-cut results were obtained with *Haemophilus*. Two closely-linked genetic markers, resistance to the antibiotics streptomycin (*str*) and cathomycin (*cat*) were used in this work. Linkage between the two was sufficient to ensure that both markers were localized on the same DNA molecule. The recipients were resistant to streptomycin, but sensitive to cathomycin and thus had the genetic constitution $str^r cat^s$, while transforming DNA was isolated from a $str^s cat^r$ donor strain. Some of the transformants thus carried the double resistance $str^r cat^r$. DNA was prepared from the recipient cells at several different times after the start of transformation, and its transforming capacities were tested on the wild-type $str^s cat^s$ strain for resistance to both antibiotics. It had been previously ascertained that both loci are cotransformed with a probability one fifth that for transmission of either one separately. Consequently, double transformation is a very likely event and it can be used as a method for titrating the str^r and cat^r alleles carried by DNA molecules from the first group of transformants.

The principal results of this experiment are summarized in Fig. 4-22A. The upper curve represents recovery of the str^r marker in the secondary transformation. This allele was originally present in the recipient cells, and its quantity simply increases exponentially with the growth of the recipient culture. The behavior of the cat^r marker, introduced into the primary transformants by the initial transforming DNA,

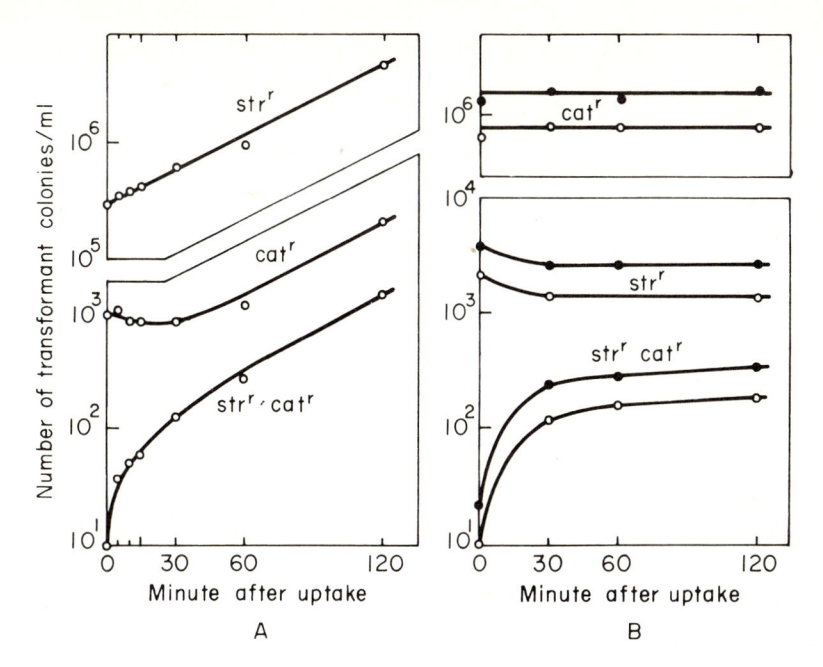

Fig. 4-22. Kinetics of formation of recombinant DNA. (A) Assay of transforming activity in recipient cell lysates for resistance to streptomycin (str^r, carried by the recipient) and cathomycin (cat^r, carried by the original transforming DNA). (B) Assay of transforming activity in starved (nongrowing) recipient cells. In this experiment, the str^r marker was carried by the original transforming DNA, and the cat^r marker was carried by the recipient. The curves connecting closed circles represent a DNA concentration twice that of the curves connecting open circles. From Voll and Goodgal (66).

is entirely different. The transforming ability of this marker remains practically constant during the first 40 minutes after initiation of transformation. The cat^r marker can therefore be quantitatively isolated from the cell into which it was originally introduced. Its quantity in the recipient remains unchanged in spite of the fact that the cells had all completed an average of one division in this period. But the most interesting fact is that $str^r cat^r$ recombinants rapidly increase from zero during this same initial period, reaching a half-maximal level after 15 minutes. This demonstrates true recombination in the primary recipient since the number of random doubly resistant transformants derived from separate molecules would be much less than that actually observed.

Formation of recombinants can apparently take place even without net DNA synthesis, suggesting a conservative mechanism of recombination. Although the number of active cat^r cistrons does not change during the first 40 minutes, the number of $str^r cat^r$ recombinants increases very sharply. This deduction was confirmed in a similar experiment on nongrowing cells which were starved for an essential nutrient (Fig. 4-22B). The results are the same in principle as those obtained in the first experiment, except that the overall quantity of markers does not increase. Here the recipient strain is cat^r and the number of cat^r markers does not change. The number of str^r markers, introduced into the recipients by the transforming DNA, decreases slightly as recombination takes place. The quantity of $str^r cat^r$ recombinants increases in the initial

40-minute period, however, to the same extent and with the same kinetics as in the growing culture. Starvation of the recipients, therefore, is not expressed by any alteration in the recombination process. We conclude that the $str^r cat^r$ recombinants were formed in the absence of marked DNA replication, and that the relevant segment of transforming DNA was introduced into the recipient chromosome by the mechanism of breakage and reunion. Of course, some repair synthesis of DNA is not excluded under these conditions.

The process of transformation and recombination in *Bacillus subtilis* was investigated by density gradient centrifugation of isotopically labeled DNA (Bodmer and Ganesan, 67). Transforming DNA was labeled simultaneously with the heavy isotope ^{15}N to alter its density, and with ^{3}H in order to make its identification easier. The DNA of the recipient cells was labeled with ^{32}P at the same time. After transformation, the recipient cells were freed of excess transforming DNA by DNase and subsequent washing. Their intracellular DNA was next extracted and subjected to CsCl density gradient centrifugation. After equilibrium had been established, the contents of the centrifuge tube were divided into several fractions and both types of radioactivity were measured in each one. Control DNA from nontransformed cells grown on ^{14}N and ^{15}N was used to determine the locations of parental molecules on the gradient and the corresponding peaks were quite distinctly separated from one another. When DNA was isolated from the transformed cells, the peak of ^{3}H radioactivity formed a shoulder between the ^{14}N and ^{15}N peaks, although it was somewhat closer to the former, which represents DNA from the recipient, than to the latter. The ^{3}H- and ^{32}P-labeled DNA's in the intermediate peak could not be separated from each other by a second centrifugation, or by thermal or alkaline denaturation. All evidence indicated that the two DNA's were associated by firm chemical bonds. The quantity of donor DNA inserted into the DNA of the recipient was found to comprise about 10% of the total. Furthermore, the molecular weight of the inserted segments was estimated to be 6×10^6 by experiments in which labeled DNA from transformed cells was subjected to ultrasonic fragmentation.

A more subtle question that may now be raised is whether one or both strands of the donor DNA is inserted into the recipient chromosome. The preferential integration of one of the two donor strands during transformation was demonstrated in the laboratory of the author (68). A strain of *Bacillus subtilis*, doubly auxotrophic for the linked markers histidine and tyrosine ($his^- tyr^-$) was transformed with hybrid DNA. The latter was prepared by melting and annealing a mixture of DNA from a strain carrying the two prototrophic markers ($his^+ tyr^+$) and from a strain carrying only one prototrophic marker ($his^+ tyr^-$), with the latter in considerable excess. Thus, the resulting DNA solution contained ($his^+ tyr^+$)/($his^+ tyr^-$) hybrid molecules together with an excess of ($his^+ tyr^-$)/($his^+ tyr^-$) DNA. If both strands of the hybrid DNA molecule are integrated into the genome of the recipient cells, one daughter cell will contain both prototrophic markers, while the other will contain only one (his^+). On tyrosine-supplemented medium, these cells should give rise to mixed colonies containing both types of segregants in approximately equal proportions. If only a single strand of the hybrid molecule is integrated, recipients should give rise to pure colonies since the initial transformants would be either ($his^+ tyr^+$)/($his^- tyr^-$) or ($his^+ tyr^-$)/($his^- tyr^-$); one of the segregants in each case would be of the parental type and thus would not be able to grow on medium supplemented with tyrosine alone. Accordingly, the transformants were spread on minimal + tyrosine plates and the resulting colonies were replica-plated onto minimal medium to test for acquisition of both prototrophic markers. Colonies whose replicas grew on minimal agar were resuspended in liquid medium supplemented with tyrosine, spread on minimal + tyrosine plates and once again replicated on minimal plates. Only 6% of the transformed cell clones, whose replicas were able to grow on minimal medium initially, proved to be mixed. The remaining 94% contained identical prototrophs of a single genotype and therefore represent the result of an event in which only one of the two DNA strands was integrated. A simple calculation revealed that the cases in which segregation occurred could be fully explained if integration of each of the two DNA strands takes place independently of the other.

This result was confirmed in pneumococcus by Fox and Allen (69), who labeled the transforming DNA with ^{32}P and the heavy isotope ^{15}N. DNA was extracted from the transformants, fragmented by sonication and centrifuged in CsCl density gradients both before and after thermal denaturation. The distribution of material showed that donor DNA is present in double-stranded molecules, but that it contributes only a single strand to the duplex. The single strand is contained in elements whose molecular weight is on the order of 1 to 2×10^6 daltons and it is chemically integrated into the recipient genome.

Thus both genetic and physical experiments show that recombination during transformation is accomplished by the insertion of donor DNA fragments into the genome of the recipient and moreover, that single strands of DNA, not double strands are inserted.

Methods have now been worked out for the separation and recovery of the two complementary strands from double-helical DNA (e.g., p. 504). The transformation of bacteria by isolated single strands of DNA can therefore be studied. The conditions for penetration of single-stranded DNA into competent *Hemophilus influenzae* cells were found by Postel and Goodgal to require a 10-minute incubation at pH 4.4, followed by incubation in neutral medium (70). Similar results were also obtained for *Bacillus subtilis* (71). Furthermore Lacks and co-workers demonstrated that transforming DNA in pneumococcus is converted to single strands after its uptake by competent cells, but prior to genetic exchange with the recipient chromosome (72). Only one strand undergoes recombination while the complementary strand is degraded to mononucleotides.

We shall now consider how transformation can be used to map the bacterial chromosome. The best object for such studies is *B. subtilis* from which a variety of metabolic mutants can be selected, much as in the case of *E. coli*. Other transformable bacteria have turned out to be much less convenient in this regard because they must be grown on enriched medium, where selection can be carried out for only a small number of markers, such as resistance to inhibitors. In addition to the dearth of suitable metabolic markers, resistance and sensitivity to inhibitors are often unsatisfactory themselves, because they commonly affect not one, but a whole group of enzymes. Results derived from complex markers of this sort have little cognitive value. Hence this exposition will be limited to mapping experiments performed with *B. subtilis*.

A general method for constructing the genetic map of *B. subtilis* by determining the order of marker replication in synchronized cell populations was suggested by Yoshikawa and Sueoka (73) and later worked out in detail by Marmur and co-workers (74). A bacterial culture is first grown to the stationary phase and at the same time, it is labeled with deuterium, the heavy isotope of hydrogen. This synchronizes replication of the chromosome since cells in the stationary phase will complete one cycle of DNA synthesis but they will not embark on the next. The culture is then transferred to fresh medium containing the normal isotope of hydrogen whereupon synchronous replication begins in a definite direction from the chromosomal origin (see p. 328ff). Aliquots of the culture are removed during the first replicative cycle and DNA is extracted from them. This DNA can then be fractionated according to density by centrifugation in a CsCl gradient. The unreplicated portion of the chromosome is most dense because it contains the maximum amount of deuterium; DNA lying between the origin and the point of replication contains only half as much deuterium as the unreplicated molecules and its density is thus significantly lower. The DNA fractions from the CsCl gradient are used to transform appropriately marked recipient cells and in this way the different genetic loci can be classified according to the density of the DNA fragment which bears them. By extracting DNA at different times during the cycle of replication, the distribution of markers between heavy and half-heavy fractions will vary, and a unique sequence can therefore be established.

The technique described above yields the order in which various markers are replicated and it is assumed that this order corresponds to their distribution on the bacterial chromosome. More refined methods are required to determine distances between markers on the *B. subtilis* genetic map, and both transformation and generalized transduction have served this purpose. These methods permit the orientation of genetic markers relative to each other within one linkage group, that is, which are usually carried by the same DNA fragment. Linkage groups occupying chromosomal fragments with molecular weights on the order of 2×10^7 daltons can be conveniently studied in one series of measurements. The principal difficulty in mapping by transformation is normalization of the number of transformants insofar as the process depends on many variables. As with conjugation, transformation frequencies are normalized by the analysis of multiple events so that unknown parameters such as the number of cells penetrated by donor DNA can be eliminated. A composite map of the *B. subtilis* chromosome, constructed on the basis of density transfer experiments, as well as on transformation and transduction, is presented in Fig. 4-23.

Fine structure genetic mapping by means of transformation is still in its early stages. Although quantitative data on various linkage groups is still meager, considerable information about the amylase (*amy*) region in *B. subtilis* is already available (75). The *amy* cistron, which specifies a protein of 48,500 daltons, is closely linked with *aro* (synthesis of aromatic amino acids), a marker used for normalization in these experiments. If the probability of transforming *aro* alone is signified by A, then the frequency with which *aro* and *amy* are cotransformed, W', is equal to 0.35 A. Consider now two deficient mutations of the *amy* cistron, amy_1^- in the recipient and amy_2^- in the donor (Fig. 4-24). In order to recover a wild type amy^+ recombinant, one crossover must occur between amy_1^- and amy_2^-. If the probability of this is W, then the probability of the multiple event involving cotransformation of *aro* and *amy* with a crossover

12. Bacterial Transformation

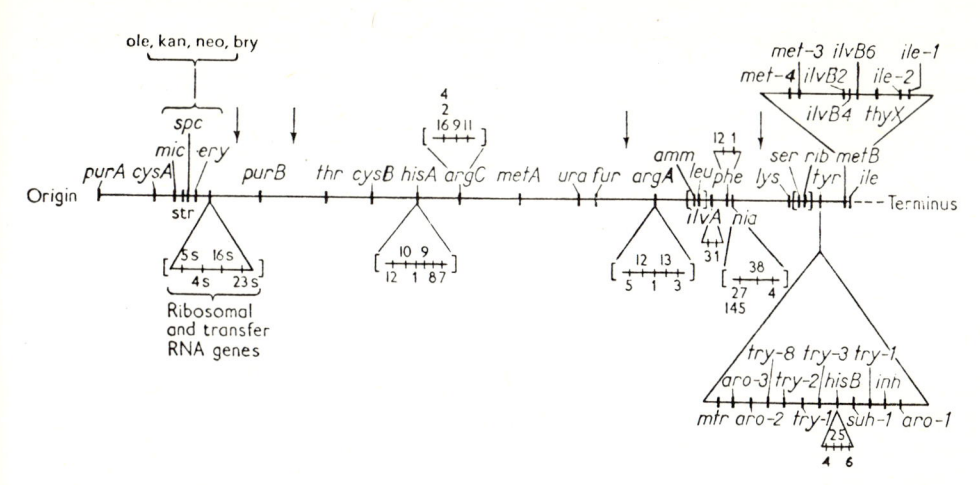

Fig. 4-23. Linkage map of the *B. subtilis* chromosome. Data were obtained from experiments on transformation, transduction, and density transfer. Linkage gaps appear in the intervals marked with vertical arrows. The markers surrounded by brackets have not been ordered relative to one another. From Dubnau *et al.* (74).

between amy_1 and amy_2 will be $W'' = W \times W'$. Sixteen different amy^- strains were examined in pairwise transformation experiments and for the two most distant mutant sites W'' was found to equal $0.045\ A$. If these two sites roughly define the boundaries of the *amy* cistron, then the dimensions of the amylase cistron are:

$$W = W''/W' = 0.045\ A/0.35\ A = 13\%$$

Recall that in the case of conjugation, the genetic dimension of the phosphatase cistron was determined to be 0.3%. The large discrepancy between the two figures indicates that genetic recombination during transformation takes place several dozen times more frequently per unit length DNA than during conjugation. This difference is most likely connected with the observation that transformation involves the integration of only one DNA strand, while recombination after conjugation demands the integration of both strands.

Independent of the work on bacterial transformation, Pritchard and others showed that in a number of organisms, such as *Aspergillus*, *Neurospora*, and yeast, the probability of multiple crossovers becomes very large within short segments of the chromosome (76). This phenomenon is referred to as *high negative interference*. The explanation for negative interference advanced by Pritchard is that during synapsis effective pairing, that is, pairing which can lead to recombination, occurs only along separate short portions of the chromosomes. It is possible that effective pairing

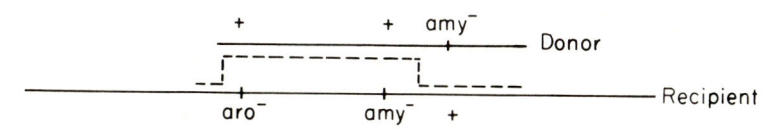

Fig. 4-24. Diagrammatic representation of the distribution of markers in a three-point cross. Plus sign indicates the presence of a functional allele. From Yuki and Ueda (75).

requires the DNA double helices to unwind over homologous stretches so as to form single-stranded loops. In that case, effective pairing could take place between one DNA strand of the first chromosome and the complementary strand of the second chromosome; the probability of genetic exchange between the two strands would be very great over that restricted region. However, extensive strand separation would be unlikely over the large chromosomal segments involved in bacterial conjugation.

The physical dimensions of a cistron may be calculated on the basis of transformation data. To this end, it is necessary to disrupt systematically the continuity of the DNA molecule and then study the effects of the procedure on recombination. In the case of conjugation, the disintegration of radioactive phosphorous serves to disrupt genetically active DNA (p. 352f), whereas in transformation, DNA can be cleaved by hydrodynamic shearing forces. Using this method, the author and his co-workers demonstrated that the average weight of the DNA fragment integrated into the *B. subtilis* chromosome during transformation is 10×10^6 daltons. This value was calculated by applying the theoretical analysis of Lacks (77) to the experimental data. It is important to note that the breaks introduced by shearing are not repaired within the cell. From the size of the integrated DNA fragment and from the genetic length of the amylase cistron we find that the molecular weight of this cistron in *B. subtilis* is at least $(0.13)(10 \times 10^6)$ or 1.3×10^6 daltons. This figure is in good agreement with theoretical predictions, as it must be, since the physical dimensions of a given cistron are invariant and thus independent of the genetic process by which the cistron is transmitted. Cistrons differ from each other only because the polypeptide chains for which they code are not all of identical size.

It is of interest that in addition to transformation by homologous DNA, *Bacillus subtilis* can also be "infected" by transformation with bacteriophage DNA. Infection occurs when competent cells are incubated in the presence of deproteinized DNA from a phage to which this species of bacteria is normally sensitive (78). This phenomenon is called *transfection*.

13. Genetic Processes in Viruses and Bacteriophages

The genetic information necessary to specify the reproduction of plant, animal, and bacterial viruses is stored exclusively in their nucleic acids. DNA was first demonstrated to be the active genetic principle of bacteriophage T2 by Hershey and Chase (79), and this result was confirmed and generalized by the discovery that single-stranded RNA carries all the genetic information required for the synthesis of tobacco mosaic virus (TMV) (Gierer and Schramm, 80). These findings have since been extended to a large number of viruses and bacteriophages, containing both single and double-stranded RNA and DNA.

Let us start by examining the evidence implicating tobacco mosaic virus RNA as a hereditary substance. When purified by repeated deproteinization with phenol and sodium dodecylsulfate, TMV RNA retains its infectivity when introduced into tobacco leaves. This material can still give rise to complete virus particles, even if protein contamination is held at less than 0.02%, and the electron microscope reveals that no intact virus particles are present. Furthermore, infective preparations are sensitive to ribonuclease, an enzyme which degrades RNA, but they are unaffected by specific antisera to TMV protein. However, the specific infectivity of purified viral RNA is less than that of native TMV particles by a factor of 200 to 300. This decrease has been attributed to the fact that free RNA is very susceptible to nuclease attack during both purification and reintroduction into the tobacco leaf.

TMV protein can be separated from its RNA core by raising the pH of a viral suspension to 10.5 or by treating it with 67% acetic acid (81). Under these conditions, the viral rods dissociate into identical protein subunits, each with a molecular weight of 17,000, and RNA. If the preparation is next incubated with RNase and dialyzed, RNA may be effectively eliminated. Purified viral protein is totally lacking in infectivity.

At pH 7.0, viral protein molecules reaggregate into small rods which superficially resemble native virus particles, although they too are noninfective. The quaternary structure of such aggregates is characteristic of TMV particles, but their assembly can stop at any stage and the length of the cylindrical protein coats appears to be randomly determined. But if reaggregation is carried out in the presence of undegraded viral RNA, the protein coat attains a length specified by the dimensions and configuration of the helically packed RNA molecules. Even more important, these spontaneously reconstituted particles possess a high degree of infectivity (Fraenkel-Conrat). The specific infectivity of viruses assembled *in vitro* can reach 80% the infectivity of native particles. The dissociation of TMV into RNA and protein is therefore a reversible process and, for all practical purposes, native and artificially reconstituted viruses are identical.

Several simple viruses consisting of a single RNA molecule and one type of protein subunit have now been successfully reconstituted *in vitro*. The self-assembly of R17, a somewhat more complicated virus has also been achieved (82). R17 is a small coliphage related to MS2 and f2 which consists of a single molecule of RNA with a molecular weight of 1.1×10^6 daltons and approximately 180 identical protein subunits, each with a molecular weight of 14,000. In addition, the protein coat contains one molecule of a second protein, called A protein, which has a molecular weight of 35,000 and is also coded for by viral RNA. Reconstitution of the phage takes place in the absence of A protein, but the resulting particles are completely devoid of infectivity and differ from native phage in sedimentation coefficient. Addition of A protein to the incubation mixture under conditions appropriate to self-assembly leads to the formation of fully infectious phage particles. The A protein is apparently required for adsorption of R17 to the bacterium. It is quite remarkable that the single molecule of A protein can spontaneously find the precise location in the phage coat to which it must attach.

The effect of mutations on tobacco mosaic virus has helped to define the relationship between viral RNA and the protein products it specifies. TMV mutants occur spontaneously and can also be induced by chemical mutagens. They differ from one another either in morphology, that is, in the size and shape of the necrotic lesions that they induce in tobacco leaves, or in the pathology of infection. Thus there are some TMV mutants that produce only local infections, confined to the leaves into which they were introduced, while others cause a generalized infection throughout the entire plant.

The only genetic experiment which can be performed with TMV consists of testing the properties of "hybrid" virus particles reconstituted from RNA and protein of different mutants. The hybrids are fully active, and when they are used to infect tobacco plants, the resulting progeny always possess the hereditary characteristics of the mutant from which the RNA was extracted, but they never exhibit any properties of the mutant contributing the protein. Unfortunately, more complicated genetic experiments cannot be performed with TMV, since these viruses have not been observed to undergo genetic recombination. In spite of this, the reconstruction data are of fundamental importance They show that all of the information necessary for synthesis of viral nucleic acid and viral protein is contained in an RNA strand with a molecular weight of 2.1×10^6, equivalent to about 6300 nucleotide residues.

TMV mutants have been produced *in vitro* by treating purified viral RNA with chemical mutagens such as nitrous acid (Gierer and Schramm, 80, 83). Several dozen TMV mutants were obtained by infecting tobacco plants with chemically altered RNA. Selection was carried out according to morphological differences in tobacco leaf lesions. The coat protein of many such mutants was purified, and after hydrolysis with the proteolytic enzyme, trypsin, they were "fingerprinted" (p. 16ff). As was expected, many of the mutant viruses possessed altered coat proteins, each one differing by at least one amino acid substitution. In a few cases, multiple mutants have been recovered whose coat protein contains several altered residues, implying that even though a number of points within the cistron specifying coat protein have been affected by the mutagen, the genetic information is still meaningful, and that functional protein is still synthesized despite differences in amino acid sequence (84).

In a significant number of mutants, however, the coat protein is not different from that of the wild type. This result indicates that TMV RNA contains more than one cistron (85). The additional proteins which are coded by TMV are probably enzymes necessary for viral replication and their number must be rather small owing to the low molecular weight of TMV RNA.

The mutagenicity of nitrous acid is attributed to its ability to deaminate guanine, cytosine, and adenine (p. 408). This compound attacks the nucleic acid chain along its entire length, and when the molecule has accumulated a sufficient number of chemical alterations, it becomes unable to synthesize a functional version of one or another virus-specific protein. The RNA molecule is said to be inactivated under these circumstances. Although some multiple mutants do survive, it has been determined that most viable mutants result from just one alteration in the RNA chain, one deamination in the case of nitrous acid. For this to be true, mutagenesis must obey first-order kinetics, and the ratio between the number of mutants, n_m, at any time t, and the maximum number of mutants, n_m^0, should be given by the equation:

$$n_m/n_m^0 = At e^{-t/\tau} \qquad (4\text{-}38)$$

where τ is a parameter which characterizes the kinetics of RNA inactivation, and A is a constant. It is evident from Fig. 4-25 that the experimental points fall on the solid curve expressed by Eq. (4-38), and that they are not consistent with the dashed curve representing second-order kinetics. A tentative calculation based on the kinetic data and mutation frequency reveals that about one fourth of the deaminations lead to viable mutants, whereas three fourths of the nucleotide alterations are lethal. This

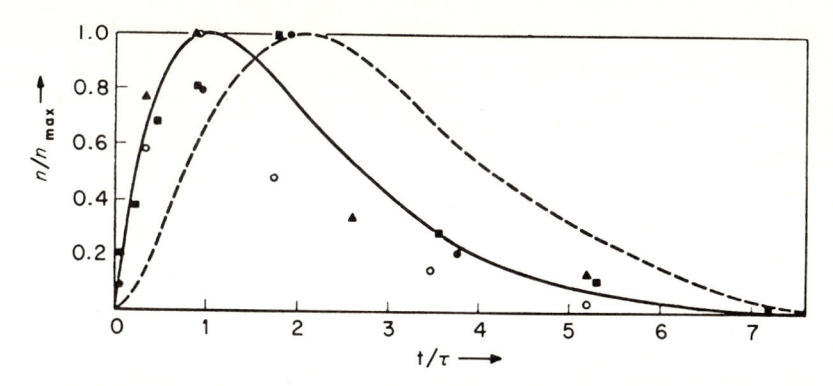

Fig. 4-25. Dependence of the concentration of tobacco mosaic virus mutants (number of necrotic lesions relative to the maximum value) on the time of incubation with nitrous acid, related to τ; ———, single-hit curve (Eq. 4-38); ------, double-hit curve. Different symbols correspond to intact TMV and purified TMV RNA from different strains and at different concentrations. From Gierer and Mundry (86).

result indicates that TMV is highly mutable, for normally one would expect a much higher ratio of lethal events.

Another method of determining the number of hits or nucleotide alterations required for lethality is to measure the kinetics of RNA inactivation and, at the same time, assay the kinetics of nucleotide base deamination. To estimate the extent of deamination, RNA chains are first completely hydrolyzed and the resulting bases are then analyzed by chromatography. Chemical changes are detectable, however, only after several percent of the bases have been deaminated. Chemical techniques are thus much less sensitive than infectivity assays. Consequently, viral RNA must be treated with the mutagen for several hours (at 20°C, pH 4, 1 M NaNO$_2$) before chemical tests can be performed, while loss of infectivity requires an exposure of only a few minutes under the same conditions. Since the reaction proceeds linearly with time, it is not difficult to calculate by extrapolation that, on the average, one to two deaminations are sufficient to inactivate one molecule of RNA.

TMV RNA molecules are also inactivated by the enzyme ribonuclease and by sonication, which physically disrupts the continuity of the nucleic acid strand. One break is enough to completely inactivate TMV RNA.

All of the foregoing experiments provide evidence that the genetic information necessary to synthesize a tobacco mosaic virus particle is borne only by its intact RNA chain. Damage to a single nucleotide residue can inactivate the nucleic acid molecule, or at least, induce a mutation in it.

Infection and formation of TMV particles in the host cell has been studied by a number of different techniques. By means of ultraviolet microscopy, it has been established that viral RNA synthesis begins soon after infection in the cell nucleus. Within 2 hours, the amount of RNA in the nuclei becomes very large, so that the total quantity of nucleic acid present increases by a factor of two. The RNA then leaves the nucleus and begins to appear in the cytoplasm. There are no mature virus particles present in the plant during the first few hours after infection, although infectious RNA molecules can be recovered from the cytoplasm and purified by preparative ultracentrifugation. A study of the free RNA molecules revealed that a double-helical *replicative form* of viral RNA is synthesized during the early stages of virus multiplication.

The synthesis of viral protein can also be traced in the infected cell. By using fluorescent antibodies specific for TMV protein (prepared by the method of Coons), it is possible to localize the synthesis of viral protein and demonstrate that it occurs in the cytoplasm. According to latest reports, TMV RNA itself serves as a template for viral protein synthesis in conjunction with ribosomes already present in the host cells.

Most of the stages in the formation of TMV particles have thus been elucidated. In a certain sense, viral reproduction is a genetic disease: the virus bypasses the genetic information of the host cell, but appropriates existing enzymes and ribosomes to synthesize its own constituents. As a result, viral proteins are produced in place of cellular proteins and viral RNA is produced instead of cellular nucleic acids. Assembly of complete virus particles apparently takes place spontaneously if external conditions are favorable and when both RNA and protein are present in sufficient quantities. The development of most plant viruses is believed to follow much the same pattern as that exhibited by the very extensively studied TMV.

TMV is not alone in yielding infective nucleic acids. Infective RNA and DNA can be extracted from a large number of plant and animal viruses. Infective DNA can also be obtained from certain bacteriophages such as λ, T1, and ΦX174, but it can be introduced into host cells only with difficulty. Intact bacteriophages are specially adapted to attack the tough bacterial cell wall, through which they must inject their DNA. The infectivity of purified phage DNA can nonetheless be tested by preparing *spheroplasts* from bacterial cells. Bacteria are treated with lysozyme, an enzyme which digests cell walls. The spheroplasts which result are viable and remain intact even though denuded of their firm supporting wall, as long as the osmotic pressure of the external medium is maintained at a sufficiently high level. Spheroplasts are permeable to purified DNA and therefore permit infection. Whole *Bacillus subtilis* cells, competent for transformation, are also penetrated and infected by phage DNA.

Bacterial viruses or *bacteriophages* have been investigated in great detail and the remainder of this section will be devoted to an exposition of their properties and their genetics. The structure of some of these viruses is quite complex, as can be seen from the schematic diagram of the DNA-containing coliphage T2, illustrated in Fig. 4-26. Let us now follow the sequence of events that occurs when normal bacterial cells are infected with intact bacteriophages (89-91). The phage is first adsorbed to specific receptors on the outer surface of the cell through its tail proteins. There it makes a small opening in the cell wall with a special lysozyme carried in its tail structure, and injects its contents into the cell interior. In large phages, injection is effected by a mechanochemical contraction of the tail proteins (Fig. 4-27). The empty protein coat or "ghost" remains on the outside of the bacterium. Cells can be readily freed of adsorbed phage "ghosts" by vigorous agitation in a blender. Individual stages of this sequence can be recorded very well in electron micrographs.

The nature of the substance introduced by the phage into bacterial cells was studied by Hershey and Chase with radioactively labeled phage particles (79). They prepared T2 phages with either ^{35}S-labeled proteins or ^{32}P-labeled nucleic acids by

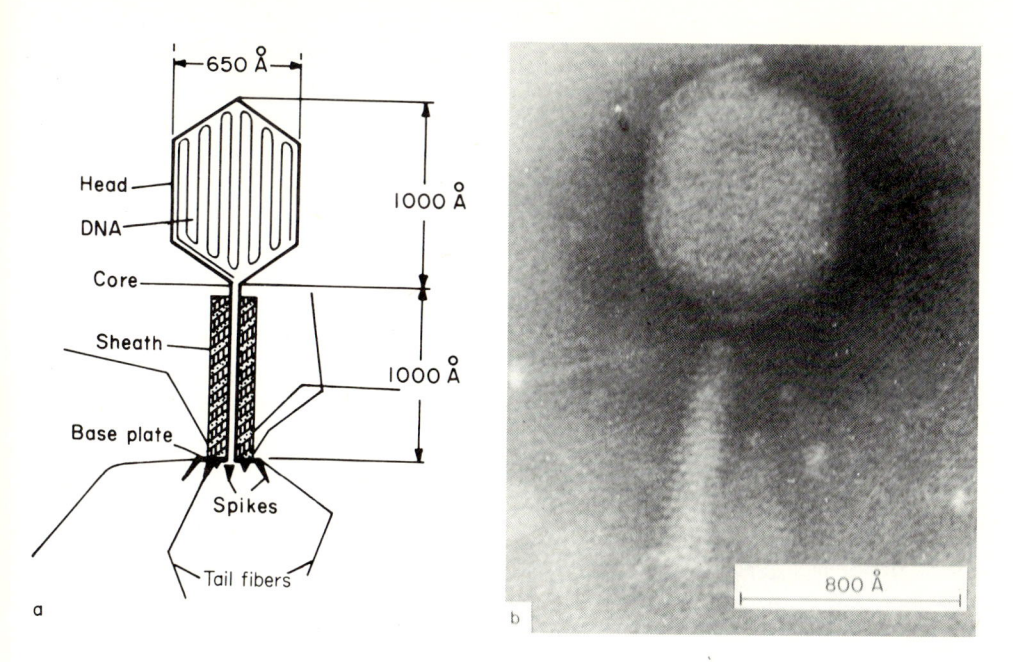

13. Genetic Processes in Viruses and Bacteriophages 371

Fig. 4-26. The structure of bacteriophage T2. (a) Diagram showing the morphological components of phage T2 and their arrangement in the intact particle. Adapted from Horne and Wildy (87). (b) Electron micrograph of phage particle embedded in phosphotungstic acid; ×400,000. From Brenner *et al.* (88).

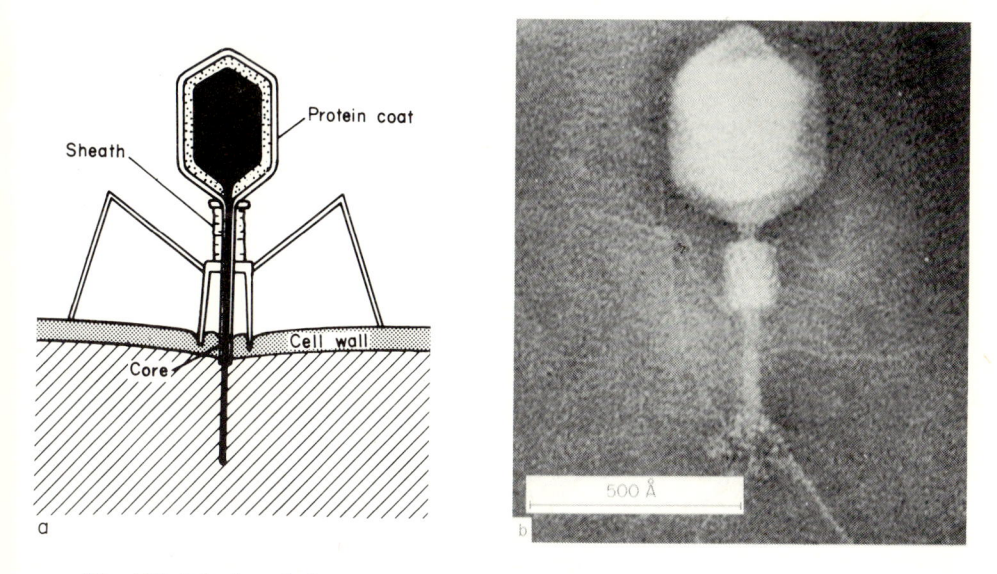

Fig. 4-27. Injection of phage DNA into the bacterial cell. (a) After adsorption to the bacterial cell wall through the spikes and fibers of the tail structure, the sheath contracts, driving the core through the wall and releasing phage DNA into the cell. From Wood and Edgar (92). (b) Electron micrograph of bacteriophage after contraction of tail sheath. Note thin inner tail core and splayed fibers of tail. From Brenner *et al.* (88).

propagating the particles on *E. coli* cells grown in the appropriate radioactive medium. After the phages were purified, they were used to infect nonradioactive *E. coli* hosts. The bacteria were then washed free of unadsorbed phages and phage ghosts and the amount of radioactive protein and DNA transferred to them was measured in separate experiments. Hershey and Chase ascertained that 70% of the total phage DNA is injected into the bacteria; losses amounting to 30% are quite understandable, for a certain number of the phage particles are inactive and, moreover, some of the phages may attach not to cells, but to cell wall fragments stemming from lysed cells. In the latter case, the phages simply inject their DNA into the medium where it is functionally inactive. Significantly, 50% of the ^{32}P label from the DNA of the infecting phages reappears in the DNA of their progeny. The small amount of protein that turns up in the infected cells does not participate in the formation of new phage particles. The conclusion drawn from these and similar experiments is that phage DNA is alone responsible for infection and that it carries all the information necessary to specify the synthesis of new enzymes and proteins which the phage needs for reproduction.

After phage T2 is adsorbed and its DNA penetrates the cell, synthesis of bacterial protein and nucleic acid is completely halted. Indeed, this can occur even without injection of phage DNA into the bacterial cell. If empty phage coats are adsorbed to *E. coli*, the cell ceases to synthesize its own components, although no virus particles are obtained. Under normal conditions, however, expression of phage DNA begins soon after its penetration into the host. The transcription of a special type of informational RNA precedes the formation of the first phage-specific proteins by a few minutes. Among the first proteins to appear are enzymes essential for the synthesis of new phage DNA. In particular, there must be enzymes to produce 5-hydroxymethylcytosine (HMC), a pyrimidine base which completely replaces cytosine in T2 DNA, but which is not elaborated by the host cell. Furthermore, many of the HMC residues are glucosylated, and a special enzyme is needed to attach glucose to the HMC hydroxyl groups. At about 6 minutes after infection, phage DNA synthesis begins and continues until the cells lyse some 15 to 20 minutes later (93). Up to 12 minutes after infection, no mature phage particles are detected if the cells are prematurely lysed. This interval is called the *eclipse period*.

Hershey followed the synthesis of phage DNA chemically, by measuring the incorporation of HMC residues. Since HMC is not present in the host cell DNA, incorporation of this base is proportional to the amount of phage DNA synthesized. Completed phage DNA molecules enter a pool in which extensive genetic exchange takes place. By minute 12 of infection, the pool contains from 30 to 50 phage genomes. Phage structural protein is also synthesized during this period, and from the twelfth minute onward, phage particles are assembled from DNA and structural proteins. The DNA pool size remains constant for 10 to 15 minutes thereafter, because the number of molecules being synthesized is equal to the number leaving the pool in completed phage particles. Between 20 and 25 minutes after the initiation of infection, the number of mature phage particles reaches 100 to 200, and the cell is lysed by a phage-specific lysozyme, liberating the mature progeny into the medium.

The development of other *E. coli* bacteriophages proceeds in an analogous fashion. However, phage λ and the T-odd phages do not contain unusual nucleotides such as hydroxymethylcytosine, so that phage DNA synthesis is initiated more rapidly after infection.

The fate of the parental phage DNA molecules after penetration was investigated by autoradiography (Levinthal and Thomas, 94). T2 phages with very "hot" DNA were prepared by labeling with high specific activity ^{32}P. The radioactive phage particles were then used to infect nonradioactive *E. coli* cells, and the progeny particles were embedded in a sensitive photographic emulsion, exposed for a number of days, and then developed. The distribution of parental ^{32}P-labeled DNA among their daughters was analyzed from the frequency of tracks in the emulsion caused by disintegrating ^{32}P nuclei. This experiment was designed specifically to determine whether the parental DNA molecule is transmitted intact to one of the progeny, that is, by a conservative mechanism, or whether it is dispersed among many of the descendants. It was established that about half of the parental DNA was inherited in the form of single fragments, each containing approximately 20% of the original ^{32}P. These fragments are conserved in successive generations. The remaining isotope was dispersed in much smaller chunks.

Stent and his co-workers studied the transmission of parental phage DNA in radioactive suicide experiments (95). The kinetics of phage inactivation were studied in the progeny of highly radioactive particles. When stored in the frozen state, phages which had received portions of the parental DNA committed suicide owing to the disintegration of ^{32}P nuclei. These experiments demonstrated that a small proportion of the progeny phages contain parental DNA fragments corresponding to between 10 and 20% of the original radioactivity. Approximately half of the parental DNA molecule could be accounted for in this way. It is significant that in the second generation the dimensions of the radioactive fragments did not change. This result agrees qualitatively with the data of Levinthal and Thomas.

Many of the experiments described so far were facilitated by development of a chromatographic technique for the preparation of chemically pure T2 phage. It is remarkable that such a large and complex structure can be chromatographed on DEAE-cellulose without inactivation and with only minimal losses of material (Hershey, Levinthal).

We shall now focus our attention on bacteriophage genetics, a field of investigation that has made many significant contributions to molecular biology. It will be helpful to initiate the discussion by considering the selection of mutants from among the T-even coliphages, T2 and T4. One class of phage mutants that was discovered very early involves differentiation according to morphology of the sterile spots or *plaques* which they form upon lysing a dense bacterial culture on the surface of nutrient agar. Bacteriophage are counted, or *titrated*, in the following way. A small portion of melted "soft agar" (0.65% agar in nutrient medium), containing bacteriophage and sensitive indicator bacteria, is poured onto the surface of a hard agar plate. An appropriate indicator strain is *E. coli* B, for it is sensitive to infection by both T2 and T4. The final concentration of bacteria is adjusted to about 10^8 cells/ml of soft agar. The phage suspension is diluted so that there are between 50 and 200 particles/plate. The soft agar overlay is allowed to harden, and the plates are incubated at 37°C. After a day or so, the hardened agar layer will appear turbid since it is filled with a "lawn" of bacteria, but there will also be a certain number of clear circular zones distributed over the plate. These regions, called plaques, are places in which phages have multiplied and lysed the indicator bacteria. Each plaque represents one phage in the original preparation. Wild-type phages yield plaques with a clear central region, about 1 to 2 mm in diameter, with a wider halo, a zone in which not all bacteria have been lysed. The halo generally is graded in density from the clear inner circle to the surrounding bacterial lawn.

From time to time, on the order of 10^{-4} per generation, altered plaques are noted. They are larger in diameter, and have sharp edges instead of indistinct halos (Fig. 4-28). Such mutants are apparently able to cause rapid lysis of the bacterial cells and

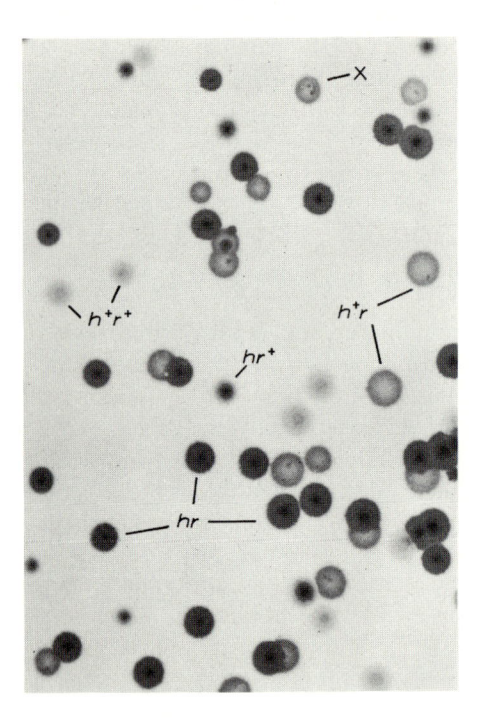

Fig. 4-28. The four types of plaques formed by phages T2h^+r^+, T2h^+r, T2hr^+, and T2hr on mixed bacterial indicators. The h^+ strains form turbid plaques and the h mutants form clear plaques on the mixed indicator strains, *E. coli* B and B/2. The r^+ plaques are characterized by fuzzy edges, the r mutant plaques by sharp edges. Adapted from Stent (91).

within the normal incubation time, the process extends farther and more fully than in the wild-type strain. If the material from such a plaque is retested, it is found that all the progeny phages have the same characteristics and it is evident that a hereditary change has taken place. Mutants of this type are designated by r, which stands for *rapid lysis*, and they all result from changes at a small number of definite genetic loci. The symbol r^+ is used to designate the wild type, which yields plaques of the kind described in the preceding paragraph. Another type of mutant produces *turbid* plaques and is designated by the symbol *tu*. Certain other more complicated differences in plaque morphology have also been observed.

A second kind of mutation which relates to a different phage locus is recovered because of differences among host cells. In any population of *E. coli* B, it is possible to find individual cells which are resistant to phage T2. As was mentioned previously, these are bacterial mutants which lack receptor groups in the cell wall for the adsorption of wild-type T2 phages. Mutants of this type are designated either B/2 or B, T2^r. Now T2 mutants can be obtained in turn which adsorb to and infect *E. coli* B/2. The phage mutants differ from their parent strain in *host range* and are designated by the letter *h*. In essence, these are mutants with an altered spectrum of lytic action. The symbol h^+ denotes wild-type behavior with respect to strain B of *E. coli*. Rapid lysis and host range mutants act like point mutations, since they can be reverted in a single step. It is not difficult, however to produce double and triple mutants with mutagenic agents, in which serveral loci are affected.

One of the most important properties of bacteriophages is the ease with which they undergo recombination. It is precisely the very high probability of genetic exchange in T-even bacteriophages which makes these organisms so indispensable in microbial genetics.

Recombination is detected in the progeny phages which result when bacterial cells are mixedly infected with two or more genetically distinct phage strains. Mixed infection is generally carried out at a high *multiplicity of infection*, that is, with a large excess of phage particles over bacteria to ensure that each cell receives at least one phage of each type. Furthermore, KCN is added to the medium in order to inhibit cellular synthetic processes. This not only provides a convenient means of synchronizing phage development, but it also prevents the DNA introduced by the first phage from evoking a response which excludes or degrades the DNA of the second phage. After infection, unadsorbed phage is removed by centrifugation or by the use of antiserum. The infected culture is then diluted into fresh nutrient medium which, in effect, eliminates the inhibition caused by the KCN. The phages develop synchronously thereafter and lyse the culture after 20 to 25 minutes' incubation. The last step is to plate the progeny phages on an agar plate specially prepared for the selection of mutants.

In the course of phage maturation, practically every phage genome participates in recombination. If the cells are infected simultaneously with three mutant strains, differing from each other at three separate loci, it is quite common to observe recombinants bearing characteristics of all three parental types (Hershey and Chase, 96). This last fact demonstrates not only that all phage DNA molecules participate in recombination, but even more, that each molecule undergoes repeated genetic exchange in the course of a single generation.

In order to study the statistics of phage recombination in greater detail, use can be made of the single-burst technique first developed by Delbrück, wherein the progeny of a single infected cell is examined. In this experiment, the cells are greatly diluted after infection, but before lysis, and aliquots are distributed among a series of tubes in such a way that no tube should contain more than one cell. In fact, a certain number of the tubes will remain empty according to the statistics of the situation. The phages mature and lyse their host cells and a complete generation can then be plated on a petri dish with appropriate indicator bacteria. This approach was used to study two-factor and three-factor crosses in phages T2 and T4 as well as others.

Visconti and Delbrück developed a statistical theory of recombination in phages to account quantitatively for the frequency of recombinants recovered from mixedly infected cells (97). The physical premises of this theory are as follows.

(1) Upon infection of the host bacterium, the viral chromosome multiplies vegetatively many times, giving rise to a pool of complete viral precursor genomes.

(2) After termination of eclipse until the time cell lysis occurs, the number of viral genomes in the pool remains constant since the rate at which the genomes are randomly and irreversibly withdrawn from the pool and incorporated into mature viruses equals the rate at which the genomes replicate.

(3) Completed phage DNA molecules in the pool of the infected cell undergo random pairwise matings with one another, leading to an exchange of genetic material by one or more crossovers.

(4) The probability of mating between any two molecules in the pool is considered equivalent and the probability of crossing-over is assumed to be the same at any point along the chromosome.

(5) From the time it is synthesized until it is withdrawn from the vegetative pool for incorporation into a mature virus particle, every DNA molecule undergoes a certain average number of matings with other molecules in the pool.

Using probability theory, it is possible to calculate the frequency with which double and triple recombinants are formed during repeated chromosomal exchange, as well as the average number of matings each chromosome undergoes. Suppose that a bacterium is infected with an equal number of two parental phages differing in the markers they carry at the x and y loci, and that the probability of recombination between x and y per mating is W_{xy}. If R_{xy} is the fraction of phages in the mating pool recombinant for these two markers, and m is the average number of mating events, or rounds of mating, per phage genome, then an increase in m by dm will bring about a corresponding increase of

$$(1/2)(1 - R_{xy})^2 W_{xy} dm$$

in the fraction of recombinants. Here $(1 - R_{xy})^2$ is simply the probability of interaction between two chromosomes nonrecombinant for x and y, that is, which have the parental genotypes. Since the input of parental phages is equal, the factor of 1/2 is required because only half of the matings occur between individuals of opposite genotype. From the above value, it is necessary to subtract

$$(1/2) R_{xy}^2 W_{xy} dm,$$

which is the fraction of previously formed recombinants converted back to parental types after further rounds of mating. The net increase in recombinants is therefore

$$dR_{xy} = (1/2)(1 - R_{xy})^2 W_{xy} dm - (1/2) R_{xy}^2 W_{xy} dm \qquad (4\text{-}39)$$

or

$$dR_{xy} = (1/2)(1 - 2 R_{xy}) W_{xy} dm \qquad (4\text{-}39')$$

Solving this differential equation, we obtain

$$R_{xy} = (1/2)(1 - e^{-W_{xy}m}), \quad (4\text{-}40)$$

an expression relating m, the average number of matings between phage genomes, R_{xy}, the experimentally determined recombination frequency, and W_{xy}, the probability of recombination per mating.

One method of evaluating m is to set up a three-factor cross in which the two parental bacteriophages differ in three genetic markers, x, y, and z. The markers are assumed to be located on a single chromosome, and the probabilities of recombination per mating between pairs of these markers are W_{xy}, W_{yz}, and W_{xz}. By analogy with Eq. (4-40), the frequency of progeny phages resulting from pairwise recombination between x and y, y and z, and x and z should be

$$R_{xy} = (1/2)(1 - e^{-W_{xy}m}) \quad (4\text{-}40)$$

$$R_{yz} = (1/2)(1 - e^{-W_{yz}m}) \quad (4\text{-}41)$$

$$R_{xz} = (1/2)(1 - e^{-W_{xz}m}) \quad (4\text{-}42)$$

These three equations contain four unknowns, m, W_{xy}, W_{yz}, and W_{xz}, but the magnitudes of W_{xy}, W_{yz}, and W_{xz} are not independent of one another. If the sequence of markers is x-y-z, the distances between them, l_{xy}, l_{yz} and l_{xz}, must be additive

$$l_{xz} = l_{xy} + l_{yx} \quad (4\text{-}43)$$

By means of Haldane's formula (Eq. 4-33, p. 351), W_{xy}, W_{yz}, and W_{xz} can be written in the following form

$$W_{xy} = (1/2)(1 - e^{-2l_{xy}}) \quad (4\text{-}43a)$$

$$W_{yz} = (1/2)(1 - e^{-2l_{yz}}) \quad (4\text{-}43b)$$

$$W_{xz} = (1/2)(1 - e^{-2l_{xz}}) \quad (4\text{-}43c)$$

An elementary transformation leads to the expression

$$W_{xz} = W_{xy} + W_{yz} - 2 W_{xy} W_{yz} \quad (4\text{-}44)$$

Now there are four simultaneous equations that can be solved for the four unknowns, m, W_{xy}, W_{yz}, and W_{xz}, after appropriate experimental values are substituted for R_{xy}, R_{yz}, and R_{xz}.

A most important result of the Visconti-Delbrück theory was the determination of m, the average number of matings per phage DNA molecule in the precursor pool for a number of different cases. Using experimental data for 3-factor crosses, m was found to be about 5 for phages T2 and T4, and between 0.5 and 1 for λ and T1. Recombination is thus a very frequent occurrence in the T-even phages, less so in certain others.

As to the actual mechanism of recombination, the experiments of Levinthal and Stent cited earlier in this section, provide support for breakage and reunion and argue against a conservative copy-choice scheme. This question was a point of dispute for many years, however, because the true molecular weight of phage DNA was not known with any degree of certainty. In other words, the integrity of the entire phage genome had not yet been confirmed and it appeared as if each particle might contain several DNA molecules. It has since been demonstrated that the DNA of phage T2 consists of a single molecule with a molecular weight of 1.3×10^8 daltons, and it is quite evident that the genome of this phage is broken up during recombination.

Delbrück suggested that fragmentation occurs as in higher organisms, that is, because of crossing-over. In higher organisms, this process takes place not on the molecular level, but on the chromosomal level and it can even be observed in the light microscope. Both strands twist around one another at synapsis, and if they originally carry the markers *AB* and *A'B'*, respectively, after breakage and reunion, they will form the new combinations *AB'* and *A'B*. Both sets of chromosomal ends are reattached simultaneously and, for that reason, recombinants of both types, usually called *reciprocal recombinants*, are recovered among the progeny in just about equal quantities.

Reciprocal recombinants are not generally found among the progeny phages stemming from a genetic cross, even when those from a single infected bacterium are analyzed. We may attribute this mainly to the fact that with phages T2 and T4, the phage genomes in the precursor pool undergo repeated mating events. Moreover, if reciprocal recombinants were indeed formed soon after infection, their initial genotypes would undoubtedly be obliterated by subsequent rounds of recombination.

The question of the mechanism of genetic exchange in phages was perhaps approached most clearly by Meselson and Weigle, who studied the relationship between the distribution of parental DNA to progeny and recombination in bacteriophage λ (98). This particular bacteriophage was chosen because it undergoes recombination less frequently than either T2 or T4, and it is thus expected that parental molecules will be less fragmented after only one generation. These authors demonstrated what had been suspected from the dispersal of T2 DNA found in earlier work: recombination requires breakage of DNA molecules and subsequent reattachment of the fragments. In the experiments to be described, mutant phages carrying two linked plaque morphology markers, *c* (*clear*) and *mi* (*minute*) were used as well as the wild type (see Fig. 4-37, p. 394, for an illustration of these plaque types); λcmi particles were propagated on cells growing in "heavy" medium containing the isotopes ^{15}N and ^{13}C, while wild-type $\lambda c^+ mi^+$ were grown up on "light" medium. The cross between the two phages can hence be written schematically as follows:

$$\lambda cmi^{15}\mathrm{N}^{13}\mathrm{C} \times \lambda c^+ mi^{+14}\mathrm{N}^{12}\mathrm{C}$$

A second experiment was performed in which the wild-type phages, $\lambda c^+ mi^+$, were grown in the presence of the heavy isotopes, and the mutant in light medium. We shall be interested in the distribution of heavy label among progeny phages with parental and recombinant genotypes.

The experiment was carried out as follows. A bacterial culture grown on normal light medium was infected by a mixture of the two types of phages at a high multiplicity and the excess parental phages were removed. After maturation and lysis, the progeny phages were centrifuged in a cesium chloride density gradient. After equilibrium was attained, the centrifuge tubes were punctured with a hollow needle and the contents were emptied dropwise into a series of tubes. In each sample, the bacteriophage were titrated in the usual way. Parental types as well as both kinds of recom-

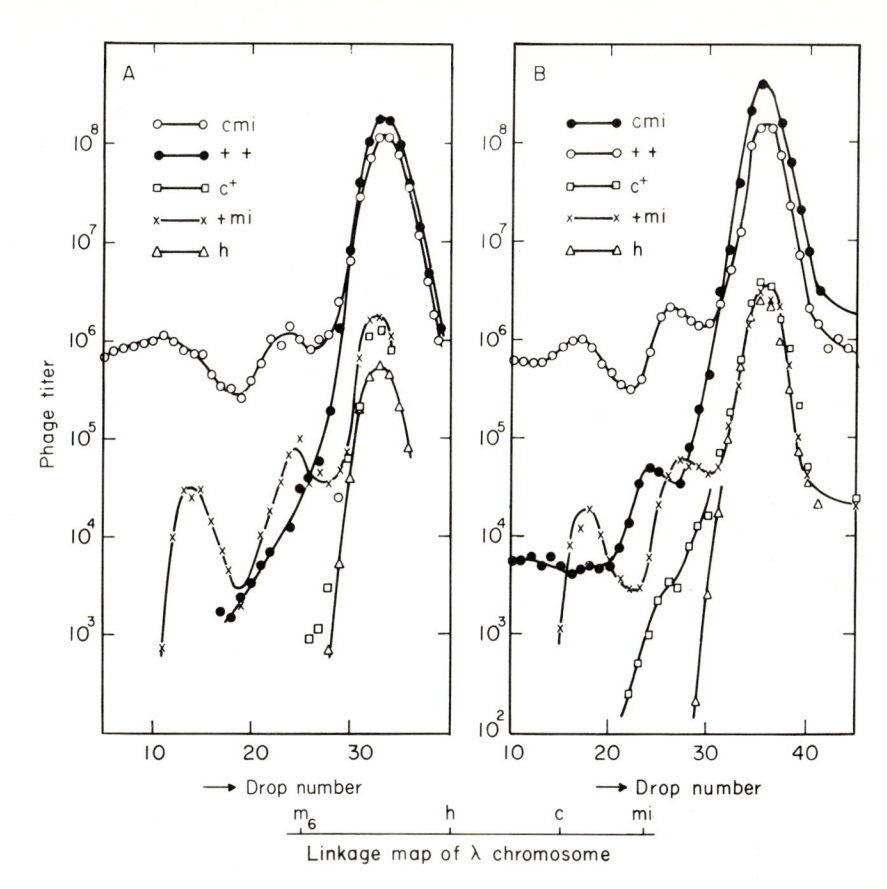

Fig. 4-29. Distribution of parental and recombinant λ phages in a density gradient. (A) Cross λcmi^{13}C^{15}N × λ++ ^{12}C^{14}N. (B) Cross λcmi^{12}C^{14}N × λ+ + ^{13}C^{15}N. The titer of the progeny phages was measured in each fraction and plotted according to genotype as shown in key; λh was added as a density and band shape reference. Inset shows position of c and mi on linkage map of the λ chromosome; the markers m_6 and mi define the extremities of the linear map. From Meselson and Weigle (98).

binants were readily differentiated by their distinctive plaque morphologies. The titer of each type was plotted on a curve as a function of position in the gradient (Fig. 4-29). The two parental phages, as well as a variety of intermediates, are all very well resolved, since particles with pure heavy DNA are 0.05 gm/cm^3 denser than those with pure light DNA. Furthermore, CsCl causes no appreciable decrease in phage viability.

Typical density gradient curves are presented in Fig. 4-29A. Curve 1 shows how phages with the genotype of the heavy parent behave after they have multiplied in light bacteria. The significance of all three peaks is quite clear and unambiguous. The first peak on the left, corresponding to the highest density, marks the position of phage with all-heavy parental DNA, which remained intact during maturation. These phages introduced their DNA into the cell but did not undergo (semiconservative) replication during maturation. Phage with pure parental DNA are observed only after multiple infection of bacterial cells with particles of the same type. The second peak corresponds to phage containing the product of semiconservative DNA replication, molecules that are half heavy and half light. Taken together, these first two peaks comprise between 1 and 2% of the entire phage population. All of the remaining material corresponds to phages with light DNA. This pattern is similar to the one obtained for semiconservative replication of DNA in bacteria (p. 264), with just one exception: among the progeny phage there are particles containing DNA that has not undergone replication. This is a peculiarity of multiple infection, where the DNA pool contains molecules which by chance remain unreplicated.

The density gradient distribution of the wild-type parental genotype (curve 2, Fig. 4-29A), which contained light constituents from the very beginning, contains one distinct peak at a density corresponding to all-light phage DNA. In Fig. 4-29B, the labeling of the two parental phages is reversed.

In order to understand the distribution of recombinants that results from mixed infection, we must take into account the following circumstance. The total number of recombinants between any two strains of λ phage reaches only 10 to 15% of the total progeny and therefore cannot materially distort the distribution of parental-type particles. The number of recombinants between the c and mi loci in this experiment is on the order of 1%. Since the c and mi loci do not occupy symmetrical positions on the λ genetic map (see inset, fig. 4-29), and since recombination requires a crossover between points c and mi, the recombinant chromosome will contain on the average 85% DNA from the donor of the c locus, and only 15% DNA from the donor of the mi locus. In the CsCl gradient, recombinants containing 85% heavy and 15% light DNA will behave approximately like the parent which was grown in the heavy medium.

The cmi^+ recombinants (curve 3, Fig. 4-29A; see also c^+mi recombinants in curve 4, Fig. 4-29B) yield three peaks consisting of heavy, half-heavy, and light DNA. The method is sufficiently accurate to permit an estimation of the density of DNA in recombinants which have not undergone replication. The heaviest peak of the recombinant curve is displaced slightly toward lower density in comparison to the maximum of all-heavy parental DNA. The content of heavy DNA in the chromosome of these recombinants is estimated to be 90%, in satisfactory agreement with the figure of 85% derived from consideration of genetic mapping experiments alone. This represents one of the few cases where it is possible to demonstrate experimentally that the genetic map distance and physical distance along the chromosome are proportional to one another. The second type of recombinants, which contain 85% light DNA from the beginning, behave very much like the parental strain that originally contained light DNA.

The basic results of this experiment may be summarized as follows. (1) In λ phage recombination, a fragment of one parental DNA molecule is joined with a fragment of a second DNA molecule derived from the second parent. (2) The fragments consist of double-stranded parental DNA. (3) Semiconservative replication of DNA next proceeds according to the Watson-Crick model. (4) Distances on the genetic map are proportional to physical distances along the DNA macromolecule which constitutes the phage chromosome.

A likely model for phage recombination is illustrated in Fig. 4-10 (p. 338). Double-stranded DNA belonging to both parental types are first cleaved internally by endonucleases. An exonuclease then successively cleaves terminal nucleotide residues from the 3'-OH end of each chain, producing single-stranded portions on the ends of the DNA fragments. The fragments which arise by this process mix together in the DNA pool. Complementary single-stranded regions of different fragments "recognize" each other and, by joining through hydrogen bonds over homologous nucleotide sequences, provide the immediate precursors of recombinant molecules. The process is completed by an enzyme with the properties of DNA polymerase which repairs all the defects remaining after the two chromosomal fragments have specifically paired. Another enzyme, ligase, closes the final single-strand nicks with phosphodiester bonds. The result is an intact duplex molecule. Anraku and Tomizawa have provided evidence for this model by showing that the introduction of fluorouracil deoxyriboside (FUdR), a specific inhibitor of DNA synthesis, retards recombination at the stage of pairing between two overlapping single-stranded segments (99). Since the chromosomal fragments are held together only by hydrogen bonds, they are susceptible to denaturation.

An important conclusion which follows from the scheme presented is that the recombinant DNA molecule is *heterozygous* for characters which fall within the overlap. The characters should segregate only during subsequent semiconservative replication of the chromosome. Such a phenomenon has in fact been observed in genetic experiments on bacteriophages. A substantial fraction of the progeny phage carries two alleles of certain chromosomal markers after mixed infection. When the phage is propagated, the characters segregate so that two different types of phages are obtained from a single initiating particle. In certain cases, this situation can be observed visually from the morphology of the plaques which are formed. For instance, r^+/r heterozygotes give rise to mottled plaques since the r segregants yield very sharply outlined, clear plaques, while the r^+ phages give turbid plaques.

Hershey and Chase found that if two T4 phage mutants, whose altered loci lie at a considerable distance from one another (on the order of twenty recombination units) are crossed, heterozygotes carrying both markers are formed in only 6% of the cases. If the mutant loci are close to one another, within two recombination units for instance, heterozygotes are formed in 75% of the cases. Levinthal broadened these observations by showing that with three markers, individuals heterozygous for the inside marker turn out to be recombinant for the two outside markers:

$$(abc) \times (a^+b^+c^+) \longrightarrow \begin{cases} (abc^+)/(ab^+c^+) \\ (a^+bc)/(a^+b^+c) \end{cases}$$

He suggested that the chromosome of a heterozygote consists of a normal double-stranded DNA molecule, but that the information carried by the two strands is different in the region of heterozygosity. A molecule of this type is called a *heteroduplex*.

Meselson confirmed the formation of heteroduplexes in λ phage crosses by means of density gradient centrifugation (100). Host cells were simultaneously infected with wild-type λc^+ particles labeled with heavy isotopes and with λc mutants grown on normal light medium. When plated on appropriate indicator bacteria the c^+ and c parental type phages give rise to turbid and clear plaques, respectively, while c^+/c heterozygotes produce easily distinguishable *mottled* plaques. Figure 4-30 shows the titer of parental and heterozygous phages resulting from the cross as a function of their density in a CsCl gradient. In particular, the left-hand peak of the curve

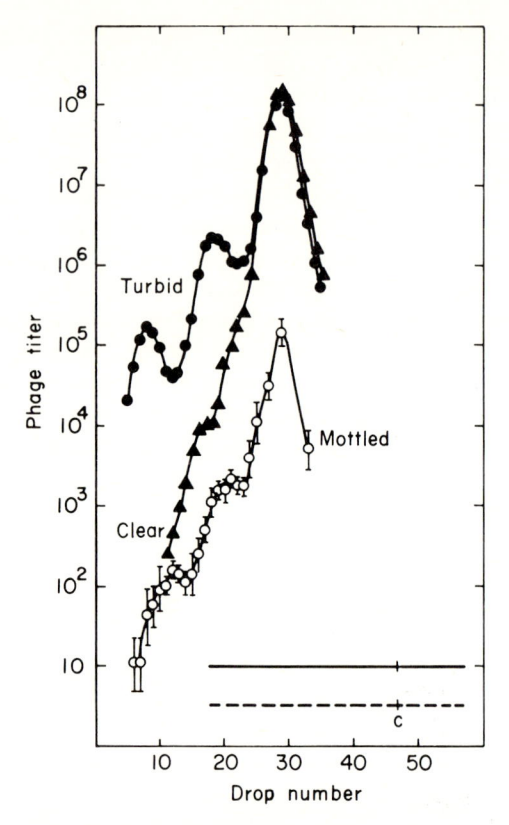

Fig. 4-30. Density gradient distribution of turbid, clear, and mottled plaque-forming progeny from a cross of $^{13}C^{15}N$-labeled λ with unlabeled λc. From Meselson (100).

labeled "mottled" corresponds to *unreplicated* heterozygotes and it has the density which would result if 3/4 of the chromosome were derived from a heavy parent and 1/4 from a light parent. This is precisely the composition expected in the heteroduplex since the c locus is situated just 1/4 of the distance from the end of the chromosome. The reciprocal heteroduplex cannot be resolved from the half-heavy parental peak.

Let us now summarize the basic phenomena of phage genetics.

(1) Coliphages T2 and T4 yield a large number of double, and a substantial number of triple, recombinants. According to the statistical calculations of Visconti and Delbrück, the average number of pairwise matings in which a phage DNA molecule participates during one generation is equal to 4 or 5 in phages T2 and T4, while for λ phage, this figure is on the order of 0.5.

(2) In addition to recombination, phage particles are formed which contain heterozygous regions and these regions are always located between two recombinant loci.

(3) In the progeny arising from fully radioactive T2 phage, individuals are recovered which carry between 10 and 20% of the parental DNA molecule. After repeated infections, the parental chromosomal fragments are conserved and dispersed no further.

(4) The mechanism of phage recombination requires breakage of a double-stranded DNA molecule and subsequent reattachment of the fragments through regions of homology.

In bacteriophage T4, investigation of the detailed or fine structure of the genetic map has confirmed and extended many of the conclusions drawn from studies of bacterial genetics. In particular, it has been determined that all genetic loci are arranged in a single linkage group, or chromosome. This corresponds well to the observation that each phage particle contains just one molecule of DNA. The proportionality of recombination frequency, W_{AB}, to map distance, Δl_{AB}, was also established in the case of T4, as long as the loci considered are not too distant from each other. As we have already mentioned, the proportionality of map distance and recombination frequency breaks down when the yield of recombinants becomes very large and approaches 50%. If recombination frequencies determined for short segments of the chromosome are summed up over the entire chromosomal length, a genetic map of 800 recombination units is obtained.

The molecular weight of T4 DNA is 1.3×10^8 daltons and it contains $(1.3 \times 10^8)/(2 \times 330) = 2 \times 10^5$ nucleotide pairs. If map units are proportional to the physical dimensions of the chromosome, a recombination distance of 1% represents $(2 \times 10^5)/800 = 250$ nucleotide pairs. One recombination unit in phages is evidently much less than the similar unit defined for bacteria (p. 354). In other words, recombination is a much more likely event in phage than in bacteria per unit length of the chromosome. The physical length of various cistrons in phage and bacteria is approximately the same, however, since the number of nucleotide residues in a cistron is directly related to the size of the protein whose synthesis it governs.

The genetic map of bacteriophage T4 is illustrated in Fig. 4-31. As in the case of the bacteria *E. coli* and *Salmonella*, the phage map consists of a single circular linkage group. The inner circle shows the positions of the classic loci, *r*, *h*, and *tu*, which were studied earliest and which we have already considered. The middle and outer circles indicate the locations of temperature-sensitive (*ts*) and amber (*am*) mutations, respectively. Both types of conditionally lethal mutants were characterized in bacteriophage T4 by Edgar and his co-workers (10, 101). Such mutants are useful in mapping studies because they cover the entire chromosome more or less uniformly. More than 40 cistrons were identified and mapped by this method.

From an inspection of the T4 genetic map, we see that there are certain regions which relate to the synthesis of phage head, tail and tail fiber proteins. There are a large number of amber mutants known in which the ability to form one of these structural components is genetically impaired. Such defective mutants can be propagated normally on permissive strains of *E. coli* carrying a suppressor gene capable of correcting the genetic error (p. 439ff). These mutants can also introduce their DNA into bacteria not carrying the suppressor, but in place of viable phage particles, phage components are produced whose nature depends on the position of the genetic block. The complete phage structure cannot be constituted, however, because the cistron altered by amber mutation does not produce an active protein product.

It was shown by Edgar and co-workers that the structural components accumulated after infection by many amber mutants of T4 are normal precursors of the blocked step and that their assembly into viable virus particles can be studied *in*

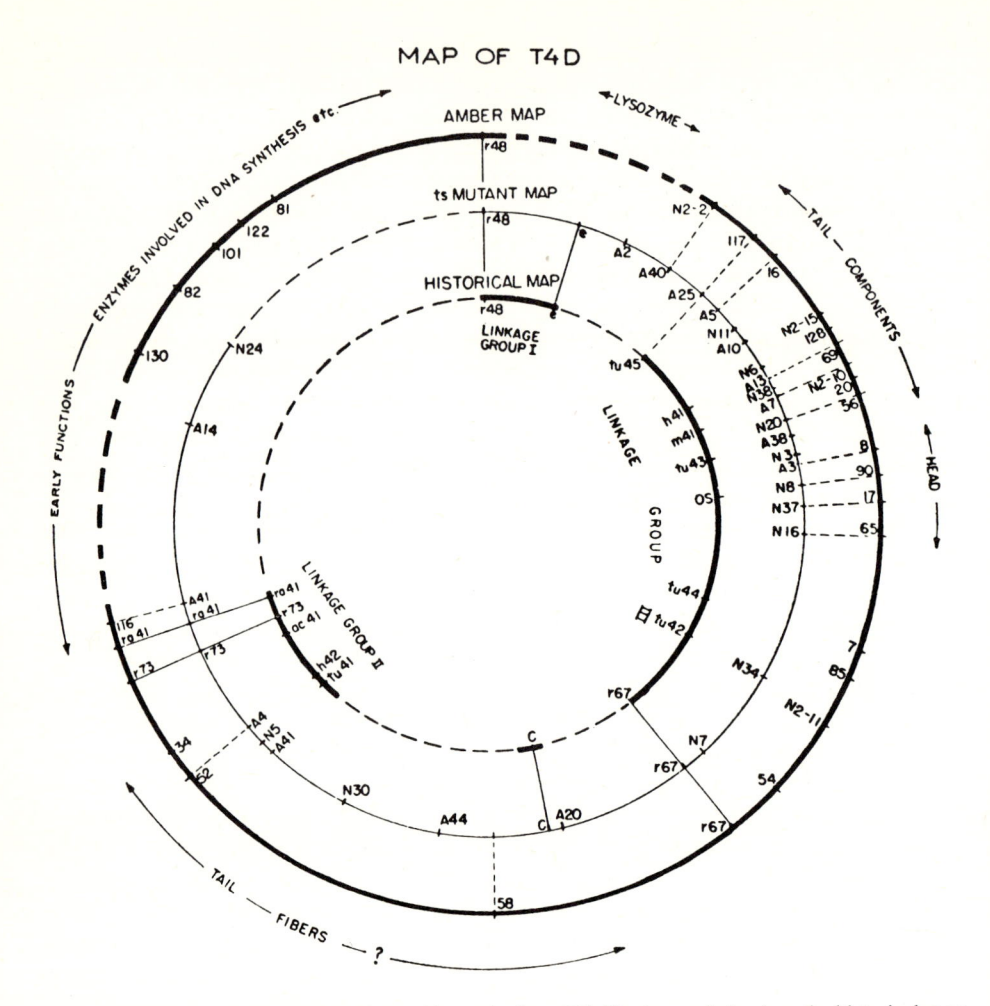

Fig. 4-31. The circular linkage map of bacteriophage T4. The inner circle gives the historical map which indicates the linkage groups inferred from early studies. The middle circle is constructed from analysis of temperature-sensitive (*ts*) mutants. The outer circle indicates the linkage of amber (*am*) mutants. A dashed radial line between the maps denotes a negative complementation test between mutants on the maps so connected. A solid radial line indicates that the marker was used to construct both maps. A solid circumferential line shows that direct linkage between the markers has been established, while a dashed circumferential line shows indirect or inferred linkage. Data of R. S. Edgar as presented by Hayes (3).

vitro (102–104). Phages with defects in different structural proteins are used to infect nonpermissive *E. coli* hosts. The cells are lysed by freezing and thawing, and the two cellular extracts are mixed together in a solution containing an optimal concentration of Mg^{2+} ions. Under these conditions, intact phage particles are assembled when the two defective phages complement one another, that is, when the first extract supplies the component(s) which the second lacks and vice versa. The formation of infective phages is easy to assay by spreading them on a petri dish seeded with a sensitive bacterial indicator strain.

These studies on the *in vitro* complementation of phage extracts, together with investigations of the components accumulated by various mutants, suggest that T4 morphogenesis occurs in a stepwise fashion and that most of the steps are under gene control. The locations of the genes contributing to this process are shown on the T4 genetic map depicted in Fig. 4-32a and the assignment of genes to particular stages in the morphogenetic pathway are given in Fig. 4-32b. The head, tail, and tail fibers of the phage are joined together into a complete virus particle only after they are first assembled independently of one another. Furthermore, the tail fibers are attached only after the tail has been attached to the head, and never to complete tails alone. The steps leading to the assembly of each of the three principle virus components appear to follow a strict sequence, although there is no evidence that this results from the time sequence in which the various gene products are synthesized. To the contrary, the order in which morphogenesis proceeds seems to be inherent in the structure of the proteins themselves and in the changes they undergo as a result of their successive interactions during the assembly process. It is of considerable interest that one of the main morphogenetic reactions, the union of finished heads (Ha) with complete tails (Ta), takes place spontaneously. This also appears to be true in the case of λ phage which can also be assembled *in vitro* from mutant extracts containing heads and tails individually (105).

Some mutations that map in the head and tail regions do not appear to affect any of the phage structural proteins. These may well be mutations in specific enzyme-like proteins which assist in the assembly of phage particles by promoting the attachment of one structural element to another. Since phage assembly *in vitro* requires neither protein nor nucleic acid synthesis, we may conjecture that the "assembly enzymes" cleave peptides from structural proteins, thereby unmasking active groups necessary for their association. In this sense, the action of these putative enzymes would resemble that of thrombin, which cleaves a peptide from fibrinogen, rendering it able to polymerize into fibrin.

The circular structure of the phage chromosome, postulated on the basis of genetic studies has been confirmed by physicochemical methods in several different cases. Sinsheimer and associates, for example, demonstrated the circularity of DNA from the small phage ΦX174 by various methods, including electron microscopy (106). Many animal viruses also contain circular chromosomes. The DNA of λ-phage can exist in either circular or linear form, depending on the circumstances. When the phage is first deproteinized, linear molecules are observed, but storage at high ionic strengths induces the formation of circular molecules (107). It has been established that there are complementary single-stranded segments at the ends of the linear DNA molecules which promote ring closure under conditions conducive to base-pairing (108). Furthermore, the replicative form of λ-DNA in the infected cell is the closed one and ring closure can be produced artificially if the phage DNA is annealed in solution. However, when the single-stranded ends are acted on by DNA polymerase, complete double-stranded molecules result and ring closure is no longer possible (109). The infectivity of λ-DNA is lost at the same time.

An even more interesting situation was found in experiments on DNA from phage T2. Although genetic studies indicate the presence of a single circular linkage group in this phage, electron micrographs of T2 DNA consistently reveal linear molecules.

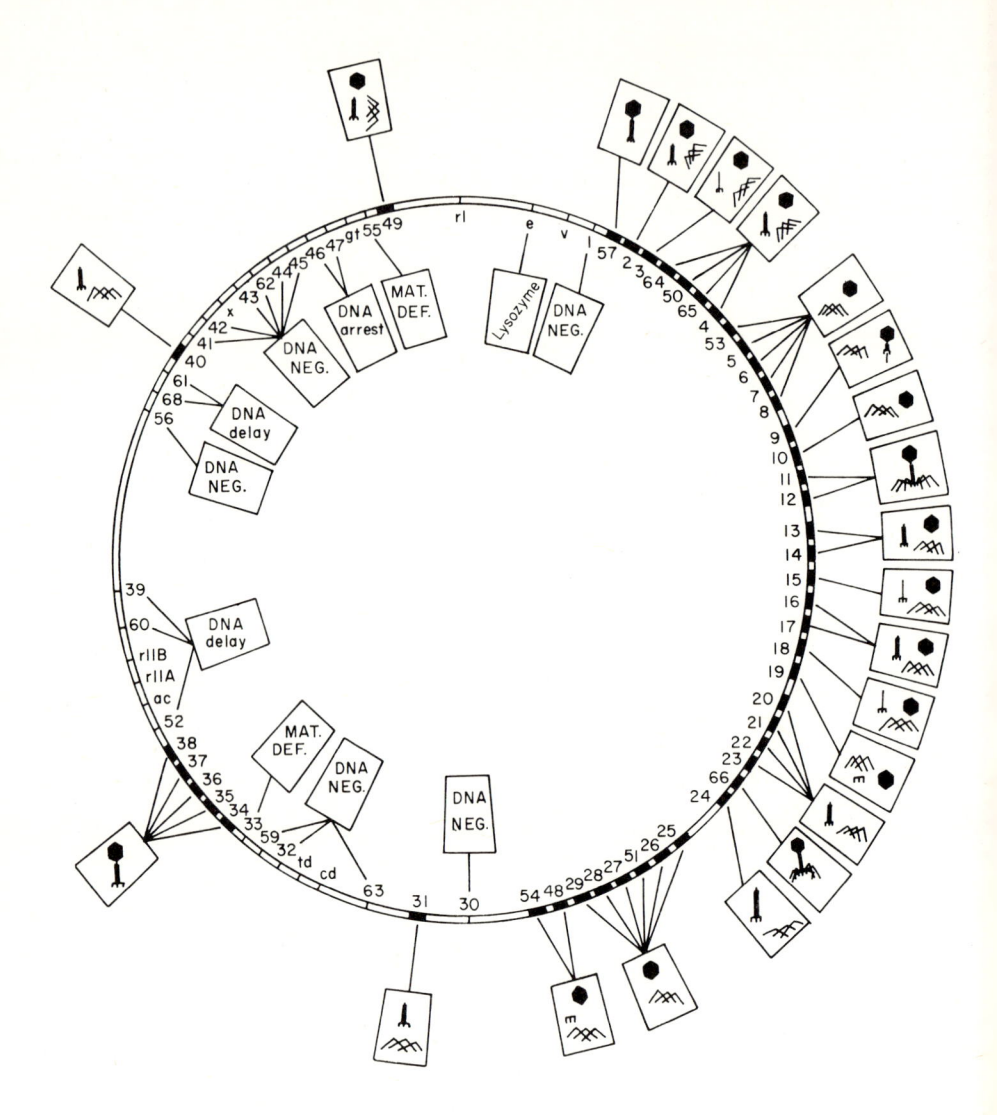

Fig. 4-32. Morphogenesis of bacteriophage T4. (a) Genetic map of phage T4 showing defective phenotypes of conditional lethal mutants. Solid black segments of the circle denote genes with morphogenetic functions. The boxes within the circle designate defects in early phage functions as follows: DNA NEG, no DNA synthesis; DNA ARREST, DNA synthesis arrested after a short time; DNA DELAY, DNA synthesis begins after some delay; MAT DEF, maturation defective, DNA synthesis is normal, but late functions are not expressed; and the late function, lysozyme. The boxed diagrams outside the circle illustrate phage structural components that can be detected in extracts of cells infected by mutants defective in genes with morphogenetic functions. In general, the function of the particular gene is necessary for formation of the component *absent* from the box. A defect in gene *11* or *12* produces complete but fragile particles that can dissociate to free heads and tails. Gene *9* mutants produce inactive particles with contracted sheaths. Heads, all tail parts, sheaths, or fibers are missing from other extracts. Adapted from Edgar and Wood (102).

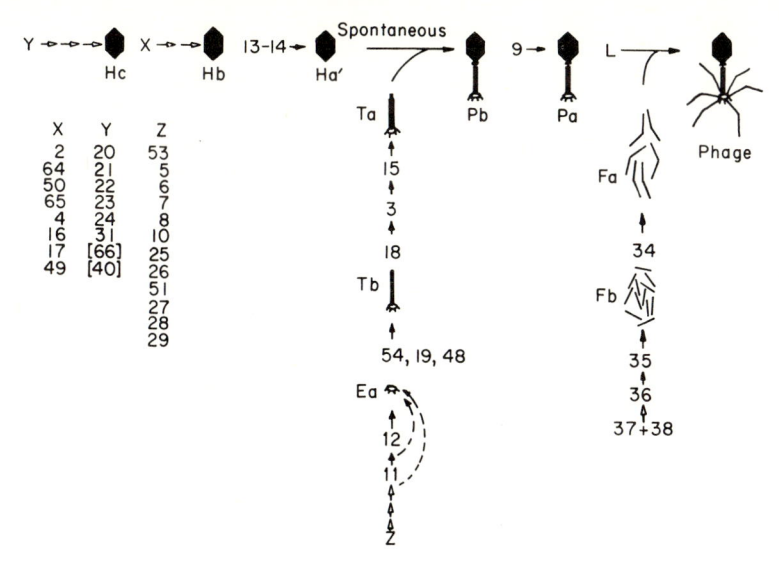

Fig. 4-32. Morphogenesis of bacteriophage T4. (b) Pathway of T4 morphogenesis. Numbers on the diagram indicate which gene controls that step in phage particle assembly. Steps indicated by filled arrows have been shown to take place *in vitro*. Intermediate forms of the head are designated Ha, Hb, and Hc; of the tail, Ta and Tb; of the tail fibers, Fa and Fb; and of the particle, Pa and Pb. The conversion of Tb to Ta appears to be polymerization of the sheath on the base plate precursor. Mutants blocked in the genes of group Y, which control the primary structure of the head, accumulate tails and fibers. Mutants defective in the genes of group X accumulate heads in addition to tails and fibers, but these heads cannot as yet be incorporated into viable particles *in vitro*. At least fifteen gene products, including those of group Z, are necessary for formation of the basic tail subunit consisting of base plate with attached core. Mutants blocked in these genes accumulate complete heads and fibers, with the exception of genes *11* and *12* which may be bypassed so as to result in defective particles. Modified from Edgar and Lielausis (104).

This paradox is explained if we imagine that the linear phage chromosomes are not all identical, but, rather, that they are circular permutations of the same closed linkage group with the ends of different molecules falling at different places (110). This would account for the apparent circularity of the genetic map. The model described was given strong support by Thomas and MacHattie, who showed that T2 DNA can form circular molecules *in vitro* upon denaturation and subsequent annealing (111). Circular molecules presumably result from the pairing of two circularly permuted strands with different termini so that the break in each strand is "covered" by the break in the other.

The fine structure of two neighboring cistrons in bacteriophage T4 was investigated by Benzer in a series of classic experiments (24, 112). Benzer selected the phage *r* region since it is especially convenient to assay. If a large number of T4 *r* mutants are selected, it is quite easy to divide them into subgroups *rI*, *rII*, and *rIII* according to their behavior in *E. coli* K12(λ). The *r* mutants identified on *E. coli* B yield the following variants on K12(λ): *rI* mutants are characterized by *r*-type plaque morphology; *rIII* mutants produce plaques similar to those of the wild-type phage; and *rII* mutants are completely unable to reproduce. Figure 4-33 shows the genetic map position of a number of *rII* mutants.

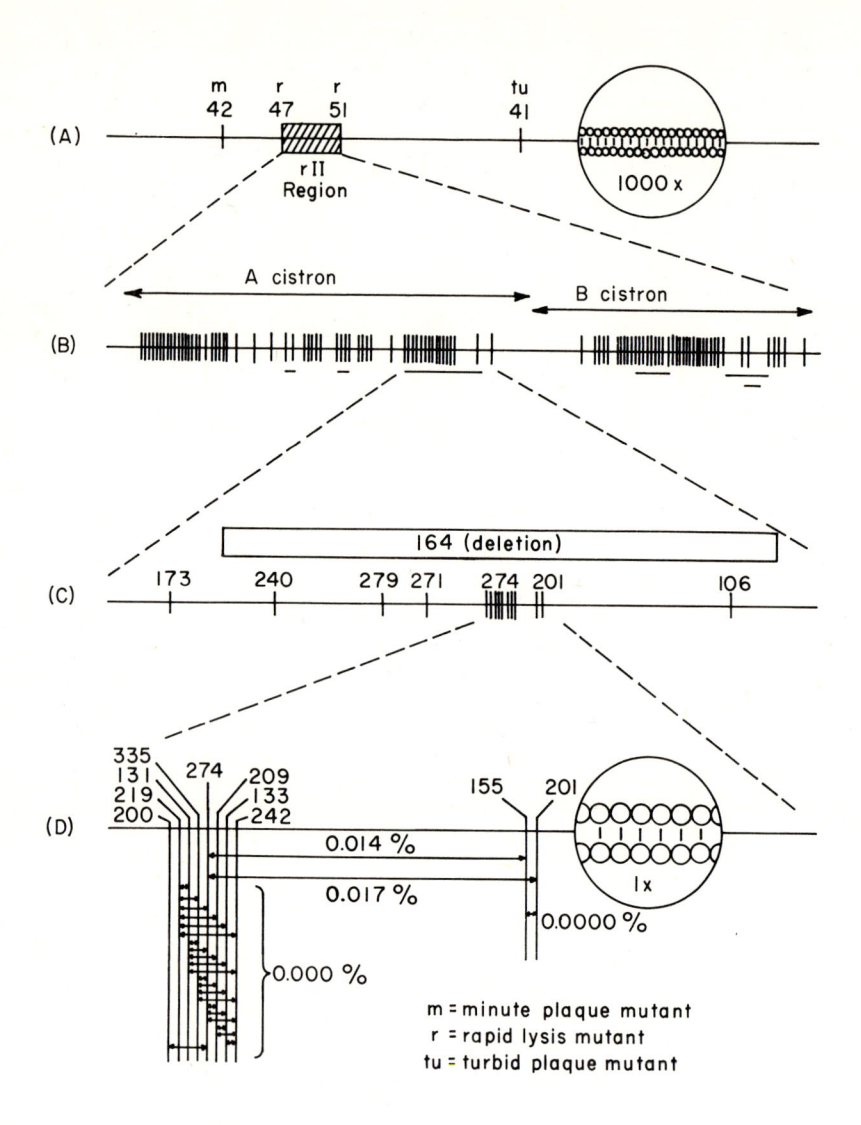

Fig. 4-33. Linkage map of the *rII* region in T4 bacteriophage. The successive drawings indicate increasing orders of magnification of the *rII* region as determined by fine structure analysis. (A) The location of the *rII* region is shown relative to several other mutants; inset shows molecular scale of Watson-Crick double helix where circles stand for nucleotide residues. (B) The locations of a number of mutants within the *rII* region. All mutants within a cistron are functionally related, whereas the *A* and *B* cistrons are functionally independent. A horizontal line represents a deletion which produces no wild-type recombinants in crosses with any mutants covered by its span. (C) A selected group of mutants within the *A* cistron is shown on larger scale. Wild-type recombinants are given by 164 (deletion) with 173 (point mutation) but not with any of the other point mutations. (D) Further magnification. The group of eight mutants on the left all have similar reversion rates and give less than $10^{-3}\%$ recombination with each other in the indicated crosses. They are assumed to be recurrences of identical mutations. From Benzer (113).

13. Genetic Processes in Viruses and Bacteriophages

The advantage of using the *rII* region in studies of this kind is the ease with which mutants can be identified. All *rII* mutations lead to the loss of capacity to synthesize one or more proteins necessary for development on K12(λ) hosts and are hence lethal with respect to this strain. In a typical mating experiment, the sensitive B strain is mixedly infected with two *rII* mutants. Progeny develop normally, and after lysis, they are scored on K12(λ). Recombinants for the *rII* region reproduce very well on the latter host, whereas both parental strains are inviable. Herein lies the great sensitivity of the method, for recombination frequencies as small as 10^{-8} can be detected. Strictly speaking, the limit of sensitivity is determined by the rate of spontaneous reversion or back mutation to the wild type. Of the more than 3000 different *rII* mutants studied by Benzer, a certain fraction turned out to be unsuitable for recombination experiments because they produced a high noise level, that is, a large quantity of spontaneous revertants. However, most of the mutants, whether spontaneous or induced by radiation or by treatment with chemical mutagens, were quite stable. There were over 2000 of these mutants, and it was possible to arrange them in linear sequence on the genetic map.

In mapping such a large number of mutations, a very methodical system of investigation had to be devised. To simplify the problem, Benzer used deletions, mutants lacking a sizable chromosomal segment in the *rII* region. Deletions were easy to recognize because, unlike point mutations, they cannot be reverted. All point mutations which fall within the region defined by a certain deletion are identified and grouped according to their one common property: they cannot yield recombinants with phage bearing that deletion. Referring now to Fig. 4-34, we see that mutants *1, 2, 3, 4,* and *5* all recombine with each other, but only *1* and *5* can recombine with the deletion *A*. Therefore, mutants *2, 3,* and *4* must lie within the region of the genetic map covered by the deletion. If several deletions can be found which overlap the entire chromosome, it is possible to group all of the mutations into subclasses, each mutant of which lies

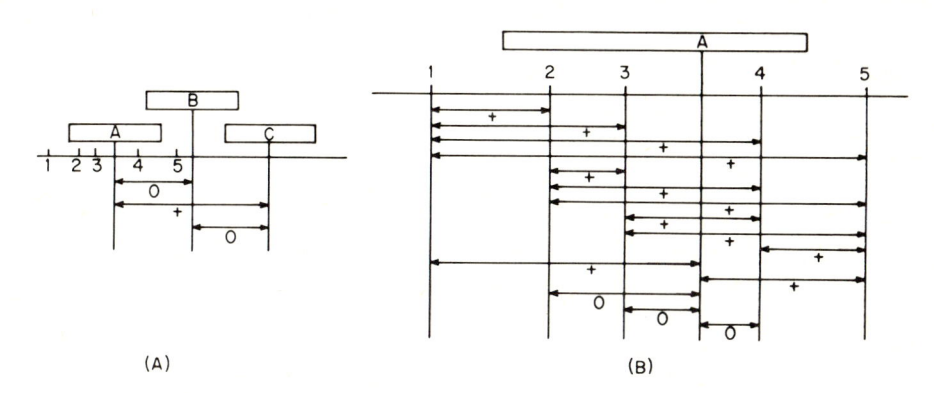

Fig. 4-34. Overlapping deletions and deletion mapping. (A) Mutants *A*, *B*, and *C* each differ from the wild type in the deletion of a portion of the genetic material. Mutants *A* and *C* can recombine with each other to produce the wild type, but neither of them can yield wild-type recombinants when crossed to *B*. (B) Deletion *A* fails to give wild-type recombinants with point mutants *2, 3,* and *4* which fall within the region of the deletion, but recombines with point mutants *1* and *5*. All of the point mutants recombine with each other. Plus sign signifies production and zero lack of production, of wild-type recombinants in a cross. From Benzer (24).

within a certain deletion. Once this is done, the problem of constructing the genetic map reduces to a much simpler task. Mutants within the segments defined by each deletion are first mapped separately, and the deletions themselves are then positioned relative to one another by measuring recombination between them. Overlapping deletions obviously do not recombine. In its logic, this procedure is reminiscent of the methods employed in determining the sequence of amino acids in a polypeptide chain. There the chain is divided up into fragments with trypsin or other proteolytic enzyme and the sequence within each fragment is determined. The complete sequence of the protein can be established when the relative alignment of all the fragments is worked out.

The distribution of point mutations throughout the *rII* region is not at all random. Some sites were represented by a single mutation, and others by a large number. Sites which readily undergo mutation were dubbed "hot spots" (114, 115). It is of great interest that various mutagenic agents (ultraviolet light, nitrous acid, hydroxylamine, and others) each have their own hot spots. This indicates that mutagenesis has a certain specificity: every mutagen attacks certain points in the cistron with higher probability than others. We shall return to the chemistry of mutagenesis in Section 15.

The entire *rII* region has a length of eight recombination units and it can be broken down into two functional subregions, designated *A* and *B*. Benzer called these two subregions *cistrons*, because they can be distinguished in a *cis-trans* test (p. 326ff). If K12(λ) is infected with two phage mutants, one impaired in the *A* region and one in the *B* region, then together they are able to reproduce because the mutations are complementary to one another. Therefore, if phages develop normally when two otherwise lethal mutations are in the *trans* configuration relative to each other (that is, on different DNA molecules), the mutations must be in different functional regions, or cistrons. The gene product lacking in one mutant is supplied by the other, and vice versa. This is also called a *complementation test*. The phages would not replicate and mature, however, if the two mutations were both *trans* and within the same cistron, since no intact copy of the affected functional region is present. The conclusion has been drawn that one cistron controls the synthesis of one polypetide chain.

According to the data of Benzer, the *A* cistron is about 5 map units long and contains $5 \times 250 = 1250$ nucleotide pairs, while the *B* cistron is 3 map units in length and therefore consists of about $3 \times 250 = 750$ nucleotide pairs. If we assume that three nucleotide pairs are necessary to code one amino acid in the corresponding protein, there should be about 400 amino acids in the *A* protein and 250 amino acids in the *B* protein, with molecular weights of 40,000 and 25,000, respectively. Unfortunately, these predictions have never been confirmed, for the *A* and *B* proteins have not yet been isolated. It is thus impossible to compare alterations in the phage DNA molecule with changes in the protein it specifies. Despite this shortcoming, investigation of the fine structure of the *rII* region in bacteriophage T4 has provided invaluable insight into the nature of the genetic material.

We are now in a better position to explain why the term *cistron* is preferred to the older word, *gene*. Traditionally, the gene concept was not rigidly defined, but included a variety of notions. The gene was at one time a unit of function, mutation, and recombination. The necessity for an operational distinction between these phenomena was recognized many year ago by Serebrovskij and Dubinin (116). As a result of his experiments, Benzer was able to distinguish and define the three elementary units required. The *cistron*, or region of genetic function, accommodates several hundred different sites capable of recombining with any of the others to restore functional activity. The gene therefore cannot be a unit of function and recombination simultaneously. The *recon*, or elementary unit of recombination, had to be defined as a separate entity, as did the *muton*, the elementary unit of mutation.

The muton is the smallest element that, when altered, leads to a mutation. In more concrete terms, it is manifested by the minimal change in nucleotide base sequence necessary to produce an amino acid replacement in the polypeptide product of the given cistron. The size of the muton can, in principle, be determined from the amount by which map distances between very closely linked markers deviate from strict additivity. Consider Fig. 4-35 in which three mutons, *1, 2*, and *3* are depicted. If the distances

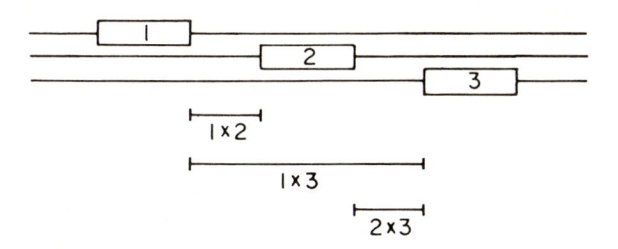

Fig. 4-35. Determining the "length" of mutation. The discrepancy between the long distance and the sum of the two short distances measures the length of the central mutation. From Benzer (24).

between the mutons are specified by the frequency of recombination among them, then $\Delta l_{13} \approx \Delta l_{12} + \Delta l_{23}$. Any difference $\Delta l_{13} - (\Delta l_{12} + \Delta l_{23})$ should yield the length of the muton. There is wide agreement at the present time that the muton probably consists of a single nucleotide residue in the case of point mutations.

The recon is the minimal distance between two point mutations which can recombine, the smallest indivisible element of the genetic material. The smallest recombination frequencies detected between mutations in the *rII* region are on the order of 0.005 to 0.01 %. If one recombination unit corresponds to 250 nucleotide pairs, then the recon should span one to two nucleotide pairs. In all likelihood, the minimum size of the recon is one nucleotide, so that this unit is very little different from the muton.

14. Lysogeny and Transduction

Certain strains of bacteria give rise to phage particles without any apparent prior infection (117). Such strains are said to be *lysogenic* and they have been shown to carry phage DNA in their own chromosomes. The phage genome is replicated in synchrony with bacterial DNA and transmitted to progeny cells along with the bacterial chromosome. In contrast to cells infected with virulent phage, no infective particles or infective nucleic acid molecules can be detected within lysogenic cells. Rather, the phage is physically integrated into the bacterial chromosome and in this state it is called a *prophage*. Integration of prophage must of course be preceded by infection of the strain with bacteriophage at some time in the history of the cell line. But from the moment lysogeny is established, the cell and all its descendents will be lysogenic, unless some event disturbs the relationship between the two chromosomes.

The very term lysogenic indicates that the prophage state is not an absolutely stable one. There is a certain probability that the prophage will be converted to the vegetative form of the phage. In this case, phage DNA is dissociated from the cellular chromosome, multiplies within the cell, and is incorporated into mature phage particles. The cell is eventually lysed and free virus particles are released into the medium. Spontaneous conversion of prophage to virulent phage can occur in any lysogenic culture, but with a frequency of only about 10^{-4} or less per generation. The release of phage particles from the lysed cell does not necessarily entail disaster for the remaining cells, since a lysogenic culture is not harmed by secondary infection with the same phage. Lysogenic cells are thus *immune* to reinfection by the phage which they carry in their genome. Of course, bacteriophage particles are adsorbed to the walls of lysogenic cells, and even inject their DNA, but new phage particles do not develop.

In certain cases, a lysogenic culture can be induced to free its prophages by ultraviolet light, x-rays, or chemical agents. In *E. coli* K12 (λ)—the symbol (λ) means that the strain is lysogenic for λ-phage—a comparatively small dose of ultraviolet light sufficient to kill about 20% of the cells, is all that is necessary to induce λ-prophage in more than 90% of the surviving cells. Induction is not a general phenomenon, however, and many lysogenic strains cannot be induced, although the lysis of individual cells does take place as the result of spontaneous conversions. A special case of induction, called *zygotic induction*, occurs when male bacterial cells containing prophage introduce their chromosome by conjugation into a nonlysogenic female cell that is not immune to the phage. In the female, the prophage enters the lytic cycle, and mature phages are produced.

None of the T phages of *E. coli*, which have been frequently mentioned in this text, can lysogenize host cells. They always produce lytic infection and are said to be *virulent*. The integration of prophage apparently requires that the phage DNA attach itself to well-defined loci in the bacterial chromosome. In order for this to occur, it is necessary to assume a certain affinity or homology between the DNA's of phage and bacterial host. This is particularly unlikely in the case of T-even phages, for their DNA contains 5-hydroxymethyl cytosine in place of cytosine and is therefore structurally dissimilar to bacterial DNA.

Lysogeny has been studied principally with phages λ, P1, P2, and Φ80, isolated from various strains of *E. coli*, and P22 isolated from *Salmonella typhimurium*. At the present time, many *temperate* bacteriophages of this type are known.

A bacteriophage manifests itself most obviously by its capacity to clear or lyse a bacterial culture, and it is on this basis that phages are generally titrated: small areas or plaques on the surface of an agar plate are cleared of indicator bacteria wherever there is an infective particle. Temperate bacteriophages are recognized by the fact that their plaques are turbid. While a certain proportion of the cells are lysed, those that are lysogenized grow normally. Thus infection of sensitive cells with temperate bacteriophages is itself a method for selecting lysogenic cells: only the survivors will harbor the prophage in their chromosomes. The frequency of lysogenization can be measured

14. Lysogeny and Transduction   393

directly, although not without difficulty. Depending on the bacterial strain employed, this figure varies from a fraction of a percent to more than half the culture.

The process of lysogenization is a complex one which depends on many external factors, including temperature and medium. In particular, conditions which inhibit protein synthesis, such as low temperature, or the presence of antimetabolites, foster the establishment of lysogeny. Fundamentally, however, lysogeny is genetically determined.

Consider the genetic properties of λ-phage, the most extensively studied of all temperate bacteriophages. The genetic map of λ is illustrated in Fig. 4-36. Mapping of the classic loci (symbols above line in Fig. 4-36) was performed by analyzing the

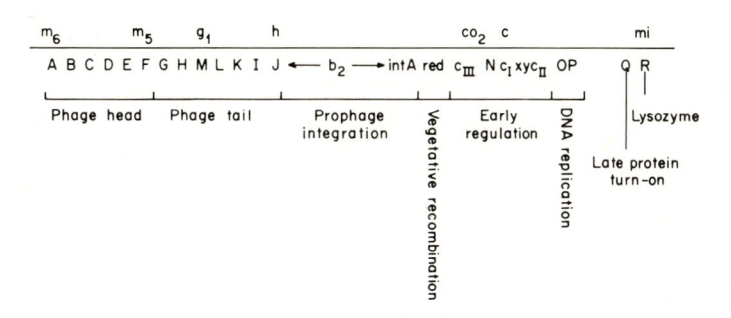

$$m_6 \quad m_5 \quad g_1 \quad h \quad \quad co_2 \quad c \quad \quad mi$$

A B C D E F G H M L K I J ← b_2 → intA red c_{III} N c_I xyc$_{II}$ OP   Q R

Phage head | Phage tail | Prophage integration | Vegetative recombination | Early regulation | DNA replication | Lysozyme, Late protein turn-on

Fig. 4-36. Linkage map of bacteriophage λ. Upper symbols indicate the classic genetic markers: m_5 and m_6, medium-sized plaque; g_1, large plaque; h, host-range mutant; c, clear plaque; mi, minute plaque. Lower symbols correspond to the current conception of the λ map with functional grouping of genes indicated below. A through R are the cistrons identified by Campbell in studies on amber and temperature-sensitive mutants of the phage; genes c_I, c_{II}, and c_{III} and N have to do with control of early functions; b_2 is a deletion interfering with lysogenization. The entire map is about 28 units in length. The λ chromosome consists of some 48,000 nucleotide pairs. From Jacob and Wollman (2) and Echols *et al.* (118).

recombinants formed after sensitive, nonlysogenic cells were multiply infected with pairs of phages bearing two or more distinct genetic markers. Mutations associated with a number of different loci cause alterations in plaque morphology, among which are *mi* or *minute*, which form tiny colonies and also possess a distinctive halo, *s* or *small*, characterized by small plaques lacking a halo, *g* which form large plaques, and m_5 and m_6, which stand for *medium* or medium-sized plaques. Mutants designated h lose their ability to adsorb to the usual *E. coli* K12 host, but are able to adsorb to and infect special indicator strains. The region of the chromosome including c_{III}, c_I, and c_{II} governs immunity to superinfection by homologous phages. Mutants in the *c* region characteristically form *clear* rather than turbid plaques. Figure 4-37 shows the morphological peculiarities of plaques formed by parental and recombinant phages in a cross where the parents differ in the three characters m_5, *c*, and *mi*.

Among the three cistrons in the *c* region, c_I is a regulatory gene governing the formation of a cytoplasmic repressor. According to current hypotheses, a repressor is elaborated under phage control which inhibits the synthesis of phage proteins and DNA, thereby preventing phage replication and maturation. Under these conditions, the prophage can become integrated into the host chromosome. For the same reason,

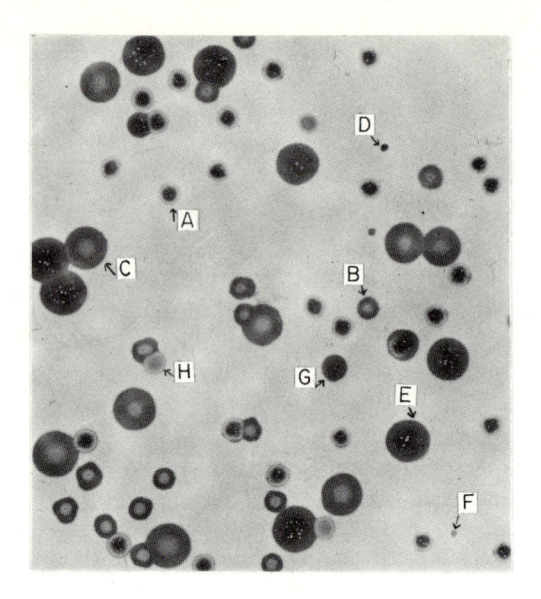

Fig. 4-37. Phenotypic differences in bacteriophage λ. Photograph shows the appearance of plaques formed by the progeny of a cross between two strains of λ phage differing in the three characters, m_5, c, and mi:

Parental types
(A) $m_5{}^+c\ mi$ (small clear)
(B) $m_5\ c^+mi^+$ (medium turbid)

Double recombinants
(G) $m_5\ c\ mi^+$ (medium clear)
(H) $m_5{}^+c^+mi$ (small turbid)

Single recombinants
(C) $m_5{}^+c^+mi^+$ (large turbid)
(D) $m_5\ c\ mi$ (minute clear)
(E) $m_5{}^+c\ mi^+$ (large clear)
(F) $m_5\ c^+mi$ (minute turbid)

From Jacob and Wollman (2)

secondary infection of the bacterial cell with the same phage type does not lead to vegetative development. Repressor is already present in the cell and it blocks synthesis of proteins under the direction of the newly introduced phage genome. The immunity of lysogenic bacteria to homologous phage therefore depends on the presence of repressor and is a dominant character. Mutations in the c_1 cistron apparently affect the nature of the repressor synthesized because they render the phage less able, or completely unable, to lysogenize the cell. At the same time, other mutations in the c_1 gene can alter the extent and specificity of immunity. Hence, this locus is frequently designated *im* for immunity. There is now convincing evidence that the repressor product of the c_1 gene is a protein (119). As we shall see in the next chapter, regulation of gene expression in phages and bacteria are similar to each other in many respects.

Susceptibility to prophage induction is also genetically determined and also depends on the c_1 locus in the phage chromosome. A mutation from ind^+ to ind^- means that the prophage can no longer be induced by ultraviolet light. If a culture lysogenic for the wild-type λ-phage, such as *E. coli* K12 (λ), is infected secondarily with the mutant λind^-, nothing changes outwardly because the prophage confers immunity on the cell. However, the wild-type prophage of such a culture ceases to be inducible by ultraviolet light. The λind^- mutation apparently causes the phage to specify an especially effective repressor which makes the prophage very stable to induction in addition to preventing development of the superinfecting mutant.

In addition to the λ-phage markers already discussed, Fig. 4-36 shows 18 cistrons, lettered A to K, which were identified by Campbell in complementation tests with amber mutants and then mapped by the methods of "marker rescue" (see p. 400ff) and standard three-factor crosses (120). The DNA of λ-phage, which has a molecular weight of 30×10^6 daltons, should contain about 50 cistrons. So far, 30 cistrons have been identified and, as a result, the functional significance of various regions of the chromosome is becoming clear. The λ-genome can be broken down into groups of cistrons that have to do with the synthesis of head protein (A to E), the synthesis of tail protein (G to J), immunity and the establishment of lysogeny (c_{III}, c_I, x, y, c_{II}), and replication of phage DNA (O and P). The R cistron specifies the synthesis of endolysin (phage lysozyme), while N and Q play a role in phage development by regulating the expression of early and late functions, respectively. Despite present knowledge of the λ-phage map, it is still incomplete and the importance of many cistrons remains to be elucidated.

The integration of phage DNA into the host chromosome is of particular interest. Integrated phage DNA is replicated along with the DNA of the host cell. During conjugation, the prophage is transferred to the female cell as part of the male chromosome and it undergoes recombination just as any other bacterial gene (Lederberg). The site of prophage attachment is located between the *bio* and *gal* regions of the bacterial genetic map (Fig. 4-38). Similar results have been obtained with other prophages; P2,

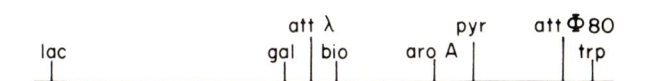

```
                    att λ              pyr      att Φ80
    lac             gal | bio     aro A          | trp
    |                 |   |          |           |
```

Fig. 4-38. Integration site for λ prophage on the bacterial chromosome. The portion of the *E. coli* genetic map surrounding *att*λ has been greatly expanded from Fig. 4-1.

for instance, occupies a position on the *E. coli* chromosome between the *xyl* (xylose utilization) and *met* (methionine biosynthesis) loci and also undergoes recombination after conjugation.

The most likely mechanism for the integration of phage DNA into the host chromosome, illustrated in Fig. 4-39, is that postulated by Campbell (121). As indicated above, the λ-chromosome is not normally a closed circle, but rather, it has two clearly defined ends. Under certain circumstances, the chromosome is able to close into a ring owing to formation of hydrogen bonds between complementary single-stranded regions at either end and to subsequent stabilization by covalent bonds. Furthermore, the λ-chromosome possesses a region designated *b2* on the map which is homologous with a certain region in the bacterial chromosome, through which the two synapse. Evidence for the role played by the *b2* locus was provided by the isolation of a λ mutant with a deletion in that region (122). The mutant is totally unable to lysogenize sensitive bacteria. Under normal conditions, however, the circular λ-chromosome is integrated into the bacterial chromosome by a crossover in the region of homology. This phenomenon is analogous to the integration of the sex factor and the F' episomes, which was discussed in an earlier section.

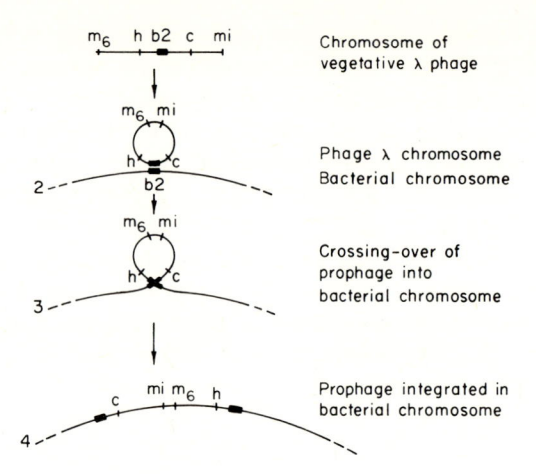

Fig. 4-39. A model for the integration of prophage λ into the bacterial chromosome. (1) Chromosome of vegetative phage λ. (2) The λ chromosome closes into a circle and synapsis occurs between it and the bacterial chromosome, presumably due to the presence in each of genetically homologous regions (heavy bars in drawing). (3) Genetic exchange takes place between the synapsed chromosomes. (4) Prophage λ integrated into the bacterial chromosome; note that the order of genetic markers is altered in comparison to vegetative λ phage. Adapted from Campbell (121).

Several specific predictions can be made from the Campbell model. First, it is clear that integration of the prophage should increase the recombination distance between the bacterial markers lying to either side of the inserted prophage by just the length of the phage chromosome. This was confirmed by recombination analysis (123). Second, when the chromosome is integrated, the circle is not broken at its normal ends, but in the *b2* region (Fig. 4-39). Therefore, the order of phage markers in the prophage is different from that in vegetative phage. This prediction was also borne out by experiment (124).

It can be assumed that during induction of the prophage, the integration process is simply reversed: the λ DNA forms a loop which is then excised from the bacterial chromosome by crossing-over.

The phenomenon of lysogeny has been studied for the most part in bacteriophages, but it has been shown that certain animal viruses can also exist in two different forms (Dulbecco, Rubin). In the vegetative state, the virus multiplies within the cell, lyses it, and infects neighboring cells. In the provirus state, free virus particles cannot be detected in the cell; their presence is known only because vegetative viruses persist in a small number of cells. The viruses have not disappeared, however, and it appears as if their DNA has formed a close association with the chromosome of the host cell. This relationship is expressed by the appearance of new cellular properties.

Dulbecco and Vogt studied the behavior of Rous sarcoma virus from birds and of polyoma virus from guinea pigs (125). Both of these viruses give rise to malignant tumors in the host organisms. When monolayer cultures of animal cells are initially infected with virus, the particles multiply vegetatively for a certain time. But viral particles eventually disappear and a second phase begins in which the viral DNA may perhaps become associated with the host cell chromosome. After this change, the culture becomes resistant to superinfection by the same virus and malignant growth starts. Disorderly cell division sets in and the clone becomes three-dimensional. Further, the cells increase in size and they lose their differentiated characteristics. This phenomenon may be analogous to lysogeny in bacteriophages. Viruses promote the same sorts of changes when they are introduced into living animals. It is distinctly possible that proviruses, established in the host cell chromosome with a loss of infectivity, constitute the pathogenic beginning of many tumors.

In 1952, Zinder and Lederberg discovered that phages are capable of mediating the transfer of genetic loci from one bacterial strain to another (126). This important new mode of genetic transfer was named *transduction*. Through the years, transduction has become one of the most important methods for studying the bacterial genetic map (127). This process is possibly the most general method of genetic exchange in bacteria, and it has already been observed in *E. coli, Salmonella, Bacillus megaterium, Bacillus subtilis*, as well as in many other bacterial species.

Transduction occurs when recipient bacteria are infected with phages that have been grown on a donor strain which carries distinct genetic markers. Although it is not necessary to use recipients lysogenic for the transducing phage, it is nonetheless of great convenience to do so. Since lysogenic recipients are immune, most of them will survive superinfection with the transducing particles. The characters transferred during transduction are wholly determined by the genotype of the donor cells.

Two major types of transduction have been distinguished, *generalized transduction* and *specialized transduction*. Generalized transduction, in which any donor locus can be transferred, has been demonstrated in *E. coli* and *Salmonella* with phages P1 and P22. Other temperate phages, such as P2 and λ, are unable to mediate this process. In specialized transduction only certain specific regions of the chromosome of the donor bacterium are borne by the phage vector. Examples are transduction of the *gal* locus by λ phage and of the *trp* locus by $\Phi 80$.

The probability that any given locus will be transferred in generalized transduction is rather small, on the order of 10^{-5} to 10^{-7} per transducing phage particle. Furthermore, only closely linked markers can be *cotransduced*, that is, transduced simultaneously. For the most part, genetic loci are transferred independently of one another, and the probability that two markers will enter the cell in two different transducing particles is simply the product of the probabilities of either event separately.

The formation of generalized transducing particles apparently unfolds in the following manner. During maturation of P1 phages within the infected cell, the bacterial chromosome is degraded to small pieces of double-stranded DNA. Fragments with a molecular weight of 60×10^6 daltons, equal in size to the phage genome, are packed into phage coats. All regions of the bacterial chromosome are assumed to have an equal probability of incorporation into such *transducing particles*. Upon lysis of the host cells, the transducing particles are released along with an excess of normal phages. They can then inject their DNA into recipient cells in the usual manner. Once the donor DNA has penetrated, recombination can take place between it and homologous region of the recipient chromosome.

Ikeda and Tomizawa elucidated this mechanism by labeling a culture of thy^- *E. coli* with bromouracil prior to infection with a virulent mutant of phage P1 (128). As phage particles were added, the bacteria were transferred to a medium containing radioactive thymine and progeny particles were harvested after maturation and lysis. Transducing particles comprise about 0.3% of the total phage particles and their position in a CsCl gradient indicates that they contain mostly bromouracil-labeled *bacterial* DNA, synthesized prior to infection. The complete absence of phage DNA in these particles was inferred from their failure to incorporate any radioactive thymine. This discovery explains why generalized transduction does not immunize the recipient cell to superinfection by homologous phages; since transducing particles are devoid of phage genes, penetration of transducing DNA in no way constitutes infection of the host cell.

Occasionally the transduced donor DNA does not undergo genetic exchange with the recipient chromosome and hence cannot be replicated, even though the genes it bears continue to be expressed. These fragments are inherited unilinearly by the descendants of the original recipient bacteria. Since the fragment is transmitted to only one of the daughters at each cell division, only a single cell of the clone possesses it at any one time. Such fragments can persist for many generations before they are finally degraded or diluted out. This phenomenon is called *abortive transduction*.

Transduction provides a means to independently verify the *E. coli* genetic map derived from studies on conjugation. Since several neighboring loci are often transduced and integrated together, a simple test for cotransduction can be used to find the relative placement of loci along short segments of the chromosome, without actually measuring recombination frequencies. If, however, attention is focused on the fine structure of a particular region, it is possible to study the map quantitatively by measuring recombination frequencies within the loci under investigation.

Ames and Hartman made use of transduction by phage P22 to establish the sequence of the nine closely linked cistrons which code for the enzymes of histidine biosynthesis in *Salmonella* (129). The fine structure of cistrons responsible for the synthesis of the enzyme tryptophan synthetase in *E. coli* was studied with P1 transduction by Yanofsky *et al.* (130).

Specialized transduction by phages λ, $\Phi 80$, and others differs in principle from generalized transduction and represents another important technique for the study of genetic processes at the molecular level. The first case of specialized transduction to be studied in detail was transfer of the *E. coli gal* region by λ phage (131). The *gal* region consists of three cistrons, two of which are essential if the cell is to utilize galactose as a sole carbon source. Induction of λ prophage leads to the formation of a lysate in which a very small proportion of the particles carry the bacterial *gal* region in addition to their own DNA. This can be readily demonstrated if the phage particles are derived from a lysogenic gal^+ donor strain and are then used to infect gal^- recipient cells. At high multiplicities of infection, the phages will transduce the gal^+ markers into the gal^- recipients with a frequency between 10^{-4} and 10^{-5} per particle. An essential condition for specialized transduction of bacterial genes is that the transducing particles be obtained by induction of a culture lysogenic for λ prophage. By contrast, vegetative λ phages cultivated in sensitive nonlysogenic *E. coli* strains cannot pick up any bacterial genes. Generalized transduction is quite different in this regard for it bears no direct relationship to the lysogenic state.

When gal^+ transductants are isolated and grown out, it is found that bacteria with the gal^- phenotype are segregated with a frequency of about 10^{-3} per cell per generation. This indicates that the *gal*⁺ transductants are actually *heterogenotes* carrying their own original *gal*⁻ marker as well as the *gal*⁺ region introduced by the phage. The transduced marker does not replace the recipient allele as it does in generalized transduction. Furthermore, when the *gal*⁺ cells obtained in this way are induced with ultraviolet light, the resulting lysate can transduce the *gal*⁺ genes at a frequency of about 50%. Such lysates are designated HFT for *high frequency transduction* as opposed to the LFT lysate from the original donor strain which mediates *low frequency transduction*. It is, of course, the HFT lysate that is particularly useful in genetic studies because it contains a large proportion of particles bearing the same region of the bacterial chromosome. A scheme for the preparation of LFT and HFT lysates is presented in Fig. 4-40.

If HFT lysates are used to infect *gal*⁻ strains at high multiplicities (a ratio of phage particles to bacteria greater than one), the resulting *gal*⁺ transductants appear to be

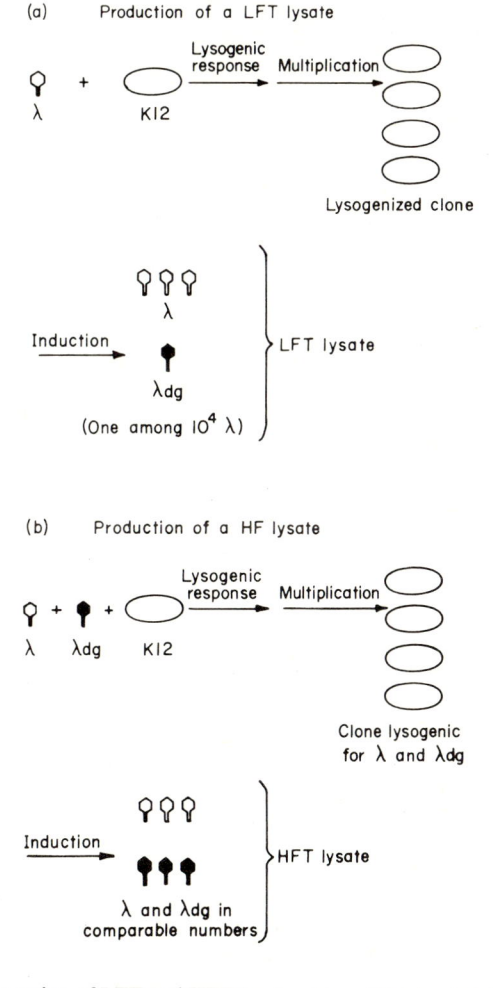

Fig. 4-40. The preparation of LFT and HFT lysates. From Weigle *et al*. (132).

normally lysogenic and themselves yield HFT lysates upon induction. At low multiplicities, however, when there are substantially fewer phage particles than bacteria and the majority of infected cells receive only a single phage, a different sort of *gal*⁺ transductant is noted. Although they are immune to superinfection by homologous phages, and lyse upon induction, no progeny phages are detected in the lysates. Bacterial cells of this type are said to be *defectively lysogenic*. It was concluded that single infection can lead to integration of the prophage into the bacterial chromosome, probably owing to the extensive homology between the *gal* regions of phage and recipient, but that the prophage is thereafter incapable of replicating autonomously. One or more functional regions of the phage chromosome vital for phage replication and maturation are apparently lost when the bacterial *gal* genes are initially incorporated into the transducing particle.

When the original LFT lysate is used to infect a recipient population at high multiplicities, it is quite probable that any cell which is infected by a *defective transducing phage* also receives a normal phage particle. The recipient becomes *doubly lysogenic* for both normal and defective prophages and, upon induction, the normal phage genome provides all the information necessary for replication of both types of prophages. Similarly, if a defective lysogen is induced and then superinfected with normal phage, a HFT lysate also results. In these two cases, normal λ acts as a *helper phage* by supplying the gene products in which the transducing particles are deficient. This explains why the HFT lysates contain comparable numbers of normal and transducing phages (Fig. 4-40). As long as recipients are infected at high multiplicities with such HFT preparations, it is likely that every infected bacterium will receive at least one normal particle in addition to a defective transducing particle. But at low multiplicities, the defective particle enters the recipient cell alone without a helper phage.

The formation of defective transducing particles can be explained very well on the basis of the Campbell model if it is recalled that such particles can arise only from prophages originally integrated into the chromosome of the donor cell (Fig. 4-41). When a lysogenic donor is induced, the prophage infrequently incorporates one or more neighboring loci from the bacterial chromosome as it is excised. Bacterial genes can be picked up from regions adjoining either end of the prophage and, as a result, a portion of the phage genome is deleted from the opposite end. The orientation of the two ends in this case is taken relative to the integrated phage, which is a circular permutation of gene sequence in the vegetative state. The incorporated bacterial genes are replicated along with the remaining portion of the phage genome and will be packaged into normal phage coats. Since they lack a sizable portion of their own chromosome, the transducing particles are expected to be incapable of autonomous replication and are generally designated as λd for *defective*, followed by the symbol for the bacterial markers which they carry. One such phage, called λd*gal* for *defective galactose-transducing*, has been discussed above. But since *att*λ, the attachment site for λ on the bacterial chromosome, adjoins both *gal* and *bio* (biotin synthesis) regions, the formation of a defective phage that transduces *bio* is also anticipated. Such a phage, called λd*bio*, has indeed been found.

The position and extent of the deletion in the genome of λ transducing phages can be determined genetically by what is called a "marker rescue" experiment (133). Bacteria were singly lysogenized with λd*gal* from a number of separate isolates. They

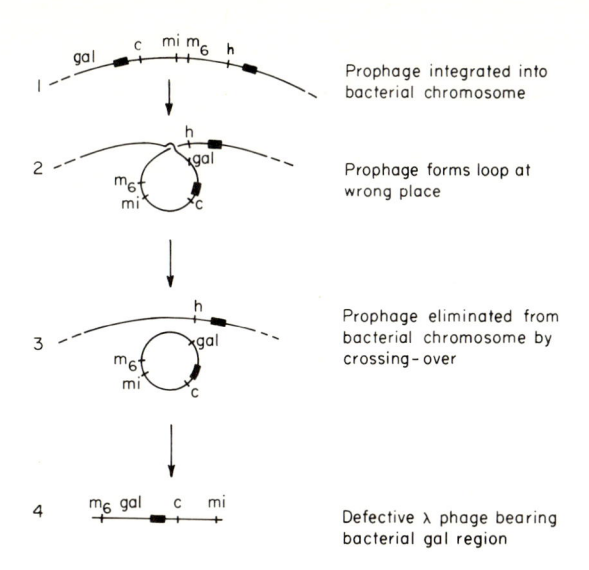

Fig. 4-41. A model for the formation of defective transducing phages. (1) The integrated prophage and surrounding region of the bacterial chromosome containing the *gal* locus. (2) A loop is formed that includes the bacterial *gal* locus, but excludes certain phage loci owing to mispairing between homologous regions of prophage and bacterial chromosome other than those responsible for prophage integration (heavy bars). (3) The prophage is excised from the bacterial chromosome by genetic exchange (crossing-over). (4) The genome of the defective λd*gal* phage now carries the bacterial *gal* locus which can be transduced into sensitive bacterial recipient cells. Adapted from Campbell (121).

were then induced and superinfected with phages bearing a series of point mutations at known positions along the left arm of the λ chromosome. Although neither λd*gal* nor the point mutants could produce a lysate on the particular bacterial strain used, recombinants between them were able to do so as long as the deletion did not overlap the point mutation. If the point mutation did fall within the deleted region, however, recombination was not possible. Thus the deletions could be quite accurately located on the λ genetic map in terms of their ability to contribute markers to the point mutants. These experiments revealed that different λd*gal* isolates lack one fourth to one third of their own genomes. The precise length of the deletion varies with the particular derivative of λd*gal* employed, as indicated in Fig. 4-42. Although the deleted region always seems to start at B, it may end at any one of several points A_1, A_2, A_3, ... A_n.

Furthermore, λd*gal* particles can be separated from wild-type λ phages by density gradient centrifugation in CsCl (134). Since the quantity of phage protein per particle is presumed to remain constant and since *E. coli* and λ DNA's have the same average base compositions, density differences must indicate differences in DNA content. This means that the amount of phage DNA lost through deletion is not exactly compensated by that acquired from the bacterial *gal* region. From the densities of several independent λd*gal* isolates, it was found that defective phage particles exhibit alterations ranging from a net gain in DNA of 8% to a net loss of 14%.

The fact that λ prophage does not always undergo a reciprocal exchange of genetic

```
        m₆           m₅      g₁     h            co₂  c              mi
    ┼─┼─┼─┼─────────┼───────┼──────┼────────────┼───┼─────────────┼
     hd₁ hd₄ hd₂ hd₆
       A₁ ┼──────────────────────────────┤B  ⎫
       A₂ ┼────────────────────────────────┤B ⎬ Extent of deletion in
       A₃ ┼──────────────────────────────────┤B ⎭ various λ dgal transducing phages
```

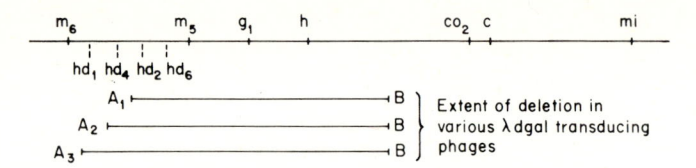

Fig. 4-42. Extent of deletion in defective galactose-transducing phages. Length of deletion shown by arrow beneath phage chromosome: "right" ends (B) fall at a well-defined point between h and c markers while "left" ends (A_n) penetrate into m_5–m_6 region to varying extents. The latter were mapped by a series of hd (host-dependent) point mutations between c and m_6. Data from Campbell (133).

material with the bacterial chromosome most likely explains the rare formation of nondefective transducing phages from λ and $\Phi 80$. These phages, designated p for plaque-forming, contain a complete phage genome as well as portions of the bacterial chromosome adjacent to the phage attachment site. Although nondefective transducing phages comprise only a small fraction of all transducing phages, they can be selected by infecting recipient cells at low multiplicities. Induction of transductants singly lysogenized with the phages will yield a lysate containing infective transducing particles, whereas induction of a culture singly lysogenic for a defective phage will lead to cell lysis without phage production. Nondefective transducing phages such as λp*bio* and $\Phi 80$p*trp* are already in wide use (p. 509).

Specialized transduction of *E. coli* bacterial genes is also mediated by $\Phi 80$ (135), a phage whose properties are very similar to those of λ. The $\Phi 80$ attachment site, called *att*$\Phi 80$, lies between *su*-C, an ochre suppressor which may be allelic with su_{III} (p. 441f), and the *trp* region, a cluster of genes relating to tryptophan biosynthesis. Phage $\Phi 80$ forms defective transducing phages carrying either *trp* or *su*-C. Besides the regions adjoining *att*$\Phi 80$, however, $\Phi 80$ can also pick up more distant loci such as *cys* B (cysteine biosynthesis) and *pyr* F (pyrimidine biosynthesis). On the *trp* side, in particular, the probability of transducing a given marker varies inversely with the distance between that marker and the prophage, as indicated in Fig. 4-43 (123). Furthermore, transducing particles carrying a given marker to the right of *att*$\Phi 80$ also carry all other loci located between the marker and the prophage. Simultaneous transduction of two markers on opposite sides of the prophage was never observed, however, indicating that one end of the DNA segment excised from the bacterial chromosome terminates within the prophage, thereby yielding defective transducing particles such as $\Phi 80$d*trp* and $\Phi 80$d*su*$_{III}$.

Finally, Beckwith and Signer have been able to insert genes of the *lac* region into the *E. coli* chromosome both immediately to the right and immediately to the left of *att*$\Phi 80$ by means of a thermosensitive F′-*lac* episome (136). Because the *lac* region in the two strains lies between *att*$\Phi 80$ and either of the normally transduceable *su*-C or *trp*

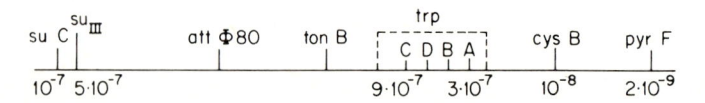

```
       su_III                           trp
    su C |           att Φ80   ton B  ┌─────┐   cys B      pyr F
     ┼───┼───────────────┼───────┼────┤C D B A├────┼──────────┼
     10⁻⁷ 5·10⁻⁷                      9·10⁻⁷ 3·10⁻⁷  10⁻⁸     2·10⁻⁹
```

Fig. 4-43. Segment of the *E. coli* chromosome surrounding *att* $\Phi 80$. The frequency with which neighboring bacterial markers are transduced by $\Phi 80$ is indicated below the line as the number of transductants per active phage. Data from Signer (123).

regions, it should also be transduceable by Φ80. This expectation was confirmed, even though in the wild-type strain, Φ80 never transduces *lac*. These results indicate that specialized transduction of a marker is a direct consequence of its proximity to the prophage on the donor chromosome. All of these observations are in accord with the Campbell model, particularly since they provide additional evidence that phage and bacterial genes are excised from the host chromosome together as a single continuous fragment of DNA.

In addition to their value in genetic studies, specialized transducing phages offer a unique opportunity for preparing specific small segments of the bacterial chromosome containing just a few cistrons. It is now relatively easy to isolate DNA of the *gal*, *bio*, *lac*, and *trp* regions from HFT lysates with little contamination by other bacterial genes. Purified gene-specific DNA obtained in this manner can be used for a variety of experimental purposes, as we shall see in Chapter V.

15. The Chemistry of Mutagenesis

One of the most important experimental techniques in genetics is artificial mutagenesis which permits the production of mutant organisms in the laboratory. The present discussion will be confined to mutagenesis in bacteria and viruses, since we are already familiar with the genetics of these two groups. Mutations can be brought about by various forms of radiant energy—ultraviolet light, x-rays and γ-rays—and by numerous chemical agents which react with DNA—nitrous acid, hydroxylamine, nitrogen and sulfur mustards, acridine dyes, and the ions of certain transition metal elements. An understanding of mutational processes is best gained from chemical mutagenesis, since its consequences are more accessible to experimental study than are the effects of radiant energy. Chemical mutagenesis was discovered simultaneously by Lobashov and Sakharov in 1936 (137) and over the years it has become an extremely valuable and widely used method.

Chemical mutagens are active both *in vivo* and *in vitro*. In the former case, a mutagen is added to a population of intact cells or viruses and its consequences are noted in the hereditary alterations borne by progeny. Insofar as concerns bacteria, the reagent must first penetrate the cell where it can act either directly on the DNA, or else modify or inhibit an enzymatic reaction essential for DNA synthesis.

The *in vitro* approach to mutagenesis offers certain advantages, because extracted nucleic acids can be subjected to direct chemical modification under controlled conditions. Mutagenesis can thus be assessed as a normal chemical reaction between two reacting substances. *In vitro* mutagenesis, however, is possible in only two instances. First, purified viral RNA, such as that of tobacco mosaic virus, can be reacted with chemical mutagens and the altered RNA then used to infect a suitable host. Viruses with mutant genotypes are observed among the progeny (138). Second, it is possible to directly alter DNA purified from transforming bacteria with chemical agents. A great variety of mutants can be recovered by this method after the modified DNA is introduced into wild-type cells (139). Although chemical mutagenesis *in vitro* is presently rather limited in scope, it will undoubtedly be of great importance in the future. Despite the intriguing possibilities which such techniques present, almost all of our knowledge of chemical mutagenesis has been gleaned from studies *in vivo*.

This discussion will be further restricted to *point mutations* which cause a single amino acid replacement in the protein product of an altered cistron. In particular, point mutations are defined by their ability to be reverted to the wild type in a single mutational step, implying that they affect just one nucleotide base pair in the DNA of the chromosome. Larger chromosomal alterations, which occasionally result from the action of radiation or chemical mutagens, will not be considered. This group includes *deletions* of extensive regions of the chromosome, *translocations* in which deleted DNA fragments are inserted at new sites on the chromosome, *duplications* of nucleotide sequences during replication and *inversions*, where the polarity of substantial segments of the chromosome undergoes a 180° change in direction. In all of these cases, mutagenic treatment causes a simultaneous change in many nucleotide residues.

Turning away from these more complex events, we shall consider the simpler and more common instances where mutations are produced by changes in single nucleotide pairs. On theoretical grounds, nucleotide pair exchanges can be classified as follows

Transition	A—T → G—C	One purine substituted by another
	G—C → A—T	One pyrimidine substituted by another
Tranversion	A—T → C—G / T—A	Purine substituted by either pyrimidine
	G—C → T—A / C—G	Pyrimidine substituted by either purine

where A as usual stands for adenine, C for cytosine, G for guanine, and T for thymine. In both transition and transversion, only one residue need be altered initially, whether by spontaneous mispairing during DNA replication, or by chemical modification. Only after the next round of replication does the altered base pair become established in both DNA strands. The base normally complementary to the altered one is replaced either because (1) it no longer forms the necessary hydrogen bonds or because (2) pairing with another base becomes energetically equivalent or more advantageous than pairing with the correct base, so that an error is very likely during replication. Thus if A is replaced by G (transition), G will pair with C instead of T in the next cycle of semiconservative DNA synthesis. Similarly, if A is replaced by C or T (transversion), they will pair with G or A, respectively, during replication, and not with T. Because of their mode of chemical action, some mutagens will produce only transitions, while others will cause both transitions and transversions. Transitions cannot be reverted by transversions in a single step, nor can the reverse take place. Important conclusions about the mechanism of chemical mutagenesis can be made from the revertability of a given point mutation with the same or another mutagen.

Bacteria and viruses can be grown on medium containing the base analogs 5-bromouracil (BU) and 2-aminopurine (AP). BU is incorporated quantitatively into the DNA of these organisms in place of thymine. The presence of BU in the nucleic acid molecules induces considerable hereditary instability (140). and the spontaneous mutation rate can reach the enormous value of 3% per generation. This is because BU has a greater tendency to form incorrect base pairs than thymine. The upper diagram in Fig. 4-44 depicts normal hydrogen bonding between adenine and the usual keto

15. The Chemistry of Mutagenesis

ADENINE   5-BROMOURACIL
(normal keto state)

GUANINE   5-BROMOURACIL
(rare enol state)

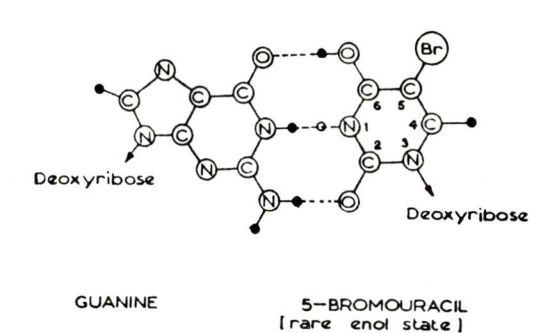

Fig. 4-44. Base pairing between 5-bromouracil and purine bases. In the normal keto state, with a hydrogen atom in the N-1 position, bromouracil binds to adenine (top). In the rare enol state, a tautomeric shift of this hydrogen atom results in specific pairing with guanine (bottom). Hydrogen atoms are represented as small black circles, and the hydrogen bonds by broken lines.

tautomer of BU when the latter simply replaces thymine. The lower diagram shows how the rarer enol form of BU can mistakenly pair with guanine. BU makes this error much more frequently than thymine because replacement of the methyl group by bromine at position C-5 of the pyrimidine ring makes the enol tautomer much more likely. Bromine is more electronegative than the methyl group and draws π-electrons away from the heterocyclic ring, thereby altering the distribution of hydrogen atoms. When BU is incorporated in place of thymine, it may pair with G in a later replication. G then pairs normally with C in the next round of DNA synthesis and the original A—T pair is replaced by a G—C pair (Fig. 4-45a). Alternatively, BU can be initially incorporated in place of cytosine due to G—BU pairing. In the next replication, BU pairs normally with adenine, and an A—T pair is eventually substituted for an original G—C pair (Fig. 4-45b). Thus by behaving like cytosine, BU can cause transitions in either direction, depending on when the mistake in base pairing occurs.

2-Aminopurine (AP), which is incorporated into bacterial DNA in place of adenine, similarly induces mistakes in base pairing (Fig. 4-46). Like adenine, AP forms two hydrogen bonds with thymine, but it can also form one hydrogen bond with cytosine even in its normal (amino) tautomeric state. Moreover, a tautomeric shift to the rare imino form renders AP capable of making two hydrogen bonds with cytosine. These last two events apparently occur quite often, since AP is a very active mutagen

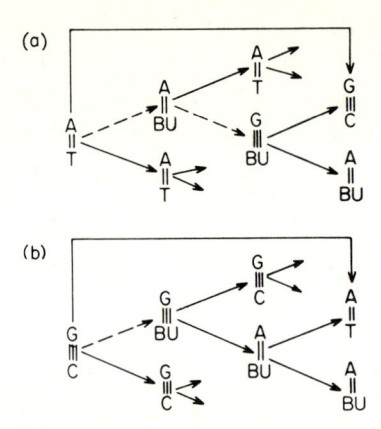

Fig. 4-45. Two types of base-pairing mistakes resulting from 5-bromouracil (BU). Base-pair transitions can occur in either direction depending on whether the erroneous pairing with guanine (G), owing to a tautomeric shift, occurs during replication (a) following correct pairing opposite adenine (A) or during its initial incorporation (b). Short lines between base pairs indicate the number of hydrogen bonds formed. Solid lines above each drawing traces the fate of altered base pair during replication. Adapted from Freese (141).

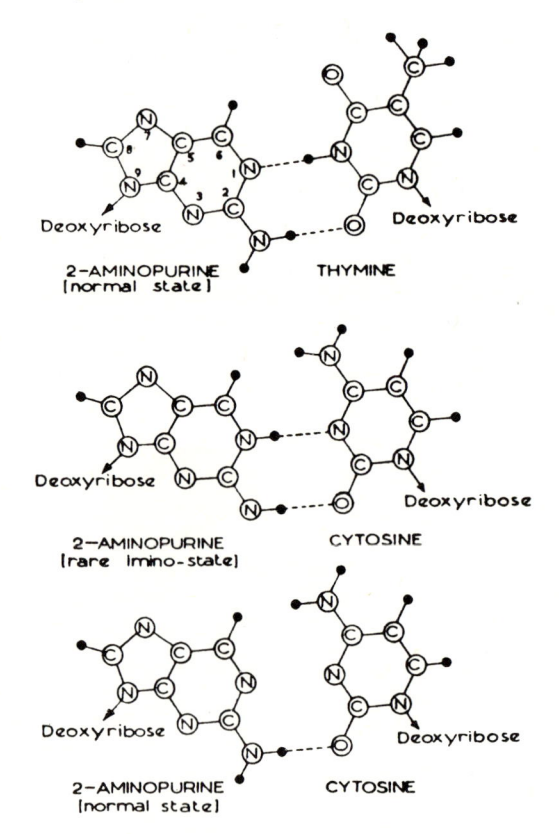

Fig. 4-46. Base pairing between 2-aminopurine and pyrimidine bases. 2-Aminopurine normally pairs with thymine by means of two hydrogen bonds (top). In the rare imino state, this base analog pairs with cytosine as a result of a tautomeric shift of a hydrogen atom to the N-1 position (middle). In the normal amino state, 2-aminopurine can also pair with cytosine by means of a single hydrogen bond (bottom). Hydrogen atoms shown as small black circles.

15. The Chemistry of Mutagenesis

even though it is incorporated into DNA in small quantities. The mechanism by which point mutations are formed in this case is just the same as in the preceding one (142); mistakes occur during replication and a new base pair is established after a subsequent replication. Amino-purine also causes transitions, but not transversions, in both directions, so that a point mutation engendered by AP can also be reverted by AP.

The reversion rate in a given cistron is almost always 10^2 to 10^3 times less than the rate of forward mutation. An auxotrophic mutation results from an alteration at any one of several hundred sites in the cistron, all of which can produce changes in the corresponding protein. In order to revert a specific mutation, however, the target for mutagenic attack has been narrowed to that residue in the DNA which was originally replaced. Even though revertants are relatively rare, they can be selected easily and at high sensitivity. Both forward and reverse mutations can be produced at the same site by mutagenic agents of the base analog type which can cause transitions in both directions. In what follows, we shall consider a ratio of 10^2 to 10^3 between the frequencies of forward and back mutation to constitute good evidence that a particular mutagen acts reversibly.

Base analogs behave mutagenically only during DNA biosynthesis, for their action depends on incorporation into the nucleic acid chain. Other mutagens act on

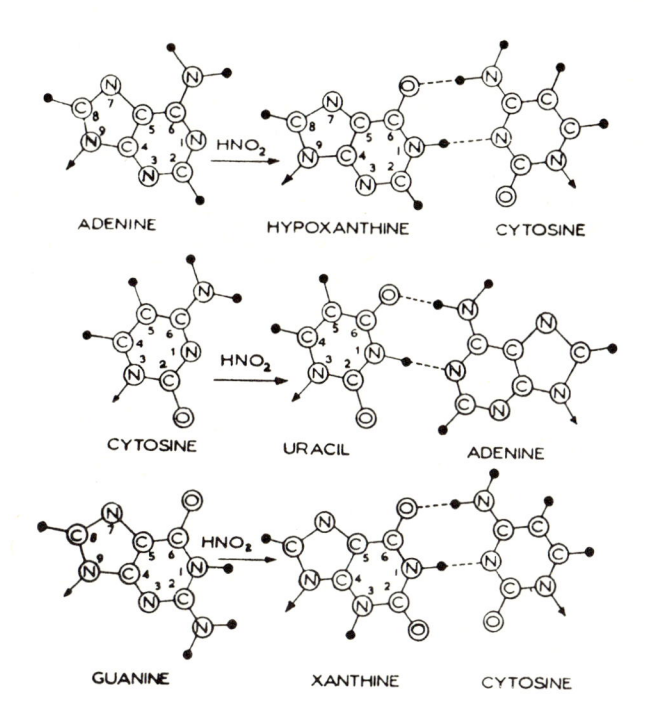

Fig. 4-47. The oxidative deamination of DNA bases by nitrous acid and its effects on base pairing. Adenine (top) is deaminated to hypoxanthine which binds to cytosine, instead of to thymine. Cytosine (middle) is deaminated to uracil which binds to adenine, instead of to guanine. Guanine (bottom) is deaminated to xanthine which continues to bind with cytosine, though with only two hydrogen bonds. Hydrogen atoms are represented as small black circles and the hydrogen bonds by broken lines.

DNA in the resting state by direct chemical attack. They can cause varied mutations in purified viral RNA and in DNA extracted from transforming bacteria, and also mutagenize free phage particles. We shall now enumerate some of the best-studied chemical mutagens of this type and their mechanism of action will be described insofar as it is known.

(1) Nitrous acid acts as a deaminating agent according to the reaction:

$$RNH_2 + HNO_2 \rightarrow ROH + N_2 + H_2O$$

In DNA, nitrous acid most readily deaminates guanine, then cytosine, and is least effective in deaminating adenine. In viral RNA, all aminated bases are attacked with equal efficiency. The hydrogen-bonding pattern of the deaminated bases is shown in Fig. 4-47. We see that nitrous acid can convert guanine to xanthine which, like guanine, pairs with cytosine, though only two hydrogen bonds are formed instead of three. Therefore, the deamination of guanine should not lead to mutations. The deamination of cytosine produces uracil, a base which can form two hydrogen bonds with adenine. Consequently, G—C pairs are replaced by A—T pairs (transition) when the DNA is replicated (Fig. 4-48). Finally, deamination of adenine produces hypoxanthine which, after tautomerization, can form two hydrogen bonds with cytosine. This leads to the substitution of A—T pairs by G—C pairs (Fig. 4-48). But since adenine is attacked most seldom, this transition is probably less frequent. Nitrous acid causes transitions in both directions, as do the base analogs, and any one of them should be able to revert a mutation evoked by any other. But since symmetry is not observed in the case of nitrous acid, this agent may not revert all transitions with equal frequency.

(2) Hydroxylamine is a very potent mutagen which is active only on isolated nucleic acids or on free phage particles. It is known for its very high selectivity (144). This mutagen reacts almost exclusively with cytosine and has negligible effects on thymine, guanine, and adenine. The reaction of hydroxylamine and cytosine is

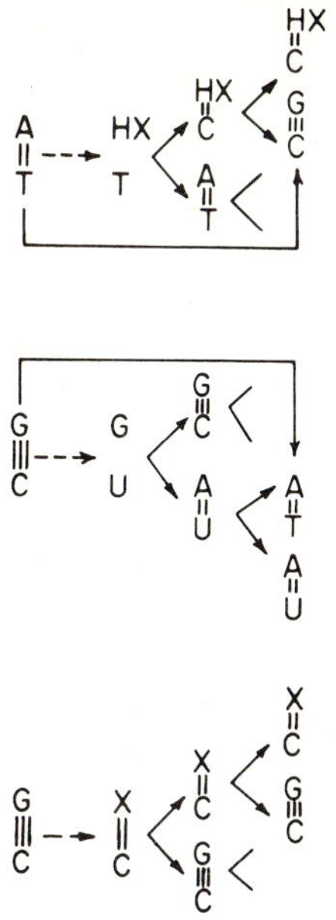

15. The Chemistry of Mutagenesis   409

Fig. 4-48. The types of base-pairing mistakes resulting from deamination with nitrous acid. Base-pair transitions in either direction can occur depending on whether adenine (A) is deaminated to hypoxanthine (HX) (top) or cytosine (C) to uracil (U) (middle). Since xanthine (X), the deamination product of guanine (G), has the same hydrogen-bonding attributes as guanine, a transition does not occur in this case (bottom). Short lines between base pairs indicate the number of hydrogen bonds formed. Solid lines above the upper two drawings trace the fate of the altered base pair during replication. Modified from Bautz-Freese and Freese (143).

After reaction, cytosine loses an absorption band between 260 and 280 mμ. One of the two tautomeric forms shown in the lower line is probably responsible for the induction of mutations. The important chemical result of this modification is the appearance of a —NH group at position N-1 in the pyrimidine ring which enables the base to pair specifically with adenine instead of guanine. This causes replacement of G—C pairs by A—T pairs after replication. Unlike nitrous acids and the base analogs, hydroxylamine induces transitions in one direction only. Mutations evoked by hydroxylamine should therefore fail to revert in a second hydroxylamine treatment. Furthermore, hydroxylamine should revert only a portion of the transitions produced by nitrous acid and the base analogs. By using combinations of hydroxylamine and the other mutagens, it is possible to determine the precise base pair substitution which occurs in any given transition.

(3) The alkylating agents dimethyl and diethyl sulfate

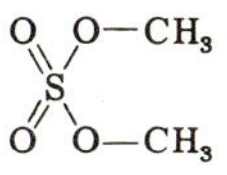

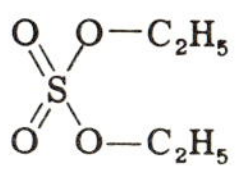

and the even more active compounds, ethyl methane sulfonate (EMS) and ethyl ethane sulfonate (EES)

attack guanine at the N-7 position. After alkylation, the deoxyriboside bond is apparently cleaved, leaving gaps in the DNA chain at locations formerly occupied by guanyl side groups. The chemical reactions which follow ethylation of guanine by EES are

Owing to an electronic rearrangement, a quaternary nitrogen is formed which destabilizes the deoxyriboside link and leads to its hydrolysis. Barring any steric restrictions, the gaps left by guanine could be filled by any one of the four bases during the next cycle of replication. Alkylating agents can thus cause transitions, transversions, and even rupture of the DNA chain weakened by depurination. These mutagens are therefore distinguished by a complex spectrum of action and a rather high toxicity. Some of the resulting mutations, in particular the transversions, should not be reverted by nitrous acid, hydroxylamine, or any of the base analogs since the latter lead only to transitions.

Another group of alkylating agents possessing even greater activity are the nitrogen mustards, one of which, β-chloroethylamine, has the following formula:

$$H_2N-CH_2-CH_2-Cl$$

Substances of this class have a very strong influence on DNA and evoke mutations both *in vivo* and *in vitro*. At the same time, they are highly toxic.

(4) Treatment with weak acid (pH 4) or heating in neutral solution removes both purine bases from DNA molecules, but not pyrimidines. Acid treatment is similar to alkylating agents in this regard, and both transitions and transversions are expected among the mutants.

(5) Several new mutagenic substances, belonging to the class of nitroso compounds, have been found to be extremely potent *in vivo*. Among the best studied representatives of this group are nitrosomethylurea:

$$\text{H}_2\text{N}-\overset{\overset{\text{O}}{\|}}{\text{C}}-\overset{\overset{\text{CH}_3}{|}}{\text{N}}-\text{NO}$$

and N-methyl-N'-nitro-N-nitrosoguanidine:

$$\text{O}_2\text{N}-\text{NH}-\overset{\overset{\text{HN}}{\|}}{\text{C}}-\overset{\overset{\text{CH}_3}{|}}{\text{N}}-\text{NO}$$

Both are extremely mutagenic and exhibit low toxicity. The use of these agents permits the production of mutations with unprecedented ease and rapidity. The formation of diazomethane:

$$\text{H}_2\text{C}\overset{\nearrow \text{N}}{\underset{\searrow \text{N}}{\|}}$$

upon decomposition of these substances probably accounts for their activity since diazomethane is a strong alkylating agent. Evidence for this mechanism is provided by the observation that *E. coli* mutants with reduced sensitivity to diazomethane simultaneously become less sensitive to the mutagenic action of nitrosoguanidine (145).

(6) The acridine dyes constitute a group of mutagens which behave in a very distinctive way. They are mutagenic for phages, and in the *rII* region of T4, acridine-induced mutations occur randomly over the genetic map. While these mutations are acridine revertable, they are not reverted by any other class of chemical mutagens. The effect of acridine dyes on bacteria is somewhat different. They are bacteriostatic and quite effectively eliminate bacterial episomes when used at low concentrations (p. 322), but they do not cause a significant number of mutations. The chemical structure of two widely used acridine dyes, proflavin and acridine orange, are given below:

Proflavin   Acridine orange

Acridine dyes are known to bind directly with DNA to form complexes of a very specific structure and this interaction may well provide the basis of their mutagenicity. Physical investigations, including x-ray diffraction analysis, of acridine-DNA complexes suggest that the acridine molecules become intercalated between adjacent base pairs in the DNA, possibly forming hydrogen bonds with surrounding bases. This disrupts the helical structure of the macromolecule by forming an extra layer perpendicular to the helix axis and results in an increase in distance between neighboring bases to 6.8 Å, exactly twice the normal axial distance between bases in native DNA (146). Compounds yielding complexes of this type with DNA are mutagenic (147) and it is apparent that there must be stringent conditions on their size, shape, and chemical structure. Mutations could result from DNA-acridine complexes by insertion or deletion of a base pair during replication. If the acridine is intercalated between two bases in the template strand, an extra base could be inserted in the new strand opposite the acridine molecule. Intercalation of an acridine molecule into the new strand could force the replicating mechanism to skip over one base in the template, causing the deletion of a base. In either case, one of the DNA strands is stretched in length by an amount exactly equal to the space occupied by a single nucleotide residue, so that the overall pattern of base pairing between the complementary strands is maintained. Heritable alterations produced by acridines are called *frame-shift mutations*, and since they have played an important role in studies on the genetic code, we shall return to them in a later section.

One compound of this class, designated ICR191, is also an effective mutagen for bacteria (148). It contains an acridine nucleus and an active group with the structure of β-chlorethylamine:

$$HN-(CH_2)_3-NH-CH_2-CH_2-Cl$$
$$OCH_3$$

Analogs of this compound possess similar mutagenic capacities.

(7) Spontaneous mutations occur at a frequency of about 10^{-7} to 10^{-8} per generation as the result of various replication errors and metabolic disorders. Both transitions and transversions have been identified among such mutations, as well as a large proportion of frame-shift mutations which can be reverted only by acridine dyes. Although spontaneous mutations are undoubtedly random phenomena, the probability with which they occur can be influenced by changes in the cell environment. Thus the frequency of spontaneous mutation in bacteria has been shown to increase linearly with the time necessary for the cells to double their mass (149).

An attempt has been made to explain spontaneous mutations on the basis of cosmic radiation, but such an interpretation does not bear up under the evidence. The background of such mutations is two orders of magnitude greater than would be expected from cosmic radiation. It follows that spontaneous mutations must be caused by various unidentified chemical reactions involving cellular DNA. It is known, for instance, that certain mutagenic agents, such as peroxides and nitrous acid, can be formed in bacteria as by-products of intermediary metabolism.

(8) All of the reactions considered thus far take place in aqueous solution and appear to be heterolytic. But reactions with free radicals can also cause mutations, as we shall see from a consideration of mutagenic effects brought about by radiant energy.

The mutagenic action of x-rays on microorganisms was discovered by Nadson and Filippov in the USSR and a similar effect on higher organisms (*Drosophila*) was first reported by Muller in the United States (150). The probability that a mutation will arise owing to irradiation with x-rays is proportional to the dose up to about 2000 r, and it does not depend on the intensity of the source (Timofeev-Ressovskij). Thus the effects of radiation are linear and the probability of mutation is proportional to the total number of energy quanta absorbed by the experimental material. These observations gave rise to the so-called *target theory*, based on the idea that mutations are caused by primary ionization within a certain small "sensitive" volume of genetic

15. The Chemistry of Mutagenesis

material, the target. Further investigation has demonstrated that target theory, and the theory of direct action of radiation as a whole, although generally correct, requires modification if it is to be applied to point mutations.

Fundamentally, radiation-induced mutagenesis is mediated by a chemical mechanism. Consider, for instance, that the probability of mutation by radiant energy is extremely dependent on the amount and nature of gasses dissolved in the cell suspension. Hence the replacement of oxygen by an inert gas such as nitrogen or argon reduces the probability of mutation several fold, whereas introduction of carbon monoxide increases the mutation rate. These observations strongly imply a chemical basis of mutation.

Ultraviolet light is an effective mutagen for bacteria, which can be subjected to radiation in a thin layer of medium. The action spectrum of ultraviolet light, that is, the probability of mutagenesis as a function of wavelength, exhibits a maximum at 260 mμ, the wavelength most strongly absorbed by nucleic acids. One of the most likely forms of photochemical damage to DNA is the production of thymine dimers as mentioned in an earlier section. In this case, a single quantum of radiant energy, $h\nu$, is not sufficient to cause ionization, but it is fully adequate to produce photochemical reactions which occur through a series of free radical intermediates. A most important fact is that mutations can be induced even when bacteria are suspended in a *previously* irradiated nutrient medium. x-Rays also cause the formation of highly reactive free radicals and excited metastable states not only in water, but in nucleic acids as well.

There is thus considerable evidence that mutagenesis by radiation proceeds via a mechanism involving free radicals. But free radicals can also be generated without radiation. They are produced, for instance, by the interaction of peroxides with reducing agents, many of which, such as the strong reductant ascorbic acid, are normally present in the mammalian cell. Therefore, hydrogen peroxide, as well as organic peroxides and hydroperoxides, can actually be mutagenic under the appropriate conditions. Finally, metal ions with alternative valence states, such as manganese, are also potentially mutagenic. In the presence of oxygen and reducing agents, manganese ions give rise to oxidation-reduction chains which continually generate free radicals. Free radicals formed as the result of homolytic reactions are chemically very active and are hence mutagenic in minute quantities. Owing to their great reactivity, free radicals attack a number of groups in DNA and are capable of causing both transitions and transversions.

As an aid to understanding spontaneous mutations, it is of interest to note that peroxides such as H_2O_2 are natural metabolites in some cells. Hydrogen peroxide is normally broken down by the enzyme catalase, but if catalase is specifically inhibited by KCN or NaN_3, H_2O_2 accumulates and begins to act as a mutagen. Thus DNA can be altered chemically by a substance usually present in the cell if the concentration of that substance surpasses permissible levels. As we have seen, mutation can be evoked by an extremely small chemical change, by a single reaction between DNA and mutagen. For this reason, we should remember that stimuli which appear insignificant in quantitative terms can provoke catastrophic results under suitable conditions.

Deductions about the mechanism of chemical mutagenesis are usually based on the study of reactions between DNA and mutagen, and on the analysis of the mutants themselves. The specificity of reversions can be particularly revealing. These methods are subject to certain limitations, however. To study the chemical effects of mutagens on DNA, a considerable amount of material must be reacted so that the alterations produced will be measurable analytically. On the other hand, a point mutation requires the alteration of just one nucleotide in 10^5 or 10^6, and the corresponding chemical event is far too small to be analyzed. In order to apply conclusions regarding nucleic acid chemistry to the actual mutational process, they must be extrapolated by several orders of magnitude.

A second and purely genetic method of investigating a mutagen is to compare its effects with those of a second, such as hydroxylamine, whose mechanism of action can be more easily interpreted. Reliable conclusions about the nature of alterations which DNA molecules undergo can generally be drawn from such studies.

Practically all the systematic data on this subject stems from studies on the *rII* region of bacteriophage T4, since mutants are easy to identify, the system is convenient to manipulate experimentally, and the genetic map has been extensively investigated. Table 4-4 presents data on both forward and reverse mutations in phage T4, obtained with a variety of different mutagens.

TABLE 4-4

Induced and Spontaneous Reversions of *rII* Mutants of Bacteriophage T4

Mutagen	Mutants tested	Base-analog revertible (%)	Base-analog nonrevertible (%)	Background of spontaneous nonrevertible mutants (%)	Ref.
2-Aminopurine	98	98	2	2	(151)
5-Bromouracil	64	95	5	2	(151)
Hydroxylamine	36	94	6	2	(152)
Nitrous acid	47	87	13	15	(151)
pH 5, 45°C	115	77	23	15	(151)
Ethyl ethane sulfonate	47	70	30	10	(153)
Proflavin	55	2	98	—	(151)
Spontaneous	110	14	86	—	(151)

Of the mutations evoked by aminopurine (AP), bromouracil (BU), hydroxylamine and nitrous acid, we see that the majority are reverted by AP and BU. This indicates that the original mutational event resulted from nucleotide transitions. To the contrary, spontaneous and proflavin-induced mutations are not revertable by base analogs for the most part. Consequently, these alterations must result from transversions, which cannot be reverted by AP or BU, or from another, more complicated event. The alkylating agent ethyl ethane sulfonate (EES) and exposure to mild acid fall between these two extremes. This corresponds with our theoretical prediction that these methods should produce both transitions, which are revertable with base analogs, and transversions, which are not. The transversions, however, should be reverted by, EES or acid treatment, and indeed, this expectation has been experimentally confirmed. Proflavin-induced mutations and most of those that occur spontaneously are in a class by themselves, for they are not reverted by either EES or acid. They can generally be reverted by a second exposure to acridine dyes, however.

On the basis of reversion studies, transition mutants can be divided into two groups, presumably according to whether the original alteration was of the G—C → A—T or the A—T → G—C type. The key to classifying transitions is the very selective and asymmetrical action of hydroxylamine which can produce only G—C → A—T substitutions. By testing the revertability of AP, BU, hydroxylamine, and nitrous acid mutations with each of the other agents in turn, a consistent pattern emerges. Despite the theoretical prediction that AP and BU should cause transitions of both types, it was found in practice that BU preferentially induces G—C → A—T replacements like hydroxylamine, while AP is more likely to evoke A—T → G—C replacements. Nitrous acid seems to be quite unselective in this regard, reverting all transition mutations produced by itself and by any other of the agents.

The work of Benzer and Freese on mutagenesis at the *rII* locus of bacteriophage T4 brought to light the curious fact that certain sites in the cistron are particularly sensitive to mutation. Such sites, called "hot spots," were found both for spontaneous mutations and for those evoked by mutagens. Most of them were specific for individual mutagenic agents, but in some cases the hot spots of two or more agents coincided. When a hot spot map was constructed showing the frequency with which each mutagen affected a particular site, it was evident that the overall mutability spectrum for every agent tested was quite distinctive.

Let us consider another example that may shed some light on the nature of hot spots. By treating phage T4 with aminopurine, two groups of *rII* mutants are obtained which differ from one another in their ability to revert by a factor of more than 100. The mutants which are easily reverted by hydroxylamine also revert with high frequency under the influence of nitrous acid, EES, and even spontaneously. But those mutants which have a low probability of reversion upon treatment with hydroxylamine are also stable with respect to the other mutagens. All of the observations can be explained if the easily reverted mutants contain erroneous G—C pairs at very reactive points in the DNA molecule.

Sites of varying reactivity in the DNA molecule are not difficult to imagine. They might arise from differences in the density of hydrogen bonds, depending on the nature of neighboring residues, or even from a difference in chemical bond stability owing to the inductive effect of π-electrons belonging to closely situated bases. One can suppose that certain combinations of neighboring bases give rise to "weak" spots which bear the brunt of attack by chemical mutagens. The existence of hot spots is probably not a property of viruses alone. Highly mutable hereditary traits were encountered long ago in the genetics of higher organisms. For example, certain genes in corn are known to mutate spontaneously with extremely high frequencies—0.1% per generation instead of the usual 0.0001%—while reversion rates reaching 1% per generation have been observed in a variety of unstable mutations of other organisms. Whether or not all of these phenomena have the same chemical basis, the study of hot spots and their susceptibility to the action of specific mutagens may provide a means for directed mutagenesis, at some future time, by which man can exert conscious control over the variability of living organisms.

Mutagenesis may be thought of as a first-order reaction between DNA and a mutagenic agent. Indeed, the fact that mutants accumulate with first-order kinetics demonstrates that only one effective encounter between DNA macromolecule and mutagen is necessary to produce a mutation, or, in more concrete terms, that only a single nucleotide residue in the DNA chain need be altered. This is also indicated by the comparatively high frequency of reversion in phages and bacteria. If mutations were to affect several nucleotides simultaneously, the probability that all of them could also be "corrected" together is extremely small.

Consider now the mutagenesis of purified transforming DNA from *Bacillus subtilis*, which, after treatment, can be introduced into competent recipient cells. This process was studied with a number of mutagens both in the laboratory of the author and in that of Freese. Mutants at the tryptophan locus were selected for convenience of observation. These mutants were blocked at the stages of tryptophan synthesis associated with the conversion of anthranilic acid to the Amadori compound, and of the latter into indolylglycerol phosphate (p. 429). Tryptophan precursors are accumulated under these conditions and since they strongly fluoresce under ultraviolet light, mutant colonies of the type sought can be readily detected on agar plates.

The action of mutagenic agents *in vitro* generally inactivates transforming DNA, since the polymer chain undergoes a variety of chemical lesions. Only in rather rare cases does an alteration occur at the cistron under study which can be established in the recipient genome as a heritable trait. We cannot separate mutagenesis from overall damage to the DNA and the kinetics of both processes must be considered together (154). Since transformation requires a double crossover between donor DNA and cellular chromosome, we must proceed as usual from the relationship between genetic map distances and recombination frequency (p. 351):

$$W = (1/2)(1 - e^{-2\Delta l \omega}) \tag{4-33'}$$

where W is the frequency of recombination between two points on the chromosome, Δl is the physical distance between them, and $1/\omega$ is a scale factor for the conversion of physical distance into genetic map distance.

Let us first consider the effects of inactivation on the integration of donor markers into the recipient genome. When a molecule of transforming DNA accumulates z inactivating or lethal, hits, its *recombination length* decreases. In other words, the segment of donor DNA which, upon integration, can introduce the given marker but not a lethal lesion, is effectively reduced. We shall consider the recombination length to be the distance from the given marker to the nearest lethal alteration (Fig. 4-49).

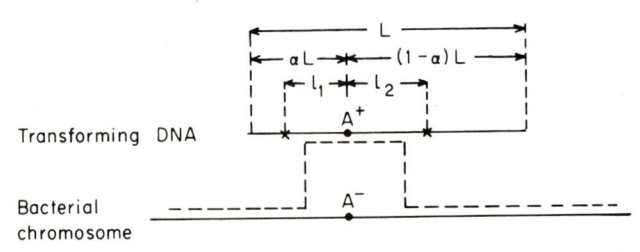

Fig. 4-49. Scheme of recombination between a molecule of transforming DNA and the bacterial chromosome containing a point mutation. The transforming DNA carries the marker A^+ while the recipient is A^-. The length of the transforming molecule is L, and the distance from the point mutation to the "left" and "right" ends of the transforming DNA are αL and $(1 - \alpha)L$, respectively. The distances between the point mutation and the nearest points of chemical inactivation to "left" and "right" are l_1 and l_2, respectively. Broken line indicates double recombination which will lead to A^+ transformant. For calculation of recombination frequency, see text. Modified from Bresler *et al.* (154).

15. The Chemistry of Mutagenesis

The probability, R, of a double crossover necessary to integrate the A^+ marker is expressed as an average of the probability W over all possible recombination distances:

$$R = \int_0^{\alpha L} W(l)\Phi(l,z)dl \int_0^{(1-\alpha)L} W(l)\Phi(l,z)dl \tag{4-45}$$

where L is the original length of the transforming molecule, αL and $(1-\alpha)L$ are the parametric coordinates of the marker, and z is the total number of lethal lesions along L. The number z increases linearly with reaction time between DNA and mutagen. The function $\Phi(l, z)$ describes the distribution of recombination lengths, that is, the distances from a given marker to the nearest lethal lesions. If the potentially lethal alterations are distributed randomly along L, then

$$\Phi(l, z) = e^{-l/\lambda}[(1/\lambda) + \delta(L-l)] \tag{4-46}$$

where $\lambda = L/z$, the average distance between two lethal hits, and δ is a function which takes on the value of either 0 or ∞. Assuming that $\lambda < \alpha L$, in other words, that the molecule sustains many lethal alterations both to the "right" and to the "left" of the given marker, we will obtain a very clear and simple formula for the kinetics of inactivation by integrating Eq. (4-45):

$$R/R_0 = 1/[(1/2\, \omega\lambda) + 1]^2 = 1/[(z'/2) + 1]^2 = 1/[(k_1 t/2) + 1]^2 \tag{4-47}$$

where R is the number of transformants resulting from mutagenized donor DNA, R_0 the same for untreated DNA, z is the number of lethal lesions per length L, and z' is the number of lethal lesions occurring in a DNA segment of length $1/\omega$, that is, per unit map distance. Since the number of lethal lesions increases linearly with time of exposure to mutagen, we can write z' as $k_1 t$.

We see that the decrease in the number of transformants with time is expressed by a very simple formula. Moreover, this theoretical relationship is in complete accord with the empirical one found by Goodgal and Rupert for inactivation of DNA with ultraviolet light. Inactivation of transforming DNA with nitrous acid, hydroxylamine, ultraviolet light, and deoxyribonuclease was also performed in the laboratory of the author, and the experimental data were in excellent agreement with Eq. (4-47).

Let us return to the question of nonlethal mutations at a given genetic locus on the transforming DNA. As the number of mutational alterations at this locus rises, the total number of lethal alterations all along the DNA molecule also rises so that the effective recombination length decreases (155). If one efficient encounter between DNA and mutagen is sufficient to produce a mutation, the number of alterations, y, will increase linearly with time: $y = k_2 t$. (If two encounters or hits were necessary, the equation $y' = k_2 t^2$ would apply, and so on.) The number of recipients, n, which have been transformed for a given marker by mutagenized donor DNA is:

$$n = Af(z)y = Af(z)k_2 t \tag{4-48}$$

where A is a constant expressing the probability of penetration and integration of the donor DNA molecule, and $f(z) = R/A$ characterizes the drop in frequency of double crossovers as a function of the number of lethal lesions sustained by the DNA molecule Dividing n by R, we obtain:

$$n/R = k_2 t = (k_2/k_1)z' \tag{4-49}$$

Hence the ratio of recipients transformed by donor DNA bearing nonlethal mutations at a given locus to the total number of recipients transformed by mutagenized DNA is a linear function of the time or of z' which is itself proportional to time. Equation (4-49) applies only when mutagenesis requires a single hit. The absolute number of transformants for the marker in question is:

$$n = A(k_2/k_1)\{z'/[(1/2\,\omega L)z + 1]^2\} = \gamma z'/[(z'/2) + 1]^2 \tag{4-50}$$

The value of z' can be evaluated independently by measuring the decrease in the number of transformants for the marker being studied along with the overall increase in mutations. Only the constant γ in Eq. (4-50) remains to be determined. But since $n_{max} = \gamma/2$, we can avoid explicit reference to γ by expressing n as a fraction of n_{max}.

The values of n obtained experimentally with a number of different mutagenic agents are plotted in Fig. 4-50 as a function of z, the average number of inactivating

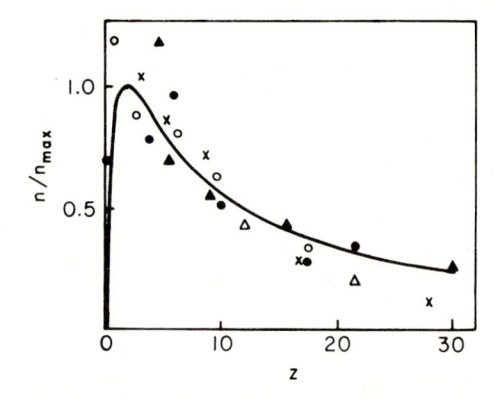

Fig. 4-50. Dependence on z of the absolute number of *trp* transformants in *B. subtilis*. The points correspond to the action of various mutagenic agents as follows: (●) NH_2OH; (○) HNO_2; (▲) N_2H_4; (△) dimethyl sulfate; (×) ultraviolet radiation. All experimental data expressed as a fraction of the maximum value of n. The solid curve is plotted according to Eq. (4-50). From Bresler et al. (156).

lesions per molecule. The fit between the data and the theoretical curve is excellent. Figure 4-51 presents the relative number of transformants for the tryptophan markers, n/R, also as a function of z. Once again there is good agreement between experiment and theory. Straight lines are obtained for all mutagens studied, attesting to the single-hit mechanism of mutagenesis. The ratio of the mutagenic and inactivation constants, k_2/k_1, can be calculated from the slopes of these lines. This ratio depends on the effectiveness of the mutagen—we see, for instance, that hydroxylamine is 20 times more effective than ultraviolet light.

The same considerations which apply to transforming DNA are also relevant to mutagenesis in bacteriophages. In phages T2 and T4, the ratio of viable to lethal mutations is a characteristic parameter with a value of between 10^{-2} and 10^{-3} for forward mutations and between 10^{-5} and 10^{-6} for reversions. The latter number confirms an earlier comment to the effect that back mutation requires a specific alteration in one nucleotide pair out of the 2×10^5 base pairs which comprise the T4 genome. But an arbitrary alteration at almost any one of the DNA residues can apparently evoke a lethal effect.

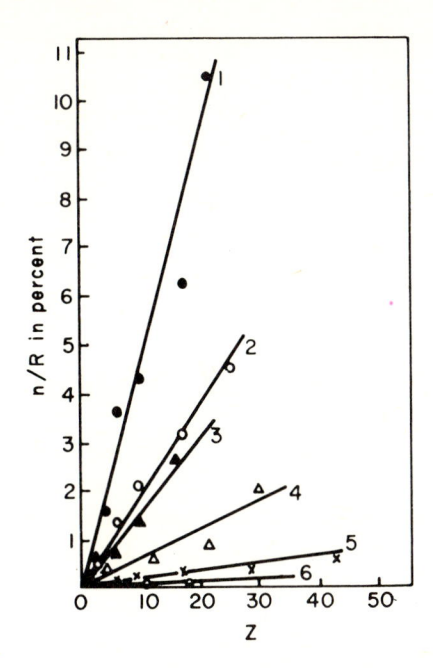

Fig. 4-51. Dependence on z of the relative number of *trp* transformants. Data relates to the treatment of transforming DNA with a number of mutagenic agents: 1 (●) NH_2OH; 2 (○) HNO_2; 3 (▲) N_2H_4; 4 (△) dimethyl sulfate; 5 (×) ultraviolet radiation; 6 (◓) pH 4.2. From Bresler *et al.* (156).

16. Mutation and the Genetic Code

In the imaginary self-replicating cybernetic machine of von Neumann, there must be a memory element in which all of the information necessary for the reproduction of an identical machine is concentrated. The memory element must be able to reduplicate or copy itself and transmit the copy to the "daughter" machine and it must contain the orders required for manufacture of all parts of the machine in proper configuration and in the proper sizes. A magnetic tape would be a likely memory element, for it can store all necessary information in the form of a code and it could actuate a mechanism—a second tape recorder, for instance—so as to reproduce this information.

The same sorts of problems also confront the cell, which is reproduced not mechanically or electronically, but by chemical synthesis. Enzymes are the working machinery of the cell, and small cytoplasmic organelles called ribosomes are its workshops. Its memory elements are chromosomes, housed in the cell nucleus, whose DNA molecules carry all the genetic information required to reduplicate the cell. Broadly speaking, this is information specifying the structure of cellular proteins, and in particular, of enzymes, and it is recorded in the polymeric DNA chain by means of a special chemical code. Investigation of the nature of this code and the means for deciphering it constitute one of the fundamental tasks of molecular biology.

The question of the code was first posed by Gamow in 1954 (157). The particular concepts that he held at that time were incorrect in their details, but the great service Gamow performed was to state the problem in a new way. Gamow reasoned as follows. Suppose that a protein is a linear polymer in which twenty different kinds of amino acids are arranged in some definite, but unknown sequence. Assume further that all of the characteristics of the protein are determined exclusively by the sequence of different amino acid residues along the polymer chain. The amino acid sequence of each protein macromolecule is therefore the property that must be encoded in the DNA molecule. But DNA consists of only four different nucleotides, so that several nucleotides are necessary to specify a single amino acids. The number of nucleotides required to code for one amino acid received a special name, the *coding ratio*. If the coding ratio is two, then the number of different doublets, or distinct nucleotide pairs which can be generated from four nucleotides is $4^2 = 16$, and this is insufficient to uniquely code all twenty amino acids. If the coding ratio is three nucleotides to one amino acid, the number of different trinucleotide combinations will be $4^3 = 64$, more than enough to specify all the common amino acids. Hence one may guess that the nucleic acid chain specifies the primary structure of the proteins whose synthesis it governs by successive trinucleotide code words or *codons*. This is called a three-letter or *triplet code*. The whole problem is somewhat similar to enciphering a text written in the normal alphabet of 26 letters by Morse code, which possesses but two different, symbols, a dot and a dash.

Gamow attempted to impose another condition on the genetic code. He proposed that there is an exact structural correspondence between the DNA and protein chains, since he believed that the nucleic acid must be a template for synthesis of the polypeptide. At the present time, we know that protein is not synthesized directly from a DNA template; the DNA does, however, serve as a template for the synthesis of messenger RNA, an exact informational replica of DNA (if we assume the equivalence of thymine in DNA with uracil in RNA) which diffuses from the cellular nucleus to the ribosomes where it directs the synthesis of protein molecules. But the participation of this RNA intermediate does not alter the essential feature of the problem, for the sequential arrangement of bases in a nucleic acid chain, whether RNA or DNA, must nonetheless be translated into protein structure.

The linear extent of a nucleotide residue along a helical nucleic acid chain is 3.4 Å, while the linear extent of an amino acid residue in a protein—actually, the length of three covalent bonds—is 3.6 Å. In order to fulfil the second requirement imposed by Gamow, it is necessary that every amino acid residue in the protein be coded by a unit of the polynucleotide chain lying directly opposite to it, that is, in one-to-one physical correspondence. To construct a three-letter code in this instance, the code words must overlap one another (Fig. 4-52a). Here every amino acid is represented by three nucleotides, but neighboring nucleotides are not independent of each other. Every one is used three times and appears in the code words of three different amino acids: the triplet UVW represents one amino acid residue. VWX the next, WXY the third, and so on. Even though the code consists of three-letter words, the ratio of the number of amino acid residues coded and the number of nucleotides which code for them is in fact unity according to this scheme. An overlapping code obviously imposes rigid conditions on possible amino acid sequences in the protein chain; no such correlations between neighboring amino acids have ever been observed and this solution was deemed unsuitable (158). Furthermore, while three code words would be affected by a single nucleotide substitution in the case of an over-

16. Mutation and the Genetic Code

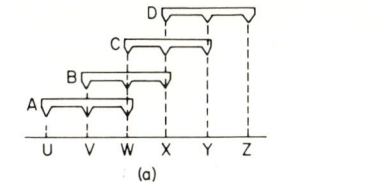

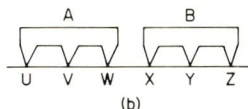

Fig. 4-52. Two types of triplet genetic codes. (a) Overlapping triplet code with a coding ratio of one. Each code letter or nucleotide base participates in the code words for three amino acids so that amino acid sequences will not be independent of each other. (b) Nonoverlapping triplet code with a coding ratio of three. Each code word is independent of all others so that a single point mutation will affect only a single amino acid.

lapping code, point mutations have been found to lead almost universally to changes in only one amino acid residue at a time.

Crick therefore proposed that the requirement for structural correspondence between nucleic acid and protein be abandoned in favor of an *adaptor hypothesis*. According to this notion, the adaptor is an intermediate molecule with at least two functional sites; one can bind a specific amino acid residue and the second must be able to specifically recognize a trinucleotide sequence in the template polynucleotide (Fig. 4-52b). The adaptor hypothesis circumvents a major difficulty encountered in the overlapping code, since structural correspondence between the two polymers is not required. A new difficulty arises now, however, for the adaptor molecules must find and identify the specific trinucleotides UVW or XYZ, but must not mistakenly bind to the intervening triplets VWX and WXY. This means either that there must be some sort of "punctuation" between trinucleotides symbolizing "correct" amino acid residues, or that the "reading" of code word sequences proceeds unidirectionally and in a stepwise fashion from a given starting point so that all the following triplets will be read in register. As we shall see, the protein chain does in fact grow progressively from one amino acid to the next in a finite period of time, and it is assumed that the adaptors with their amino acids are aligned on the polynucleotide template in order from the point at which chain growth is initiated.

In genetic studies on bacteriophage T4, Crick, Brenner, and colleagues (159) experimentally confirmed the triplet code and indirectly demonstrated that reading of the code words must take place in one direction along the polynucleotide template from a specific starting point so that successive code words are read in register. This achievement was made possible by a special type of mutation that can arise spontaneously or under the influence of acridine dyes. In particular, Crick and Brenner studied proflavin-induced mutations in the *B* cistron of the phage T4 *rII* region. Such mutants differ in two important ways from most other point mutants. First, they can be reverted only spontaneously or by other acridines, but not by mutagens which cause direct chemical alterations of bases in the DNA. Second, they act according to the all-or-nothing principle by completely eliminating the functional activity of a cistron. In contrast, many revertable point mutations only affect the activity of a protein quantitatively without totally impairing its function. These are the so-called *leaky mutants* which are still viable despite alterations in a given gene product. Thus *r*-type mutants of phage T4 can grow on *E. coli* B, albeit with altered plaque morphology, but not on strain K12(λ). However, there are leaky *rII* mutants that are capable of growing on K12(λ), although with *r*-type plaque morphology, once again indicating a functional difference between them and wild-type phages. Among the *rII* mutants induced with acridines, however, leakiness is never observed.

The behavior of acridine mutants implies that at least one entire cistron is devoid of functional activity. Although these might at first be thought to result from large deletions, studies of reversion and recombination do not bear this out, for they act like point mutations. A likely explanation is that acridine dyes cause the insertion or deletion of a single nucleotide base pair during DNA replication. We shall indicate these two possible mutations by + and −, respectively. The hypothesis was verified by analyzing spontaneous reversions of the proflavin-induced *rII* mutations. Of the 20 revertants isolated by their ability to grow on K12(λ), 18 turned out to be double mutants, since, when crossed with the wild-type phage, the *rII* recombinants recovered were of two different types. One class was identical to original *rII* mutations because they did not subsequently recombine with them. The second class, however, consisted of distinct nonleaky *rII* mutants which all mapped within the *B* cistron. In fact, all of the mutations which corrected or suppressed the original *rII* mutants were found to occupy a region equal to about one tenth of the total cistron length, extending to either side of the original sites.

This result can be explained as follows with the aid of Fig. 4-53. If the first mutation is assumed to be of the type designated +, then the second has to be −. In the original mutant, the entire *B* cistron is out of register from the point at which the mutation occurred to the end because the insertion of an extra base (frame 3, line b, Fig. 4-53) breaks the succeeding "text" down into incorrect nucleotide triplets and will in all probability lead to the appearance of a premature nonsense or chain-terminating codon in the subsequent portion of the message. The second mutation removes a residue in frame 6 (see line c) so that the code is read improperly between frames 3 and 6, but starting with frame 6, the normal reading frame is restored and all triplets beyond that point are expressed just as they are in the wild type. A single point mutation of this *frame-shift* type therefore disrupts a significant portion of the putative *B* protein and it is very unlikely that such a protein would be active, thus explaining the nonleaky phenotype of the proflavin-induced lesions. Strictly speaking, reversion does not lead to the restoration of the wild-type cistron in terms of nucleic

	Frame of codon number										
	1	2	3	4	5	6	7	8	9	10	
(a)	ABC	ABC	ABC	ABC	ABC	ABC	ABC	ABC	ABC	ABC	Wild type
(b)	ABC	ABC	A*AB	CAB	CAB	CAB	CAB	CAB	CAB	CAB	+ Mutant
(c)	ABC	ABC	A*AB	CAB	CAB °ABC	ABC	ABC	ABC	ABC	+− Revertant
(d)	ABC	ABC	A*AB	CAB	C*CA	BCA	BCA	BCA	BCA	BCA	++ Mutant
(e)	ABC	ABC	A*AB	CAB	C*CA	BCA	BCA	B*AC	ABC	ABC	+++ Revertant

Fig. 4-53. The mechanism of frame-shift mutation. (*a*) The frames numbered 1 to 10 correspond to a sequence of ten successive code words. (*b–e*) Asterisks (*) mark positions in the polynucleotide chain where a base has been inserted, while the symbol (°) indicates the deletion of a base. See text for details.

acid sequence, for the region between the two mutations remains "out of step." But if the distorted segment of the nucleic acid chain is not greater than one tenth of the total cistron, corresponding to about twenty triplets, the synthesis of the *B* protein takes place and the product is functional. Furthermore, the integrity of the polypeptide chain is apparently not affected, although the amino acid sequence may be completely changed in the interval coded by the nucleotide sequence between the two shifts in reading frame. The altered amino acids evidently fall outside the functionally active region of the protein.

Revertants to the second class of mutants, of − genotype, were next isolated. Upon mating these new strains with the wild type, *rII* recombinants of two types resulted once again. And again, some were identical to the parent strain, and some mapped at distinct sites. The latter, which must have the + genotype, are all located near to the − lesions in their parents. The new + revertants were themselves reverted again, producing a third generation of independent nonleaky *rII* mutants, this time of the − type. All in all, 80 point mutations were selected for study in this way. They all bore the *r* phenotype, but they could be divided into two unambiguous groups, scored either + or −, depending on the number of stages necessary for their isolation.

Recombination between + and − mutations can lead to wild-type behavior because the original reading frame for the triplet code is restored with the exception of the segment between the two mutations. But recombination between two + mutants or two − mutants never produces wild type phages, for the insertion or deletion of two nucleotide pairs does not restore the original reading of trinucleotide code words (line d, Fig. 4-53). When the triple mutants + + + and − − − were constructed, however, they formed plaques on K12(λ). This result was of great importance, for it proved that the genetic code consists of three-letter words (or a multiple of three if more than one base is inserted or deleted during acridine mutagenesis). Three single nucleotide insertions or deletions would return the reading frame to its initial state outside of the region defined by the two outermost frame-shift mutations (line e, Fig. 4-53).

Streisinger and collaborators have recently shown directly that all the properties of frame-shift mutations deduced from purely genetic experiments are correct (160). He and his colleagues isolated a number of proflavin-induced mutations in the cistron of phage T4 governing the synthesis of lysozyme. This enzyme is responsible for releasing phage particles at the end of the lytic cycle by digesting the bacterial cell wall. Although single mutations of this type lead to inactive lysozyme, the strains carrying them could be isolated by lysing the bacteria in which they developed with external lysozyme. When certain mutants of this kind were mated, their recombinants were able to synthesize lysozyme and form plaques on indicator bacteria. Phages bearing pairs of frame-shift mutations in the lysozyme cistron were distinguished from wild-type phages on the basis of plaque type. Wild-type and a number of mutant lysozymes were isolated and hydrolyzed with trypsin, and the resulting peptides were separated chromatographically. Where peptide differences were noted, the amino acid sequences of the peptides in question were compared with those of the wild type. It was found that in the mutant peptides, sequences of two to five residues were altered. This at once confirmed the genetic experiments of Crick, Brenner, and colleagues, and further, the amino acid alterations were consistent with the trinucleotide code word assignments deduced independently from experiments *in vitro* (p. 436). The importance of Streisinger's experiments was that the codons were verified *in vivo*, in a living bacterial cell. This work also demonstrated that the nucleotide code is read in sequence from a definite starting point and permitted the authors to conclude that translation of the polynucleotide messenger proceeds from the 5'- to the 3'-end.

All of these experiments, as well as a number of others that we will not be able to treat, makes the triplet code without punctuation appear very likely. Moreover, the code is undoubtedly *degenerate*; in other words, several different nucleotide triplets code for the same amino acid. We noted earlier that it is possible to compose 64 different triplets from 4 nucleotides, while there are in all only 20 amino acids commonly found in proteins. Degeneracy may be inferred from the fact that in $+-$ *rII* and lysozyme mutants of phage T4, a substantial portion of the cistron contains the "wrong" nucleotide sequence, yet they lead to the synthesis of functionally active protein. If 44 out of the 64 possible trinucleotide combinations were meaningless and did not code for any amino acid, the probability of obtaining functional protein in these cases would be vanishingly small. Another fact in favor of the degenerate code is the large variation in the $(G + C)/(A + T)$ nucleotide base ratio among different species of bacteria (Belozersky and Spirin, 161). This ratio can fall anywhere between 0.45 and 2.8, depending on the species. Yet the amino acid composition of total protein from these same bacterial species differs very little from one to another despite the sharp divergence in their DNA base ratios (Sueoka, 162). If the code is identical in all these organisms, the only reasonable explanation of the above findings is provided by code degeneracy which allows for species differences in nucleotide composition without denying the existence of overall regularities in the genetic systems of diverse bacteria.

Although a universal code was assumed in the foregoing, a direct test of the extent to which the genetic code differs in different bacteria was of great importance. This question was studied in the laboratory of Levinthal by the means of the following experiments (163). When a strain of *E. coli* carrying the F'-*lac* episome is mixed with cells of *Serratia marcescens*, the episome is transferred to *Serratia* recipients with a frequency of nearly 100%. Even though the difference in the DNA composition of these two species is significant—*E. coli* DNA contains 50% (G + C) and *Serratia* DNA, 58% (G + C)—the foreign DNA replicates very well in *Serriatia* cells. But is the genetic information introduced on the *E. coli* episome expressed in *Serratia*? Ability to synthesize the bacterial enzyme alkaline phosphatase was used to settle this question.

Both *E. coli* and *Serratia* synthesize phosphatases which are similar in their enzymatic properties, but different in their amino acid compositions and in their immunological and physicochemical behavior. For instance, the isoelectric point of *E. coli* phosphatase is pH 4.5, while that of *Serratia* phosphatase is pH 6.2. If mixed together, the two phosphatases are quite easy to separate by starch gel electrophoresis and they give indisputably different fingerprints after hydrolysis with trypsin. Now the *E. coli* F'-*lac* episome also carries the *pho* (phosphatase) cistron, and when the episomal culture is mated with wild-type *Serratia*, cells are formed with information for both phosphatases. The *Serratia pho* cistron is situated on the cellular chromosome while the *E. coli pho* cistron resides on the introduced episome. It was found that such cells can synthesize both phosphatases simultaneously and in approximately equal quantities. Proof of this remarkable phenomenon was provided by electrophoresis, immunology, and fingerprints of tryptic hydrolyzates. The genetic code is thus read the same way in these two bacterial strains, at least. This result has been generalized by other methods of investigation.

Consider now a general approach to deciphering the genetic code which makes use of both genetic and biochemical investigations. First, a variety of bacterial or viral strains are selected which carry mutations in a given cistron. The position of each mutation on the fine structure genetic map is ascertained by recombination analysis. Next, the amino acid sequence of the protein whose synthesis is governed by that cistron is worked out, both for wild type and mutants, so as to define the position and the nature of the amino acid replacements brought about by each mutation. If the mutagen used to produce the mutation has a well-defined specificity of action, the nature of the nucleotide alteration in the cistron can also be deduced. By comparing amino acid replacements at a given site in the protein chain evoked by a single nucleotide base change, and by analyzing how the DNA must have been altered to produce such replacements in the protein, we can in principle decipher the genetic code. One example of this procedure has already been considered in our discussion of the effect of frame-shift mutations of phage T4 lysozyme. In that case, the mutagen had a very distinctive mode of action. In general, however, the specific nature of nucleotide base substitutions is hard to determine with certainty and the problem is usually more difficult than this simple protocol would suggest.

A study which served as a prototype for subsequent investigations of this type is the classical work of Ingram on normal and abnormal human hemoglobins (164, 165). The hemoglobin cistron could not, of course, be analyzed genetically, but Ingram's approach to protein structure analysis paved the way for similar studies in microorganisms where genetic investigations are possible.

A number of hereditary diseases of the red blood cells, or erythrocytes, have been known for many years. In one of them, called *sickle-cell anemia*, certain conditions cause the erythrocytes to assume a sickled, rather than an ellipsoidal, shape. If this occurs, hemolysis frequently follows and oxygen transport throughout the body is thereby impaired. The ways in which this disease is inherited suggests that it is a genetic defect resulting from a single mutation. If an individual is heterozygous for the trait, that is, if he obtains a normal gene from one parent and a mutant gene from the other, he develops normally and his erythrocytes do not sickle *in vivo*, although they can be shown to sickle under conditions of low oxygen tension. But if two heterozygous parents both transmit the sickle-cell gene to a child so that he is homozygous for the trait, sickle cells can be observed in his bloodstream and he will most likely die at an early age owing to severe hemolytic anemia.

The question arises why the sickle-cell trait did not disappear in the process of evolution insofar as it leads to the death of homozygous progeny. Since this disorder is particularly widespread in Africa and other regions where malaria is endemic, it was suggested by Allison that possession of the sickle-cell trait, even in heterozygous form, may confer resistance to malaria upon the carrier (166). According to this hypothesis, the abnormal erythrocytes of individuals with the sickle-cell trait are less easily parasitized by the protozoans which cause malaria than normal erythrocytes.

Where the incidence of malaria is very high, there would hence be a selective advantage in carrying the sickle-cell gene heterozygously. This is perhaps the reason why the number of sickle-cell carriers has not decreased, but has increased, from generation to generation.

An explanation for sickle-cell anemia was first advanced by Linus Pauling and associates (167). He proposed that the mutation affects one of the two genes governing the synthesis of hemoglobin, the major protein component of erythrocytes, whose function it is to transport molecular oxygen to all cells of the body. Pauling and his co-workers studied and compared the properties of normal and mutant hemoglobins and noted that the electrophoretic mobility of sickle-cell hemoglobin (Hb S) differs from that of normal hemoglobin (Hb A). This abnormality does not cause sickle-cell hemoglobin to lose its ability to bind oxygen, for the binding constants of Hb S and Hb A were found to be approximately the same. However, Hb S is substantially less soluble than Hb A and it aggregates within the red blood cells when its concentration is sufficiently high, accounting for the appearance of sickle-shaped cells in persons homozygous for the trait. The intracellular precipitation of Hb S renders the erythrocyte more susceptible to hemolysis, and anemia results.

Hemoglobins from individuals suffering with a variety of blood diseases were subsequently found to differ electrophoretically from Hb A. Further study led to the identification of a great many abnormal hemoglobins including some that cause no ill effects in addition to those associated with serious disorders. The name "molecular diseases" given to such abnormalities by Pauling was clever, but not very accurate. It is the human organism which is afflicted because of alterations in the structure of a specific protein that carries out a vital function.

Ingram elucidated the chemical nature of mutant hemoglobins by analyzing their amino acid sequences. He first separated the α- and β-polypeptide chains which make up native hemoglobin and then digested them with trypsin and fingerprinted the resulting peptides by two-dimensional electrophoresis and chromatography. The α-chains of both normal hemoglobin A and mutant hemoglobin S were identical, but fingerprints of β-chains from the two proteins revealed a difference in just one peptide out of 26 (see Fig. 1-3, p. 17). When this peptide was analyzed in greater detail, it became clear that the difference reduced to the replacement of a single amino acid out of the 146 residues comprising the β-chain. Comparative amino acid sequences for several kinds of hemoglobin are presented in Table 4-5. They explain why electrophoretic differences were noted between normal and mutant molecules; in Hb S, the normal acidic residue glutamic acid is replaced by the uncharged residue valine, while in Hb C, a normal hemoglobin with an alteration at the same site, it is replaced by the basic amino acid lysine.

The existence of mutant forms of hemoglobin provides excellent evidence for the presence of cistrons in higher organisms which govern the synthesis of specific polypeptide chains. The "one gene–one enzyme" principle enunciated by Beadle and Tatum for microorganisms can be applied to human hemoglobins in the slightly modified form, "one cistron–one polypeptide chain." Hemoglobin is not suited to studies on the genetic code, however, for it is difficult to perform the necessary genetic experiments on higher organisms.

Correlation of the position of amino acid sequence alterations in proteins with map positions of the structural gene mutations that provoke them has been attempted

TABLE 4-5

N-Terminal Amino Acid Sequences in the β-Chain of Normal and Abnormal Human Hemoglobins[a]

Hemoglobin	Physiology	Terminal amino acid sequence
Hb A	Normal	H_2N-Val-His-Leu-Thr-Pro-*Glu*-Glu-Lys- (+ + − − +)
Hb S	Sickle-cell anemia	H_2N-Val-His-Leu-Thr-Pro-*Val*-Glu-Lys- (+ + − +)
Hb C	Normal, but present in excess in thalassemia	H_2N-Val-His-Leu-Thr-Pro-*Lys*-Glu-Lys- (+ + + − +)

[a] From Ingram (164).

with bacteria and viruses in many laboratories with the aim of establishing the *colinearity* of genetic map and gene product. One of the first experiments of this sort was carried out by Levinthal, Garen, and Rothman with enzyme alkaline phosphatase in *E. coli* (9, 168).

Alkaline phosphatase is one of the perhaps 2000 different proteins that the bacterial cell can produce. Its apparent function is to supply inorganic phosphate when its concentration is insufficient in the growth medium by cleaving orthophosphate from organic phosphomonoesters. In order to study the chemical structure of phosphatase, it is necessary to employ bacteria which can synthesize this enzyme in large amounts. There are two ways of accomplishing this. Phosphatase synthesis is repressed in the presence of excess inorganic phosphate, and in this case the enzyme comprises no more than 0.03% of the total *E. coli* protein. Its synthesis is greatly stimulated, however, if phosphate ions are removed from the medium. At the maximum synthetic level, alkaline phosphatase constitutes as much as 5 to 6% of the total cellular protein. Another means of increasing the cellular output of the enzyme is to use special strains which lack normal phosphatase regulation. These so-called *constitutive* strains elaborate a large quantity of the enzyme even in the presence of inorganic phosphate. Both wild type and mutants can be grown under these conditions so as to secure the maximum amount of protein from them. When the bacteria have reached an appropriate density, the enzyme is extracted and purified by chromatography on DEAE-cellulose.

Physical and chemical characterization demonstrates that alkaline phosphatase has a molecular weight of 80,000 and that it consists of two identical subunits held together by ionic bonds. Each subunit is a single polypeptide chain containing 380 amino acid residues. Hydrolysis with trypsin yields 35 peptides which can be separated by electrophoresis and chromatography (fingerprinting).

We have already seen that a wide variety of mutants with lesions in the phosphatase cistron can be selected quite readily (p. 309). Such mutants either synthesize an enzyme with reduced activity, and sometimes with altered physicochemical properties as well, or else they synthesize no active enzyme at all (*pho*⁻ phenotype). Phosphatase from the first group of mutants generally cross-reacts immunologically with antibodies specific to the wild-type enzyme. Protein of this sort is called *cross-reacting material* or CRM for short. Phosphatase CRM is also found in certain *pho*⁻ mutants, although it is completely absent in others. The CRM-forming strains all resulted from alterations in the phosphatase structural gene, since they contained demonstrably different enzyme molecules. The non-CRM-formers were also assumed to carry mutations in this gene because both kinds of mutants cluster together within a small region of the *E. coli* genetic map, just 0.3 recombination units in length. By recombination analysis, the map positions of the various *pho* mutations were established with an accuracy of two to three nucleotide pairs. Levinthal and associates also obtained several *pho*⁺ revertants by back mutation of *pho*⁻ strains. Some of them yielded phosphatase identical to that of the wild type, while another group produced active, but altered, enzymes, differing from the wild type in electrophoretic mobility or fingerprint.

In the course of these studies, two closely linked *pho* mutations were shown to cause alterations in the same tryptic peptide. The map distance between *E26*, a mutant that produced an enzyme of low activity, and *U24*, a CRM⁻ mutant, was 0.005 recombination units, equivalent to one sixtieth of the entire structural gene. Tryptic fingerprints of the phosphatase made by *E26* lacked a single peptide, but no chemical studies could be carried out with *U24* since it did not synthesize a phosphatase CRM. However, a revertant of *U24* that synthesized an active, but electrophoretically altered, phosphatase was found to map within 0.0003 recombination units of the original forward mutation in *U24*. Due to the very great proximity of forward and reverse mutations, the protein of the *U24* revertant was used to identify the location of the original alteration. As predicted, the tryptic fingerprint of this protein lacked the same peptide as *E26*, demonstrating that the close linkage between genetic sites in *E26* and *U24* is paralleled by close linkage of the amino acids controlled by these sites. This was a first step on the way to proving the colinearity of a cistron with its polypeptide product.

Another CRM⁻ mutant at the *pho* locus was found to revert to *pho*⁺ as a result of mutation at a second site separated by 0.14 recombination units, that is, by 50% of the length of the cistron, from the site of the forward mutation. Furthermore, the forward mutation mapped at the same place as *E26*, and as might have been expected, the revertant protein lacked the same tryptic peptide as *E26*. But the position of another peptide on the fingerprint was shifted simultaneously, presumably due to the reverse mutation. Analysis of the amino acid composition of this peptide in both normal and revertant proteins revealed a single amino acid change from alanine to valine.

The fact that an alteration at one point in the polypeptide chain, which leads to loss of enzymatic function, can be corrected or suppressed by a second amino acid change at a distant point is indeed intriguing. At first glance, this finding would appear to contradict our assumptions regarding the active site of the enzyme, but such a conclusion is premature. The active site may contain functional groups which, although quite distant from each other along the polypeptide backbone, are brought together by the specific way the chain folds. Thus the substitution of an amino acid with a charged side group for one with an uncharged side group can be reflected in an altered tertiary structure, and the change may be enough to disrupt the sensitive balance of forces that creates the active site. A second alteration that restores the

16. Mutation and the Genetic Code

original tertiary structure may then occur at a completely different site in the polypeptide. The ability to repeatedly alter the enzyme through mutation without inactivating it provides new opportunities for the study of enzymatic mechanisms and their dependence on the configuration of the protein macromolecule. Alkaline phosphatase appears to be particularly flexible in this respect, since many of the functionally active revertant proteins are clearly different in primary structure.

The correspondence between genetic map and protein structure was also investigated by Yanofsky and his co-workers, using the enzyme tryptophan synthetase from *E. coli* (169, 170). Tryptophan synthetase is actually a complex of two proteins, denoted A and B, which can be separated chromatographically. Only the complex possesses the enzymatic activity necessary to synthesize tryptophan in the *E. coli* chromosome, although, judging from genetic evidence, the two cistrons are probably adjacent to one another.

Tryptophan synthetase catalyzes the final step in a series of reactions leading to the biosynthesis of tryptophan. In microorganisms, the last stages of this pathway are the following:

[Anthranilic acid] → [Amadori compound]

[Tryptophan] ← [Indolyl-3-glycerol phosphate]

The main role of tryptophan synthetase is to join indolylglycerol phosphate with L-serine, but this enzyme can also catalyze two other related reactions:

Indolyl-3-glycerol phosphate $\rightleftharpoons$ (± glycerol phosphate) Indole; Indolyl-3-glycerol phosphate + serine → Tryptophan + glycerol phosphate; Indole + serine → Tryptophan

The reaction between indole and serine occurs by means of a very complicated mechanism known to require the presence of the coenzyme pyridoxal phosphate.

Yanofsky and co-workers obtained many *E. coli* strains with mutations in the tryptophan synthetase cistrons, several of which completely lacked either the A or the B protein. Mutants were also found that synthesized enzymatically inactive protein of the same immunological specificity as native tryptophan synthetase. Finally, mutants were isolated in which only one of the three catalytic activities normally associated with the wild-type protein remained: the ability to convert indole and serine to tryptophan. The active site of the enzyme was altered in such a way that its functions were limited, but not eliminated. These mutants are easy to distinguish from the wild type because they accumulate the precursor, indolylglycerol phosphate, and require free indole for growth. For this reason, they were chosen for further study.

After several such mutants were selected and mapped by transduction with phage P1, they all turned out to belong to the *A* cistron. Tryptophan synthetase was next extracted from wild-type and mutant cells, and further separated into A and B subunits. Each mutant A protein was fingerprinted in order to identify the altered tryptic peptide and the amino acid composition and primary structure of that peptide was determined by chemical techniques.

In most revertable point mutants, it was found that a single amino acid had been replaced by another at a unique site in the polypeptide chain of the A protein. In addition, these studies demonstrated that there is a very good correlation between the genetic map positions of tryptophan *A* mutations and the corresponding positions in the A protein at which amino acid replacements occur. The complete sequence of roughly one third of the 280 amino acids comprising the A protein was worked out and within this segment, the order of several amino acid replacements is the same as the order of mutational sites in the *A* gene that caused them. Furthermore, this system permitted Yanofsky and his associates to verify that the distance between two mutations on the genetic map is proportional to the distance between altered amino acid residues in the polypeptide (Fig. 4-54). The fundamental postulate of colinearity of the gene and the protein for which it codes was thus proven.

Of particular interest was the discovery that a specific residue in the polypeptide chain can be replaced by at least two different amino acids as the result of different mutational events. For example, the glycine residue at position 47 of the wild-type tryptophan synthetase A protein is replaced in one mutant (*A23*) by arginine, and in a second (*A46*) by glutamic acid. By recombination between these two mutants, it was possible to recover bacteria with wild-type tryptophan synthetase containing glycine at position 47. This provides evidence that mutants *A23* and *A46* did not arise by alteration of the same mutable site, since no wild-type recombinants could have been formed under such circumstances. Moreover, the frequency of recombination between *A23* and *A46* indicates that the distance separating the two mutations is equal to 1/1000 of the length of the *A* gene. Because this gene codes for 280 amino acids, the minimal recombination distance must correspond to the dimensions of a single nucleotide, assuming a triplet code. From *in vitro* studies, the most likely codons for arginine and glutamic acid in the present case are GAA and AGA, respectively, both derivable by a single nucleotide replacement from the glycine codon, GGA (see p. 436). Not only are these assignments in accord with the possibility of *intracodon*

16. Mutation and the Genetic Code

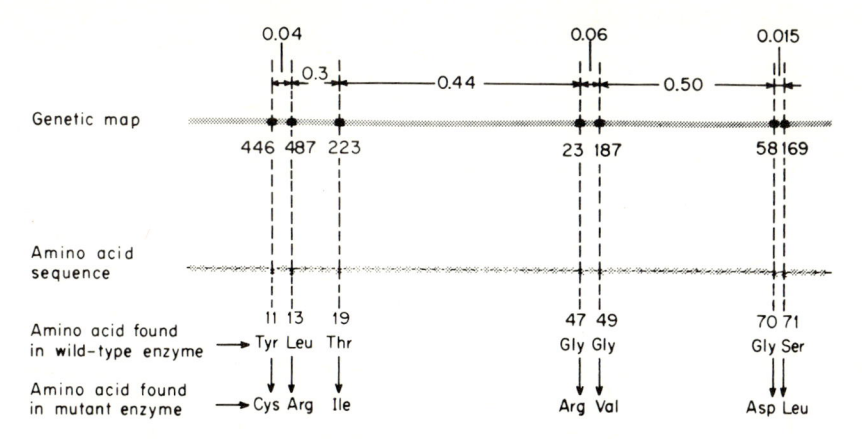

Fig. 4-54. Colinearity of the gene and its protein product. The genetic map for one fourth of the tryptophan synthetase *A* gene is presented on the top line with map distances between various tryptophan synthetase mutations. The corresponding segment of the tryptophan synthetase A protein is depicted on the lower line showing the amino acids normally present in the protein as well as the amino acid replacements resulting from each mutation. The numbers in the amino acid sequence refer to their position in the 75-residue segment studied. Adapted from Yanofsky *et al.* (170).

recombination, but they also indicate that the smallest mutable site, like the elementary recombination unit, consists of just one base pair in the structural gene.

Spontaneous or induced reversions of mutants *A23* and *A46* can result in further alterations at position 47 of the tryptophan synthetase A protein, providing a means for investigating codon assignments *in vivo*. The A protein of some enzymatically active revertants of *A23* contain glycine at the altered site, suggesting a return to the original wild-type coding unit, while others contain serine, threonine, or isoleucine. Protein function was restored in *A46* revertants by replacement of glutamic acid with glycine, alanine, or valine. Furthermore, alanine and valine are never found at position 47 among the A proteins isolated from *A23* revertants, nor is serine, threonine, or isoleucine found at that position in the A proteins of *A46* revertants. The simplest explanation of these findings is that each amino acid alteration results from a single nucleotide change. Assuming this, one can deduce that (1) no serine, threonine, or isoleucine codon can be derived from the *A46* glutamic acid codon by a single nucleotide substitution and (2) no valine or alanine codon can be derived from the *A23* arginine codon by a single nucleotide substitution. Thus the series of amino acid changes imposes specific restrictions on the nucleotide composition and sequence of the coding units involved. Figure 4-55 shows how the amino acid replacement data can be related to codons proposed on the basis of *in vitro* studies (p. 436). An arrow designates a change that can be caused by a single mutational event, and the codons assigned in each case demonstrate that all interconversions can be accounted for by single nucleotide replacements. The codon assignments are also consistent with the specificity of the mutagens employed to bring about the observed alterations. This work provides strong *in vivo* confirmation of the genetic code determined *in vitro*. Unfortunately, our knowledge of the chemical changes in DNA resulting from mutation are still insufficient to permit complete elucidation of the code solely from experiments of this sort.

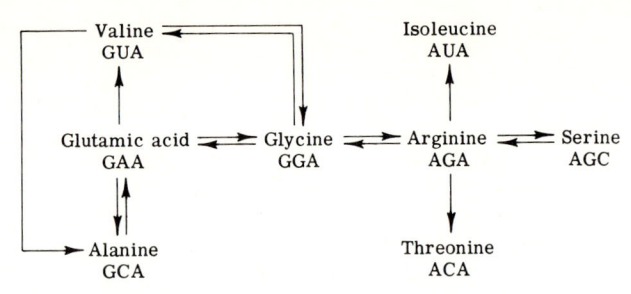

```
        Valine  ⇌         Isoleucine
         GUA              AUA
          ↑                ↑
          │                │
Glutamic acid ⇌ Glycine ⇌ Arginine ⇌ Serine
    GAA         GGA         AGA       AGC
     ↕                       ↓
     │                       │
    Alanine              Threonine
     GCA                   ACA
```

Fig. 4-55. Amino acid replacements at position 47 in the tryptophan synthetase A protein with likely codon assignments. The wild-type protein has glycine at position 47. The A proteins with valine, alanine, serine, threonine, or isoleucine are functional, while A proteins containing either arginine or glutamic acid are inactive. Each arrow indicates that the amino acid substitution can occur as the result of a single mutational event or, in other words, a single nucleotide base change. Adapted from Yanofsky *et al.* (171).

Somewhat more progress on the chemistry of nucleotide base changes has been made with the RNA of tobacco mosaic virus. The RNA of TMV carries the genetic information of the virus and is exclusively responsible for its infectivity. A large number of TMV mutants have been produced by reacting purified viral RNA with nitrous acid and the resulting amino acid alterations in the TMV coat protein have been analyzed (Wittmann and Wittmann-Liebold, 172). Coat protein mutants usually contain one, sometimes two, and, rarely, three amino acid replacements after mutagenesis. Since multiple alterations recovered at the same time are almost always located at some distance from each other, they most likely result from separate mutational events and therefore provide no evidence for an overlapping code. From the known action of nitrous acid on nucleotide bases, a set of logical conclusions can be drawn regarding the nature of certain codons by analyzing allowable amino acid replacements in TMV coat protein.

To characterize the ways in which codons can be changed by nitrous acid treatment, Wittmann constructed special "octet" diagrams of the following kind:

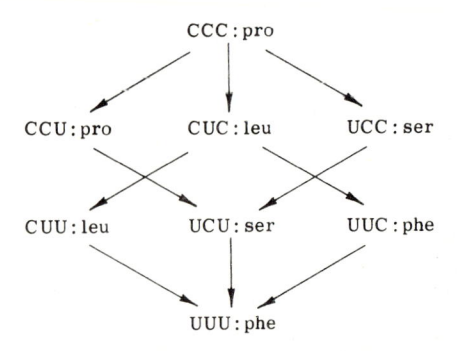

```
                    CCC : pro
                   ╱    ↓    ╲
            CCU : pro  CUC : leu  UCC : ser
                  ╲  ╳    ╳  ╱
            CUU : leu  UCU : ser  UUC : phe
                   ╲    ↓    ╱
                    UUU : phe
```

This scheme portrays all of the codon interconversions which could occur by deamination of cytosine with nitrous acid, starting with codons containing cytosine, or combinations of cytosine and uracil. It also includes the currently accepted codon assignments. An important implication of this diagram is that the four amino acids shown are not freely exchangeable with one another. In fact, the only allowed amino acid replacements should be Pro → Ser, Pro → Leu, Ser → Phe and Leu → Phe if a single cytosine residue in one of the above trinucleotides is deaminated to uracil by nitrous acid. The amino acid replacements that are actually induced in TMV coat protein by nitrous acid adhere to this scheme quite closely. When Wittmann started his analysis of TMV mutants, the composition of many codons was known, but the sequence of nucleotide bases within them remained to be determined. From the known compositions and the specificity of nitrous acid, he was able to predict the sequence of a number of codons from amino acid replacement data.

17. Direct Methods for Deciphering the Genetic Code; General Properties of the Code

Although study of mutants has made an important contribution to deciphering the genetic code, the fundamental achievements in this endeavor have resulted from a more direct biochemical approach. The basic idea of the biochemical method has been to use synthetic polyribonucleotides of a defined composition and structure as templates for the production of simple polypeptides. Such model systems have provided a Rosetta Stone for unraveling the genetic code.

The essential tool for biochemical investigations of the genetic code is the cell-free system, a preparation of enzymes and ribosomes extracted from bacterial or mammalian cells, in which the synthesis of protein can be followed by measuring the incorporation of ^{14}C-labeled amino acids. Cell lysis, whether brought about mechanically, enzymatically, or by the use of detergents, leads to a crude extract, containing ribosomes, various kinds of RNA, soluble proteins, and low-molecular-weight substances of the cell cytoplasm. Protein synthesis *in vitro* can be studied directly in the crude extract by introducing radioactively labeled amino acids and studying their incorporation into the protein fraction which is precipitated with hot trichloroacetic acid (TCA). The cell-free system permits the investigator to intervene actively in the process and, in particular, to separate all of the elements participating in protein synthesis for individual study. The necessary and sufficient components for protein synthesis in a cell-free system are the following:

(1) Ribosomes, small cytoplasmic organelles, are sedimented from the crude extract by centrifugation at 100,000 g, washed free of contaminants, and stored in the frozen state.

(2) Soluble enzymes from the 100,000 g supernatant fluid, which, for the most part, are neither purified further nor identified. These are generally kept in the presence of a reducing agent such as β-mercaptoethanol.

(3) A set of all twenty common amino acids, any or all of which may be labeled radioactively, depending on the aims of the experiment.

(4) A set of all transfer RNA's normally present in the cell; these are the adaptors we spoke of on page 421.

(5) A source of chemical energy, usually provided by ATP and GTP (adenosine and guanosine triphosphate). A system to regenerate ADP and GDP is also essential in order to sustain the required level of energy donors. The transfer of high energy phosphate groups from phosphoenolpyruvate by the enzyme pyruvate kinase is

frequently selected for phosphorylation of the nucleotides, although creatine phosphate and creatine phosphokinase can also serve in this capacity.

(6) Certain ions, among which are Mg^{2+}, used at a concentration of about 10^{-2} M, and K^+ or $(NH_4)^+$ at about 10^{-1} M.

(7) Template or messenger RNA (mRNA), carrying information necessary to specify the structure of the polypeptides, must also be added.

Cell-free systems have been successfully prepared from bacteria, the simplest of all organisms, as well as from the cells of higher organisms. Among the latter, the cell-free system from rabbit reticulocytes deserves special note. Reticulocytes are very unusual cells, for 80% or more of the protein they synthesize is of just one kind: hemoglobin. Reticulocytes are the precursors of red blood cells (erythrocytes) and their concentration in the blood fluid can be greatly increased by injecting the animal with phenylhydrazine. This compound produces anemia and the compensatory manufacture of reticulocytes in the bone marrow is accelerated. If these cells are extracted from the blood of the treated animal and placed in a medium containing a complete set of amino acids and a certain amount of blood serum, the reticulocytes synthesize hemoglobin even outside the body. Furthermore, by studying cell-free systems from reticulocytes, we can very easily test the results of environmental influences on the protein product owing to the great specialization of the protein-synthesizing machinery. The hemoglobin produced can be extracted, purified chromatographically, and investigated as desired.

Ribosomes prepared from bacterial cells always contain a certain quantity of messenger RNA which is gradually used up and degraded. When this happens, usually after 10 to 15 minutes, protein synthesis rapidly comes to a halt. Nirenberg and Matthaei were the first to discover that the introduction of synthetic polynucleotides of a known structure into such an exhausted system results in the synthesis of simple polypeptides (173). The triplet codon, UUU, was deciphered at once in this way. Polyuridylic acid (Poly U) stimulates the synthesis of polyphenylalanine in the cell-free system and it was concluded that UUU is the code word for the amino acid phenylalanine. Polyphenylalanine is characterized by extremely low solubility, but it can nonetheless be isolated in quantities sufficient for chemical investigation. This method provided a direct approach to deciphering the genetic code.

In subsequent investigations (Nirenberg *et al.*; Ochoa and co-workers, 174), polynucleotides of different compositions, containing mainly uracil with varying amounts of adenine, cytosine, and guanine, were also tested. These copolymers were synthesized with the enzyme polynucleotide phosphorylase (see p. 237) and the sequence of residues along the chain was random. The idea behind this work was that such polyribonucleotides should contain sequence information necessary for the synthesis of at least short stretches of polyphenylalanine. At the same time, other amino acids will also be incorporated if the triplets, containing adenine, cytosine, and guanine in different combinations with uracil, constitute "meaningful" code words. For instance, we can estimate the probability that given nucleotide triplet combinations, such as CUU will occur relative to the expected number of UUU triplets in poly UC, and compare them with the actual frequency with which the corresponding amino acids, here leucine and phenylalanine, are incorporated. If the two frequencies are consistent with one another over many polynucleotides with

17. Direct Methods for Deciphering the Genetic Code; General Properties of the Code

varying composition, the code word assignment is verified. By this procedure, codons for several amino acids were deduced.

The next step in this direct analysis of the code was to introduce polyadenylic acid (poly A) and copolymers of adenine with other nucleotides into the cell-free system. Poly A itself codes for polylysine, and through studies of adenine-rich copolymers, the number of triplets which could be identified was extended to 47 out of the 64 possible combinations.

A number of conclusions were drawn from these data: (1) The code is extremely degenerate; for certain amino acids, six alternative codons were found. (2) The experimental data did not contradict the postulate of a triplet code in any way. (3) The question of whether the same code applies universally to all living organisms was given a limited test by adding synthetic template polynucleotides to cell-free systems prepared from material other than *E. coli*. It was shown that when supernatant enzymes, tRNA's, and ribosomes were extracted from rat liver, rabbit reticulocytes, and human placenta, all incorporated the same amino acids under the direction of a given template as did the bacterial system.

In the experiments described, it was possible to determine the *composition* of the codons, but the *sequence* of nucleotides within them remained uncertain, insofar as the templates were random synthetic copolymers. At first appearances, it seemed that it would be a complicated task to discover the precise chemical structure of the triplet codons, but the problem was eventually solved.

As mentioned previously, Crick had predicted that amino acids must be aligned on the polynucleotide template by means of special adaptor molecules. It has been found that transfer RNA fulfils this function. Although the nature of tRNA will be discussed in greater detail in Chapter V, it will suffice to note here that each amino acid has at least one, and often more, chemically distinct tRNA adaptors. However, all tRNA molecules have the common property of being able to bind their specific amino acids through an aminoacyl bond to the 3'-ribose carbon atom of the terminal nucleotide. Somewhere within each tRNA chain there is a specific sequence of three nucleotides called the *anticodon* which pairs specifically with a corresponding *codon* in the template. It is quite natural to suppose that the codon and anticodon are complementary and that they bind to each other through hydrogen bonds, observing the same rules as apply to base pairing in the Watson-Crick model of DNA.

Although skeptical views were expressed concerning the stability of complexes held together by only three complementary base pairs, the weight of evidence demonstrates that this is quite possible. In one confirmatory experiment, the synthetic polynucleotide poly U was chemically attached to DEAE-cellulose and the resulting sorbent was used to chromatograph a mixture of unfractionated tRNA's. As it turned out, tRNA specific for phenylalanine alone was bound to the column. Since this system is quite simple and contains no ribosomes, enzymes, or energy source, there is no doubt that the principle of base complementarity is observed and that three hydrogen-bonded base pairs are anough to form a stable complex (175). Even the mononucleotide AMP is specifically adsorbed to poly U, demonstrating that hydrogen bonds formed by one complementary base pair are sufficient for limited stability. Experiments were also undertaken to chemically modify the tRNA anticodon with a variety of different agents. Despite the imperfections of these efforts, they all indicated that the anticodon is a triplet of nucleotides complementary to the codon, and that no additional apparatus is necessary to bind the two together. Once this had been recognized, a new approach to the coding problem was developed.

Nirenberg and Leder (176) showed that the code could be studied by synthesizing all 64 possible trinucleotide code words individually. These synthetic triplets could be bound to ribosomes and the capacity of the complex to bind aminoacyl-tRNA's was tested in turn. In these experiments, the triplets usually bind one specific type of tRNA, charged with its corresponding amino acid. No polypeptide synthesis occurs under these circumstances, of course, since the "template" codes only a single amino acid. This is a very sensitive test for recognizing complementarity between codon and tRNA, however, since radioactive amino acids can be used. This method permitted the determination of practically all codons in a very short time (177). In fact, the proper designation of a certain number of codons can be used to completely decipher all others in conjunction with the experimental data on amino acid replacements in the mutant proteins of TMV, tryptophan synthetase, and a number of others.

A complete set of codon assignments is presented in Table 4-6. We see that in

TABLE 4-6

The Genetic Code[a,b]

First nucleotide	Second nucleotide				Third nucleotide
	U	C	A	G	
U	Phe	Ser	Tyr	Cys	U
	Phe	Ser	Tyr	Cys	C
	Leu	Ser	Ochre	Nonsense	A
	Leu	Ser	Amber	Trp	G
C	Leu	Pro	His	Arg	U
	Leu	Pro	His	Arg	C
	Leu	Pro	Gln	Arg	A
	Leu	Pro	Gln	Arg	G
A	Ile	Thr	Asn	Ser	U
	Ile	Thr	Asn	Ser	C
	Ile	Thr	Lys	Arg	A
	Met[c]	Thr	Lys	Arg	G
G	Val	Ala	Asp	Gly	U
	Val	Ala	Asp	Gly	C
	Val	Ala	Glu	Gly	A
	Val[c]	Ala	Glu	Gly	G

[a] Best allocation of the 64 codons. The evidence used to produce this table comes mainly from *E. coli*, although it is likely that the genetic code is very similar or identical in other organisms.

[b] Adapted from Crick (178).

[c] At the 5'-end of the template RNA, this codon is believed to serve as an initiator of polypeptide synthesis; in that case it codes for N-formylmethionine.

eight cases, where entire boxes are filled by one amino acid, the nature of the third nucleotide is indifferent. Other boxes attest to differences in codon expression depending on whether the third nucleotide is a purine or a pyrimidine. The data presented in Table 4-6 are well established, since they have undergone very rigid verification in recent years. As mentioned in an earlier section, Khorana and his co-workers synthesized a series of regular ribopolynucleotides of high molecular weight in which two, three, or four nucleotides are periodically repeated along the polymer chain. The polypeptides synthesized on such templates are also found to have periodically repeating amino acid sequences and the amino acids themselves correspond to the trinucleotide designations given in Table 4-6 (179). This experiment represents the fullest and most direct confirmation of the triplet code and the code word assignments.

Although in Chapter I glutamine and asparagine were considered as derivatives of glutamic acid and aspartic acid, respectively, we see that they have their own codons and, accordingly, that they are inserted into the polypeptide chain as distinct residues. Two other important amino acids, cystine and hydroxyproline, have no codon assignments, however. These substances are formed only secondarily after the protein has been synthesized: cystine by spontaneous oxidation of two cysteine residues and hydroxyproline by enzymatic oxidation of proline.

Among the many interesting consequences of the amino acid code, let us touch for a moment on degeneracy. The occurrence of several triplets that code for one and the same amino acid must be the result of an evolutionary process. It is only natural that, during the course of evolution, mutations have affected the structure of tRNA and in particular, of its anticodon. If each anticodon in the tRNA corresponds to a complementary codon in the mRNA template, then mutations in the anticodon might be stabilized by the appearance of a reciprocal change in codon structure. We might suppose that in the majority of cases, the codons for a particular amino acid will change in such a way that two nucleotides in the triplet will remain unaltered while the third varies. This rule finds support in Table 4-6; moreover, it is evident that the nucleotide in the third position varies most frequently.

Indeed, the existence of multiple tRNA's for several amino acids has been verified. Purification of *E. coli* tRNA's by countercurrent distribution, for instance, has revealed the presence of four distinct leucine-specific fractions which manifest the anticipated differences in ability to recognize the numerous leucine codons during polypeptide synthesis directed by mRNA's containing varying proportions of C, U, and G (180). One might be tempted at this point to assume that the anticodon of each tRNA species can recognize only a single codon according to the strict standards of complementarity specified by the Watson-Crick model. There is evidence to the contrary, however, derived from experiments on the binding to ribosomes of purified yeast and *E. coli* tRNA stimulated by trinucleotides of defined sequence (181). It was observed that certain aminoacyl-tRNA's behave ambiguously, recognizing several different codons for the amino acid to which they correspond, while others are much more selective in the codons with which they can pair.

The pattern of multiple codon recognition by a single tRNA can be explained on the basis of the *wobble hypothesis* advanced by Crick (182). From the nature of the codon assignments, Crick deduced that pairing of the first two bases in the codon with corresponding bases in the anticodon obeyed standard rules of complementarity, that is, A pairs with U, G pairs with C. While standard pairing can also occur in the third position, a certain steric flexibility or wobble would permit the formation of alternative base pairs provided the alignment of juxtaposed polynucleotide chains is not too different from the standard one. The most likely possibilites for pairing at the third position are presented in Table 4-7. The table includes inosine because it is a

TABLE 4-7

Pairing at the Third Position in the Codon[a]

Base in the anticodon	Bases recognized in the codon
U	A or G
C	G
A	U
G	U or C
I	U or C or A

[a] From Crick (182).

relatively common constituent of tRNA and has already been tentatively identified in the anticodon of several tRNA's whose structures are known. One example will suffice to demonstrate the implications of this scheme. While the wobble hypothesis does not predict exactly how many different types of tRNA will be found for any amino acid, it does set a lower limit. If an amino acid is coded for by all four bases in the third position (Pro, Thr, Val, Ala, . . . , see Table 4-6), there must be at least two corresponding tRNA's. Furthermore, the nucleotides in the two anticodons that pair with the third position in the codon must be either U and G, or C and I, in order to cover all four possible code words for the amino acid in question. Experiments on the binding of purified aminoacyl-tRNA's to specific trinucleotides, as described above, lend considerable support to this theory.

A second interesting aspect of the code is the occurrence of *nonsense codons*, which do not correspond to any amino acid. Three such codons have been found: UAG (amber), UAA (ochre), and UGA. Synthesis of the polypeptide chain ceases when these triplets are encountered in a synthetic template *in vitro*. There is good evidence that the same thing occurs with natural messenger RNA's *in vivo*, and it may well be that nonsense codons are a normal form of punctuation that give a signal for peptide chain termination. We shall return to the expression of amber and ochre triplets in special suppressor strains later in this section.

Besides three nonsense codons, it is possible that, because of code degeneracy, not all codons are used by every living organism. In other words, a given organism may have a set of tRNA's which correspond not to all the possible parallel codons, but to only a portion of them. If this were true, some of the triplets might be read as nonsense in one species, but as sense in another if the appropriate tRNA is present. All of the evidence available at present argues against this supposition, however, and there is every reason to believe that the genetic code is universal. This explains why polypeptides are synthesized *in vitro* when the tRNA's and supernatant enzymes are derived from one organism and the information, that is, the mRNA, from another.

17. Direct Methods for Deciphering the Genetic Code; General Properties of the Code

One very interesting result of nonsense mutations is the incidence of special mutations that suppress them. In bacteria, for instance, there are reversions which do not map at the site of the original mutation. If the primary lesion occurs at one specific locus, inactivating a vitally important enzyme, a second mutation at a completely different and even distant region of the chromosome can correct the first error and synthesis of active enzyme recommences. What is most intriguing in this phenomenon is that the two mutations affect distinct genetic loci. We may assume that the first alters a structural gene coding the amino acid sequence of an enzyme, but the function of the second, suppressor mutation requires further explanation.

A mechanism for the behavior of suppressor genes was proposed by Brody and Yanofsky (183). Let us assume that a mutation has caused the replacement of a single nucleotide residue in a particular triplet of a cistron, changing it from sense to nonsense. The cell now cannot synthesize the protein governed by that cistron at all. A suppressor mutation must then represent a change in the way the cell reads the code, probably affecting one of the multiple tRNA's for one amino acid. We further suppose that the altered tRNA has an anticodon complementary to one of the nonsense triplets not normally read as sense by the given organism. As a result, the mutant or suppressor tRNA reads the nonsense mutation meaningfully and even though the genetic information is defective, we obtain an active protein molecule.

Let us illustrate the notion of suppression on a concrete example. Garen and his co-workers discovered a whole series of *E. coli* mutants which completely lacked the capacity to synthesize alkaline phosphatase (184). In fact, the cell extracts of these mutants did not even contain any inactive protein with the immunological specificity of phosphatase. In most *pho⁻* mutants, however, proteins are formed which cross-react with antibodies to phosphatase; for this reason, they are called CRM-formers where CRM is an abbreviation for cross-reacting material. Garen showed that mutants which do not synthesize CRM form a group that can be corrected by a single suppressor mutation called su_I. This can be explained if it is assumed that the suppressor mutation causes the cell to synthesize a new kind of tRNA whose anticodon is complementary to a specific, and normally unsuitable, triplet in the structural gene. By analyzing the primary structure of phosphatases produced in bacteria carrying su_I, Garen and Weigert demonstrated that the suppressed proteins differed by a single amino acid from wild-type enzyme (185). In two of the mutants studied, glutamine was replaced by serine and in a third, tryptophan was replaced by serine.

Bacterial suppressors were also found to affect the expression of certain mutations in the bacteriophages T4 (Stretton and Brenner, 186) and f2 (Zinder and co-workers, 187). As already noted on page 307f, there is an extensive group of mutations in phage T4 known as amber mutations. Such mutants are unable to mature in normal *E. coli* B hosts, but multiply very well in special strains of *E. coli* K12. It turned out that a number of mutants of this strain, such as *E. coli* CR63, harbor suppressor genes in their own chromosomes which correct amber mutations in the viral genome. Even more remarkable, the very same bacterial suppressor strain, su_I, which was used by Garen to restore synthesis of alkaline phosphatase, is also suitable for the correction of amber mutations in bacteriophage T4.

Certain T4 amber mutations map in the cistron specifying head structural protein. When *E. coli* B is infected with such phages, no functional head protein is formed because the genetic information in this cistron is defective. However, short polypeptide fragments with some resemblance to phage head protein are produced and their size depends on the particular phage mutant employed. Hence, every amber head protein mutant manufactures a characteristic protein fragment. These polypeptides are apparently the products of protein synthesis that is initiated normally, but that is terminated prematurely when a nonsense amber triplet is encountered in the template.

Phage head protein fragments were extracted from cells infected with a number of these mutants, and their structure was compared with that of normal phage head protein. Although the structural formula of this polypeptide is still not completely worked out, it was possible to draw a number of important conclusions. First, by comparing many mutants, it was determined with very great accuracy that the length of the polypeptide fragment, that is, the distance from the N-terminus where synthesis starts to the point where synthesis breaks off, is proportional to the genetic map distance from the beginning of the cistron to the site of the amber mutation determined, as usual, by recombination analysis (188). These experiments therefore provide excellent quantitative proof of the complete correspondence between two linear sequences, in other words, of the colinearity of the polynucleotide cistron and the polypeptide chain which it specifies. Second, it was possible to establish which amino acids are substituted in the phage head protein when the bacterial suppressor gene is present in the host. Synthesis of the mutant polypeptide fragments always terminates immediately before a glutamine or tryptophan residue. Under the influence of su_1, serine is inserted at this point, and chain synthesis continues. The agreement between these observations and those of Garen on the action of su_1 in the synthesis of phosphatase is indeed remarkable.

The influence of bacterial suppressor genes on the development of the small RNA phage f2 has been studied in less detail. A series of f2 mutants was found which were unable to synthesize phage coat protein, but which could be corrected by the suppressors of *E. coli*. It was shown that synthesis of phage protein in the presence of su_1 resulted in the substitution of serine for the glutamine residue which occurs at the same position in wild-type coat protein.

These brilliant experiments can be explained on the basis of codon designations presented in Table 4-6. The amber triplet has been assigned the sequence UAG (189). When the trinucleotide UpApG was tested for its ability to bind tRNA to *E. coli* ribosomes, no interactions were detected. According to many considerations, however, it is likely that this triplet is read as serine in the suppressor mutant, *E. coli su*$_1$ (185). The amber sequence UAG is very similar to those which normally code for glutamine, CAG, and tryptophan, UGG, and it can be obtained from them by a single nucleotide replacement. In accord with several lines of evidence, it may be imagined that UAG is a nonsense triplet in wild-type *E. coli* but that it codes for serine in su_1 owing to a mutation in the anticodon of a serine-specific tRNA. This would occur if the normal serine anticodon CGA is mutated to CUA, the amber anticodon, with a replacement of G by U in the second position. Therefore, it is quite natural that mutations of the type:

$$\begin{matrix} \text{CAG (glutamine)} \\ \text{UGG (tryptophan)} \end{matrix} \longrightarrow \text{UAG (amber)}$$

in the codons of both bacteria and phage cistrons are corrected by one and the same suppressor tRNA, which reads the nonsense triplet as a meaningful code word. In addition, it has been found that in the *E. coli* suppressor strains su_{II} and su_{III}, glutamine and tyrosine, respectively, are inserted at the place of the amber mutation (190).

A second class of suppressible mutations has been found both in bacteria and in phage T4. These are the ochre mutations which are corrected by another set of suppressors, although at much lower efficiency than the amber suppressors with amber mutations. The ochre suppressors are also less specific, since they suppress amber mutations in addition to ochre mutations. By all criteria, ochre mutations also cause premature chain termination and give rise to incomplete polypeptide fragments rather than complete proteins. Judging from amino acid replacements in revertants from this kind of mutation and from the fact that it is connected to the amber codon by a single base change, ochre has been assigned the triplet UAA (189). Finally, a third group of suppressible nonsense mutations with the codon UGA has been identified (191). Suppressors of UGA act only on this nonsense codon and do not correct either UAG or UAA.

The nature of genetic suppression has been confirmed by *in vitro* synthesis of bacteriophage proteins under the direction of natural mRNA's. In one case, the extract of a bacterial strain carrying the amber suppressor su_I was shown to contain an altered tRNA capable of inserting serine into phage coat protein at the site of an amber mutation, thus proving that the amber codon UAG could be read by a tRNA molecule (192). This result was extended to su_{II} and su_{III} in the following experiments. When RNA is isolated from *E. coli* after infection with normal bacteriophage T4 and then added to a complete cell-free system containing ribosomes and supernatant from uninfected bacteria, a phage-specific enzyme called lysozyme is produced in measurable quantities (193). This enzyme is a particularly suitable object for such studies because it is not coded for or made by the uninfected bacterium. The appearance of lysozyme in this system has all the attributes of *de novo* synthesis and it must therefore be specified by the added phage mRNA. This is the first instance in which an enzymatically active protein has been successfully produced *in vitro*. If the cell-free system is now programed with RNA from cells infected with T4 lysozyme amber mutants, *in vitro* enzyme synthesis does not occur when the supernatant fraction is derived from su^- bacteria. But when tRNA's isolated from bacterial strains carrying any of the amber suppressors su_I, su_{II}, or su_{III} are added to the system, active lysozyme is synthesized from the defective mRNA templates (194). This result demonstrates that suppression of amber mutations is mediated by tRNA in all three amber suppressor strains.

Conclusive evidence for the participation of mutationally altered tRNA's in genetic suppression has come from structural studies on allelic tRNA's from normal and suppressor strains of bacteria (195). While little doubt remained that the suppressor product was a tRNA molecule, the precise function of the suppressor gene was still unknown. The experiment to be described below demonstrated that the su_{III} gene in *E. coli* is the structural gene for one of the normal tyrosine tRNA's and that the mutation of wild-type su_{III}^- to amber suppressor su_{III}^+ results from a single base change in the anticodon of this tRNA permitting recognition of the UAG amber codon. This was deduced from a determination of the complete nucleotide sequences of the two tRNA's.

The tRNA product of the su_{III} gene comprises only about 15% of the total tyrosine tRNA in the cell and it was therefore necessary to enrich for this component so as to facilitate its extraction and study. The solution decided on was an ingenious one. Since su_{III} maps very near the attachment site for phage Φ80, it was possible to obtain defective transducing phages carrying either the wild-type su_{III}^- allele (Φ80dsu_{III}^-) or the su_{III}^+ suppressor allele (Φ80dsu_{III}^+). Infection of each bacterium with several such transducing phages introduced multiple copies of the su_{III} gene into the cell and, accordingly, the amount of su_{III} tRNA was greatly increased since many DNA templates were available from which it could be transcribed.

Soon after su_{III}^- E. coli were infected with phage Φ80dsu_{III}^+ suppressor activity appeared in the cells and the overall amount of tyrosine tRNA present rose sharply. Furthermore, this tyrosine tRNA was able to recognize the amber codon in ribosomal binding experiments. Finally, specific hybridization of tyrosine tRNA's with DNA from Φ80dsu_{III}^+ showed that this phage does contain a structural gene for tyrosine tRNA, pointing to the conclusion that the suppressor gene itself specifies a tyrosine tRNA.

To determine whether there were structural differences in the products of the two su_{III} alleles, tyrosine tRNA was extracted and purified from cells multiply infected with either of the defective phages Φ80dsu_{III}^- and Φ80dsu_{III}^+. After complete nucleotide sequences had been worked out in the two cases, it was found that su_{III}^- and su_{III}^+ tyrosine tRNA's differed only in the nucleotide occupying position 36 of the chain within the anticodon. The sequence of anticodon (underlined) and surrounding regions are:

$$\ldots \text{AGAC}\underline{\text{UGU}}\text{AAA } \Psi \text{CU} \ldots \quad \text{in } su_{III}^-$$
$$\ldots \text{AGAC}\underline{\text{UCU}}\text{AAA } \Psi \text{CU} \ldots \quad \text{in } su_{III}^+$$

where Ψ stands for pseudouracil. Wild-type su_{III}^- tRNA contains the anticodon GUA which, according to the wobble hypothesis, can recognize either tyrosine codon, UAU or UAC. The suppressor mutation changes the anticodon to CUA which uniquely recognizes UAG, the amber codon.

Taken together, the studies recounted above yield a highly consistent explanation of suppression and of its relationship to the genetic code.

The current table of codons has now undergone verification by many methods, both biochemical and genetic. In particular, we have seen in the last two sections how the pattern of amino acid replacements in missense, nonsense, and frame-shift mutations, and in intracodon recombination has been used to substantiate at least certain code word assignments. A complete correlation of data obtained *in vitro* and *in vivo* lends considerable reliability and persuasiveness to the entire conception of the code. The reasons why this specific code developed in the course of evolution is not known, but its general applicability to organisms of all degrees of complexity is truly remarkable.

In conclusion, let us consider the special role of the two codons AUG and GUG. When these codons are present at the 5'-end of the RNA message, they serve to initiate synthesis of the protein chain. Under these conditions they code a very specific amino acid derivative, *N*-formylmethionine:

$$\begin{array}{c} \text{CH}_3 \\ | \\ \text{S} \\ | \\ \text{CH}_2 \\ | \\ \text{CH}_2 \\ | \\ \text{O}=\text{C}-\text{N}-\text{C}-\text{COOH} \\ |\ |\ | \\ \text{H}\text{H}\text{H} \end{array}$$

If these codons are encountered at an internal position in the template RNA strand, then AUG codes methionine and GUG, valine.

The initiating capacity of these codons was ascertained using synthetic RNA templates of defined sequence (196). When the codons are present in the polynucleotide chains, polypeptide synthesis begins with an *N*-formylmethionyl residue, but if they are absent, synthesis can begin at any point along the polynucleotide sequence. Moreover, presence of the initiating triplets increases the rate of polypeptide synthesis by a factor of ten. Polypeptides synthesized *in vitro* from natural template RNA's are also found to contain *N*-formylmethionine as their *N*-terminal residues (197). In the experiments of Zinder, for instance, RNA of bacteriophage f2 was used to stimulate the synthesis of phage coat protein in an *E. coli* cell-free system. *N*-Formylmethionine was determined to be the terminal residue of all protein chains synthesized. When phage protein is formed *in vivo*, however, the N-terminal position of the protein chain is occupied by alanine, and with a free amino group, at that. Since alanine is the second residue in phage coat protein, there must be an enzyme in the cell which specifically cleaves formylmethionine from the N-terminal position. In *E. coli*, a large proportion of cellular proteins have unformylated methionine as an N-terminal residue. This indicates that the *N*-formyl groups attached to the terminal methionine residues in the newly formed protein chains are enzymatically deformylated at some later time. Although initiation of protein chains with *N*-formylmethionine was established beyond doubt in bacteria, evidence of this mechanism in higher organisms is still lacking.

References

1. G. W. Beadle and E. L. Tatum, *Proc. Natl. Acad. Sci. U.S.* **27**, 499 (1941); see also R. P. Wagner and H. K. Mitchell, "Genetics and Metabolism," 2nd Ed., Wiley, New York, 1964.
2. F. Jacob and E. L. Wollman, "Sexuality and the Genetics of Bacteria," Academic Press, New York, 1961.
3. W. Hayes, "The Genetics of Bacteria and their Viruses," 2nd Ed., Blackwell, Oxford, 1968.
4. J. D. Gross, *in* "The Bacteria" (I. C. Gunsalus and E. Y. Stanier, eds.), Vol. 5, p. 1. Academic Press, New York, 1964.
5. "Genetika i Selektsiya Mikroorganizmov" (S. I. Alikhanyan, ed.). Nauka, Moscow, 1964.
6. A. L. Taylor and C. D. Trotter, *Bacteriol. Rev.* **31**, 332 (1967).
7. M. Demerec and N. Ohta, *Proc. Natl. Acad. Sci. U.S.* **52**, 317 (1964); K. E. Sanderson, *Bacteriol. Rev.* **31**, 354 (1967).
8. B. D. Maling and C. Yanofsky, *Proc. Natl. Acad. Sci. U.S.* **47**, 551 (1961).
9. A. Garen, C. Levinthal, and F. Rothman, *J. Chim. Phys.* **58**, 1068 (1961).

10. R. S. Edgar and R. H. Epstein, *Science* **134**, 327 (1961); R. S. Edgar, R. P. Feynman, S. Klein, I. Lielausis, and C. M. Steinberg, *Genetics* **47**, 179 (1962).
11. S. Benzer and S. P. Champe, *Proc. Natl. Acad. Sci. U.S.* **48**, 1114 (1962); A. Garen, S. Garen, and R. C. Wilhelm, *J. Mol. Biol.* **14**, 167 (1965); E. R. Signer, J. R. Beckwith, and S. Brenner, *J. Mol. Biol.* **14**, 153 (1965); E. Gallucci and A. Garen, *J. Mol. Biol.* **15**, 193 (1966); B. M. Ohlsson, P. F. Strigini, and J. R. Beckwith, *J. Mol. Biol.* **36**, 209 (1968).
12. J. Lederberg and E. M. Lederberg, *J. Bacteriol.* **63**, 399 (1952).
13. B. D. Davis, *J. Am. Chem. Soc.* **70**, 4267 (1948); B. D. Davis, *Proc. Natl. Acad. Sci. U.S.* **35**, 1 (1949); J. Lederberg, *in* " Methods in Medical Research " (J. H. Comrie, Jr., ed.), Vol. 3, p. 5. Year Book Publ., Chicago, 1950.
14. S. E. Luria and M. Delbrück, *Genetics* **28**, 491 (1943).
15. J. Lederberg and E. L. Tatum, *Nature* **158**, 558 (1946); E. L. Tatum and J.Lederberg, *J. Bacteriol.* **53**, 673 (1947).
16. T. F. Anderson, E. L. Wollman, and F. Jacob, *Ann. Inst. Pasteur* **93**, 450 (1957).
17. A. Garen and P. D. Skaar, *Biochim. Biophys. Acta* **27**, 457 (1958); S. D. Silver, *J. Mol. Biol.* **6**, 349 (1963).
18. W. Hayes, *Cold Spring Harbor Symp. Quant. Biol.* **18**, 75 (1953); L. L. Cavalli-Sforza, J. Lederberg, and E. M. Lederberg, *J. Gen. Microbiol.* **8**, 89 (1953); L. L. Cavalli-Sforza and J. L. Jinks, *Genetics* **54**, 87 (1956).
19. T. Loeb and N. D. Zinder, *Proc. Natl. Acad. Sci. U.S.* **47**, 282 (1961).
20. Y. Hirota, *Proc. Natl. Acad. Sci. U.S.* **46**, 57 (1960).
21. R. Curtiss, *Ann. Rev. Microbiol.* **23**, 69 (1969).
22. F. Jacob and E. A. Adelberg, *Compt. Rend.* **249**, 189 (1959); see also F. Jacob, P. Schaeffer, and E. L. Wollman, *Symp. Soc. Gen. Microbiol.* **10**, 67 (1960).
23. J. Marmur, R. Rownd, S. Falkow, L. S. Baron, C. Schildkraut, and P. Doty, *Proc. Natl. Acad. Sci. U.S.* **47**, 972 (1961).
24. S. Benzer, *in* " The Chemical Basis of Heredity " (W. D. McElroy and B. Glass, eds.), p. 70. Johns Hopkins Press, Baltimore, Maryland, 1957.
25. J. Cairns, *J. Mol. Biol.* **6**, 208 (1963).
26. J. Cairns, *Cold Spring Harbor Symp. Quant. Biol.* **28**, 43 (1963).
27. M. Abe and J. Tomizawa, *Proc. Natl. Acad. Sci. U.S.* **58**, 1911 (1967); B. Wolf, A. Newman, and D. A. Glaser, *J. Mol. Biol.* **32**, 611 (1968).
28. E. Cerdá-Olmedo, P. C. Hanawalt, and N. Guerola, *J. Mol. Biol.* **33**, 705 (1968).
29. E. Freese and E. B. Freese, *Radiation Res. Suppl.* **6**, 97 (1966).
30. R. Okazaki, T. Okazaki, K. Sakabe, K. Sugimoto, and A. Sugino, *Proc. Natl. Acad. Sci. U.S.* **59**, 598 (1968).
31. A. A. Blinkova, S. E. Bresler, and V. A. Lanzov, *Z. Vererbungslehre* **96**, 267 (1965).
32. F. Jacob and S. Brenner, *Compt. Rend.* **256**, 298 (1963); see also F. Jacob, S. Brenner, and F. Cuzin, *Cold Spring Harbor Symp. Quant. Biol.* **28**, 329 (1963).
33. A. Ryter and F. Jacob, *Ann. Inst. Pasteur* **107**, 384 (1964).
34. J. Lederberg, *J. Cellular Comp. Physiol.* **45**, Suppl. 2, 75 (1955).
35. O. H. Siddiqi, *Proc. Natl. Acad. Sci. U.S.* **49**, 589 (1963).
36. S. E. Bresler, V. A. Lanzov, and A. A. Blinkova, *Genetics* **56**, 117 (1967).
37. W. Szybalski, *Abhandl. Deut. Akad. Wiss. Berlin* No. 4, p. 1 (1964).
38. S. B. Zimmerman, J. W. Little, C. K. Oshinsky, and M. Gellert, *Proc. Natl. Acad. Sci. U.S.* **57**, 1841 (1967); B. Weiss and C. C. Richardson, *Proc. Natl. Acad. Sci. U.S.* **57**, 1021 (1967).
39. J. Scaife and J. D. Gross, *Genet. Res.* **4**, 328 (1963).
40. J. Scaife and J. D. Gross, *Abhandl. Deut. Akad. Wiss. Berlin* No. 4, p. 65 (1964).
41. D. Freifelder, *J. Mol. Biol.* **34**, 31 (1968).
42. F. T. Hickson, T. F. Roth, and D. R. Helinski, *Proc. Natl. Acad. Sci. U.S.* **58**, 1731 (1967).
43. J. R. Beckwith, E. R. Signer, and W. Epstein, *Cold Spring Harbor Symp. Quant. Biol.* **31**, 393 (1966).
44. A. J. Clark and A. D. Margulies, *Proc. Natl. Acad. Sci. U.S.* **53**, 451 (1965); P. Howard-Flanders and L. Theriot, *Genetics* **53**, 1137 (1965).
45. E. L. Wollman and F. Jacob, *Compt. Rend.* **240**, 2449 (1955); E. L. Wollman and F. Jacob, *Ann. Inst. Pasteur* **95**, 641 (1958).

References

46. E. L. Wollman, F. Jacob, and W. Hayes, *Cold Spring Harbor Symp. Quant. Biol.* **21**, 141 (1956).
47. See, for instance, A. Garen, *Symp. Soc. Gen. Microbiol.* **10**, 239 (1960).
48. J. B. S. Haldane, *J. Genet.* **8**, 299 (1919).
49. C. Verhoef and P. G. de Haan, *Mutation Res.* **3**, 101 (1966); P. G. de Haan and C. Verhoef, *Mutation Res.* **3**, 111 (1966).
50. A. D. Hershey, M. D. Kamen, J. W. Kennedy, and H. Gest, *J. Gen. Physiol.* **34**, 305 (1951).
51. G. S. Stent and C. R. Fuerst, *J. Gen. Physiol.* **38**, 441 (1955); C. R. Fuerst, F. Jacob, and E. L. Wollman, *Compt. Rend.* **243**, 2162 (1956).
52. F. Jacob and E. L. Wollman, *Symp. Soc. Exptl. Biol.* **12**, 75 (1958).
53. P. J. Driskell-Zamenhof and E. A. Adelberg, *J. Mol. Biol.* **6**, 483 (1963).
54. S. D. Silver, E. E. M. Moody, and R. C. Clowes, *J. Mol. Biol.* **12**, 283 (1965).
55. S. Falkow and R. V. Citarella, *J. Mol. Biol.* **12**, 138 (1965).
56. C. Levinthal, *Brookhaven Symp. Biol.* **12**, 76 (1959).
57. F. Griffeth, *J. Hyg.* **27**, 113 (1928).
58. O. T. Avery, C. M. Macleod, and M. McCarty, *J. Exptl. Med.* **79**, 137 (1944).
59. A. W. Ravin, *Advan. Genet.* **10**, 61 (1961); P. Schaeffer, in "The Bacteria" (I. C. Gunsalus and R. Y. Stanier, eds.), Vol. 5, p. 87. Academic Press, New York, 1964.
60. R. Pakula and W. Walczak, *J. Gen. Microbiol.* **31**, 125 (1963).
61. R. D. Hotchkiss, in "The Chemical Basis of Heredity" (W. D. McElroy and B. Glass, eds.), p. 321. Johns Hopkins Press, Baltimore, Maryland, 1957.
62. M. S. Fox and R. D. Hotchkiss, *Nature* **179**, 1322 (1957); M. S. Fox and R. D. Hotchkiss, *Nature* **187**, 1002 (1960); J. Spizizen, *Proc. Natl. Acad. Sci. U.S.* **44**, 1072 (1958).
63. S. E. Bresler, M. I. Mosevitskij, and A. L. Timkovskij, *Dokl. Akad. Nauk. SSSR* **149**, 721 (1963).
64. R. D. Hotchkiss and J. Marmur, *Proc. Natl. Acad. Sci. U.S.* **40**, 55 (1954).
65. M. S. Fox, *Nature* **187**, 1004 (1960); M. S. Fox, *Proc. Natl. Acad. Sci. U.S.* **48**, 1043 (1962).
66. M. J. Voll and S. H. Goodgal, *Proc. Natl. Acad. Sci. U.S.* **47**, 505 (1961).
67. W. F. Bodmer and A. T. Ganesan, *Genetics* **50**, 717 (1964).
68. S. E. Bresler, R. A. Kreneva, V. V. Kushev, and M. I. Mosevitskij, *Z. Vererbungslehre* **95**, 288 (1964).
69. M. S. Fox and M. K. Allen, *Proc. Natl. Acad. Sci. U.S.* **52**, 412 (1964).
70. E. H. Postel and S. H. Goodgal, *J. Mol Biol.* **28**, 247 (1967).
71. M. D. Chilson and B. D. Hall, *J. Mol. Biol.* **34**, 439 (1968).
72. S. Lacks, B. Greenberg, and K. Carlson, *J. Mol. Biol.* **29**, 327 (1967).
73. H. Yoshikawa and N. Sueoka, *Proc. Natl. Acad. Sci. U.S.* **49**, 559 (1963).
74. D. Dubnau, C. Goldthwaite, I. Smith, and J. Marmur, *J. Mol. Biol.* **27**, 163 (1967).
75. S. Yuki and Y. Ueda, *Nippon Idengaku Zasshi* **43**, 121 (1968).
76. R. H. Pritchard, *Genet. Res.* **1**, 1 (1960).
77. S. Lacks, *J. Mol. Biol.* **37**, 179 (1968).
78. B. E. Reilley and J. Spizizen, *J. Bacteriol.* **89**, 782 (1965).
79. A. D. Hershey and M. Chase, *J. Gen. Physiol.* **36**, 39 (1952).
80. A. Gierer and G. Schramm, *Nature* **177**, 702 (1956).
81. H. Fraenkel-Conrat, "Design and function at the Threshold of Life: The Viruses." Academic Press, New York, 1962.
82. J. W. Roberts and J. E. Argetsinger Steitz, *Proc. Natl. Acad. Sci. U.S.* **58**, 1416 (1967).
83. A. Gierer, *Symp. Soc. Gen. Microbiol.* **10**, 248 (1960).
84. A. Tsugita and H. Fraenkel-Conrat, *J. Mol. Biol.* **4**, 73 (1962).
85. H. G. Wittmann, *Proc. 5th Intern. Congr. Biochem., Moscow,* 1961 **1**, 240 (1963).
86. A. Gierer and K. W. Mundry, *Nature* **182**, 1457 (1958).
87. R. W. Horne and P. Wildy, *Virology* **15**, 348 (1961).
88. S. Brenner, G. Streisinger, R. W. Horne, S. P. Champe, L. Barnett, S. Benzer, and M. W. Rees, *J. Mol. Biol.* **1**, 281 (1959).
89. M. H. Adams, "Bacteriophages," Wiley (Interscience), New York, 1959.
90. "The Viruses" (F. M. Burnet and W. M. Stanley, eds.), Vols. 1–3. Academic Press, New York, 1959.
91. G. S. Stent, "Molecular Biology of Bacterial Viruses," Freeman, San Francisco, California, 1963.

92. W. B. Wood and R. S. Edgar, *Sci. American* **217**, No. 1, 60 (1967).
93. H. Uchida and G. S. Stent, *J. Mol. Biol.* **2**, 251 (1960); D. Pratt, G. S. Stent, and P. D. Harriman, *J. Mol. Biol.* **3**, 409 (1961).
94. C. Levinthal and C. A. Thomas, *Biochim. Biophys. Acta* **23**, 453 (1957).
95. G. S. Stent and N. K. Jerne, *Proc. Natl Acad. Sci. U.S.* **41**, 704 (1955); G. S. Stent, G. H. Sato, and N. K. Jerne, *J. Mol. Biol.* **1**, 134 (1959).
96. A. D. Hershey and M. Chase, *Cold Spring Harbor Symp. Quant. Biol.* **16**, 471 (1951).
97. N. Visconti and M. Delbrück, *Genetics* **38**, 5 (1953); A. D. Hershey, *Cold Spring Harbor Symp. Quant. Biol* **23**, 19 (1958).
98. M. Meselson and J. J. Weigle, *Proc. Natl. Acad. Sci. U.S.* **47**, 857 (1961).
99. N. Anraku and J. Tomizawa, *J. Mol. Biol.* **11**, 501 (1965); J. Tomizawa and N. Anraku, *J. Mol. Biol.* **11**, 509 (1965).
100. M. Meselson, *in* "Heritage From Mendel" (R. A. Brink, ed.), p. 81. Univ. of Wisconsin Press, Madison, Wisconsin, 1967.
101. R. H. Esptein, A. Bolle, C. M. Steinberg, E. Kellenberger, E. Boy de la Tour, R. Chevalley, R. S. Edgar, M. Susman, G. Denhardt, and A. Lielausis, *Cold Spring Harbor Symp. Quant. Biol.* **28**, 375 (1963).
102. R. S. Edgar and W. B. Wood, *Proc. Natl. Acad. Sci. U.S.* **55**, 498 (1966).
103. J. King, *J. Mol. Biol.* **32**, 231 (1968).
104. R. S. Edgar and I. Lielausis, *J. Mol. Biol.* **32**, 263 (1968).
105. J. J. Weigle, *Proc. Natl. Acad. Sci. U.S.* **55**, 1462 (1968).
106. W. Fiers and R. L. Sinsheimer, *J. Mol. Biol.* **5**, 424 (1962); A. K. Kleinschmidt, A. Burton, and R. L. Sinsheimer, *Science* **142**, 961 (1963).
107. H. Ris and B. L. Chandler, *Cold Spring Harbor Symp. Quant. Biol.* **28**, 1 (1963).
108. A. D. Hershey and E. Burgi, *Proc. Natl. Acad. Sci. U.S.* **53**, 325 (1965); A. D. Kaiser and R. B. Inman, *J. Mol. Biol.* **13**, 78 (1965).
109. H. B. Strack and A. D. Kaiser, *J. Mol. Biol.* **12**, 36 (1965).
110. G. Streisinger, R. S. Edgar, and G. H. Denhardt, *Proc. Natl. Acad. Sci. U.S.* **51**, 775 (1964).
111. C. A. Thomas and L. A. MacHattie, *Proc. Natl. Acad. Sci. U.S.* **52**, 1297 (1964).
112. S. Benzer, *Proc. Natl. Acad. Sci. U.S.* **41**, 344 (1955); S. Benzer, *Proc. Natl. Acad. Sci. U.S.* **47**, 403 (1961).
113. S. Benzer, *Brookhaven Symp. Biol.* **8**, 3 (1956).
114. S. Benzer and E. Freese, *Proc. Natl. Acad. Sci. U.S.* **44**, 112 (1958).
115. E. Freese, *J. Mol. Biol.* **1**, 87 (1959).
116. A. S. Serebrovskij and N. P. Dubinin, *Usp. Eksperim. Biol.* **4**, 235 (1929).
117. G. Bertani, *Advan. Virus Res.* **5**, 151 (1958).
118. H. Echols, L. Pilarski, and P. Y. Cheng, *Proc. Natl. Acad. Sci. U.S.* **59**, 1016 (1968).
119. M. Ptashne, *Proc. Natl. Acad. Sci. U.S.* **57**, 306 (1967).
120. A. Campbell, *Virology* **14**, 22 (1962).
121. A. Campbell, *Advan. Genet.* **11**, 101 (1962).
122. G. Kellenberger, M. L. Zichichi, and J. Weigle, *J. Mol. Biol.* **3**, 399 (1961).
123. E. R. Signer, *J. Mol. Biol.* **15**, 243 (1966).
124. E. Calef and G. Licciardello, *Virology* **12**, 81 (1960); A. Campbell, *Virology* **20**, 344 (1963).
125. R. Dulbecco and M. Vogt, *Proc. Natl. Acad. Sci. U.S.* **46**, 1617 (1960).
126. N. D. Zinder and J. Lederberg, *J. Bacteriol.* **64**, 679 (1952).
127. P. E. Hartman, *in* "The Chemical Basis of Heredity" (W. D. McElroy and B. Glass, eds.), p. 408. Johns Hopkins Press, Baltimore, Maryland, 1957; A. Campbell, *in* "The Bacteria" (I. C. Gunsalus and R. Y. Stanier, eds.), Vol. 5, p. 49. Academic Press, New York, 1964.
128. H. Ikeda and J. Tomizawa, *J. Mol. Biol.* **14**, 85 (1965).
129. B. N. Ames and P. E. Hartman, *Cold Spring Harbor Symp. Quant. Biol.* **28**, 349 (1963).
130. C. Yanofsky and E. S. Lennox, *Virology* **8**, 425 (1959); C. Yanofsky and I. P. Crawford, *Proc. Natl. Acad. Sci. U.S.* **45**, 1016 (1959).
131. M. L. Morse, E. M. Lederberg, and J. Lederberg, *Genetics* **41**, 142 (1956); M. L. Morse, E. M. Lederberg, and J. Lederberg, *Genetics* **41**, 758 (1956).
132. J. Weigle, M. Meselson, and K. Paigen, *Brookhaven Symp. Biol.* **13**, 125 (1959).
133. A. Campbell, *Virology* **9**, 293 (1959).

References

134. J. Weigle, M. Meselson, and K. Paigen, *J. Mol. Biol.* **1**, 379 (1959); J. Weigle, J. Mol. *Biol.* **3**, 393 (1961); G. Kayajanian and A. Campbell, *Virology* **30**, 482 (1966).
135. A. Matsushiro, *Virology* **19**, 475 (1963).
136. J. R. Beckwith and E. R. Signer, *J. Mol. Biol.* **19**, 254 (1966).
137. M. E. Lobashov, *Tr. Leningrad. Obshchestva Estestvoispitat.* **66**, 346 (1937); V. V. Sakharov, Biol. Zh. **4**, 107 (1936).
138. H. Schuster and G. Schramm, *Z. Naturforsch.* **13b**, 697 (1958).
139. R. M. Litman, *J. Chim. Phys.* **58**, 997 (1961); S. E. Bresler and D. A. Perumov, *Biokhimiya* **27**, 927, (1962).
140. S. Zamenhof, *Ann. N.Y. Acad. Sci.* **81**, 784 (1959).
141. E. Freese, *J. Mol. Biol.* **1**, 87 (1959).
142. E. Freese, *Proc. 5th Intern. Congr. Biochem., Moscow,* 1961 **1**, 204 (1963).
143. E. Bautz-Freese and E. Freese, *Virology* **13**, 19 (1961).
144. E. Freese, E. Bautz, and E. Bautz-Freese, *Proc. Natl. Acad. Sci. U.S.* **47**, 845 (1961).
145. E. Cerdá-Olmedo and P. C. Hanawalt, *Mol. Gen. Genet.* **101**, 191 (1968).
146. L. S. Lerman, *Proc. Natl. Acad. Sci. U.S.* **49**, 94 (1963).
147. A. Orgel and S. Brenner, *J. Mol. Biol.* **3**, 762 (1961).
148. B. N. Ames and H. J. Whitfield, Jr., *Cold Spring Harbor Symp. Quant. Biol.* **31**, 221 (1966).
149. A. Novick and L. Szilard, *Proc. Natl. Acad. Sci. U.S.* **36**, 708 (1950).
150. G. A. Nadson and G. S. Filippov, *Vestn. Rentgenol. i Radiol.* **3**, 95 (1925); H. J. Muller, *Science* **66**, 84 (1927); see also N. P. Dubinin, "Problems of Radiation Genetics," Oliver & Boyd, Edinburgh and London, 1964.
151. E. Freese, *Brookhaven Symp. Biol.* **12**, 63 (1959).
152. E. Freese, E. B. Freese, and E. K. F. Bautz, *J. Mol. Biol.* **3**, 133 (1961).
153. E. B. Freese, *Proc. Natl. Acad. Sci. U.S.* **47**, 540 (1961).
154. S. E. Bresler, V. L. Kalinin, and D. A. Perumov, *Biopolymers* **2**, 135 (1964).
155. S. E. Bresler and D. A. Perumov, *Dokl. Akad. Nauk SSSR* **158**, 967 (1964).
156. S. E. Bresler, V. L. Kalinin, and D. A. Perumov, *Mutation Res.* **5**, 1 (1968).
157. G. Gamow, *Nature* **173**, 318 (1954); for a general review of this topic, see C. R. Woese, "The Genetic Code." Harper & Row, New York, 1967.
158. S. Brenner, *Proc. Natl. Acad. Sci. U.S.* **43**, 684 (1957).
159. F. H. C. Crick, L. Barnett, S. Brenner, and R. J. Watts-Tobin, *Nature* **192**, 1227 (1961).
160. G. Streisinger, Y. Okada, J. Emrich, J. Newton, A. Tsugita, E. Terzaghi, and M. Inouye, *Cold Spring Harbor Symp. Quant. Biol.* **31**, 77 (1966).
161. A. N. Belozersky and A. S. Spirin, *Nature* **182**, 111 (1958).
162. N. Sueoka, *Proc. Natl. Acad. Sci. U.S.* **47**, 1141 (1961).
163. E. R. Signer, A. Torriani, and C. Levinthal, *Cold Spring Harbor Symp. Quant. Biol.* **26**, 31 (1961).
164. V. M. Ingram, *Biochim. Biophys. Acta* **36**, 402 (1959).
165. J. A. Hunt and V. M. Ingram, *Biochim. Biophys. Acta* **42**, 409 (1960).
166. A. C. Allison, *Brit. Med. J.* **1**, 290 (1954).
167. L. Pauling, H. A. Itano, S. J. Singer, and I. C. Wells, *Science* **110**, 543 (1949).
168. C. Levinthal, A. Garen, and F. Rothman, *Proc. 5th Intern. Congr. Biochem., Moscow,* 1961 **1**, 196 (1963).
169. D. R. Helinski and C. Yanofsky, *Proc. Natl. Acad. Sci. U.S.* **48**, 173 (1962); U. Henning and C. Yanofsky, *Proc. Natl. Acad. Sci. U.S.* **48**, 183 (1962); U. Henning and C. Yanofsky, *Proc. Natl. Acad. Sci. U.S.* **48**, 1497 (1962); D. R. Helinski and C. Yanofsky, *J. Biol. Chem.* **238**, 1043 (1963).
170. C. Yanofsky, B. C. Carlton, J. R. Guest, D. R. Helinski, and U. Henning, *Proc. Natl. Acad. Sci. U.S.* **51**, 266 (1964).
171. C. Yanofsky, J. Ito, and V. Horn, *Cold Spring Harbor Symp. Quant. Biol.* **31**, 151 (1966).
172. H. G. Wittmann and B. Wittmann-Liebold, *Cold Spring Harbor Symp. Quant. Biol.* **28**, 589 (1963).
173. M. W. Nirenberg and J. H. Matthaei, *Proc. Natl. Acad. Sci. U.S.* **47**, 1588 (1961); see also J. H. Matthaei, O. W. Jones, R. G. Martin, and M. W. Nirenberg, *Proc. Natl. Acad. Sci. U.S.* **48**, 666 (1962).
174. M. W. Nirenberg, O. W. Jones, P. Leder, B. F. C. Clark, W. S. Sly, and S. Pestka, *Cold Spring Harbor Symp. Quant. Biol.* **28**, 549 (1963); J. F. Speyer, P. Lengyel, C. Basilio, A. J. Wahba, R. S. Gardner, and S. Ochoa, *Cold Spring Harbor Symp. Quant. Biol.* **28**, 559 (1963).

175. S. Erhan, L. G. Northrup, and F. R. Leach, *Proc. Natl. Acad. Sci. U.S.* **53**, 646 (1965).
176. M. Nirenberg and P. Leder, *Science* **145**, 1399 (1964).
177. M. Nirenberg, P. Leder, M. Bernfield, R. Brimacombe, J. Trupin, F. Rottman, and C. O'Neal, *Proc. Natl. Acad. Sci. U.S.* **53**, 1161 (1965); M. Nirenberg, T. Caskey, R. Marshall, R. Brimacombe, D. Kellogg, B. Doctor, D. Hatfield, J. Levin, F. Rottman, S. Pestka, M. Wilcox, and F. Anderson, *Cold Spring Harbor Symp. Quant. Biol.* **31**, 11 (1966).
178. F. H. C. Crick, *Cold Spring Harbor Symp. Quant. Biol.* **31**, 1 (1966).
179. S. Nishimura, D. S. Jones, E. Ohtsuka, H. Hayatsu, T. M. Jacob, and H. G. Khorana, *J. Mol. Biol.* **13**, 283 (1965); S. Nishimura, D. S. Jones, and H. G. Khorana, *J. Mol. Biol.* **13**, 302 (1965); H. G. Khorana, H. Büchi, H. Ghosh, N. Gupta, T. M. Jacob, H. Kössel, R. Morgan, S. A. Narang, E. Ohtsuka, and R. D. Wells, *Cold Spring Harbor Symp. Quant. Biol.* **31**, 39 (1966).
180. T. P. Bennett, J. Goldstein, and F. Lipmann, *Proc. Natl. Acad. Sci. U.S.* **53**, 385 (1965).
181. D. Söll, D. S. Jones, E. Ohtsuka, R. D. Faulkner, R. Lohrmann, H. Hayatsu, H. G. Khorana, J. D. Cherayil, A. Hampel, and R. M. Bock, *J. Mol. Biol.* **19**, 556 (1966); D. A. Kellogg, B. P. Doctor, J. E. Loebel, and M. W. Nirenberg, *Proc. Natl. Acad. Sci. U.S.* **55**, 912 (1966).
182. F. H. C. Crick, *J. Mol. Biol.* **19**, 548 (1966).
183. S. Brody and C. Yanofsky, *Proc. Natl. Acad. Sci. U.S.* **50**, 9 (1963).
184. A. Garen and O. Siddiqi, *Proc. Natl. Acad. Sci. U.S.* **48**, 1121 (1962).
185. M. G. Weigert and A. Garen, *J. Mol. Biol.* **12**, 448 (1965).
186. A. O. W. Stretton and S. Brenner, *J. Mol. Biol.* **12**, 456 (1965).
187. G. W. Notani, D. L. Engelhardt, W. Konigsberg, and N. D. Zinder, *J. Mol. Biol.* **12**, 439 (1965).
188. A. S. Sarabhai, A. O. W. Stretton, S. Brenner, and A. Bolle, *Nature* **201**, 13 (1964).
189. S. Brenner, A. O. W. Stretton, and S. Kaplan, *Nature* **206**, 994 (1965).
190. M. G. Weigert, E. Lanka, and A. Garen, *J. Mol. Biol.* **14**, 522 (1965).
191. D. Zipser, *J. Mol. Biol.* **29**, 441 (1967); J. F. Sambrook, D. P. Fan and S. Brenner, *Nature* **214**, 452 (1967).
192. M. R. Capecchi and N. G. Gussin, *Science* **149**, 417 (1965); D. L. Engelhardt, R. E. Webster, R. C. Wilhelm, and N. D. Zinder. *Proc. Natl. Acad. Sci. U.S.* **54**, 1791 (1965).
193. W. Salser, R. F. Gesteland, and A. Bolle, *Nature* **215**, 588 (1967).
194. R. F. Gesteland, W. Salser, and A. Bolle, *Proc. Natl. Acad. Sci. U.S.* **58**, 2036 (1967).
195. H. M. Goodman, J. Abelson, A. Landy, S. Brenner, and J. D. Smith, *Nature* **217**, 1019 (1968).
196. R. E. Thach, T. A. Sundararajan, K. F. Dewey, J. C. Brown, and P. Doty, *Cold Spring Harbor Symp. Quant. Biol.* **31**, 85 (1966).
197. J. M. Adams and M. R. Capecchi, *Proc. Natl. Acad. Sci. U.S.* **55**, 147 (1966); R. E. Webster, D. L. Engelhardt, and N. D. Zinder, *Proc. Natl. Acad. Sci. U.S.* **55**, 155 (1966).

Chapter V

RNA Function

1. Introduction

Ribonucleic acid (RNA) is the second component of living tissue which performs a cybernetic function, in this case by providing and supporting a flow of information in the cell. Its role is perhaps even more complicated than that of deoxyribonucleic acid (DNA). While DNA stores the information necessary for the growth and reproduction of each cell, different types of RNA participate directly in this process, supplying working components of the protein-synthesizing machinery. Accordingly, the quantity of DNA in the cell is invariant, and even in complex and highly differentiated organisms, all cells contain the same amount of this substance. By contrast, the quantity of RNA per cell varies substantially, depending on the rate of protein synthesis. The latter is influenced in its turn by external conditions of growth such as the carbon source, temperature, and nutrient supplements.

The synthesis of each specific enzyme or protein is determined by a special kind of RNA which contains the necessary information in the form of a chemical code akin to that of DNA. But in addition to carrying information about protein structure, RNA also plays a role in controlling the rate of synthesis of each protein. The mechanism by which protein synthesis is regulated depends in large measure on interactions between the cell and its surrounding medium. In this chapter we shall consider the feedback loop which permits the cell to initiate the synthesis of just the enzyme or enzymes it needs to cope with a particular change in the environment. We should emphasize at this point that the number of informational RNA molecules of a given type present in the cell at any one time is a material reflection of the operation of this mechanism. The result will be the rapid synthesis of one or more enzymes for which there is a need and the inhibition or decline in the synthesis of others not as essential for efficient growth.

Although nucleotide sequences in RNA are copied or transcribed in a complementary fashion from segments of the DNA, the base composition of total RNA does not necessarily correspond to that of DNA. But while the two are not identical, there is, however, a correlation between the base compositions of RNA and DNA extracted from a single bacterial species. In light of what has been said, this is not surprising: DNA composition is constant, but the number and type of RNA molecules transcribed from various cistrons change with the particularlar needs of the cell.

There are three principal classes of RNA in every cell: *ribosomal* (rRNA), *template* or *messenger* (mRNA), and *transfer* (tRNA). Ribosomal RNA is the nucleic acid component of ribosomes. It is a stable substance of high molecular weight, comprising from 75 to 85% of the total RNA in bacterial cells. There are special genetic loci or cistrons in the bacterial chromosome which govern the synthesis of ribosomal RNA molecules. While it is known that ribosomes play a fundamental role in protein synthesis, the precise manner in which ribosomal RNA functions is not yet clear. Since about 65% of the weight of bacterial ribosomes consists of RNA, and the remaining 35% of protein, it is assumed that rRNA behaves in part as a structural material. The great stability of rRNA was shown in experiments of the following design. When a bacterial culture is grown in the presence of "heavy" isotopes and then transferred to "light" medium, isotopically labeled rRNA is transmitted to progeny cells without alteration, and can be detected in its original form even in the fourth and fifth generations. The isotopic composition of the inherited RNA does not change in the least, although all of the new components synthesized by the growing cells contain light constituents.

In all cells, regardless of their origin, the participation of ribosomes in protein synthesis is obligatory. Ribosomes have even been found in the nuclei of cells from higher organisms where they mediate the synthesis of specific nuclear proteins. As with other cellular organelles, ribosomes separate a particular chemical reaction in space from others taking place concurrently in the same milieu. For example, amino acids, the building blocks of protein, can take part in many cellular processes since they are present in the cytoplasm as soluble compounds. The reactions leading to protein synthesis must therefore be separated spatially from oxidative or degradative reactions. This explains why distinct organelles are necessary for protein synthesis. Generally speaking, they must also be accessible only to essential substances in the cytoplasm such as amino acids and the compounds which mediate energy transfer, adenosine triphosphate (ATP) and guanosine triphosphate (GTP).

A second type of RNA, called template or messenger RNA, carries genetic information from the chromosomal DNA to the sites of protein synthesis, that is, to the ribosomes where it becomes bound. Such RNA molecules apparently act as a template which directs the assembly of polypeptide chains. Messenger RNA is rapidly metabolized in bacterial cells throughout exponential growth; its synthesis requires 20 to 30 seconds and it is generally degraded in the course of a few minutes. The steady-state concentration of mRNA in the cell is on the order of 3 to 8% of the total cellular RNA. The number of mRNA molecules corresponding to a given protein is variable, depending on the requirements of the cell and on the rate at which each protein must be made.

Besides high-molecular-weight rRNA and mRNA, there is a low-molecular-weight RNA called transfer RNA (tRNA) present in the cell. This third type of RNA has a special function completely distinct from those of either previous class. It covalently binds amino acids and transports them to the sites of protein synthesis. Its role as an adaptor molecule in the actual assembly of the polypeptide chain was considered in the preceding chapter. Transfer RNA comprises between 10 and 20% of the total cellular RNA.

Finally, monoribonucleotides and their phosphorylated derivatives such as ATP, GTP, and others play an important role in the cellular economy. They are widely distributed and often behave as coenzymes in addition to their participation in energy metabolism. We shall not be concerned with monoribonucleotides in this chapter since the reader can find a detailed exposition on this topic in any textbook of biochemistry.

2. Transfer RNA and the Activation of Amino Acids

Since all types of RNA are implicated in protein synthesis, let us begin our discussion with transfer RNA (tRNA), actually a class of similar compounds that transport different amino acids to the sites of protein synthesis and there participate in the very first stages of polypeptide chain formation (1). Transfer RNA is highly soluble and is further distinguished by a rather low molecular weight, on the order of 25,000 to 30,000 daltons. Most tRNA molecules in the cell are attached to amino acids which they acquire after the amino acid is first raised to a higher energy level, or *activated*, in a reaction with ATP. This corresponds to the observation that polypeptide chain synthesis requires the presence of covalently bonded complexes between amino acids and tRNA, not free amino acids. Without doubt, tRNA structure must be specially suited for the transfer of amino acids to growing polypeptide chains. In fact, tRNA apparently serves as an adaptor for the proper positioning of the amino acid during the assembly of proteins on a nucleic acid template.

One of the principal functions of tRNA is to incorporate cellular amino acids into a chemically labile compound from which they can then be transferred to the growing polypeptide on the ribosomes under energetically favorable conditions. The role of tRNA exemplifies a difficulty with which the cell must continually cope. Many essential metabolites, which either penetrate the cell membrane from the external medium or are elaborated inside the cell itself, must be protected against undesired side reactions that could prevent them from fulfilling more useful ends. Therefore, amino acids are activated specifically for protein synthesis when they are temporarily sequestered by tRNA molecules and are not simply burned up in the cell as an energy source. If one or more of the amino acids are present in excess, a certain proportion remains in the free state and these molecules can be oxidized via the Krebs cycle, subjected to transamination, or degraded in some other manner. In this case, the cellular economy is well served, since the excess amino acids which cannot be used in protein synthesis are instead employed by the cell for other purposes.

The reactions which lead to the formation of a covalent bond between amino acid and tRNA have been extensively studied. However, the chemistry of the final and decisive stage in protein synthesis, the formation of the peptide bond and cleavage of tRNA from the growing end of the chain, has not yet been completely elucidated.

In the first step of this sequence, ATP and amino acid react with elimination of pyrophosphate

$$\text{ATP} + \text{amino acid} \rightarrow \text{AMP-amino acid} + \text{pp}$$

The amino acid and AMP are united by a mixed anhydride bond between the carboxyl group of the former and the phosphoric acid group attached to the 5'-C of AMP:

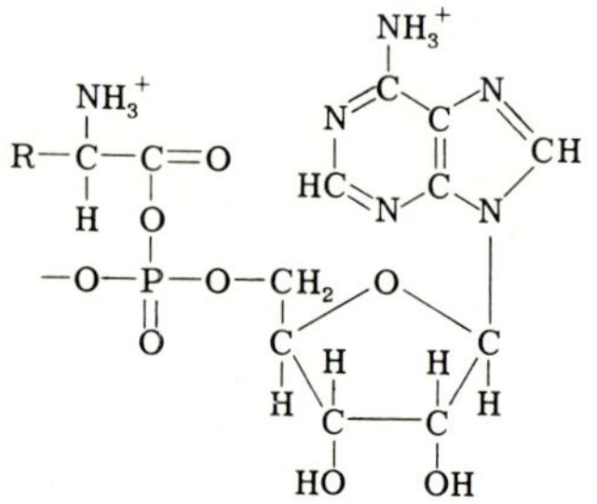

The compound which results from this reaction is an *aminoacyl adenylate* while the reaction itself is called *amino acid activation*.

To each of the twenty amino acids there corresponds a specific enzyme which catalyzes the formation of acyl-phosphate bonds. Such activating enzymes have been found in all cells and tissues in which they have been sought, from the simplest of bacteria to the highly differentiated cells of plants and animals. These enzymes are found among the so-called "pH 5-enzymes," normally soluble proteins which can be precipitated in weakly acidic medium (Zamecnik and Hoagland). Phenylalanine, isoleucine, and tyrosine activating enzymes from *Escherichia coli* have been extensively purified. The enzyme specific for tyrosine activation has been found to have a molecular weight of 160,000 daltons. Many other such enzymes have been obtained in enriched preparations with specificity for only one amino acid, that is, free from homologous enzymes which activate other amino acids. Even chemically very similar amino acids such as valine and leucine are activated by distinct enzymes.

In the second reaction of this sequence the aminoacyl group is transferred to the free 3'-hydroxyl group of a specific tRNA molecule. Now there exist at least as many different types of tRNA as there are amino acids. Actually the number is greater, between 30 and 40, if we take the degeneracy of the code into account. Each amino acid is joined to its corresponding tRNA in a reaction catalyzed by the same specific activating enzyme which participated in the first reaction. For this reason, these enzymes are called *aminoacyl-tRNA synthetases*.

All of the tRNA molecules taking part in the "mobilization" of amino acids contain three terminal residues in common, regardless of their specificity:

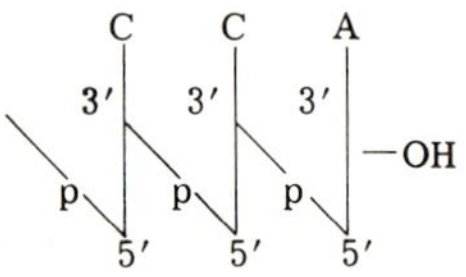

2. Transfer RNA and the Activation of Amino Acids

Guanine frequently occupies the 5'-terminal position at the other end of the molecule. The "universal" 3'-terminal trinucleotide —C—C—A can be cleaved by enzymes present in the pH 5 fraction other than the activating enzymes. After cleavage of the terminal residues, the corresponding mononucleotides, AMP and CMP, appear in the medium and tRNA loses its ability to accept amino acids. If the nucleoside triphosphates ATP and CTP are added back to the solution, the terminal —C—C—A sequence is restored to the tRNA under the influence of another enzyme present in the pH 5 fraction. After purification, this enzyme was shown to catalyze the attachment of all three terminal residues with the simultaneous elimination of pyrophosphate.

The reaction by which activated amino acid and tRNA are joined is in itself reversible. If the reaction is written in two steps, with an intermediate aminoacyl adenylate-enzyme complex, it is clear that the reaction can be displaced toward cleavage of the amino acid by adding an excess of AMP, or even better, of AMP and pyrophosphate together:

$$\text{ATP} + \text{amino acid} + \text{enzyme} \rightleftharpoons \text{enzyme(AMP-amino acid)} + \text{pp}$$
$$\text{enzyme(AMP-amino acid)} + \text{tRNA} \rightleftharpoons \text{tRNA-amino acid} + \text{enzyme} + \text{AMP}$$

In the presence of excess AMP and pyrophosphate, the original mixture of free amino acids is regenerated. This prediction was fully confirmed by Lipmann and his co-workers.

Feldmann and Zachau (2) determined the nature of the bond between amino acid and tRNA. The carboxyl group of the amino acid is esterified to the hydroxyl group of 3'-C of the terminal adenylic acid residue:

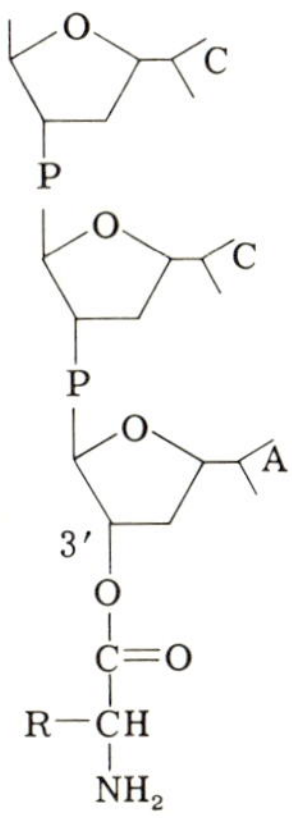

This was proved by the fact that the terminal ribosyl group in tRNA can no longer be oxidized by periodic acid (HIO_4) after reaction of tRNA and amino acid. This reagent attacks the terminal ribosyl group of ribonucleic acid chains only where there are two neighboring hydroxyl groups in the β-position.

The specificity of tRNA molecules with respect to the amino acids they can accept is pronounced, but not absolute. Very close chemical analogs can sometimes be incorporated instead of the normal amino acids. Thus fluorophenylalanine replaces phenylalanine if the former compound is present in excess. Insofar as concerns the species specificity of tRNA and amino acid activating enzymes, the situation is rather complicated. Detailed investigations performed by Benzer and others (3–6) demonstrated that species specificity is absent or very weakly expressed in the tRNA's for certain amino acids, such as valine, but very marked for others, such as arginine and tyrosine. The binding of amino acids to the tRNA's of bacteria, yeast, and higher organisms under the influence of homologous and heterologous aminoacyl-tRNA synthetases is illustrated in Table 5-1.

TABLE 5-1

Species Specificity of Amino Acid Transfer-RNA and Aminoacyl-tRNA Synthetases

Source of aminoacyl-tRNA synthetase	Amino acid tested	Relative amino acid acceptor activity of tRNA from			Ref.
		E. coli	Yeast	Rat liver	
E. coli	Arg	1.00	0.02	—	(3)
Yeast		0.01	1.00	—	
Rabbit liver		1.18	0.02	—	
E. coli	Tyr	1.00	0.10	—	(3)
Yeast		0.05	1.00	—	
Rabbit liver		0.01	1.12	—	
E. coli	Tyr	1.00[a]	0.0[a]	—	(4)
Yeast		0.0[a]	1.00[a]	—	
E. coli	Lys	1.00	0.24	—	(3)
Yeast		0.73	1.00	—	
E. coli	Leu	1.00	0.05	0.01	(5)
Yeast		0.02	1.00	1.13	
Rat liver		0.01	1.22	1.00	
Rat kidney		0.01	—	1.25	
Ox liver		0.02	—	0.74	
Pigeon liver		0.07	—	1.08	
Frog liver		0.02	—	0.89	

[a] Purified tRNA; all other assays performed on unfractionated tRNA mixtures.

An important question is how each tRNA molecule "recognizes" its specific amino acid. Direct recognition is excluded since the end of the chain which binds the amino acid is exactly the same in all types of tRNA. The process of recognition must therefore be mediated by the enzyme which catalyzes the reaction between tRNA and amino acid. This explains why more than twenty specific activating enzymes are needed to charge all types of tRNA with their respective amino acids. These proteins therefore participate in the transfer of genetic information insofar as they must ensure the proper match between amino acid and tRNA adaptor which in turn contains the anticodon governing placement of the tRNA on the template. This substantially complicates the problem of translating the genetic code into protein structure. Any of these enzymes, for instance, can be altered by mutation as can any of the tRNA's. Thus in addition

2. Transfer RNA and the Activation of Amino Acids

to the structural gene itself, there are upward of 40 individual mutable factors—at least twenty activating enzymes and at least twenty tRNA's—which participate in protein synthesis. The entire system therefore has an enormous number of degrees of freedom.

The participation of tRNA in the final stages of peptide bond formation is beyond question (7). However, one important supporting observation merits attention. The antibiotic puromycin inhibits the growth of many microorganisms by specifically interfering with protein synthesis [Yarmolinsky and De La Haba (8)]. Investigation revealed that puromycin acts neither on amino acid activation nor on the transfer of activated amino acids to tRNA, but that it does stop the third stage, elongation of the protein chain by peptide bond formation. The structure of the antibiotic, which was confirmed by complete synthesis, mimics that of aminoacyl-tRNA:

Puromycin   Aminoacyl-tRNA

Because of its structural similarity to the aminoacyl portion of the tRNA molecule, puromycin can replace tRNA at the end of the peptide. There it is incorporated as the terminal residue, preventing further chain growth and producing incomplete polypeptides at the same time.

An experiment providing direct proof that tRNA participates in protein synthesis as an adaptor was performed by Lipmann and his colleagues (9). Transfer RNA molecules were enzymatically charged with cysteine and then subjected to hydrogenation under mild conditions on the solid catalyst, Raney nickel. Under these circumstances, the sulfhydryl group is split from cysteine, turning it into alanine without disturbing the ester bond between tRNA and the amino acid. An abnormal combination results, with alanine bound to cysteine-specific tRNA:

The question now is whether this compound will behave as cysteine or alanine during assembly of the polypeptide chain on the RNA template. The synthetic template poly UG was selected to test this in the cell-free system according to Nirenberg and Matthaei. This polynucleotide stimulates the incorporation of cysteine but does not contain any alanine codons. It was found that when alanine is derived from cysteine already bound to cysteine-specific tRNA, it is incorporated extremely well during *in vitro* polypeptide synthesis under the direction of poly UG. When messenger RNA-ribosome complexes extracted from rabbit reticulocytes were "tricked" in this way, it was possible to synthesize hemoglobin α-chains in which alanine replaced specific cysteine residues (10). This case is particularly clear-cut since alanine was inserted into at least one tryptic peptide which normally contains no alanine. Therefore, the specific tRNA molecule dictates the position of its attached amino acid in the polypeptide chain.

The role of tRNA in coding can also be demonstrated by making use of degeneracy. Since there are several codons for many amino acids, there must also be several tRNA varieties with corresponding anticodons, although according to the "wobble" hypothesis, certain anticodons can recognize more than one codon. Let us take leucine and its specific tRNA's as an example. Five different tRNA's with leucine acceptor activity have been isolated from a mixture of tRNA's by the method of countercurrent distribution. One of them was found to be necessary for leucine incorporation when poly UG served as a template and a second when poly UC was employed. This experiment shows that degeneracy of the code is associated at least in part with a multiplicity of adaptor tRNA's (11).

As final confirmation of the significance of degeneracy, it was necessary to show experimentally that during protein synthesis different codons in the template can represent the same amino acid and that they are filled by different adaptors. In other words, if only one of the five leucine-specific tRNA's is charged with radioactive leucine, and the remaining ones with unlabeled leucine, then in the process of protein synthesis radioactive leucine should occupy only certain well-defined sites in the polypeptide chain, corresponding to only one of the six parallel leucine codons.

Surprisingly, this experiment was not successful for a long time. Labeled leucine was incorporated indiscriminately into all positions occupied by leucine residues. We now understand that this is at least partly due to the "wobble" in codon-anticodon recognition (see p. 437). Nonetheless, measurable differences between the various adaptor tRNA's were still anticipated. The question was resolved when it was discovered that the activating enzymes present in the cell-free system were able to catalyze transacylation, the transfer of an amino acid from one molecule of tRNA to another [Yamane and Sueoka (12)]. When this circumstance was clarified, measures were taken to prevent the undesired reaction. By adding a large excess of unlabeled leucine to the cell-free system, it was possible to saturate the corresponding aminoacyl-tRNA synthetases and thereby avert transacylation. Using this method, Benzer and his co-workers obtained the predicted result in a study of hemoglobin synthesis in ribosomal preparations from rabbit reticulocytes (13). One variety of leucine-specific tRNA extracted from *E. coli*, after charging with radioactive amino acid, introduced labeled leucine into only one place during synthesis of the hemoglobin α-chain. Other types of leucine-specific tRNA apparently insert their amino acids into several other sites during polypeptide synthesis.

3. Ribosomes and the Synthesis of Proteins

All cellular protein synthesis takes place in association with ribosomes, compact subcellular organelles composed of RNA and structural protein. Ribosomes are ubiquitous in living organisms and they can be extracted from almost any kind of cell. Depending on the tissue of origin, ribosomal particles possess characteristic sedimentation coefficients of either 70 S (bacteria, blue-green algae, chloroplasts of green plants) or 80 S (yeast, fungi, cytoplasm of plants and animals), and molecular weights ranging from 2.5 to 4×10^6 daltons. As can be seen from Fig. 5-1, they approximate prolate ellipsoids in shape with a major axis between 250 and 350 Å and minor axes between 170 and 250 Å. Bacterial ribosomes are believed to be homogeneous, but this may not be true for higher organisms. In animal and plant cells, cytoplasmic ribosomes are not free, but bound to the membranes of the endoplasmic reticulum. When necessary, the particulates can be liberated from cell membranes with detergents which break up and dissolve the lipoprotein matrix.

That protein synthesis *in vivo* requires the participation of ribosomes has been demonstrated in *E. coli* (14b). Depending on nutrition and growth rate, normal bacterial cells contain from 5,000 to 10,000 70 S ribosomal particles. The structural integrity of these particles is contingent on the presence of magnesium ions. When the bacteria are placed in a medium lacking Mg^{2+} ions, existing ribosomes are degraded and no new ones are formed. As a result of this artificially induced ribosome deficiency, protein synthesis comes to a complete halt. Moreover, Mg^{2+}-starved cells cannot support lytic infection by bacteriophages since phages are unable to develop in the absence of protein synthesis. The introduction of magnesium ions into the medium quickly restores the normal internal organization of the cytoplasm: ribosomes are formed, protein synthesis soon resumes, and the cells once again become susceptible to phage infection.

Most of our knowledge about the structure and function of ribosomes derives from studies *in vitro* and the remainder of this section will be devoted to a discussion of experiments carried out with cell-free extracts or with purified ribosomes. Detailed investigations of how the cellular protein-synthesizing apparatus operates and, in particular, how the ribosome functions in this process, were made possible by the discovery that amino acids are incorporated into polypeptide chains in cell lysates as well as in intact cells.

In the simplest experiments on protein synthesis *in vitro*, concentrated suspensions of bacterial or mammalian cells are first disrupted by some suitable method such as osmotic shock, grinding with abrasive powder, or repeated freezing and thawing. Cell membranes, cell walls, and remaining whole cells are then removed from the lysate by low speed centrifugation. The resulting cell-free extract consists of a complex mixture of ribosomes, messenger RNA, soluble proteins, small cellular metabolites including amino acids, and a variety of specific tRNA molecules. Polypeptide synthesis can be readily detected in such a system by measuring the incorporation of radioactively labeled amino acids into the fraction precipitated by hot trichloroacetic acid (TCA).

Although the requirements for protein synthesis *in vitro* have already been discussed in Chapter 4 (p. 433f.), it will be well to review some of the more important

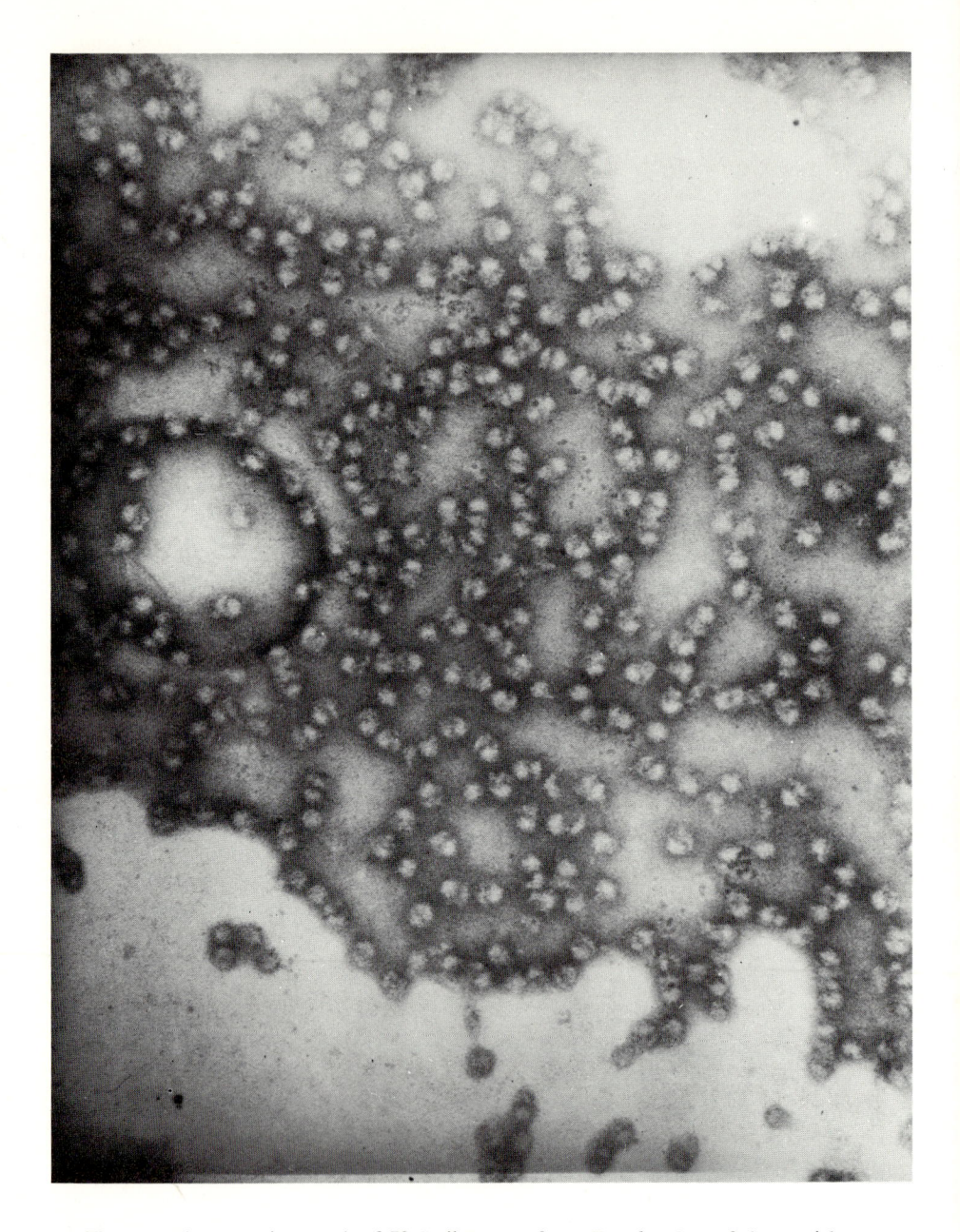

Fig. 5-1. Electron micrograph of 70 S ribosomes from *E. coli*. Most of the particles present show two unequal subunits, some of which may be seen separated on the left-hand side. Particles were negatively stained with phosphotungstic acid; ×200,000. From Huxley and Zubay (14a).

properties of cell-free amino acid incorporation here. The first *in vitro* systems were quite inefficient and they could not sustain protein synthesis for extended periods. Newly formed polypeptides rarely exceeded a few thousandths, or at best, a few hundredths, of a percent of the protein already present in the extract and after incubation at 37°C for 10 to 20 minutes, amino acid incorporation rapidly declined and then stopped. Nonetheless, it was still possible to show that the conversion of labeled amino acids to TCA-insoluble form was irreversible, indicating that genuine chemical synthesis had taken place or, in other words, that amino acids had been joined together by peptide bonds into a polypeptide chain. The fact that all twenty amino acids were necessary for this to occur ruled out an exchange reaction and strongly implied that net synthesis was taking place.

The low efficiency of the *in vitro* system may be accounted for in part by recognizing that the cell extract is 10^3 to 10^4 times more dilute than the cytoplasm of the cell; for this reason, certain cell components necessary for protein synthesis may not be present at sufficient concentrations. The ability of various substances to stimulate amino acid incorporation was therefore tested. The addition of excess tRNA was found to enhance the performance of the cell-free system considerably, as did a source of chemical energy in the form of adenosine triphosphate (ATP). Of particular interest was the discovery that catalytic amounts of guanosine triphosphate (GTP) were specifically required for the transfer of amino acids from aminoacyl-tRNA to the growing polypeptide chain. It was also convenient to supply an enzymatic phosphorylating system to continually regenerate ATP and GTP from ADP and GDP.

Finally, the duration of polypeptide synthesis *in vitro* was greatly increased by supplementing the system with high-molecular-weight polyribonucleotides (15). Natural and synthetic RNA's were both effective to this end. From studies on the polypeptide products resulting from the use of simple synthetic polynucleotides of known composition it soon became evident that these compounds served as templates for amino acid assembly during protein synthesis. The initial cell lysate always contained a certain amount of natural RNA templates, collectively called messenger RNA or mRNA, but nucleases present in the extract degraded them in a relatively short time, rendering them ineffective in stimulating amino acid incorporation. The fact that hemoglobin synthesis continued for so long in extracts from mammalian reticulocytes without the addition of polynucleotides was attributed to the unusual stability of hemoglobin-specific mRNA.

Significant advances in the understanding of protein synthesis resulted from fractionation of the complex cell-free system into its components (Zamecnik, Lipmann, and co-workers). In one of the earliest experiments of this type, ribosomes were sedimented from the soluble extract by centrifugation at 100,000 *g*. Neither the ribosome fraction nor the supernatant alone could stimulate amino acid incorporation, but this capacity was restored when the two were mixed together once again. Ribosomes and supernatant enzymes were therefore to be numbered among the many essential participants in protein synthesis along with tRNA, mRNA, amino acids, and a source of energy.

Before discussing further fractionation of the cell-free incorporating system and the investigation of intermediates in polypeptide synthesis, let us first touch briefly on the properties of bacterial ribosomes. Bacterial ribosomes are of special interest since they have been used most extensively in such studies. At the relatively high Mg^{2+} concentrations ($\approx 10^{-2}$ M) required for *in vitro* protein synthesis, the majority of bacterial ribosomes sediment at 70 S. But when the concentration of Mg^{2+} ions is reduced to about 10^{-4} M, 70 S ribosomes reversibly dissociate into two nonidentical subunits with sedimentation coefficients of 30 S and 50 S. In fact, there is now some evidence indicating that the dissociated state is the normal one for ribosomes in the cell and that 70 S particles are assembled specifically for the purpose of protein synthesis (16). Each 30 S subunit contains a single molecule of structural RNA (rRNA) with a molecular weight of 5.6×10^5 daltons and a sedimentation coefficient of 16 S. In addition, there are some twenty structural proteins present in the smaller subunit that together make up about 35% of its total mass. There are two RNA molecules in the 50 S subunit which have molecular weights of 3×10^4 daltons (5 S) and 1.1×10^6 daltons (23 S). The 50 S subunit contains from 30 to 40 distinct structural proteins. The proportion of RNA and protein is roughly the same in both ribosomal subunits—two-thirds RNA and one-third protein by weight. These facts about ribosome structure are summarized in Fig. 5-2.

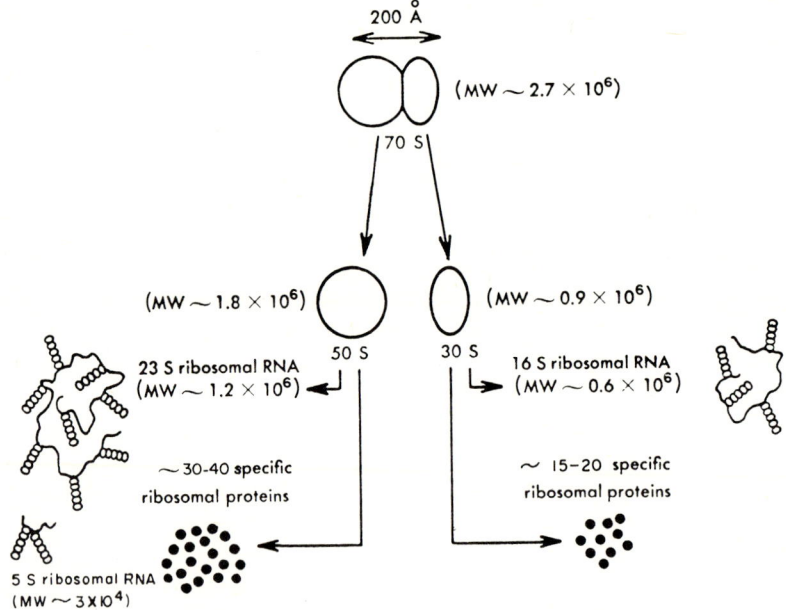

Fig. 5-2. Structure of the *E. coli* ribosome. The 70 S ribosome and its 30 S and 50 S subunits are identified by their sedimentation constants, measured in svedbergs; 16 S and 23 S are the sedimentation constants of the smaller and larger ribosomal RNA molecules. Each subunit also contains a large number of different structural proteins which are intimately associated with the rRNA in the intact particle. From Watson (17), p. 326.

An intermediate complex consisting of ribosome, mRNA template, tRNA, and incomplete polypeptide chain can be isolated from cell-free extracts actively engaged in protein synthesis. This complex has been widely studied in recent years. The dissociation of 70 S ribosomes at low Mg^{2+} concentrations liberates tRNA and mRNA from the complex and partially releases a compound containing both the growing polypeptide chain and a molecule of tRNA. The tRNA molecule is esterified to the

3. Ribosomes and the Synthesis of Proteins

terminal carboxyl group of the polypeptide (see Fig. 5-3, p. 462). This substance, called *peptidyl-tRNA*, can be completely extracted from *E. coli* ribosomes with a solution of sodium dodecylsulfate after prior incubation in the presence of ^{14}C-labeled amino acids (18). Peptidyl-tRNA has also been obtained from ribosomes of rabbit reticulocytes by both Gierer and von Ehrenstein. It is not difficult to show that these intermediates contain both polynucleotide and polypeptide moieties by subjecting them to electrophoresis at pH 4.5. At this pH most proteins are uncharged and therefore relatively immobile, but nucleic acids bear one negative charge per residue and move rapidly toward the anode. Peptidyl-tRNA also migrates toward the anode, although somewhat more slowly than pure polynucleotides. The nature of these intermediates can also be demonstrated by treatment with 0.001 N base (pH 11) which cleaves the ester bond linking the polypeptide with tRNA. The two components can then be identified separately.

Gilbert studied the peptidyl-tRNA intermediate formed during synthesis of polyphenylalanine on the synthetic template, polyuridylic acid (poly U) (19). Using ^{32}P-labeled tRNA and ^{14}C-labeled phenylalanine to accurately quantitate the results, he concluded that ribosomes participating in protein synthesis each bind one molecule of tRNA which is itself attached to the end of the growing polypeptide chain. In other words, at any given moment each 70 S ribosome takes part in the synthesis of a single protein chain and every incomplete protein is anchored to the ribosome through a bound tRNA molecule. Besides peptidyl-tRNA, each ribosome also binds a molecule of aminoacyl-tRNA bearing the next amino acid residue to be inserted into the peptide. Aminoacyl-tRNA is bound less firmly to the ribosome than peptidyl-tRNA and it can be exchanged with homologous, isotopically labeled tRNA, or even washed off completely with buffer at 10^{-2} M Mg^{2+}.

When the inhibitor puromycin is added to a cell-free system, protein synthesis rapidly ceases and short, newly formed polypeptide chains are released from the ribosomes. These are not peptidyl-tRNA molecules, however, but polypeptides with puromycin covalently bound to their C-terminal ends. This is consistent with our previous observation that puromycin is structurally analogous to the aminoacyl end of tRNA and evidently, a specific antagonist of it (p. 455).

Some of the enzymes involved in peptide bond synthesis can be dissociated from the ribosomes and fractionated on DEAE-Sephadex [Lipmann and co-workers (20)]. Two enzymes, T and G, have been isolated and found to catalyse separate reactions; their sedimentation coefficients are 5.1 S and 5.5 S, respectively (21). T has been further separated into two fractions, stable T$_s$ and unstable T$_u$, which may, however, be subunits of a single enzyme. Enzyme T is essential for formation of the complex between aminoacyl-tRNA and the ribosome, a step requiring the dephosphorylation of one molecule of GTP. Purified enzyme G alone acts as a GTPase in the presence of washed ribosomes. In the active complex, GTP hydrolysis appears to be coupled with the translocation of peptidyl-tRNA and it was shown that under optimal conditions, a mole of peptide bonds is synthesized at the expense of two moles of GTP converted to GDP (22). Additional enzymatic steps are of course entirely possible; it is already certain, for instance, that the final peptidyl transferase reaction is catalyzed by an enzyme firmly bound to the 50 S ribosomal subunit.

Our current knowledge about the mechanism of protein synthesis is summed up by the scheme presented in Fig. 5-3. Although much of it is still speculative, we can be

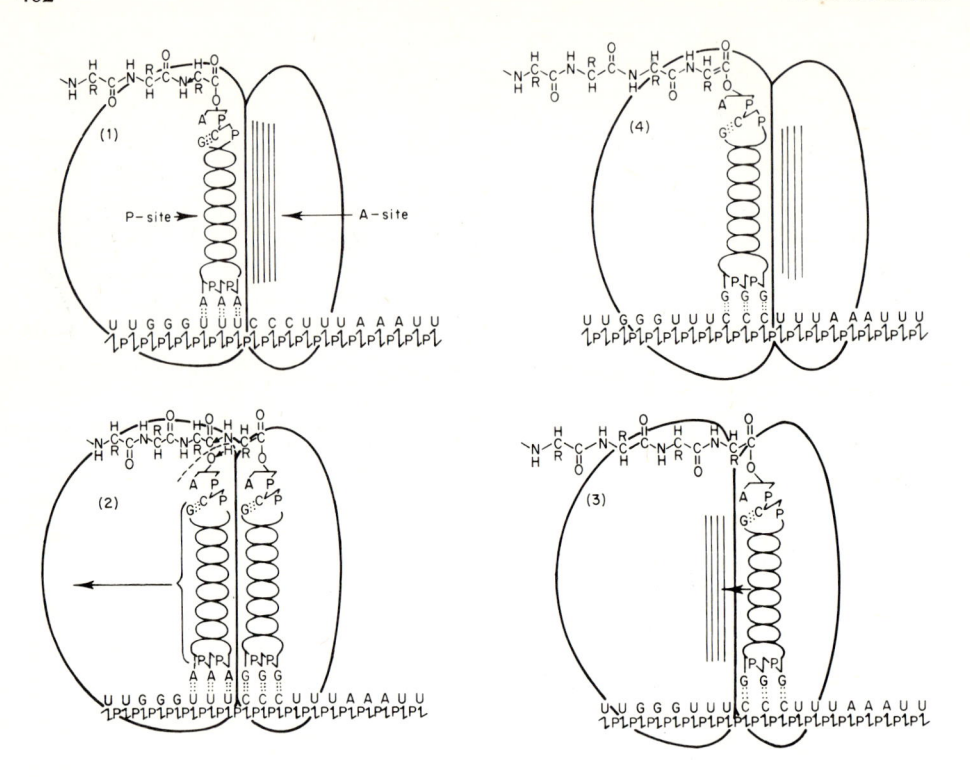

Fig. 5-3. The peptide condensation cycle. A ribosome with peptidyl-tRNA at its P-site (1) binds aminoacyl-tRNA to its A-site (2) via specific anticodon-codon interaction with the mRNA; the peptidyl residue is next condensed with the aminoacyl-tRNA (3), and the lengthened peptidyl-tRNA undergoes translocation from A-site to P-site (4), preparing the ribosome for another cycle. Adapted from Lipmann (23, p. 177).

sure that the peptide condensation cycle consists in essence of four stages. In the first stage, the growing polypeptide chain in the form of peptidyl-tRNA occupies a specific site on the 50 S subunit which we will call the *peptidyl-* or *P-site* and the mRNA template is bound specifically to the 30 S subunit. If chain growth is halted before completion of the protein and the 70 S ribosomes are dissociated by reducing the magnesium ion concentration to 10^{-4} M, the mRNA is released from the complex, but a substantial fraction of the nascent peptidyl-tRNA remains bound to the 50 S particles. This result provides evidence that there exists a special site for peptidyl-tRNA on the 50 S subunit.

Isolated 50 S subunits, however, bind neither peptidyl-tRNA nor aminoacyl-tRNA, although isolated 30 S subunits spontaneously form complexes with both of these compounds in the presence of mRNA containing a codon complementary to the tRNA anticodon. The complex of peptidyl-tRNA and mRNA with the 30 S subunit is quite stable and that with the aminoacyl-tRNA relatively labile. From this it may be concluded that there is a second site for tRNA located on the 30 S subunit and we will call it the *aminoacyl-* or *A-site*. Here the anticodon of the appropriate aminoacyl-tRNA is apparently "recognized" by the mRNA triplet specifying the next amino acid residue in the protein chain. Addition of 50 S subunits renders the complex more stable and provides a second site for the proper alignment of incoming and outgoing tRNA's. Without these two sites, it would be difficult to envision how peptide bond synthesis on the mRNA template could take place.

The second stage of the cycle involves binding of the aminoacyl-tRNA corresponding to the next amino acid in the polypeptide chain to the A-site on the 30 S subunit by specific interaction between tRNA anticodon and mRNA codon. This completes formation of the complex which can carry out peptide bond synthesis. The A-site can also be occupied by a molecule of free, nonaminoacylated tRNA, although free tRNA is not generally present in the cell under normal conditions.

In the third stage of the cycle, the peptidyl moiety is condensed with the aminoacyl-tRNA and a peptide bond is formed. The peptide is transferred from the P-site to the A-site in this step and at the same time, the tRNA at the P-site is cleaved from the peptide and then released from the ribosome. Recent data suggest that transfer of the incomplete polypeptide from the outgoing tRNA to the aminoacyl group of the incoming tRNA molecule is catalyzed by a protein provisionally called peptidyl transferase which is integrated into the structure of the 50 S ribosomal subunit (24). Alternatively, one of the supernatant factors discovered by Lipmann may become tightly bound to the ribosome at this stage so as to promote the next reaction.

The fourth and last stage consists of *translocation* of the lengthened peptidyl-tRNA chain from the A-site to the P-site. Simultaneously, the mRNA is advanced by one nucleotide triplet, thus exposing a new triplet at the A-site which codes for the next amino acid in the chain. Translocation of peptide and neucleotide chains require the hydrolysis of one molecule of GTP, probably catalyzed by supernatant factor G, which has been shown to possess GTPase activity. Both translocation and mRNA advancement must be energy-requiring mechanochemical processes based on conformational transformations in specific ribosomal proteins. The energetics and mechanism of this putative reaction remain among the most interesting problems outstanding in the study of ribosome function. After the fourth stage, the protein(s) that accomplish the mechanochemical reaction return to their original state and the cycle begins anew.

Two vital steps in protein synthesis are *initiation* and *termination* of the polypeptide chain (p. 438ff). Initiation requires the presence of a special tRNA which transports N-formylmethionine and which is bound by the codons AUG or GUG (25). Natural mRNA's such as RNA from phage f2 and synthetic messengers such as random poly AUG and poly UG both stimulate the formation of a specific complex of the 30 S subunit with N-formylmethionyl-tRNA when protein initiation factors are also added to the reaction mixture. Other aminoacyl-tRNA's will not bind to the 30 S subunits at the low Mg^{2+} concentrations thought to prevail within the cell even if the appropriate codon is present. This complex is similar to the one formed by peptidyl-tRNA and the 30 S subunit already described. If a 50 S subunit is next bound to the initiating complex, the N-formylmethyionyl-tRNA is shifted to the P-site, leaving the A-site free for occupancy by an aminoacyl-tRNA, and the first peptide bond in the chain can be synthesized.

Formation of the initiation complex occurs in this manner *in vitro* at a magnesium ion concentration of 3×10^{-3} M, presumed to correspond to actual physiological

464   V. RNA Function

conditions. To initiate polypeptide synthesis on artificial messengers not containing the specific codons AUG and GUG, it is necessary to use Mg^{2+} concentrations on the order of 10^{-2} M or higher. Although this overcomes the problem of initiation, it simultaneously raises the frequency of errors in translation of the mRNA. However, after initiation has occurred under these conditions, the magnesium ion concentration can be reduced to 5×10^{-3} M without interfering with further chain growth.

Although little is known about the final stages of polypeptide synthesis, it is clear that chains can terminate at the nonsense codons UAG, UAA, and UGA. Furthermore, these codons must transmit a chemical signal and cannot act simply by stopping

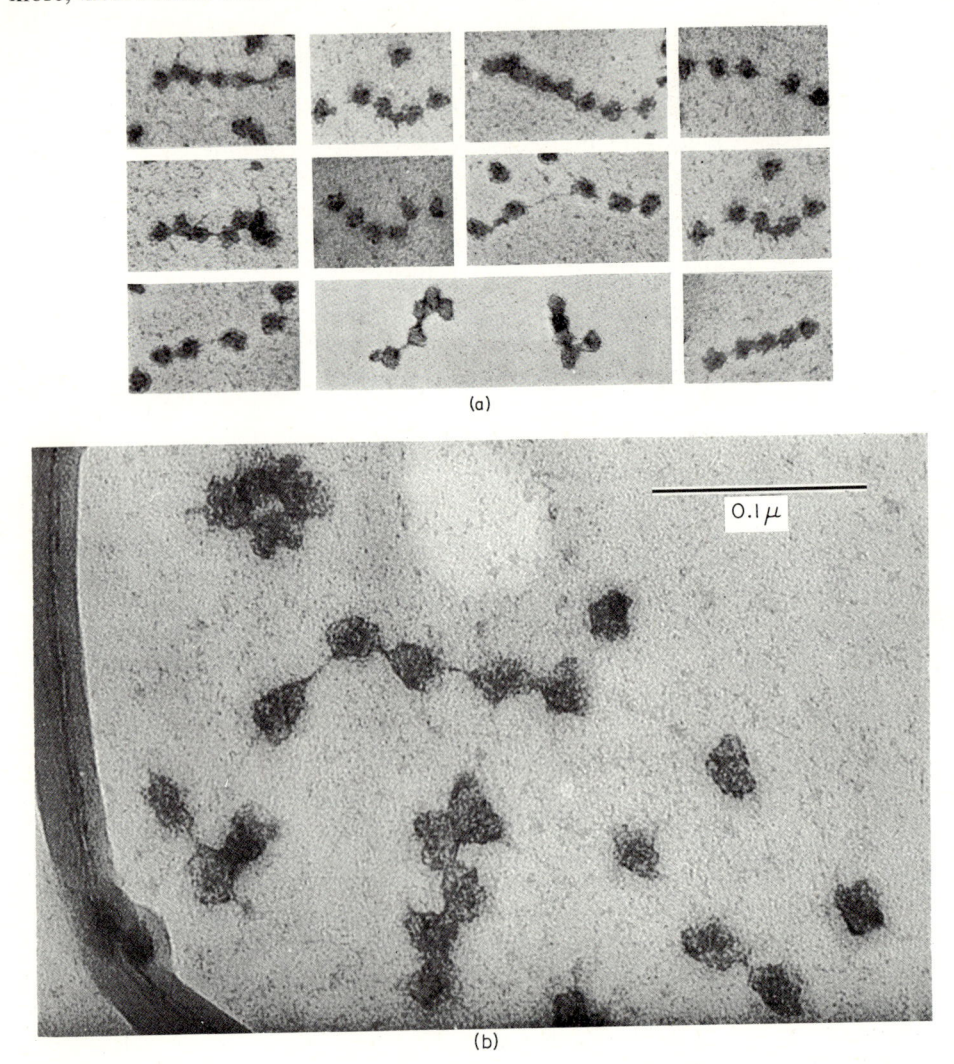

Fig. 5-4. Polyribosomes from rabbit reticulocytes. Particles stained in positive contrast with uranyl acetate. (a) Shows the polysomes in linear array with the ribosomes separated by a thin strand; ×100,000. (b) Micrograph showing three pentamers in various configurations; ×280,000. From Slayter *et al.* (29).

protein synthesis (26). Deficiency in an aminoacyl-tRNA required to continue chain propagation can bring protein synthesis to a halt at some other codon, but it does not cause either termination or cleavage of the ester bond in peptidyl-tRNA. Nonsense codons do provoke this reaction. In addition, a supernatant protein has been isolated that can specifically cleave the necessary ester link in a substrate consisting of the ribosome–mRNA–peptidyl-tRNA complex after chain growth has been halted by one of the three nonsense triplets (27). While this protein is required for termination, it is not known what other components are required for the reaction and, in particular, the existence of a terminating tRNA which can recognize nonsense codons has not been ruled out.

After the central role of ribosomes in protein biosynthesis had been understood for some time, it was discovered in a number of laboratories (Lipmann, Rich, Gierer, and their colleagues, 28), that several ribosomes can be associated with one molecule of messenger RNA at the same time. These ribosomal aggregates, called *polyribosomes* or *polysomes*, are observed both in living cells and in cell-free systems, wherever protein synthesis is taking place. By centrifugation in sucrose gradients, it was shown that polysomes containing up to twenty or more 70 S ribosomes can be extracted from lysed cells, and that they retain their functional activity during the isolation procedure. Excellent photographs of polysomes have been obtained with the electron microscope (Slayter *et al.*, 29) and some of them are presented in Fig. 5-4.

By adding radioactively labeled ribosomes to a cell-free system, it has been possible to demonstrate that "old" ribosomes are exchanged for "new" in the polysomal complexes and that this turnover is functionally associated with protein synthesis (30). The mRNA strand, which can serve as template for one or several proteins, contains from 1,000 to 10,000 nucleotide residues with an end-to-end length between 1 and 10 μ. It is natural to suppose that ribosomes move along the template, reading out the coded information for protein structure as it goes. In this way, up to a dozen or more ribo-

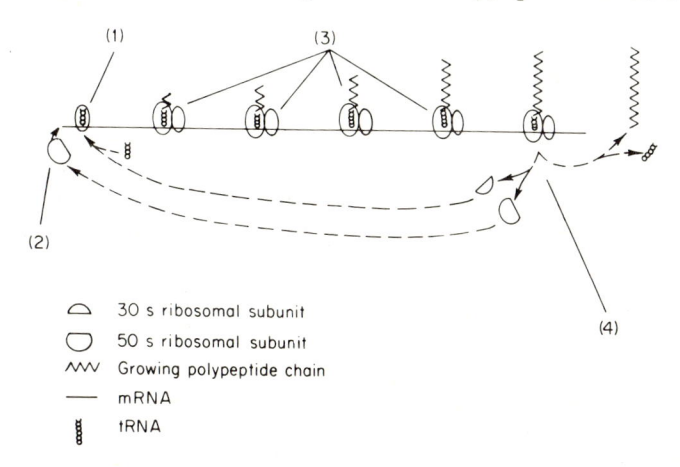

Fig. 5-5. Schematic representation of polyribosome during protein synthesis. (1) 30 S ribosomal subunit combines with mRNA and *N*-formylmethionyl-tRNA to form initiating complex. (2) A 50 S ribosomal subunit joins the complex, permitting protein synthesis to begin. (3) Polypeptide chain is lengthened as ribosome moves along the mRNA template from left to right. (4) Protein synthesizing complex dissociates upon completion of polypeptide chain.

somes can carry out polypeptide synthesis at the same time on a single template. Within a small time interval, a portion of the 70 S monomers complete their work and slip off the end of the template as 30 S and 50 S subunits, while other subunits assemble to begin synthesis of a new protein chain at the opposite end (Fig. 5-5). The exchange of 30 S and 50 S partners during polypeptide synthesis has been shown by experiments with density-labeled ribosomes. The relative motion of ribosomes and mRNA template must be attributed to a mechanochemical process which perhaps involves a configurational change within the ribosome, although there is as yet no concrete evidence for this.

An important problem in the general scheme of protein synthesis is the direction of motion of the ribosomes along the mRNA, or, in other words, the direction in which the code is read by the protein synthesizing apparatus. This has recently been demonstrated by investigating the nature of the short oligopeptides synthesized on small synthetic templates (31). Thus the hexanucleotide template A_3U_3:

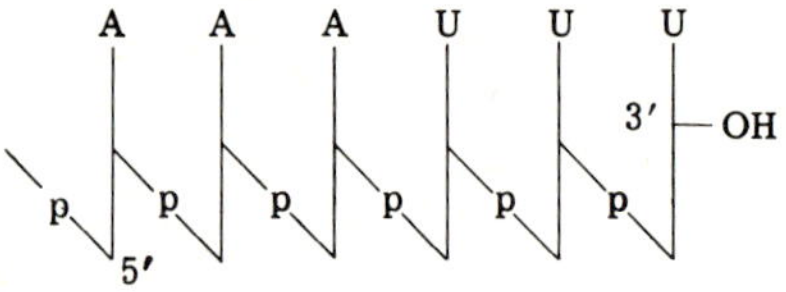

directs the synthesis of the dipeptide lysyl-phenylalanine:

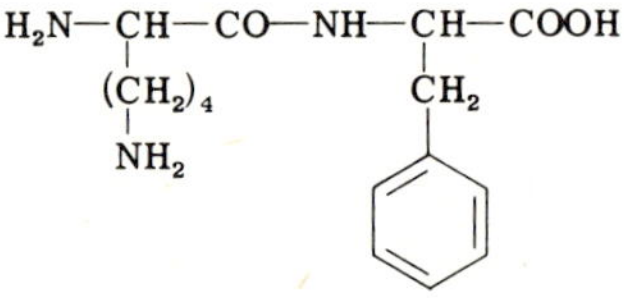

in the cell-free system. The dipeptide phenylalanyl-lysine was never made, however. As will be seen in the next section, synthesis of the peptide chain always starts from the N-terminal end and moves toward the C-terminal residue. Therefore, the *in vitro* experiment described above unambiguously shows that the polynucleotide coding sequence is read from the 5'-end to the 3'-end of the template. This direction is the same as that in which RNA is transcribed from DNA.

Ribosomes affect not only the rate and presumably, the direction, of protein synthesis, but also the way mRNA codons are read by their anticodons in tRNA molecules. By changing the conditions under which polypeptide synthesis occurs on a synthetic template, it is possible to provoke erroneous translation of the genetic code. The first intimation of this was provided by Gorini and his co-workers (32), who observed that coding mistakes occurred in the presence of streptomycin, a metabolic inhibitor that specifically acts on ribosomes. A polyuridylic acid (poly U) template normally stimulates the incorporation of phenylalanine into polyphenylalanine in the presence of ribosomes from a strain of *E. coli* sensitive to the drug. When streptomycin is added, however, substantial amounts of leucine, isoleucine, tyrosine, and serine are incorporated in addition to phenylalanine. Ribosomes obtained from streptomycin-resistant mutants of *E. coli* were insensitive to the inhibitor *in vitro*, and no misreading occurred in its presence. Subsequently, it was shown that low and high temperatures, high concentrations of Mg^{2+} ions (3×10^{-2} M), mildly alkaline conditions, the presence of polyamines and alcohols, and other drugs similar to streptomycin all lead to errors in coding (33–36). The nature of some of the induced errors is indicated in Table 5-2.

TABLE 5-2

Errors of Amino Acid Incorporation in Cell-Free Systems from Bacteria

Source	Agent inducing error	Polynu- cleotide	Amino acid normally incorporated	Amino acids misincorporated in presence of agent[a]	Ref.
Escherichia coli	High Mg^{2+}	Poly U	Phe	Leu > Ile > Ser > Tyr	(33, 34)
	Low temperature	Poly U	Phe	Leu > Ile > Ser > Tyr	(33)
	High pH	Poly U	Phe	Leu > Ile > Ser	(34)
	Ethanol	Poly U	Phe	Leu, Ile	(35)
	Aminoglycoside antibiotics such as streptomycin	Poly U	Phe	Ile > Ser > Tyr > Leu	(36)
	Ethanol	Poly C	Pro	Thr, Leu	(35)
	Aminoglycoside antibiotics such as streptomycin	Poly C	Pro	Ser > Thr > His >> Ala >> Leu	(36)
	Aminoglycoside antibiotics such as streptomycin	Poly A	Lys	Val > Arg	(36)
Bacillus stearother- mophilus	High Mg^{2+}	Poly U	Phe	Leu	(37)
	Low temperature[b]	Poly U	Phe	Leu	(37)

[a] Order corresponds to magnitude of effect.
[b] These bacteria normally grow at 65°C; assay carried out at 37°C.

Similar results were obtained by Weinstein and Friedman in studies on ribosomes derived from the thermophilic bacterium, *Bacillus stearothermophilus*, which grows optimally at 65°C (37). At the optimum temperature for growth, extracted ribosomes incorporate only phenylalanine under the direction of a poly U template; but at 37°C, leucine is "incorrectly" incorporated in addition to the "correct" polymerization of phenylalanine. This effect of temperature is opposite to that found with ribosomes from mesophiles.

By studying the nature of the coding errors induced by various abnormal environmental conditions, it is clear that the amino acid substitutions are not by any means random. Different conditions lead to identical errors. This becomes more understandable when we note that the codons for isoleucine (AUU), leucine (CUU), tyrosine (UAU), and serine (UCU) are distinguished from that of phenylalanine (UUU) by just one letter per triplet. Under abnormal conditions, the fidelity of translation is reduced and the complementarity between two bases in the codon and anticodon appears sufficient for binding tRNA to the template. This situation can most likely be attributed to conformational changes in the ribosomes themselves induced by streptomycin, nonoptimal temperatures, high pH, and so on. In particular, conformational changes at or near the A-site on the 30 S ribosomal subunit could perturb the structure of the template RNA by altering distances between neighboring bases. Binding of complementary codon and anticodon triplets would become impossible and a transition would be made to binding through doublets. Whether or not this explanation is correct, it is evident that ribosomes actively participate in protein synthesis and that they are capable of influencing the recognition of codon and anticodon in particular. Unfortunately, very little more is known about the relationship of ribosome structure to its function.

Studies on the dissociation of structural proteins from ribosomal subunits have yielded valuable information on ribosome assembly and on the function of the individual components. These relatively insoluble proteins can be separated from their rRNA without denaturation by exposing the ribosomes to high concentrations of monovalent cations such as Li^+ and Cs^+. A partial dissociation of one third to one half of the ribosomal proteins has been performed with concentrated ($\approx 5\ M$) solutions of CsCl in several laboratories (38). Even though a substantial portion of the ribosomal protein is split off in CsCl, the resulting ribonucleoprotein cores retain a compact structure as judged from their sedimentation behavior. The remaining ribosomal proteins can be removed from the protein-deficient cores with a mixture of $4\ M$ urea and $2\ M$ LiCl. Ribosomal RNA is precipitated by this treatment while the proteins remain in solution.

Investigation by polyacrylamide gel electrophoresis (see Fig. 1-53, p. 117) indicates that most of the ribosomal proteins are different from one another and that their evident heterogeneity does not merely result from aggregation. This has been confirmed by the isolation of many of these proteins in pure form. After partial hydrolysis, all ribosomal proteins that have been examined yield distinctive sets of peptides.

The complete reconstitution of ribosomal subunits takes place if purified rRNA and the necessary ribosomal proteins are mixed together under carefully controlled conditions. In particular, Nomura and his colleagues showed that 30 S ribosomal particles with full functional activity can be obtained when totally dissociated components from native *E. coli* 30 S subunits are incubated together at 43°C in an appropriate buffer (39). Reconstitution is considerably less efficient at lower temperatures. Particles are apparently formed as the result of spontaneous *self-assembly* when the urea and salts required to solubilize dissociated ribosomal proteins are removed by dialysis. It is supposed that each protein is bound to a strictly determined site on the ribosomal RNA. These remarkable results leave little doubt that ribosomes *in vivo* are also formed by self-assembly and that their structure represents an equilibrium configuration with minimum free energy.

The interaction between ribosomal proteins and rRNA is very specific. Reconstitution of active *E. coli*-type 30 S subunits occurs only from *E. coli* 16 S rRNA and the proteins originally derived from native 30 S subunits. Proteins from 50 S subunits do not associate with 16 S rRNA to produce active particles. Furthermore, use of fragmented *E. coli* 16 S or 23 S rRNA does not lead to the formation of 30 S subunits, nor does 16 S rRNA from unrelated organisms, in the presence of *E. coli* 30 S proteins. If certain ribosomal proteins are withheld from the mixture, however, self-assembly can take place, although full functional activity may not be restored. Artificial ribosomes deficient in one specific protein can thus be used to examine the way in which protein synthetic activity is altered by deletion of that component. The function of the missing protein can then be inferred.

Using such techniques, Traub and Nomura have established that in bacteria a mutant ribosomal protein is responsible for the acquisition of resistance to the antibiotic streptomycin (40). The mutation to streptomycin resistance was previously known to alter the ribosomes since the antibiotic has different effects on *in vitro* amino acid incorporation by ribosomes from streptomycin-sensitive and streptomycin-resistant cells. But not until a specific 30 S protein from resistant ribosomes produced resistant "hybrid" 30 S subunits when reassociated with 16 S rRNA from a sensitive strain was it known that the mutation affected a protein rather than the rRNA itself. In the future, similar studies can be expected to pinpoint the ribosomal components responsible for such varied functions as the binding of aminoacyl-tRNA and peptidyl-tRNA, and the movement of mRNA along the ribosomal surface.

4. The Rate and Direction of Protein Synthesis

Experiments on the kinetics of cellular processes have been used to solve many basic problems in molecular biology. When we study biological phenomena, particularly their biochemical aspects, we generally deal with *steady-state* processes. This means that cellular reactions occur at constant rates or, at least, that changes in rate are very slow, with a time lag on the order of the generation time of the cells.

Consider now how we may investigate the kinetics of individual cellular processes in a bacterial culture growing under steady-state conditions. A pulse of radioactively labeled metabolite is administered to the culture for a certain time interval and then removed by dilution or centrifugation. While it is present, the labeled component is incorporated into proteins, nucleic acids, or other cellular constituent, depending on its nature. If the time of contact between the cell culture and the radioactive component is large, then the substances into which the component is incorporated become uniformly labeled. But if the time of exposure is small compared to the generation time, the distribution of the radioactive component may be far from uniform. Using the latter technique, frequently called *pulse-labeling*, we can study transient processes in the cell and measure their kinetic constants without materially disrupting the overall steady-state growth of the culture.

A second method of determining the kinetics of cellular synthetic processes consists of upsetting steady-state growth of the cells by conjugation, introduction of inhibitors, phage infection, or enzyme induction (p. 486ff). Suppose that an F^- bacterial cell cannot produce protein of a given type. Within 3 to 5 minutes after the introduction of a functional copy of the cistron from a Hfr donor during conjugation, the recipient cell begins to synthesize the corresponding protein at a normal rate. The processes which occur in these few minutes before synthesis of the protein is fully established are of a transient, rather than a steady-state, nature. An investigation into these processes yields valuable information about the rates of individual stages in protein synthesis. As another example, we know that bacteriophage infection is followed by a period of several minutes in which all synthetic processes in the cell are reorganized. A cell which is penetrated by phage DNA begins to synthesize new enzymes, phage DNA, and then structural proteins from which progeny phage particles are assembled.

The central questions of protein synthesis to which kinetic techniques have been applied include the following: Does synthesis of a protein macromolecule occur by successive addition of amino acids from one end of the chain to the other, or by the simultaneous assembly of all residues? If synthesis is unidirectional, at which end is it initiated? How long does the synthesis of one protein macromolecule require?

Hemoglobin was the first protein to be extensively studied in this way. This substance invited investigation because of its important biological role, the ease with which it may be purified, and the fact that it is the only protein synthesized in significant quantities by reticulocytes. The manner in which hemoglobin is assembled was studied with radioactively labeled amino acids (41). It had already been established that the N-terminal residues (those bearing free α-amino groups) of both α- and β-chains of hemoglobin is valine. The kinetic experiment was set up in the following manner. A preparation of washed reticulocyte ribosomes was incubated in medium with a complete set of amino acids, including ^{14}C-labeled valine. After a relatively short period of synthesis, the newly formed protein was extracted and its total radioactivity was determined. The radioactivity of N-terminal valine was also measured by reacting the polypeptide chains with FDNB or phenylisothiocyanate, after which the end groups were cleaved and purified chromatographically. Now, a molecule of rabbit hemoglobin consists of 670 amino acid residues, 46 of which are valine. Four of the valine residues occupy N-terminal positions, one in each of four polypeptide chains which together comprise a hemoglobin molecule. Hence terminal valine makes up 8.7% of the total valine in the protein. In a control experiment in which hemoglobin was synthesized by reticulocytes under steady-state conditions in a medium containing ^{14}C-valine, the fraction of N-terminal valine was found to be 8.6%, in excellent agreement with the analytical data. When reticulocyte ribosomes were incubated for a very short time in the presence of ^{14}C-valine, the results seemed to be paradoxical: labeled hemoglobin was obtained, but none of the N-terminal valine residues were radioactive. This result can be better understood by consulting the model for protein synthesis illustrated in Fig. 5-6.

Let us assume that ribosomes contain templates upon which polypeptide chains gradually increase in length from one end to the other. Since synthesis of a complete polypeptide chain requires an appreciable period of time, the templates are always partially occupied by growing, but not yet completed, molecules. On the average, the templates are 50% occupied by growing chains under steady-state conditions, although the number of completely free templates at any particular time is equal to zero for all practical purposes.

This model was confirmed by an experiment in which cells were first permitted to reach the steady-state on a medium containing ^{14}C-valine. Ribosomes were then extracted and incubated for 10 minutes in nonradioactive medium. In this case, labeled N-terminal groups account for 13% of the ^{14}C-valine released from the ribosomes in complete hemoglobin molecules. From this figure, it can be estimated that the hemoglobin molecules finished during the final 10-minute incubation *in vitro* were already 63% complete before the ribosomes were extracted. It is clear that if the

4. The Rate and Direction of Protein Synthesis

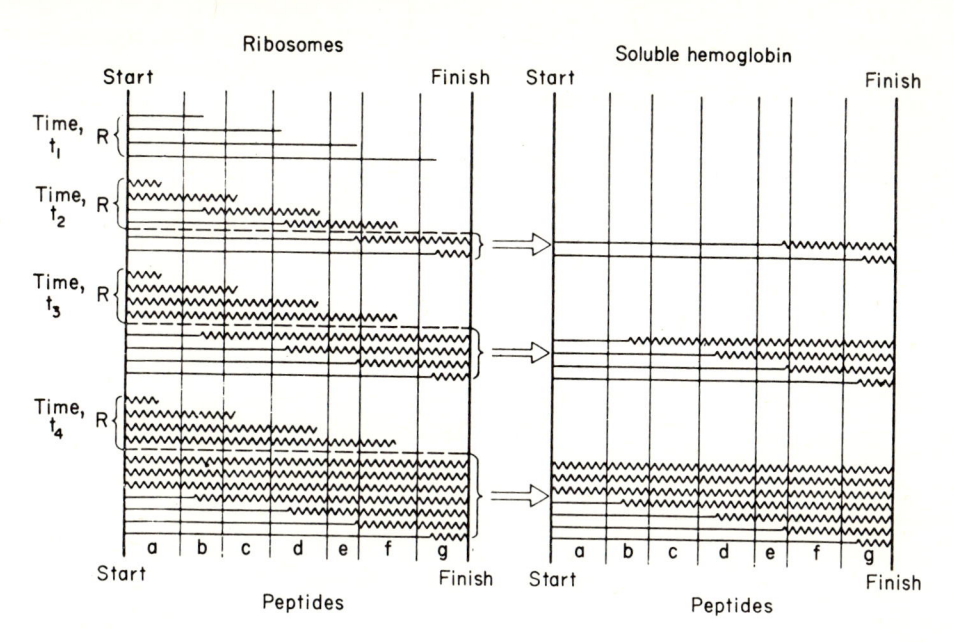

Fig. 5-6. Model of sequential synthesis of hemoglobin chains on the ribosome. The straight lines represent unlabeled polypeptide chains; the zigzag lines represent radioactively labeled polypeptide chains formed after the addition of radioactive amino acids at time t_1. The letter R indicates unfinished chains attached to the ribosomes at each time. At time t_2, for instance, the top two completely zigzag lines represent polypeptide chains formed completely from amino acids during the time interval between t_1 and t_2. The middle two lines represent chains which have grown during the time interval but have not yet reached the finish line and are therefore still attached to the ribosomes. The bottom two chains represent those which have crossed the finish line, left the ribosomes, and are to be found mixed with other molecules of soluble hemoglobin in the cell. From Dintzis (42).

in vitro incubation time is divided into two equal intervals, molecules which leave the template in the first 5 minutes are nearest to completion and naturally contain the most ^{14}C-valine. Molecules leaving the template during the second 5 minutes contain less ^{14}C-valine, since they were further from completion upon removal of the labeled amino acid. But terminal valine should account for a larger percent of the total label in this case. This prediction was substantiated experimentally when it was found that 16% of the ^{14}C-valine occupied terminal positions in the second sample. This means that the corresponding macromolecules were 45% complete just prior to incubation in nonradioactive medium. The average extent to which the templates are occupied under steady-state conditions of protein synthesis is estimated from these data to be 50%, in full agreement with the predictions of the model.

The experiments described above were confirmed and supplemented by the even more convincing results of Dintzis (42), who also studied the kinetics and polarity of hemoglobin synthesis in intact reticulocytes. The medium contained ^{14}C-leucine throughout the incubation and consequently, all 32 leucine residues per hemoglobin molecule were labeled uniformly with this isotope. A pulse of high specific activity ^{3}H-leucine was then administered to the reticulocytes for various periods of time ranging from 15 seconds to several minutes. At the end of the ^{3}H-leucine pulse, the cells were lysed by osmotic shock and fully synthesized hemoglobin molecules were purified. The α- and β-chains were separated by ion exchange chromatography and then subjected to trypsin hydrolysis. The 35 peptides which resulted were resolved from one another by the method of fingerprinting, each spot was eluted into individual tubes, and each of the tryptic peptides was assayed for both ^{3}H and ^{14}C radioactivity. The ^{14}C radioactivity gave the number of leucine residues in the given peptide. Tritium radioactivity showed the degree of penetration of leucine into the peptide during the selected time interval.

The results which Dintzis obtained were extremely clear. All of the tryptic peptides could be arranged in linear order according to the increase in relative pulse label (^{3}H:^{14}C ratio), as can be seen in Fig. 5-7. The pattern of incorporation is in

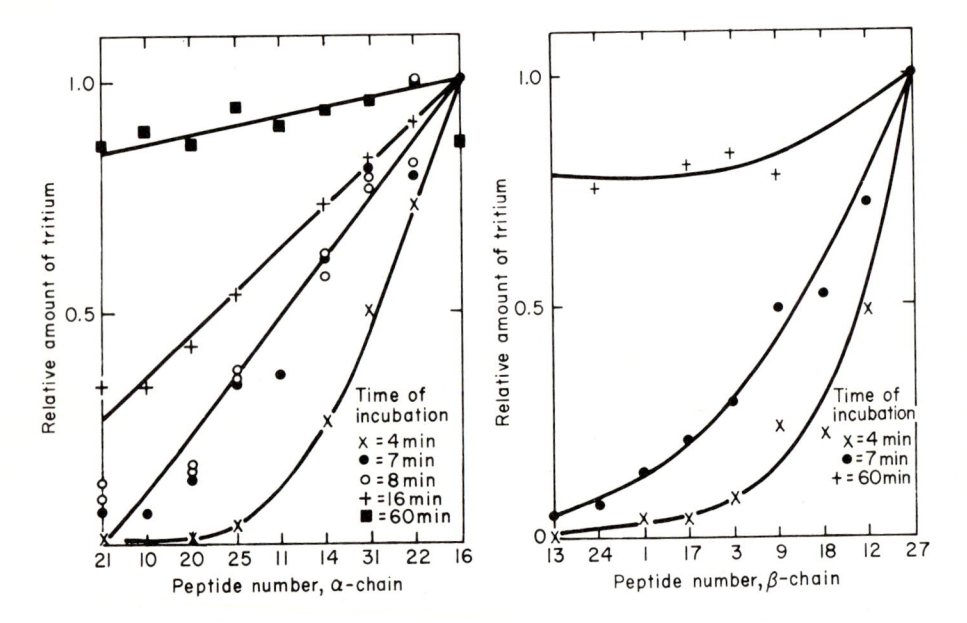

Fig. 5-7. Distribution of ^{3}H-leucine among tryptic peptides of soluble rabbit hemoglobin after various times of incubation at 15°C. The gradient in radioactive label establishes the sequence of peptides in the hemoglobin chain and demonstrates that the protein molecules are synthesized sequentially from one end to the other. From Dintzis (42).

complete agreement with the model depicted in Fig. 5-6. During a given time interval, all growing chains are extended by a certain number of amino acid residues. A fraction of the molecules on the template will be finished during the time interval, and those closest to completion will contain tritium label only in the final peptides. As the interval is increased, label appears in peptides nearer to the starting point of the chain, although the terminal peptides are always more densely labeled.

These experiments prove that the protein chain grows by successive addition of amino acid residues from one end of the molecule to the other. If growth of the chain had occurred from many points, it would have been impossible to arrange all the peptides in a unique sequence according to increases in relative ^{3}H radioactivity.

Furthermore, from the known primary structure of hemoglobin, it was demonstrated that protein synthesis begins with the N-terminal residue and proceeds toward the C-terminal residue. Finally, the sequence of peptides established by this work coincides precisely with that derived for hemoglobin by purely chemical techniques.

The method developed by Dintzis makes it possible in principle to ascertain the correct sequence of peptides from any limited protein hydrolyzate by combining pulse-labeling techniques with fingerprint analysis. The experiment also yields several important kinetic constants. The longest time required for the incorporation of radioactivity into all peptides of the completed chain permits one to find the synthetic time of a single polypeptide chain. From the net rate of hemoglobin production, we may easily calculate that one cell simultaneously synthesizes between 1000 and 2000 molecules. To summarize, each one of the 1500 to 2000 ribosomes per reticulocyte cell produces one molecule of protein every 1.5 minutes.

Let us now turn to the more difficult case of protein synthesis in bacterial cells. We must first briefly digress on the protein and nucleic acid "economy" of an *E. coli* cell. Protein amounts to 60% of the bacterial dry weight, while DNA constitutes about 5% and RNA from 10 to 25%, depending on the growth rate. Two thirds of the cellular protein, or 40% of the cell weight, consists of soluble proteins situated in the cytoplasm. About 10% of the protein is found in the cell wall and membrane, and another 5 to 15% is located in the structural proteins of cytoplasmic particulates, mainly ribosomes. If an average *E. coli* protein is assumed to have a molecular weight of 50,000, a single cell contains on the order of a million protein molecules, while the number of different kinds of proteins may reach 1000 to 2000. On the average, there should be 500 to 1000 molecules of each kind, although certain proteins are undoubtedly represented by much fewer copies. These figures must hence be taken as very approximate.

Up to 85% of the total bacterial RNA is located in 70 S ribosomes, of which there are from 5000 to 10,000 per cell. Ribosomes can be isolated by centrifugation from an aqueous solution at 100,000 g in the course of 1 to 2 hours. These particles can also be very effectively purified and studied by centrifugation from a thin layer into a sucrose density gradient. A concentration gradient is first established in a centrifuge tube prior to sedimentation, usually running from 5% sucrose at the top to 20 or 30% at the bottom. The sample is layered over the top and the gradient is centrifuged. During the run, components of the sample sediment into the sucrose at different rates depending on their size and shape. Owing to the great stability of the sucrose gradient to convection, the distribution of particles will be maintained when the centrifuge is stopped, and sequential layers of the gradient fluid can then be separated and analyzed individually.

Roberts and his colleagues designed kinetic experiments to determine the rate of bacterial protein synthesis *in vivo*, using sucrose gradient analysis (43). They followed the incorporation of radioisotopes into ribosomes and their subsequent release into the cytoplasm as complete proteins. Since the cellular proteins were all precipitated together with trichloroacetic acid (TCA) for analysis of radioactivity, no information was available about the synthesis of specific types of proteins. In addition, bacterial cells are capable of doubling their weight every 30 minutes so that protein synthesis must proceed much more rapidly than in reticulocytes, requiring on the order of several seconds rather than several minutes as in the case of hemoglobin.

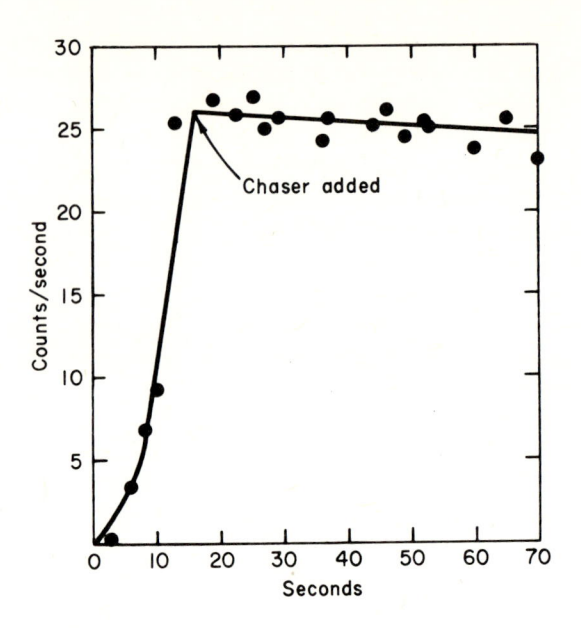

Fig. 5-8. Kinetics of incorporation of $^{35}SO_4^{2-}$ into the protein of whole *E. coli* cells. Radioactive label was added at zero time and a nonradioactive $^{32}SO_4^{2-}$ "chaser" at 16 seconds. Note prompt incorporation of the tracer and rapid cessation of incorporation after addition of excess nonradioactive sulfate. From McQuillen *et al.* (43).

The kinetics of $^{35}SO_4^{2-}$ incorporation into total protein were first measured by precipitating aliquots of the culture with hot TCA (Fig. 5-8). In order to limit the incorporation of ^{35}S during a short pulse, an excess of the sulfur-containing amino acids, methionine and cysteine, as well as unlabeled SO_4^{2-}, were added to the culture after 16 seconds to dilute the ^{35}S radioactivity. The incorporation of ^{35}S stops abruptly within a few seconds after the nonradioactive "chaser" is added. The size of the sulfate pool in *E. coli* must therefore be insignificant and, moreover, penetration of the cell membrane and diffusion through the cytoplasm do not at all limit the kinetics of sulfur incorporation.

A more detailed analysis of events taking place in the cell was carried out in the following manner. An *E. coli* culture was first grown on enriched medium at 37°C, and then transferred to a medium without sulfur. After 30 minutes of starvation, the cells exhaust their reserve pool of sulfur entirely, and growth comes to a halt. Radioactive ^{35}S was then introduced into the culture in the form of $Na_2^{35}SO_4$. After 15 seconds, the bacterial suspension was rapidly cooled to 0°C in order to stop, or greatly reduce, the incorporation of labeled sulfur into proteins, and the cells were collected by centrifugation. They were next broken open by freezing and thawing in the presence of lysozyme and the cell juices were freed of unlysed cells, cell membranes, and other insoluble matter by another round of centrifugation. The soluble extract was then centrifuged through a sucrose gradient for 75 minutes at 37,000 rpm, as described above. The gradients were fractionated at the end of the run, and the radioactivity and absorption at 260 mμ were measured in each sample. From the results presented in Fig. 5-9a, we see

4. The Rate and Direction of Protein Synthesis

that the ribosomes sediment quite far into the gradient while soluble proteins remain in the upper layers of the centrifuge tube. After short pulses of radioactivity such as employed in this experiment, the peptide-bound radioactivity is almost evenly divided between the ribosome and soluble fractions.

In order to demonstrate that the ribosome-bound polypeptides, labeled with radioactive sulfur, actually represent precursors of cellular proteins, a nonradioactive chaser was added to another portion of the same culture and incubated a further 15 seconds. The nonradioactive chaser prevents the further incorporation of radioactivity,

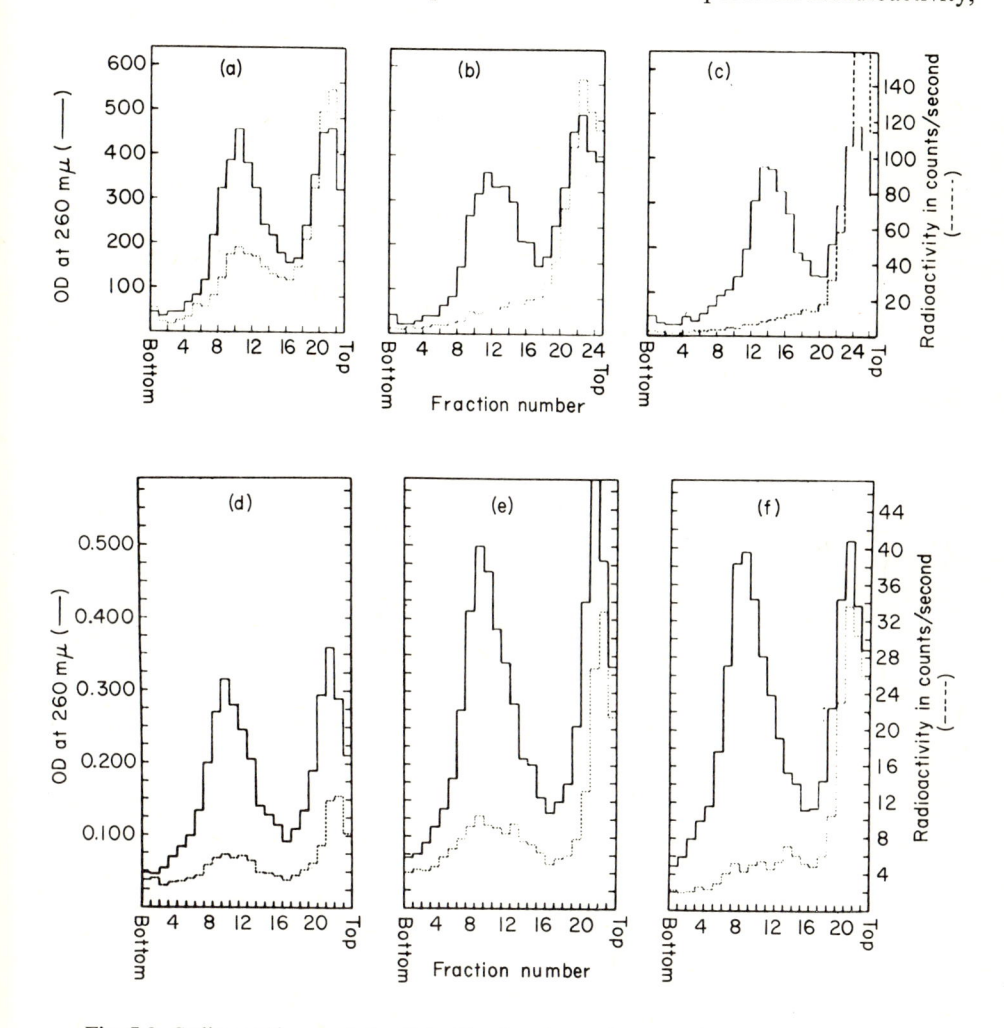

Fig. 5-9. Sedimentation analysis of the kinetics of protein synthesis in *E. coli*. A. Cell extracts prepared after (a) 15 seconds' incubation with $^{35}SO_4^{2-}$, (b) 15 seconds' incubation with $^{35}SO_4^{2-}$ followed by 15 seconds' incubation with ^{32}S chaser, and (c) 15 seconds' incubation with $^{35}SO_4^{2-}$ followed by 120 seconds with ^{32}S chaser. B. Cell extract prepared after (d) 5 seconds' incubation with ^{14}C-amino acids, (e) 12 seconds' incubation, and (f) 12 seconds' incubation with ^{14}C-amino acids followed by 120 seconds' incubation with ^{12}C-amino acids. In both cases, note transfer of radioactivity from 70 S to 85 S region to nonsedimenting region. From McQuillen et al. (43).

but permits labeled precursors in the cell to mature into final products. Such a technique is therefore very useful in establishing precursor-product relationships. An extract of this culture was prepared and fractionated on a sucrose gradient. The results are presented in Fig. 5-9b. The sucrose gradient patterns convincingly show that the radioactivity contained in polypeptides is transferred from the ribosomes to the soluble fraction during the chase in nonradioactive medium. If the chaser is added for 120 seconds after a 15-second pulse of radioactive sulfur, the results are even more dramatic since the ribosomes are almost completely liberated of radioactivity which is found in the soluble fraction instead (Fig. 5-9c).

Additional control experiments were performed in order to determine that the TCA-precipitable soluble proteins labeled in a short pulse are truly representative of normal cellular proteins. The decisive proof was provided by co-chromatography of pulse-labeled protein with nonradioactive soluble protein from *E. coli* on DEAE-cellulose. The radioactive peaks in the effluent curve were found to coincide precisely with peaks of protein nitrogen. Therefore, the pulse-labeled products do indeed correspond to a typical mixture of normal soluble proteins.

In subsequent kinetic experiments, ^{35}S label was replaced by a mixture of amino acids, uniformly labeled with the radioactive isotope of carbon, ^{14}C. A complete mixture of radioactive amino acids was prepared as follows. The unicellular alga, *Chlorella*, was grown under conditions such that $^{14}CO_2$ of high specific radioactivity was its sole carbon source. Total protein was then extracted from the algal cells, purified, and hydrolyzed with 6 N HCl. A uniformly labeled mixture of most usual amino acids results, though a few of them are destroyed in the course of acid hydrolysis.

When the ^{14}C-labeled amino acids were incorporated into polypeptides during short pulses of 5 and 12 seconds, the results are very similar to those of the experiments in which ^{35}S was used (Fig. 5-9, d and e). After 5 seconds, most of the radioactivity is associated with the ribosomes; this situation prevails also after 12 seconds, but more of the radioactivity has left the ribosomes, apparently in the form of soluble proteins whose synthesis has been completed. The labeled soluble fraction continues to increase with time, but after 20 seconds the radioactivity of the ribosomes does not increase further. Kinetics of this sort directly indicate that soluble proteins must pass through an intermediate ribosome-bound stage. After a 30-second chaser in the presence of nonradioactive amino acids, an even greater proportion of radioactivity has been transferred to the soluble fraction as in the experiments with ^{35}S (Fig. 5-9f).

From the rate at which radioactivity is released from the ribosomes, the time necessary to synthesize an individual protein molecule has been evaluated. This figure is less than 5 seconds and probably quite close to 3 seconds. While it is difficult to measure this number with greater accuracy, it is clear that when radioactive precursors are added to a bacterial culture, they are first incorporated into 70 S ribosomes up to a certain limiting level of activity which is reached in about 5 seconds. Thereafter, the radioisotope is transferred from the ribosomes into the soluble protein fraction at the same rate as new isotope is fixed. The increase in the total radioactivity of the ribosomes with time corresponds to the gross rate of synthesis of soluble protein, but the limiting activity corresponds to the amount of protein synthesized in 3 seconds. The most likely time of synthesis for an individual molecule of protein in bacterial cells is therefore between 3 and 5 seconds. More recent data for the growth rate of individual polypeptide chains yield a value of 15 to 20 residues per second.

If it is assumed that each ribosome is engaged in the synthesis of one protein molecule at all times, the 5000 to 10,000 ribosomes in the cell are completely sufficient to account for bacterial growth. Let us calculate the gross rate of protein synthesis in an *E. coli* cell growing with a doubling time of 30 minutes. The relative rate of growth in this case is about 0.1 % per second, and the total number of protein molecules synthesized per second must be about 1000 if the cell contains a million proteins with an average molecular weight of 50,000. If the time necessary to synthesize one polypeptide chain is taken as 3 seconds, the synthetic apparatus of the cell is able to produce between 1500 and 2000 protein molecules per second. Thus there is even a small reserve capacity available.

It was also important to establish that protein chains are built up by successive addition of amino acid residues in the bacterial cell, as was proved by Schweet and Dintzis for hemoglobin synthesis in reticulocytes. In the experiments of Goldstein and Brown (44), a short pulse of radioactive leucine was administered to an auxotrophic culture of *E. coli* limited by that amino acid. After the pulse, total protein was extracted from the cells and digested with carboxypeptidase, an enzyme that splits amino acids from the C-terminal end of the polypeptide chain. Carboxypeptidase cleaves residues sequentially from the end of the molecule, but the process comes to a halt when the enzyme encounters cysteine, proline, or certain other amino acids. If the percent of leucine cleaved from the chain is measured as a function of time for which the cells are exposed to radioactivity, then it turns out that carboxypeptidase liberates all of the ^{14}C-leucine only when the pulse time is very short in comparison with the time required to synthesize a complete polypeptide chain. In the experiment described, this was accomplished by incubating the cells at 0°C. At this temperature, the protein synthetic time was on the order of 50 seconds in contrast to the 3-second synthetic time found at 37°C. It was concluded that protein chains in bacteria also grow in a stepwise fashion from the N-terminus toward the C-terminal residue.

5. Messenger RNA and Protein Synthesis

That protein synthesis takes place on ribosomes and that the finished protein molecules are released from the ribosomes into the cytoplasm have been demonstrated by numerous experiments both in living cells (Dintzis, Roberts and collaborators) and in cell-free preparations (Zamecnik, Schweet, Lipmann, Tissières, Watson and colleagues). The experiments of Jacob and other investigators showed that the synthesis of a specific enzyme by the cell requires the presence of an undamaged structural gene carrying information for the given protein.

It follows from the juxtaposition of these two facts that there must be an uninterrupted flow of information from the chromosome to the ribosomes. What then is the material agent which mediates this information transfer? Some time ago it was thought that this substance was the high-molecular-weight structural RNA of ribosomes. Such RNA would have to serve as a template for protein synthesis and, in one way or another, its structure would have to reflect that of the cellular DNA.

There are many facts that argue against a role in information transfer for the structural RNA of ribosomes. First of all, the relative proportions of nucleotide bases in the DNA of several different bacterial species varies within an extremely wide range, with (G + C)/(A + T) ratios running from 0.5 to 2.8 (Belozersky and Spirin, (45)). However, the base ratios of total RNA from the same species varies within much narrower limits; in all, the maximum difference in the (G + C)/(A + T) fraction is only 20% although the variation is correlated with that of the DNA to some extent. It was therefore necessary to inquire whether there might not exist a special kind of informational RNA more closely related in base composition to DNA, and comprising only a small proportion of the total cellular RNA, which could serve as a template for protein synthesis.

A second inconsistency in the hypothesis that ribosomal RNA (rRNA) plays a major role in information transfer arose from the experiments of Davern and Meselson on the stability of rRNA (46). When microbial cells are grown on a medium containing the heavy isotopes ^{13}C and ^{15}N, their ribosomal RNA is passed on to successive generations in an unaltered form when the cells are resuspended in light medium. That is, intact heavy rRNA is detected after three to four cell divisions in the light medium, but no rRNA molecules with an intermediate isotopic composition were ever found. This attests to the great metabolic stability of rRNA.

But most of the evidence available indicated that the template for protein synthesis must undergo a rather quick metabolic turnover. The quantity in which many types of protein molecules can be produced by the bacterial cell is subject to rapid regulation within broad limits and, in particular, the synthesis of certain proteins can start and stop within very short times after changes in the composition of the growth medium. Under conditions of induction, for instance, the overall production of two extensively studied *E. coli* proteins, alkaline phosphatase and β-galactosidase, can increase by a factor of 1000. At maximum levels of induction, these enzymes comprise 5 to 6% of the total protein, as compared with a few hundredths of a percent under normal conditions. The synthesis of β-galactosidase is induced by the replacement of glucose with lactose as sole carbon source. Within 1 or 2 minutes after addition of inducer, significant β-galactosidase activity can be detected in the cells and a high rate of synthesis soon becomes established. When lactose is diluted out of the medium, however, synthesis of β-galactosidase comes to a rapid halt. The highly stable structural RNA of ribosomes could not be expected to react to changes in the cell environment with such rapidity.

A similar phenomenon is observed upon mating. If an *E. coli* Hfr strain carrying the *lac*$^+$ marker is mated with a *lac*$^-$ female, and the Hfr males are subsequently killed, then the synthesis of the enzyme β-galactosidase begins in the recipients of the *lac*$^+$ marker within a few minutes after conjugation (47). Since the time necessary for the appearance of induced enzymes is short compared with the generation time of the culture, the number of ribosomes synthesized by the cell during that period is insignificant. Therefore it is logical to suppose that a ribosome can expedite the synthesis of any protein depending on the "orders" which it receives from the chromosome. These experiments provide support for the idea that there exists an unstable intermediate compound, called messenger RNA, which is specified by the cellular DNA and which can serve as a template for the synthesis of protein.

A further argument against rRNA as mediator of information transfer in the cell is the fact that infection of *E. coli* with T-even bacteriophages leads to the immediate cessation of ribosomal RNA synthesis. After a lag of several minutes, the synthesis of a number of new phage-specific proteins begins in place of the hundreds of proteins normally made by the cell. In order for this transition to occur, the first templates for phage protein must be made within an interval on the order of a minute or two.

All of the evidence cited above implies the existence of a special kind of RNA whose function it is to carry information from the chromosome to the ribosomes where protein synthesis takes place. Such RNA must be unstable, it must have a short lifetime, and it must be continually synthesized and degraded. Moreover, the template or messenger RNA must reflect the nucleotide composition and sequence of DNA in its own structure.

Volkin and Astrachan discovered a class of RNA similar to DNA in nucleotide composition by exposing bacteriophage-infected *E. coli* cells to radioactive precursors for short intervals of time (48). In these experiments, cells infected with phage T2 were incubated in medium containing ^{32}P-labeled phosphate for 1 to 3 minutes. They were then broken open and the total RNA was extracted and deproteinized. The base ratio was determined on the total RNA as well as on the radioactive fraction, and the two sets of figures were found to be quite different. The base ratio of the labeled RNA synthesized during the short pulse corresponded to that of bacteriophage DNA, substituting uracil for thymine. An analogous result was obtained when *E. coli* cells were infected with phage T7, whose DNA composition is completely different from that of T2. These experiments constituted the first direct evidence for the existence of messenger RNA.

The next important step in the study of messenger or informational RNA came from the experiments of Brenner, Jacob, and Meselson (49). They used infection with bacteriophage T4 to "turn on" the synthesis of new proteins in *E. coli* host cells. Bacteria were grown on heavy medium containing components labeled with the heavy isotopes ^{13}C and ^{15}N. The density of bacterial nucleic acids and proteins in such a culture was higher by 0.05 to 0.1 gm/cm^3 than in the culture grown on normal light medium, whose components contained ^{12}C and ^{14}N. After a certain time, heavy cells were transferred to light medium into which radioactive phosphorus was introduced and the cells were infected with phage T4. Within 7 minutes after infection, the culture was lysed and their ribosomes were centrifuged in a CsCl density gradient. The presence of Mg^{2+} ions at a concentration of 3×10^{-2} M was required to stabilize the ribosomes. The distribution of ribosomes in the CsCl gradient showed that almost all of them bore the original heavy label. This means that despite the diversion of the cellular synthetic apparatus to the production of new bacteriophage proteins, the ribosomes participating in this process were synthesized prior to infection. Moreover, the ribosomal fraction contained radioactive ^{32}P, incorporated into phage-specific mRNA produced *after* infection. This experiment supports the concept that ribosomes are universal and nonspecific sites of protein synthesis, programed by messenger RNA transcribed from the cellular or phage chromosome. In attaching to preformed ribosomes, newly formed phage-specific mRNA diverts them to the synthesis of bacteriophage proteins for which the cell initially contains no information.

Messenger RNA from bacteria not infected with bacteriophage has also been extracted and studied. The basic approach, used by Gros, Watson, Spiegelman, and others (50–52), is to label the cells with a radioactive pulse of ^{14}C-uracil or ^{32}P-phosphate. The optimal duration of the pulse depends on the generation time of the culture. If the generation time is on the order of an hour, a pulse time of about 20 to 30 seconds is usually employed. For very long generation times, the length of the pulse is extended proportionately.

The first method used to isolate mRNA was centrifugation of crude cell extracts in sucrose gradients at low concentration of Mg^{2+} ions (about 10^{-4} M). Presence of a stable density gradient in the centrifuge tube prevents convectional mixing and permits the separation of cell components according to sedimentation rate. The protocol of a typical experiment is as follows: (1) an *E. coli* culture is incubated in the presence of ^{14}C-uracil or $^{32}PO_4^{2-}$ for 20 to 30 seconds; (2) the cells are washed free from the radioactive medium with tris buffer, pH 7.5, containing 10^{-4} M Mg^{2+}; (3) from 0.5 to 1.0 gm (wet weight) of cells are broken open by grinding with alumina, freezing and thawing in the presence of lysozyme, or by some other method, and the suspension is centrifuged for 15 minutes at 10,000 rpm to remove cellular debris; (4) deoxyribonuclease is added to a final concentration of 5 $\mu g/ml$ in order to eliminate DNA; (5) 0.2 ml of the suspension from step 4 is layered onto the top of a preformed 5 to 20% sucrose gradient containing 10^{-4} M Mg^{2+} in a 5-ml centrifuge tube; (6) the gradient is centrifuged for 2 to 4 hours in a swinging-bucket rotor of a preparative ultracentrifuge at 35,000 rpm; and (7) after centrifugation, the bottom of the tube is pierced with a hollow needle and its contents are collected dropwise and divided into a series of fractions.

After analysis of ultraviolet absorption (at 260 mμ) and radioactivity in each fraction, the data are plotted as illustrated in Fig. 5-10B. Peaks of ultraviolet-absorbing material mark the positions of ribosomal subunits with sedimentation coefficients of 30 S and 50 S, while a small amount of 70 S ribosomes are also present. This is a typical distribution of components in a cell extract made at low Mg^{2+} concentration. The radioactivity is concentrated in a broad zone with a maximum at about 16 S, although about 10% of the label sediments at 70 S. The 16 S peak is not detected optically because it contains very little material in comparison with other components in the gradient. It was estimated, in fact, that mRNA comprises between 3 and 5% of the total *E. coli* RNA. For comparison, a similar gradient pattern run at high Mg^{2+} concentration is presented in Fig. 5-10A. In this case, about half of the mRNA is found in ribosomes or polysomes where protein synthesis is believed to take place. If Mg^{2+} ions are added to a suspension containing a mixture of 30 S and 50 S subunits as well as labeled mRNA, reconstructed 70 S ribosomes are obtained with which 50% of the radioactivity is associated.

If the crude extract is deproteinized with phenol and sodium dodecylsulfate after step 4, and centrifuged in a sucrose gradient for 10 hours, two major peaks of ultraviolet-absorbing material corresponding to ribosomal RNA are observed at 16 S and 23 S, as well as a peak at 4 S which contains tRNA (Fig. 5-10C). In this case, the radioactivity is broadly distributed between 16 S and 4 S components with a maximum between 8 S and 10 S. This indicates that the pulse-labeled mRNA is characterized by molecular weights of several hundred thousand.

It soon became evident that mRNA is extremely sensitive to degradation by ribonuclease present in the crude cell extract. If the cells are broken open in the presence of phenol and sodium dodecylsulfate so as to achieve immediate denaturation of proteins, then mRNA is distributed very widely throughout the gradient with a maximum at about 20 S to 30 S and a tail extending to 40 S or 45 S. The molecular

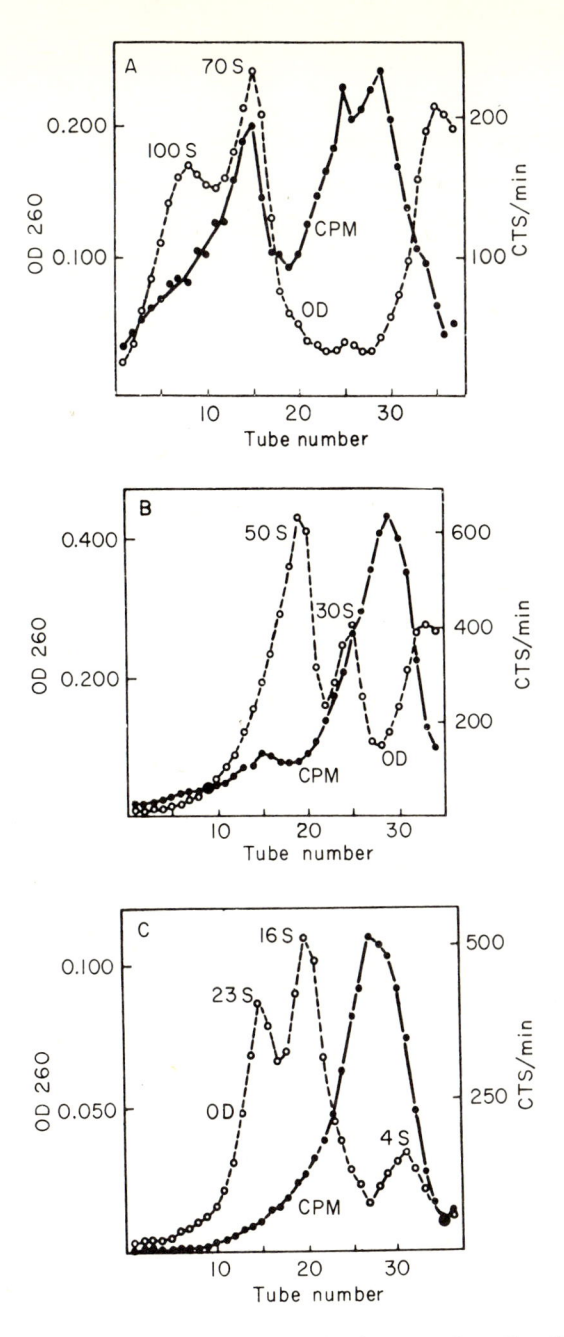

Fig. 5-10. Sucrose gradient sedimentation analysis of extracts from *E. coli* cells grown on ^{14}C-uracil for 20 seconds. (A) Extract at 10^{-2} M Mg^{2+} showing the newly synthesized mRNA associated in part with intact 70 S and 100 S ribosomes; the remainder sediments in a broad peak between the ribosomes and 4 S tRNA. (B) Extract at 10^{-4} M Mg^{2+} in which the ribosomes dissociate into 30 S and 50 S subunits. Almost all the labeled RNA has been liberated from the ribosomes and forms a broad peak between the 30 S and 4 S components. (C) Extract purified with phenol to eliminate protein. The mRNA sediments more slowly than the peaks of ribosomal structural rRNA with sedimentation constants of 16 S and 23 S. From Gros *et al.* (53).

weight of undegraded mRNA therefore extends to several millions, which strongly implies that the templates are polycistronic (54).

Sedimentation analysis has also been used to study messenger RNA from cells infected with phage T2 (52). ^{32}P-labeled phosphate was added to the culture from 3 to 8 minutes after infection and the cells were subsequently lysed. Sucrose gradient centrifugation of extracts from the infected cells again reveals the presence of a distinct radioactive peak. After deproteinization, the heterogeneous radioactive fraction sediments with a maximum at about 16 S. Even though the time of exposure to radioactive label was relatively long in these experiments, all of the radioactivity was found in mRNA. It was concluded that ribosomal RNA is not synthesized in phage-infected cells. By contrast, if a pulse of comparable duration is used to label uninfected cells, a very large fraction of the total radioactivity if found in ribosomes.

In studies on base composition, ^{32}P-labeled mRNA is subjected to alkaline hydrolysis and the resulting mixture of nucleotides is separated chromatographically (Watson, Hiatt, Spiegelman). Results of several such analyses are presented in Table 5-3. Note particularly the figures for *Bacillus megaterium* in which the compositions of

TABLE 5-3

NUCLEOTIDE BASE RATIOS OF RNA AND DNA IN BACTERIA AND PHAGE

Source of nucleic acid	Nucleic acid	Purines / pyrimidines	A + T(U) / G + C	Ref.
Escherichia coli	DNA	0.96–1.04	0.92–1.00	(50)
	rRNA	1.30	0.88	
	tRNA	1.12	0.64	
	mRNA	1.09	0.95	
Bacteriophage T2	DNA	0.98	1.83	(50)
	mRNA	1.06	1.68	
Pseudomonas aeruginosa	DNA	1.00	0.56	(55)
	RNA (total)	1.21	0.88	
	mRNA	1.03	0.72	
Bacillus megaterium	DNA	1.00	1.63	(55)
	RNA (total)	1.19	0.86	
	mRNA	1.05	1.30	

DNA and ribosomal RNA are very different. It is quite clear in this case that pulse-labeled mRNA resembles DNA in composition and not rRNA.

If mRNA is an exact replica of one or more functional regions of the DNA (cistrons), there must be, in addition to identity of base composition, complete correspondence in nucleotide sequence between mRNA and various portions of the DNA molecule. It should thus be possible to form a Watson-Crick double helix from mRNA and the complementary strand of DNA. To prove the complete homology in sequence between mRNA and DNA, the method of Marmur and Doty was used to anneal a mixture of phage T2 DNA with mRNA isolated from T2-infected bacteria (Spiegelman). Phage DNA was labeled with ^{32}P and mRNA with ^{14}C- or ^{3}H-uracil. The solution was heated for 15 minutes at 90°C and then slowly cooled over a period of 3 to 4 hours. The annealed mixture was then centrifuged to equilibrium in a CsCl density gradient.

In the resulting gradient distribution, labeled RNA formed a band lying close to, but not coinciding with, that of DNA. This band also contained a certain amount of DNA which was detected by the presence of ^{32}P label. It appeared that the intermediate structure, labeled with both ^{32}P and ^{14}C, consisted of DNA-mRNA duplex hybrids. A number of control experiments demonstrated conclusively that specific hybridization between the two chains had occurred. Thus an unheated mixture of phage DNA and mRNA yielded no such hybrid band. Replacement of phage T2 DNA with that of phage T5 also produced no doubly labeled complexes, demonstrating the specificity of binding between the two polymers. Even though the average base composition of T5 DNA is very close to that of T2, it hybridizes neither with T2 DNA nor with T2-specific mRNA. Furthermore, the hybrid DNA–mRNA duplexes are not degraded by either ribonuclease or deoxyribonuclease, nor by the combination of the two enzymes.

Similar results were obtained by Gros and his colleagues for the hybridization of purified *E. coli* mRNA with homologous and heterologous DNA (Fig. 5-11). If the mRNA contains radioactive phosphorous, ^{32}P label appears in the optical density peak representing DNA after annealing. When DNA from another bacterial strain is used, none of the labeled mRNA is found in the DNA peak for there are few or no complementary nucleotide sequences in the two heterologous nucleic acids.

Specific annealing between DNA and RNA provides a means for rapidly isolating phage mRNA by adsorption to homologous DNA embedded in an agar matrix (p. 283). When a column of this sort is charged with DNA from bacteriophage T4, it is possible to prepare relatively large amounts of mRNA from *E. coli* cells infected with that phage (Bautz, 56). The retrieval of as much as 0.1 mg of phage-specific mRNA from 5 mg of total RNA permitted further study by conventional physicochemical techniques. The mRNA was found to be heterogeneous in molecular weight with some fractions sedimenting as rapidly as 35 S. By measuring the optical density as a function of temperature, the single-stranded molecules exhibited a significant hypochromic effect (18%), indicating the presence of secondary structure. The mean temperature for the disappearance of the hypochromic effect fell at 53°C and the average helix content was computed to be close to 40%.

The best of the methods for extracting specific mRNA molecules consists of hybridizing them with DNA on nitrocellulose membrane filters (Gillespie and Spiegelman, 57). The DNA of transducing phages is especially good for isolating specific bacterial mRNA's since they bear only a small portion of the bacterial genome. In this technique, DNA is first denatured and adsorbed to nitrocellulose filters. The bound DNA is then annealed with mRNA in a small volume of buffer at 60°C over a period of about 24 hours. The filter is next treated with RNase in order to remove unannealed RNA and the RNase is inactivated with iodoacetic acid. Up to 90% of the specifically hybridized mRNA in the duplexes can finally be washed from the surface of the filter with a small amount of dilute buffer at 95°C. The selectivity of hybridization is so great under these conditions that background adsorption in the presence of noncomplementary RNA is less than 0.001%.

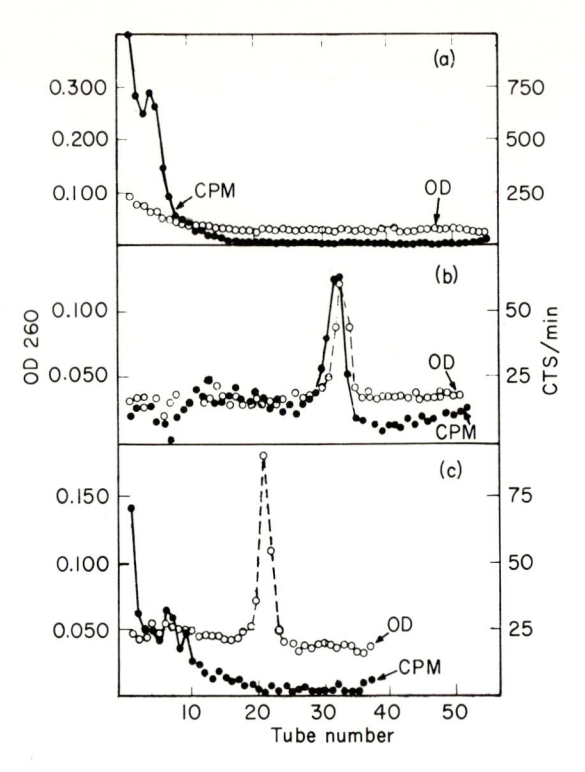

Fig. 5-11. The formation of hybrid double helices consisting of DNA and mRNA as the result of denaturing and renaturing the two components together in solution. mRNA was prepared from E. coli cells after a 20-second exposure to ^{32}P for selective labeling of their constituent nucleotides. DNA was prepared from E. coli and Bortadella bronchiseptica, denatured by heating to 100°C, mixed with ^{32}P-labeled mRNA, and cooled slowly. The renatured solutions were then centrifuged for 5 days in a cesium chloride density gradient and were subsequently fractionated and analyzed for ultraviolet-absorbing material and radioacvivity. (A) ^{32}P pulse-labeled RNA from E. coli alone. (B) ^{32}P pulse-labeled RNA annealed with E. coli DNA. (C) ^{32}P pulse-labeled E. coli RNA annealed with DNA from Bortadella bronchiseptica. The optical density curve is attributable almost exclusively to DNA, the radioactivity to mRNA. Note that E. coli mRNA forms molecular hybrids with homologous DNA from E. coli but not with the heterologous DNA from an unrelated strain of bacteria. From Gros et. al. (50).

Specific hybridization of mRNA with DNA was used to show that mRNA molecules are transcribed from only one of the two complementary DNA strands. Of course this experiment required prior separation of the two DNA strands and a method for doing this was worked out by Szybalski and associates (see p. 504). Then the two complementary molecules could be tested individually for their capacity to entrap mRNA on a membrane filter. Although only a single strand is transcribed within the confines of a single cistron, it is not necessarily the same strand which is copied from cistron to cistron. Thus during infection with bacteriophages T4 and λ, mRNA for phage "early proteins" is transcribed from one of the DNA strands, mRNA for the "late proteins" mainly from the other.

The rate at which mRNA is metabolized in bacterial cells has been measured in several different ways. It is evident that if mRNA is synthesized rapidly, it must also be degraded rapidly. In order to make certain of this, a culture of *E. coli* cells was incubated with ^{14}C-uracil for 20 seconds and then transferred to a medium containing unlabeled uracil. Five minutes later, the cells were harvested and an extract analyzed by sucrose gradient centrifugation. It was determined that the peak of radioactive mRNA had almost completely disappeared after the 5-minute chase. This time interval was sufficiently large to permit the degradation of most mRNA molecules into fragments which were subsequently used for the synthesis of other types of RNA. Kinetic measurements on the formation and degradation of labeled polysomes in *E. coli* indicated that the average lifetime of a messenger RNA molecule was on the order of 10 minutes (58). In another experiment, *Bacillus subtilis* cells were subjected to a short pulse of ^{14}C-uracil in order to selectively label the mRNA. Actinomycin D, a specific inhibitor of DNA-dependent RNA synthesis, was then added to the medium. Since this inhibitor prevents the synthesis of all kinds of RNA in the bacterial cell, it was possible to measure the kinetics with which polymerized ^{14}C-uracil was degraded to acid-soluble fragments (59). The decay of mRNA is exponential with a half-life between 1 and 2 minutes. From these results, an average mRNA template can be estimated to direct the synthesis of approximately 20 to 40 polypeptide chains before degradation. An analogous experiment performed on cells of higher organisms showed that pulse-labeled RNA breaks down much more slowly over a period of several hours.

Further evidence that mRNA carries information for the synthesis of proteins was provided by experiments on protein synthesis *in vitro*. The early cell-free systems employed by Zamecnik, Hoagland, and Lipmann were deficient in many respects: The maximal level of polypeptide synthesis was only $3 \times 10^{-3}\%$ of the protein present, and amino acid incorporation declined and stopped after 10 to 15 minutes. What this system lacked was a source of new mRNA to replace that degraded during incubation, leaving ribosomes without a template for protein synthesis. In order to rectify these deficiencies and intensify protein synthesis *in vitro*, Watson demonstrated that the addition of purified *E. coli* mRNA to a complete cell-free system stimulated amino acid incorporation by 100 to 200%. A similar stimulation occurs if the cell-free preparation is not freed of DNA and if all four nucleoside triphosphates necessary for RNA synthesis (ATP, GTP, UTP, and CTP) are also added. No stimulation is observed when one of the triphosphates is omitted, or when the extract is treated with deoxyribonuclease. Newly synthesized polyribonucleotides are apparently transcribed from DNA *in vitro* and themselves go on to serve as templates for polypeptide synthesis.

An even greater stimulatory effect was obtained by Wood and Berg (60), who extracted and purified RNA polymerase, an enzyme from *E. coli* which catalyzes the DNA-dependent synthesis of RNA from nucleoside triphosphates. When Berg added this enzyme along with purified phage or bacterial DNA and all four nucleoside triphosphates to a complete cell-free system, a large stimulation of amino acid incorporation was noted. Not only does peptide synthesis continue for 60 to 70 minutes in this case, but the total amount of protein produced increases by a factor of ten over the system in which mRNA synthesis does not take place. The transcription of mRNA from DNA in the incubation mixture permits the continuous manufacture of new templates for polypeptide synthesis to replace those which have been degraded.

The RNA of viruses and RNA-containing bacteriophages itself serves as mRNA for the synthesis of proteins, as demonstrated in cell-free incorporating systems. One of the best examples is RNA from the phage f2 which Zinder and colleagues showed is a messenger RNA for phage coat protein as well as two other proteins needed by the phage.

Although the cell-free system contributed greatly to the understanding of protein synthesis and the genetic code, it was not known for many years whether complete protein molecules with native structure could be synthesized by this method, or whether the products were simply nonfunctional polypeptide fragments. This question was resolved by the use of RNA from f2 phage as a template in a cell-free system derived from *E. coli* (Zinder *et al.*, 61). Phage coat protein was recovered as a major reaction product and the two other proteins coded by the viral RNA were synthesized in smaller quantities (62). The similarity of the major protein product to native phage coat protein was verified by its chromatographic behavior on DEAE-cellulose and by the distribution of certain radioactive amino acids in its fingerprints. The lysozyme specific to phage T4 has also been synthesized in a cell-free system from uninfected *E. coli* upon addition of mRNA extracted from T4-infected cells (63). This is the first instance in which a functionally active enzyme has been synthesized entirely *in vitro*.

The synthesis of proteins in higher organisms is also mediated by mRNA. Mirsky and his co-workers isolated a ribonucleic acid component from the nuclei of calf thymus cells which was similar in all respects to bacterial mRNA. Its synthesis is relatively rapid and its base composition reflects that of DNA from the same cells. Significantly, this type of RNA is located in the nuclei where it is continuously transcribed from the cellular DNA and transferred to the cytoplasm for use in protein synthesis. A number of other investigators have used the pulse-labeling technique to study mRNA in a wide variety of animal tissues. Pulse times on the order of 10 to 20 minutes are usually selected for these investigations. The absolute quantities of mRNA in the tissues of higher oranisms is of course substantially greater than in bacteria. Furthermore, the tissue specificity exhibited by mRNA from higher organisms demonstrates that mRNA from physiologically different cells is copied from different functional regions of the DNA (64).

6. Regulation of Protein Synthesis

One of the most complex problems of molecular biology is how the cell regulates the number of enzyme molecules it produces (65). Chemical reactions in the cell are delicately balanced and there are many mechanisms which automatically regulate the rates of metabolic processes in accord with the requirements of the cell. We saw in Chapter 2 that some of these mechanisms operate at the level of enzyme activity through specific activation and inhibition. The phenomena of selective permeability and active transport across membranes also play a significant role in regulation. In this section, we shall investigate another aspect of regulation, the mechanism determining the rate at which enzymes are synthesized by the cell.

Consider, for example, the way in which synthesis of enzymes leading to the formation of the amino acid arginine is regulated in bacterial cells. The more of these enzymes there are in the cell, the more the arginine that will be produced per unit time. If the bacteria are grown on a medium liberally supplied with arginine, however, endogenous

6. Regulation of Protein Synthesis

synthesis of the amino acid is superfluous and, consequently, there is no need for enzymes which catalyze arginine formation. There is a feedback mechanism in the cell through which arginine, the end product of the enzymatic reactions, governs the rate of synthesis of almost all of the enzymes in its own biosynthetic pathway. Arginine thus acts as a *repressor of enzyme synthesis* in this case. Repression of enzyme synthesis by the end products of the reactions they catalyze is a very general phenomenon, most often associated with the control of anabolic (biosynthetic) enzymes. One should not confuse the term *repression* with *feedback inhibition* in which the end product of a biosynthetic pathway inhibits the activity of the first enzymes in the sequence.

We have already noted that the bacterial cell is capable of synthesizing more than 1000 different enzymes. When such cells are grown on an enriched medium which supplies most metabolites that the cell would normally have to synthesize itself, the synthesis of the majority of biosynthetic enzymes may be repressed to a greater or lesser extent. The potential capacity of the cell to synthesize each of its proteins is in general significantly higher than that realized during normal cell growth. Thus when the cell is supplied exogenously with most of its metabolic needs, the quantity of each protein actually synthesized amounts to no more than several hundredths or a few tenths of a percent of the total cellular protein. But when some vital substance does not suffice, a regulatory mechanism automatically comes into play to correct the imbalance. In the presence of inorganic phosphate, for instance, *E. coli* cells do not make the enzyme alkaline phosphatase. But if the bacteria are raised on a medium lacking inorganic phosphate, they respond by synthesizing phosphatase since this enzyme is capable of hydrolyzing organic phosphate esters. In this case, phosphatase synthesis is derepressed because its normal repressor is absent. The enzyme is synthesized by the cells in abnormally large amounts, sometimes reaching 5% of the total protein, presumably at the expense of other proteins.

Formation of the enzyme β-galactosidase which catalyzes the hydrolysis of the disaccharide lactose (4-glucose-β-D-galactoside) into glucose and galactose

is also regulated, but in a somewhat different way. Here the substrate of the enzyme, or a compound related to it, *induces* the synthesis of the enzyme. Induction is most common among enzymes catalyzing catabolic reactions whereby various carbon sources are utilized for respiration and fermentation. The existence of compounds that repress and induce the formation of enzymes provides direct evidence that regulatory mechanisms are present.

At one time it was believed that no β-galactosidase is synthesized when the medium contains glucose or other carbon source more accessible than lactose. According to this older theory, introduction of the substrate into the medium causes the cell to undergo heritable alterations leading to the appearance of the capacity to synthesize β-galactosidase. The same sort of explanation was used in regard to other examples of induction, such as synthesis by penicillin-resistant bacteria of the enzyme penicillinase which degrades the antibiotic and thereby protects the cell. Since the formation of penicillinase always requires the presence of penicillin, this process was also thought by some to be wholly determined by presence of the drug. Thus microbiology remained the citadel of Lamarckism long after this theory had been abandoned in other biological fields.

All doubts about the nature of induced enzyme synthesis were dispelled, however, when more accurate experiments demonstrated that cells capable of adaptively synthesizing enzymes, such as β-galactosidase and penicillinase, usually produce them in very small quantities even in the absence of inducer. This indicates that the cell contains the genetic information required for synthesis of the inducible protein, whether or not it is expressed. In other words, the chromosome carries a cistron in which the chemical structure of the enzyme protein is coded. Owing to the regulatory mechanisms, however, residual enzyme synthesis occurs slowly and the rate of enzyme formation sharply increases only in the presence of a specific substrate or inducer.

Furthermore, the enzyme β-galactosidase acts on numerous β-D-galactoside substrates, even though the majority of them are unable to induce enzyme synthesis. A typical example of this sort is phenyl-β-D-galactoside

which, although an excellent substrate for the enzyme, is unable to serve as an inducer of its synthesis. On the other hand, there are substances which are very good inducers of β-galactosidase synthesis, but which are not themselves substrates of the enzyme and represent only remote analogs of its substrates. One such compound is isopropyl-β-D-thiogalactoside (IPTG), whose structure is

while certain other analogous compounds, including methyl-β-D-thiogalactoside (TMG), behave in a similar fashion.

The effects of β-galactosidase induction on the growth of E. coli are depicted in Fig. 5-12. At the beginning of the experiment, the culture received galactose as a sole carbon source. The initial galactose was exhausted within an hour and growth stopped at that time. The culture was then divided into two portions, one of which received methyl-β-D-thiogalactoside, an inducer but not a substrate of the enzyme, while the

6. Regulation of Protein Synthesis

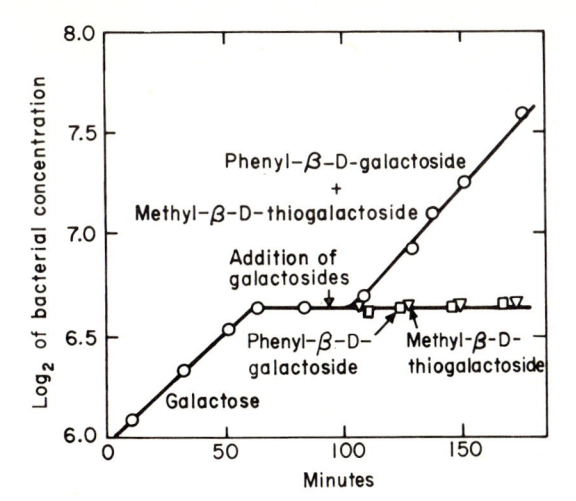

Fig. 5-12. Growth of *E. coli* on galactose and the indicated galactosides demonstrating the independence of inducer and substrate functions. Phenyl-β-D-glactoside is a substrate of the enzyme β-galactosidase but not an inducer, while methyl-β-D-thiogalactoside is an inducer but not a substrate. From Hogness (66).

other received phenyl-β-D-galactoside, a substrate of the enzyme but not an inducer. In neither case did growth of the culture resume. Only when both substances were added together was the culture restored to exponential growth.

Genetic experiments have made a significant contribution to understanding how β-galactosidase synthesis is regulated in *E. coli*. Two types of mutation relating to β-galactosidase synthesis were noted quite early. The first class of mutants was unable to synthesize active enzyme, apparently due to a defect in the genetic information determining its primary structure. The cistron governing the synthesis of β-galactosidase in the *lac* region is designated by the letter z. A mutation in which capacity to synthesize the enzyme is lost can be represented schematically as $z^+ \to z^-$. When the z^- strain is known to produce an enzymatically inactive protein that is serologically similar to β-galactosidase, it is designated z^-_{CRM}, where CRM stands for *cross-reacting material*.

The second type of mutant, by contrast, was capable of synthesizing β-galactosidase at maximal rates even in the absence of inducer. These were called *constitutive* strains to emphasize that no exogenous compound was necessary for enzyme synthesis. Mutations of this sort do not map at the z locus, but rather at a separate locus designated by the letter i, so that a mutation to constitutivity can be written as $i^+ \to i^-$. It is apparent that i^- strains have lost the capacity to regulate β-galactosidase synthesis since the enzyme is always produced, regardless of the needs of the cell. The i cistron is therefore called a *regulator gene*.

A further complication was soon discovered in the *lac* region. Besides the z and i cistrons, an additional cistron was recognized, denoted by y, which determines the synthesis of a special type of protein called galactoside permease. The synthesis of permease is induced by the very same substances which induce the synthesis of β-galactosidase. Galactosidase permease is responsible for the active transport of lactose and other galactosides and thiogalactosides across the cell membrane. In the presence of permease, galactoside concentration within the cell may be many times greater than in the medium. The formation of permease can be followed independently of β-galactosidase by measuring the rate at which radioactive ^{35}S-labeled IPTG penetrates the cell.

The independence of the z and y cistrons was determined by complementation tests in bacteria diploid for the *lac* region, of the type z^+y^-/z^-y^+. Diploid cells can be formed by conjugation or sexduction, but the former are less satisfactory because the merozygotes segregate after a few cell divisions. Stable heterogenotes can be obtained, however, if the relevant locus can be isolated on an episome. If a F'-z^-y^+ episome is introduced into a z^+y^- cell to form a heterogenote of the type $z^+y^-/$F'-z^-y^+, both β-galactosidase and permease are synthesized, demonstrating that the z^- and y^- mutations belong to separate functional units. Analysis of recombinants showed that the z and y cistrons are adjacent to one another in the *lac* region, not far from the *i* gene which controls their synthesis.

The kinetics of β-galactosidase induction are illustrated in Fig. 5-13a. When the

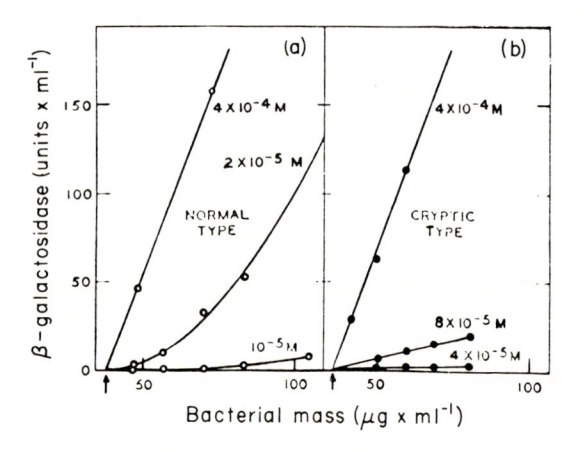

Fig. 5-13. Kinetics of induced β-galactosidase synthesis in (a) normal *E. coli* and (b) the permeaseless "cryptic" mutant type. The inducer employed was IPTG. The absence of permease results not only in decreased sensitivity of the cells to inducer, but in a qualitative change in the kinetics of induction as well. From Herzenberg (67).

concentration of inducer is sufficiently large ($\approx 4 \times 10^{-4}$ M), the induction mechanism begins to operate very soon after inducer is added to the medium. After the first permease molecules are formed, the intracellular inducer concentration increases rapidly. From that point on, the reaction is self-sustaining and the steady-state rate of β-galactosidase synthesis is established within a few minutes. The new rate can exceed that prior to induction by a factor of 1000 or more and if synthesis is maintained at this high level for long periods, β-galactosidase can comprise up to 6% of the total cellular protein. When the inducer concentration is less than 4×10^{-4} M, a longer time is required for the cells to reach steady-state synthesis of the enzyme. Nonetheless, once it is reached, the maximal steady-state rate of enzyme formation is just the same as when higher inducer concentrations are used. The role of the permease is thus to raise the intracellular inducer concentration to a level which permits the maximum rate of β-galactosidase synthesis, independent of inducer concentration in the external medium. Differences in external inducer concentrations are reflected only in the time necessary to establish steady-state conditions. The intracellular feedback mechanism here apparently works by the all-or-none principle, just as the relay in a thermostat; synthesis is either close to zero or else it proceeds at the maximum rate.

In cells with the genetic constitution $i^+z^+y^-$, β-galactosidase is synthesized upon induction, but inducer penetrates the cell only very slowly since it must depend on diffusion and not on active transport. Therefore, the steady-state rate of β-galactosidase synthesis becomes established rather slowly in these circumstances and it is a function of extracellular inducer concentration (Fig. 5-13b). β-Galactosidase synthesis in the $i^+z^+y^-$ mutants loses its biological significance to a considerable degree. In the presence of a galactoside such as lactose, the cells begin to synthesize the enzyme required for utilization of lactose as a carbon source. But the concentration of lactose within the cell is no higher than in the external medium since its penetration into the cell is limited by diffusion. Unless the lactose concentration is very high, the y^- bacteria behave as *lac*$^-$ even though they produce β-galactosidase. For this reason such strains are known as *cryptic mutants*.

Before returning to the mechanism of regulation, it will be useful to consider some of the chemical properties of β-galactosidase. By chromatography on DEAE-cellulose, it is possible to isolate the enzyme with a purity of 99% or higher, even when it comprises only 0.1% of the total cellular mass. A unit of enzyme activity has been defined which corresponds to the hydrolysis of 1 mμM of *o*-nitrophenyl-β-D-galactoside per minute at 28°C, pH 7.1 and 0.1 M sodium phosphate, when substrate is present in excess. The colorless substrate yields the bright yellow compound nitrophenol upon hydrolysis, making kinetic measurements very simple.

The molecular weight of the enzyme was determined with the method of equilibrium dialysis by measuring the amount of inhibitor bound by a given quantity of protein. The resulting figure was 135,000 daltons per active site. In fact, the purified protein turned out to be a tetramer of the basic functional unit with a molecular weight of 540,000. With the aid of highly purified preparations, it was found that one unit of enzymatic activity corresponds to 3×10^{-9} gm of enzyme, or 3.5×10^9 protein molecules, assuming that each protein macromolecule contains four active sites. Under the conditions of the assay, its turnover number is 1.7×10^5 moles of substrate per mole of enzyme per minute. From these figures, it was calculated that uninduced *E. coli* cells contain approximately one molecule of enzyme apiece; when the synthesis of β-galactosidase is fully induced, each cell produces up to 1000 enzyme molecules.

The appearance of protein after enzyme induction could be explained either by *de novo* synthesis from amino acids or by activation of a preexisting high-molecular-weight precursor called a *zymogen*. In order to resolve this question, Hogness, Cohn, and Monod (68) performed the following experiment. *Escherichia coli* cells were grown on a medium containing a limiting amount of sulfate as $^{35}SO_4^{2-}$. After all the radioactive sulfur had been taken up by the cells, a large excess of unlabeled sulfate was added to the culture together with the inducer methyl-β-D-thiogalactoside. Growth resumed and the formation of induced enzyme began. After a certain time interval, β-galactosidase was isolated from the cells and its radioactivity was measured. The enzyme protein contained no labeled sulfur, so that it could not have arisen from a polypeptide precursor synthesized prior to induction. Hence, β-galactosidase must have been synthesized *de novo* from amino acids after the inducer was added. Rotman and Spiegelman came to the same conclusion using ^{14}C-labeled amino acids (69). These experiments also showed that protein turnover, that is, rapid degradation and resynthesis of proteins, does not occur to a significant extent. If the inducer is withdrawn the synthesis of β-galactosidase quickly ceases, but the protein molecules produced up to that time are stable.

The next logical question is whether or not mRNA templates for β-galactosidase synthesis exist in the cell prior to induction. If galactosidase-specific mRNA is formed only after addition of inducer, its synthesis must be quite rapid, since steady-state enzyme synthesis is established within a few minutes. Pardee (70) used a uracil-requiring mutant of *E. coli* to investigate this problem. When uracil or phosphate was withheld from the medium, no β-galactosidase synthesis was observed upon introduction of the inducer. It was concluded that RNA synthesis is indeed necessary for the induced formation of the enzyme. On the other hand, cellular DNA synthesis is not required for induction and specific inhibitors of DNA synthesis fail to alter induced synthesis of the enzyme in any way.

We know that the *E. coli* cell contains a small number of β-galactosidase molecules even before induction. Thus the chromosome has structural information available for enzyme synthesis, but galactosidase-specific mRNA is apparently very scarce. If there were just one template in the cell at all times and if one polypeptide monomer of the β-galactosidase molecule were produced from it every 5 to 10 seconds, there would be 100 or so β-galactosidase tetramers formed per doubling time. In fact, the average number of enzyme molecules in uninduced cells is on the order of one or two. Soon after induction, the rate of β-galactosidase synthesis increases by a factor of 1000 and in order to guarantee the new rate of protein synthesis, specific mRNA templates must be formed. This explains why Pardee found that RNA synthesis is necessary for enzyme induction.

Mutation at the *i* locus has no effect on β-galactosidase structure. The enzymes synthesized by induced and constitutive strains are identical by the criteria of chromatography, electrophoresis, and immunology. They further possess the same Michaelis constants, demonstrating that their catalytic properties are identical, too. Consequently, the *i* gene does not influence the quality of β-galactosidase synthesized, but rather the quantity in which it is produced. The regulatory function of the *i* gene must therefore be exerted through an intermediate substance for whose synthesis it contains the necessary genetic information. The synthesis of the *i* gene product is constitutive, that is, not subject to regulation itself.

We shall now consider two alternative hypotheses concerning the nature of the *i* gene product and the role that it must play in the bacterial cell.

1. Assume that wild-type bacteria continuously elaborate a certain substance called *repressor*, which hinders the formation of mRNA templates for protein synthesis. The structure and synthesis of this endogenous repressor is determined by the i^+ allele of the regulator gene. When inducer molecules are introduced from the outside, they inactivate the repressor by their interaction with it. As a result, the synthesis

of galactosidase-specific mRNA is no longer prevented. In the constitutive i^- strain, active repressor is not synthesized, owing to a mutational alteration in the regulator gene. The proper mRNA template is hence formed continuously by the structural gene for β-galactosidase and protein synthesis occurs at maximal rates. This mode of regulation can be called repression at the level of transcription. By modifying the assumptions somewhat, a similar argument can be made for repression at the level of protein synthesis or translation, whereby the repressor would determine whether or not a specific mRNA molecule, once synthesized, can be read on the ribosome. The evidence is against translational regulation, however.

2. According to the second hypothesis, inducer directly stimulates the synthesis of β-galactosidase mRNA. The i^- genotype possesses the capacity to synthesize an *endogenous inducer* essential for switching on enzyme synthesis. The wild-type strain, lacking the ability to produce endogenous inducer, requires an external inducer instead to trigger β-galactosidase synthesis. Note carefully that in the first hypothesis it is assumed that the i^+ gene is the functional allele producing the repressor and that the i^- gene is inactive. The second hypothesis, however, predicts that the i^- gene is functionally active, producing an endogenous inducer, while the i^+ allele is assumed to be inactive. This provides a means for distinguishing the two hypotheses, since the allele which specifies an active product should be dominant in diploid zygotes.

Dominance at the i locus was determined in a series of ingenious experiments performed by Pardee, Jacob, and Monod (47). Parental strains with the genotypes i^+z^+ and i^-z^-, neither of which synthesizes β-galactosidase constitutively, were mated in reciprocal crosses. The synthesis of β-galactosidase was then followed in the resulting merozygotes during the 1 to 2 hours before recombinants were formed. When the i^-z^- genes are transferred from the Hfr donor to i^+z^+ recipients, β-galactosidase is not synthesized to any appreciable extent in the absence of inducer, suggesting that the i^+ allele is dominant to i^-. The merozygotes are inducible, however, just as are the female parents. But when the i^+z^+ genes are transferred to a i^-z^- female, β-galactosidase synthesis begins a few minutes after entry of the *lac* region and at rates characteristic of constitutive strains. The rate of enzyme formation falls to zero after an hour, however, although addition of inducer restores it to its initial rate. This is interpreted to mean that the i^+ gene produces a cytoplasmic repressor which prevents the z^+ gene from functioning. When the i^+ allele is transferred into the cytoplasm of an i^- recipient which initially contains no repressor, an appreciable interval of time elapses before enough repressor is synthesized to turn off the structural gene. This result was confirmed in stable heterogenotes of the type i^-z^+/F'-i^+z^-, which are inducible for enzyme synthesis. The genetic configuration of this diploid underscores the fact that repressor is transmitted through the cytoplasm, because the episomal i^+ gene represses activity of the bacterial z^+ structural gene even when they are oriented *trans* relative to one another. Only homogenotes of the type i^-/F'-i^- are constitutive.

It is natural at this point to inquire as to the chemical nature of the repressor specified by the i gene. While some early experiments were taken to indicate that repressor formation was unaffected by specific inhibitors of protein synthesis, there is now conclusive evidence that the *lac* repressor is a protein. The discovery of temperature-sensitive mutants (p. 307) in the i gene strongly implied that this locus codes the structure of a protein (71). Even more persuasive evidence was provided by the identification of amber mutations in the *i* gene which can be restored by the amber suppressors, su_I and su_{III} (72). Since suppressor mutations act by correcting nonsense codons in mRNA templates during translation into amino acid sequences, the repressor must be a protein. Gilbert and Müller-Hill have recently reported a successful assay for *lac* repressor which uses equilibrium dialysis against radioactively labeled inducer (IPTG). With this method, they demonstrated beyond doubt that the *i* gene product is a protein with the properties expected of a repressor (73). Uninducible itself, the molecule has a molecular weight of between 150,000 and 200,000, and it occurs in about 10 copies per gene. Repressor protein was isolated from wild-type cells and from specially selected strains designated i^t (tight-binding) in which the *i* gene product has a higher than normal affinity for IPTG. The absence of a substance binding inducer was shown in extracts from strains carrying an unsuppressed amber i^- allele, deletions in the *i* locus, or the i^s gene. The i^s (super-repressor) allele appears to produce a repressor that has no affinity for the inducer and therefore permanently shuts off the synthesis of β-galactosidase even when the structural gene is perfectly sound.

Very little is known about the structural peculiarities of inducers which permit them to bind with endogenous repressor, although such interactions are, naturally, of very great interest. In the case of β-galactosidase, presence in the inducer of the galactopyranoside ring appears to be essential, although the inducer need not be a close structural analog of the substrate. Thus the most effective inhibitors of β-galactosidase are seldom good inducers and vice versa. This is not surprising, since the enzyme inhibitor must recognize the active site on the enzyme, while the inducer must possess affinity for the active site of the repressor molecule so as to inactivate it.

To account for the influence of repressor on protein synthesis, Jacob and Monod (74) proposed that there exists another genetically determined element called the *operator* which is necessary for the transcription of structural genes in the *lac* region. The repressor molecules would thus interact with the operator to prevent transcription of mRNA. According to this hypothesis, there should be mutations in the operator which reduce its affinity for repressor, thereby leading to constitutive enzyme synthesis. Such mutations should also be independent of genetic alterations at the *i* locus and they should not map in that gene. Mutants of this type, designated o^c for *operator constitutive*, were indeed found among inducible heterogenotes of the configuration $i^+z^+/F'-i^+z^+$ (75). Since there are two copies of the i^+ allele in these strains, it would be very unlikely to find constitutives due to altered repressor, for both *i* genes would have to be affected simultaneously. All of the o^c mutations map in a small segment of the *lac* region between the *i* and *z* cistrons, although they are most closely linked with *z*.

The o^c mutations are dominant over both i^s, the super-repressor, and o^+, the wild-type operator, and permit constitutive synthesis of β-galactosidase, permease, and a third enzyme, galactoside acetylase (designated *a* on the genetic map) whose synthesis is coordinated with that of the other two proteins. In o^+/o^c heterogenotes, the o^c mutation is said to be *cis*-dominant because it affects only the structural genes which are located on the same chromosome. Thus in heterogenotes of the type $i^+o^cz^+y^-/F'-i^+o^+z^-y^+$, β-galactosidase is synthesized constitutively because the bacterial chromosome carries both o^c and the z^+ structural gene. In the same strain, permease synthesis is inducible, since the y^+ cistron is located on the episome, *cis* to an operator

(o^+) that mediates wild-type regulation, but *trans* relative to o^c. To stress the independence of regulation in the two chromosomes, z^+/z^-_{CRM} heterogenotes were constructed. While the z^+ gene gives rise to normal β-galactosidase, the z^-_{CRM} allele produces a protein (CRM or immunologically cross-reacting material) which, though enzymatically inactive, possesses the same immunological specificity as normal β-galactosidase. When the two structural genes are present in a heterogenote homozygous for o^+, both kinds of protein are synthesized upon induction. But in a heterogenote with the genetic constitution $i^+o^cz^+y^+/F'\text{-}i^+o^+z^-_{CRM}y^+$, β-galactosidase and permease are produced constitutively, while the inactive CRM protein is synthesized only after induction.

The extraction and purification of *lac* repressor has enabled Gilbert and Müller-Hill to directly confirm the binding of repressor to DNA from the *lac* region (76). Repressor was labeled with ^{14}C-amino acids to facilitate identification of its complex with *lac*-specific DNA obtained from the specially constructed transducing phage Φ80 d*lac* (p. 402f). Since the sedimentation coefficient of this DNA is 20 S to 30 S, and that of the repressor is 7 S, the DNA:repressor complex can be easily distinguished in the ultracentrifuge. The complex is characterized by a dissociation constant on the order of 10^{-12} to 10^{-13} M. Recall for comparison that dissociation constants for the most strongly bound enzyme-substrate complexes do not generally exceed 10^{-5} to 10^{-6} M. The exceptionally low value in the case of the DNA:repressor complex testifies to a very large enthalpy (or entropy) of binding. When such strong binding prevails, it is easy to calculate that 10 to 20 molecules of repressor per cell are quite sufficient to maintain the operator in the bound state as long as inducer is absent.

In these same experiments it was shown that the DNA:repressor complex can be dissociated by adding the inducer isopropyl-β-D-thiogalactoside to the solution. A particularly convincing control was the demonstration that if DNA was obtained from a bacterial strain bearing the o^c (operator constitutive) mutation, the binding of repressor to it could not be detected by sedimentation analysis. These data provide impressive confirmation of the model for genetic regulation advanced by Jacob and Monod.

Still another genetic element, a controlling site called the *promoter* (*p*), has been identified in the *lac* region. Consistent with the proposal of Jacob, Ullman, and Monod (77), the promoter is believed to be the initiation point for transcription of the structural genes of the *lac* region and it is possibly the site at which RNA polymerase becomes attached to the DNA. The promoter is located between the *i* gene and the operator, probably contiguous with the latter (Fig. 5-14). The existence of this genetic element was proved by the isolation of distinctive mutants which differ from the wild type in that the maximal rates of synthesis of β-galactosidase, permease, and transacetylase are coordinately reduced (78). All such mutants occupy a region of the DNA whose length corresponds to just a few nucleotide residues. Furthermore, the promoter mutations are *cis*-dominant, that is, they reduce only the rate of expression of genes on the same chromosome, and they do not alter the function of either operator or *i* gene. The promoter site is thought to be the initial substrate in the enzyme-substrate complex formed with RNA polymerase, the enzyme that transcribes mRNA from DNA. Mutations in the promoter region might therefore reduce the affinity of the site for the polymerase. The operator, placed immediately after the *p* region, but before the structural gene for β-galactosidase, is the site of regulator action. When the operator is occupied by repressor, the latter could block the progress of RNA polymerase along the DNA into the structural genes, even if the enzyme is bound to the promoter. An inducer such as lactose or some other β-galactoside changes the structure of the repressor, most likely due to an allosteric reaction, and as a result, the repressor is dissociated from the operator and the polymerizing enzyme can transcribe the z, y, and a genes.

The *lac* region thus contains four kinds of genetic elements: closely linked and probably contiguous *structural genes* (z, y, a), specifying the synthesis of β-galactosidase, galactoside permease, and galactoside acetylase; an *operator* (o) which is situated near to the end of the cluster of structural genes and which controls the expression of all three of them coordinately in conjunction with the product of the regulator gene; a *promoter* (p), the site of attachment of RNA polymerase to the DNA and of initiation of mRNA transcription; and a *regulator gene* (i). The orientation of these loci on the genetic map is shown in Fig. 5-14. This unit of coordinately expressed structural genes

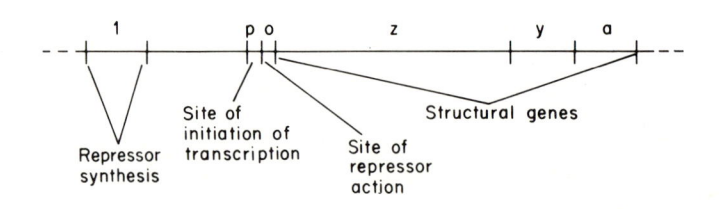

Fig. 5-14. Genetic linkage map of the *lac* operon in *E. coli*. Map shows the relative lengths of the individual elements and their position with respect to one another. The repressor, elaborated by the *i* gene, interacts with the operator *o* region in such a way as to control initiation by the promoter *p* of coordinate transcription of the operon, which contains the structural genes *z* for β-galactosidase, *y* for galactoside permease, and *a* for galactoside transacetylase. Adapted from Ippen *et al*. (78).

along with the controlling sites o and p is called an *operon*. The regulator gene, which is outside the operon, produces a cytoplasmic repressor that blocks the transcription of mRNA from the entire operon by binding to the operator. If no mRNA is produced, the proteins specified by the structural genes cannot be synthesized. In order for these genes to be expressed, an inducer must associate with and inactivate the repressor so that the latter can no longer interact with the operator. Released from repression in this way, the operator permits transcription of specific messenger RNA. The entire process is depicted schematically in Fig. 5-15A, p. 498.

The coordinated expression of gene clusters, each of whose components determines a separate stage in a single biochemical pathway, is not in any way limited to the *lac* operon. Similar phenomena have been observed among other groups of enzymes governing the utilization of sugars. Moreover, an apparently related mechanism of control applies to clusters of genes determining enzymes for the synthesis of pyrimidines and of several amino acids. Such biosynthetic pathways may contain as many as ten separate enzymes, all under coordinate control. But in these cases, the enzymes are not induced, but *coordinately repressed* by the end product of the pathway. In particular, a whole group of enzymes can be repressed by a single end product which is neither a substrate nor a product of the intermediate enzymes in the pathway, and which may even possess little or no structural similarity with its precursors.

An example of the coordinate repression of biosynthetic enzymes is provided by the histidine operon of *Salmonella typhimurium*. In this species of bacteria, the amino acid histidine is synthesized in ten sequential enzymatic steps, and the genes which control the structures of the nine necessary enzymes are all grouped together in one region of the chromosome. All of these enzymes are repressed coordinately and in approximately the same ratio when histidine is present in the medium (79). The products of the intermediate reactions do not contribute to repression of enzyme formation, however, as demonstrated with mutants in which the activity of one of the enzymes has been impaired. In such mutants, the genetic block leads to the accumulation of intermediates in the biosynthetic pathway, but they are not able to bring about a halt in the synthesis of the enzymes themselves. Although fully constitutive mutants have not been obtained in the case of the histidine enzymes, there is good evidence for existence of an operator region at the extreme end of the gene cluster. Presumptive operator constitutive mutants have been isolated in this region whose enzyme levels are considerably higher under conditions of repression than they are in the wild-type strain. In diploids, constitutive behavior is found only for those markers which are *cis* relative to the operator mutation. The position of the operator determined genetically corresponds with independent evidence on the direction in which the *his* operon is transcribed.

What all inducible and repressible systems have in common is that the expression of a group of functionally related cistrons, probably contiguous to one another along the chromosome, is regulated by a single operator located at one extremity of the gene sequence. The operon is thus a unit in which transcription of several cistrons is coordinated. The repressor acts by combining with the operator in a specific way so as to block transcription and prevent the flow of information to the ribosome where protein synthesis takes place. A general scheme for coordinate induction and repression was formulated by Jacob and Monod (65). Its main features are presented in Fig. 5-15. In the case of induction (Fig. 5-15A), the repressor directly interacts with the operator to block mRNA transcription; inducer binds the repressor so as to inactivate it and transcription of the operon can then proceed. Interpretation of coordinate repression, however, requires us to modify somewhat our conception of the product of the regulator gene. In repression (Fig. 5-15B), the regulator gene is assumed to specify an *aro-repressor* which cannot combine directly with the operator unless it is activated by the end product of the biosynthetic pathway or some derivative of it, called the *co-repressor*, to yield functional repressor. There is now evidence that the real co-repressor for regulation of the histidine operon in *Salmonella* is not histidine itself, but histidyl-tRNA. The small molecules which either activate or inactivate the product of the regulator gene are known generally as *effectors*. The interaction between effector and the product of the regulator gene may well be an allosteric one. In induction, the effector would act as an allosteric inhibitor, while in repression, the biosynthetic end product would be an allosteric activator of the apo-repressor.

In addition to the genetic evidence indicating that induction and repression exert their regulatory effects at the level of transcription of the genetic message, there is also biochemical evidence to support this conclusion. Both the *gal* and *lac* regions in *E. coli*

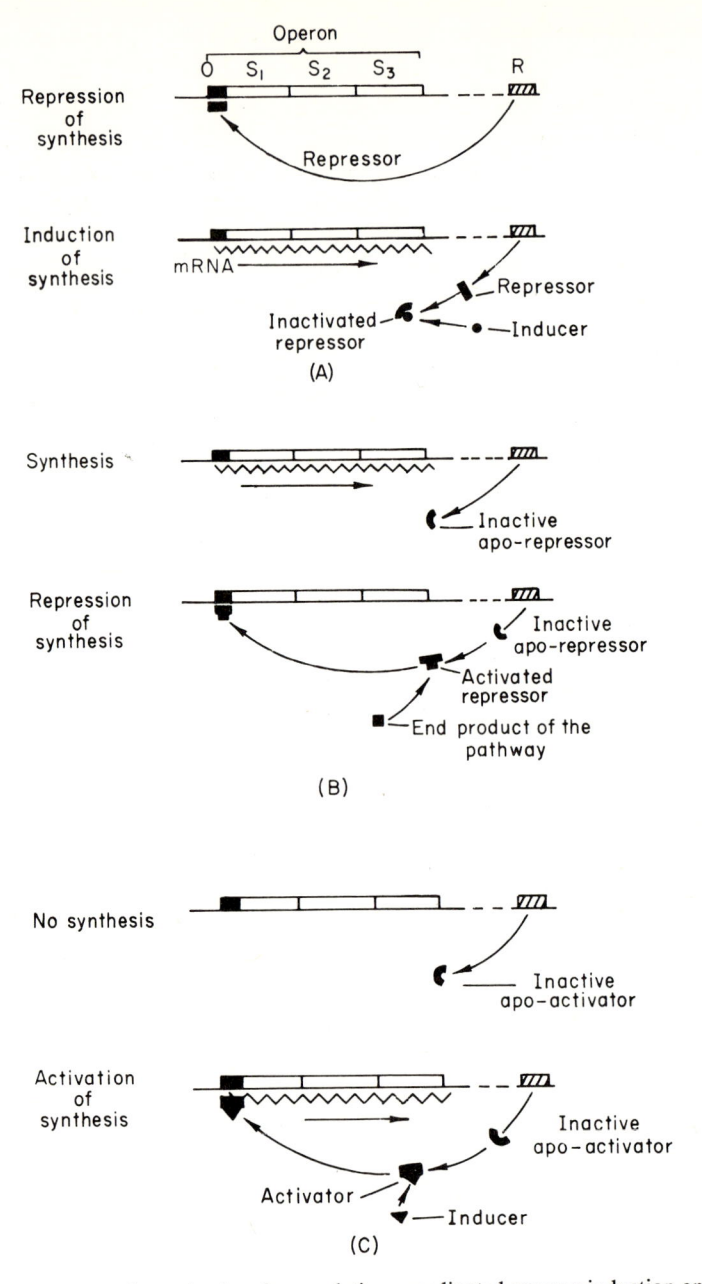

Fig. 5-15. The genetic mechanism for regulating coordinated enzyme induction and repression. S_1, S_2, and S_3 are three contiguous structural genes determining the synthesis of three enzymes serving a single metabolic pathway. O is the operator, the site at which transcription of genetic information of S_1, S_2, and S_3 into messenger RNA is regulated. R is the regulator gene. (A) In *coordinated enzyme induction* under *negative control*, the product of R is a specific repressor which, by combining with O, prevents transcription of the structural genes and, therefore, synthesis of the enzymes they specify. The function of inducer is to inactivate the repressor, thus releasing the operator from repression. (B) In *coordinated enzyme repression*, the product of R is an inactive apo-repressor which is specifically activated by the end product of the pathway. (C) In *positive control* of *coordinated enzyme induction*, the R gene elaborates an apo-activator which, in conjunction with an inducer forms an activator that switches on transcription of the operator upon interacting with O. Adapted from Hayes (80), p. 613.

contain inducible genes, and the specific mRNA transcribed from them should hybridize with corresponding segments of bacterial DNA, isolated as previously described from transducing phages or episomal particles. To test this hypothesis, DNA was extracted from the transducing phage λdgal and from the F'-lac episome and immobilized in agar gels. RNA prepared from bacteria both before and after induction of the respective locus was then passed through the gel. A large increase in complementary RNA was observed after induction, demonstrating that this process does lead to the production of specific *gal* or *lac* mRNA (81). In an analogous experiment with immobilized DNA from the defective phage Φ80d*trp* which carries the *E. coli* tryptophan locus, it was shown that the quantity of RNA that hybridizes with the tryptophan-specific DNA drops appreciably when the enzymes of this pathway are repressed by addition of excess tryptophan to the medium (82).These experiments leave no doubt that induction and repression of enzyme synthesis is regulated on the level of transcription of mRNA from the structural genes.

Furthermore, the mRNA is not transcribed from each cistron of an operon separately, but in the form of a long *polycistronic* message. To demonstrate this, Martin extracted mRNA from two strains of *Salmonella typhimurium* (83). One strain was derepressed for the histidine enzymes and was therefore assumed to be producing histidine-specific mRNA at a high rate. The other strain bore a large deletion in the middle of the histidine operon and hence served as a control. When both extracts were centrifuged in a sucrose density gradient, the first yielded an excess of mRNA with a sedimentation coefficient of 34 S, tentatively identified as the histidine-specific mRNA. Its molecular weight was estimated to be 4×10^6 daltons, equivalent to 1.3×10^4 nucleotide residues and of just about the size necessary to accommodate the nine cistrons of the histidine operon.

The repressor theory of genetic regulation has been applied with considerable success to the relationship between prophage and bacterial cell in lysogenic cultures. Recall that temperate phages such as λ can either multiply vegetatively within the bacterial cell, or else their DNA can become stably integrated into the bacterial chromosome as a prophage, where it is replicated along with the cellular DNA. One of the peculiarities of the prophage state is that it confers immunity upon the lysogenized cell to superinfection by the same temperate bacteriophage.

Conjugation between lysogenic and nonlysogenic cells has shed considerable light on the nature of immunity. If a male lysogenic for phage λ is mated with a nonlysogenic female, the prophage enters the vegetative state as soon as it penetrates the female cell. After a certain interval the zygote lyses, releasing infective phage particles. This phenomenon is called *zygotic induction*. When the male is nonlysogenic and the female lysogenic, or when both are lysogenic, zygotic induction does not occur. Note that the occurrence of zygotic induction is not dependent on the genetic constitution of the zygotes, for even though one chromosome bore prophage and the other did not in both cases described, the results were quite different. Zygotic induction must therefore be determined by the cytoplasm of the female strain. From this observation it was proposed that in lysogenic bacteria there is a cytoplasmic immunity substance or repressor which prevents the expression of superinfecting phages as well as of the prophage itself. When the prophage is introduced into a nonlysogenic cell whose cytoplasm contains no repressor, its genes are derepressed and the prophage multiplies vegetatively. The extremely low level of RNA in lysogenic cells capable of hybridizing with λ-DNA, and the appearance of considerable amounts of such RNA upon prophage induction, indicates that λ-repressor interferes with transcription of the λ-genome (84).

The similarities between lysogeny and repression were confirmed by the isolation of a number of mutants affecting immunity and lysogenization. Besides its structural genes, the λ-chromosome carries the c_I locus which is necessary for the maintenance of lysogeny and which is also responsible for immunity against superinfecting phages. Certain mutations of the c_I gene prevent the conversion of vegetative phage to prophage so that lytic infection always results. Such c_I^- mutants were assumed to be incapable of synthesizing active repressor that under normal conditions prevents the expression of phage genes and leads to the lysogenic state. These mutations are analogous to i^- mutations in the *lac* region. Other strains with mutations at the c_I locus, called *ind*$^-$, are able to lysogenize bacteria in the normal way, but they cannot be induced by ultraviolet light. The *ind*$^-$ mutants, however, are sensitive to zygotic induction, showing that they must continue to synthesize repressor. These phage mutants apparently produce an altered repressor molecule which is more resistant to the inducing effects of ultraviolet irradiation. The *ind*$^-$ mutant can be compared to the i^s super-repressor mutant in the *lac* region. Finally, virulent mutants, denoted λ_{vir}, have been obtained which are insensitive to repressor and are hence capable of multiplying vegetatively in lysogenic cells. These are quite similar to *lac* operator constitutive mutations.

The λ-repressor protein has recently been isolated by Ptashne (85). Special techniques were employed to increase the amount of λ-repressor synthesized to about 10% of the total protein produced by infected cells. These methods included inhibition of the synthesis of cellular proteins with massive doses of ultraviolet light, inhibition of the synthesis of most phage proteins by the use of special mutants, and infection of each bacterium with 30 to 35 phage particles so as to maximize the number of copies of the c_I gene present. Extracts from cells infected under these conditions yielded a single prominent peak of newly synthesized radioactive protein when chromatographed on DEAE-cellulose. This protein was not made in cells infected with amber mutants of the c_I gene, but was synthesized in modified form by temperature-sensitive mutants in the same gene. The λ-repressor was found to be an acidic protein with a molecular weight of about 30,000.

As with the *lac* system, the ability of the λ-repressor protein to bind with λ-phage DNA was examined by ultracentrifugal analysis (86). Labeled repressor was obtained from a phage bearing the *ind*$^-$ mutation in its c_I gene which apparently causes synthesis of a more stable repressor than that produced by the wild-type allele. Since repressor sediments at about 2.8 S, its complex with λ-DNA, which has a sedimentation coefficient of 33 S, is quite easy to discern by sucrose gradient analysis (Fig. 5-16). Repressor is bound at only two sites on the entire λ-genome. The dissociation constant in this case is estimated to be roughly 10^{-9} to 10^{-10} M. As a control, the ability of DNA from the phage mutant *imm*^{434} to bind repressor was also tested. This mutant differs from the wild type in that it bears the immunity region and hence the c_I gene of phage 434 although all of its other genes are identical to those of λ-phage. Phages with the *imm*^{434} region do not make and are not sensitive to λ-repressor, and it would therefore be likely that DNA from these mutants would not bind with λ-repressor. From Fig. 5-16

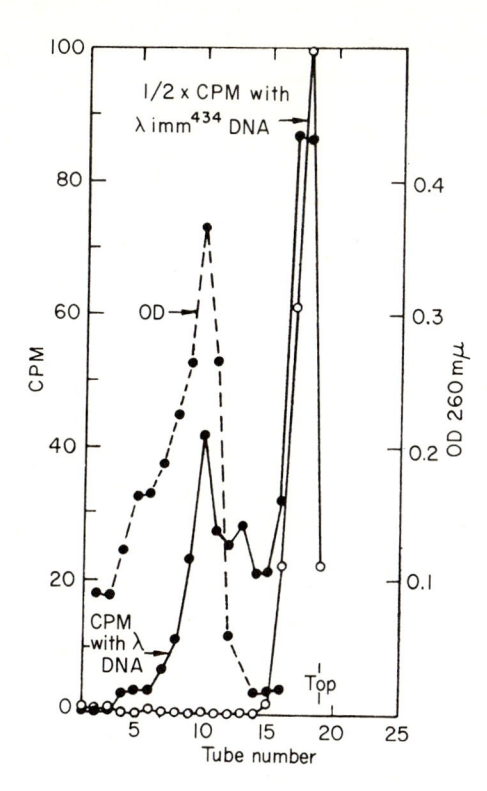

Fig. 5-16. Specific binding of the λind^- repressor to λ-DNA. ^{14}C-labeled λind^- repressor was mixed with 100 μg of λ or λimm^{434} DNA; the mixture was incubated for 5 minutes at 37°C and then centrifuged for several hours on a linear sucrose gradient. Separate fractions were assayed for ultraviolet absorption to locate the DNA and for radioactivity to locate the labeled repressor. From Ptashne (86).

it is apparent that this prediction is fully borne out. Furthermore, when cells are infected with a λ-mutant containing an unsuppressed amber mutation in its c_I gene, none of the proteins in the extract can form a complex with λ-DNA.

The influence of λ-repressor on the enzymatic transcription of mRNA from λ-DNA *in vitro* was tested by Echols and co-workers (87). They found that transcription of phage DNA by purified RNA polymerase was indeed inhibited several fold by λ-repressor.

All of the results cited represent a triumph for the theory of genetic regulation advanced by Monod and Jacob (65). However, data has accumulated on systems in which regulation of protein synthesis diverges in one or more aspects from the Monod-Jacob model. In particular, there appear to be alternative solutions to almost all of its major postulates. But rather than compromising the theory, these new findings serve to broaden and extend it, illustrating the way in which similar basic mechanisms have developed to meet specific metabolic needs in the course of evolution.

Let us first examine the postulate that functionally related genes are grouped together in polycistronic operons. This is of course a rational way for the genes of one pathway to be distributed on the chromosome, but it does not always occur. Thus the eight cistrons whose enzyme products are necessary for the biosynthesis of arginine in *E. coli* are scattered over the entire genome. Four of the *arg* genes are completely unlinked while the remaining four are tightly clustered. Even though there is clearly no operon containing all the *arg* cistrons, they are nonetheless subject to control by a single regulatory gene. In most *E. coli* strains, this gene apparently codes for an aporepressor which is activated by arginine so as to exert parallel repression of the *arg* genes (88). One operator is thought to be associated with each unlinked gene and at least one with the clustered genes.

The organization of genetic loci into operons could have arisen as the result of major chromosomal rearrangements such as translocations and deletions. During such events, chromosomal genes can be reshuffled and, if the outcome is useful to the cell, or at least not harmful, the mutation is fixed by natural selection. That such rearrangements can occur is supported by experimental evidence. For example, Jacob, Ullman, and Monod isolated mutant strains of *E. coli* bearing a deletion which extends from the *z* gene on one side to a cistron for purine biosynthesis on the other (89). The *i* gene and the *lac* controlling sites are excised by this deletion and the *z* gene is inactivated so that the cells cannot synthesize β-galactosidase. As a result, the remaining genes of the *lac* operon which code for permease (*y*) and acetylase (*a*) are fused to the purine operon and their synthesis becomes subject to purine regulation, that is, they are repressed by purines rather than induced by galactosides. This sort of chromosomal rearrangement is obviously of no value to the cell, but one can imagine that useful rearrangements also occur. Since evolution is a slow process, it is not at all surprising that while certain related genes are found to be widely separated from one another on the chromosome, others have been shifted into close proximity and have been organized so as to increase the efficiency of the cell.

The exclusively negative character of regulation is a second feature of the Monod-Jacob theory requiring modification. In the original formulation, regulation was envisioned to occur via a repressor synthesized under the control of a regulator gene, whether the structural genes were inducible or repressible. Although this appears to be the case for *lac* and a number of other systems, regulation of the enzymes participating in the metabolism of certain sugars (arabinose, rhamnose) appears to be *positive* (see Fig. 5-15C, p. 498). This means that the regulatory genes specify the synthesis of an *activator protein* that stimulates expression of an operon.

The best example of positive control is the arabinose (*ara*) operon in *E. coli* which contains three structural genes coding for three enzymes in the pathway for arabinose catabolism and which is inducible by arabinose. The operon lies between the leucine and threonine loci on the *E. coli* chromosome. Closely linked to the structural genes is the *C* gene which exerts coordinate regulation over the three genes of the operon. The *C* gene is thought to specify an inactive apo-activator protein capable of binding arabinose as co-activator to produce the activator (90). The mechanism of control was elucidated by the study of mutations in the *C* gene. In mutants designated C^- none of the three enzymes is synthesized although their structural genes remain intact. The recovery of C^- amber mutations showed that the product of the gene is necessary for expression of the operon and provides strong evidence that this product is a protein. Diploid strains constructed with the episome F'-*ara*, which carries the *ara* operon and the *C* gene, demonstrated that the wild-type C^+ allele is dominant to C^- since the *ara* enzymes remain inducible. Constitutive mutants of the C gene, C^c, were also obtained in which all three enzymes of the arabinose operon were synthesized in the absence of arabinose, the putative co-activator. In diploids, C^c is dominant to C^- as expected, but recessive to C^+, suggesting that the activator protein is composed of several subunits and that mixed multimers are produced in the C^+/C^c diploids which cannot serve as activators in the absence of arabinose. These observations provide further evidence that the C gene product is a protein activator whose absence prevents induction of the arabinose operon.

The vegetative development of λ-phage from induction or infection to the completion of mature virus particles is subject to regulation by viral genes at a number of successive stages. The appearance of viral gene products roughly fall into early, intermediate, and late periods of the vegetative cycle. Since separate groups of genes are expressed at different times during phage development, this system provides a very simple model for differentiation. At least three principal regulatory genes have been distinguished in λ-phage. Gene c_1 codes for λ-repressor which exerts negative control over expression of the entire viral genome. Mutations in the N gene produce multiple defects in both early functions such as DNA replication and the synthesis of two early proteins, exonuclease and an antigenic protein called β, and in late functions such as the synthesis of phage coat and tail proteins (Fig. 5-17). Mutations in gene Q affect

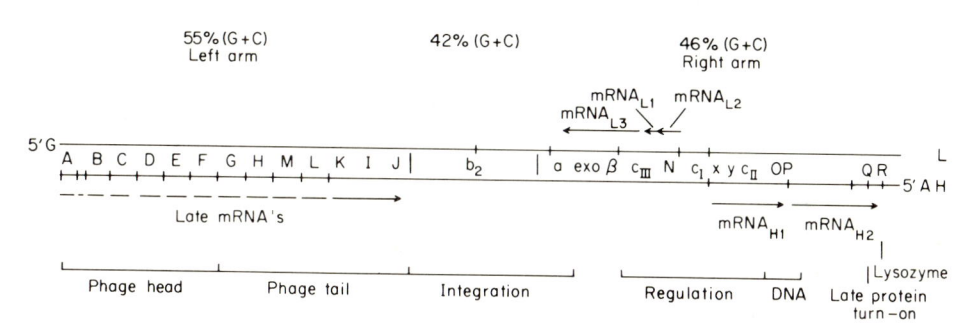

Fig. 5-17. The vegetative genetic map of λ-phage. The functional clustering of λ-genes is indicated below the map. Above the map are the base compositions in percent $G + C$ of both arms of λ-DNA and of the central b_2 region. 5'-G and 5'-A identify the 5'-terminal nucleotides and the polarity of the L and H strands, which differ from each other in density and can be separated by centrifugation in CsCl. The arrows indicate direction and strand from which various species of mRNA are transcribed as well as their approximate localization relative to the λ-map. The distribution of cytosine-rich clusters along the DNA is indicated by the symbol (─┼──┼─). Adapted from Taylor et al. (93); Echols et. al. (87); and Kourilsky et al. (94).

the expression of late functions subsequent to DNA replication. Both N and Q are thought to exert positive regulation over the transcription of groups of phage genes.

The role of N has been elucidated by studying its influence on the synthesis of the early proteins, exonuclease and β, whose structural genes are distinct from the N gene and map just to the left of it (91). Specifically, amber mutations in N block the formation of these two proteins when the cells do not carry a suppressor, while temperature-sensitive N mutants fail to produce them when the cells are grown at high temperature. Thus the product of the N gene must be a protein which is itself required for the synthesis of exonuclease and β-antigen. The N protein may therefore exercise positive control over the expression of the relevant structural genes.

The time sequence in which the various regions of the λ-chromosome are transcribed can be measured with considerable precision. This is done by hybridizing λ-specific mRNA synthesized at different times after induction or infection with DNA fragments corresponding to well-defined regions of the chromosome according to the method of Gillespie and Spiegelman (p. 483). In particular, phage mRNA can be hybridized (1) with the entire phage genome, (2) with DNA from either the left or right arms of the chromosome individually, or (3) with either of the two complementary DNA strands individually.

In order to separate the two DNA strands, denatured λ-DNA is annealed with the synthetic polynucleotide poly IG. Of the two strands, one contains an excess of cytosine-rich clusters (indicated by short vertical bars in Fig. 5-17) which preferentially bind G-rich polynucleotides. After centrifugation in a CsCl density gradient, two well-separated bands are recovered. The strand binding an excess of poly IG, denoted by the letter H, is found in the denser band. The less dense band contains the complementary or L strand.

Fractionation of λ-DNA into left and right half-molecules is performed by shearing the DNA in a hydrodynamic gradient after which the fragments are separated by chromatography on methylated albumin-kieselgur columns (92). The strands can be "dissected" even further by using DNA's from defective λ-mutants with deletions over well-defined segments of the chromosome. The failure of λ-specific mRNA to hybridize with either DNA strand bearing such a deletion provides good evidence that the mRNA is complementary to the region deleted. The particular strand copied can be determined by hybridization with separate L and H strands from normal λ-DNA. In addition, it is easy to see that the use of DNA from the defective transducing phage λd*gal* whose left arm from A to J is deleted and replaced by DNA from the bacterial chromosome is almost equivalent to shearing λ-DNA into half-molecules. The DNA from λd*gal* can be used to test the hybridization of mRNA's specifically with the right arm of the chromosome. These methods provide an accurate means for determining which genes and which DNA strand is being transcribed at any given time in phage development.

Using the techniques described above, Szybalski and his co-workers examined the mRNA synthesized under the direction of the λ-genome prior to prophage induction and at various times after the vegetative cycle had begun (93). In noninduced *E. coli* cells lysogenic for λ-phage, the great preponderance of mRNA is transcribed from the bacterial chromosome. Nonetheless, from 0.03 to 0.06% of the mRNA labeled in a 2-minute pulse of ^{3}H-uridine is complementary with λ-DNA. Although the amount of λ-specific mRNA is very small, these values are significantly higher than for mRNA from nonlysogenic cells. The λ mRNA made before induction is preferentially transcribed from the L strand and it is probably the product of the c_I gene since repressor synthesis is constitutive in lysogenic bacteria.

If prophages are induced by treating lysogenic cells with the antibiotic mitomycin C or, if the lysogen synthesizes a temperature-sensitive repressor, simply by raising the temperature to 42°, much more phage mRNA is produced. The proportion of phage mRNA rises to 0.5 to 2.0% of the total mRNA in early stages of phage development and from 6 to 12% in later stages, but no higher since the bacterial chromosome continues to be transcribed. All of the initial phage-specific mRNA hybridizes with the right arm of the chromosome and up to 90% is transcribed from a segment of the L strand to the left of c_1. The remainder is complementary to a region on the H strand to the right of c_1. Since this mRNA pattern is unchanged if cellular protein synthesis is inhibited by chloramphenicol or if the phage carries an unsuppressed polar or nonsense mutation in either the N or x gene, the portions transcribed early were inferred to be within a-N and x-O, respectively. Further transcription would then require the synthesis of a protein specified by genes in either or both of these two segments.

Most of the late mRNA, synthesized about 60 minutes after induction, hybridizes with the H strand in the left arm which contains the cistrons for phage coat and tail proteins. This was confirmed by isolation of mRNA following induction of $\lambda dgal$, a defective phage with a deletion covering the A-J region. All the early mRNA's from this strain are normal, but the late mRNA's which hybridize with the left arm of the chromosome are not formed. In addition, a substantial fraction of the late mRNA specifically hybridizes with the right arm of the H strand. The right arm of the chromosome is known to contain at least one late gene, R, the structural gene for phage endolysin (lysozyme). Infection of the cells with virulent forms of λ phage that are insensitive to or unable to synthesize repressor lead to much the same conclusions.

The nature of messenger RNA's formed early after induction has been examined in greater detail by fractionating them according to size before hybridization (94). When heat-induced lysogens are pulse-labeled with ^{3}H-uracil between 3 and 4 minutes after induction, five types of early phage-specific mRNA can be distinguished, none of which is present in the noninduced cell. All of these mRNA's are produced before the onset of DNA replication and they all hybridize with the right arm of the λ-chromosome. Their molecular weights were estimated by first separating the pulse-labeled extracts on sucrose gradients or polyacrylamide gels, and then hybridizing the individual fractions with various kinds of λ DNA. The approximate sizes of the early mRNA's were found to be

mRNA$_{L1}$	30,000 daltons
mRNA$_{L2}$	225,000 daltons
mRNA$_{L3}$	1,700,000 daltons
mRNA$_{H1}$	900,000 daltons
mRNA$_{H2}$	2,000,000 daltons

where the subscripts H and L denote the DNA strand to which the mRNA is complementary. Due to their large size, mRNA$_{L3}$, mRNA$_{H1}$ and mRNA$_{H2}$ are assumed to be transcribed from large polycistronic operons. Together, these five mRNA's represent about two thirds of the right arm of the λ-chromosome.

If cellular protein synthesis is inhibited by the addition of chloramphenicol at the time of induction, then mRNA$_{L3}$ and mRNA$_{H2}$ are not produced, though the other three mRNA's are. Thus the transcription of mRNA$_{L1}$, mRNA$_{L2}$, and mRNA$_{H1}$ does not require protein synthesis and it must be under the direct control of λ repressor. The significance of mRNA$_{L1}$ is as yet unknown, although it appears to be too small to code for a protein. The transcription of mRNA$_{L3}$ and mRNA$_{H2}$ requires a substance produced by at least one of the other three mRNA's. This substance is most likely a protein whose structural gene is in the N region since nonsense mutations in N also prevent formation of mRNA$_{L3}$ and mRNA$_{H2}$.

One of the mRNA's has been characterized further by studying a polar mutation in the x gene which is defective for the x-O functions and fails to replicate DNA. This mutant synthesizes mRNA that hybridizes almost exclusively with the L strand from the right arm of λ-DNA. A fraction of this mRNA is of very high molecular weight and has been identified as mRNA$_{L3}$. Its transcription site has been localized by using the DNA of phage λd*bio*, a biotin-transducing phage in which the *a*, *exo*, and β-genes are deleted. The high-molecular-weight mRNA$_{L3}$ hybridizes much less with this DNA than wild-type DNA, indicating that it is transcribed from the region deleted in λd*bio*. Conversely, mRNA$_{L3}$ is not synthesized after induction of λd*bio*, although another high-molecular-weight mRNA, designated mRNA$_{H2}$, is transcribed from the H strand of the defective phage DNA. Since (1) mRNA$_{L3}$ fails to hybridize with DNA bearing a deletion to the left of N and covering the genes for exonuclease and β-antigen and since (2) the synthesis both of mRNA$_{L3}$ and of those two proteins responds to mutations in the *N* gene and to chloramphenicol in a similar fashion, it may be concluded that mRNA$_{L3}$ is transcribed from the region shown in Fig. 5-17.

The positions of the other early mRNA's have also been confirmed by hybridization with DNA from appropriate deletion mutants. Thus mRNA$_{H1}$ corresponds to the x-y-c_{II}-O operon, and mRNA$_{L1}$ and mRNA$_{L2}$ cover the *N* gene and possibly c_{III}. It seems likely that mRNA$_{H2}$ is transcribed from the vicinity of the *Q* gene (Fig. 5-17).

A rather detailed picture of the control of early mRNA synthesis in λ phage emerges from these experiments. The λ-repressor specifically interacts with two operators on either side of the c_I or immunity region. One operator adjoins the left end of the c_I gene and controls transcription of gene *N* into mRNA$_{L2}$ from right to left. The other operator is located at the right end of c_I and controls transcription of mRNA$_{H1}$ from left to right along the x-y-c_{II}-O operon. The placement of the two operators has been verified by studies on λ*vir* (virulent), a mutant strain which develops even in the presence of λ-repressor. The λ*vir* strain contains multiple mutations (v_1, v_2, and v_3) which map at the left (v_2) and right (v_1 and v_3) extremities of the c_I region. As was shown by binding experiments *in vitro*, DNA from λ derivatives carrying either v_2 or v_1 and v_3 exhibits a markedly lower affinity for repressor than wild-type DNA (95). Such operator mutations presumably render the adjacent operons at least partially constitutive *in vivo*. After the two operons under direct control of the repressor are transcribed, one or more of their protein products is required to initiate transcription of mRNA$_{L3}$ and mRNA$_{H2}$, the first of which probably specifies the early proteins, exonuclease and β-antigen. It is hoped that this kind of analysis can be extended in the future to include phage operons necessary for the late functions in the vegetative cycle. For the present we can only conclude that the hierarchy of controls suggested by these experiments contributes substantially to the efficiency of phage development.

The third major postulate of the Monod-Jacob theory states that regulation occurs at the level of chromosomal transcription. An examination of *polarity* in the expression of some operons, however, indicates that regulation of protein synthesis may not be limited to transcription alone. *Polar mutations* are defined as lesions in an operon that lower or abolish the expression of genes distal to the mutated gene, taking the operator as origin. In the *lac* and tryptophan operons of *E. coli*, and in the tryptophan and histidine operons of *Salmonella typhimurium*, certain mutations both inactivate the cistron in which they occur and reduce the activity of all genes distal to the mutation, even though these genes are not themselves damaged. Moreover, the extent to which the expression of distal genes is reduced depends on the position of the polar mutations within a given gene. Mutations of this class are much more strongly polar when they are placed at the operator-proximal as opposed to the operator-distal end of a single gene. From studies on the nature of mutations in various operons, it has been determined that nonsense and frame-shift mutations are in general polar, while missense mutations are not. Since the first two classes of mutations interfere with protein synthesis, polar effects must arise during translation of the messenger RNA.

Interpretations of polarity have been based on the assumption, for which there is presently good physical evidence, that all genes within an operon are transcribed onto one polycistronic messenger RNA molecule. This explains very simply how a single operator can control the transcription of an entire set of genes. If mRNA molecules could be transcribed from several points within the operon, it would be very difficult to formulate a hypothesis accounting for their joint control by an operator situated at a unique position on the chromosome.

On the basis of these considerations, similar molecular mechanisms of polarity have been postulated independently by Yanofsky and Ito and by Martin and coworkers (96). During protein synthesis, according to this theory, ribosomes can become associated with the polycistronic mRNA only at the operator end. When the ribosome encounters a nonsense codon, chain termination occurs and the ribosome slips along the subsequent portion of the mRNA molecule without participating in protein synthesis. If the ribosome is dissociated from the mRNA before reaching the end of the cistron, none of the cistrons distal to that containing the nonsense codon will be translated. But if the ribosome encounters the chain-initiating region of another cistron before dissociating, polypeptide synthesis may resume. Whether or not it actually does so depends in part on the efficiency of the chain-initiation signal of the particular cistron. Thus polarity results not only from a finite probability that the ribosome will dissociate from the mRNA over a region which it is not translating (a probability that increases with the physical distance between nonsense codon and the next chain-initiation region), but also from the properties of the specific chain-initiation apparatus. An efficient initiator reorients the ribosome and permits it to resume translation. If the initiator is weak, however, the ribosome may fail to begin synthesis of a polypeptide chain and will continue to slip along the messenger until it either dissociates or encounters yet another chain-initiating region. Both of these factors together contribute to the diminishing extent of translation of cistrons in an operon, the farther they lie from the operator.

The validity of this explanation was generally confirmed by the discovery of strong polar effects *in vitro* when RNA from phage f2 bearing an amber mutation near the beginning of its first cistron was used as a template in a cell-free system from *E. coli* (97). By contrast, an amber mutation towards the end of the first cistron exhibited practically no polarity with regard to *in vitro* translation of the phage mRNA.

The incidence of polarity therefore indicates that regulation of genetic expression could occur during translation of information from mRNA into an amino acid sequence. It might, for instance, explain the observation that in the *lac* operon, β-galactosidase, the enzyme nearest the operator, is synthesized in much greater quantities than acetylase at the distal end of the operon. However, the expression of operons in the cell does not as a rule lead to decreasing amounts of distal gene products unless mutations are present in some of the genes. Furthermore, it can be argued that the mechanism postulated above lacks the specificity that usually characterizes genetic regulation.

Nonetheless, there are increasing indications that the tempo of mRNA transcription is set by the rate at which mRNA is translated into protein structure. The simplest hypothesis to account for such phenomena was formulated by Stent (98), who suggested that ribosomes attach to mRNA during its synthesis as soon as it emerges from the RNA polymerase and begin to translate it before transcription is completed. This process is illustrated in Fig. 5-18. The advance of the ribosome along the template

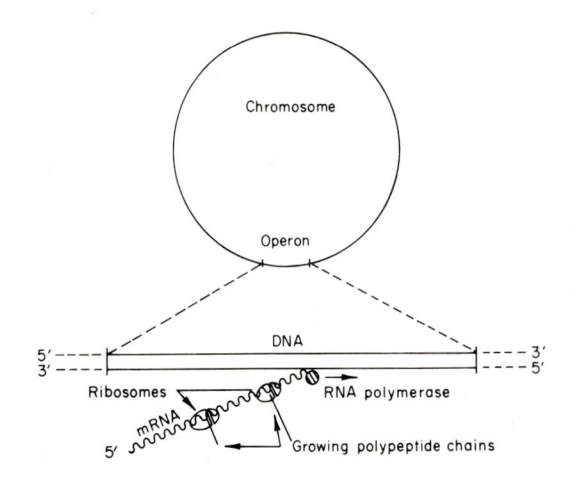

Fig. 5-18. Schematic representation of suggested coupling between translation and transcription. Ribosomes engage mRNA during its synthesis, forming a polyribosomal aggregate closely associated with DNA template. Translation starts immediately and the relative motion of mRNA and ribosomes during protein synthesis pulls the newly formed ribopolynucleotide away from the DNA-polymerase complex. After Stent (98).

could then directly influence the rate at which RNA polymerase transcribes the mRNA from the genome if the relative motion of mRNA and ribosomes in protein synthesis actively pulls the newly made RNA away from the DNA-polymerase complex. Among the many arguments which can be adduced in favor of this scheme is the important fact that the chemical direction in which mRNA is synthesized is the same as the chemical direction in which mRNA is translated during protein synthesis, namely, from the 5'-terminus of the polynucleotide. An advantage of this mechanism would be that the intracellular level of mRNA would not exceed the number of ribosomes available for its translation.

Experimental evidence for the postulated link between translation and transcription has been obtained by Imamoto and Yanofsky (99). They investigated the properties of a variety of polar nonsense and frame-shift mutations distributed throughout the five cistrons of the tryptophan (*trp*) operon in *E. coli*. When this operon is derepressed by starvation for tryptophan, the enzymes coded by genes containing polar mutations are not synthesized and those operator-distal to the mutation are formed in reduced amounts. The nature of the *trp*-specific mRNA produced by these mutants was examined to determine whether translational polarity was associated with alterations in transcription. Accordingly, polar mutants were incubated with radioactive uridine for short periods soon after derepression in order to label *trp*-specific mRNA. After extraction, the mRNA was hybridized with DNA from a series of Φ80pt transducing phages which carried either the entire *trp* operon or specific segments of it (Fig. 5-19).

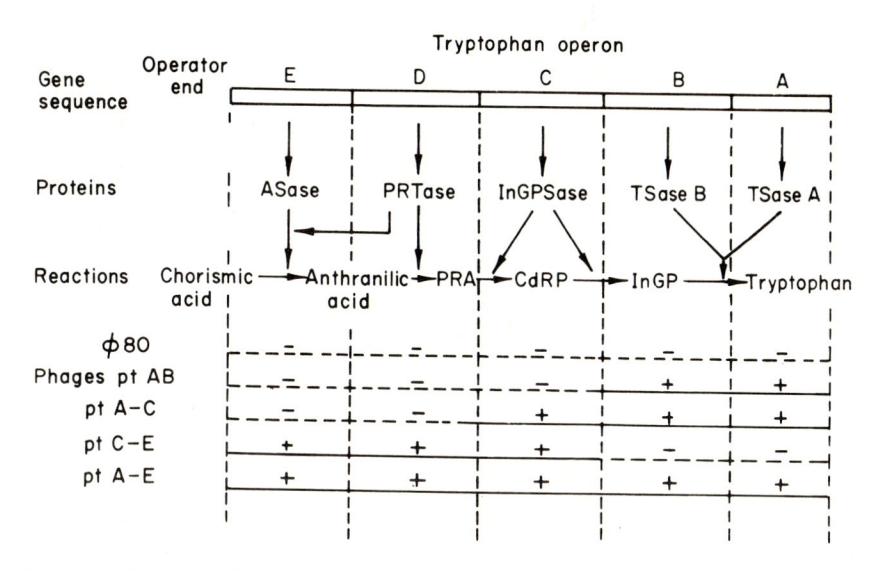

Gene sequence	Operator end	E	D	C	B	A
Proteins		ASase	PRTase	InGPSase	TSase B	TSase A
Reactions	Chorismic acid →	Anthranilic acid →	PRA →	CdRP →	InGP →	Tryptophan
Φ80 Phages pt AB	−	−	−	−	+	+
pt A–C	−	−	−	+	+	+
pt C–E	+	+	+	−	−	
pt A–E	+	+	+	+	+	

Fig. 5-19. Segments of the tryptophan operon carried by various Φ80pt transducing phages. Each region indicated by + is carried by the particular Φ80pt. Each one of these phages is a source of highly specific DNA bearing precise regions of the tryptophan operon. From Imamoto and Yanofsky (99).

Using the various *trp* DNA's, it was possible to assay which *trp* genes were transcribed into mRNA by bacteria with different polar mutations in the operon. Strong polarity mutants were consistently found to contain less total *trp*-mRNA than the wild type strain and most of the *trp*-mRNA molecules formed in each mutant were smaller than normal. Specific hybridization with DNA from pt phages such as pt$_{AB}$ and pt$_{A-C}$ indicated that the *trp*-mRNA from these strains was deficient in the segments corresponding to genes which are operator-distal to the mutated gene. Furthermore, the size of the *trp*-mRNA produced by various polar mutants increased in proportion to the distance between operator and mutation. There is no evidence that the *trp*-mRNA's found in the polar mutants stem from selective degradation of larger molecules. Since nonsense and frame-shift mutations cause premature termination of polypeptide synthesis, the above results imply that the halt in translation evokes a corresponding halt in transcription in the vicinity of a nonsense codon. Of course, such events do not always occur because the polar mutation can sometimes be bypassed, in which case both translation and transcription continue.

Although the experiments cited here indicate that transcription is influenced by translation, study of regulation at the level of translation has only just begun. It is quite possible that translational controls are of fundamental importance, particularly in the cells of higher organisms where the messenger RNA appears to be much more stable than in bacterial cells.

References

1. M. B. Hoagland, M. L. Stephenson, J. F. Scott, L. I. Hecht, and P. C. Zamecnik, *J. Biol. Chem.* **231**, 241 (1958); J. Preiss, P. Berg, E. J. Ofengand, F. H. Bergmann, and M. Dieckmann, *Proc. Natl. Acad. Sci. U.S.* **45**, 319 (1959).
2. H. Feldmann and H. G. Zachau, *Biochem. Biophys. Res. Commun.* **15**, 13 (1964).
3. S. Benzer and B. Weisblum, *Proc. Natl. Acad. Sci. U.S.* **47**, 1149 (1961).
4. B. P. Doctor, J. E. Loebel, and D. A. Kellogg, *Cold Spring Harbor Symp. Quant. Biol.* **31**, 543 (1966).
5. R. Rendi and S. Ochoa, *J. Biol. Chem.* **237**, 3707 (1962).
6. T. Yamane, T.-Y. Cheng, and N. Sueoka, *Cold Spring Harbor Symp. Quant. Biol.* **28**, 569 (1963).
7. E. B. Keller and P. C. Zamecnik, *J. Biol. Chem.* **221**, 45 (1956); D. Nathans and F. Lipmann, *Proc. Natl. Acad. Sci. U.S.* **47**, 497 (1961); G. von Ehrenstein and F. Lipmann, *Proc. Natl. Acad. Sci. U.S.* **47**, 941 (1961).
8. M. B. Yarmolinsky and G. L. De La Haba, *Proc. Natl. Acad. Sci. U.S.* **45**, 1721 (1959).
9. F. Chapeville, F. Lipmann, G. von Ehrenstein, B. Weisblum, W. J. Ray, Jr., and S. Benzer, *Proc. Natl. Acad. Sci. U.S.* **48**, 1086 (1962).
10. G. von Ehrenstein, B. Weisblum, and S. Benzer, *Proc. Natl. Acad. Sci. U.S.* **49**, 669 (1963).
11. B. Weisblum, S. Benzer, and R. W. Holley, *Proc. Natl. Acad. Sci. U.S.* **48**, 1449 (1962).
12. T. Yamane and N. Sueoka, *Proc. Natl. Acad. Sci. U.S.* **51**, 1178 (1964).
13. B. Weisblum, F. Gonano, G. von Ehrenstein, and S. Benzer, *Proc. Natl. Acad. Sci. U.S.* **53**, 328 (1965).
14a. H. E. Huxley and G. Zubay, *J. Mol. Biol.* **2**, 10 (1960).
14b. B. J. McCarthy, *Biochim. Biophys. Acta*, **55**, 880 (1962).
15. M. W. Nirenberg and J. H. Matthaei, *Proc. Natl. Acad. Sci. U.S.* **47**, 1588 (1961).
16. D. Schlessinger, G. Mangiarotti, and D. Apirion, *Proc. Natl. Acad. Sci. U.S.* **58**, 1782 (1967).
17. J. D. Watson, "Molecular Biology of the Gene." Benjamin, New York, 1965.
18. S. E. Bresler, N. N. Vasilyeva, R. A. Grayevskaya, S. V. Kirillov, and E. M. Saminskij, *Biokhimiya* **29**, 353 (1964); S. Bresler, R. Grajevskaja, S. Kirilov, E. Saminski, and F. Shutov, *Biochim. Biophys. Acta* **123**, 534 (1966).
19. W. Gilbert, *J. Mol. Biol.* **6**, 374 (1963); F. O. Wettstein and H. Noll, *J. Mol. Biol.* **11**, 35 (1965).
20. J. E. Allende, R. Monro, and F. Lipmann, *Proc. Natl. Acad. Sci. U.S.* **51**, 1211 (1964); J. Lucas-Lenard and F. Lipmann, *Proc. Natl. Acad. Sci. U.S.* **55**, 1562 (1966); F. Lipmann, *Science* **164**, 1024 (1969).
21. A. Parmeggiani, *Biochem. Biophys. Res. Commun.* **30**, 613 (1968).
22. J. M. Ravel, R. L. Shorey, C. W. Garner, R. C. Dawkins, and W. Shive, *Cold Spring Harbor Symp. Quant. Biol.* **34**, 321 (1969); H. Weissbach, N. Brot, D. Miller, M. Rosman, and R. Ertel, *Cold Spring Harbor Symp. Quant. Biol.* **34**, 419 (1969); A. Skoultchi, Y. Ono, J. Waterson, and P. Lengyel, *Cold Spring Harbor Symp. Quant. Biol.* **34**, 437 (1969); J. Lucas-Lenard, P. Tao, and A.-L. Haenni, *Cold Spring Harbor Symp. Quant. Biol.* **34**, 455 (1969).
23. F. Lipmann, in "International Symposium on Regulatory Mechanisms in Nucleic Acid and Protein Biosynthesis, Luntern, 1966" (V. V. Konigsberger and L. Bosch, eds), p. 177. Elsevier, Amsterdam, 1967.
24. B. E. H. Maden, R. R. Traut, and R. E. Monro, *J. Mol. Biol.* **35**, 333 (1968).
25. H. P. Ghosh and H. G. Khorana, *Proc. Natl. Acad. Sci. U.S.* **58**, 2455 (1967); C. Guthrie and M. Nomura, *Nature* **219**, 232 (1968).
26. M. S. Bretscher, *J. Mol. Biol.* **34**, 131 (1968).
27. M. R. Capecchi, *Proc. Natl. Acad. Sci. U.S.* **58**, 1144 (1967).

References

28. G. J. Spyrides and F. Lipmann, *Proc. Natl. Acad. Sci. U.S.* **48**, 1977 (1962); J. R Warner, P. M. Knopf, and A. Rich, *Proc. Natl. Acad. Sci. U.S.* **49**, 122 (1963); A. Gierer, *J. Mol. Biol.* **6**, 48 (1963).
29. H. S. Slayter, J. R. Warner, A. Rich, and C. E. Hall, *J. Mol. Biol.* **7**, 652 (1963).
30. A. Rich, J. R. Warner, and H. M. Goodman, *Cold Spring Harbor Symp. Quant. Biol.* **28**, 269 (1963).
31. R. E. Thach, M. A. Cecere, T. A. Sundararajan, and P. Doty, *Proc. Natl. Acad. Sci. U.S.* **54**, 1167 (1965).
32. L. Gorini and E. Kataja, *Proc. Natl. Acad. Sci. U.S.* **51**, 487 (1964); J. Davies, W. Gilbert, and L. Gorini, *Proc. Natl. Acad. Sci. U.S.* **51**, 883 (1964).
33. W. Szer and S. Ochoa, *J. Mol. Biol.* **8**, 823 (1964).
34. M. Grunberg-Manago and M. Dondon, *Biochem. Biophys. Res. Commun.* **18**, 517 (1965).
35. A. G. So and E. W. Davie, *Biochemistry* **3**, 1165 (1964).
36. J. Davies, L. Gorini, and B. D. Davis, *Mol. Pharmacol.* **1**, 93 (1965).
37. S. M. Friedman and I. B. Weinstein, *Proc. Natl. Acad. Sci. U.S.* **52**, 988 (1964).
38. K. Hosokawa, R. Fujimura, and M. Nomura, *Proc. Natl. Acad. Sci. U.S.* **55**, 198 (1966); M. I. Lerman, A. S. Spirin, L. P. Gavrilova, and V. F. Golov, *J. Mol. Biol.* **15**, 268 (1966); T. Staehelin and M. Meselson, *J. Mol. Biol.* **16**, 245 (1966).
39. P. Traub and M. Nomura, *Proc. Natl. Acad. Sci. U.S.* **59**, 777 (1968).
40. P. Traub and M. Nomura, *Science* **160**, 198 (1968).
41. J. Bishop, J. Leahy, and R. Schweet, *Proc. Natl. Acad. Sci. U.S.* **46**, 1030 (1960).
42. H. M. Dintzis, *Proc. Natl. Acad. Sci. U.S.* **47**, 247 (1961).
43. K. McQuillen, R. B. Roberts, and R. J. Britten, *Proc. Natl. Acad. Sci. U.S.* **45**, 1437 (1959).
44. A. Goldstein and B. J. Brown, *Biochem. Biophys. Acta* **53**, 438 (1961).
45. A. N. Belozersky and A. S. Spirin, *Nature* **182**, 111 (1958).
46. C. I. Davern and M. Meselson, *J. Mol. Biol.* **2**, 153 (1960).
47. A. B. Pardee, F. Jacob, and J. Monod, *J. Mol. Biol.* **1**, 165 (1959).
48. E. Volkin and L. Astrachan, in "The Chemical Basis of Heredity" (W. D. McElroy and B. Glass, eds.), p. 686. Johns Hopkins Press, Baltimore Maryland, 1957.
49. S. Brenner, F. Jacob, and M. Meselson, *Nature* **190**, 576 (1961).
50. F. Gros, W. Gilbert, H. H. Hiatt, G. Attardi, P. F. Spahr, and J. D. Watson, *Cold Spring Harbor Symp. Quant. Biol.* **26**, 111 (1961).
51. S. Spiegelman, *Cold Spring Harbor Symp. Quant. Biol.* **26**, 75 (1961); S. Spiegelman, B. D. Hall, and R. Storck, *Proc. Natl. Acad. Sci. U.S.* **47**, 1135 (1961).
52. B. D. Hall and S. Spiegelman, *Proc. Natl. Acad. Sci. U.S.* **47**, 137 (1961).
53. F. Gros, H. Hiatt, W. Gilbert, C. G. Kurland, R. W. Risebrough, and J. D. Watson, *Nature* **190**, 581 (1961).
54. S. Spiegelman and M. Hayashi, *Cold Spring Harbor Symp. Quant. Biol.* **28**, 161 (1963).
55. M. Hayashi and S. Spiegelman, *Proc. Natl. Acad. Sci. U.S.* **47**, 1564 (1961).
56. E. K. F. Bautz, *Proc. Natl. Acad. Sci. U.S.* **49**, 68 (1963).
57. D. Gillespie and S. Spiegelman, *J. Mol. Biol.* **12**, 829 (1965).
58. G. Mangiarotti and D. Schlessinger, *J. Mol. Biol.* **29**, 395 (1967).
59. C. Levinthal, A. Keynan, and A. Higa, *Proc. Natl. Acad. Sci. U.S.* **48**, 1631 (1962); C. Levinthal, D. P. Fan, A. Higa, and R. A. Zimmermann, *Cold Spring Harbor Symp. Quant. Biol.* **28**, 183 (1963).
60. W. B. Wood and P. Berg, *Cold Spring Harbor Symp. Quant. Biol.* **28**, 237 (1963).
61. D. Nathans, G. Notani, J. H. Schwartz, and N. D. Zinder, *Proc. Natl. Acad. Sci. U.S.* **48**, 1424 (1962).
62. E. Viñuela, M. Salas, and S. Ochoa, *Proc. Natl. Acad. Sci. U.S.* **57**, 729 (1967).
63. W. Salser, R. F. Gesteland, and A. Bolle, *Nature* **215**, 588 (1967).
64. B. J. McCarthy and B. H. Hoyer, *Proc. Natl. Acad. Sci. U.S.* **52**, 915 (1964).
65. F. Jacob and J. Monod, *J. Mol. Biol.* **3**, 318 (1961).
66. D. S. Hogness, in "Biophysical Science" (J. L. Oncley, F. O. Schmitt, R. C. Williams, M. D. Rosenberg, and R. H. Bolt, eds.), p. 256. Wiley, New York, 1959.
67. L. A. Herzenberg, *Biochim. Biophys. Acta*, **31**, 525 (1959).

68. D. S. Hogness, M. Cohn, and J. Monod, *Biochim. Biophys. Acta* **16**, 99 (1955).
69. B. Rotman and S. Spiegelman, *J. Bacteriol.* **68**, 419 (1954).
70. A. B. Pardee, *Proc. Natl. Acad. Sci. U.S.* **40**, 263 (1954).
71. A. Novick, E. S. Lennox, and F. Jacob, *Cold Spring Harbor Symp. Quant. Biol.* **28**, 397 (1963).
72. S. Bourgeois, M. Cohn, and L. E. Orgel, *J. Mol. Biol.* **14**, 300 (1965); B. Müller-Hill, *J. Mol. Biol.* **15**, 374 (1966).
73. W. Gilbert and B. Müller-Hill, *Proc. Natl. Acad. Sci. U.S.* **56**, 1891 (1966).
74. F. Jacob and J. Monod, *Cold Spring Harbor Symp. Quant. Biol.* **26**, 193 (1961).
75. F. Jacob, D. Perrin, C. Sanchez, and J. Monod, *Compt. Rend.* **250**, 1727 (1960).
76. W. Gilbert and B. Müller-Hill, *Proc. Natl. Acad. Sci. U.S.* **58**, 2415 (1967).
77. F. Jacob, A. Ullman, and J. Monod, *Compt. Rend.* **258**, 3125 (1964).
78. K. Ippen, J. H. Miller, J. Scaife, and J. Beckwith, *Nature* **217**, 825 (1968).
79. B. N. Ames and B. Garry, *Proc. Natl. Acad. Sci. U.S.* **45**, 1453 (1959); B. N. Ames, B. Garry, and L. A. Herzenberg, *J. Gen. Microbiol.* **22**, 369 (1960); J. R. Roth, D. F. Silbert, G. R. Fink, M. J. Voll, D. Anton, P. E. Hartman, and B. N. Ames, *Cold Spring Harbor Symp. Quant. Biol.* **31**, 383 (1966).
80. W. Hayes, "The Genetics of Bacteria and their Viruses." Blackwell, Oxford, 1964.
81. G. Attardi, S. Naono, J. Rouvière, F. Jacob, and F. Gros, *Cold Spring Harbor Symp. Quant. Biol.* **28**, 363 (1963).
82. F. Imamoto, N. Morikawa, K. Sato, S. Mishima, and T. Nishimura, *J. Mol. Biol.* **13**, 157 (1965).
83. R. G. Martin, *Cold Spring Harbor Symp. Quant. Biol.* **28**, 357 (1963).
84. W. S. Sly, H. Echols, and J. Adler, *Proc. Natl. Acad. Sci. U.S.* **53**, 378 (1965).
85. M. Ptashne, *Proc. Natl. Acad. Sci. U.S.* **57**, 306 (1967).
86. M. Ptashne, *Nature* **214**, 232 (1967).
87. H. Echols, L. Pilarski, and P. Y. Cheng, *Proc. Natl. Acad. Sci. U.S.* **59**, 1016 (1968).
88. G. A. Jacoby and L. Gorini, *J. Mol. Biol.* **39**, 73 (1969).
89. F. Jacob, A. Ullman, and J. Monod, *J. Mol. Biol* **13**, 704 (1965).
90. D. Sheppard and E. Engelsberg, *Cold Spring Harbor Symp. Quant. Biol.* **31**, 345 (1966); D. E. Sheppard and E. Engelsberg, *J. Mol. Biol.* **25**, 443 (1967).
91. C. M. Radding and H. Echols, *Proc. Natl. Acad. Sci. U.S.* **60**, 707 (1968).
92. D. S. Hogness and J. R. Simmons, *J. Mol. Biol.* **9**, 411 (1964).
93. K. Taylor, Z. Hradecna, and W. Szybalski, *Proc. Natl. Acad. Sci. U.S.* **57**, 1618 (1967).
94. P. Kourilsky, L. Marcaud, P. Sheldrick, D. Luzatti, and F. Gros, *Proc. Natl. Acad. Sci. U.S.* **61**, 1013 (1968).
95. M. Ptashne and N. Hopkins, *Proc. Natl. Acad. Sci. U.S.* **60**, 1282 (1968).
96. C. Yanofsky and J. Ito, *J. Mol. Biol.* **21**, 313 (1966); R. G. Martin, D. F. Silbert, D. W. E. Smith, and H. J. Whitfield, Jr., *J. Mol. Biol.* **21**, 357 (1966).
97. D. L. Engelhardt, R. E. Webster, and N. D. Zinder, *J. Mol. Biol.* **29**, 45 (1967).
98. G. S. Stent, *Proc. Roy. Soc. (London)* **B164**, 181 (1966).
99. F. Imamoto and C. Yanofsky, *J. Mol. Biol.* **28**, 1 (1967).

Author Index

Numbers in parentheses are reference numbers and indicate that an author's work is referred to although his name is not cited in the text. Numbers in italics show the page on which the complete reference is listed.

A

Abe, M., 329(27), *444*
Abelson, J., 441(195), *448*
Adams, J. M., 443(197)
Adams, M. H., 370(89), *445*
Adelberg, E. A., 325, 354(53), *444*, *445*
Adler, J., 500(84), *512*
Akinrimis, E. O., 244(47), *296*
Alberts, B. M., 247(50), *296*
Alberty, R. A., 142(18), *198*
Alden, R. A., 98(91), 99(91), *126*
Aldoshin, V. G., 32(27), 68(27), 69, *124*
Allende, J. E., 461(20), *510*
Allison, A. C., 425, *447*
Ames, B. N., 398, 412(148), *446*, *447*, 497(79), *512*
Anderer, F. A., 22, *124*
Anders, M., 254(69), *297*
Anderson, F., 436(177), *448*
Anderson, T. F., 319, *444*
Anfinsen, C. B., 7(2), 73, *124*, *125*, 132, 134(11), 135, 140, *198*
Anraku, N., 381, *446*
Anton, D., 497(79), *512*

Apgar, J., 210(4), 213, 285, 286(4), *295*
Apirion, D., 460(16), *510*
Argetsinger Steitz, J. E., 367(82), *445*
Asadourian, A., 54(57), *125*
Astbury, W. T., 195, *199*
Astrachan, L., 479, *571*
Attardi, G., 479(50), 482(50), 484(50), 499(81), *511*, *512*
Aurisicchio, A., 282(108), *298*
Avery, O. T., 355, *445*

B

Baltimore, D., 259(80), *297*
Bamford, C. H., 26(21a), 28, 34(32), *124*, *125*
Barclay, R. K., 216(11), *295*
Barnett, L., 371(88), 421(159), *445*, *447*
Baron, L. S., 326(23), *444*
Barrell, B. G., 286(119), *298*
Basilio, C., 434(174), *447*
Bautz, E. K. F., 414(152), *447*, 483, *511*
Bautz-Freese, E., 408(144), 409, *447*
Beadle, G. W., 300, *443*
Beaudreau, G., 259(81), *297*

Becker, A., 257(64), *297*
Becker, Y., 259(80), *297*
Beckwith, J. R., 307(11), 339(43), 402, *444, 447*, 495(78), 496(78), *512*
Beighton, E., 195, *199*
Bello, J., 54(88), 96(88), *126*
Belozerskij, A. N., 213, *295* 424, *447*, 478, *511*
Bennett, T. R., 437(180), *448*
Benzer, S., 307(11), 326, 391(88), 387, 388, 389, 390(114), 391, *444, 445, 446*, 454, 455(9), 456(10, 11), *570*
Berends, F., 136(12), *198*
Berg, P., 254(72), 297, 451(1), 485, *570, 571*
Berger, A., 40, *125*
Bergmann, F. H., 451(1), *510*
Bernfield, M., 436(177), *448*
Bernhard, S. A., 148(19), *198*
Bertani, G., 391(117), *446*
Billeter, M. A., 259(78), *297*
Birshtein, T. M., 65(68), *125*
Bishop, J., 470(41), *511*
Biszku, E., 105(96), *126*
Blake, C. C. F., 54(86), 96(86), *126*
Bleake, R. D., 294(133), *298*
Blinkova, A. A., 332(31), 333, 337(36), *444*
Blout, E. R., 35(33), 46(47), 47(48), 48, 54, 55, 56, 58, 59(33, 61), 63(66, 67), *125*
Blow, D. M., 54(90), 98(90), *126*
Bluhm, M. M., 86, *126*
Bock, R. M., 294(131), *298*, 437(181), *448*
Bodmer, W. F., 362, *448*
Bodo, G., 79(75), 81(75), 83(80), 84, 86(81), 87(75), 89, *125, 126*
Boedtker, H., 221(18), 222(18), 223(181), 246(18), 288(18), 289, 291, *296, 298*
Bogdanova, E., 294, *298*
Bolle, A., 383(101), 440(188), 441(193, 194), *446, 448*, 486(63), *511*
Bollum, F. J., 248(53), *296*
Bolton, E. T., 283, *298*
Borst, P., 259(78), *297*
Boshart, G. L., 28(23), *124*
Botts, D. J., 148(19), *198*
Bourgeois, S., 494(72), *512*
Boy de la Tour, E., 383(101), *446*
Boyes-Watson, J., 110, *126*
Bradbury, J. H., 44(44), 45(44), *125*
Bragg, J. K., 65(68), *125*
Brandenburg, D., 31(25), *124*
Branson, H. R., 32(28), 34(28), *124*
Braunshtejn, A. E., 156, *198*
Bremer, H., 31(25), *124*
Brenner, S., 107(99), *126*, 307(11), 333, 371, 411(147), 420(158), 421, 439, 440(188, 189), 441(189, 191, 195), *444, 445, 447, 448*, 479, *511*

Bresler, S. E., 32(27), 59(62, 63), 60, 68(27), 69(27), 70(62, 63), 71, 74, *124, 125*, 290(122), *298*, 332(31), 333(31), 337(36), 358, 359, 362(68), 403(139), 416(154), 417(155), 418, 419, *444, 445, 447*, 461(18), *510*
Bretscher, M. S., 465(26), *510*
Brimacombe, R., 436(177), *448*
Brinkoff, O., 31(25), *124*
Britten, R. J., 118(105), *126*, 473(43), 474(43), 475(43), *571*
Brody, S., 439, *448*
Brohult, S., 103(95), *126*
Brot, N., 461(22), *510*
Brown, B. J., 477, *511*
Brown, H., 21(17), *124*
Brown, J. C., 443(196), *448*
Brown, J. F., Jr., 165(30), *199*
Brown, L., 34(32), *125*
Brownlee, G. G., 286(119), *298*
Büchi, H., 239(38), *296*, 437(179), *448*
Buerger, M. J., 80(79), *126*
Bunville, L. G., 227(27), 231(29), *296*
Burdon, R. H., 259(78), *297*
Burgi, E., 385(108), *446*
Burma, D. P., 264(70), *297*
Burton, A., 385(106), *446*
Byrd, C., 251(57), *296*

C

Cacere, M. A., 466(31), *511*
Cairns, J., 273, *297*, 328, 330, *444*
Calef, E., 396(124), *446*
Campbell, A., 395, 396, 397(127), 400(133), 401(134), 402, *446, 447*
Canfield, R. E., 20, *124*
Capecchi, M. R., 441(192), 443(197), *448*, 465(27), *510*
Carlson, K., 363(72), *445*
Carlton, B. C., 429(170), 431(170), *447*
Caskey, T., 436(177), *448*
Caspar, D. L. D., 106(97), 108(97), *126*
Cavalli-Sforza, L. L., 320, *444*
Cerdá-Olmedo, E., 331(28), 411(145), *444, 447*
Chamberlin, M., 254(67, 74), *297*
Champe, S. P., 307(11), 371(88), *444, 445*
Chance, B., 154, 166, *198, 199*
Chandler, B. L., 293, *298*, 385(107), *446*
Changeux, J.-P., 166(32, 33), 168, 170(33), 173(33), 182(39), *199*
Chapeville, F., 455(9), *510*
Chargaff, E., 213, 282(112), *295, 298*
Chase, M., 366, 370, 375, *445, 446*
Cheng, P. Y., 393(118), *446*, 501(87), 503(87), *512*

Cheng, T.-Y., 454(6), *510*
Cherayil, J. D., 437(181), *448*
Chevalley, R., 383(101), *446*
Chilson, M. D., 363(71), *445*
Christensen, H. N., 174(36), 176, *199*
Citarella, R. V., 354(55), *445*
Clark, A. J., 340, *444*
Clark, B. F. C., 294(128), *298*, 434(174), *447*
Clowes, R. C., 354(54), *445*
Cochran, W., 80(79), *126*
Cohen, C., 63(66), *125*
Cohen. J. A., 136(12), *198*
Cohn, M., 491, 494(72), *572*
Colvill, A. J. E., 259(77), *297*
Connors, P. G., 294(131), *298*
Corey, R. B., 32(28, 29, 30,) 33, 34(28), 35(30, 34, 35), 37, *124*, *126*, 215, *295*
Corradini, P., 7(1), *124*
Cox, J. M., 54(84), 95(84), *126*
Cox, R. A., 284(118), *298*
Cozzarelli, N. R., 248(54), *296*
Craig, L. C., 121, *126*
Cramer, F., 294(130), *298*
Crawford, I. P., 398(130), *446*
Crestfield, A. M., 133(78), 134, *198*
Crick, F. H. C., 80(79), *126*, 214(7, 8), 216, 243(98), 244, *295*, *296*, 421, 436, 437, *447*, *448*
Crothers, D. M., 225, 227, 228, *296*
Cruickshank, P. A., 28(23), *124*
Cullis, A. F., 92(83), 95(83), *126*
Curtiss, R., 324(21), 325(21), *444*
Cuzin, F., 333(32), *444*

D

Darnell, J. E., 259(80), 280, *297*
Davern, C. I., 478, *571*
Davidson, B., 38(37), 64, *125*
Davidson, E., 110(100), *126*
Davie, E. W., 467(35), *571*
Davies, D. R., 54(76), 79(76), 91(76), 93(76). *126*, 239(39), 240(43), 242(45b), 243(48), 244, *296*
Davies, J., 466(32), 467(36), *511*
Davis, B. D., 310, *444*, 467(36), *511*
Davis, B, J., 116(103), *126*
Davison, P. E., 268, 274(98), *297*
Dawkins, R. C., 461(22), *510*
de Haan, P. G., 351, 352(49), *445*
De La Haba, G. L., 455, *510*
Delbrück, M., 312, 314, 317, 376, *444*, *446*
DeLorenzo, F., 73(71), *125*
de Lozé, C., 54(57), *125*
Demerec, M., 304(7), *443*

Denhardt, G. H., 383(101), 387(110), *446*
Denkewalter, R. G., 31(26b), *124*
Deutscher, M. P., 248(54), *296*
Dewey, K. F., 443(196), *448*
Dickerson, R. E., 54(76), 79(76, 77), 80, 89(77), 91(76), 92(77), 93(76), *126*
Dieckmann, M., 451(1), *510*
DiMarzio, E. A., 65, *125*
Dintzis, H. M., 79(75), 81(75), 83(80), 84(80), 86(81), 87(75), 89(80), *125*, *126*, 471, 472, *511*
Dixon, M., 127(1), *198*
Doctor, B. P., 294(128), *298*, 436(177), 437(181), *448*, 454(4), *510*
Doerfler, W., 282(110), *298*
Dondon, M., 467(34), *511*
Doty, P., 41(42), 43, 44, 45(44), 46(46), 47(48, 49), 48(48), 49, 50, 53(56), 54, 55, 61(42), 62, *125*, 221(17, 18), 222, 223, 225, 227, 246, 269, 270, 271, 272(92), 273, 274(100), 275(100), 276(100), 278, 279(100), 288, *295*, *296*, *297*, 326(23), 443(196), *444*, *448*, 466(31), *511*
Drenth, J., 100(92), *126*
Driskell-Zamenhof, P. J., 354(53), *445*
Dubinin, N. P., 390, 412(150), *446*, *447*
Dubnau, D., 364(74), 365, *445*
Dulbecco, R., 396, *446*
Dutton, G., 232(30), *296*
Dvorkin, G. A., 224, 294(126), *296*, *298*

E

Echols, H., 393, *446*, 500(84), 501, 503(91), *512*
Edgar, R. S., 307(10), 371, 383, 384(102, 104), 386, 381(110), *444*, *446*
Edman, P., 13, *124*
Edsall, J. T., 32(30), 35(30), 105, *124*, *126*
Eigner, J., 225, 227, 273(95), *296*, *297*
Elliott, A., 26(21a), 28, 34(32), *124*, *125*
Elödi, P., 105(96), *126*
Emlich, J., 423(160), *447*
Engelgardt, V. A., 186, 191, *199*
Engelhardt, D. L., 439(189), 441(192), 443(197), *448*, 507(97), *512*
Engelsberg, E., 502(90), *512*
Englander, J. J., 233(32), *296*
Englander, S. W., 233(32), *296*
Englund, P. T., 248(54), *296*
Epstein, C. J., 73(71), *125*
Epstein, J., 52, *125*
Epstein, R. H., 301(10), 383(10, 101), *444*, *446*
Epstein, W., 339(43), *444*
Erhan, S., 435(175), *448*
Eriksson-Quensel, I.-B., 103(95), *126*
Ertel, R., 461(22), *510*

Estabrook, R. W., 166(31), *199*
Evans, A., 254(69), *297*
Evans, A. H., 141(16), *198*
Everett, G. A., 210(4) 213, 285, 286(4), *295*

F

Falkow, S., 279(103), *297*, 326(23), 354(55), *444*, *445*
Fan, D. P., 441(191), *448*, 485(59), *571*
Fasman, G. D., 35(33), 38(37), 58(33), 59(33, 61), 64, *125*, 244(47), *296*
Faulkner, R. D., 437(181), *448*
Feldmann, H., 453, *510*
Felsenfeld, G., 240, *296*
Fernández-Morán, H., 174(34), *199*
Feuer, G., 191, *199*
Feughelman, M., 216, *295*
Feynman, R. P., 307(10), 383(10), *444*
Fiers, W., 385(106), *446*
Filippov, G. S., 412, *447*
Fink, G. R., 499(79), *512*
Fisher, H. F., 74(73), 75, *125*
Fitts, D. D., 61(65), *125*
Flessel, C. P., 244(46), *296*
Flory, P., 225, *296*
Fox, M. S., 365, 360, 363, *445*
Fraenkel-Conrat, H., 22(18), 106(98), *124*, *126*, 367(81), 368(84), 445
Freese, E., 218(21), *295*, 331(29), 390(114, 115), 406, 407(142), 408(144), 409(14), 151, 152), *444*, *446*, *447*
Freese, E. B., 331(29), 414(152, 153), *444*, *447*
Freifelder, D., 274(98), *297*, 337(41), *444*
Frenkel, J., 268, *297*
Frenkel', S. Ya., 59(63), 60, 70(63), 72, *125*
Freundlich, M., 166(32), *199*
Fresco, J. R., 221(18), 222(18), 223(18), 244(46), 246(18), 247(50), 288(18), 294(133), *296*, *298*
Friedman, S. M., 467(37), *511*
Friess, S. L., 148(19), *198*
Fuerst, C. R., 352(51), *445*
Fujimura, R., 468(38), *511*
Fukuda, K., 31(25), *124*
Furth, J. J., 254(66, 69), *297*

G

Gajtskhoki, V. S., 185(43), *199*
Gallucci, E., 307(11), *444*
Gamow, G., 419, *447*
Ganesan, A. T., 362, *445*
Gardner, R. S., 434(174), *447*
Garen, A., 305, 307(11), 318, 348, 349, 427, 439, 440(185), 441(190), *443*, *444*, *445*, *447*, *448*

Garen, S., 307(11), *444*
Garner, C. W., 461(22), *510*
Garry, B., 497(79), *512*
Gavrilova, L. P., 290(122), 294(126), *298*, 468(38), *511*
Gefter, M., 257(64),
Geiduschek, E. P., 227(27), 231(29), 254(73), 259, 274(97), 282(108), *296*, *297*, *298*
Gellert, M., 259(62), *297*, 337(38), *444*
Gest, H., 352, *445*
Gesteland, R. F., 441(193, 194), *448*, 486(63), *511*
Ghosh, A., 166(31), *199*
Ghosh, H. P., 437(179), *448*, 463(25), *510*
Giacomoni, D., 281(106), *298*
Gibbs, J. H., 65, *125*
Gierer, A., 366, 368, 369, *445*, 465, *511*
Gilbert, W., 461, 466(32), 479(50), 481(53), 482(50), 484(50), 494(73), 495, *570*, *511*, *512*
Gillespie, D., 283, *298*, 483, *511*
Gish, D. T., 32(18), *124*
Givol, D., 73(71), *125*
Glaser, D. A., 329(27), *444*
Goaman, L. C. G., 54(84), 98(84), *126*
Goldberger, R. F., 73(71), *125*
Goldmann, M., 254(66), *297*
Goldstein, A., 477, *511*
Goldstein, J., 437(180), *448*
Goldthwaite, C., 364(74), 365(74), *445*
Golov, V. F., 468(38), *511*
Golub, E. I., 224(19), *296*
Gomatos, P. J., 292, *298*
Gonano, F., 456(13), *510*
Goodgal, S. H., 360, 361, 363, *445*
Goodman, H. M., 441(195), *448*, 465(30), *511*
Goodman, M., 26(21b), *124*
Gorbachev, L. P., 224(19), *296*
Gorini, L., 466, 467(36), 502(88), *511*, *512*
Goulian, M., 253, *297*
Gray, W. R., 13, *124*
Grabar, P., 122, 123, *126*
Grayevskaya, R. A., 461(18), *510*
Graziosi, F., 282(108), *298*
Green, M. H., 258, 281(107), *297*, *298*
Greenberg, B., 363(72), *445*
Greenberg, D. M., 152, *198*
Griffeth, F., 355, *445*
Gros, F., 479, 481, 482(50), 484, 499(81), 503(94), 505(94), *511*, *512*
Gross, E., 18(12), *124*
Gross, J. D., 301(4), 337, 339, *443*, *444*
Grossman, L., 244(47), *296*
Grunberg-Manago, M., 236, 237(37), *296*, 467(34), *511*
Guerola, N., 331(28), *444*
Guest, J. R., 429(170), 431(170), *447*

Gulland, J. M., 227(25, 26), 228, *296*
Gupta, N. K., 239(38), *296*, 437(179), *448*
Gussin, N. G., 441(192), *448*
Guthrie, C., 463(25), *510*
Gutte, B., 31, *124*

H

Haber, E., 134(11), 135(10), *198*
Haenni, A.-L., 461(22), *510*
Haldane, J. B. S., 351, *445*
Hall, B. D., 363(71), *445*, 479(51, 52), 482(52), *511*
Hall, C. E., 45, 46(46), *125*, 255, *297*, 464(29), 465(29), *511*
Hall, Z. W., 218(14), *295*
Hamilton, L. D., 214(9), 216(11), *295*
Hampel, A., 294(131), *298*, 437(181), *448*
Hanawalt, P. C., 331(28), 411(145), *444, 447*
Hanby, W. E., 26(21a), 28, 34(32), *124, 125*
Hanson, J., 195, *199*
Happey, F., 34(32), *124*
Hargitay, B., 189(47), *199*
Harker, D., 54(88), 98(88), *126*
Harpst, J. A., 224(21), *296*
Harriman, P. D., 372(93), *446*
Harrington, W. F., 71(69), *125*
Hart, R. G., 54(76), 79(76), 91(76), 93(76), *126*
Hartley, B. S., 13, *124*
Hartman, P. E., 397(127), 398, *446*, 497(79), *512*
Hartsuck, J. A., 98(89), *126*, 164(29), *198*
Haruna, I., 259(81), *297*
Haselkorn, R., 221(18), 222(18), 223(18), 246(18), 259(88), 288(18), *296, 297*
Hatfield, D., 436(177), *448*
Hayashi, M., 258, 281(107), 282(109), *297, 298*, 482(54, 55), *511*
Hayashi, M. N., 258, 282(109), *297, 298*
Hayatsu, H., 437(179, 181), *448*
Hayes, W., 301(3), 320, 342(46), 384, *443, 444, 445*, 498, *512*
Heason, J. Z., 148, *198*
Hearst, J. E., 262(85), *297*
Hecht, L. I., 451(1), *510*
Helinski, D. R., 337(42), 429(169, 170), 431(170), *444, 447*
Henderson, R., 54(90), 98(90), *126*
Henning, U., 429(169, 170), 431(170), *447*
Hermans, J. Jr., 42, *125*
Hershey, A. D., 269, *297*, 352, 366, 370, 375, 376, 385(108), *445, 446*
Herzenberg, L. A., 490, 497(79), *511, 512*
Hiatt, H. H., 479(50), 481(53), 482(50), 484(50), *511, 571*
Hickson, F. T., 337(42), *444*
Higa, A., 485(59), *511*

Hill, A. V., 187(45), 188, *199*
Hirota, Y., 322(20), *444*
Hirs, C. H. W., 14(10), 19, *124*
Hirschmann, K., 31(26b), *124*
Hirth, L., 293(128), *298*
Hoagland, M. B., 451(1), *510*
Hogness, D. S., 282(110), *298*, 489, 491, 504(92), *511, 512*
Holland, I. B., 259(81), *297*
Holley, R. W., 210(4), 213, 285, 286(4), *295*, 456(11), *510*
Holly, F. W., 31(26b), *124*
Holmes, K. C., 294(128), *298*
Holtzer, A. M., 44(44), 45(44), *125*
Hooper, C. W., 214(9), 216(11), *295*
Hopkins, N., 506(95), *512*
Horn, P., 72, *125*
Horn, V., 432(171), *447*
Horn, R. W., 107, *126*, 371(88), *445*
Hosokawa, K., 468(38), *511*
Hotchkiss, R. D., 141(16), *198*, 356, 357, 360, *445*
Howard-Flanders, P., 340, *444*
Hoyer, B. H., 283(115, 116), *298*, 486(64), *511*
Hradecna, Z., 282(113), *298*, 503(93), 504(93), *512*
Hubbard, R. W., 8(5), *124*
Hunt, J. A., 425(165), *447*
Hurwitz J., 254(66, 69), 257(64), *297*
Huxley, H. E., 195, 197, *199*, 458, *510*

I

Ikeda, H., 397, *446*
Imahori, K., 53, 55, *125*
Imamoto, F., 499(82), 508, 509, *512*
Ingram, V. M., 16(11), 17, *124*, 425, 427, *447*
Inman, R. B., 252(58), *297*, 385(108), *446*
Inouye, M., 423(160), *447*
Ippen, K., 495(78), 496, *512*
Itano, A., 426(167), *447*
Ito, J., 432(171), *447*, 507, *512*
Ivanov, I. I., 129(2), *198*

J

Jacherts, D., 293(125), *298*
Jacob, F., 301(2), 319(16), 323, 235, 333, 334(33), 335, 342(46), 346, 348, 352(51), 353, 393, 394, *443, 444, 445*, 166(32), *199*, 478(47), 479, 486(65), 493(71), 494(75), 495, 497, 499(81), 501, 502, *511, 512*
Jacob, T. A., 31(26b), *124*
Jacob, T. M., 258(74), *297*, 437(179), *448*

Jacoby, G. A., 502(88), *512*
Jansonius, J. N., 100(92), *126*
Jansz, H. S., 136(12), *198*
Jernberg, N., 31(26c), *124*
Jerne, N. K., 373(95), *446*
Jinks, J. L., 320(18), *444*
Johnson, L. N., 97, *126*
Jones, D. S., 437(179, 181), *448*
Jones, O. W., 434(173, 174), *447*
Jordan, D. O., 204(3), 227(25, 26), 228, *295, 296*
Josse, J., 248(55, 56), 249, 250, 262, *296*
Jovin, T. M., 248(54), *296*

K

Kaiser, A. D., 248(56), 249(56), 250(56), 262(56), *296*, 385(108, 109), *446*
Kalckar, H. M., 188, *199*
Kalinin, V. L., 416(154), 418(156), 419(156), *445, 447*
Kamen, M. D., 352, *445*
Kaplan, S., 440(189), 441(189), *448*
Karkas, J. D., 282(112), *298*
Karlson, R. H., 35(33), 58(33), 59, *125*
Kartha, G., 54(88), 96(88), *126*
Kasha, M., 220(16), *295*
Kataja, E., 466(32), *511*
Katchalsky, A., 189, *199*
Katsoyannis, P. G., 26(21), 31(25), 73(71), *124, 125*
Kayajanian, G., 401(134), *447*
Kazakova, T. B., 185(43), *199*
Kellenberger, E., 383(101), *446*
Kellenberger, G., 395(122), *446*
Keller, E. B., 355(7), *510*
Kellogg, D. A., 436(177), 437(181), *448*, 454(4), *510*
Kelly, R. B., 248(54), *296*
Kendrew, J. C., 54(76), 76(76, 77), 81(75), 83(80), 84(80), 86(81), 87, 89(80), 91, 92, 93, 95(85), *126*
Kennedy, J. W., 352, *445*
Kenner, G. W., 26(21b), *124*
Kcynan, A., 485(59), *511*
Khorana, H. G., 204(3), 239, 251, 258(74), *295, 296, 297*, 437(181), *448*, 463(25), *510*
Kierkegard, L., 294(131), *298*
Kim, S.-H., 294(132), *298*
King, J., 384(103), *446*
Kirillov, S. V., 461(18), *510*
Kirkwood, J. G., 61(65), *125*
Kis, H., 385(107), *446*
Kiselev, N., 294(126), *298*
Kitai, R., 21(16, 17), *124*

Kitajgorodskij, A. A., 80(79), *126*
Kittel, C., 182(39), *199*
Klein, S., 307(10), 383(10), *444*
Kleinschmidt, A. K., 293, *298*, 385(106), *446*
Klostermeyer, H., 31(25), *124*
Klug, A., 106(97), 108(97), *126*, 294(128), *298*
Knight, C. A., 22(18), *124*, 259(78), *297*
Knopf, P. M., 465(28), *511*
Koekoek, R., 100(92), *126*
Koenig, D. F., 54(86), 96(86), *126*
Kössel, H., 437(179), *448*
Konigsberg, W., 439(187), *448*
Koreneva, L. G., 224(19), *296*
Kornberg, A., 248(51, 52, 56), 248(54), 249(56), 250(56), 252(52, 58), 253, 260, 262(50), *296, 297*
Koshland, D. E., Jr., 161, 163, 164, *198*
Kourilsky, P., 503, 505(94), *512*
Krasna, A. I., 224(21), *296*
Kraut, J., 98(91), 99(91), *126*
Kreneva, R. A., 362(68), *445*
Kröger, H., 264(70), *297*
Kubinski, H., 282(111), *298*
Kudner, R., 282(112), *298*
Kuhn, W., 189, *199*
Kumar, A., 239(38), *296*
Kurland, C. G., *481*(53), *511*
Kushev, V. V., 362(68), *445*
Kushner, V. P., 59(63), 60, 70(63), *125*

L

Labanauskas, M., 294(131), *298*
Lacks, S., 363, 366, *445*
Landy, A., 441(195), *448*
Lane, D., 277, *297*
Lang, D., 293(125), *298*
Langridge, R., 214(9), 216(11), 244(47), 292, 294(133), *295, 296, 298*
Lanka, E., 441(190), *448*
Lanzov, V. A., 332(31), 333(31), 337(36), *444*
Leach, F. R., 435(175), *448*
Leach, S. J., 52, *125*
Leahy, J., 470(41), *511*
Leder, P., 434(174), 436(177), *447, 448*
Lederberg, E. M., 308, 320(18), 398(131), *444, 446*
Lederberg, J., 308, 310, 318, 320(18), 336, 397, 398(131), *444, 446*
Leggett Bailey, J., 7(4), 12(4), 13(4), *124*
Lehman, I. R., 218(147), 257(61), 275(101), *295, 297*
Lehninger, A. L., 180, 182, 183, 184, *199*
Lengyel, P., 434(174), *447*, 461(22), *510*

Lennox, E. S., 398(130), *446*, 493(71), *512*
Lerman, L. S., 411(146), *447*
Lerman, M. I., 468(38), *571*
Leslie, I., 203, *295*
Levin, J., 436(177), *448*
Levinthal, C., 267, 268, *297*, 305, 354(56), 373, 424, 427, *443*, *445*, *446*, *447*, 485(59), *511*
Licciardello, G., 396(124), *446*
Lielausis, A., 383(101), *446*
Lielausis, I., 384(104), 387, *446*
Lifson, S., 189(47), *199*
Lindblow, C., 244(47), *296*
Linderström-Lang, K. U., 39, 40, 41, *125*, 163, *198*
Lipmann, F., 437(180), *448*, 455(7), 461, 462, 465, *510*, *511*
Lipscomb, W. N., 98(89), *126*, 164(29), *198*
Lipsett, M. N., 244(96), *296*
Lipson, H., 80(79), *126*
Litman, R. M., 403(139), *447*
Litt, M., 221(18), 222(18), 223(18), 246(18), 288(18), *296*
Littauer, U. Z., 284(118), *298*
Little, J. W., 259(62), *297*, 337(38), *444*
Liu, A. K., 20, *124*
Lobashov, M. E., 403, *447*
Loeb, T., 322, *444*
Loebel, J. E., 437(181), *448*, 454(4), *570*
Lohrmann, R., 437(181), *448*
Low, B. W., 32(30), 35(30), *124*
Lucas-Lenard, J., 461(20, 22), *570*
Ludwig, M. L., 98(89), *126*, 164(29), *198*
Lumry, R., 151, *198*
Lundberg, R. D., 47(99), 48, 49, 50, *125*
Luria, S. E., 280, *297*, 312, 314, 317, *444*
Luzatti, D., 503(94), 505(94), *572*
Luzzati, V., 111, *126*, 234, 293, 294(127), *296*, *298*
Lyn, G., 257(64), *297*
Lyubimova, M. N., 186, 191, *199*

M

McCarthy, B. J., 283(116), *296*, 457(14b), 486(64), *570*, *511*
McCarthy, M., 355, *445*
Macleod, C. M., 355, *445*
McGill, B. B., 223, *296*
MacHattie, L. A., 387, *446*
McQuillen, K., 473, 474, 475, *511*
Maden, B. E. H., 463(24), *570*
Madison, J. T., 210(4), 213, 285, 286(4), *295*
Mahler, H. R., 232(30), *296*
Mair, G. A., 54(86), 96(86), *126*

Maling, B. D., 305(8), *443*
Mandel, M., 279(103), *297*
Mandelkern, L., 225, *296*
Mandell, J. D., 269, *297*
Mangiarotti, G., 460(16), 485(58), *510*, *511*
Marcaud, L., 503(94), 505(94), *512*
Marcker, K. A., 294(128), *298*
Margulies, A. D., 340, *444*
Marmur, J., 269(90), 270, 271(90), 272(92), 273(95, 96), 274(100), 275(100), 276(100), 277, 278, 279(100, 103), *297*, 326(23), 360, 364, 365(74), *444*, *445*
Marquisee, M., 210(4), 213, 285, 286(4), *295*
Marshall, R., 436(177), *448*
Martin, R. G., 434(173), *447*, 499, 507, *512*
Maslova, R. N., 233, *296*
Masson, F., 234(34), *296*
Mathis, A., 234(34), *296*
Matsushiro, A., 401(135), *447*
Matthaei, J. H., 434, *447*, 459(15), *570*
Matthews, B. W., 54(90), 98(90), *126*
Mehrotra, B. D., 232(30), *296*
Meinhofer, J., 31(25), *124*
Mekshenkov, M. I., 224(19), *296*
Melnikova, M. P., 185, (43)*199*
Menten, M. L., 142(17), 145, *198*
Mcrrifield, R. B., 30(24a, 24b), 31(24a, 24b, 26c,) *124*
Merrill, S. H., 210(4), 213, 285, 286(4), *295*
Meselson, M., 262(84), 264, 265, 266, *297*, 378, 379, 381, 382, 399(132), 401(134), *446*, *447*, 468(38), 478, 479, *511*
Michaeli, I., 189(47), *199*
Michaelis, L., 142, 145, *198*
Michelson, A. M., 204(3), 242(45a), *295*, *296*
Miles, H. T., 241, *296*
Miller, D., 461(22), *510*
Miller, J. H., 495(78), 496(78), *512*
Mills, D., 259(81), *297*
Mishima, S., 499(82), *512*
Mitchell, H. K., 300(1), *443*
Miyazawa, J., 273(94), *297*
Miyazawa, T., 55, 56, *125*
Moelwyn-Hughes, E. A., 137, 150, *198*
Moffitt, W., 61(65), *125*
Monod, J., 166(32, 33), 168, 170(33), 173, *198*, *199*, 478(47), 486(65), 491, 493 494,(75), 495, 497, 501, 502, *511*, *512*
Monro, R. E., 461(20), 463(24), *510*
Moody, E. E. M., 354(54), *445*
Moohr, J. W., 254(73), *297*
Moon, M. W., 251(57), *296*
Moore, S., 8, 19(13), 20(14), *124*, 133(7, 8), 134(7), *198*
Morales, M. F., 148(19), 192, *198*, *199*
Morgan, R., 437(179), *448*

Morikawa, N., 499(82), *512*
Morris, S. J., 294(128), *298*
Morse, M. L., 398(131), *446*
Mosevitskij, M. I., 290(122), *298*, 358(63), 354(63), 362(68), *445*
Muirhead, H., 54(84), 92(83), 95(83, 84), *126*
Muller, H. J., 412, *447*
Müller-Hill, B., 494(72, 73), 495, *512*
Mundry, K. W., 369, *445*

N

Nadson, G. A., 412, *447*
Nakamoto, T., 254(68), *297*
Naono, S., 499(81), *512*
Narang, S. A., 437(179), *448*
Nathans, D., 455(7), 486(61), *510*, *511*
Natta, G., 7, *124*
Nejfakh, S. A., 185, *199*
Neilands, J. B., 127(1), *198*
Némethy, G., 75(74), 76, *125*
Neurath, H., 138(14), *198*
Newman, A., 329(27), *444*
Newton, J., 423(160), *447*
Nicolaieff, A., 111(101), *126*
Nirenberg, M. W., 434(173), 436(177), 437(181), *447*, *448*, 459(15), *510*
Nishimura, S., 258(74), *297*, 437(179), *448*
Nishimura, T., 499(82), *512*
Nishizuka, Y., 461(20), *510*
Nobbs, C. L., 95(85), *126*
Noll, H., 461(19), *510*
Nomura, M., 117(104), *126*, 463(25), 468(38), 469, *510*, *511*
Norland, K. S., 35(33), 58(33), 59(33), *125*
North, A. C. T., 54(86), 92(83), 95(83), *126*
Northrup, L. G., 435(175), *448*
Notani, G. W., 439(187), *448*, 486(61), *511*
Novick, A., 412(149), *447*, 493(71), *512*
Nutt, R. F., 31(26b), *124*

O

Ochoa, S., 236, 264(70), *296*, *297*, 434, *447*, 454(5), 467(33), 486(62), *510*, *511*
Ofengand, E. J., 451(1), *510*
Ohlsson, B. M., 307(11), *444*
Ohta, N., 304(7), *443*
Ohtsuka, E., 234(38), 251(57), *296*, 437(179, 181), *448*
Okada, Y., 423(160), *447*
Okazaki, R., 331(30), *444*
Okazaki, T., 331(30), *444*
Okuda, T., 31(25), *124*

Olivera, B. M., 257(61), *297*
O'Neal, C., 436(177), *448*
Ono, Y., 461(22), *510*
Ooi, T., 164, *198*
Oosterbaan, R. A., 136(12), *198*
Opara-Kubinska, Z., 282(111), *298*
Orgel, A., 411(147), *447*
Orgel, L. E., 494(72), *512*
Ornstein, L., 116(103), *126*
Ortiz, P. J., 236(36), *296*
Oshinsky, C. K., 259(62), *297*, 337(38), *444*
Ozawa, T., 164, *198*

P

Paigen, K., 399(132), 401(134), *446*, *447*
Pakula, R., 356, *445*
Paradies, H. H., 294(128, 129), *298*
Pardee, A. B., 478(47), 492, 493, *511*, *512*
Parmeggiani, A., 461(21), *510*
Parrish, R. G., 79(75), 81(75), 87(75), *125*
Pauling. L., 32(28, 29, 30), 33, 34, 35(30, 34, 35), 37, *124*, *125*, 215, *295*, 426, *447*
Penswick, J. R., 210(4), 213, 285, 286(4), *295*
Perrin, D., 494(75), *512*
Perumov, D. A., 403(139), 416(154), 417(155), 418(156), 419(156), *445*, *447*
Perutz, M. F., 92(83), 95(84), 110(100), *126*
Pestka, S., 434(174), 436(177), *447*, *448*
Phillips, D. C., 54(76, 86), 79(75, 76, 77), 81(75), 87(75), 89(77), 91(76), 92(77), 93(76), 96(86), 97, *126*
Pilarski, L., 393(118), *446*, 501(87), 503(87), *512*
Pochon, F., 242(45a), *296*
Podelski, T. R., 182(39), *199*
Poglazov, B. F., 129(2), *198*
Pratt, D., 372(93), *446*
Preiss, J., 451(1), *510*
Printz, M. P., 233(31), *296*
Pritchard, R. H., 365, *445*
Ptashne, M., 394(119), *446*, 500(86), 501, 506(95), *512*
Ptitsyn, O. B., 65(68), *125*
Pullman, B., 218(13), *295*

Q

Quiocho, F. A., 98(89), *126*, 164(29), *198*

R

Radding, C. M., 503(91), *512*
Raj Bahndary, U. L., 294(131), *298*

Author Index

Ravel, J. M., 461(22), *510*
Ravin, A. W., 355(59), *445*
Rawitscher, M. A., 231(29), *296*
Ray, W. J., Jr., 455(9), *510*
Raymond, S., 116(103), *126*
Reeke, G. N., 98(89), *126*, 164(29), *198*
Rees, M. W., 371(88), *445*
Reilley, B. E., 366(78), *445*
Rendi, R., 454(5), *510*
Rice, S. A., 221(17), 222, 223, *295*, *296*
Rich, A., 220(16), 239(39, 40), 240(43), 243, 244(47), 245, 294(132), *295*, *296*, *298*, 464(29), 465(29, 30), *511*
Richards, E. G., 244(46), *296*
Richards, F. M., 131, 132, *198*
Richardson, C. C., 248(52), 252, 257(63), *296*, *297*, 337(38), *444*
Rideal, E. K., 51, *125*
Ris, H., 293, *298*
Risebrough, R. W., 481(53), *511*
Roberts, J. W., 367(82), *445*
Roberts, R. B., 118(105), *126*, 473(43), 474(43), 475(43), *511*
Roberts, R., 51, *125*
Robertson, J. M., 80(79), *126*
Roholt, O. A., 152, *198*
Rosenheck, K., 53(56), 54, 62, *125*
Rosman, M., 461(22), *510*
Rossmann, M. G., 92(83), 95(83), *126*
Roth, J. R., 497(79), *512*
Roth, T. F., 337(42), *444*
Rothman, F., 305, 427, *443*, *447*
Rotman, B., 492, *512*
Rottman, F., 436(177), *448*
Rouvière, J., 499(81), *512*
Rownd, R., 273(96), *297*, 326(23), *444*
Ryabchenko, N. I., 235(35), *296*
Ryle, A. P., 21(16), *124*
Ryter, A., 334(33), *444*

S

Saenger, W., 294(130), *298*
Sagik, B. P., 281(107), *298*
Sajgó, M., 105(96), *126*
Sakabe, K., 331(30), *444*
Sakharov, V. V., 403, *447*
Salas, M., 486(62), *511*
Salser, W., 441(193, 194), *448*, 486(63), *511*
Sambrook, J. F., 441(191), *448*
Saminskij, E. M., 32(27), 68(27), 69(27), *124*, 461(18), *510*
Sanchez, C., 494(75), *512*
Sander, C., 244(47), *296*
Sanderson, K. E., 304(7), *443*

Sanger, F., 12, 14, 21(17), *124*, 286, *298*
Sarabhai, A. S., 440(188), *448*
Sarkar, P. K., 244(46), *296*
Sarma, V. R., 57(86), 96(86), *126*
Sarnat, M., 259(77), 282(108), *297*, *298*
Sasisekharan, V., 239(41), *296*
Sato, G. H., 373(95), *446*
Sato, K., 499(82), *512*
Scaife, J., 337, 339, *444*, 495(78), 496(78), *512*
Schachman, H. K., 101(93), *126*
Schaeffer, P., 325(221), 33(54), *444*, *445*
Schellman, J. A., 71(69), *125*, 163, *198*,
Scheraga, H. A., 40(40), 42, 51(52), 52, 75(74), 76, *125*, 225(24), *296*
Schildkraut, C. L., 248(52), 252(52, 58), 273(95, 96), 274(100), 275, 276, 279, *296*, *297*, 326(23), *444*
Schlessinger, D., 460(16), 485(58), *510*, *511*
Schlimme, E., 294(130), *298*
Schnàbel, E., 31(25), *124*
Schneider, M. C., 259(78), *297*
Schramm, C. H., 26, *124*
Schramm, G., 366, 368, 403(138), *445*, *447*
Schuster, H., 403(138), *447*
Schwartz, J. H., 486(61), *411*
Schweet, R., 470(41), *511*
Scott, J. F., 451(1), *510*
Scouloudi, H., 89(82), *126*
Seeds, W. E., 216(11), *295*
Sela, M., 135(10), *198*
Serebrovskij, A. S., 390, *446*
Sgaramella, V., 239(38), *296*
Sheehan, J. C., 28(23), *124*
Sheldrick, P., 503(94), 505(94), *512*
Shemyakin, M. M., 156, *198*
Sheppard, D., 502(90), *512*
Shipp, W., 259(79), *297*
Shive, W., 461(22), *510*
Shore, V. C., 54(76), 79(76, 77), 89(77), 91(76), 92(77), 93(76), *126*
Shorey, R. L., 461(22), *510*
Shugayeva, N. V., 213(6), *295*
Shutov, F., 461(18), *510*
Siddiqi, O. H., 336, 439(184), *444*, *448*
Sigler, E. B., 54(90), 98(90), *126*, 239(41), 294(131), *296*, *298*
Signer, P. R., 307(11), 339(43), 396(123), 402(123), 424(163), *444*, *446*, *447*
Silbert, D. F., 497(79), 507(96), *512*
Silver, S. D., 318, 354(54), *444*, *445*
Simmons, J. R., 504(92), *512*
Simmons, N. S., 63, *125*
Singer, S. J., 426(167), *447*
Sinsheimer, R. L., 230, 253, *296*, *297*, 385, *446*
Sjöstrand, F. S., 181, *199*
Skaar, P. D., 318, *444*

Skoultchi, A., 461(22), *510*
Slayter, H. S., 255, *297*, 464, 465, *511*
Sly, W. S., 434(174), *447*, 500(84), *512*
Smith, D. W., F. 507(96), *512*
Smith, I., 364(74), 365(74), *445*
Smith, J. D., 441(195), *448*
Smith, L. F., 21(16), *124*
Smyth, D. G., 20, *124*
Snell, E. E., 156, *198*
Snyder, L., 259(77), *297*
So, A. G., 467(35), *511*
Söll, D., 437(181), *448*
Solomon, A. K., 179(38), *199*
Spackman, D. H., 8, *124*
Spahr, P.-F., 479(50), 482(50), 484(50), *511*
Speyer, J. F., 434(174), *449*
Spiegelman, S., 258, 259(81), 280, 281(107), 282(109), 283, *297*, *298*, 479, 482(52, 54, 55), 483, 492, *511*, *512*
Spirin, A. S., 213, 287, 290, 294(126), *295*, *298*, 424, *447*, 468(38), 478, *511*
Spitkovskij, D. M., 235(35), *296*
Spizizen, J., 356, 366(78), *445*
Spyrides, G. J., 465(28), *511*
Sroka, W., 31(25), *124*
Staehelin, T., 468(38), *511*
Stahl, F. W., 262(84), 264, 265, 266, *297*
Stein, W. H., 8, 19(13), 20(14), *124*, 131(3), 133(7, 8), 134(7), *198*
Steinberg, C. M., 307(10), 383(10, 101), *444*, *446*
Steitz, T. A., 98(89), *126*, 164(29), *198*
Stent, G. S., 352(51), 372(93), 373, 390(91), 394, *445*, *446*, 508, *512*
Stephenson, M. L., 451(1), *510*
Stewart, J. M., 31(26c), *124*
Stodolsky, M., 282(108), *298*
Stokes, A. R., 216(11), *295*
Stanley, W. M., 22(18), *124*
Strack, H. B., 385(109), *446*
Strandberg, B. E., 54(76), 79(76,77), 89(77), 91(76), 92(77), 93(76), *126*
Straub, F. B., 191, *199*
Streisinger, G., 371(88), 387(110), 423(160), *445*, *446*, *447*
Stretton, A. O. W., 439, 440(188, 189), 441(189), *448*
Strigini, P. F., 307(11), *444*
Stumpf, P. K., 127(1), *198*
Sturtevant, J. M., 231(29), *296*
Sueoka, N., 269(90), 271(90), 272, *297*, 364, 424, *445*, *447*, 454(6), 456, *510*
Sugimoto, K., 331(30), *444*
Sugino, A., 331(30), *444*
Sundararajan, T. A., 443(196), *448*, 466(31), *511*
Susman, M., 383(101), *446*

Suzuki, K., 31(25), *124*
Svedberg, T., 103, *126*
Svensson, H., 113, *126*
Swain, C. G., 165(30), *199*
Swen, H. W., 100(92), *126*
Szent-Györgyi, A., 191, *199*
Szent-Györgyi, A. G., 63(66), *125*
Szer, W., 467(33), *511*
Szabolcsi, G., 105(96), *126*
Szilard, L., 412(149), *447*
Szybalski, W., 260(83), 282(111, 113), *297*, *298*, 338, *444*, 503(93), 504, *512*

T

Talmud, D. L., 74, *125*
Tanaka, J., 53, *125*
Tanford, C., 23, 52, *124*, *125*
Tao, P., 461(22), *510*
Tatum, E. L., 300, 318, *443*, *444*
Taylor, A. L., 302, 303(6), 323, *443*
Taylor, C. A., 80(79), *126*
Taylor, K., 282(113), *298*, 503, 504(93), *512*
Terzaghi, E., 423(160), *447*
Thach, R. E., 443(196), *448*, 466(31), *511*
Theriot, L., 340, *444*
Thiéry, J., 182(39), *199*
Thoma, J. A., 163(27), 164(27), *198*
Thomas, C. A., 267, 273(94), *297*, 373, 387, *446*
Tilak, M., 31(25), *124*
Timasheff, S. N., 293(127), *298*
Timkovskij, A. L., 358(63), 359(63), *445*
Tinoco, I., Jr., 219, *295*
Tiselius, A., 113, *126*
Tochini-Valentini, G. P., 282(108), *298*
Tometsko, A., 31(25), 73(71), *124*, *125*
Tomizawa, J., 329(27), 381, 397, *444*, *446*
Torriani, A., 424(163), *447*
Traub, P., 117, *126*, 468(39), 469, *571*
Traut, R. R., 463(24), *510*
Trotter, C. D., 302, 303(6), 323, *443*
Trotter, I. F., 34(32), *125*
Trupin, J., 436(177), *448*
Tsejtlin, P. I., 235(35), *296*
Ts'o, P. O. P., 244(47), *296*
Tsugita, A., 22, *124*, 368(84), 423(160), *445*, *447*
Tu, L., 117(104), *126*
Tung, Y., 182(39), *199*
Tuppy, H., 140, *198*
Turovskij, V. S., 185(43), *199*

U

Uchida, H., 372(93), *446*
Ueda, Y., 364(75), 365, *445*

Author Index

Ullman, A., 495, 502, *512*
Umbarger, H. E., 166(32), *199*
Urnes, P., 41(42), 55, 61(42), 62, *125*

V

Vanyushin, B. F., 213(6), *295*
Varga, S. L., 31(26b), *124*
Varshavsky, Ya, M., 233, *296*
Vasilyeva, N. N., 461(18), *510*
Veber, D. F., 31(26b), *124*
Verhoef, C., 351, 352(49), *445*
Vinograd, J., 262, *297*
Viñuela, E., 486(62), *511*
Visconti, N., 376, *446*
Vitali, R. A., 31(26b), *124*
Vithayathil, P. J., 131, 132, *198*
Vogt, M., 396, *446*
Volkin, E., 479, *511*
Voll, M. J., 360, 361, *445*, 497(79), *512*
von der Haar, F., 294(130), *298*
von Ehrenstein, G., 455(79), 456(10, 13), *510*
von Hippel, P. H., 233(31), *296*

W

Wada, A., 45(45), 46(47), *125*
Wagner, R. P., 300(1), *443*
Wahba, A. J., 434(174), *441*
Walczak, W., 356, *445*
Wang, Y. T., 116(103), *126*
Warner, J. R., 464(29), 465(28, 29, 30), *511*
Warner, R. C., 264(70), *297*
Waterson, A. P., 107(99), *126*
Waterson, J., 461(22), *510*
Watson, H. C., 79(77), 89(77), 95(85), *126*
Watson, J. D., 214(7, 8), 216, 243(48), 244, *295, 296,* 460, 479, 481(53), 482(50), 484, *510, 511*
Watts-Tobin, R. J., 421(159), *447*
Webb, E. C., 127(1), *198*
Weber, H., 239(38), *296*
Webster, R. E., 441(192), 443(197), *448,* 507(97), *512*
Weibull, C., 195, *199*
Weigert, M. G., 439, 440(185), 441(190), *448*
Weigle, J. J., 378, 379, 385(105), 395(122), 399, 401(134), *446, 447*
Weill, J. D., 264(70), *297)*
Weinstein, I. B., 467(37), *511*
Weintraub, L., 116(103), *126*
Weisblum, B., 454(3), 455(9), 456(10, 11, 13), *510*
Weiss, B., 257(63), *297,* 337(38), *444*
Weiss, S. B., 254(68,73), 257, 282(108), *297, 298*
Weissbach, N., 461(22), *510*

Weissmann, C., 259(78), *297*
Wells, I. C., 426(167), *447*
Wells, R. D., 437(179), *448*
Wetlaufer, D. B., 63(66), *125*
Wettstein, F. O., 461(19), *510*
White, F. H., Jr., 133(9), 135(10), *198*
Whitfield, H. J., Jr., 412(148), *447,* 507(96), *512*
Wilcox, M., 436(177), *448*
Wildy, P., 107(99), *126,* 371, *445*
Wilhelm, R. C., 307(11), 441(192), *444, 448*
Wilkins, M. H. F., 214(9), 216(11), *295*
Will, G., 92(83), 95(83), *126*
Williams, C. A., Jr., 122, 123, *126*
Wilson, H. R., 80(79), *126,* 214(9), 216(11), *295*
Witkop, B., 18(12), *124*
Wittmann, H. G., 22(19), *124,* 368(85), 432, *445, 447*
Wittmann-Liebold, B., 22(19), *124,* 432, *447*
Witz, J., 111(101), *126,* 234(34), 293(127), *296, 298*
Woese, C. R., 419(157), *447*
Wolf, B., 329(27), *444*
Wollman, E. L., 301(2), 319(16), 323, 325(22), 335, 342, 344, 346, 348, 352(51), 353, 393, 394, *343, 344, 345*
Wolthers, B. G., 100(92), *126*
Wood, W. B., 371, 384(102), 386, *446,* 485, *511*
Woodward, R. B., 26, *124*
Wright, C. S., 98(91), 99, *126*
Wyckoff, H. W., 79(75), 81(75), 83(80), 84(80), 87(75), 89(80), *125, 126*
Wyman, J., 166(33), 168, 170(33), 173(33), *199*

Y

Yagi, K., 164, *198*
Yamane, T., 454(6), 456, *510*
Yang, J. T., 46(47), 47(48), 48(48), 61(65), *125,* 244(46), *296*
Yankeelov, J. A., Jr., 163(27), 164(27), *198*
Yankofsky, S. A., 280(105), *298*
Yanofsky, C., 305, 398(130), 429, 431, 432, 439, *443, 446, 447, 448,* 507, 508, 509, *512*
Yarmolinsky, M. B., 455, *510*
Yoshikawa, H., 364, *445*
Young, J., 22(18), *124*
Yuki, S., 364(75), 365, *445*
Yuryev, V. A., 129(2), *198*

Z

Zabel, R., 31(25), *124*
Zachau, H. G., 453, *510*
Zahn, H., 31(25), *124*

Zahn, R. K., 293(125), *298*
Zalut, C., 31(25), *124*
Zamecnik, P. C., 451(1), 455(7), *510*
Zamenhof, S., 404(140), *447*
Zamir, A., 210(4) 213, 285, 286(4), *295*
Zichichi, M. L., 395(122), *446*
Zimm, B. H., 65(68), *125*, 224, 225, 227, 228, *296*

Zimmermann, R. A., 485(59), *511*
Zimmerman, S. B., 259(62), *297*, 337(38), *444*
Zinder, N. D., 322, 397, 439, 441(192), 443(197), *444*, *446*, *448*, 486, 507(97), *511*, *512*
Zipser, D., 441(191), *448*
Zubay, G., 458, *510*
Zwick, M., 189(47), *199*

Subject Index

A

Absorption spectrum,
  of DNA, 219
  of protein, 51
Acetone, as protein precipitant, 118–119
Acetylcholine, 182
Acridine dyes,
  elimination of sex (F) factor by, 320, 322, 411
  interaction with DNA, 411
  mechanism of mutagenesis by, 411–412, 421–422
  as mutagens, 411–412
  structural formulas, 411
Acridine mustard, as mutagen, 412
Acrylamide, 116
Actin, 26, 105, 191–192, 195, 197
  extraction of, 191–192
  interaction with myosin, 26, 105, 192
  molecular weight, 105, 192
Actinomycin D, as inhibitor of DNA-dependent RNA synthesis, 259, 485
Activating enzyme, *see* Aminoacyl-tRNA synthetase
Activation energy,
  for active transport, 175
  in enzymatic reactions, 149–152
Activator,
  allosteric, 167, 173
  in positive regulation of enzyme synthesis, 498, 502–503
Active site, *see under* Enzyme(s)

Active transport, 128, 174–186
  against concentration gradient, 174–175, 177
  of amino acids, 179
  energetics of, 176–177, 179
  as enzymatic process, 175, 178
  functional role of, 174–175
  of glycerol, 178
  hypothesis accounting for, 175–177
  intracellular, 182–186,
    regulation of, 185–186
  "ion pump" and, 175, 179, 182
  membrane and, 174–175
  and permeases, 178,
  phospholipids in, 179
Actomyosin, 105, 184, 186, 191–192, 195
Actomyosin gels, 191–193, 195
Adaptor hypothesis, 421
  evidence for, 455–456
Adenine,
  action of mutagens on, 408–410
  hydrogen-bonding capacities, 215, 217, 243, 244–245
  structural formula, 206
Adenosine, 204
Adenosine diphosphate (ADP), 179, 185
Adenosine monophosphate (AMP), 167
Adenosine triphosphatase (ATPase), 179, 186
Adenosine triphosphate (ATP), 176, 177, 179, 182–185, 186, 187, 188, 189, 192, 193, 195, 255, 450, 451, 452, 453, 459
  regulation of synthesis of, 182–185

Adenovirus, structure of, 107
Agar gel,
  chromatography of nucleic acids in, 233, 283, 483, 499
  for growth of bacteria, 308, 311
  in immunoelectrophoresis, 122–123
Alanine,
  codons for, 436
  structural formula, 4
Albumin, see Serum albumin; Ovalbumin
Alkaline phosphatase,
  and colinearity of gene with protein product, 426–429
  fine structure genetic mapping of, 348–350
  genetic regulation of synthesis of, 487
  mutants of, in E. coli, 304, 305, 309, 310, 348–350, 427–429, 439
  properties of, 26, 105, 427
  and suppression, 439
  and universality of genetic code, 424–425
Alkylating agents,
  in modification of proteins, 133, 134, 136, 160
  as mutagens, 409–411
Allele, definition of, 345
Allelic genes, 304, 341, 345
Allosteric effectors, 167, 168, 182, 497
  activators, 167, 173
  inhibitors, 167, 172, 173, 174
Allosteric interactions, in regulation of protein synthesis, 497
Allosteric proteins, 167–168, 169, 172–173, 182
  subunit interactions in, 168, 169, 171
Allosteric sites, 167, 168, 172
  interaction between, 168
Allostery, 166–174
  enzyme activation and, 167, 172
  enzyme inhibition and, 167, 172–173
  structural basis of, 167, 168, 173–174
  theory of, 168–174
Alpha-helix, see α–Helix
Amber mutations, see under Mutations
Amino acid analyzer, 8–10
Amino acids,
  abbreviations for, 4–5
  acidic, 3
  activation of, for protein synthesis, 451–453
  analysis of composition, 7–10
    automatic, 8–10
  basic, 4, 9
  carboxyanhydrides of, 26–29, 139
  codons for, 436
  C-terminal, 3, 11, 12, 14, 18, 132
  D-, 6, 47–50, 58, 148
  detection of specific, 18
  dimethylamides of, 27–28
  DNP-, 12–13
  DNSCl derivatives of, 13
  ion-exchange chromatography of, 7–9
  L-, 6, 34, 47–50, 58, 148, 179
  list of, 4–5
  N-terminal, 3, 11, 12–14, 18
  pK values of, 9
  PTH-, 13–14
  radioactive, preparation of, 476
  replacement
    in human hemoglobin, 17, 426, 427
    in tobacco mosaic virus coat protein, 368, 432–433
    in tryptophan synthetase, 430–431, 432
    suppression and, 439–441
  R_f values of, 12
  sequence analysis, 10–23
  sequences,
    methods of determining, 10–23
    in proteins, 20–22
  stereoisomeric configurations of, 6
  structural formulas, 4–5
  sulfur-containing, 5, 7
  unusual, 4–6
D-Amino acid oxidase, 164
Aminoacyl adenylates, 452, 453
Aminoacyl-tRNA, 452–453, 459, 461–463, 465
  formation of, 452–453
  nature of bond in, 453
  in protein synthesis, 455–456, 461–463
  ribosomal (A) site for, 462, 463, 468
Aminoacyl-tRNA synthetase, 452, 456
  molecular weight of, 452
  species specificity of, 453–454
  specificity of, 454–455
p-Aminobenzoic acid (PABA), 140–141, 311
α-Aminobutyric acid, 6
Amino group, 3, 8–10
2-Aminopurine,
  hydrogen-bonding capacities of, 406
  as mutagen, 405–407, 414–415
α-Aminovalerianic acid, 5
Ammonium sulfate in fractional precipitation of proteins, 118
β-Amylase, proposed mechanism of action, 162–163
Anabolic reaction, 128, 487
Anion exchangers, 12, 115, 209
Anomalous dispersion, see under Optical rotatory dispersion
Anthranilic acid, 429
Antibiotics,
  actinomycin D, 259, 485
  cathomycin, 356, 360–361
  mitomycin C, 504
  penicillin, 310, 488
  puromycin, 455, 461

Subject Index

resistance to, 140–141, 311, 312, 328, 341, 356–358, 360–361, 469
  streptomycin, 311, 328, 466–467
  sulfanilamide, 140–141
Antibodies, 122, 123, 129
Anticodon, 435–438, 442, 462–463
  definition, 435
Antigen, 122, 123
Antimetabolites, 140–141
Apo-activator, 498, 502
Apo-repressor, 497, 498
Arabinose metabolism in *E. coli*,
  clustering of structural genes for, 502
  genetic regulation of, 502–503
Arginine, 9, 15, 16
  biosynthesis in *E. coli*, 327
    genetic regulation of, 486–487, 501–502
  codons for, 436
  residues, cleavage by trypsin at, 15
  structural formula, 4
Arrhenius equation, 149, 150, 164
Asparagine,
  codons for, 436
  structural formula, 3
Aspartate kinase, 167
Aspartic acid, 29, 51
  codons for, 436
  structural formula, 4
Association-dissociation phenomena,
  in DNA, 274
  in proteins, 24–25, 103–108
  in ribosomes, 460
Asymmetric carbon atom, 6, 41, 42, 49, 57, 61, 221
Asymmetric synthesis, of polymer, 6–7
Autoradiography,
  analysis of chromosome replication by, 264, 267–268, 328–329, 330, 373
  analysis of DNA molecular weight by, 224
  in studies of DNA dispersion, 267–268, 373
  track, 267, 373
Auxotrophic mutants,
  definition, 308
  methods of selecting, 308–310
Axial ratio, of macromolecules, 70, 72, 77, 108 109, 111, 192
8-Azaguanine, 261
Azotobacter vinelandii, 236, 256

B

Bacillus megaterium, 281
  base ratios of nucleic acids from, 482
Bacillus natto, 279
Bacillus stearothermophilus, ribosomes of, 467
Bacillus subtilis,
  bacteriophages of, 281

competence of, in transformation, 359–360
DNA of, 275, 276, 279, 282
genetic analysis of, 309, 363–366
mutagenesis of transforming DNA from, 416–419
mRNA of, 485
transformation of, 359–360, 362, 363, 364
Bacteria,
  active transport in, 175
  auxotrophic, 308
  basis of variability in, 312–318
  chemical composition, 202, 473
  genetic analysis in,
    advantages of, 299–300, 311
    by conjugation, 301, 318–320, 340–355
    selection of recombinants in, 340–341
    by transduction, 301, 397–403
    by transformation, 300–301, 363–366
  haploid nature of, 320
  mutants of, 304–312
    selection of, 308–312
  prototrophic, 308
  radioactive "suicide" in, 332–333, 352–354
  sexual differentiation in, 318–321, 326
Bacterial virus, *see* Bacteriophage
Bacteriophage, 370–403
  chromatography of, 373
  chromosome of,
    circularity of, 385
    circular permutations of, 387
  defective, 400–403
  dimensions of, 230, 371
  DNA of,
    content of, 203
    as genetic material, 202, 254, 366, 370–372
    integrity of, 377–378
    molecular weight of, 224, 230, 269, 378, 383, 395
  genetic analysis of, 373–383, 387–390, 393–395
    complementation test, 390
    DNA pool and, 376
    fine structure, 387–391
    heterozygotes and segregation, 381–382
    marker rescue in, 395, 400–401, 402
    mixed infection and, 376, 378
    multiple rounds of mating, 376–377, 382
    reciprocal recombination and, 377
    single burst experiment, 375–376
    three-factor crosses, 375–377
    Visconti-Delbrück theory, 376
  "ghosts," 370
  infection by, 370–373
    adsorption, 370
    DNA pool and, 372, 376

DNA synthesis during, 372
eclipse period, 372
effects on host cell metabolism, 372, 478–479
injection of DNA, 370–372
lysis, 372
maturation, 372
nature of "early" proteins, 372
infectious nucleic acids from, 202, 254, 259, 366–368, 370
lysozyme of, 370, 372, 395, 423–424, 441, 505
male-specific, 322
morphogenesis of, 383–385, 386, 387
morphological components of, 371, 385, 386, 387
multiplicity of infection (MOI), 375
mutants of, 374–375, 383–384, 387–389, 393–395
    amber, 383, 384
    conditional lethal, 383, 384
    host-range, 375
    plaque-morphology, 374–375, 378, 393
    temperature-sensitive, 383, 384
osmotic shock of, 267–268
plaque, 373–375
    definition of, 373
    morphology of, 374–375, 381, 394
    mottled, 381–382
prophage state, 312, 391–393
protein comonents of, 371, 372, 383–385, 386, 387
radioactive "suicide" and, 373
receptors for, on bacterial surface, 311, 322, 370
recombination in, 375–383, 400–401
replication of DNA from, 268, 373
resistance to, 311–312, 328
mRNA production by, 372, 479, 504–506
RNA-containing,
    assembly of, 367
    attachment of, to pili, 322
    f2, 322, 367, 439, 440, 442, 486, 507
    MS2, 367
    proteins of, 367
        in vitro synthesis of, 441, 486, 507
    Qβ, 259
    R17, 367
    replication of RNA of, 259
    RNA of, as mRNA, 441, 485–486
temperate,
    definition of, 392
    strains of, see under specific strain
titration of, 373–374
transduction by, 397–403
virulent,
    definition of, 392
    strains of, see under specific strain

Bacteriophage λ, 392–396, 398–402, 499–501, 503–506
attachment site of, on bacterial chromosome, 395
attachment site deletions in, 395
chromosome map of, 503
circular DNA of, 385
    electron micrograph of, 292
    molecular weight of, 395
defective mutants of, 400, 504
    lysogenic state and, 400
    mapping extent of deletion in, 400–401, 402
    properties of, 400–401
"early" proteins of, 503
genetic analysis of, 378–380, 381–382, 393–395
genetic map of, 393
heterogenotes mediated by, 399
heterozygotes in, 381–382
hybrid, with immunity region of phage 434, 500–501
hybridization of mRNA and DNA of, 504
immunity and, 392, 393–395, 499–501
induction of prophage, 392, 394–395, 499–501
integration of, 392, 395–396
morphogenesis, in vitro, 385
mutants of, 393–395, 500–501, 503–506
    affecting regulation of transcription, 500–501, 503–506
    affecting immunity, 393–394, 500–501
    amber, 395
    host-range, 393
    noninducible, 394–395, 500–501
    plaque-morphology, 378, 393, 394
    temperature-sensitive, 504
    virulent, 500, 506
recombination in, 378–381
    mechanism of, 380–381
    with bacterial chromosome, 395–396
regulation of function in, 394–395, 499–501, 503–506
repressor of, 499–501, 503–506
    interaction with DNA, 500, 501, 506
sequence of gene expression in, 503–506
separation of DNA strands of, 282, 504
transducing particles of
    defective, 400–401, 402
        λdgal, 400–401, 402, 499, 504
        λdbio, 400, 506
    formation of, 398–401
    nondefective, 401–402
        λpbio, 402
transcription of DNA of, 282–283, 484, 504–506
zygotic induction of, 392, 499
Bacteriophage P1, 392, 397, 398, 430

Subject Index 529

acquisition of bacterial genes by, 397–398
  transduction by, 329, 398, 430
Bacteriophage P2, 392, 397
Bacteriophage P22, 392, 397, 398
Bacteriophage Φ80, 392, 398, 402–403
  attachment site of, on bacterial chromosome, 402
  transducing particles of,
    defective, 402
      Φ80d*lac*, 403, 495
      Φ80sd*su*, 442
      Φ80d*trp*, 402, 499
    nondefective, 402
      Φ80p*trp*, 402, 509
Bacteriophage, ΦX174,
  circular DNA of, 231, 253–254, 385,
    molecular weight of, 230
  inactivation of DNA of, by ^{32}P, 352
  molecular weight and size of, 230
  replicative form of, 230–231, 253–254, 282
  single-stranded DNA of, 230–231, 235, 251, 282
  spheroplast assay for infectivity of, 254
  transcription of DNA of, 282
  in vitro synthesis of infectious DNA of, 252–254
Bacteriophage S13, 231
Bacteriophage T-even,
  genetic analysis in, 373–378, 381, 382, 387–390
  glucosylation of DNA of, 205, 372
  hydroxymethylcytosine in DNA of, 205, 213, 372
  mutants of, 374–375, 383–384, 387–389
  replication of, 372
  rounds of mating in, 375, 376–377
  structure of, 371
Bacteriophage T2,
  autoradiography of chromosome of, 267–268, 373
  circularly permuted chromosome of, 387
  DNA of,
    base ratio of, 482
    electron micrograph of, 293
    as genetic material, 370–372
    molecular weight, of, 269, 378
  molecular weight of, 230
  mRNA of, 482, 483
    base ratio of, 482
Bacteriophage T4,
  correlation of genetic and physical distances in, 383
  DNA, molecular weight of, 224, 383
  electron micrograph of, 371
  fine structure genetic analysis of, 387–390
    deletions and, 389–390
  genetic map of, 383, 384, 386

  circularity of, 383
  length of, 383
  head protein of, 439–440
    and colinearity of gene with protein product, 440
  heterozygotes in, 381
  lysozyme of,
    mutants of, 423–424, 441
    in vitro synthesis of, 441, 486
  morphogenesis of, 383–385, 386, 387
    in vitro complementation and, 384–385
  morphogenetic map of, 386
  mutants of,
    amber, 383, 384, 439, 440
    assembly, 383–385, 386, 387
    conditional lethal, 383, 384, 385
    frameshift, 421–424
    r, 374–375, 387
      phenotypes of, 387
    r_{II}, 387–390, 421–423
      complementation test and, 390
      genetic analysis of, 389–390, 421–423
      length of cistron, 390
    temperature-sensitive, 218, 383, 384
  transcription of DNA of, 484
Base analogs, as mutagens, 404–407
Base pairing in DNA, 385–387
  principle of complementarity and, 214–216
  replication errors affecting, 216–218, 247, 405–407
  specificity of, 217–218, 232–233
  between unusual nucleotides, 260
  Watson-Crick model for, 214–216
Base ratio,
  of bacterial DNA, 213, 424, 482
  of bacterial RNA, 213, 482
  of bacterial mRNA, 482
  of bacteriophage DNA, 230, 482
  and buoyant density, 271–272
  and Chargaff's rule, 213, 230
  correlation of DNA,
    with amino acid composition of protein, 424
    with base ratio of mRNA, 479, 482
    with base ratio of total RNA, 213, 478
  definition of, 213
  and degeneracy of genetic code, 424
  of DNA synthesized *in vitro*, 248
  of plant and animal DNA, 213
  variations in, 213, 424
Beta-galactosidase *see* β-Galactosidase
Boltzmann distribution, 103, 263
Boltzmann equation, 66
Bond,
  anhydride, 27, 452
  conjugated, 32, 51, 156
  coordination, 134, 136, 153, 154, 155, 291

covalent, 24, 25, 26, 31, 32, 134, 153
disulfide, *see* Disulfide bonds
electrostatic, *see* Electrostatic bond
high-energy, 177, 182, 187, 188
hydrogen, *see* Hydrogen bonds
hydrophobic, *see* Hydrophobic interactions
ionic, *see* Electrostatic bond
peptide, 3, 24, 32–33, 53
van der Waals, *see* van der Waals interactions
Bradykinin, chemical synthesis of, 31
Breakage and reunion, *see* Recombination, genetic, mechanisms of
5-Bromouracil,
 as density label, 265, 329
 hydrogen-bonding capacities of, 405
 as mutagen, 404–405, 406, 414–415
 as thymine substitute in DNA, 260–261, 329
Brownian motion, 24, 278

C

Calf thymus DNA, 219, 222, 223, 224, 228, 235, 271
Calorimeter, adiabatic, 68
Campbell model,
 for formation of transducing particle, 400–403
 for prophage integration, 396
Capsid, viral, 107
Carbobenzoxylation of amino acids, 29
Carbodiimides,
 in chemical synthesis of polypeptides, 28, 30
 in chemical synthesis of RNA, 238
α-Carbon atom, 3, 6, 41, 43, 57
Carbonic anhydrase, 154–155
Carbonyl group, 32, 158
Carboxyanhydrides of amino acids, 26, 27, 29, 139
 formation of, 27
 reactivity of, 27–28
Carboxyl group, 3, 8, 9, 10, 46, 51, 73, 107, 111–112, 153, 190–191, 452, 453
 effect of ionization of, on protein structure, 46, 51–53, 73
Carboxymethylcellulose (CMC), *see under* Cellulose
Carboxypeptidase,
 in C-terminal amino acid sequence analysis, 14, 18, 132, 477
 and "induced fit" hypothesis of enzyme action, 164
 specificity of, 14, 477
 x-ray diffraction analysis of, 97, 98, 100, 164
Casein, 4, 7
Catabolic reaction, 128, 487

Catalase, 51, 127, 413
Catalysis,
 concerted, 98, 158, 160, 165
 enzymatic, 127–165, 166–174
  advantages over nonenzymatic, 127–128, 164–165
 inorganic, 127–128
 nonenzymatic, 127–128, 158, 160, 164–165
 nucleophilic attack in, 158
Cathomycin, as marker in transformation, 356, 360–362
Cation exchanger, 8–9, 15, 115, 131
Cell division, and DNA replication, 334
Cell-free system, *see* Protein synthesis, *in vitro*
Cellulose, ion-exchange,
 characteristics of derivatives,
  carboxymethyl (CM)-, 115
  diethylaminoethyl (DEAE)-, 115
  ECTEOLA, 115
 uses of,
  carboxymethyl (CM)-, 15
  diethylaminoethyl (DEAE)-, 115, 211, 239, 267, 373, 427, 435, 491, 500
Centrifugation, *see* Sedimentation analysis
Cesium chloride, in density gradient centrifugation, 262
Chain termination, *see under* Protein, synthesis
Chain termination mutations, *see* Mutations, nonsense
Chargaff's rule, 213, 230
Chemical mutagenesis, *see under* Mutagenesis
Chemical potential, 168–169, 170
2-Chloroethanol, 32, 39, 70
β-Chloroethylamine, 410
Chloroform, 39, 43, 44, 45, 47, 56, 58, 59, 294
p-Chloromercuribenzene sulfonate (PCMBS), 84, 90
p-Chloromercuribenzoate (PCMB), 85, 136, 146, 189
Chloroplast DNA, 200
Chromatography,
 of amino acids, 8–9, 12, 13, 14
 anion exchangers for, 12, 115, 209
 cation exchangers for, 8–9, 15, 115, 131
 cellulose derivatives for, 15, 115
 DNA-agar gels in, 233, 283, 483, 499
 ion-exchange, 8–9, 12, 14, 15, 16, 115–116, 131, 209, 211, 239, 267, 373, 427, 435, 491, 500
 resins for, 8, 12, 131
 method of Bolton and McCarthy, 233, 283, 483, 499
 methylated albumin-kieselguhr (MAK), 269, 280, 281, 282, 504
 molecular exclusion on dextran gels, 116, 280
 of mononucleotides, 209

Subject Index

of nucleic acids, 233, 269, 280, 281, 282, 283, 435, 483, 499, 504
paper, 12–13, 14, 16
of peptides, 16
of phages, 115–116, 267, 373
of polynucleotides, 211, 239
of proteins, 15, 115, 131, 427, 491, 500
Chromophore, 51, 219
Chromosome,
  bacterial,
    circularity of, 321, 329, 330
    genetic and physical correlations in, 353–355
    methods of mapping, 340–353, 363–366
    replication of, 328–334
    transfer of, in conjugation, 318–325, 332–333
  bacteriophage,
    circularity of, 231, 383–387
    genetic and physical correlations in, 380, 383
    length of, 383
    methods of mapping, 376–377, 389–390, 400–401
    replication of, 373
  circularly permuted, 387
  distance, 350–352, 376–377
    genetic and physical correlations and, 353–355, 380, 383
  functional unit of, 300, 326–327, 390
  map, *see* Genetic map
  mapping, *see* Genetic mapping
  recombination of, *see* Recombination
  replication of, *see* Replication
  single-stranded DNA molecules as, 230–231
Chymotrypsin,
  active site of, 98, 136–139, 140, 158
  in amino acid sequence determination, 18, 19, 22
  esterase activity of, 136, 158–159
  inhibition of, 136, 147–148
  mechanism of, 157–159, 164
  modification of, 139
  specificity of, 18, 139
  x-ray diffraction analysis of, 97, 98, 100
Chymotrypsinogen, 73, 138
  activation of, 138, 139
  amino acid sequence of, 138
Cis-trans test, *see* Complementation test
Cistron, 300, 390–391
  average size of, 300, 383
  definition of, 326–327, 390–391
  molecular dimensions of, 354–355, 366
Clustering of functionally related genes, 496–497, 501–502
Co-activator, 502–503

Coat protein,
  of bacteriophage, 367, 441, 486, 507
  of tobacco mosaic virus, 22, 106–107, 367–368, 432–433
Code, genetic, *see* Genetic code
Coding ratio,
  from analysis of frameshift mutations, 421–423
  from cistron length, 354–355
  definition, 420
Codons(s), 420, 435
  amber, 438–442
  binding of tRNA to ribosomes by trinucleotide, 436, 442, 463
  chain-initiating, 442–443, 463
  chain-terminating, 422, 439–442, 464
  for N-formylmethionine, 442–443, 463
  interaction of, with anticodon, 435–436, 442, 462–463
    suppression and, 439
    "wobble" hypothesis concerning, 437–438
  list of, 436
  methods of determining nucleotide composition of, 434–435
  methods of determining nucleotide sequence of, 436–437
  misreading of, 466–468
  nonsense, 422, 438–442, 464–465
  ochre, 438, 441
  suppression of nonsense, 439–441
  synthetic trinucleotide, 436
  UGA, 438, 441
  in vivo confirmation of assignments, 423–424, 430–433
Coenzymes, 134–135, 156–157
Cohesive ends of phage chromosome, 385
Colinearity of gene and protein product,
  evidence from bacteria, 426–430, 431
  evidence from bacteriophage, 440
Collagen, 227
Competence, for transformation,
  factors determining, 356
  of recipient bacteria, 355–356, 358–360
Competitive inhibition,
  for active transport, 178
  in enzymatic reactions, 140–141, 146–148
  in transformation, 357–358
Complementarity of bases in DNA, *see* Base pairing in DNA
Complementation, interallelic, 326–327, 390
Complementation test,
  in bacteria, 327
    by heterogenotes mediated by substituted sex (F′) factor, 327, 490, 493, 494–495
    by conjugal merozygotes in rec^- strains, 327
  in bacteriophage, 390, 395

use of, in determining character dominance, 327–328
use of, in mapping, 390
Conditional lethal mutants, 307–308, 383, 384
    definition of, 307
    value of, 306–307
β-Conformation of polypeptide chain
    cross-β, 35, 38, 39, 54, 63, 64–65, 98, 99, 100
    dimensions characterizing, 36
    discrimination of, from α-helix, 35, 38, 39, 41, 55–56, 60–61, 64
    pleated sheet, 35, 36–37, 38, 78, 195
        antiparallel, 35, 37, 38
        parallel, 35, 37
    in polypeptides, 38–39, 56, 59, 63
    in proteins, 38, 54, 56, 63, 64–65, 72, 78, 97–100, 195
Conjugation, bacterial,
    chromosome transfer in, 318–325
        DNA as material agent in, 318–319
        DNA synthesis and, 332–333
        mechanisms of, 322–324, 332
    discovery of, 318
    electron micrograph of, 319
    in $F^+ \times F^-$ crosses, 320, 322, 325
    genetic mapping in, 340–353
    heterogenote formation in, 327
    Hfr-mutants and, 320–321
    in $Hfr \times F^-$ crosses, 321, 322
    interspecific, 326
    segregation of chromosome following, 320
    sexduction and, 325
    sexual differentiation for, 320
    species susceptible to, 301
    stable diploids in, 327
    stages of, 318–320
    transfer of sex (F) factor in, 320, 324, 332
    zygotic induction and, 392, 499
Constitutive mutants, *see under* Mutants
Contractile proteins, 129, 182, 184, 186, 191–193, 195
Control mechanisms, *see* Regulation
Coordinated enzyme induction, 496–497, 498, 502–503
Coordinated enzyme repression. 496–497, 498
Coprecipitation, in protein purification, 118
Copy-choice, *see* Recombination, genetic, mechanism of
Co-repressor, in regulation of protein synthesis, 497, 498
Cotransduction of linked genes, 397, 398
Cotton effect, 63–64
Coulomb forces, 69, 74, 189
Countercurrent distribution, fractionation of macromolecules by, 119, 120–121, 210, 456

Counterion, 8, 113, 179, 290
Cresol, 35, 39, 45, 56
Cristae, of mitochondria, 180, 182, 183
Crossing-over,
    multiple, 345, 347, 350–351, 365–366
    nonreciprocal, 336
    recombination and, 334–336
    reciprocal, 334–336
Cross-reacting material (CRM), 428, 489
Crystal,
    properties of, 61, 80–81
    of proteins, 83–85, 110–111
    of tRNA, 294–295
Crystallization,
    as criterion of protein purity, 115, 119
    intramolecular,
        in DNA, 231, 277–278
        in proteins, 24, 43
    isomorphous replacement and, 84–85, 295
    water of, 110–111
C-terminal amino acid residue of protein, 3, 11, 12, 14, 15, 18, 30, 132
    determination of, 14, 18
    and direction of protein synthesis, 466, 473
C-terminal sequence determination, with carboxypeptidase, 14, 18, 132, 477
Cyanogen bromide, 18
Cysteine, 98
    codons for, 436
    cross-linkage between residues of, in protein, 7, 15, 25, 69–74
    in formation of disulfide bonds, 7, 14, 18
    structural formula, 5
Cystine, 5, 7
Cytochromes, 51, 184
Cytochrome *c*, 139–140
    species differences in, 140
Cytoplasm, 128, 174, 200, 457
Cytoplasmic inheritance, 201
Cytosine,
    action of mutagens on, 368, 407, 408–409
    hydrogen-bonding capacities of, 215, 217
    modified residues of, 205
    structural formula, 206
    ultraviolet absorption by, 219

D

DEAE-cellulose, *see under* Cellulose
Debye-Hückel diffuse layer, 113, 114, 240
Defective bacteriophage, *see* Bacteriophage, defective
Deletions,
    bridging two operons, 502
    induced by acridines, 411–412, 422
    in r_{II} region of bacteriophage T4, 389
    use of, in mapping, 389–390

Subject Index

Denaturation,
  of DNA, 221–224, 227–229, 231–236, 273–278
  and helix-coil transition, 38, 42, 65–69, 70–73, 231–236
  of proteins, 26, 42, 47, 59–60, 70–74, 112, 133
    with organic solvents, 26, 64–65
    thermal, 42, 59–60, 65–69, 70–71, 72–73
    with urea, 32, 73, 112, 133
  of RNA, 284–285, 288–289
Density, buoyant, of nucleic acids, 263, 367, 271–272
Density gradient centrifugation,
  analysis
    of bacteriophage heterozygotes by, 381–382
    of bacteriophage recombination by, 378–380
    of complementarity between DNA and mRNA, 281–283, 482–483
    of DNA base composition by, 269–271
    of DNA denaturation by, 274
    of DNA:DNA hybrids by, 279–280
    of DNA heterogeneity by, 271–272
    of DNA integration in transformation by, 362, 363
    of genetic map in transformation by, 364
    of replication by, 254, 262–267, 329
    of ribosome stability by, 479
    of transducing phage by, 398, 401
  description of method and theory, 262–264
  identification of DNA:RNA hybrids by, 280–281
  of intact bacteriophage, 379, 401
  isolation of growing point of replicating chromosome by, 329
  sedimentation equilibrium in CsCl, 262–264
  separation of DNA strands by, 254, 282, 504
Density labeling, 262, 264–267, 329
Deoxyribonuclease (DNase), 209, 249, 256, 355, 480, 483
Deoxyribonucleic acid (DNA),
  antiparallel nature of strands in helix, 214, 249–250, 331
  base pairing in, 214–216, 260, 385–387
    erroneous, 216–218, 405–407
    principle of complementarity and, 214–216
    specificity of, as basis of structure, 217–218, 232–233
  base ratios of, 213, 230, 248, 269–272, 424, 478, 479, 482
  and genetic code, 424
  Chargaff's rule, 213, 230
  cellular content of, 202–203
  chemical structure of, 204, 206
  chromatography of, 233, 269, 280, 281, 282, 283 504

circular molecules of, 231, 330, 337, 385–387
  supercoil structure, 231
denaturation of, 227–229, 273–278
  thermal, 221–224, 231–236
denatured, properties of, 228–229
density of, and base composition, 271–272
density gradient sedimentation of, 262–267
deproteinization of, 209
double-helical structure of, 214–216
effect of decay of incorporated ^{32}P on, 269, 332–333, 352–354, 373
effect of electrostatic charge and pH, 227–228, 231–232
effect of mutagenic agents on, 403–416
electron microscopy of, 252, 257, 292, 385
enzymatic synthesis of, 247–251, 252–254
extranuclear, 200
flexibility of double-stranded, 223–225
  and statistical segment, 223, 225, 286
  and "stiff random coil" configuration, 224
fractionation of, 269, 273
fragmentation of, 224, 268–269, 273, 357
as genetic material, 201–203, 254, 299, 449
glucosylation of, 205, 372
helical structure of, 214–217
helix-coil transition in, 221–223, 227–229, 231–236, 273–278
heteroduplex, 381–382
heterogeneity of, 209, 269–273
hybridization of, 273–284, 482–485, 499, 504–506, 509
hydrodynamic properties of, 222, 224–227
hydrogen bonding in, 214–215, 232–233, 260, 263, 274, 278
hypochromic effect in, 219, 222, 223, 227, 230, 233
infectivity of, 202, 254, 366, 370
isotope effects in, 232–233
length of, 329
macromolecular structure of, 213–216, 218–230
melting point and base composition, 269–271
model of three-dimensional structure, 216
molecular weight of,
  in *E. coli* and various phages, 200, 209, 224, 230, 268, 269, 329, 331–332, 354, 357, 358, 378, 383, 395
  methods of determining, 223–227
    by autoradiography, 224
    by intrinsic viscosity, 224–227, 229
    by light scattering, 223–224
    by sedimentation analysis, 224–227, 229
"nearest-neighbor" analysis of, 248–251
nucleotide components, of 204–206
  minor, 204–205
nucleotide sequence analysis of, 209

optical properties of, 219–223
photochemical action of ultraviolet light on, 339–340
polyelectrolyte behaviour of, 227–228, 229, 235, 286
purification of, 204, 209
recombination and, *see* Recombination, genetic
renaturation of, 273–278
replication of, *see* Replication
single-stranded, 230–231, 235
    of phage ΦX174, 230–231, 253–254
    properties of, 230
    reaction with formaldehyde, 230, 274
    replicative form of, 230–231, 253–254, 258
    in transformation, 363
    in vitro synthesis of infectious, 252–254
strand fractionation,
    in alkaline CsCl, 282
    by CsCl density gradient, 254, 282, 504
    on MAK, 281, 282, 504
strand selection in transcription, 259, 281–283, 484
strand separation during replication, 262–266, 329–331
structure, 218–230, 269–278
    general properties of, 200
    primary, 206, 209
    secondary, 213–230
of substituted sex (F') factor, 337
synthetic, 238–239, 251–252, 255, 258
synthetic polyribonucleotides as models of, 239–245
synthesis *in vitro*, 247–255, 265–267
    of infectious ΦX174 DNA, 252–254
    with synthetic templates, 251–252
    without template, 255
    template requirements for, 248
template for RNA synthesis, 255–259, 280–283
transcription of, *see* Transcription of RNA
transfer of, in conjugation, 318–319
transformation by, *see* Transformation
ultraviolet absorption by, 219–221
unusual bases in, 204–205, 259–261
Watson-Crick model of structure of, 214–218
x-ray diffraction analysis of, 213–214
Deoxyribose, 204, 208, 239
Deproteinization of nucleic acids, 200, 201, 209, 286, 366, 481
Dialysis, 116, 155, 491, 494
Diazomethane, 411
Dichloroacetic acid, 26, 32, 39, 42, 43, 44, 47, 58, 59
Dichloroethane, 38, 39, 45

Dichroism,
    infrared, 55–56
    ultraviolet, 220–221
Dicyclohexylcarbodiimide, 28, 238
Diffraction, *see* X-ray diffraction analysis
Diffusion, 174
    chemical, 175–177
    exchange, 176, 177
    of macromolecules, 102
Diffusion coefficient, 102
Diisopropylphosphofluoridate (DFP), 136, 146
Dimerization, 105
    of ribonuclease, 133, 134
    of thymine residues in DNA, 340
1-Dimethylaminonaphthalene-5-sulfonyl chloride (DNSCl), 13
    for determination of N-terminal amino acid, 13
Dimethylformamide, 27, 28, 32, 44, 45
    in protein structure analysis, 71, 72
N^2-Dimethylguanosine, 208
Dinitrofluorobenzene, *see* 1-Fluoro-2,4-dinitrobenzene
2,4-Dinitrophenol, 179, 183
Dioxane, 27, 28, 32, 45, 50, 70, 71, 294
Diplococcus pneumoniae, transformation of, 355, 357–358, 360, 363
Diploid state, 326–328
Dipolar ion, 9, 12
Dipole transition moment, 45, 53, 54, 61, 219–220
Dispersion force, 77, 218, 232
Disruption of cells, methods for, 480
Dissociation constant, 8, 111, 494, 500
Disulfide bonds, 7, 18
    behavior during protein denaturation and renaturation, 73, 133–134
    cleavage of,
        by oxidation, 14
        by reduction, 14, 71–72, 73, 105, 133
    interchain, 15
    intrachain, 15
    and protein structure, 25, 26, 40, 69–74, 100
    in insulin, 14–15, 71
    in ribonuclease, 14, 20, 22, 71, 72–73, 133–135
DNA, *see* Deoxyribonucleic acid
DNA-dependent RNA polymerase, *see* RNA polymerase, DNA dependent
DNA polymerase, 218, 248–255, 259, 265–267, 331, 337, 338
    bacterial, 248–251
    bacteriophage, 218, 251
    mammalian, 251
    molecular weight of, 248
    reaction catalyzed by, 248, 251, 259
        direction of synthesis in, 251, 331

Subject Index

role of, in DNA replication, 248–251, 252–255
DNase, *see* Deoxyribonuclease
DNP-amino acids, 12–13
  isolation and characterization, 12–13
  reaction of formation, 12
DNSCl derivatives of amino acids, 13
Dominance, definition of, 327–328
Drude equation, 61

E

E. coli, *see Escherichia coli*
Edman technique, *see* N-terminal amino acid residue of protein, determination of
Effective residue rotation,
  definition of, 57
  and secondary structure of polypeptides and proteins, 57–59, 61–65
Electron microscopy, in determination of macromolecular structure, 45, 46, 107, 180, 196–197, 252, 257, 271, 292, 293, 294, 319, 337, 371, 385, 458, 464
π-Electrons, 51, 156, 220
Electrophoresis, 112–114
  diagnostic uses of, 113, 123
  free boundary, 123
  of human blood plasma, 113, 122–123
  immuno-, 122–123
  paper, 16, 116, 123, 209
  of peptidyl-tRNA, 461
  polyacrylamide gel, 116–117, 468
  of proteins and polypeptides, 112–113, 116–118
  starch gel, 116, 424
  theory of, 113–114
  two-dimensional, 286
  zone, 116–117
Electrophoretic mobility, 114
Electrostatic bond, 26, 69, 74, 77, 134, 189–190, 291
  energy of, 77
Electrostatic charge,
  effect on molecular dimensions and conformation,
    in DNA, 227–228
    in flexible polyelectrolytes, 188–190
    in polypeptides, 38, 46–47, 58
    in proteins, 74, 77, 111–112
    in RNA, 239–240, 289–291
  proteins and, 111–114
  relation to electrophoretic mobility, 114
Ellipsoid, equivalent, 108, 109, 111
Enantiomorph, 6, 7, 50
End-group analysis, *see* C-terminal amino acid residue of protein; N-terminal amino acid residue of protein

Endonucleases, 210–212, 337
Endopeptidase, 15
Endoplasmic reticulum, 201, 457
Entropy, 168–169, 171
  of activation, 149–150, 151
  of formation of enzyme-substrate complex, 150–151, 163
  of formation of hydrophobic bonds, 76
  of helix-coil transition, 65–69, 70–71, 231–232
Enzyme(s),
  activation energy and, 150, 153
  active site of, 130, 132–134, 157–161, 163–165, 167, 168
  and active transport, 178
  chemical modification of, 136–139
  role and organization of, 130, 139–142
  x-ray diffraction analysis of, 97–100
  allosteric transformation in, 166–168, 172–173
  as catalysts, 127–128, 129–142
    advantages of, 164–165
    metathesis and, 129–130
  conformational changes in, 98, 164
  effect of pH on, 136–137, 152–153
  induction of, *see* Induction, of enzyme synthesis
  inhibition, 136–139, 140–141, 146–149, 155–156, 172–173
    as probe of active site, 136–139, 140–141, 155–156
  kinetics, 142–153, 168–172
    activation energy and, 149, 150
    conditions for study, 142
    effect of pH on, 152–153
    effect of substrate concentration on, 143–144
    inhibition and, 146–149, 172–173
    Lineweaver-Burk equation and, 144, 145, 147
    Michaelis constant, 144, 147–148, 150–151, 152–153
    Michaelis-Menten equation and, 144, 145, 146, 149, 168, 169, 171–172
    rate constant, 144, 148, 149–150
    relation between rate and equilibrium constants, 144
    steady-state treatment of, 143–144
    theory of, 142–149
      allostery and, 168–172
      thermodynamics of, 149–152
  mechanism of action, 130, 153–165
    acid-base, 160
    concerted, 158, 160
    double displacement hypothesis, 161–163
    "induced fit" hypothesis, 98, 162, 163–164

metal ions and, 154–156
nucleophilic attack and, 158
pyridoxal phosphate and, 158
prosthetic groups of, 134–135
regulation of activity of, 165–168, 172–174
   feedback inhibition and, 165, 167, 172
repression of, see Repression of enzyme synthesis
specificity of, 15, 127, 139, 157, 163–164
structure of, in relation to function, 96–100, 130, 131–134, 136–142, 163–164
zymogens and, 139, 491
Enzyme-substrate complex, 142–146, 157–165
formation of, 144, 150–151, 152, 153–155, 161, 163, 165
substrate activation and, 149–150, 153, 156–157, 163–164, 165
x-ray diffraction analysis of, 96–98,
Episome,
definition of, 324
integration of, 337–339
replication of, 333–334
sex (F) factor as, 324
substituted sex (F′) factor as, 325
temperate phage as, 339
Equilibrium constant, 144, 147–148, 150–151, 171, 172
Equilibrium dialysis, 491, 494
Erythrocytes, 92, 175, 178, 425
Escherichia coli,
autoradiograph of chromosome, 330
bacteriophages of, 370–403
base ratios of nucleic acids from, 482
chemical constituents of, 473
chromosomes, growing points of, 329
circularity of chromosome, 321, 329, 330
conjugation in, 318–325
correlation of genetic and physical distances in, 354
DNA content of, 203, 329
fluctuation test of mutation in, 312–318
genetic analysis in, 340–355
genetic map of, 303
haploid nature of, 320
kinetics of protein synthesis in, 473–477
mating types of, 320, 325
recombination in, 334–339
regulation of enzyme synthesis in, 427, 478, 487–496, 497–498, 501–503, 508–510
ribosomes of, 280–281, 450, 457, 458, 459–469
mRNA of, 479–482
sex (F) factor of, 320–326
selection of mutants of, 308–312
substituted sex (F′) factors of, 325–326, 334, 337, 354

suppressor tRNA's of, 441–442
suppressors of, 307–308, 439–442
transduction of, 397–403
Esterase, 129, 136,
activity,
of chymotrypsin, 136, 158–159
of trypsin, 136, 158–159
mechanism of action, 136
Ethyl ethane sulfonate (EES), as mutagen, 410
Ethyl methane sulfonate, (EMS), as mutagen, 410
Ethylene dichloride, 45
Exonuclease, 210–212, 337, 338
of λ phage, 503, 504, 506

F

F (fertility) factor, see Sex (F) factor of *E. coli*
F⁺ donor, see *Escherichia coli*, mating types of
F′ factor, see Substituted sex (F′) factor of *E. coli*
F⁻ recipient, see *Escherichia coli*, mating types of
F-duction, see Sexduction
Feedback inhibition, 167, 487
allostery and, 167–173
Feedback loops,
in biological regulatory systems, 165–166, 185, 449
in regulation of biosynthesis, 167, 487, 491
Fibrin, 26, 106, 139
Fibrinogen, 25, 26, 75, 105, 106, 139
Fibrous proteins, 25, 56, 78
Fingerprinting, 368, 424, 426, 427, 430, 472, 486
in amino acid sequence determination, 16–18
Flavin adenine dinucleotide (FAD), 135
Flexibility of nucleic acid molecules, 223–225, 286
statistical segment as measure of, 223, 225, 286
Flory–Mandelkern equation, 225, 229, 284, 286
Fluctuation test, 312–318
Fluorodeoxyuridine, 381
1–Fluoro-2,4-dinitrobenzene (FDNB), 12–13, 178, 187
for determination of N-terminal amino acid, 12–13
Fluorophenylalanine, 453
5–Fluorouracil, 260–261
Folic acid, 140–141
Formamide, 44, 45
Formic acid, 39, 56
N-Formylmethionyl-tRNA, 442–443
Fractional precipitation, 131, 452
in purification of proteins, 118–119
Fractionation
of nucleic acids, 204, 210, 269, 273, 456

of proteins, 112–113, 115–123
Frameshift mutations, 421–424
  and confirmation of codon assignments, 423–424
  and establishment of triplet code, 421–423
  induction of, by acridine dyes, 421–422
Free energy,
  of activation, 149–150, 151, 152
  and base pairing in DNA, 218
  of formation of enzyme-substrate complex, 150–151, 152
  of formation of hydrophobic bonds, 76
  in helix-coil transition, 66–68
Free radicals, as mutagens, 413
Frictional coefficient, 102, 108
  relation to molecular size and shape, 108
  for spheres, 108
Frictional ratio, for ellipsoids of revolution, 108

G

Galactose metabolism, transduction of genes for, by λ phage, 398–401
β-Galactosidase,
  chemical and physical properties of, 491–492
  genetic regulation of, 487–496
  inducers of,
    isopropyl-β-D-thiogalactoside (IPTG), 488, 490, 494, 495
    lactose, 487
    methyl-β-D-thiogalactoside (TMG), 488–489, 491
  induction of, 487, 490–491, 496
  mutations of, 489–495
  polar mutants of, 506–507
  substrates of,
    lactose, 487
    o-nitrophenyl-β-D-galactoside, 491
    phenyl-β-D-galactoside, 488–489
  synthesis of, 490, 491–492, 496
  synthesis of mRNA for, 492, 496
Galactoside transacetylase, 494, 496
Galactoside permease, 489–490, 494–495, 496
Gaussian distribution, 264, 271
Gene(s), 300, 426
  definition of, 390–391
  clustering of functionally related, 496–497, 501–502
  as determinants of protein structure, 300
  dominant, 327–328
  mapping of, see Genetic mapping
  mutation of, see Mutation(s)
  organization into operons, 496–497, 501–502
  recessive, 328
  regulatory, 325, 489, 496, 497, 498, 502
  structural, 325, 496, 498
  suppressor, 307–308, 383, 439
  mechanism of action, 439–442
  tRNA products of, 441–442
  wild type, 301, 308
Gene-enzyme relationships, 300, 426
Genetic analysis,
  advantages of bacteria in, 299–300, 341
  of bacteriophage, 373–383, 387–390
  of E. coli, methods of, 340–355
  by transduction, 397–403
  by transformation, 363–366
  molecular scale of, 354–355
Genetic code,
  acridine-induced frameshift mutations and, 421–424
  adaptor hypothesis and, 421, 435, 455
  anticodon, 435–438, 439, 442, 462–463
  assignment of nucleotide sequences, 434–437
  biochemical analysis of, 433–443
  chain-terminating mutations and, 422, 440
  coding ratio and, 354–355, 420, 423
  codon, 435–438, 442
    definition of, 420
  deciphering of, 434–437
    with defined polynucleotides, 258, 437, 466
    with defined trinucleotides, 436, 442
    frameshift mutations and, 423–424
    with random polynucleotides, 434–435
  in vivo confirmation of, 423–424, 430–433
  degeneracy of, 424, 435, 437–438, 452
  doublet, 420
  evolution and, 437
  general nature of, 202, 419–421
  genetic analysis of, 425–433
  information flow and, 202, 419
  list of codons for, 436
  misreading of, 466–468
  nonsense triplets, 438
  overlapping, 420–421
  punctuation of, 421
  reading frame for, 421–423
  suppression and, 439–442
  theoretical aspects of, 419–421
  translation of, 455–456, 461–465
  triplet, 420, 421–423
  universality of, 424–425, 435, 438
  "wobble" hypothesis and, 437–438
Genetic distances,
  in bacteriophage, 383
  in physical units, 353–354, 383
  in recombination units, 345, 347–351, 354, 383
  relationships between 351–352, 354, 383
  in time units, 343, 344
Genetic exchange in bacteria, modes of, 300–301
Genetic fine structure analysis, 304, 340

in bacteriophage, 387–391
by conjugation, 348–350
in *E. coli*, 340, 348–350
resolving power of, 390, 428, 430–431
by transduction, 398
by transformation, 364–365
Genetic map, 300, 301
of *Bacillus subtilis*, 365
of bacteriophage λ, 393
of bacteriophage T4, 384, 386, 388
circular, 303, 383
colinearity of, with protein product, 426–430, 431, 440
of *E. coli*, 303, 323, 349
fine structure, 349, 388
length of, 354, 383
Genetic map unit,
in bacteria, 343–347, 364–365
in bacteriophage, 383
definition of, 344–345, 350–352
interpretation of, in physical terms, 353–355, 366, 383, 390
Genetic mapping,
by complementation test, 390
by conjugation, 340–353
by gradient of transmission, 343–344
by interrupted mating, 342–343
by marker rescue, 395, 400–401, 402
by ^{32}P disintegration, 352–353
by recombination analysis, 344–352, 376–377
by time of transfer, 342–343
by transduction, 329, 364, 365, 397–398, 430
by transformation, 363–366
by use of deletions, 389–390
Genetic markers, 300
in *E. coli*, 301–303
and selection of mutants, 308–312
and selection of recombinants, 340–341
in transformation, 356, 363
Genetic recombination, *see* Recombination, genetic
Globular proteins, 23, 25, 100,
Bresler-Talmud theory of, 74–75, 100
hydration of, 109–110
structure of, 69–78, 108–113
x-ray diffraction analysis of, 78–101
Globulins, electrophoretic separation of, 113, 123
Glucosylation of bacteriophage DNA, 205, 372
Glutamic acid, 29, 51
codons for, 436
structural formula, 4
Glutamic dehydrogenase, 174
Glutamine,
codons for, 436
structural formula, 3

Glyceraldehyde-3-phosphate dehydrogenase, 26
Glycine,
codons for, 436
structural formula, 4
Gradient elution, 15, 116, 211, 282
Guanidine, 26, 32
Guanine,
action of mutagens on, 408–410
hydrogen-bonding capacities of, 215, 217
structural formula, 206
Guanosine triphosphatase (GTPase), 461, 463
Guanosine triphosphate (GTP), 450, 451, 459
in peptide bond formation, 459, 461, 463

H

Haemophilus influenzae, transformation of, 358, 360–362, 363
Heat production, during muscle contraction, 187–188
Heavy atoms, in x-ray diffraction analysis, 84–85, 90
Helical structure,
of DNA, 214–217
of RNA, 287, 288–289
of tRNA, 233, 285
of synthetic polyribonucleotide complexes, 239–245
of tobacco mosaic virus, 106, 288
α-Helix,
dimensions of, 33–34, 45
discrimination of, from β-conformation, 35–38, 39, 41, 55–56, 60–61, 64
evidence for, 33–34
methods of estimating, 34, 39, 41–43, 53–54, 57–58, 61–65, 68–69
in polypeptides, 34, 35, 38–39, 41–50, 53–56, 58–64
in proteins, 24, 54, 61–62, 65, 69, 70–72, 78, 79, 91, 96–100
sense of rotation,
effect of stereoisomerization on, 47–50
left, 35, 43, 58, 61–62
right, 34–35, 43, 58
Helix-coil transition,
in DNA, 221–223, 227–229, 231–236, 273–278
and base composition, 269–271
heat of, 232
hydrodynamic changes associated with, 227–229, 235
isotope effects and, 233
reversibility of, 273–278
strand separation and, 233–235, 273, 277
temperature of, 221, 231–232, 235, 236, 269–271
thermodynamics of, 231–232
in polynucleotides, 245–246

Subject Index

in polypeptides, 38–39, 42–43, 45–47, 48, 58–59
in proteins, 41, 42, 59–60, 65–69, 70–73
  cooperativity of, 38, 42, 46, 65–68
  effect of solvent on, 70, 71
  entropy and, 66–69, 70–71
  heat of, 68–69
  temperature of, 24, 41, 42, 68–69, 70
  thermodynamics of, 65–69
Helper phage, 400
Heme, 51, 79, 93, 94, 95, 96
  proteins containing, 51, 79
Hemoglobin, 7, 15, 51, 69, 79, 100, 113, 130
  abnormal human, 17, 426, 427
    amino acid replacements in, 17, 426, 427
  allostery and, 169, 170, 171
  association with O_2, 170, 171
    conformational changes and, 95–96
  dissociation of, 15, 103–104
  fractionation of α- and β-chains of, 15, 93, 472
  function of, 92
  hydration of, 109–111
  model, 94
  molecular weight of, 92
  N-terminal amino acid residues of, 470
  sickle-cell, 17, 18, 113, 425–426, 427
  structure of, 92–95
    quaternary, 95–96, 164
  subunits of, 15, 92, 103–104
    contact between, 95–96
  synthesis of, *in vitro*, 434, 456, 470–473
  tryptic digests of, 17, 18, 472
    alignment by pulse-labeling, 472
    fingerprinting, 17, 18
  x-ray diffraction analysis of, 92–96, 109–111, heavy metal ions and, 85
Hershey–Chase experiment, 366, 370–372
Heteroduplex, 381–382
Heterogeneity, molecular,
  of nucleic acids, 269–273
    estimation by density gradient sedimentation, 271–272
    estimation by temperature of helix-coil transition, 269–271
  of ordinary linear polymers, 1, 101–102
  of proteins, 1–2, 101–102
Heterogenote, 327–328
  definition of, 327
  formation by sexduction, 327, 490, 493, 494–495
  formation by transduction, 399
  in complementation tests, 327, 490, 493, 494–495
Heterozygote,
  in bacterial crosses, 318–328, 334–350

  in bacteriophage crosses, 381–382
Hfr donor bacteria, 320–326
  discovery of, 320
  properties of, 320–325
  selection of, 324–325
  types of, 321, 323, 324, 341, 343
HFT, *see* Transduction, high frequency
High frequency of recombination, *see Hfr* donor bacteria
Histidine,
  at active site
    of chymotrypsin, 98, 137–139, 157–159
    of proteolytic enzymes, 98
    of ribonuclease, 132–133, 134, 160–161
    of trypsin, 136–139, 157–160
  biosynthesis of,
    genetic regulation of, 497, 499
    operon for, 497
  codons for, 436
  structural formual, 4
Homogeneity of proteins, 1–2, 101–102
  criteria for, 122
"Hot spots," mutational, 390, 415
Hybridization of nucleic acids,
  of DNA with DNA, 273–280
    annealing process in, 273–274, 278
    interspecific, 279–280, 283
  of DNA with RNA, 280–284, 482–485, 499, 504–506, 509
    annealing process in, 280, 483
    competition experiments and, 281, 284
    in demonstration that mRNA is transcribed, 482–485, 499, 509
    in demonstration that rRNA is transcribed, 280–281
    in demonstration that tRNA is transcribed, 281
    for detection of DNA:RNA complementarity, 280–284, 499, 509
    for determination of which DNA strand is transcribed, 281–283, 484–485, 504–506
    resistance of complex to DNase and RNase, 280, 281, 483
  methods of analysis, 279, 280, 283–284
    agar gel chromatography, 283
    density gradient sedimentation, 279, 280–281
    membrane filter assay, 283–284, 483
Hydration,
  of hydrodynamic particles, 109–111
  of protein crystals, 84–85, 110–111
  of protein molecules 23, 109–111
Hydrodynamic constants, 108–111
Hydrodynamic properties,
  of nucleic acids, 224–229
  of proteins, 44–45, 108–111

Hydrogen bonds,
  in DNA,
    determining specificity of base pairing, 214–218
    determining specificity of replication, 248–251, 261–262
    dimensions of, 215
    energy of, 218
    estimation by isotopic exchange, 232–233, 285
    and helix-coil transition, 231–236, 273–274, 278
    hypochromic effect and, 219
  in polynucleotides, 239–245
  in polypeptides, 31–41
    dimensions of, 33
    effect on infrared absorption, 55–56
    effects of solvents on, 32, 38–39, 42–43, 44–45
    energy of, 24, 32, 68
    estimation by isotopic exchange, 39–41
    hypochromic effect and, 53–54
    between peptide groups, 31–35, 36, 37
  in proteins, 40–41, 71, 78, 131
    between amino acid side chains, 51–53, 73
    determining secondary structure of, 24–26, 74, 100
    and helix-coil transition, 65–68
  in RNA, determining secondary structure of, 233, 285–286, 287, 288–289
Hydrogen bromide, 29
Hydrogen peroxide, as mutagen, 413
Hydrophilic amino acid side chains, 74, 77
Hydrophobic amino acid side chains, 74, 76, 97
Hydrophobic interactions,
  energy of, 76–77
  in nucleic acids, 218, 232–233
  in polypeptides, 35, 38
  in proteins, 25, 26, 70, 74–77, 92, 96–100
Hydroxylamine, as mutagen, 408–409
N-Hydroxylysine, 6
5-Hydroxymethyl cytidine, 205, 213, 372
Hydroxyproline, 5, 10, 14, 24, 25
Hyperchromic effect, 221
Hypochromic effect,
  in DNA, 219, 221–223, 227, 230, 271
  in polynucleotides, 239–240
  in polypeptides, 53–54
  in proteins, 52, 54, 59–60, 64
  in RNA, 284–285, 288

I

Immunity of lysogenic bacteria, 392, 393–394, 499–500
  genes for, in λ bacteriophage, 393, 394–395, 500
  mechanism of, 393–394, 499–501
  repressor and, 499–501
Immunoelectrophoresis, 122–123
In vitro protein synthesis, *see under* Protein
Index of refraction, 57, 58
Indole, 429
Indolylglycerol phosphate, 429–430
"Induced fit" hypothesis of enzyme specificity, 98, 163–164
Induced mutations, *see under* Mutation(s)
Inducers and regulation of protein synthesis, 487–489, 491, 492–493, 494, 496, 497–498
Induction,
  of enzyme synthesis, 488–489, 490–491, 496–497, 498, 502–503
    coordinated, 496–497, 498
    early theories of, 487–488
    mechanism of, 496–497, 498, 502–503
  of prophage,
    by chemical agents, 392, 504
    mechanism of, 394–395, 499–501
    mutants insensitive to, 394–395, 500, 501
    spontaneous, 392
    by ultraviolet light, 392, 500
    zygotic, 392, 499
Infection, bacteriophage, *see* Bacteriophage, infection by
Infectivity,
  of bacteriophage DNA, 202, 254, 366, 370
  of viral RNA, 202, 259, 288, 289, 366–368
Information, genetic,
  flow of, in cell, 201–202, 449, 477
  storage of, 201–202, 299, 499
Infrared dichroism, 55–56, 59, 78
Infrared spectroscopy,
  absorption attributable to C=O bonds, 55, 56
  absorption attributable to N—H bonds, 55, 56
  as means of distinguishing α- and β- conformation in protein, 56
Inhibition of enzymes, 136–139, 140–141
  allosteric, 172–174
  competitive, 146–148
  feedback, 167, 487
  noncompetitive, 148–149
Initiation,
  of DNA synthesis, 333
  of protein synthesis, 442–443, 463–464
Initiator of DNA synthesis, 333
Injection of phage DNA, mechanism of, 370–372
Inosine,

hydrogen-bonding capacities of, 242, 244
  as minor base in tRNA, 208, 213, 438
  structural formula, 208
Insulin, 7, 14, 51, 75, 85, 101, 119
  amino acid sequence of, 21
  chemical synthesis of, 31
  difference spectrum of, 52
  disulfide bonds of, 7, 14, 40, 71
  isotope exchange and, 40–41, 71
  molecular weight of, 101
  ordered structure in, 62, 65
  oxidation of, 14, 71
  species differences in, 21
  titration curve of, 52
Interrupted mating, *see under* Genetic mapping
Intrinsic viscosity,
  and DNA molecular weight, 224–227
  and DNA structure, 222, 224–229, 235–236
  and polypeptide molecular weight, 44–45
  and protein structure, 70, 108–109, 192
  and RNA structure, 289–290
  of poly-γ-benzyl-L-glutamic acid, 43–45
Invertase, 136, 137, 145, 146
Iodoacetamide, 4, 134, 136
Ion pump, 175, 179, 182
Ion-exchange chromatography, *see under* Chromatography
Ionic bond, *see* Electrostatic bond
Ionic strength, 47, 69, 240, 289, 291
  definition of, 114
Isoelectric point, 112–113
  definition of, 112
  table of, 104–105
Isoenzymes, 131
Isoleucine,
  codons for, 436
  structural formula, 4
Isomorphous replacement, *see* X-ray diffraction analysis
Isotope effects, 42, 232–233
Isotopic exchange,
  in DNA, 232–233
  estimation of protein secondary structure by, 39–41, 54, 64
  in insulin, 40–41
  in polypeptides, 39–40
  in ribonuclease, 40–41
  in tRNA, 233, 285

K

Keratin, 25, 56, 77–78
Kinetics,
  of chromosome transfer in conjugation, 342–344
  of enzymatic catalysis, 142–153
  of enzyme induction, 490–491
  of metabolic processes, methods of study, 469–470
  of mutagenesis, 305–306, 368–369
  of protein synthesis, 470–477
  of radioactive "suicide," 332–333, 352–354
  of mRNA synthesis, 480
Kornberg enzyme, 248–255, 259, 265–267

L

Lactose utilization in *E. coli*, 487–496
  enzymes mediating,
    β-galactosidase, 487–496
    galactoside permease, 489–490, 494–495, 496
    galactoside transacetylase, 494, 496
  genetic analysis of, 489–491, 493, 494–495
    cis-dominance and, 494, 495
  genetic regulation of, 488–489, 490–491
    function of *i* gene in, 492–493, 494
    operator for, 494–495
    promoter for, 495–496
    repressor and, 493–494, 495, 496
  mapping of genes for, 346–348
  mutations affecting, 489–495, 507–508
  operon for, 496
Langmuir adsorption isotherm, 168, 169, 358
Lanthionine, 6
Leucine,
  codons for, 436
  structural formula, 4
LFT, *see under* Transduction
Light scattering,
  and molecular weight of nucleic acids, 224
  and molecular weight of proteins, 44, 101
  relation to radius of gyration, 45
Lineweaver-Burk equation, 144
  plot of, 145, 147
Linkage, genetic
  in bacteriophage, 383, 393
  and conjugation, 321, 342–344, 346–350
  and transduction, 397–398, 403
  and transformation, 358–360, 362–363, 364–365
Lipids, 174, 179, 182
Lipoprotein, in cellular membranes, 174, 175, 181, 182, 185
Locus, genetic, 301
Luria-Delbrück fluctuation experiment, 312–318
Lysine, 9, 15, 16, 29
  codons for, 436
  specificity for cleavage by trypsin at residues of, 15
  structural formula, 4
Lysogenic bacteria, 391–392
  defectively, 400–401

doubly, 400
induction of prophage in, 392, 394–395
nature of immunity in, 393–394
phage production by, 392, 398–399
relation between prophage and bacterial chromosome, 392, 395–396
superinfection of, 394–395, 400, 401
Lysogenization,
mechanism of, 393, 395–396
frequency of, 393
Lysogeny, 391–397
and formation of transducing particles, 398
mutations affecting, 393–395
Lysozyme,
hen egg white, 54, 62, 254
active site of, 96
amino acid sequence of, 20
model of, 97
x-ray diffraction analysis of, 96
phage λ, 395
phage T4, 372
frameshift mutations of, 423–424
in vitro synthesis and suppression, 441

M

Magnesium ion,
and configuration of RNA in solution, 289–291
effects on protein synthesis *in vitro*, 463–464, 467
requirement for, by DNA and RNA polymerases, 251, 255
reversible dissociation of ribosomes and, 460
in stability of ribosomal structure, 457
MAK, *see under* Chromatrography
Map, genetic, *see* Genetic map
Mapping, genetic, *see* Genetic mapping
Mapping function, 351
Marker, genetic, *see* Genetic markers
Marker rescue, 395, 400–401, 402
Mating types in bacteria, 320, 325
Mean residue rotation, 57
Mechanochemical proteins, 129, 182, 184, 186, 192–193, 195
Mechanochemical reactions, 129, 182–183, 185, 186–195, 463, 466
Media for bacterial growth,
complex or enriched, 308
solid, 308
synthetic, 308
Membrane,
active transport across, 128, 174–186
as capacitor, 182
chemical diffusion across, 175–177

composition of, 174
electrical excitability of, 179–182
function of, 174, 183–185
mitochondrial, 174, 181, 182–185
structure of, 174, 181
Mercaptalbumin, 104
β-Mercaptoethanol, 73, 133, 433
Merodiploid, 328, 336
Merozygote, 332, 334, 336, 343, 346
Messenger RNA (mRNA),
amount of, in bacterial cell, 450, 480
attachment to ribosome, 461, 462, 480, 481
bacteriophage, 479, 482, 484, 504–506
base ratios of, 482
direction of transcription, 466, 508
direction of translation, 424, 466, 473, 477
discovery of, 479
evidence for, 477–486
function of, 450, 477–479
hybridization with DNA, 281–283, 482–484, 499
sequence homology and, 482–483
specificity of, 483
isolation from bacterial cells, 480–482
lability of, 480
in mammalian cells, 486
metabolic instability of, 450, 478–479, 485
polycistronic, 482
transcription from operon, 499, 505, 507, 508–510
translation on ribosome, 506–508
properties required of, 479, 483
purification and characterisation of, 483
rate of degradation, 485
sedimentation analysis of, 480, 481, 482
size of, 465, 480, 483
synthesis of,
in bacterial cells, 479–482, 484
coupling to protein synthesis, 508–510
during phage infection, 281, 479, 482–484, 504–506
regulation of, 497–499
in vitro, 485
synthetic polynucleotides as, 434–437, 456, 459, 461, 466–467
as template for protein synthesis, 420, 450, 459,485–486
transcription of, 482–484
translation of, *see* Protein synthesis
Metabolic pathways, regulation of, 486–510
Metabolic processes, regulation of, 128, 165–174, 182–186, 486–487
Metal ions,
in enzyme function, 135, 154–156
in protein function, 79
Metathesis, 129–130

Subject Index

Methionine, 18
  cleavage at residues of, by cyanogen bromide, 18
  codons for, 436
  N-formyl, and initiation of protein synthesis, 442–443, 463–464
  structural formula, 5
5–Methylcytosine, 204, 213
1-Methylguanosine, 205
1-Methylinosine, 208
N–Methyl-N′–nitro–N–nitrosoguanidine, as mutagen, 331, 411
Michaelis constant, 31, 144, 147–148, 150–151, 152–153
Michaelis-Menten equation, 144, 145, 146, 149, 168, 169, 171–172
Mitochondria, 128–129, 174, 182–185
  conformational changes of, 182–185
  electron micrograph of, 180
  in regulation of metabolic processes, 182–185
  structure of, 181, 182
Moffit equation, 61
  b_o and helix content, 61–63
  determination of b_o and λ_o, 61
Monod-Jacob model of genetic regulation, 497–498
  and clustering of genes in operon, 501–502
  and polar mutations, 506–510
  and positive control, 502–503
Morphogenesis of bacteriophage T4, 383–385, 386, 387
Morse potential curve, 269
Mottled plaque, 381–382
mRNA, see Messenger RNA
Multiple rounds of mating in bacteriophage, 376–377, 382
Multiplicity of infection, 375
Muscle, 175, 182, 186–197
  contraction of, 186–188, 193–195
    actomyosin gels in study of, 191–193, 195
    anisotropy and, 190–191, 194
    energetics of, 187–189
    inhibitors and, 189
    physical mechanism of, 189–191, 193–194
    thermodynamics of, 191, 193
  heat production in, 187–188
  efficiency of, 187
  mechanochemical properties of, 186–195
  molecular structure of, 195, 196–197
  proteins of, 186, 192
  synthetic models of, 189–191
  thermal relationships in, 187–188
Muscle fiber, 193–197
  electron micrographs of, 196, 197
  molecular architecture of, 195, 196–197
Mutability spectrum,
  definition of, 415
  "hot spots" and, 390, 415
  interpretation of, 415
Mutagenesis, 304–306, 403–419
  in bacteria, 305
  in bacteriophage, 390, 411, 414–415, 418
  chemical, 306, 403–416
  kinetics of, 305–306, 368–369, 416–419
  radiation and, 305, 403, 412–413
  specificity of, 390, 414–415
  in tobacco mosaic virus, 367, 368–369, 403
  in vitro, 368–369, 403, 416–419
  in vivo, 403–416
Mutagenic agents,
  chemical, 306, 403–412
    acridine dyes, 411–412
    alkylating, 409–410, 411
      ethyl ethane sulfonate (EES), 410, 414, 415
      ethyl methane sulfonate (EMS), 410
      nitrogen mustards, 410
    base analogs, 404–407
      2-aminopurine, 405–407, 414, 415
      5–bromouracil, 217, 261, 404–405, 406, 414, 415
    free radicals as, 413
    hydroxylamine, 408–409
    N-methyl-N′-nitro-N-nitrosoguanidine, 410–411
    nitrous acid, 368, 407, 408, 409
    reacting directly with DNA, 408–413
    requiring DNA replication, 403–407
    weak acid, 410
  genetic studies of mechanism of, 414–415
  radiation as, 403
  ultraviolet light as, 305–306, 413
  x-rays as, 412–413
Mutants,
  acridine-induced, 421–422
  antibiotic-resistant, 141–142, 310–311, 312, 328, 341, 356–358, 360–362
  auxotrophic, 308
  bacterial, 304–312
  bacteriophage, 307–308, 373–375, 393–395
  conditional lethal, 307–308, 383
  constitutive, 427, 489, 492, 493, 494, 497, 503
  "cryptic," 490–491
  host-range, 375, 393
  "leaky," 421
  phage-resistant, 311–312, 314, 317, 328, 375
  plaque-morphology, 374–375, 378, 393
  r, 374–375, 387
  r_{II}, 387–390, 414–415, 421–423
  recombination-deficient (rec^-), 324, 340
  selection and characterization of, 306–307, 308–312

temperature-sensitive, 307, 334, 383, 384, 402–403, 504
Mutation(s),
  as alteration in DNA molecule, 403, 415–416
  amber, 308, 383, 384, 439–442
  anticodon, 439–442
  base substitution errors and, 216–218
  as basis of heritable variation, 312
  chain-terminating, 422, 440, 507
  conditional lethal, 307–308, 383
  deletion, 306, 389–390, 404
  duplication, 404
  effect on protein function, 304–305, 310, 425–426, 428–429, 430–431
  effect on protein structure, 101, 140, 141, 425–433
  frameshift, 412, 421–424
  "hot spots" and, 390, 415
  induced, 304–306, 390, 404–416
  inversion, 306, 404
  lethal, 304, 306–307
  Luria-Delbrück fluctuation test and, 312–317
  molecular basis of, 216–217, 304, 403–416
  muton and, 391
  nonsense, 422, 438–441, 507
  nucleotide base pair changes and, 404, 405, 408, 409, 414, 415
  ochre, 308, 441
  operator, 494–495, 497, 500, 506
  point, 306, 391, 404
    as alteration of single base pair, 247, 404, 415–416, 420–421, 431, 441
    reversion of, 404
  polar, 506–510
  promoter, 495
  rate of, 218, 306, 310, 314, 317
  replication errors and, 218, 247, 405–407
  reversion of, 306, 309–310, 404, 407, 414–415, 421–423
  size of, 391
  spontaneous, 217–218, 247, 304, 312–317, 412
  suppressible, 307–308, 439–441
  suppressor, 307–308
    and genetic code, 439–442
  transition, 404, 405, 407, 408, 409, 410, 412, 413, 414–415
  translocation, 306, 404
  transversion, 404, 410, 412, 413, 414
  UGA, 441
Muton,
  definition of, 391
  minimum size of, 391, 431
Myofibril, 196, 197
Myoglobin, 51, 65, 79–92, 93, 100, 101
  combination with O_2, 170, 172
  function of, 79
  heavy atoms and, 84–86, 90
  helix content of, 55, 65, 79, 91
  models of, 80, 81, 84, 89
  molecular weight of, 79
  structural features of, 79–80, 88–89, 91–92
  x-ray diffraction analysis of, 54, 79–92, 93
Myosin, 25, 26, 75, 105, 187, 191–192, 195, 197
  ATPase activity of, 188
  extraction of, 191–192
  interaction with actin, 26, 105, 192
  molecular weight of, 105, 187, 192
  phosphorylation of, 188–189, 193

N

$n \to \pi^*$ transition, 221, see $n \to \pi^*$ Transition
N-terminal amino acid residue of protein, 3, 11, 12–14, 18, 470
  determination of, 12–14
  direction of protein synthesis and, 466, 470–473
  purity of proteins and, 122
N-terminal amino acid sequence determination, 13–14
Naladixic acid, 332
"Nearest neighbor" frequency analysis,
  in DNA, 248–251
  in RNA, 256
Nerve excitation, 179–182
Neuron, 175, 182
Nicotinamide adenine dinucleotide (NAD), 135, 185, 255
  reduced (NADH), 160
Ninhydrin, 9, 10, 16
  reaction with amino acids, 10
Nitrobenzene, 27, 28
Nitrogen mustard, as mutagen, 410
Nitrophenol, 189, 491
Nitrosoguanidine, see N-Methyl-N'-nitroso-N-nitroguanidine
Nitrosomethylurea, as mutagen, 410–411
Nitrothiophenol, 189
Nitrous acid, as mutagen, 368–369, 407, 408, 409
Nonsense mutations, see under Mutations
Normalization of genetic data, 345–347, 350–351, 365–366
Novobiocin, see Cathomycin
Nucleases,
  DNase, 209, 249, 256, 355, 480
  endonuclease, 210–212, 337
  exonuclease, 210–212, 337, 338
  phosphodiesterase, 210–212, 275, 276, 279
  RNase, 131–134, 160–161, 210–212, 280, 281

Nucleic acids, *see also* Deoxyribonucleic acid; Ribonucleic acid
  as genetic material, 202–203, 254, 259, 288, 289, 299, 366–368, 370–372, 499
  cellular content of, 202–203
  chemical structure of, 204–213
  components of, 204–205, 206, 208
    minor base, 204, 205, 208
    nomencalture of, 204
  configuration of, 223–229, 284–292
  differences between DNA and RNA, 204, 208, 286
  hybridization of, 273–284
  methods for analysizng structure of,
    electron microscopy, 292, 293, 294
    hydrodynamic properties, 223–229
    optical properties, 218–223
    sequence of nucleotides, 209–213, 286
    x-ray diffraction, 213–214, 293–295
  primary structure of, 204, 206–207, 209–213, 285–286
  secondary structure of, 213–229, 284–292
Nucleophilic attack, 158, 165
Nucleoprotein, 200
Nucleoside triphosphates, 204
Nucleosides, 204
Nucleotide bases,
  tautomeric shifts in, 216–217, 228, 405, 406, 408, 409
  unusual, 204–208, 259–261
Nucleotide sequence,
  methods of determining, 209–213, 286
    minor bases and, 210
    specific nucleases and, 211–212
  of alanine-specific tRNA, 213
  of 5 s ribosomal RNA, 286
Nucleotide triplets, *see* Codon(s)
Nucleotides, 204
  structure of, 204–206
Nucleus, 128, 174, 200, 203, 255

O

Ochre mutations, *see* Mutations, ochre
Octet diagram, and genetic code, 432–433
Oligodeoxyribonucleotides, 251–252, 258
One gene–one enzyme hypothesis, 300, 426
Operator,
  and control of transcription, 496, 497–499
  as control element in protein synthesis, 494–496, 497, 498, 502
  definition of, 494
  interaction of repressor with, 494, 495, 498, 506
  mutations of, 494, 495, 497, 500, 506
Operon,

clustering of related genes in, 496–497, 501–502
definition of, 496
elements of,
  operator, 496
  promoter, 496
  structural genes, 496
fusion of, 502
histidine, 497
lactose, 496
  map of, 496
negative control of, 498, 502
polarity of transcription from, 506–510
polycistronic mRNA from, 499, 505, 507, 508–510
positive control of, 498, 502–503
regulation of function in, 496–498
regulator genes and, 496, 497, 498, 502–503
tryptophan, 508–510
Optical activity,
  and helical structure of DNA, 220–223
  and secondary structure of polypeptides, 41, 47, 56–59
  and secondary structure of proteins, 60–61
  influence of chemical environment on, 57
Optical density, 219
Optical rotation, 41–43, 56–59, 60–65
  and secondary structure of DNA, 220–223
  and secondary structure of polypeptides, 41–43, 46, 47–50, 56–59
  and secondary structure of proteins, 60–65, 69, 70, 71
Optical rotatory dispersion (ORD), 61–65, 69, 221–223, 224
  and Cotton effect, 63–64
  and secondary structure of DNA, 221–223, 224
  and secondary structure of polypeptides, 61–64
  and secondary structure of proteins, 61–65, 69
  anomalous, 61, 63–65, 221–223, 224
  Drude equation for, 61
  Moffit equation for, 61
Ordinary linear polymers, 1, 23, 24, 101, 109
Oriented films,
  of nucleic acids, 220–221
  of polypeptides, 38, 39, 55–56, 59
Ornithine, 5
Osmotic shock, 267–268
Ovalbumin, 72
  denaturation of, 59, 60, 70
  disulfide bonds of, 70
  helix content of, 62, 65, 69, 70
  optical rotation and, 69, 70
Oxidative phosphorylation, 128, 179, 182, 183
  regulation of, 183–185

P

π-electrons, *see* π-Electrons
π→π* transition, *see* π→π* Transition
PABA, *see* p-Aminobenzoic acid
Papain, 18, 97, 100
Paper chromatography, *see under* Chromatography
Paper electrophoresis, *see under* Electrophoresis
Paramyosin, 54, 61, 62, 63
Partial specific volume, 45
Partition coefficient, 120–121
Pasteur effect, 185
Pauling-Corey helix, *see* α-Helix
PCMB, *see* p-Chloromercuribenzoate
Penicillin, 310, 488
    in selection of auxotrophic mutants, 310
Penicillinase, 488
Pepsin, 4, 7, 18
Peptide,
    bond,
        as structural feature of proteins, 3, 32
        biosynthesis of, 459, 461–465
        chemical synthesis of, 26–28
        ultraviolet absorption maximum of, 53
    group,
        interatomic distances in, 32–33
            partial double-bend character and, 32–33
            planar configuration of, 33–34
    separation of, 16–18
    stepwise degradation of, 14
    tryptic, 16–18
Peptidyl-tRNA, 460–461, 462–463, 465
    identification of, 461
    ribosomal (P) site for, 462, 463
Performic oxidation, 14
Periodic acid, 453
Permease
    in active transport, 178
    galactoside, 489–490
Phage, *see* Bacteriophage
Phenol,
    catalytic activity of, 164–165
    use of, in deproteinization of nucleic acids, 200, 201, 209, 286, 366, 481
Phenotypic lag, 320
Phenylalanine, 51
    codons for, 436
    structural formula, 5
Phenylisothiocyanate (PTC), 13, 178
    for determination of N-terminal amino acid, 13
    for determination of N-terminal amino acid sequence, 13–14
Phenylthiohydantoin-amino acid, *see* PTH-amino acids

Phosphodiester bond, 204, 206, 210
Phosphodiesterase, 210–212, 275, 276, 279
Phosphoglucomutase, 164
Phosphoglyceraldehyde dehydrogenase, 26, 105, 119
Phosphomonoesterase, 212, 427
Phosphorylase b, 167–168
Phosphoserine, 4, 7
Pili, bacterial, 318, 322
pK, 8–9, 51, 52, 107, 111–112, 137, 153
    of amino acids, 9
    definition of, 8
Plaque, bacteriophage, 373
β-Pleated sheet conformation, 35, 36–37
    in polypeptides, 38
    in proteins, 78, 195
Pneumococcus, transformation of, 355, 357–358, 360, 363
Polar mutations, 506–510
Polarization of electronic orbitals, in enzymatic catalysis, 153–155
Polyacrylamide gel electrophoresis, 116–117, 468
Poly-O-acetyl-L-serine, 39, 59
Poly-O-acetyl-L-threonine, 39
Poly-DL-alanine, 39–40
Poly-L-alanine, 39, 56
Poly-β-benzyl-L-aspartate, 35, 58–59, 61
    left helix in, 35, 58–59
Poly-γ-benzyl-L-glutamate, 34, 38, 39, 42–43, 44–45, 47–50, 58–59, 63
    effect of solvent on structure of, 43, 44
    helicity of, 34, 42–43, 45
    molecular conformation in, 45
Polycistronic mRNA, *see under* Messenger RNA
Polydeoxyribonucleotides,
    chemical synthesis of, 238–239, 251, 258
    enzymatic synthesis of, 255, 265–267
    poly dAT, 255, 265–267, 269, 271, 273
    as templates for replication, 251–252, 265–267
    as templates for transcription, 258
Polydispersity,
    of DNA, 269–273
    of proteins, 101–102
Polyelectrolyte, 42, 46–47, 58, 77, 111, 189–190, 227–228, 229, 235, 239–240, 286, 289–290
Poly-L-glutamic acid, 38, 42, 45, 46, 53, 58
    effect of pH on, 46
    electron micrograph of, 45
Poly-L-homoserine, 39
Poly-L-leucine, 34, 38
Poly-L-lysine, 38, 47, 58, 64
    α→β transition in, 38, 64

Subject Index

Polymerase,
  DNA, 218, 248–255, 259, 331
  DNA-dependent RNA, 255–258
Poly-L-methionine, 63
Poly-γ-methylglutamic acid, 34
Polynucleotide chain, 204, 206
  direction of, 204, 211
Polynucleotide ligase, 231, 239, 253, 254–255, 331, 332, 337, 338
Polynucleotide phosphorylase, 237
Polypeptide chain, 2
  addition of amino acids to, 30, 47–50
  cleavage with chymotrypsin, 18
  cleavage with cyanogen bromide, 18
  cleavage with trypsin, 15
  α-helical conformation of, 32–34
  β-pleated sheet conformation of, 35, 36–37
Polypeptides,
  chemical synthesis of, 26–31
    initiation, 27
    protection of amino groups, 30
    protection of side groups, 29–30
    removal of protecting groups, 29, 30
  helix-coil transition of, 38–39, 42–43, 45–47, 48, 58–59
  as model compounds, 24, 26, 31
  secondary structure of, 31–69
    β-conformation, 38–39, 56, 59, 63
    cross-β in, 39
  effect of charge on, 38, 46–47, 58
  effect of solvent on, 38, 39, 42–43, 44–45, 47, 48, 58, 59
  effect of stereoisomerization on, 47–50
  α-helix in, 34, 35, 38–39, 41–50, 53–56, 58–64
  α→β transition in, 56
Polyphenylalanine, 34
Polyribonucleotides,
  enzymatic synthesis of, 236–237
  formation of complexes between, 239–247
  and genetic code, 258, 434–437
  helix-coil transition in, 245–246
  poly A, 237, 239–245, 246, 435, 467
  poly AU, 246
  poly C, 220, 237, 243, 246, 467
  poly G, 237, 243
  poly I, 237, 243–245, 246
  poly rT, 237, 245
  poly U, 237, 239–245, 246, 434, 461, 466–467
  as templates for polypeptide synthesis, 434–437, 456, 459, 461, 466–467
Polyribosomes, 465–466, 508
  electron micrograph of, 464
  and protein synthesis, 465–466
Poly-L-serine, 39
Polysome, *see* Polyribosomes

Polystyrene,
  in chemical synthesis of polypeptides, 30–31
  in chromatographic media, 8
Poly-L-threonine, 39
Porphyrin, 51, 79, 95, 139
Proflavin,
  and induction of frameshift mutations, 421
  structural formula, 411
Proline, 10, 14, 24, 25, 69, 100
  and disruption of protein secondary structure 24, 69
  codons for, 436
  structural formula, 5
Promoter, 495–496
Prophage, 391–396, 398–403
  attachment site for, on bacterial chromosome 395–396, 400
  defective, 400–403
  definition of, 391
  excision, 396
    incorporation of bacterial genes during, 398–403
  model for, 400–401
  immunity, 392, 393–394
    mechanism of, 393–394, 499–501
  induction of, 392, 394–395, 504
    mechanism of, 394–396, 499–501
    zygotic, 392, 499
  integration, 391, 395–396
  recombination with bacterial chromosome, 395–396, 400–403
Prosthetic groups, of enzymes, 134–135
Proteases, 18, 129, 136
  structure of, 97–100
Protein,
  absorption spectrum of, 51, 53, 219
  alkylation of, 133, 134, 136, 160
  amino acid composition of, 7–10
  amino acid sequence analysis of, 10–23
  chromatography of, 15, 115–116, 131, 427, 491, 500
  complete chemical synthesis of, 30–31
  contractile, 129, 182, 184, 186, 191–193, 195
  criteria of purity, 122
  cross-linking of, 7, 14, 25, 69–74
  denaturation of, 26, 42, 47, 59–60, 64–69, 70–74, 112, 133
  dimerization of, 105, 133, 134
  Edman degradation of, 7
  electrochemical properties of, 111–114
  electrophoresis of, 112–113, 116–118, 122–123
  elementary analysis of, 7
  enzymatic activity of, 127, 129–174
  fibrous, 25, 56, 78
  filamentous, 25

fingerprinting of, 16–18
globular, 23, 25, 69–78, 100, 108–113
　Bresler-Talmud theory of structure, 74–75, 100
helix content of, 54, 61–62, 65, 69, 79, 96–100
helix-coil transition in, 41, 42, 59–60, 65–69, 70–73
heterogeneity of, 1–2, 101–102
hydration of, 109–111
hydrolysis of, 7, 10, 15, 18
hydrodynamic properties of, 44–45, 108–111
molecular weight determination,
　by chemical means, 101
　by light scattering, 45
　by sedimentation, analysis 101–103
molecular weight, table of, 104–105
N-terminal amino acids of, 12–14
polyelectrolyte effects in, 77, 111
renaturation of, 73–74, 133–134, 135
ribosomal, 117, 468–469
separation and purification of, 115–123
structural, 75, 128
structure of, 23–26, 32–35, 69–78, 79–101
　effect of solvent on, 26, 56, 64–65, 70, 71
　general properites of, 23–26, 31–38, 74–78, 100, 112
　primary, 10–23, 24, 26
　　as determinant of secondary and tertiary structure, 73–74, 134, 142–143, 420
　　principles of analysis, 10–12
　quaternary, 25, 26, 92–96, 103–108, 130, 131
　secondary, 24–26, 40–42, 51–53, 54, 59–60, 62, 63, 69, 74, 79–80, 88–90, 91, 95, 96–101
　　β-conformation in, 38, 56, 63, 72, 78, 97, 100, 195
　　cross-β in, 38, 54, 63, 64–65, 98, 99, 100
　　α-helix in, 54, 61–62, 65, 69, 70–72, 78, 79, 91, 96–100
　　$\alpha \rightarrow \beta$ transition in, 56, 78, 195
　tertiary, 25, 26, 69–78, 91–95, 96–101, 108–112
synthesis,
　activation of amino acids for, 451–453
　codon-anticodon recognition and, 462–463, 468
　dependence on mRNA, 459, 485–486
　direction of, 424, 466, 473, 508
　enzymes mediating, 461, 463
　errors in, 466–468
　hydrolysis of GTP and, 461, 463
　initiation of, 463–464, 507
　　codons specifying, 442–443, 463
　　N-formylmethionyl-tRNA and, 442–443, 463

　intermediate complex in, 460
　kinetics of, 470–477
　mechanism of, 461–465
　polyribosome cycle and, 465–466
　rate of, 473, 476–477
　regulation of, 487–510
　ribosomes and, 457–468, 478
　and RNA synthesis, 485, 508–510
　as a sequential process, 421–423, 466, 472–473, 477
　termination of, 464–465
　translocation and, 463
　in vitro, 433–434, 457–459, 470–473, 485–486
　　functional products from, 441, 486
　　requirements for, 433–434, 459
　in vivo, 473–477
　viral, 22, 106–108, 367–368, 371, 372, 383–385, 386, 387, 432–433, 439–440, 441, 486, 503, 507
　x-ray diffraction analysis of, 79–101
Proteolytic enzymes, 15, 18, 97–100, 129, 136
Prototroph, definition of, 308
Provirus, 396–397
Pseudomonas aeruginosa, base ratios of nucleic acids from, 482
Pseudouracil, 205, 442
PTH-amino acids,
　isolation and characterization, 13–14
　reaction of formation, 13
Pulse-labeling,
　in sequencing peptide fragments, 473
　in study of DNA replication, 329, 332
　in study of metabolic processes, 469
　in study of protein synthesis, 470–477
　in study of mRNA synthesis, 479–482
Purines, as nucleic acid components, 204
Puromycin, as inhibitor of protein synthesis, 455, 461
　mechanism of action, 455
Pyridine, 16, 45
Pyridoxal phosphate, 156–157, 430
Pyrimidines, as nucleic acid components, 204
Pyrophosphate, 248, 452, 453

Q

Quartz crystal, 61
Quaternary structure, of proteins, 25, 26, 92–96, 103–108, 130, 131
　allostery and, 168–174

R

r_{II} mutants of bacteriophage T4, 387–390, 414–415, 421–423

Subject Index

complementation between, 390
genetic analysis of, 387–390
R_f, definition of, 12, 16
Racemic mixtures, 6, 48, 49, 50
Radiation, as mutagenic agent, 305–306, 412–413
Radioactive "suicide"
in bacteria, 332–333, 352–354
in bacteriophage, 352, 373
Radius of gyration, 45, 223
Random coil configuration, 23, 32, 38, 39, 42, 45, 47, 48, 53, 58, 59, 61, 227, 228–229, 234, 287
and RNA in solution, 286–287, 293
"stiff," and DNA in solution, 223–224, 225
Recessiveness, 328
Recombination, genetic
between adjacent base pairs, 391, 430–431
in bacteriophage, 375–383
theory of, 376–377
in conjugation, 319, 340–352
DNA synthesis in, 337, 338, 361–362
effective pairing in synapsis, 337, 365–366
in formation of transducing particle, 400–401
intracodon, 430–431
intragenic, 348–349, 364–365, 387–390
mathematical analysis of,
in bacteria, 350–352
in bacteriophage, 376–377
mechanisms of, 334–340, 345, 347
breakage and reunion, 335, 336–337, 360–363, 377, 378–381
copy-choice, 335, 336
repair synthesis and, 337–340, 362
methods of selection in, 340–341
molecular models for, 335, 338
multiple crossovers in, 345, 347, 350–351, 365–366
mutants defective in, 324, 340
nonreciprocal, 336, 378, 401–402
normalization of data in, 345–347, 349–350, 364
radiation repair mechanism and, 339–340
reciprocal, 334–336, 378
recon and, 391
between sex (F) factor and bacterial chromosome, 337–339
synapsis and, 319, 337, 365
in transduction, 398–403
in transformation, 360–363
unit of,
in *E. coli*, 344–351, 354
in bacteriophage T4, 383
Recombination analysis,
in large segments of chromosome, 340, 342–344, 364
in short segments of chromosomes, 340, 345, 346, 365, 398
Recombination frequency,
additivity principle and, 345, 350–351, 377
and size of muton, 391
definition of, 345–347, 350–352
in bacteriophage crosses, 376–377, 383
in conjugation, 345–350
in transformation, 365–366
relation to crossover frequency, 350–351
relation to map distance, 350–352, 383
relation to physical distance, 354, 383
Recon,
definition of, 391
minimum size of, 391, 430–431
Reconstitution of viruses, *see under* Virus
Regulation
of protein synthesis, 449–450, 486–510
at level of translation, 507–510
biochemical evidence for, 497–499, 504–506
genetic analysis of, 489–491, 493, 494–495, 497, 500, 502–503
in λ bacteriophage, 499–501, 504–506
in *E. coli*, 427, 478, 487–496, 501–503, 508–510
in *Salmonella*, 497
negative control systems in, 498, 502, 503
polar mutations and, 506–508
positive control systems in, 498, 502–504
synthesis of mRNA and, 497–499, 504–506, 507–510
operon and 496, 497
of DNA synthesis, 333–334
Regulator gene, 325, 489, 496, 497, 498, 502–504
definition of, 497
properties of, 492–493, 497, 498, 502–503
protein products of, 493–494, 500–501, 503–504
Renaturation,
of DNA, 273–278
of proteins, 73–74, 133–134, 135
of RNA, 288–289
Repair of radiation damage, 339–340
Replica-plating,
description of method, 308–309
for isolation of auxotrophic mutants, 308–309
for isolation of F' strains, 325–326
for isolation of *Hfr* donors, 324–325
for investigating mutational basis of variability, 317–318
in recombination analysis, 347, 363
Replication, 216–218, 261–268, 328–334
accuracy of, 217–218, 248–251, 405–407
autoradiography in study of, 264, 267–268, 328–329, 330, 373

and cell division, 334
of circular chromosome, 329
  in vitro, 252–254
conjugation and, 332–333
direction of, 252, 329, 331
dispersive, 264, 268, 373
errors of, 216–218, 247
initiation points for, 329, 332, 333
polynucleotide ligase and, 253, 254–255, 331 332
replicon hypothesis of, 333–334
semiconservative, 256, 261–262, 264–267, 380
simultaneous copying of both strands in, 331–332
strand separation and, 262, 329–331
of synthetic oligodeoxyribonucleotides, 251–252
in vitro, 247–251, 252–255, 265–267
template requirement for, 248
Replicative form,
  of single-stranded DNA viruses, 230–231, 253–254, 258
  of single-stranded RNA viruses, 259, 370
Replicator, in regulation of DNA synthesis, 333
Replicon hypothesis, 333–334
Repression of enzyme synthesis, 487
  coordinated, 496–497, 498
  by end-product of biosynthetic pathway, 487, 496–497
  mechanism of, 497, 498
Repressor,
  of bacteriophage λ induction, 499–501
    effect on transcription of λ DNA *in vitro*, 501
    properties and nature of, 500–501
  of lactose operon, 492–494
    effect of inducer on, 492, 494, 495, 496
    as protein, 493–494, 497, 498, 500–501
    interaction with DNA, 495, 496, 500, 501, 506
    interaction with inducer, 494, 495, 496
  in regulation of enzyme synthesis, 487, 492–494, 497, 498
  super-,
    of bacteriophage λ function, 500–501
    of lactose operon, 494, 500
Resistance,
  to antimetabolites, 141–142, 310–311, 312, 328, 341, 356–358, 360–362, 488
  to bacteriophages, 311–312, 316–318, 328, 375
Reticulocytes, 434
  synthesis of hemoglobin in extracts of, 456, 459, 470–473
Reversion of mutations, 306, 309–310, 404, 407, 414–415, 421–423

Revertants, 310, 389, 422, 423
Ribonuclease, pancreatic,
  A, 131,
    chemical synthesis of, 31
  active site of, 96–97, 132–134, 160–161
  alkylation of, 132–134, 160–161
  chemical modification of, 132–133, 139, 160–161
  chromatography of, 131
  controlled degradation of, 131, 132
  denaturation of, 42, 53, 71, 72–73, 133–134, 135
  dimer formation by, 133, 134
  disulfide bonds in, 15, 22, 71, 72–73, 133–134
    reduction of, 73, 133, 135
  helix content of, 54, 65, 71, 97
  hydrogen bonds in, 40–41, 51–53, 73
  isotopic exchange and, 40–41
  mechanism of, 160–161
  molecular weight of, 104
  optical rotation and, 62, 65
  oxidation of reduced, 73, 133–134, 135
    reactivation by, 73, 133–134, 135
  purification by fractional precipitation, 118
  reconstitution of, 131, 132–134, 135
  resistance of DNA to, 209
  resistance of DNA:RNA complexes to, 280, 281, 483
  S, 131
  S-peptide from, 31, 131
  S-protein, 131
    chemical synthesis of, 31
  specificity of, 131, 211–212
  structure,
    primary, 19, 20, 21
    secondary, 40–41, 54, 62, 65, 71, 97
    tertiary, 72–73, 97
  structure-function studies, 131–134
  x-ray diffraction analysis of, 96–97
Ribonuclease(s), 210–212,
  specificity of, 212
  T_1-, specificity of, 211–212
Ribonucleic acid (RNA),
  alkaline hydrolysis of, 208
  alteration of properties by unusual bases, 260–261
  base ratios of, 213, 482
  cellular content of, 203, 450
  chemical structure of, 204, 207
  components of, 204
    minor, 205, 208
  configuration of, in solution, 287, 293
  deproteinization of, 209, 284
  double-stranded, 259, 292, 370
  flexibility of, 286–287, 293
    statistical segment and, 286–287

Subject Index

fractionation of, 210
function of, 201, 449
as genetic material of viruses, 202, 259, 288, 289, 366–368, 370
helix-coil transition in, 284–285, 287, 288–291
hybridization of, with DNA
    for detection of RNA:DNA complementarity, 280–284, 483, 499
    methods of detection, 280, 283, 284, 483
    during transcription *in vitro*, 256
infectivity of, 202, 259, 288, 289, 366–368
messenger, *see* Messenger RNA
"nearest neighbor" sequence analysis of, 256
nucleotide sequence analysis of, 209–213, 286
polyelectrolyte behavior of, 239–240, 289–290
purification of, 209
replication of viral, 259, 370
replicative form of, 259, 370
ribosomal, *see* Ribosomal RNA
structure of, 287–295
    primary, 209–213, 285
    secondary, 285–286, 287–292
    tertiary, 287, 293, 294–295
synthesis of, *in vitro*, 236–238
    from DNA template, 255–259, 280–283
synthetic, 236–247
tobacco mosaic virus, *see* Tobacco mosaic virus, RNA of
transcription of, *see* Transcription
transfer, *see* Transfer RNA
viral, as mRNA, 370, 441, 485–486
x-ray diffraction analysis of, 292, 293–295
Ribose, 204, 208, 239
Ribosomal RNA (rRNA),
amount of, in bacterial cell, 286, 450
base ratio of, 482
electron micrograph of, 294
genetic loci for, in bacteria, 281, 450
hybridization with DNA, 281
molecular weight of, 281, 286, 460
primary structure of 5s, 286
secondary structure of, 287
sedimentation coefficient of, 286
unsuitability of, as informational intermediate, 477–478
x-ray analysis of, 293–294
Ribosomes, 128, 174, 201, 286, 450
aminoacyl-tRNA (A) site of, 462–463, 468
association with mRNA, 460–461, 462
association with tRNA, 460–461, 462, 463
components of, 450, 460
    protein, 460, 468–469
    RNA, 281, 286–287, 460, 468–469
controlled degradation of, 468
electron micrograph of, 458
and fidelity of translation, 466–468
function of, 450
metabolic stability of, 450, 479
number of, in bacterial cell, 457, 473
peptidyl-tRNA (P) site of, 462–463
and protein synthesis, 457–468, 470–477, 478, 479
reconstitution of, 468–469
    specificity of, 468
separation by centrifugation, 459, 473
size and shape, 457, 458
structure of, 457, 459–460
subunits of, 281, 286, 458, 459, 461, 462, 465–466, 480, 481
reversible dissociation of, 460
Rigid macromolecules, 24, 25, 44, 46, 227
RNA, *see* Ribonucleic acid
RNA polymerase, DNA-dependent
bacterial, 256, 257
DNA template for, 255–258
mammalian, 255–256
promoter, as attachment site for, 495
reaction catalyzed by, 255–256
RNA synthetase, 259
RNase, *see* Ribonuclease

S

Salmonella, 279, 326, 392, 396
circular linkage group of, 304
histidine biosynthesis in, 497, 499, 506–507
Salting out of protein, *see* Fractional precipitation
Schiff bases, 156
Sedimentation analysis,
in density gradients, 262–264
by equilibrium, 103
in sucrose gradients, 118, 473
by velocity, 102–103
Sedimentation coefficient
correction to standard conditions, 103
definition of, 102
of proteins, table of, 104–105
relation to molecular weight
    of nucleic acids, 224–227, 229, 286
    of proteins, 102
Segregation of bacterial chromosome, 320
Selection
of mutants, 301, 306–307, 308–312
    bacterial, 308–312
    bacteriophage, 307–308, 373–375, 393–395
    color tests in, 309
    penicillin method, 310
    replica-plating and, 308–309
of recombinants,
    bacterial, 319, 320, 340–341, 342, 343, 346, 348–350, 356

bacteriophage, 375–376, 379, 381, 387–388, 390, 394
   methodology of, 340–341
  of revertants, 309–310
Selective media, 309, 341
Semiconservative replication, 256, 261–262, 264–267, 380
Sephadex, 115, 116
Serine, 429
  at active site of chymotrypsin, 136, 138, 157–159
  at active site of trypsin, 136, 138, 157–159
  codons for, 436
  structural formula, 4
Serratia marcescens, transfer of *E. coli* F′ factor to, 326, 354, 424–425
Serum albumin, 104, 113, 119, 122–123, 219
  bovine, 51, 54, 62, 219
  denaturation of, 70–72
  disulfide bonds in, 25, 70, 71, 72
  horse, 69
  human, 70–72, 119
  secondary structure of, 54, 62, 69
  tertiary structure of, 70–72
  ultraviolet absorption spectrum of, 51
Serum proteins, 25
  fractionation of, by electrophoresis, 113, 122–123
Sex (F) factor of *E. coli*,
  chemical nature of, 320, 354
  as determinant of mating type, 320, 322–324
  elimination of, 320, 322
  as episome, 324
  functions of, 321–322
  integration with bacterial chromosome, 322–323
    determination of chromosomal origin by, 322–323, 332
    multiple sites for, 323, 324–325, 332
    by recombination, 324, 339
  properties of, 320–325
  replication of, 322, 328
    autonomous, 324, 332, 333
  size of, 354
  states of, 324
  transmission of, 320–322, 328, 334
Sexduction, 325, 326, 490
  interspecific, 326, 354, 424–425
Sexuality in bacteria, 318–321, 326
  genetic determination of, 320
Shear, effect of on DNA molecules, 268–269
Shigella dysenteriae, 279, 326
Sickle-cell anemia, 17, 425–426, 427
Silk fibroin, 25
Single-burst experiment, 375–376
Solvation, *see* Hydration

Solvent,
  effect on molecular configuration,
    in polypeptides, 38, 39, 42–43, 44–45, 47, 48, 58, 59
    in proteins, 26, 56, 64–65, 70, 71
  in hydrodynamic particles, *see* Hydration
  method of variation of,
    in studies of polypeptide secondary structure, 43, 47, 48, 58, 59
    in studies of protein secondary structure, 64–65
    in studies of protein tertiary structure, 70, 71
Specific activity, of protein, definition, 115
Specific rotation,
  definition, 56
  of DNA, 221
  of polypeptides and proteins, 41, 47, 56, 57
Spheroplast, 254, 370
Spontaneous mutation, *see under*, Mutation(s)
Statistical segment, 223, 225, 286–287
Stereoisomers, effect on polypeptide configuration, 47–50
Stokes' law, 108
Streptococcus, transformation of, 352
Streptomycin,
  and misreading of genetic message, 466–467
  resistance to,
    as genetic marker in transformation, 356–358, 360–361
    as recessive trait, 328
    ribosomal basis of, 328, 469
    as selected marker in conjugation, 341
Structural gene, 325, 496, 498
Structure of macromolecules, methods of determining
  discrimination of α- and β-conformation in proteins, 35–38, 39, 41, 55–56, 60–61, 64
  electron microscopy, 45, 46, 107, 180, 196–197, 252, 257, 271, 292, 293, 294, 319, 337, 371, 385, 458, 464
  frictional coefficient, 108
  hypochromic effect, 52, 53–54, 59–60, 64, 219, 221–223, 227, 230, 239–240, 271, 284–285, 288
  infrared spectroscopy, 55–56, 59, 78
  isotopic exchange, 39–41, 54, 64, 232–233, 285
  optical rotation, 41–43, 46, 47–50, 56–59, 60–65, 69, 70, 71, 220–223
  optical rotatory dispersion, 61–65, 69, 221–223, 224
  solvent variation, 43, 47, 48, 58, 59, 64–65, 70, 71
  viscosity, 43–47, 70, 108–109, 192, 222, 224–229, 235–236, 289–290

Subject Index

ultraviolet spectroscopy, 50–54, 219–221
x-ray diffraction analysis, 79–101, 293–294
Substituted sex (F′) factor of *E. coli*,
 circularity of DNA in, 337
 as episome, 325
 F′-*ara*, 325, 502–503
 F′-*gal*, 325
 F′-*lac*, 325, 334, 337–339, 354, 402–403, 424, 490, 499
  temperature-sensitive, 334, 402–403
 F′-*pro*, 325
 in genetic analysis, 327–328, 402–403, 490, 493, 494–495
 integration into bacterial chromosome, 325, 334
  by recombination, 337–339
 properties of, 325
 replication of, 325, 328
  autonomous, 326, 333–334
 selection of, 325–326
Subtilisin, 18, 131
 active site of, 98, 99
 x-ray diffraction analysis of, 98–100
Subunit structure,
 of proteins, 7, 25, 92, 103–108, 130
  and allostery, 168
 of ribosomes, 286, 460
 of viruses, 106–107, 367
Sucrose gradient,
 in centrifugation, 118, 465, 473, 476, 480, 482
 in electrophoresis, 117
Sulfanilamide, 140–141, 311, 356
Sulfhydryl groups, 85, 104, 455
 alkylation of, 14
 and enzymatic activity, 98, 105, 133, 135, 136, 146, 178, 179
 oxidation of, 7, 14, 73, 105, 133
 reduction of, 14, 71–72, 73, 133
Superinfection, 392, 393–394, 398, 401
Suppression, 307–308, 439–443
 intergenic, 307–308, 439
 intragenic, 428–429
 mechanism of, 439–441
Suppressor,
 of amber mutation, 308, 439–441
 gene, 307–308, 439, 441
 mutation, 307–308, 439, 441
 nature of, 439
 of ochre mutation, 308, 441
 as tRNA molecule, 439–443
  evidence for, 441–443
 of UGA mutation, 441
Svedberg unit, 102
Svedberg equation, 102
Synapsis, *see* Recombination, genetic

T

Target theory, 412–413
Tautomeric shifts,
 in enzymatic catalysis, 156–157
 mutation hypothesis based on, 216–218, 404–407
 in nucleotide bases, 216–217, 228, 405, 406, 408, 409
Temperate bacteriophages, *see under*, Bacteriophage
Temperature of helix-coil transition, *see under* Helix-coil transition
Temperature-sensitive mutants, *see under* Mutants
Template,
 in synthesis of DNA from DNA, 248–255
 in synthesis of protein from RNA, 450, 459, 485–486
 in synthesis of RNA from DNA, 255–259
 synthetic polyribonucleotides as, 434–437, 459, 466–467
Tertiary structure of proteins, 25, 26, 69–78, 79–80, 91–95, 96–101, 108–112
Tetrahedral intermediate, 158
Tetrahydrofuran, 27
Thermodynamics,
 of enzymatic reactions, 149–152
 of helix-coil transition,
  in nucleic acids, 231–232, 269–271
  in proteins, 65–69
Thermolysin, 18
Thiol groups, *see* Sulfhydryl groups
Three-factor cross, 375–377
Threonine,
 codons for, 436
 structural formula, 4
Threonine deaminase, 172–173
Thrombin, 106
Thymine,
 dimerization of, in DNA, 340
 hydrogen-bonding capacities of, 215, 217
 structural formula, 206
Thyroxine, 4
Titration,
 of bacteriophage, 373–374
 electrochemical, 111–112, 227–228
TMV, *see* Tobacco mosaic virus
Tobacco mosaic virus,
 amino acid replacements in protein of, 368, 432–433
 amino acid sequence of protein of, 22
 dissociation of, with acetic acid, 106, 367
 hybrid particles of, 360
 lesions caused by, 367, 368
 molecular weight of protein of, 106, 367
 mutagenesis of, 368–369, 432

physiology of infection by, 369–370
reassembly of, 106–107, 367
reconstitution of protein coats, 106–107, 367
RNA of,
    as genetic material, 366–367, 368, 369
    infectivity of, 288, 289, 366–368
    helix-coil transition in, 288–291
    replicative form of, 370
    as mRNA, 370
    size and configuration, 287–291
    structure of, 106–107, 287–288, 367
Toluene-p-sulfonyl chymotrypsin, 98
Transcription of RNA,
    dependence on translation, 508–510
    from DNA template, 255–259, 280–283
        in λ bacteriophage, 484, 504–506
        complementarity of product to template, 256
        direction of synthesis, 258, 466, 508
        inhibition by DNase, 256
        initiation of, 495
        strand selection in, 258–259, 281–283, 504–506
    from RNA template, 259
    from synthetic oligodeoxyribonucleotide, template, 258
Transducing bacteriophage particle,
    defective, 400–402
    gene-specific DNA from, 403, 499, 504, 509
    HFT preparations of, 399–400
    models for formation of,
        generalized, 397–398
        specialized, 400–401
    nondefective, 401–402
Transduction, 397–403
    abortive, 398
    discovery of, 397
    generalized, 397–398
    high frequency (HFT), 399–400
    integration of chromosomal fragments in, 397–398, 400–403
    of linked markers, 397, 398
    low frequency (LFT), 399–400
    specialized, 398–403
    species susceptible to, 301, 397
Transfection, 366
Transfer RNA (tRNA),
    as adaptor, 435, 451, 455–456
    alanine-specific, 210–213, 285
        primary structure of, 213
        secondary structure of, 285
    and amino acid activation, 451–453
    aminoacyl-, 452–453, 459, 461–463, 465
    amount of, in bacterial cells, 284, 451
    base ratio of, 482
    binding of,
        to ribosomes, 460–462
        to trinucleotide-ribosome complexes, 436
    cysteine-specific, modification of, 455–456
    degeneracy and, 437–438, 452, 456
    electron micrograph of, 294
    N-formylmethionyl-, and initiation of protein synthesis, 442–443, 463–464, 465
    histidyl-, as co-repressor of histidine biosynthesis, 497
    leucine-specific, degeneracy and multiple forms of, 437, 456
    molecular weight of, 210, 281, 284, 451
    minor base components of, 205, 208
    peptidyl-, 460–461, 462–463, 465
    and protein synthesis, 451, 455–456, 460–464
    purification of, 210, 456
    specificity of acceptor activity, 453, 454–455
    structure of,
        anticodon site in, 286, 442
        base-paired regions of, 284–285
        primary, 209–213, 285
        secondary, 233, 284–286, 294–295
        tertiary, 294–295
    suppressor, 439–442
        mutation at anticodon of, 441–442
        as product of suppressor gene, 440–441
        in vitro suppression by, 441
    transcription of, 281
    unusual nucleotide bases in, 205, 208
    "wobble" hypothesis and, 437–438, 456
    x-ray diffraction analysis of, 294–295
Transformation, 301, 355–366
    by bacterial DNA, 202, 276, 355
    competence factors in, 356
    competence of recipient bacteria for, 355–356, 358–360
    competitive inhibition in, 357–358
    discovery of, 355
    dispensibility of DNA replication in, 361–362
    DNA binding in, 358
    DNA integration in, 360–363
    DNA uptake in, 355–358
    effect of DNA concentration on, 357–358, 359
    evidence for single DNA strand incorporation in, 362–363
    genetic mapping by, 363–366
    of linked genes, 358–359, 360–361, 362–363, 364–365
    recombination in, 360–363, 365–366
        by breakage and reunion, 362–363
    with single-stranded DNA, 363
    species susceptible to, 301, 354
    specificity of, 356
Transforming DNA, 355–363

Subject Index

binding of, 358
hybrid strands in, 362–363
inactivation of, 416–418
molecular size of, 358, 366
single-stranded, 363
$n \to \pi^*$ Transition, 221
$\pi \to \pi^*$ Transition, 51, 220–221
Transition mutation, 404, 415
Translation of nucleotide sequence into amino acid sequence, see under Protein
Translocation,
   of genetic material by F' factor, 402–403
   mutations, 306, 404
   as step in protein synthesis, 463
Transpeptidation, 161
Transport,
   active, see Active transport
   by diffusion, 174–175
Transversion mutation, 404, 414
Trichoroacetic acid (TCA), 433, 457, 473
Trifluoroacetic acid, 39
Trinucleotides, synthetic,
   as codons, 436
   use of, in deciphering genetic code, 436, 442
Triplets, nucleotides, see Trinucleotides, synthetic
Trypsin,
   active site of, 136–139, 140, 158
   denaturation of, 59–60, 65, 72
   esterase activity of, 136, 158–159
   mechanism of action, 157–160, 164
   peptidase activity of, 160
   specificity of, 15, 139
   use of, in amino acid sequence determination, 15–19, 22
Trypsinogen, 138
   activation of, 138, 139
   amino acid sequence of, 138
Tryptophan, 14, 51
   biosynthesis,
      genetic regulation of, 499, 509
         transcription of specific mRNA and, 499
      operon for, 509
      and translational regulation, 508–510
   codon for, 436
   structural formula, 5
Tryptophan synthetase, 130
   in establishment of colinearity of gene and protein product, 429–430, 431
   mutant forms of, 304, 305, 430–431
   reaction catalyzed by, 429
   in studies on genetic code in vivo, 430–431, 432
Turnover number, 144, 491
Tyrosine,
   codons for, 436
   hydrogen bond formation by hydroxyl groups of, 51–53, 73
   structural formula, 5

U

Ultracentrifugation, see Sedimentation analysis
Ultraviolet light,
   absorption of,
      by nucleic acids, 219–221
      by proteins and polypeptides, 51–54
   elimination of sex (F) factor by, 320, 322
   induction of prophage by, 392, 500
   as mutagen, 305–306, 413
   photochemical action of, on DNA, 339–340
Ultraviolet dichroism, 220–221
Ultraviolet spectrum,
   of DNA, 219
   of serum albumin, 51
Unit cell, 81, 84, 85, 86, 110
Unshared electron pairs, 154
Unusual nucleotide bases, see under Deoxyribonucleic acid; Transfer RNA
Uracil,
   hydrogen-bonding capacities of, 239–241, 242–244
   structural formula, 207
Urea,
   as denaturing agent, 26, 73, 112, 133
   effect on structure of proteins, 32, 112
   effect on structure of RNA, 211
UV, see Ultraviolet light

V

Valine,
   codons for, 436
   structural formula, 4
van der Waals interactions,
   and hydrophobic bonds,
      in DNA, 218, 232–233
      in polypeptides, 35, 38
      in proteins, 25, 26, 70, 74–77, 100
   energy of, 76–77
Variation of solvent, method of, 43, 47, 48, 58, 59, 64–65, 70, 71
Virus,
   animal, 396–397
   bacterial, see Bacteriophage
   infectivity of nucleic acids from, 202, 254, 259, 288, 289, 366–368, 370
   plant, 366–370
   provirus state, 396–397
   reconstitution of, 106–107, 367
   structural organization of, 106–107, 371
   symmetry of, 107–108
   tobacco mosaic, see Tobacco mosaic virus
Visconti-Delbrück theory, 376–377

Viscosity,
  dependence on molecular weight,
    in nucleic acids, 224–227, 229
    in proteins, 44–45
  intrinsic, definition of, 108
  of protein solutions, 23
  relation to DNA configuration, 222, 224–229, 235–236
  relation to polypeptide configuration, 43–47
  relation to protein configuration, 70, 92, 108–109
  relation to RNA configuration, 289–290

W

Watson-Crick model of DNA structure, 214–218, 259
  application to synthetic polyribonucleotide complexes, 239
  evidence for, 220, 234, 249–251, 256, 264
Wild-type strain, definition of, 301, 308
"Wobble" hypothesis, 437–438, 442, 456

X

X rays, as mutagenic agents, 412–413
X-ray diffraction analysis,
  crystals and, 80–82
  electron density map and, 85–88, 91, 92, 93
  in detection of ordered structure, 61, 65, 79, 91, 96–100
  for determination of solvation, 110–111
  of DNA, 213–214
  Fourier synthesis in, 83, 85–88, 90
  isomorphous replacement and, 84–86, 90, 92, 96
  pattern resulting from, 82
  phase problem and, 83–86
  principles of, 80–88
  protein crystals and, 83–85, 110–111
  of proteins, 65, 79–101, 110–111, 195
    carboxypeptidase, 97, 98, 100
    chymotrypsin, 97, 98
    egg-white lysozyme, 96
    hemoglobin, 92–96, 110–111
    myoglobin, 79–92, 93
    papain, 97, 98
    ribonuclease, 96–97
    subtilisin, 97, 98–100
  of RNA, 292
  RNA crystals and, 294
  of tRNA, 294–295
  in study of active site of enzymes, 96–100
  of synthetic polynucleotides, 239–245
  of synthetic polypeptides, 34, 38, 45, 61
  of viruses, 287–288
X-ray scattering, low angle, 293
  of DNA, 234
  of RNA, 293

Y

Yeast,
  structure of alanine-specific tRNA of, 213
  tRNA of, 454

Z

Zinc, 101, 154
Zone electrophoresis, 116–117
Zone sedimentation, 118
Zwitterion, 9, 12
Zygote, 318–320, 326–327, 332, 334–336
  formation of, in conjugation, 318–319, 334
  segregation of, 320, 336
Zygotic induction, 392, 499
  mechanism of, 499–501
Zymogen, 139, 491